Petrology of Sedimentary Rocks

Sam Boggs, Jr.
University of Oregon

Macmillan Publishing Company
New York

Maxwell Macmillan Canada
Toronto

Maxwell Macmillan International
New York Oxford Singapore Sydney

Dedicated to Theodore (Ted) R. Walker,
who taught me some things about sedimentary petrology worth thinking about

and

to my wife, Sumiko,
whose patience and understanding made the writing of this book possible.

Editor: Robert A. McConnin
Production Editor: Stephen C. Robb
Art Coordinator: Peter A. Robison
Photo Editor: Chris Migdol
Cover Designer: Robert Vega
Production Buyer: Pamela D. Bennett

This book was set in Times Roman by The Clarinda Company and was printed and bound by R. R. Donnelley & Sons Company. The cover was printed by Lehigh Press, Inc.

Copyright © 1992 by Macmillan Publishing Company, a division of Macmillan, Inc.

Printed in the United States of America

All rights reserved. No part of this book may be reproduced or transmitted in any form or by any means, electronic or mechanical, including photocopy, recording, or any information storage and retrieval system, without permission in writing from the Publisher.

Library of Congress Cataloging-in-Publication Data
Boggs, Sam.
Petrology of sedimentary rocks / Sam Boggs, Jr.
    p. cm.
  Includes bibliographical references and index.
  ISBN 0-02-311790-7
  1. Rocks, Sedimentary. I. Title.
QE471.B6818 1992
552'.5—dc20
                                                        91-28804
                                                         CIP

Macmillan Publishing Company
866 Third Avenue
New York, NY 10022

Macmillan Publishing Company is part of the Maxwell Communications Group of Companies.

Maxwell Macmillan Canada, Inc.
1200 Eglinton Avenue East, Suite 200
Don Mills, Ontario M3C 3N1

Printing: 1 2 3 4 5 6 7 8 9   Year: 2 3 4 5

Preface

The aim of this book, echoing F. J. Pettijohn in the third edition of his classic textbook *Sedimentary Rocks*, is to tell the user something about sedimentary rocks rather than about sedimentation. Therefore, this book emphasizes the characteristics of sedimentary rocks rather than sedimentary processes. It focuses on the compositional, textural, and structural characteristics of sedimentary rocks and the various ways that we can use these properties as tools for interpreting provenance and diagenetic history. Aspects of depositional environments are also considered, particularly with respect to the chemical/biochemical sedimentary rocks; however, no attempt is made to provide comprehensive coverage of depositional environments.

This book is aimed at advanced undergraduate and graduate students. Some of the chapters are long and some of the topics are covered in greater detail than may be typical for undergraduate texts. The book provides something for both undergraduates and graduates, and it is complete enough in itself so that instructors need not supply extensive reference material from other sources. This textbook is divided into four parts. Part 1 is introductory, and it furnishes some insight into the depositional basins and tectonic settings in which sedimentary rocks accumulate. Part 2 covers the siliciclastic sedimentary rocks and includes discussion of their composition, classification, petrographic characteristics, provenance, and diagenesis. Part 3 discusses the carbonate rocks. It describes the special characteristics of limestones and dolomites and discusses their classification, origin, and diagenesis. Part 4 deals with the remaining chemical/biochemical sedimentary rocks (evaporites, cherts, iron-rich sedimentary rocks, phosphorites) and carbonaceous sedimentary rocks such as coals and oil shales.

Although the petrographic microscope remains the fundamental tool for studying particle composition of sedimentary rocks, no modern work on sedimentary petrology can rely on petrographic microscopy alone as a means of characterizing sedimentary rocks. Geologic study still begins in the field, and field study of sedimentary textures, structures, and facies sequences is a fundamental part of petrologic analysis. In the laboratory, petrographic study is being increasingly supplemented by a variety of techniques for doing chemical analyses of sedimentary rocks, such as electron microprobe analysis of individual minerals, as well as isotope analysis. Cathodoluminescence microscopy is gaining importance as a tool for both diagenetic studies and provenance interpretation, X-ray diffraction techniques provide mineralogic data on fine-grained sedimentary rocks, and the scanning electron microscope is finding wide application in a variety of studies such as analysis of quartz-grain surface textures and clay-mineral textures and studies of diagenetic alteration. This book attempts to present a balanced view of the characteristics of sedimentary rocks as revealed by these various analytical techniques.

The discipline of sedimentary petrology is extremely broad, covering study of the compositions, characteristics, and origins of sediments and sedimentary rocks. Researchers are becoming increasingly specialized, and few individuals are able to extend their research to the entire field of sedimentology. Therefore, this book draws heavily on the published work of numerous researchers. Even though this book is heavily referenced, the literature on sedimentary petrology is so extensive that only a fraction of the available reference material is used. Therefore, I may have omitted some references that readers may consider important. For convenience to the reader, references are collected at the end of the book and a list of additional readings is provided at the end of each chapter.

I wish to thank those individuals who critically reviewed the entire manuscript or various chapters of the manuscript for their constructive comments. These reviewers include Bruce L. Bartleson, Western State College of Colorado; John A. Campbell, Fort Lewis College, Colorado; Adrian Cramp, University College, Swansea, United Kingdom; Ralph Hunter, U.S. Geological Survey, Menlo Park, California; Brian K. McKnight, University of Wisconsin, Oshkosh; Karl A. Mertz, Jr., consultant, Newcastle, California; Robert Q. Oaks, Jr., Utah State University; R. Lawrence Phillips, U.S. Geological Survey, Palo Alto, California; M. Dane Picard, University of Utah. I am also greatly indebted to numerous individuals who supplied much-needed photographs and specimens. These individuals are acknowledged within the text. Cindy S. Shroba, University of Oregon, proofread the entire manuscript.

Contents

Chapter 3
Sedimentary Structures   79

Chapter 4
Sandstones: Composition   126

Chapter 5
Sandstones: Classification and Petrography

Chapter 6
Conglomerates

PART 1
Principles

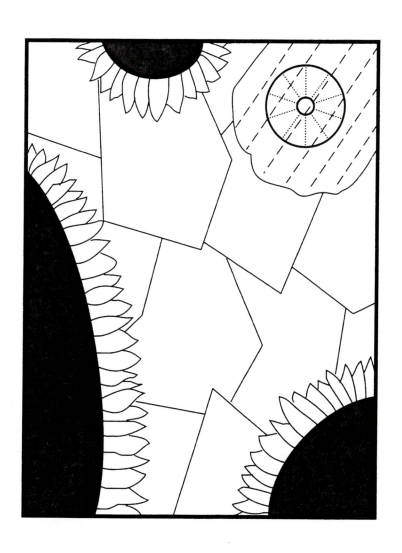

Chapter 1
Origin, Classification, and Occurrence of Sedimentary Rocks

1.1 Introduction

Sedimentary rocks form at low temperatures and pressures at the surface of Earth owing to deposition by water, wind, or ice. By contrast, igneous and metamorphic rocks form mainly below Earth's surface, where temperatures and pressures may be orders of magnitude higher than those at the surface, although volcanic rocks eventually cool at the surface. These fundamental differences in the origin of rocks lead to differences in physical and chemical characteristics that distinguish one kind of rock from another. Sedimentary rocks are characterized particularly by the presence of layers, although layers are also present in some volcanic and metamorphic rocks, and by distinctive textures and structures. Many sedimentary rocks are also distinguished from igneous and metamorphic rocks by their mineral and chemical compositions and fossil content.

Sedimentary rocks cover roughly three-fourths of Earth's surface. They have special genetic significance because their textures, structures, composition, and fossil content reveal the nature of past surface environments and life forms on Earth. Thus, they provide our only available clues to evolution of Earth's landscapes and life forms through time. These characteristics of sedimentary rocks are in themselves reason enough to study sedimentary rocks. In addition, many sedimentary rocks contain minerals and fossil fuels that have economic significance. Petroleum, natural gas, coal, salt, phosphorus, sulfur, iron and other metallic ores, and uranium are examples of some of the extremely important economic products that occur in sedimentary rocks.

Many different terms are used to describe the study of sedimentary rocks, including stratigraphy, sedimentation, sedimentology, and paleontology. This book deals with sedimentary petrology, which is that particular branch of study concerned especially with the composition, characteristics, and origins of sediments and sedimentary rocks. The book focuses particularly on the physical, chemical, and biological characteristics of the principal kinds of sedimentary rocks; however, it is concerned also with the relationship of these properties to depositional conditions and provenance. I have attempted, where appropriate, to identify major problems and concerns regarding the origin of particular kinds of sedimentary rocks or particular properties of these rocks. Where controversy surrounds the origin, as for the origin of dolomites and iron-formations, different points of view are examined.

In this opening chapter of the book, I give a brief, generalized discussion of the origin, classification, occurrence, and study of sedimentary rocks. I treat in somewhat greater detail the tectonic setting of sediment accumulation. The composition of siliciclastic sedimentary rocks, in particular, is strongly influenced by tectonic provenance and the kinds of depositional basins and depositional conditions present in the tectonic setting. Therefore, it seems appropriate in this opening chapter to examine tectonic setting and basin architecture as a framework for discussion in succeeding chapters. Chapters 2 and 3 examine the sedimentary textures and structures that are common to many kinds of sedimentary rocks. Chapters 4 and 5 describe the characteristics of sandstones, Chapter 6 discusses conglomerates, and Chapter 7 describes the characteristic features of shales or mudrocks. The extremely important topic of sediment provenance is discussed in Chapter 8, followed in Chapter 9 by discussion of diagenesis of siliciclastic sedimentary rocks. Chapters 10–13 deal with the chemical/biochemical and carbonaceous sedimentary rocks. Chapter 10 describes limestones, Chapter 11 discusses dolomites, and Chapter 12 examines the diagenesis of these carbonate rocks. Chapter 13 describes the characteristics of evaporites, cherts, phosphorites, and iron-rich sedimentary rocks and discusses some of the controversial aspects of their origin. The final chapter of the book, Chapter 14, discusses the organic-rich, carbonaceous sedimentary rocks such as oil shales and coals.

1.2   Origin and Classification of Sedimentary Rocks

As mentioned, all sedimentary rocks originate in some manner by deposition of sediment through the agencies of water, wind, or ice. They are products of a complex sequential series of geologic processes that begin with **formation of source rocks** through intrusion, metamorphism, volcanism, and tectonic uplift. Physical, chemical, and biologic processes subsequently play important roles in determining the final sedimentary product. **Weathering** causes the physical and chemical breakdown of source rocks, leading to concentration of resistant particulate residues (mainly silicate mineral and rock fragments) and formation of secondary minerals such as clay minerals and iron oxides. At the same time, soluble constituents such as calcium, potassium, sodium, magnesium, and silica are released in solution. Soluble constituents are constantly carried from weathering sites in surface waters that discharge ultimately into the ocean. **Explosive volcanism** may also contribute substantial quantities of particulate (pyroclastic) debris, including feldspars, volcanic rock fragments, and glass.

In time, particulates are removed from the land by erosion and undergo **transportation** by water, wind, or ice to depositional basins at lower elevations. Within depositional basins, transport of particulates eventually stops when the particles are **deposited** below wave base. Soluble constituents delivered to basins by surface waters, or added to ocean water by water–rock interactions along midocean spreading ridges, may eventually accumulate in basin waters in concentrations sufficiently high to cause their removal by inorganic processes. In many cases, however, precipitation of dissolved constituents is aided in some way by biologic processes. Also, plant or animal organic residues, which wash in from land or originate within the depositional basins, may be deposited along with land-derived detritus or chemical/biochemical precipitates.

After deposition of particulate sediment or chemical/biochemical precipitates, burial takes place as this sediment is covered by successive layers of younger sediment. The increased temperatures and pressures encountered during burial bring about **diagenesis** of the sediment, leading to solution and destruction of some constituents, generation of some new minerals in the sediment, and eventually consolidation and lithification of the sediment into sedimentary rock.

This highly generalized sequence of sedimentary processes leads to generation of four fundamental kinds of constituents—terrigenous siliciclastic particles, chemical/biochemical constituents, carbonaceous constituents, and authigenic constituents—that in various proportions make up all sedimentary rock.

1. Terrigenous Siliciclastic Particles. The processes of terrestrial explosive volcanism and rock decomposition owing to weathering generate gravel- to mud-size particles that are either individual mineral grains or aggregates of minerals (rock fragments or clasts). The minerals are mainly silicates such as quartz, feldspars, and micas. The rock fragments are clasts of igneous, metamorphic, or older sedimentary rock that are also composed dominantly of silicate minerals. Further, fine-grained secondary minerals, particularly iron oxides and clay minerals, are generated at weathering sites by recombination and crystallization of chemical elements released from parent rocks during weathering. These land-derived minerals and rock fragments are subsequently transported as solids to depositional basins. Because of their largely extrabasinal origin and the fact that most of the particles are silicates, we commonly refer to them as terrigenous siliciclastic grains, although some pyroclastic particles may originate within depositional basins. These siliciclastic grains are the constituents that make up common sandstones, conglomerates, and shales.

2. Chemical/Biochemical Constituents. Within depositional basins, chemical and biochemical processes may lead to extraction from basin water of soluble constituents to form minerals such as calcite, gypsum, and apatite, as well as the calcareous and siliceous tests, or shells, of organisms. Some precipitated minerals may become aggregated into silt- or sand-size grains that are moved about by currents and waves within the depositional basin. Carbonate ooids and pellets are familiar examples of such aggregate grains. There is no commonly accepted group name, analogous to the term siliciclastic, for precipitated minerals and mineral aggregates; they are referred to here simply as chemical/biochemical constituents. These constituents are the materials that make up intrabasinal sedimentary rocks such as limestones, cherts, evaporites, and phosphorites.

3. Carbonaceous Constituents. The carbonized residues of terrestrial plants and marine plants and animals, together with the petroleum bitumens, make up a third category of

sedimentary constituents. **Humic** carbonaceous materials are the woody residues of plant tissue and are the chief components of most coals. **Sapropelic** residues are the remains of spores, pollen, phyto- and zooplankton, and macerated plant debris that accumulate in water. They are the chief constituents of cannel coals and oil shales. **Bitumens** are solid asphaltic residues that form from petroleum through loss of volatiles, oxidation, and polymerization.

4. Authigenic Constituents. Minerals precipitated from pore waters within the sedimentary pile during burial diagenesis constitute a fourth category of constituents. These

TABLE 1.1
Classification of sedimentary rocks

Composition	Group name	Particle size	Principal constituents	Main rock types
<~15% Carbonaceous residues — <50% Terrigenous siliciclastic grains	or siliciclastic rocks	>2 mm	Rock fragments	Conglomerates and breccias
		1/16–2 mm	Silicate minerals and rock fragments	Sandstones
		<1/16 mm	Silicate minerals	Shales (mudrocks)
>50% Chemical–biochemical constituents	Chemical–biochemical rocks	Variable	Carbonate minerals, grains; skeletal fragments	Carbonate rocks (limestones and dolomites)
			Evaporite minerals (sulfates, chlorides)	Evaporites (rock salt, gypsum, anhydrite)
			Chalcedony, opal, siliceous skeletal remains	Siliceous rocks (cherts and related rocks)
			Ferruginous minerals	Ironstones and iron formations
			Phosphate minerals	Phosphorites
>~15% Carbonaceous residues	Carbonaceous rocks	Variable	Siliciclastic or chemical–biochemical constituents: carbonaceous residues	Sapropelites (oil shales) / Impure coals
			Carbonaceous residues	Humic coals / Cannel coals / Solid hydrocarbons (bitumens)

secondary, or authigenic, constituents may include silicate minerals such as quartz, feldspars, clay minerals, and glauconite and nonsilicate minerals such as calcite, gypsum, barite, and hematite. They may be added during burial to any type of sedimentary rock but are never the dominant constituents of sedimentary rocks.

Depending upon the relative abundance of siliciclastic, chemical/biochemical, and carbonaceous constituents, we recognize three fundamental types of sedimentary rocks (Table 1.1): siliciclastic (terrigenous) sedimentary rocks, chemical/biochemical sedimentary rocks, and carbonaceous sedimentary rocks. As shown in Table 1.1, each of these major groups of sedimentary rocks can be further subdivided on the basis of grain size and/or mineral composition. Thus, the siliciclastic sedimentary rocks are divided by grain size into conglomerates/breccias, sandstones, and shales (mudrocks), each of which can be classified on a still finer scale on the basis of composition. The chemical/biochemical sedimentary rocks are divided by composition into carbonates, evaporites, cherts, ironstones and iron-formations, and phosphorites. Carbonaceous sedimentary rocks may be separated by composition into sapropelites (oil shales and impure coals), coals, and bitumens.

Although we recognize many types of sedimentary rocks on the basis of composition and grain size, only three of these rock types are volumetrically important. As discussed in greater detail below, shales (mudrocks), sandstones, and limestones make up the bulk of all sedimentary rocks in the rock record. The compositions, textures, and structures of sandstones and limestones make them particularly important as indicators of past depositional conditions. Therefore, I have placed major emphasis in this book on these two important groups of rocks.

1.3   Distribution of Sedimentary Rocks in Space and Time

Sedimentary rocks and sediments range in age from Precambrian to modern. The age of metamorphism of the oldest known metamorphosed sedimentary rocks has been determined by radiometric methods to be about 3.8 billion years (Windley, 1977). The ages of deposition of these sedimentary rocks must be still older, perhaps as much as 4 billion years or more. The first rocks that formed on Earth were probably basic volcanic rocks. Sedimentary rocks began to form once Earth's atmosphere and oceans had developed owing to degassing of Earth's interior.

The area of Earth's surface covered by sedimentary rocks has increased progressively with time as the area of volcanic rocks has been successively reduced by erosion (Fig. 1.1). Sedimentary rocks now cover about 80 percent of the total land area of Earth (Ronov, 1983). They also cover most of the floor of the ocean, above a basement of volcanic rocks. According to Ronov, sedimentary rocks make up about 11 percent of the volume (9.5 percent of mass) of Earth's crust and 0.1 percent of the volume (0.05 percent of mass) of the total Earth. Average thickness of Earth's sedimentary shell is 2.2 km, but thickness varies widely in different parts of the continents and ocean basins.

Most of the volume of sedimentary rocks of Earth's crust (about 70 percent) is concentrated on the continents, which make up about 29 percent of Earth's surface (Ronov, 1983). About 13 percent of sedimentary rocks occur on the continental shelf and continental slope, which together make up about 14 percent of Earth's surface. Approx-

FIGURE 1.1
Percent of continents covered by
most important groups of rocks as
a function of age. (*After Ronov, A.
B., 1983, Am. Geol. Inst. Reprint
Ser.: V. Fig. 17, p. 31, reprinted
by permission.*)

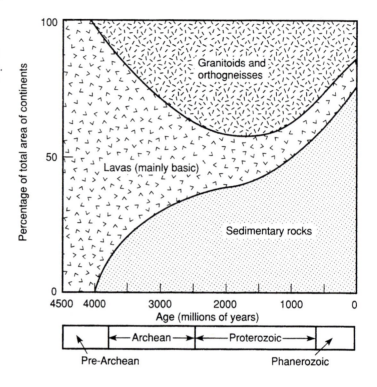

imately 17 percent of the total volume of sedimentary rocks occurs on the floors of the oceans, which constitute about 58 percent of Earth's surface.

As mentioned, the rocks that make up Earth's sedimentary shell are mainly shales, sandstones, and carbonate rocks. Past estimates, by different workers, of the relative proportion of these rock types in the total sedimentary pile have varied significantly. Recent estimates by Ronov (1983), on the basis of data obtained by direct measurement of the distribution of the most important rock types, suggest that shales make up about 50 percent of the sedimentary rocks on the continents, sandstones 24 percent, carbonate rocks 24 percent, evaporites about 1 percent, and siliceous rocks (cherts) about 1 percent. In this tabulation, Ronov has apparently lumped iron-rich sedimentary rocks with carbonate rocks, possibly under the assumption that the iron-rich rocks formed by alteration of siderites (iron carbonates). Phosphorites and carbonaceous sedimentary rocks are omitted from the tabulation because their overall volume is quite small compared to that of the other sedimentary rocks. Conglomerates are not mentioned; they are probably included with sandstones.

The distribution of sedimentary rock types by age is shown in Figure 1.2. Note that the relative volume of preserved shale per unit age has not changed significantly since early/middle Precambrian time. Also, the volume of sandstone of various ages is fairly constant, although the proportion of different sandstone types (graywackes, arkoses, quartzitic sands) has changed somewhat through time. The most notable changes in volume of preserved sediment per unit age are the marked decrease in iron-rich sedimentary rocks (jaspilites) after late Precambrian time and the significant increase in carbonate rocks and evaporites after the Precambrian.

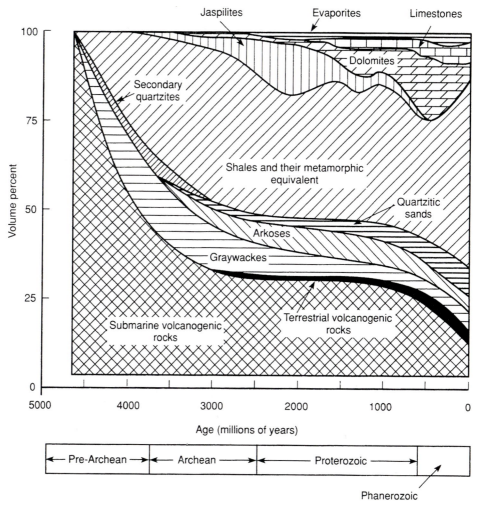

FIGURE 1.2
Volume percent of sedimentary rocks as a function of age. (*After Ronov, A. B., 1983, Am. Geol. Inst. Reprint Ser.: V. Fig. 19, p. 33, reprinted by permission.*)

1.4   *Recycling of Sedimentary Rocks*

Figure 1.3 depicts the volume of sedimentary rocks graphed as a function of age of the rocks. This graph shows a very strong trend of increasing mass of sedimentary rock per unit time from the Precambrian into the Cenozoic. This trend reflects both rates of sedimentation and rates of erosion. Keep in mind that the volume of older sedimentary rocks has been progressively reduced through time by erosion. Thus, the volume of sediments shown for a given age in Figure 1.3 does not represent the total volume of sediment deposited during that period of time. Rather, it is the preserved remnant of that original volume.

FIGURE 1.3
Relative volume of sedimentary rock on the continents per unit of age. Crosshatched areas denote metamorphosed equivalents of the sedimentary rocks of late Proterozoic age. (*After Ronov, A. B., 1983, Am. Geol. Inst. Reprint Ser.: V. Fig. 8, p. 14, reprinted by permission.*)

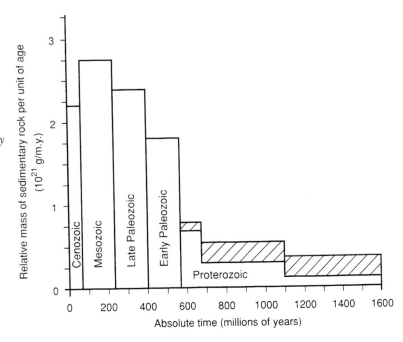

The particles that made up the first sedimentary rocks that formed on Earth were derived by erosion of basic volcanic rocks. Through time, the area of Earth's surface covered by sedimentary rocks increased as the area covered by basic volcanic rocks decreased (Fig. 1.1). Some of these early-formed sedimentary rocks were ultimately uplifted after burial and lithification to become the source rocks for a new generation of sedimentary rocks. These sedimentary rocks, in turn, were subsequently uplifted and exposed to become the source rocks for a still younger generation of sedimentary rocks, and so on. The constituents that make up younger sedimentary rocks have thus been recycled through the processes of uplift, weathering, and erosion. The number of times that sedimentary rock of a particular type has been recycled is a function of both the tectonic setting of the rocks within a continental mass and the relative susceptibility of the rocks to destruction by weathering and erosion. Tectonic setting and climate govern the intensity of the weathering/erosion process; rock type determines the relative ease of destruction. In general, evaporites are the most soluble and most easily destroyed sedimentary rocks. Limestones are next, dolomites are third, and shales, sandstones, and volcanogenic sediments are fourth (Garrels and Mackenzie, 1971). Owing to the greater susceptibility of evaporite rocks to destruction, Garrels and Mackenzie suggest that such rocks may have been recycled up to 15 times in the last three billion years. Carbonate rocks have been recycled about 10 times and shales and sandstones five times.

Garrels and Mackenzie suggest two possible models to account for recycling of sedimentary rocks through time. The **constant-mass model** assumes an early degassing of Earth. All the water of the hydrosphere and atmosphere was presumably released at this time, along with all the CO_2, HCl, and other acid gases that could react with primary igneous rock to form sedimentary rock. The total volume of sedimentary rock was thus created very early in Earth's history. Since that time, no completely new sediment has been created because no new acid gases have been released to create them. Through time, these early-formed sediments have been recycled owing to erosion and destruction by metamorphism, with concomitant recycling of CO_2 and HCl. The **linear-accumulation model** assumes that water, CO_2, and HCl are being continuously degassed from Earth's interior at a linear rate. New sedimentary rocks have thus continued to form through time by breakdown of primary igneous rock. Therefore, the mass of sediments has grown linearly through time from zero to the currently existing mass. This model represents the extreme opposite conditions to those assumed in the constant-mass model. It is possible, of course, that the real recycling process may have combined elements of the two models. That is, an early high rate of degassing may have been followed by a continuously decreasing, possibly irregular, rate of degassing.

In any case, either model could account for the volume of preserved sediment that now exists. What is important in studying the petrology of sedimentary rocks is to keep in mind that the recycling process has brought about some important changes in sedimentary rocks through time. For example, the mineralogy of siliciclastic sedimentary rocks must have been affected through time as chemically and mechanically less stable minerals and rock fragments were selectively destroyed as they moved through several cycles of uplift, weathering, erosion, transportation, deposition, and diagenesis—moving sediments toward a state of greater compositional maturity. Textural properties such as shape, roundness, and grain size must also have been affected by multiple cycling, resulting, for example, in enhanced rounding of detrital grains. Recycling of sediments has also produced changes through time in the bulk chemical composition of sedimentary rocks, particularly in the amounts of major elements such as Fe, Mn, Ca, Mg, K, Na, and Si. The patterns of chemical change as a function of time are complex and not easily generalized; readers are referred to Garrels and Mackenzie (1971) and Ronov (1983) for details of some of these changes.

1.5   Tectonic Setting of Sediment Accumulation

1.5.1   Introduction

The physical, chemical, and biological properties of sedimentary rocks are strongly influenced by the nature of sediment source areas (provenance) and the conditions of the depositional environment. The characteristics of source areas and depositional environments, in turn, are the result of the tectonic and geologic history of the region in which the sediments accumulate. For example, source-rock types are intimately related to the regional tectonic setting; e.g., volcanic source rocks originate mainly within magmatic-arc settings, plutonic igneous rocks are more characteristic of continental-block provenances, and metamorphic and sedimentary source rocks typically occur in orogenic belts characterized by collision tectonics. Furthermore, the topographic expression and relief of source areas are controlled by uplift and deformation. Similarly, such aspects of the depositional environment as basin size and geometry, water depth, proximity to source areas, and rate of basin subsidence are influenced by the position of the depositional environment within the regional tectonic framework.

Tectonism, through its influence on provenance and depositional environments, thus exerts an important, indirect control on sedimentation patterns and sedimentary rock characteristics. We will examine more closely the nature of this relationship in appropriate sections of the book.

1.5.2   Geosynclines and Tectonics

From about the 1860s to the 1960s, geological thought regarding the relationship of tectonics and sedimentation centered around the geosynclinal theory. This theory was first presented by an American geologist, James Hall, in 1857 and was published by Hall in 1859. Hall and subsequent early American geologists such as Dana (1873) conceived of geosynclines as relatively narrow, elongate sediment-filled troughs, located along the margins of continents. Shallow-marine deposits putatively accumulated in these troughs to great thickness as a result of continued subsidence of the geosyncline. Ideas about the cause of deep subsidence differed. For example, Hall attributed subsidence to sediment loading, whereas Dana believed that subsidence was caused by downbuckling of Earth's crust owing to lateral compression generated by shrinking of Earth.

European geologists accepted the general principles embodied in the geosynclinal theory; however, early European writers such as Haug (1900) and Suess (1909) regarded the sediment in geosynclines to be principally deep-marine deposits and the geosynclines to lie in intercontinental positions rather than immediately along the margins of continents. Deep-sea trenches such as those found in the western Pacific region were considered by these workers to be examples of modern geosynclines. Thus, a fundamental difference of opinion between American and European geologists about the nature and origin of geosynclines developed very early. See Dott (1974, 1978, 1979) for discussion of the evolution of the geosynclinal concept and the specific terminology used for geosynclines.

1.5.3   Plate Tectonics and Seafloor Spreading

Although the geosynclinal concept guided geological thinking for nearly one hundred years, by the 1960s an increasing number of geologists were raising serious doubts about some aspects of the geosynclinal theory (Coney, 1970). While some geologists were still arguing about geosynclines, a quiet geological revolution was under way that was soon to have a profound effect on every aspect of geological thought. Spearheaded by geologists such as Harry Hess, Robert Dietz, and J. Tuzo Wilson, the concept of seafloor spreading and plate tectonics emerged in the late 1950s and early 1960s. Although the plate tectonics theory is now familiar to all geologists, the concept of spreading ridges, moving crustal plates, and subduction zones forced some dramatic new ideas about tectonics and sedimentation upon a reluctant generation of geologists weaned on geosynclinal concepts. Eventually, it became necessary for these geologists to either abandon the geosynclinal concept in favor of plate tectonics or attempt to meld and reconcile geosynclinal dogma with the new concept of global tectonics. As Dott (1974) mentions, some geologists argued that the geosynclinal theory is dead or dying and that the term itself should be abandoned. Others have sought to rationalize essentially every detail of the older geosynclinal concept with the new plate tectonics theory. It is not my intention here to get caught up in the arguments about whether or not we should retain the geosynclinal concept and geosyncline terminology. Readers who wish to pursue this line of inquiry further, or who wish simply to develop a fuller understanding of the geosynclinal concept, may consult Aubouin (1965), Coney (1970), Dott (1974, 1978, 1979), Dott and Shaver (1974), Dickinson (1971, 1974a, 1974b), Mitchell and Reading (1986), and Reading (1982).

What is important at this time is for sedimentologists to develop models that can relate global tectonic settings to patterns of sedimentation. Sedimentary basins are now commonly classified in terms of (1) the type of crust on which the basins rest, (2) the positions of the basins with respect to plate margins, and (3) for basins lying close to a plate margin, the type of plate interactions occurring during sedimentation (Dickinson, 1974a; Miall, 1990, p. 501). Several classifications of sedimentary basins that take into account these criteria have been proposed; for example, Dickinson (1974a), Dickinson and Suzcek (1979), Bally and Snelson (1980), and Mitchell and Reading (1986). See also Bally (1984), Klein (1987), and Ingersoll (1988).

Mitchell and Reading (1986) suggest that we can group tectonic settings for sediment accumulation into six basic types:

1. **Interior basins, intracontinental rifts,** and **aulacogens** (including thermally related rifts and collision-related rifts)
2. **Passive** or **rifted continental margins** (Atlantic-type margins comprising a shelf, slope, and rise)
3. **Oceanic basins** and **rises** (the deep-ocean floor and smaller basins associated with midocean spreading ridges and rises, but excluding transform-fault–related settings)
4. **Subduction-related settings** (convergent plate boundaries characterized by a trench, arc-trench gap, and volcanic arc)
5. **Strike-slip/transform-fault–related settings**
6. **Collision-related settings** (resulting from closure of an oceanic or marginal basin)

These tectonic settings are illustrated in Figure 1.4. Figure 1.4A shows source regions and depositional basins associated with continental blocks, including intracratonic rift basins and rifted margin settings. Oceanic spreading ridges and transform-fault systems are also depicted in this figure. Figure 1.4B illustrates collision-related settings, and Figure 1.4C shows the complex system of source areas and depositional basins typical of magmatic-arc (subduction-related) settings. These tectonic settings are discussed more fully below.

1.5.4   Interior Basins, Intracontinental Rifts, and Aulacogens

Ovate Downwarps

Some interior basins are relatively large circular to ovate downwarps within the interiors of more or less stable cratonic massifs or shields. North American examples of such basins include the large Paleozoic Michigan, Illinois, and Williston basins as well as the smaller Mesozoic–Tertiary basins of the central and southern Rocky Mountains (Fig. 1.5). Examples from other continents include the Mesozoic–Tertiary Chad Basin and similar basins of West Africa, the Eyre Basin of Australia, and numerous basins on the Russian platform. Some interior basins of this type are filled with siliciclastic or carbonate marine sediment deposited from epicontinental seas; others contain nonmarine sediment. Many of these basins contain important oil- and gas-producing formations. The origin of interior basins of this type is poorly understood, and it is difficult to relate their development to plate tectonic events. They have been variously suggested to originate as a result of (1) subsidence owing to the presence of ancient underlying rift systems, (2) cooling and subsidence following a thermal event, as intrusion of dense material in the mantle, (3) mantle phase changes, (4) mantle hotspots, and (5) shallow subduction.

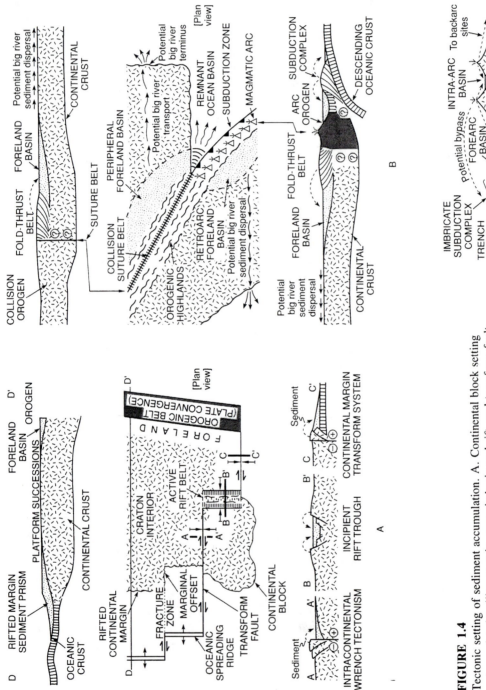

FIGURE 1.4

Tectonic setting of sediment accumulation. A. Continental block setting and associated trailing margin, oceanic basin and rift, and transform-fault settings. B. Collision-related settings, including uplifted subduction complex. C. Subduction-related settings in magmatic arcs. (*After Dickinson, W. R., and C. A. Suczek, 1979, Am. Assoc. Petroleum Geologists Bull., v. 63. Fig. 5, p. 2174; Fig. 6, p. 2175; Fig. 7, p. 2177; reprinted by permission of AAPG, Tulsa, Okla.*)

FIGURE 1.5
Interior basins of the central and southern Rocky Mountains. 1. Folds and fault blocks. 2. Uplifts with Precambrian basement rocks in higher parts. 3. Basins with Paleocene sediments. 4. Basins with Eocene sediments deposited over Paleocene sediments. 5. Areas of lake deposits, mainly of Eocene age. (*From King, P. B., Evolution of North America. Copyright © 1959. Fig. 66, p. 114, reprinted by permission of Princeton University Press, N.J.*)

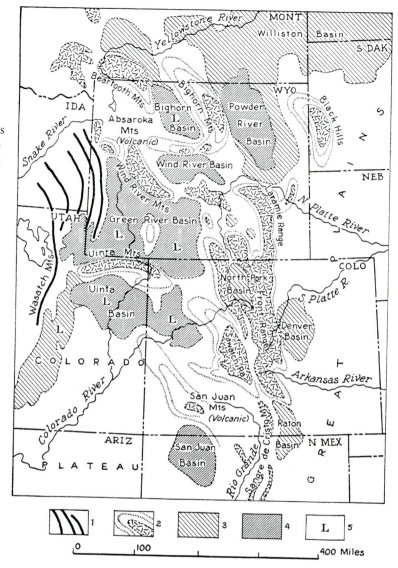

Rifts and Aulacogens

Narrow, fault-bounded rift valleys constitute a second type of interior basin. These basins range in size from small grabens a few kilometers wide to gigantic rifts such as the East African Rift System, which is nearly 3000 km long and 30–40 km wide. Most cratonic rift systems are believed to originate in response to a thermal event that causes rifting and spreading. Thermally related rifts include grabens of various types and origins, so-called failed rifts, and aulacogens. **Aulacogens** are long, narrow troughs, with thick sediment fills, that extend into continental cratons at a high angle from fold belts (Burke, 1977; Fig. 1.6). The term **failed rift** is applied to rifts striking from ocean basins into continents. Failed rifts are presumably the failed arms of triple-point junctions. One arm of a trilete spreading rift system is believed to stop spreading after a few million years, before or shortly after development of oceanic crust. The rifts making up the other two arms

FIGURE 1.6
Aulacogen north of the Black and
Caspian seas on the Russian
platform. *(After Burke, K., 1977,
Aulacogens and continental
breakup: Ann. Rev. Earth and
Planetary Sciences, v. 5. Fig. 5,
p. 393, reprinted by permission of
Blackwell Scientific Publications,
Oxford.)*

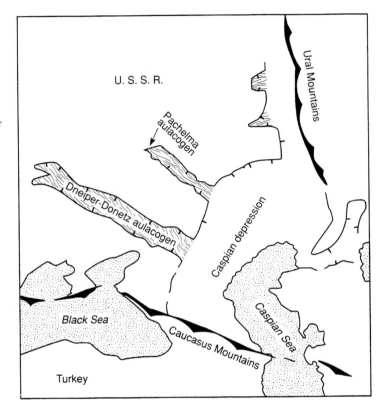

continue to spread, eventually causing separation of the continent and development of an
ocean. Putatively, aulacogens develop from failed rifts owing to subsequent deformation,
probably related to ocean closing. Not all failed-rift systems develop into aulacogens.

Many smaller rift basins that develop within continents are not directly related to
failed rift systems. The numerous grabens in the Basin and Range Province of the western
United States provide a good example of such basins. Crustal extension has been proposed
as a mechanism for developing these graben-type basins.

Aulacogens and other thermally related rifts are believed to go through a series of
stages in their development, beginning with an initial doming or prerift stage caused by
a thermal event. This stage is followed in turn by (1) an initial rifting stage with accompanying early sedimentation, (2) a stage of slow to rapid subsidence during which a thick
sediment fill can accumulate, and (3) a succeeding stage during which downwarping
degenerates or ceases. In the case of aulacogens, an additional stage occurs that involves
deformation (possibly owing to continental collision) and generation of a fold belt.
Hoffman et al. (1974) propose several possible models for the evolution of aulacogens and
related orogenic belts, as shown in Figure 1.7.

Collision-related Rifts

Burke (1978) suggests that some rifts may have been caused by collision resulting from
continental impact. The Rhine graben, extending northward across the Jura Alps, and the

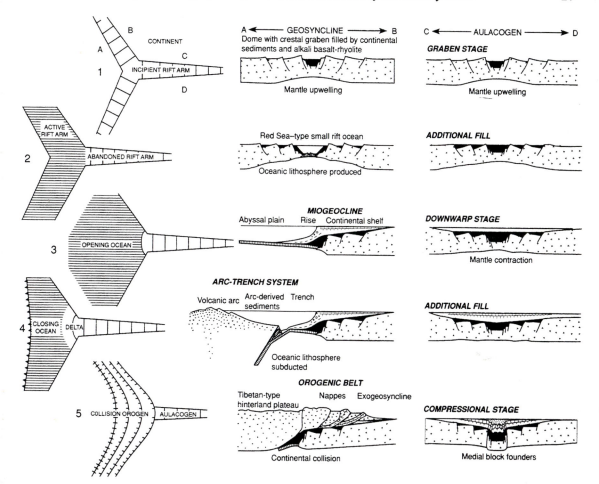

FIGURE 1.7

Models for evolution of aulacogens and related orogenic belts. Model 1. A three-armed, radial rift system accompanied by crustal doming; a strike-slip component (transform fault) may be important along any of the rift arms. Model 2. Two rift arms spread to produce a narrow ocean similar to the Red Sea; rifting of third arm is insufficient for continental separation. Model 3. Spreading of two active rift arms produces a large ocean basin. Downwarping and terrace-rise sedimentation occur along new aseismic continental margins. Third arm fails and remains as a transverse trough located at reentrant on continental margin. The failed rift arm evolves from incipient rift to broad downwarp (intracratonic basin). Model 4. Ocean is closed by subduction along a trench, producing an adjacent magmatic arc. The process of ocean closure can take many paths, one of which is shown here. Model 5. Closing of ocean ultimately results in continental collision and development of collision orogen. Abandoned rift arm is preserved as an aulacogen located where orogen makes reentrant into its foreland. Aulacogen is further loaded with peripheral-foreland sediments derived from advancing orogen and its medial block founders, resulting in final stage of compressional deformation and faulting. (*After Hoffman, P., et al., 1974, Aulacogens and their relation to geosynclines: Soc. Econ. Paleontologists and Mineralogists Spec. Pub. 19. Fig. 10, p. 52, reprinted by permission of SEPM, Tulsa, Okla.*)

(Lake) Baikal rift, in southern Siberia, are suggested to have originated in this way owing to collision in the Alpine and Himalayan mountains. These rifts presumably formed as a result of tensional forces that developed following ocean closing and the resultant orogeny. Sediments of collision-related rifts are distinguished from those of aulacogens because they date back only to the time of collision, not to the time of continental rupture.

Sedimentary Rocks in Rift Basins

A variety of depositional environments may characterize rift basins, depending upon the size of the basins, their position within a continent, and their stage of development. Thus, the sediments of some rift basins that developed well within continental margins may include nonmarine sandstones, shales, conglomerates, and evaporites of fluvial, swamp, or lacustrine origin, as well as associated volcanic rocks. Rift systems connected to the open ocean may accumulate thick sequences of sediments in marginal-marine and marine settings. These sedimentary sequences may include carbonates, evaporites, deltaic sands and muds, shelf sands and muds, and deeper-water turbidites and pelagic deposits. The thick sequence of sediments in rift basins commonly includes both favorable source rocks and reservoir rocks for petroleum, as well as good stratigraphic and structural traps for petroleum. The sedimentary beds of rift basins may also serve as hosts for stratiform ore deposits of various types.

1.5.5   Passive Continental Margin Settings

Passive continental margins are defined as those margins that have little or no seismic or volcanic activity. They are sometimes also referred to as Atlantic-type margins, rifted

FIGURE 1.8
Present-day passive and active continental margins. Also shown are submarine basins and continental areas that include Mesozoic–Cenozoic mountain ranges and intramontane sedimentary basins (gray zones). (*From Bally, W. W., and S. Snelson, 1980, Realms of subsidence, in Miall, A. D., ed., Facts and principles of world petroleum occurrence: Canadian Soc. Petroleum Geology Mem. 6. Fig. 3, p. 13, reprinted by permission.*)

CONTINENTAL MARGINS

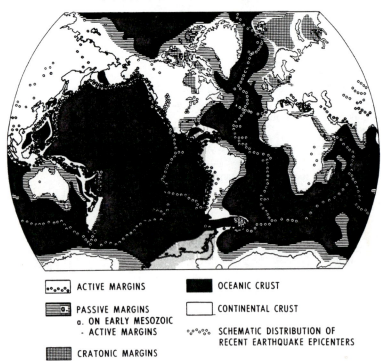

- ╍╍╍ **ACTIVE MARGINS**
- ▪░▪ **PASSIVE MARGINS**
  a. ON EARLY MESOZOIC - ACTIVE MARGINS
- ▓▓▓ **CRATONIC MARGINS**
- ■ **OCEANIC CRUST**
- □ **CONTINENTAL CRUST**
- °°°°° **SCHEMATIC DISTRIBUTION OF RECENT EARTHQUAKE EPICENTERS**

margins, and divergent margins. They form parts of lithospheric plates that are moving away from ocean spreading ridges, and they lack the distinctive island arcs, trenches, and associated basins that characterize margins undergoing active subduction. With the exception of those margins surrounding the Pacific, most of the world's present continental margins are passive margins (Fig. 1.8).

Sediments accumulate in a variety of subsettings on passive continental margins, which include the continental shelf and slope and the continental rise at the foot of the slope. Many passive continental margins, such as the Atlantic margin of North America, are marked by the presence of offshore basins separated by platforms (Fig. 1.9). Sediment thickness in these basins may exceed 15 km, whereas thickness of sediments on the

FIGURE 1.9
Sedimentary basins on the passive continental margin of eastern North America. F.Z. = fracture zone. (*After Sheridan, R. E., 1974, Atlantic continental margin of North America, in Burk, C. A., and C. L. Drake, eds., The geology of continental margins. Fig. 12, p. 403, reprinted by permission of Springer-Verlag, New York.*)

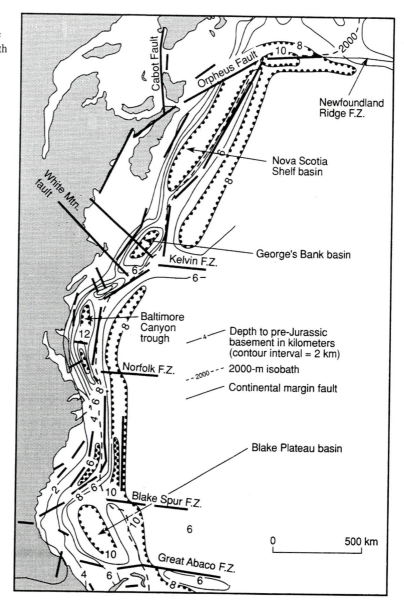

platforms is significantly less. Passive continental margins are created owing to rifting caused by spreading from midocean ridges. The origin of the basin and platform systems on these margins is also related to rifting processes. Thick prisms of sediment may accumulate on passive continental margins as a result of continued subsidence of the margin. The causes of subsidence are not well understood, although several possible mechanisms have been proposed. These suggested mechanisms include (a) **deep crustal metamorphism,** which causes subsidence owing to increase in density of lower crustal rocks, (b) **crustal stretching** and **thinning,** in which subsidence is the result of extension of the crust, increase in heat flow and crustal density, and (c) **sediment loading** (Mitchell and Reading, 1986).

Siliciclastic sediment deposited on most passive margins is derived exclusively from the continent and carries a continental-block provenance signature (Dickinson and

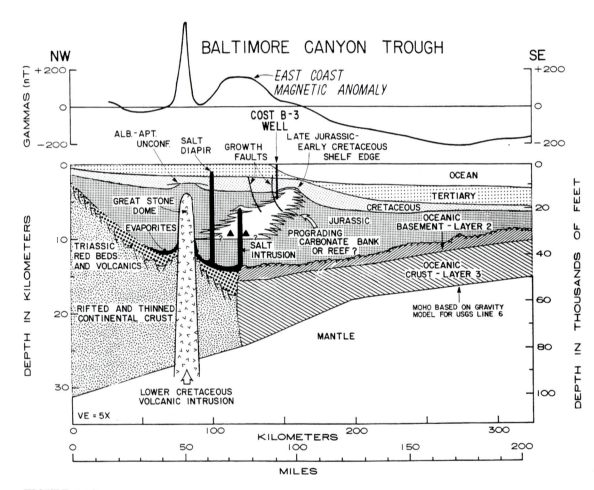

FIGURE 1.10

Interpretative section across the Atlantic continental margin of North America in the vicinity of Baltimore Canyon trough. Based on geophysical data and drill-hole information. (*After Grow, J. A., 1981, Structure of the Atlantic margin of the United States, in Geology of passive continental margins: Am. Assoc. Petroleum Geologists Educ. Course Note Ser. No. 19. Fig. 13, p. 3–20, reprinted by permission of AAPG, Tulsa, Okla.*)

Suzcek, 1979). An exception to this general provenance pattern in the modern ocean is the northwest European continental margin, where major offshore islands also supply sediment. The processes of sediment transport on continental shelves is still poorly understood and controversial (Boggs, 1987) but may include transport by tidal currents, storm-generated waves and currents, turbidity currents, and major ocean (surface) currents. Movement of sediments into deeper water is accomplished largely through turbidity currents and nepheloid-layer transport.

Sediment that accumulates on passive margins thus includes siliciclastic sands, gravels, and muds deposited in marginal-marine environments (deltas, tidal flats, beaches, estuaries) and on continental shelves; carbonate sands, muds, and reefs generated on carbonate platforms; and hemipelagic muds and turbidites deposited in deeper water in continental-slope and -rise environments. Some continental-rise turbidites may subsequently be reworked by deep bottom currents (contour currents) to produce contourites, which are typically better sorted, better laminated, and less regularly graded than turbidites. Many ancient passive-margin deposits also include evaporites, which appear to have formed during an early stage of rifting and continental separation as ocean water entered subaerial rift basins (Burke, 1975). An example of the types of sediment prisms formed on passive continental margins is shown in Figure 1.10.

1.5.6   Ocean Basins and Rises

The deep-ocean floor encompasses a variety of environments for deposition of dominantly fine-grained sediment. Sediment accumulates on broad, nearly flat abyssal plains, on the tops of seamounts, rises, and ridges, and in local basins associated with rises and ridges (Fig. 1.11). Major ocean basins are created by continental rifting. Rifting initially produces narrow, elongate depressions, which may fill with evaporites, nonmarine siliciclastic deposits, and associated volcanic rocks. Continued rifting results eventually in formation of a fully developed midocean ridge/deep-ocean floor basin system as the ocean basin opens to its fullest extent. Local, (normal) fault-bounded basins may form more or less parallel to ridge crests owing to tensional forces operating during the spreading process.

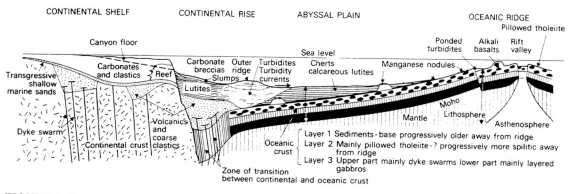

FIGURE 1.11

Generalized cross section across the western Atlantic showing the depositional setting and deposits typical of the deep-ocean basin/ridge environment. (*After Dewey, J. F., and J. M. Bird, 1970, Mountain belts and the new global tectonics: Jour. Geophys. Research, v. 75. Fig. 6, p. 2632, reprinted by permission.*)

Sediments that accumulate in deep-ocean environments are mainly pelagic clays, biogenic oozes, turbidites, and minor authigenic deposits such as manganese nodules and crusts. Pelagic clays, probably consisting mainly of windblown dust, are nearly ubiquitous in deep-ocean environments, where sedimentation rates are low. Siliceous biogenic oozes occur in those parts of the deep ocean where surface productivity of siliceous organisms is high. Diagenesis and lithification of these deposits produces deep-water cherts. Calcareous biogenic oozes deposited above the calcium carbonate compensation depth (CCD) may survive to produce deep-water pelagic limestones (chalks). Deep-ocean basins adjoining passive continental margins may receive fine turbidite sediments that pass across continental rises onto abyssal plains. Siliciclastic turbidites have a continental-provenance signature; carbonate turbidites may form adjacent to carbonate platforms. Turbidites may also accumulate locally in small basins located along ridges or rises (ponded turbidites) or on the abyssal plain at the foot of rises. These turbidites consist of resedimented pelagic biogenic oozes or clay derived from ridges or rises within the deep-ocean basin. Manganese nodules and crusts accumulate in deep-ocean environments, where rates of sedimentation are so low that the nodules are not buried by pelagic clays or oozes.

Sediments deposited in the deep-ocean environment may eventually be subducted and consumed during an episode of ocean closing. Alternatively, they may be offscraped in trenches during subduction to become part of a subduction complex. In general, sedimentary rocks at the base of the sediment pile in the deep ocean become progressively older away from midocean ridges (Fig. 1.11).

1.5.7   Subduction-related Settings

Subduction-related settings are a prominent feature of the seismically active margins of the modern Pacific Ocean (Fig. 1.8). Figure 1.4C is a highly generalized sketch illustrating the fundamental tectonic character of subduction-related arc settings characterized by a deep-sea trench, an active volcanic arc, and an arc-trench gap separating the two. Figure 1.12 shows additional details of the forearc regions of subduction-related settings. Subduction may take place in an oceanic arc system, where one ocean plate is thrust beneath another, or in a continental-margin arc, where an oceanic plate descends beneath a continental margin and the volcanic arc is located landward of the continental margin. Note the difference between the active-margin setting illustrated in Figure 1.4C and the passive-margin (rifted-margin) setting shown in Figure 1.4A.

The most important depositional sites in subduction-related settings normally include a deep-sea trench, forearc basins that lie within the arc-trench gap, and backarc, or marginal, basins that lie behind the volcanic arc in some arc-trench systems, or so-called retro-arc basins, that lie on continental crust behind fold-thrust belts in some continental-margin arcs. Thinner deposits also accumulate on platforms separating forearc or backarc basins and on the tops of associated ridges and rises. Siliciclastic sediment deposited in these settings may be derived mainly from volcanic sources in undissected volcanic arcs or from both volcanic and continental sources in dissected arcs (Dickinson and Suczek, 1979). Deposits that accumulate within the trench may include both terrigenous sediment derived from the arc and oceanic sediment offscraped from a subducting plate.

The Japan arc-trench system provides a well-studied example of a modern subduction-related system and illustrates many of the features typical of such systems.

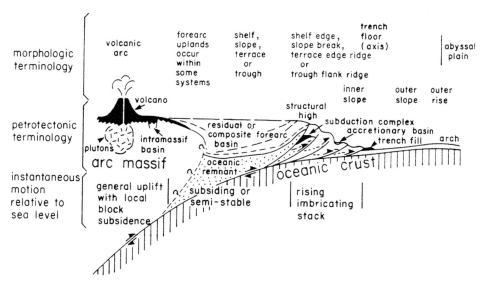

FIGURE 1.12

Model of a forearc region. (*After Dickinson, W. R., and D. R. Seeley, 1979, Structure and stratigraphy of forearc regions: Am. Assoc. Petroleum Geologists Bull., v. 63, Fig. 2, p. 5, as modified slightly by Miall, A. D., 1990, Principles of sedimentary basin analysis, 2nd ed. Fig., p. 524, reprinted by permission of Springer-Verlag, New York.*)

Figure 1.13 shows the geographic position of Japan within the Western Pacific arc-trench system. Japan lies near the triple-point junction of three major plates: the Eurasian plate, the Pacific plate, and the Philippine Sea plate. A structural profile across the forearc and backarc in the northern Honshu region (Fig. 1.14) illustrates the variety of depositional settings associated with the arc. Quaternary sediments deposited around Japan include sands, gravels, and muds on the continental shelf; hemipelagic muds in trenches, on the forearc and backarc continental slope, and on the tops of local ridges and rises; turbidites in the trenches, forearc basins, and backarc basins; biogenic oozes and clays in parts of both the forearc and backarc; and disseminated volcanic ash and pumice, plus distinctive ash layers, in both the forearc and backarc. Much the same Quaternary-age facies occur in both the Japan backarc and forearc regions (Boggs, 1984), although sandy turbidites are somewhat more common in the backarc and biogenic oozes and clays are more abundant in the forearc. Seismic data show the presence of a subduction complex, or accretionary prism, within both the Japan Trench and the Nankai Trough. Presumably, this subduction complex is made up of oceanic sediment scraped off the Pacific and Philippine Sea plates and accreted to the inner trench walls. The composition of this complex has not been determined by deep drilling.

1.5.8   Strike-Slip/Transform-Fault Settings

Strike-slip faults are particularly prominent features of ocean spreading ridges, where faulting produces major offsets in ridge systems. They also form the transform boundaries between some major crustal plates, and they occur on continental margins and within

FIGURE 1.13

The location of Japan with respect to major plate boundaries and trenches of the western Pacific. *(From Boggs, S., 1984, Quaternary sedimentation in the Japan arc-trench system: Geol. Soc. America Bull., v. 95, Fig. 1, p. 669.)*

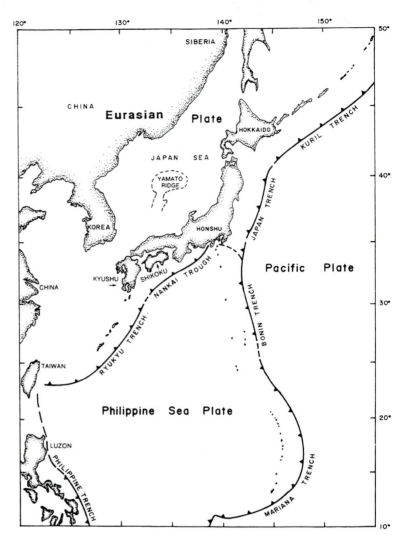

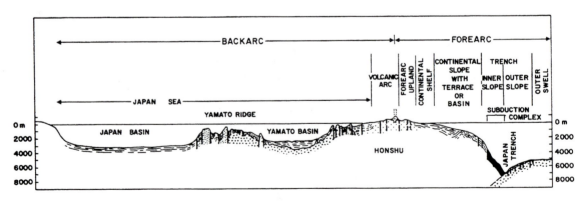

FIGURE 1.14

Structural profile across the Japan forearc and backarc in the vicinity of northern Honshu. *(From Boggs, S., 1984, Quaternary sedimentation in the Japan arc-trench system: Geol. Soc. America Bull., v. 95, Fig. 2, p. 670.)*

continents on continental crust. Some of these settings for transform faults are shown schematically in Figure 1.4A. Movement along strike-slip faults can, depending upon the shapes and orientations of the faults (Fig. 1.15), produce a variety of pull-apart basins. Most basins formed by strike-slip faulting are small (a few tens of kilometers across) and commonly show evidence of significant local syndepositional relief, such as the presence of fault-flank conglomerate wedges (Miall, 1990, p. 552).

Modern examples of strike-slip basins include the Gulf of California and numerous small basins in the southern California borderland (continental-margin basins), the Dead Sea basin (on land), and the Yallahs basin off Jamaica (in the ocean within the boundary zone between the North American plate and the Caribbean plate). Because strike-slip basins occur in a wide variety of depositional settings, ranging from the deep-ocean floor to continental interiors, no distinctive facies assemblages or provenance signatures are characteristic of the deposits of strike-slip basins. Sediments in many of these basins tend to be quite thick, owing to high sedimentation rates, and not uncommonly are marked by numerous localized facies changes. Continued strike-slip movement may offset and displace the basins considerable distances from their source areas.

FIGURE 1.15
Types of strike-slip fault patterns and resulting basins that form owing to fault movements. A, B, C. Braided, or anastomosing, fault patterns. D. Fault terminations. E. En echelon, or side-stepping, faults. (*After Reading, H. G., 1980, Characteristics and recognition of strike-slip fault systems, in Ballance, P. F., and H. G. Reading, eds., Sedimentation in oblique-slip mobile zones: Internat. Assoc. Sedimentologists Spec. Pub. 4, Fig. 3, p. 12.*)

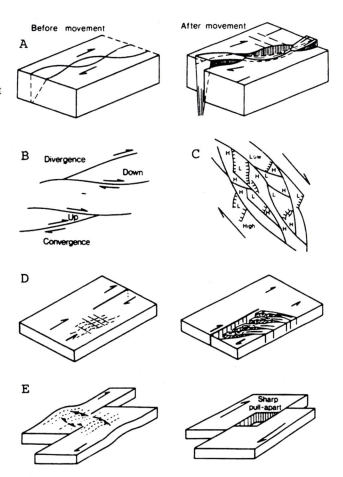

1.5.9   *Collision-related Settings*

Closing of an ocean or marginal basin ultimately leads to collision between continents or active arc systems. Collision creates both compressional and shearing forces. Shearing forces generate strike-slip movements that may be manifested considerable distances away from the collision zone. These movements can create strike-slip basins of the types described above. Compressional forces cause uplift of mountain systems along the collision suture belt, forming fold-thrust belts and associated **foreland basins** (Fig. 1.4B). These foreland basins develop between the fold-thrust belts and the craton, or continental interior. Foreland basins in the Indo-Gangetic trough south of the Himalayas are good examples of foreland basins created by continental collision. Other examples include the many foreland basins recognized in the Alpine-Mediterranean region (Fig. 1.16). Figure 1.17 shows examples of three tectono-stratigraphic styles in the U.S. Appalachian foreland basin. Foreland basins may also be created behind fold-thrust belts associated with active subduction (Fig. 1.4B), such as the basins that developed east of the Rocky Mountains in North America during Jurassic to Cretaceous time. For more information on foreland basins, see the special volume on foreland basins edited by Allen and Homewood (1986).

Owing to the irregular shapes of continents and island arcs and the fact that these land masses tend to approach each other obliquely during collision, portions of an old ocean may remain unclosed after collision occurs. These surviving embayments are called remnant basins (Fig. 1.4B). The Bay of Bengal, lying between India and the Sundra outer arc (Fig. 1.18), is a modern example of a remnant basin that escaped closure when India collided with Tibet during the Tertiary.

Some foreland basins are isolated from the ocean and receive only nonmarine sediments; others have an ocean connection. A variety of depositional environments, including alluvial fan, fluvial, deltaic, and marine environments, may thus occur in foreland basins. Deposits of such basins may include sands, gravels, and muds of either nonmarine or marine origin, turbidites, carbonates, and evaporites. Remnant basins appear to be characterized especially by turbidite sedimentation. Siliciclastic sediments may

FIGURE 1.16
Foreland basins associated with the western Mediterranean and Alpine-Carpathian fold belt system. (*From Mitchell, A. H. G., and H. G. Reading, 1986, Sedimentation and tectonics, in Reading, H. G., ed., Sedimentary environments and facies, 2nd ed. Fig. 14.65, p. 515, reprinted by permission of Blackwell Scientific Publications Limited, Oxford.*)

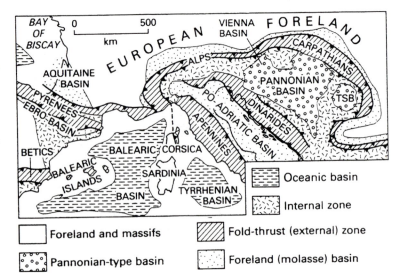

FIGURE 1.17

Three tectono-statigraphic styles (clastic wedges) in the Appalachian foreland basin. These styles developed in response to different mechanisms of foreland basin construction. (*From Tankard, A. J., 1986, On the depositional response to thrusting and lithospheric flexure: Examples from the Appalachian and Rocky Mountain basins: Internat. Assoc. Sedimentologists Spec. Pub. 8. Fig. 5, p. 375, reprinted by permission of Blackwell Scientific Publications Limited, Oxford.*)

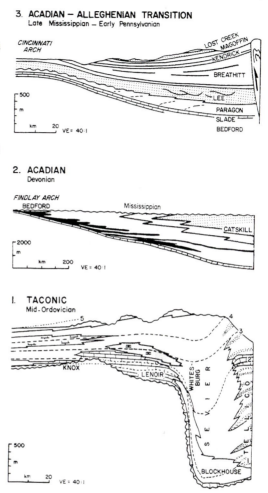

be derived both from continental interiors and from orogenic highlands located along the suture belt. Depending upon the types of plates involved in collision, a wide variety of source rocks for sediments may be present in collision-related settings. Dickinson and Suczek (1979) recognize three main types of collision-related provenances: (1) subduction complexion provenances, which may include chert, argillite, greenstone, and turbidites, (2) collision orogen provenances, typically composed of sedimentary and metasedimentary rocks, and (3) foreland uplift provenances, which may include sedimentary rocks and exposed plutonic basement rocks.

1.6   Study of Sedimentary Rocks

1.6.1   Field Study

Geologists can gain indispensable clues to geologic history through field studies of stratification styles, bedding characteristics, and sedimentary structures. It is in the field also that samples are collected for all subsequent laboratory analyses. Depending upon the objectives of the investigation, field studies of sedimentary rocks can range from simple

FIGURE 1.18
The Ganges-Indus peripheral basin
and the Bay of Bengal intrasuture
embayment. (*From Miall, A. D.,
1990, Principles of sedimentary
basin analysis, 2nd ed. Fig. 9.67,
p. 580, reprinted by permission of
Springer-Verlag, New York.*)

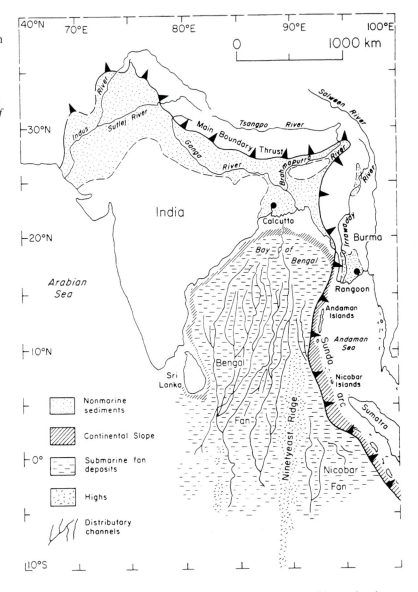

reconnaissance descriptions of principal rock types to detailed geophysical investigations.
The more common types of field studies performed by a single investigator or small
groups of investigators include mapping the distribution of formations or rock types
(geologic mapping), determining lateral and vertical changes in lithofacies or biofacies,
measuring the thickness of rock units, describing textural properties and sedimentary
structures, measuring the orientation of directional sedimentary structures such as cross-
bedding, identifying mineral components of rocks, studying modern sedimentation processes such as sediment transport and deposition, and collecting samples for later laboratory analysis. Several books are available that describe the various methods of field
study and mapping. See, for example, Ahmed and Almond (1983), Barnes (1981),
Bouma (1969), Compton (1962), Moseley (1981), and Tucker (1982).

Sampling is a critically important aspect of field work because the value of interpretations based on laboratory analyses of sedimentary rocks is uniquely dependent upon the sampling technique used in the field. Exhaustive laboratory analysis of samples is useless if samples are not representative of the unit investigated. In fact, improperly collected samples are worse than useless because they can lead to erroneous conclusions. Griffith (1971) philosophized that sampling is like religion: "Everybody is for it, but few seem to practice it." Field sampling is performed to obtain samples for laboratory analysis of many different rock properties, such as chemical and mineral composition, isotopic composition, grain size, grain shape, and fossil content. Sampling methods vary with the intended purpose of the analysis and the type of rock being sampled. For example, the technique used to collect samples from a vertical exposure of heterogeneous, layered, consolidated sedimentary rock will differ from the method used to sample the surface layer of a relatively homogeneous, unconsolidated modern beach or fluvial deposit. An investigator may be concerned primarily with differences between beds in the layered rocks but interested only in spatial variations within the surface layer of the unconsolidated deposits. A paleontologist may take spot samples from several different beds in a layered sequence if the object is to establish differences in fossil assemblages from bed to bed but may take a channel sample across all the beds if concerned only with the total assemblage of fossils in the beds. It is hardly possible to overemphasize the importance of using the proper sampling technique for a given situation. The problem of selecting the proper technique is compounded if the investigator at the time of sample collection has only a hazy notion of the intended purpose of the samples.

Detailed discussion of the theory and practice of sampling is beyond the scope of this book. The serious student or investigator will, however, make a special effort to become familiar with sampling methods before beginning a sampling project. Information about sampling methods is available in several publications, including Cheeney (1983), Griffith (1967, 1974), Krumbein and Graybill (1965), Koch and Link (1970), and Till (1974). A simple, concise summary of the sampling problem is given in Blatt et al. (1980, p. 7–9).

1.6.2 Laboratory Study

Geologists tend to think of the petrographic microscope as the primary tool for petrologic study; however, petrographic microscopy is only one of the many techniques available for laboratory study of sedimentary rocks. Some techniques, such as sieve and pipette analyses for sediment grain-size determination, are time-tested methods that have been around for several decades. Other techniques, such as electron microscopy of very small particles, X-radiography of sedimentary structures, thermoluminescence studies of carbonates, grain-shape studies involving computer-assisted Fourier analysis, and chemical analysis of sediments by techniques such as X-ray fluorescence and ICP (inductively coupled argon plasma emission spectrometry) are comparatively recent developments. It is not feasible here to attempt description of these various methods of studying sedimentary rocks. Available books that provide detailed information about some of these methods include Carver (1971a), Griffith (1967), Hutchinson (1974), Milner (1962), Mueller (1967), Tickell (1965), and Tucker (1988). Many issues of the *Journal of Sedimentary Petrology* also contain notes and full-length articles describing special methods for measuring or analyzing sedimentary rock properties.

1.6.3   *Basin Analysis*

In recent years, it has become popular to refer to detailed stratigraphic and sedimentologic analysis of depositional systems as **basin analysis**. As suggested in Figure 1.19, basin analysis may include aspects of magnetostratigraphy, seismic stratigraphy and sequence stratigraphy, and radiometric age dating, as well as more conventional stratigraphic and petrologic analysis, including provenance study. Such comprehensive analysis is rarely possible for an individual investigator but is becoming increasingly important in larger research efforts. For additional information on basin analysis, see Klein (1987), Ingersoll (1988), Kleinspehn and Paola (1988), and Miall (1990).

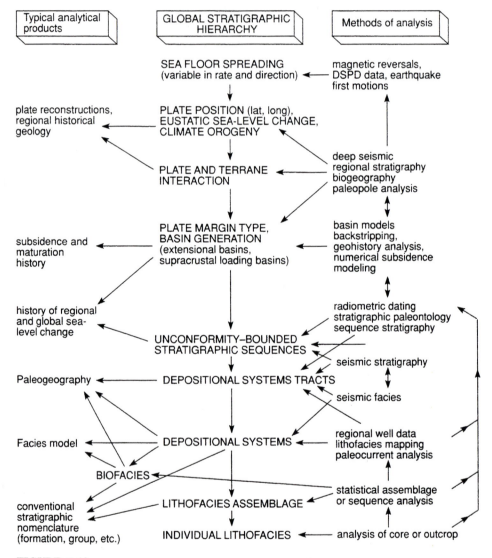

FIGURE 1.19

Flow sheet illustrating the various kinds of studies involved in basin analysis. (*After Miall, A. D., 1990, Principles of basin analysis, 2nd ed. Fig. 1.1, p. 6, reprinted by permission of Springer-Verlag, New York.*)

Additional Readings

Allen, P. A., and P. Homewood, eds., 1986, Foreland basins: Internat. Assoc. Sedimentologists Spec. Pub. No. 8, Blackwell Scientific Publications, Oxford, 453 p.

Bally, A. W., A. B. Watts, J. A. Grow, W. Manspeizer, D. Bernoulli, C. Schreiber, and J. M. Hunt, 1981, Geology of passive continental margins: Am. Assoc. Petroleum Geologists Educ. Course Note Ser. No. 19, individually paginated.

Bott, M. H. P., ed., 1976, Sedimentary basins of continental margins and cratons: Elsevier, Amsterdam, 314 p.

Conybeare, C. E. B., 1979, Lithostratigraphic analysis of sedimentary basins: Academic Press, New York, 555 p.

Dickinson, W. R., 1974, Tectonics and sedimentation: Soc. Econ. Paleontologists and Mineralogists Spec. Pub. 22, Tulsa, 204 p.

Dott, R. H., and R. H. Shaver, eds., 1974, Modern and ancient geosynclinal sedimentation: Soc. Econ. Paleontologists and Mineralogists Spec. Pub. 19, Tulsa, 380 p.

Doyle, L. J., and O. H. Pilkey, eds., 1979, Geology of continental slopes: Soc. Econ. Paleontologists and Mineralogists Spec. Pub. No. 27, 374 p.

Garrels, R. M., and F. T. McKenzie, 1971, Evolution of sedimentary rocks: W. W. Norton, New York, 397 p.

Kleinspehn, K. L., and C. Paola, eds., 1988, New perspectives in basin analysis: Springer-Verlag, New York, 453 p.

Leggett, J. K., ed., 1982, Trench-forearc geology: Sedimentation and tectonics on modern and ancient active plate margins: Blackwell Scientific Publications, Oxford, 576 p.

Miall, A. D., 1990, Principles of sedimentary basin analysis, 2nd ed.: Springer-Verlag, New York, Berlin, 668 p.

Milner, H. B., 1962, Sedimentary petrography, v. I, 4th ed.: George Allen & Unwin, London, 643 p.

Ramberg, I. B., and E. R. Neumann, eds., 1978, Tectonics and geophysics of continental rifts: D. Reidel, Dordrecht, 444 p.

Ronov, A. B., 1983, The Earth's sedimentary shell: AGI Reprint Ser.: V, American Geological Institute, 80 p.

Stanley, D. J., and G. T. Moore, eds., 1983, The shelfbreak: Critical interface on continental margins: Soc. Econ. Paleontologists and Mineralogists Spec. Pub. No. 33, 467 p.

Tucker, M., ed., 1988, Techniques in sedimentology: Blackwell Scientific Publications, Oxford, 394 p.

PART 2
Siliciclastic Sedimentary Rocks

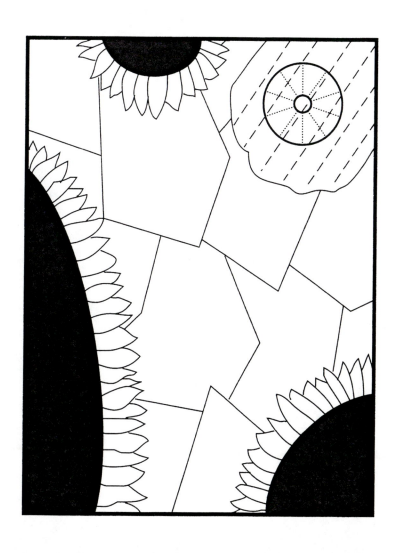

Chapter 2
Sedimentary Textures

2.1 Introduction

Few subjects in the field of sedimentology have been researched more thoroughly than sedimentary textures. The number of published articles, books, and book chapters that deal in some way with textures of siliciclastic sedimentary rocks must now number several hundred. This strong interest in sedimentary textures has apparently arisen out of the conviction of many workers that sedimentary texture is a valuable tool for environmental analysis. The size, shape, and arrangement (fabric) of siliciclastic grains have been examined and reexamined over a period of decades in an effort to establish through empirical and experimental studies the validity of this assumption. Unfortunately, this goal of environmental interpretation remains elusive, and many problems still beset investigators who attempt to use sedimentary texture as a tool for environmental analysis. Nonetheless, texture is a fundamental attribute of siliciclastic sedimentary rocks. Along with other properties, it helps to characterize and distinguish them from other types of rocks and aids in their correlation. Furthermore, the texture of sedimentary rocks affects such derived properties as porosity, permeability, bulk density, electrical conductivity, and sound transmissibility. These derived properties are of particular interest to petroleum geologists, hydrologists, and geophysicists.

Sedimentary texture encompasses three fundamental properties of sedimentary rocks: grain size, grain shape (form, roundness, and surface texture, or microrelief, of grain surfaces), and fabric (grain packing and orientation). Grain size and shape are properties of individual grains. Fabric is a property of grain aggregates. The characteristics of each of these properties are explored in this chapter.

2.2   Grain Size

2.2.1   Grain-size Scales

Natural siliciclastic particles range in size from clay to boulders. Because of this wide range of sizes, the most useful grade scales for expressing particle size are logarithmic or geometric scales that have a fixed ratio between successive elements of the series. The grade scale most widely used by sedimentologists is the **Udden-Wentworth scale** (Wentworth, 1922). Each value in this scale is either two times larger than the preceding value or one-half as large, depending upon the sense of direction (Table 2.1). The Udden-Wentworth scale extends from <1/256 mm (0.0039 mm) to >256 mm and is divided into four major size categories (clay, silt, sand, and gravel). Some of these major size categories can be further subdivided, as shown in Table 2.1.

Although the Udden-Wentworth scale adequately expresses the wide range of particle sizes found in natural sediments and sedimentary rocks, it does not lend itself especially well to the purposes of graphical plotting and statistical calculations. Because the magnitude of each size class in the scale is different, and because many of the size classes are in fractions of millimeters, the scale is difficult to work with when graphing. This problem can be avoided in part by plotting the logarithm to base 10 of the millimeter sizes. This procedure yields even-size divisions, but the divisions have fractional values. The phi (ϕ) scale is a logarithmic scale to base 2 that overcomes this problem of fractional size classes by allowing grain-size classes to be expressed in integers. The scale is based on the relationship

$$\phi = -\log_2 S \qquad (2.1)$$

where ϕ is phi size and S is grain size in millimeters. Equivalent phi and millimeter sizes are shown in Table 2.1. Note that increasing absolute value of negative phi numbers indicates increasing millimeter size, whereas increasing positive phi numbers indicate decreasing millimeter size.

2.2.2   Measuring Grain Size

Methods

Several methods for measuring the grain size of siliciclastic particles are available, with the choice of methods depending upon the sizes of the particles and their state of consolidation (Table 2.2). Some of these measurement techniques have been used for several decades; others are comparatively new. The principal methods of interest are described briefly below.

Unconsolidated Sediment

Older, conventional techniques for measuring the grain size of unconsolidated **sandy sediment** include **sieving** (Folk, 1974, p. 33–35; Ingram, 1971) and **sedimentation methods** that involve measuring the fall time of particles through water in a settling tube (Galehouse, 1971). Fall time can be equated empirically to particle diameter. More recently, automated settling tubes (so-called rapid sediment analyzers) have been developed that allow grain size of sandy sediment to be measured rapidly and easily. The weight of sediment accumulating in a pan at the bottom of the settling tube, or change in pressure of the water column as sediment settles, is automatically measured as a function of time. The resulting data are simultaneously recorded on an *X-Y* plotter or chart recorder

TABLE 2.1

Udden-Wentworth grain-size scale for sediments and the equivalent phi (φ) scale.

	U.S. Standard sieve mesh	Millimeters		Phi (φ) units	Wentworth size class
GRAVEL		4096		−12	
		1024		−10	Boulder
		256	256	− 8	
		64	64	− 6	Cobble
		16		− 4	Pebble
	5	4	4	− 2	
	6	3.36		− 1.75	
	7	2.83		− 1.5	Granule
	8	2.38		− 1.25	
SAND	10	2.00	2	− 1.0	
	12	1.68		− 0.75	
	14	1.41		− 0.5	Very coarse sand
	16	1.19		− 0.25	
	18	1.00	1	0.0	
	20	0.84		0.25	
	25	0.71		0.5	Coarse sand
	30	0.59		0.75	
	35	0.50	½	1.0	
	40	0.42		1.25	
	45	0.35		1.5	Medium sand
	50	0.30		1.75	
	60	0.25	¼	2.0	
	70	0.210		2.25	
	80	0.177		2.5	Fine sand
	100	0.149		2.75	
	120	0.125	⅛	3.0	
	140	0.105		3.25	
	170	0.088		3.5	Very fine sand
	200	0.074		3.75	
MUD — SILT	230	0.0625	1/16	4.0	
	270	0.053		4.25	
	325	0.044		4.5	Coarse silt
		0.037		4.75	
		0.031	1/32	5.0	
		0.0156	1/64	6.0	Medium silt
		0.0078	1/128	7.0	Fine silt
		0.0039	1/256	8.0	Very fine silt
CLAY		0.0020		9.0	
		0.00098		10.0	Clay
		0.00049		11.0	
		0.00024		12.0	
		0.00012		13.0	
		0.00006		14.0	

as a cumulative curve. When the curve is properly calibrated, grain size can be read from it. The latest development in rapid sediment analysis is to feed the output from the analyzer directly into a microcomputer, which, with appropriate software, digitizes the data. The computer then calculates grain-size statistics and produces various kinds of grain-size graphs or charts (Poppe et al., 1985). Geologists disagree about whether sieving or settling tube methods give the best results, where best presumably refers to the greatest usefulness in interpreting depositional environments (see, for example, Tanner, 1983).

TABLE 2.2
Methods of measuring sediment grain size

Type of sample	Sample grade	Method of analysis
Unconsolidated sediment and disaggregated sedimentary rock	Boulders Cobbles Pebbles	Manual measurement of individual clasts
	Granules Sand Silt ___	Sieving, settling tube analysis, image analysis
	Clay	Pipette analysis, sedimentation balances, photohydrometer, Sedigraph, laser-diffractometer, electro-resistance (e.g., Coulter counter)
Lithified sedimentary rock	Boulders Cobbles Pebbles	Manual measurement of individual clasts
	Granules Sand Silt ___	Thin-section measurement, image analysis
	Clay	Electron microscope

Most methods for measuring the grain size of **fine-size sediment** (fine silt and clay) are based in some way upon **Stokes' Law**

$$D = \frac{\sqrt{C}}{\sqrt{V}} \tag{2.2}$$

where D is particle diameter in centimeters, V is settling velocity, and C is a constant that equals $(\rho_s - \rho_f)g/18\mu$, in which ρ_s is the density of the settling grains, ρ_f is density of the fluid, μ is viscosity of the fluid, and g is gravitational acceleration.

The standard, conventional way of measuring particle size on the basis of Stokes' Law is by **pipette analysis** (Galehouse, 1971). Fine, unconsolidated sediment is stirred into a suspension in a measured volume of distilled water in a settling tube. Uniform-size aliquots of this suspension are withdrawn with a pipette at specified times, evaporated to dryness in an oven, and weighed. These weight data can then be used in a modified version of Stokes' Law to calculate particle diameter

$$D = \frac{\sqrt{x/t}}{\sqrt{C}} \tag{2.3}$$

where x is the depth in centimeters to which particles have settled in a given time (withdrawal depth), t is the elapsed time in seconds, and $x/t = V$ (settling velocity).

Because pipette analysis is such a laborious and time-consuming process, several new automated techniques for measuring the grain size of fine-grained sediment have been developed, most of which are also based on Stokes' Law. These methods require the use of sophisticated (and expensive) equipment that may not be available in every sedimentology laboratory.

Sedimentation balances are types of automated settling tubes for fine sediment that work on much the same principle as rapid sediment analyzers for sandy sediment. A **photohydrometer** (or hydrophotometer) is a special type of automated settling tube that empirically relates changes in intensity of a beam of light passed through a column of suspended sediment to particle settling velocities and thus to particle size (Jordan et al., 1971). The **Sedigraph** determines particle size by measuring the attenuation of a finely collimated X-ray beam as a function of time and height in a settling suspension (Stein, 1985; Jones et al., 1988). A **laser-diffraction size analyzer** operates on the principle that particles of a given size diffract light through a given angle, the angle increasing with decreasing particle size (McCave et al., 1986). **Electro-resistance size analyzers** such as the Coulter counter or Electrozone particle counter measure grain size on the basis of the principle that a particle passing through an electrical field maintained in an electrolyte will displace its own volume of the electrolyte and thus cause a change in the field. These changes are scaled and counted as voltage pulses, with the magnitude of each pulse being proportional to particle volume (Swift et al., 1972; Muerdter et al., 1981).

These various methods of measuring the grain size of fine-grained unconsolidated sediment may give slightly different results because they are based on somewhat different principles. For a comparison of the results of size analysis by four of these techniques (photohydrometer, Sedigraph, laser, electro-resistance) see Singer et al. (1988).

Consolidated Sedimentary Rock

The grain size of consolidated sedimentary rocks that cannot be adequately disaggregated cannot be measured by the various techniques described above. The conventional technique for measuring sand- and coarse silt-size grains in consolidated rocks is measurement in thin sections by use of a petrographic microscope fitted with an ocular micrometer (Textoris, 1971). The grain size determined in this way is the section diameter of randomly oriented grains, which is commonly smaller than the maximum diameter of the grains. This phenomenon is sometimes referred to as the **corpuscle effect** (Burger and Skala, 1976), meaning that grains cut marginally in a thin-section plane have smaller apparent diameters. Thin-section measurements do not yield the same results as sieve analysis, which measures the intermediate diameter of grains. Therefore, thin-section grain-size measurements are commonly corrected in some way to make them agree more closely with sieve data (Friedman, 1962; Burger and Skala, 1976).

A more sophisticated (and expensive) method of measuring the size of particles in thin section is now available using so-called **image analysis** (Ehrlich et al., 1984; Mazzullo and Kennedy, 1985; Schäfer and Teyssen, 1987). With this system, a TV camera with a special viewing tube mounted on a petrographic microscope ''sees'' the boundary of grains as pixel units. Pixels are electronically digitized arrays or squares defined by two spatial coordinates (X,Y). A conversion factor is used to convert the data from square pixels to square micrometers, from which the diameter of the particle is calculated. The problem of the corpuscle effect encountered with microscopic measurement is also present with image analysis. Burger and Skala (1976) and Schäfer and Teyssen (1987) describe techniques to correct for the corpuscle effect by Monte Carlo simulation. Image analysis is especially useful for measuring the size of grains in consolidated rocks, where the grains cannot be measured by sieving or settling tube methods, but the method can be applied equally well to measurement of loose grains in unconsolidated sediment (Mazzullo and Kennedy, 1985).

2.2.3 Reducing and Displaying Grain-size Data

Graphical Methods

The measurement techniques described in Table 2.2 yield large amounts of data that must be reduced in some way before meaningful comparison can be made between different sediment samples. The techniques for reducing and presenting grain-size data include both graphical and statistical or quasistatistical methods. Graphical presentation of grain-size data commonly involves plotting the data on bivariate diagrams in which either individual weight percent of each grain-size class or cumulative weight percent is plotted against phi size. **Histograms** (Fig. 2.1A) are bar diagrams constructed by plotting individual weight percent (frequency) along the ordinate and the phi size of each size class along the abscissa. Such diagrams provide an easily visualized pictorial representation of the grain-size distribution, but their shape is affected by the phi-size intervals selected for plotting (1/4ϕ, 1/2ϕ, etc.). A **frequency curve** is similar to a histogram except that the bar diagram is replaced by a smooth curve (Fig. 2.1B). Crude frequency curves can be constructed by connecting the midpoints of each size class in a histogram with a line. More-accurate frequency curves can be constructed by a technique described by Folk (1974) involving graphical differentiation of the cumulative curve. A **cumulative curve** is generated by plotting cumulative weight percent frequency against phi size. Cumulative curves can be constructed that use either an arithmetic scale (Fig. 2.1C) or a log-probability scale (Fig. 2.1D) for the ordinate. Plotting a cumulative curve on a probability ordinate yields a straight line if the grain-size values have a log-normal distribution (explained below).

Mathematical Methods

General Statement.   Although graphical plots provide a convenient visual method for evaluating the grain-size distribution of a given sample, comparison of large numbers of such plots can be very cumbersome. Also, the average grain size and sorting of samples cannot be determined very accurately by visual inspection of grain-size graphs. To avoid these difficulties, mathematical methods that permit statistical treatment of grain-size data can be used to derive parameters that describe grain-size distributions mathematically.

Average Grain Size.   Three mathematical measures can be used to describe the average size of grains in a sediment sample. The **mode** is the most frequently occurring particle size in a population of grains. The modal diameter corresponds to the diameter of grains represented by the peak of a frequency curve or the steepest point (inflection point) of a cumulative curve. The **median size** represents the midpoint of the grain-size distribution. Half of the grains by weight in the sample are larger than the median, and half are smaller. The median corresponds to the 50 percentile diameter on the cumulative curve (Fig. 2.2). The **mean size** is the arithmetic average of all the particle sizes in a sample. The true arithmetic mean size of grains in a sediment sample cannot be calculated from the data obtained by most grain-size measurement techniques, other than manual measurement of individual clasts, because such measurements do not yield the sizes of individual particles. An approximation of the mean size can be calculated from Formula 1, Table 2.3.

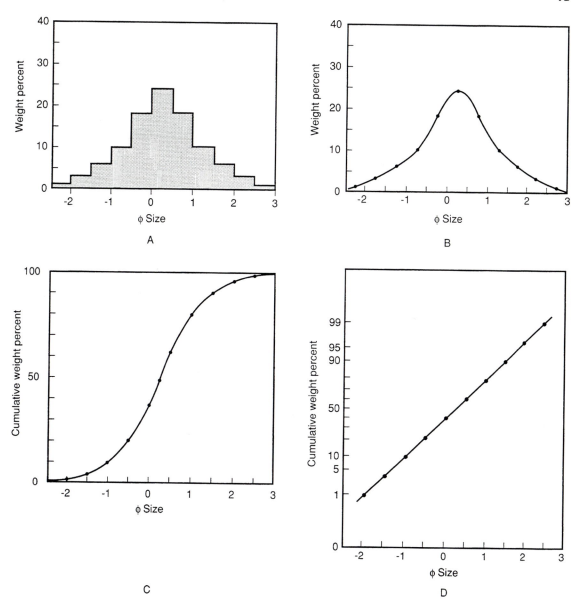

FIGURE 2.1
A hypothetical, nearly log-normal grain-size distribution plotted as a histogram (A), a frequency curve (B), a cumulative curve with an arithmetic ordinate (C), and a cumulative curve with a log-probability ordinate (D).

The graphic mean size (Table 2.3) is obtained by calculating the average of the 16th, 50th, and 84th percentile diameters determined from the cumulative curve (Fig. 2.2).

Grain-size Sorting.   The mathematical expression of sorting is standard deviation. In conventional statistics, one standard deviation encompasses approximately the central 68 percent of the area under the frequency curve (Fig. 2.3). The conventional formula for

TABLE 2.3

Formulas for calculating grain-size statistical parameters by graphical methods

Graphic mean	$M_z = \dfrac{\phi_{16} + \phi_{50} + \phi_{84}}{3}$	(1)
Inclusive graphic standard deviation	$\sigma_i = \dfrac{\phi_{84} - \phi_{16}}{4} + \dfrac{\phi_{95} - \phi_5}{6.6}$	(2)
Inclusive graphic skewness	$SK_i = \dfrac{(\phi_{84} + \phi_{16} - 2\phi_{50})}{2(\phi_{84} - \phi_{16})} + \dfrac{(\phi_{95} + \phi_5 - 2\phi_{50})}{2(\phi_{95} + \phi_5)}$	(3)
Graphic kurtosis	$K_G = \dfrac{(\phi_{95} - \phi_5)}{2.44(\phi_{75} - \phi_{25})}$	(4)

Source: Folk, R. L., and W. C. Ward, 1957, Brazos River bar: A study in the significance of grain-size parameters: Jour. Sed. Petrology, v. 27, p. 3–26.

calculating standard deviation cannot be used with grain-size data; however, a formula for calculating the approximate standard deviation of a grain-size distribution by graphical-statistical methods is given in Table 2.3. Formula 2 in Table 2.3 yields standard deviation expressed in phi units (phi standard deviation). Verbal terms for sorting corresponding to various values of graphic phi standard deviation are given below, after Folk (1974):

Phi standard deviation	*Sorting*
$<0.35\phi$	very well sorted
0.35 to 0.50ϕ	well sorted
0.50 to 0.70ϕ	moderately well sorted
0.70 to 1.00ϕ	moderately sorted
1.00 to 2.00ϕ	poorly sorted
2.00 to 4.00ϕ	very poorly sorted
$>4.00\phi$	extremely poorly sorted

Skewness is an additional measure of grain-size sorting that reflects sorting in the tails of the distribution. When plotted as a frequency curve, the grain-size distributions of most natural sediments do not yield a perfect bell-shaped curve such as the idealized curve

FIGURE 2.2

Calculating percentile values from a cumulative curve.

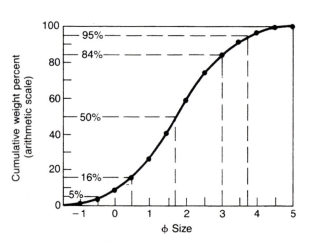

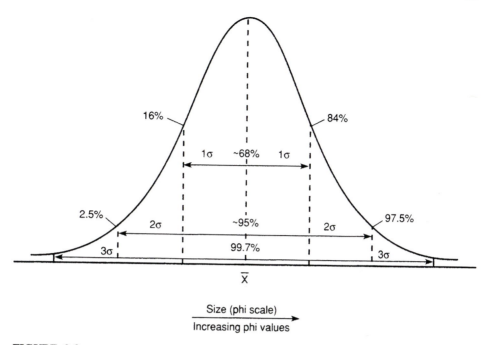

FIGURE 2.3

Frequency curve for a normal distribution showing the relation of standard deviation (σ) to the population mean (X). One standard deviation (1σ) accounts for approximately 68 percent of the area under the curve, two standard deviations (2σ) account for approximately 95 percent, and three standard deviations (3σ) account for 99.7 percent.

shown in Figure 2.4A. That is, they do not exhibit a normal, or log-normal, grain-size distribution (even distribution of sizes about the mean size). Instead, they display an asymmetrical or skewed distribution. When an excess of fine particles is present in a sample, the frequency curve has a fine "tail" and the grain-size distribution is said to be fine-skewed, or positively skewed (fine sediment has positive phi values) (Fig. 2.4B). When a coarse tail is present, the grain population is coarse-skewed, or negatively skewed (Fig. 2.4C). The numerical value of skewness (+ or −) can be calculated by using the graphic skewness formula shown in Table 2.3. The more this numerical value deviates from zero, the greater the skewness.

The sharpness or peakedness of a grain-size frequency curve is referred to as **kurtosis**. Sharp-peaked curves are said to be leptokurtic; flat-peaked curves are platykurtic. Sharp-peaked curves indicate better sorting in the central portion of the grain-size distribution than in the tails, and flat-peaked curves indicate the opposite. A formula for calculating graphic kurtosis is given in Table 2.3.

The mean size, standard deviation, skewness, and kurtosis of a grain-size distribution can be calculated directly, without reference to cumulative curve plots, by the **moment method**. This method of deriving grain-size statistical parameters has been known since the 1930s but was not used extensively until computers became readily available to facilitate the involved computations. These computations involve multiplying a weight (weight frequency in percent) by a distance (from the midpoint of each size grade

FIGURE 2.4

Hypothetical frequency curves of normal and skewed grain-size populations. (*After Friedman, G. M., and J. E. Sanders, 1978, Principles of sedimentology, Fig. 3.18, p. 75. © 1978, John Wiley & Sons, Inc. Reprinted by permission of John Wiley & Sons, Inc., New York.*)

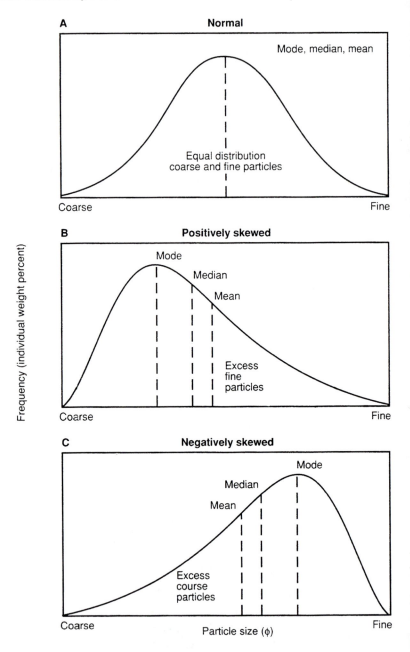

to the arbitrary origin of the abscissa). Equations for computing moment statistics are given in Table 2.4, and a sample computation form using $1/2\phi$ size classes is given in Table 2.5.

2.2.4   Significance of Grain-size Data

Over the past several decades, many investigators have tried to use grain-size data in environmental analysis. For example, Friedman (1967, 1979) made use of two-

TABLE 2.4

Formulas for calculating grain-size statistical parameters by the moment method

Mean (1st moment)	$\bar{x}_\phi = \dfrac{\Sigma fm}{n}$	(1)
Standard deviation (2nd moment)	$\sigma_\phi = \sqrt{\dfrac{\Sigma f(m - \bar{x}_\phi)^2}{100}}$	(2)
Skewness (3rd moment)	$Sk_\phi = \dfrac{\Sigma f(m - \bar{x}_\phi)^3}{100\,\sigma_\phi{}^3}$	(3)
Kurtosis (4th moment)	$K_\phi = \dfrac{\Sigma f(m - \bar{x}_\phi)^4}{100\,\alpha_\phi{}^4}$	(4)

where f = weight percent (frequency) in each grain-size grade present
m = midpoint of each grain-size grade in phi values
n = total number in sample; 100 when f is in percent

component grain-size variation diagrams in which one statistical parameter is plotted against another; e.g., skewness versus standard deviation. This method putatively allows separation of the plots into major environmental fields such as beach environments and river environments. Passega (1964, 1977) developed a graphical approach to environmental analysis that makes use of what he calls *C-M* and *L-M* diagrams. Three properties of grain-size distributions are used in these diagrams: (1) the grain diameter (*C*) corresponding to the first percentile on the cumulative curve, to measure the coarsest grains in a deposit, (2) the median grain diameter (*M*), and (3) the percentile finer than 0.031 mm (*L*) to measure the fine tail of the size distribution. When these values are plotted as either a *C-M* or *L-M* diagram, most samples from a given environment will theoretically fall within a specific environmental field of the diagram. Several workers (Visher, 1969; Sagoe and Visher, 1977; Glaister and Nelson, 1974; Middleton, 1976) have suggested that the shapes of grain-size cumulative curves plotted on log-probability paper have environmental significance. Such curves typically display two or three straight-line segments rather than the single straight line predicated for a normally distributed population. These curve segments are interpreted by the above workers to represent subpopulations of grains transported simultaneously by different transport modes; that is, traction, saltation, and suspension. Differences in curve shapes and the locations of truncation points of the curve segments allegedly allow discrimination of sediments from different environments. Dias and Neal (1990) used the polymodality of grain-size distributions, as displayed in grain-size frequency curves, to evaluate the origin and distribution of continental shelf sediment.

Although successes have been reported by using these methods for typing and differentiating sediments from several modern environments (e.g., beach, river, dune), skepticism about the reliability of such methods has been growing (Folk, 1977; Picard, 1977; Reed et al., 1975; Tucker and Vacher, 1980; Sedimentation Seminar, 1981; Vandenberghe, 1975). Some researchers have reported success in grain-size environmental analysis by using more sophisticated multivariate statistical techniques such as factor analysis and discriminant function analysis (Klovan, 1966; Greenwood, 1969; Chambers and Upchurch, 1979; Taira and Scholle, 1979; Stokes et al., 1989); however, these techniques have not been widely applied. Furthermore, some of the assumptions used in these statistical methods have also come under criticism (Forrest and Clark, 1989).

More recent approaches to interpretation of grain-size data also involve some fairly advanced statistical methods. For example, several papers have appeared in print since the

TABLE 2.5

Form for computing moment statistics using $1/2\phi$ size classes

Class interval (ϕ)	m Midpoint (ϕ)	f Weight %	fm Product	m − x̄ Deviation	(m − x̄)² Deviation squared	f(m − x̄)² Product	(m − x̄)³ Deviation cubed	f(m − x̄)³ Product	(m − x̄)⁴ Deviation quadrupled	f(m − x̄)⁴ Product
0–0.5	0.25	0.9	0.2	−2.13	4.54	4.09	−9.67	−8.70	20.60	18.54
0.5–1.0	0.75	2.9	2.2	−1.63	2.66	7.71	−4.34	−12.59	7.07	20.50
1.0–1.5	1.25	12.2	15.3	−1.13	1.28	15.62	−1.45	−17.69	1.63	19.89
1.5–2.0	1.75	13.7	24.0	−0.63	0.40	5.48	−0.25	−3.43	0.16	2.19
2.0–2.5	2.25	23.7	53.3	−0.13	0.02	0.47	0.00	0.00	0.00	0.00
2.5–3.0	2.75	26.8	73.7	0.37	0.13	3.48	0.05	1.34	0.02	0.54
3.0–3.5	3.25	12.2	39.7	0.87	0.76	9.27	0.66	8.05	0.57	6.95
3.5–4.0	3.75	5.6	21.0	1.37	1.88	10.53	2.57	14.39	3.52	19.71
>4.0	4.25	2.0	8.5	1.87	3.50	7.00	6.55	13.10	12.25	24.50
Total		100.0	237.9			63.65		−5.53		112.82

Source: McBride, E. F., Mathematical treatment of size distribution data, in Carver, R. E., ed., Procedures in sedimentary petrology, Table 2, p. 119. © 1971, John Wiley & Sons, Inc. Reprinted by permission of John Wiley & Sons, Inc., New York.

late 1970s dealing with the so-called **log-hyperbolic distribution**. The hyperbolic distribution was formally introduced to geologists by Barndorff-Nielsen (1977) and subsequently discussed and amplified by Bagnold and Barndorff-Nielsen (1980) and Barndorff-Nielsen et al. (1982). The principle involved in log-hyperbolic distributions is discussed briefly below.

Most geologists assume that the sizes of grains in natural sediments tend toward a normal distribution when the logarithm of grain size is plotted as a frequency curve. Thus, we say that sediments have a log-normal distribution, or a log-normal probability density function. This assumption is made even though we know that most natural sediments have grain-size distributions that exhibit some degree of skewness. If a truly log-normal population of grains is plotted on double log (log-log) paper so that the logarithms of both the grain size and frequency are plotted, or the logarithm to base *e* or base 10 of the frequency is plotted against phi size on arithmetic scales, the resulting curve is a parabola with continuously inward-curving tails. Barndorff-Nielsen (1977) and Bagnold and Barndorff-Nielsen (1980) maintain that when the size distribution of natural sediment, such as dune sand, is plotted on log-log plots and a curve is fitted to these data, the resulting curve actually has the shape of a hyperbola, whose straight-line asymptotes or "tails" extend to include the smallest measurable frequencies. They suggest that "granular size distributions occurring in nature do not conform to the 'normal' probability function, but to a different, 'hyperbolic,' probability function" (Bagnold and Barndorff-Nielsen, 1980). They further suggest that not only is the hyperbolic probability function a better descriptor of natural grain populations than the normal probability function, but that it holds the potential to discriminate between sediments from different environments.

Some workers who have attempted to use the log-hyperbolic function in environmental analysis, e.g., Christiansen et al. (1984), Christiansen (1984), and Vincent (1986), report success with this method. On the other hand, Fieller et al. (1984) and Wyrwoll and Smyth (1985) reported no apparent gain over the normal probability function in the ability to characterize the depositional environment of sediments by using the parameters of the log-hyperbolic distribution. Wyrwoll and Smith point out a factor that has been mentioned by many other workers—that sediment-sizing techniques such as sieving tend to sort sediments by shape as well as size. As a result, extremes of grain-size distributions may be as much an artifact of the measuring technique as factors arising from the mechanics of sediment transport within a particular environment. Thus, the log-hyperbolic distribution is unlikely to be any better than the log-probability distribution for describing an instrumentally skewed population.

Another relatively new approach to environmental interpretation on the basis of grain size data involves **entropy analysis**. Entropy analysis attempts to compare and contrast size-frequency data to enhance differences between samples. Ideally, this method allows determination of the optimal number of grain-size class intervals (e.g., in a histogram) and the class-interval widths to maximize information content (Full et al., 1984). Earlier applications of entropy analysis involved univariate parameters (Full et al., 1984); more recently, multivariate techniques have been used to extend the theory (Forrest and Clark, 1989). For a brief description of the use of entropy analysis to determine optimum class-interval widths, see the discussion under Fourier analysis in Section 2.3.5 of this chapter. Entropy analysis has not been widely applied to grain-size studies, but is not likely to provide a panacea for environmental interpretation.

And so it goes! The eternal optimism and the persistent skepticism about the environmental significance of grain-size distributions seem to keep pace with each other—and the techniques for interpreting grain-size data appear to demand increasingly more-sophisticated statistical applications. Apparently fed up with all this, Ehrlich (1983) suggests that perhaps the time has come to stop plaguing generations of students with complex grain-size techniques that are never seriously used. He suggests that students' time would be more profitably engaged in tatting!

2.3   Grain Shape

2.3.1   Methods of Expressing Shape

Numerous parameters or measures have been suggested to describe the shapes of particles (see review by Barrett, 1980). As used herein, particle **shape** is taken to encompass all aspects of the external morphology of particles, including form, roundness, and surface texture. **Form** refers to the gross, overall morphology or configuration of particles. Most measures of form consider the three-dimensional shape of the grains. **Roundness** is a measure of the sharpness of the corners of a grain and is commonly measured in two dimensions only. **Surface texture** refers to microrelief features, such as scratches and pits, that appear on the surfaces of clastic particles, particularly particles that have undergone transport. Changes in form or roundness brought about by abrasion during sediment transport or solution or cementation during diagenesis can affect surface texture by creating new grain surfaces. Thus, the three aspects of shape can be thought of as constituting a hierarchy, where form is a first-order property, roundness a second-order property superimposed on form, and surface texture a third-order property superimposed on both the corners of a grain and the surfaces between the corners (Barrett, 1980; Fig. 2.5). These three shape properties of grains are independent parameters. One property can vary without necessarily affecting the others, although dramatic changes in form or roundness are likely to also affect surface texture.

2.3.2   Particle Form

The gross morphology or overall shape of particles has proven to be a difficult parameter to quantify with great precision and accuracy. Several measures of form were proposed by early workers, including flatness, elongation, and sphericity (Barrett, 1980). Particular emphasis has been given to the sphericity of grains, that is, the degree to which the shape of grains approaches the shape of a sphere (Wadell, 1932). Practical determination of sphericity involves measurement of the three orthogonal axes of particles and calculation of a sphericity value on the basis of relative lengths of these axes. The assumption is made that the more nearly equal the lengths of the three axes, the more nearly the particle approaches the shape of a sphere. Mathematical relationships generated for the purpose of calculating sphericity include those for determining intercept sphericity (ψ_I) (Krumbein, 1941)

$$\psi_I = \sqrt[3]{\frac{D_S D_I}{D_L^2}} \tag{2.4}$$

and maximum projection sphericity (ψ_p) (Sneed and Folk, 1958)

$$\psi_p = \sqrt[3]{\frac{D_S^2}{D_L D_I}} \tag{2.5}$$

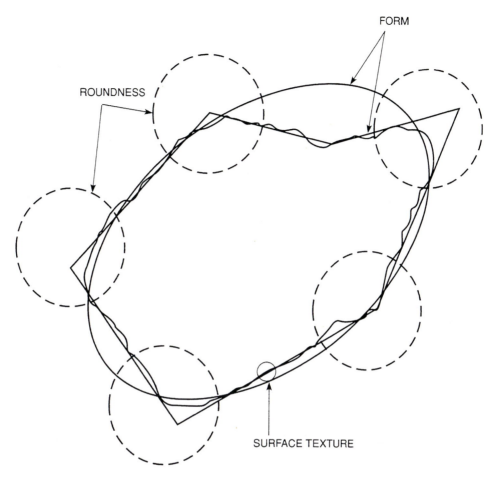

FIGURE 2.5
The hierarchal relationship of form, roundness, and surface texture. The heavy, solid line is the particle outline. (*From Barrett, P. J., 1980, The shape of rock particles, a critical review: Sedimentology, v. 27. Fig. 2, p. 293, reprinted by permission of Elsevier Science Publishers, Amsterdam.*)

where D_S refers to the length of the short particle axis, D_I the length of the intermediate axis, and D_L the length of the long axis. Sneed and Folk suggest that maximum projection sphericity is the better sphericity measure because it better expresses the behavior of particles as they settle in a fluid or are acted on by fluid flow. It is not necessarily a better sphericity measure for particles deposited in other ways (e.g., glacial deposits).

Some widely used alternative expressions of particle form are provided by the shape classifications of Zingg (1935) and Sneed and Folk (1958). Zingg's shape classification is derived by plotting on a bivariate diagram the ratio of the intermediate to long particle axis versus the ratio of the short to intermediate particle axis (Fig. 2.6A). Four shape fields, or four types of particle shapes, are identified: roller, bladed, oblate, and equant. This shape measure is somewhat similar to sphericity; note, however, that a particle with a particular value of (intercept) sphericity can fall in more than one of Zingg's shape fields

(Fig. 2.6B). Sneed and Folk (1958) classify form on a triangular diagram in which ratios of particle axes are plotted in such a way as to create ten form fields (Fig. 2.7). End-member particle shapes are compact, platy, and elongated. As in the case of Zingg's shape fields, note that lines of equal (maximum projection) sphericity plotted on this diagram can cross several form fields. Additional form parameters have been proposed by other workers (e.g., Winklemolen, 1982; see also the review in Barrett, 1980); however, none of these form measures has received widespread acceptance.

2.3.3   Particle Roundness

Several methods for measuring and expressing particle roundness have been proposed (reviewed by Dobkins and Folk, 1970; Barrett, 1980). Wadell (1932) is credited with introducing the most widely used mathematical expression for roundness. He defined roundness as the arithmetic mean of the roundness of the individual corners of a grain in the plane of measurement (a two-dimensional measure). The roundness of individual corners is given by the ratio of the radius of curvature of the corners to the radius of the maximum-size circle that can be inscribed within the outline of the grain in the plane of measurement. The degree of Wadell roundness (R_W) is thus expressed as

$$R_W = \frac{\Sigma(r/R)}{N} = \frac{\Sigma(r)}{RN} \tag{2.6}$$

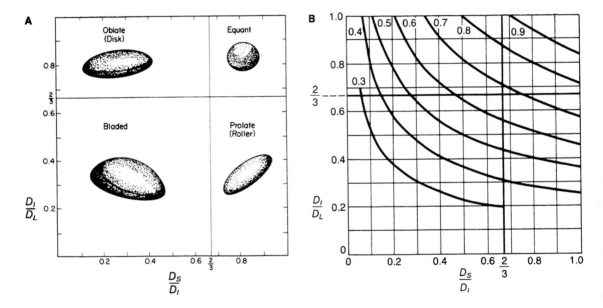

FIGURE 2.6
A. Classification of shapes of pebbles after Zingg (1935). B. Relationship between intercept sphericity and Zingg shape fields. The curves represent lines of equal sphericity. (A, from Blatt, H., G. Middleton, and R. Murray, Origin of sedimentary rocks, 2nd ed., 1980, Fig. 3.20, p. 80. Reprinted by permission of Prentice-Hall, Englewood Cliffs, N.J. B, after Pettijohn, F. J., Sedimentary rocks, 1975. Fig. 3.19, p. 54, reprinted by permission of Harper & Row, New York.)

FIGURE 2.7

Classification of pebble shapes after Sneed and Folk. The symbol V refers to the adjective very (e.g., very platy, very bladed, very elongated). (*After Sneed, E. D., and R. L. Folk, 1958, Pebbles in the Lower Colorado River, Texas, a study in particle morphogenesis: Jour. Geology, v. 66. Fig. 2, p. 119, reprinted by permission of University of Chicago Press.*)

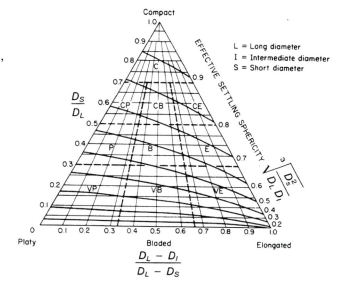

where r is the radius of curvature of individual corners, R is the radius of the maximum inscribed circle, and N is the number of corners.

Measuring the values of r and R in loose grains or in thin section is a very laborious process—so much so that most previous investigators have not been willing to spend the necessary time. Boggs (1967a) describes a fairly rapid method for obtaining these parameters by using an electronic particle-size analyzer. Also, simpler roundness measures have been proposed that require only that the radius of the sharpest corner be measured and divided by the radius of the inscribed circle (Dobkins and Folk, 1970). Nonetheless, measuring the roundness of numerous grains is very time-consuming. Therefore, many investigators resort to visual estimates of grain roundness aided by the use of visual estimation charts. Several such charts have been proposed; however, the roundness scale of Powers (1953; Fig. 2.8) appears to be most widely used. Although visual estimates of

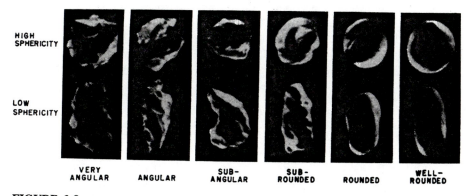

FIGURE 2.8

Grain-size images for estimating the roundness of sedimentary particles. (*After Powers, M. C., 1953, A new roundess scale for sedimentary particles: Jour. Sed. Petrology, v. 23. Fig. 1, p. 118, reprinted by permission of the Society of Economic Paleontologists and Mineralogists, Tulsa, Okla.*)

grain roundness can be done quite rapidly, reproducibility of results, even by the same operator, can be quite low.

2.3.4 Significance of Form and Roundness

In spite of the considerable preoccupation that geologists have displayed in measuring the form and roundness of natural particles, these parameters have not proven to be reliable guides to the provenance and transport histories of siliciclastic sediment. Sphericity does have an influence on the settling velocity of particles in a fluid and departure of a grain from a spherical shape causes a decrease in settling velocity (Komar and Reimers, 1978). Also, sphericity affects the transportability of particles moving by traction; particles of more-equant shape tend to be transported preferentially over disc- or blade-shaped particles. Sphericity appears to be largely an inherited property that is modified only slightly by transport processes. In spite of the relationship between sphericity and particle transportability, few successes have been reported in using sphericity to differentiate sediments from different depositional environments or to identify different sources of sediments.

Much the same thing can be said about roundness. In a series of interesting experiments in the late 1950s, Kuenen (1959, 1960) demonstrated experimentally that sand-size quartz grains undergo relatively little rounding by stream transport but are effectively rounded during moderately short distances of wind transport. Kuenen found wind transport to be 100 to 1000 times more effective than water transport in rounding sand-size grains. Unfortunately, the problems of grain recycling make these findings less useful in environmental studies than they might otherwise be. Although well-rounded grains may signify an episode of wind transport in the history of the grains, that episode may not necessarily have been the last. Some success has been reported in discriminating depositional environments on the basis of grain roundness, especially by using multivariate statistical techniques (e.g., Sahu, 1982). On the whole, however, grain roundness has not proven to be a highly reliable indicator of depositional environment.

The failure of sphericity and roundness to serve as dependable guides to source and depositional environments rests in part on the fact that many natural variables interact to produce the characteristics of a particular deposit. A single parameter such as roundness or sphericity may not be a sensitive enough measure to identify these variables. A more important reason may be, as suggested by Barrett (1980), that even the best of the commonly used procedures for determining sphericity and roundness are limited by observational subjectivity and low discriminating power. For example, measurement of the three axes of a particle does not provide a unique characterization of particle form.

2.3.5 Other Methods for Analyzing and Quantifying Two-dimensional Particle Shape

General Statement

Owing to the inability of sphericity and roundness parameters to delineate the shapes of particles with a high degree of accuracy, some geologists have sought more exact, mathematical methods for characterizing particle shape. Several such methods are possible (Clark, 1981). Most workers have focused on measuring particle shape by Fourier analysis, which appears to work well for describing the regular shapes of most natural particles. Highly irregular or strongly embayed particles are less amenable to Fourier analysis. An alternate method, based on use of the so-called fractal dimension of particles, has been proposed by some workers for analysis of such particles.

Fourier Analysis

General Principles.   The fundamentals of Fourier analysis of particle shape were introduced to geologists in the papers of Schwarcz and Shane (1969) and Ehrlich and Weinberg (1970). The problem of finding an exact method for describing, and regenerating, the (two-dimensional) outline of a grain can be attacked by cutting and unrolling the outline of a grain, as shown in Figure 2.9. When unrolled in this way, it can be seen that the outline of the grain is a periodic function, somewhat resembling the shape of a sine wave.

FIGURE 2.9

Method of "unrolling" a grain outline to produce a periodic wave. A. Grain outline showing radii measured from grain center to points on the grain perimeter. B. Unrolled grain outline as constructed from radii measurements. Note that the form of the unrolled grain is a crude sine wave.

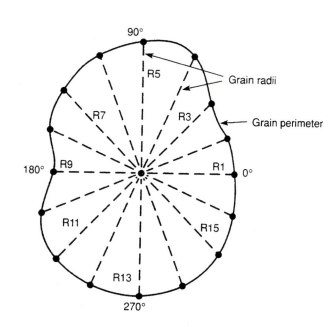

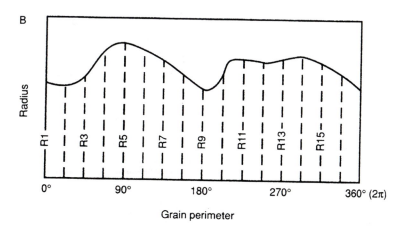

The simplest periodic function is given by

$$y = A \sin (\omega x + \varphi) \tag{2.7}$$

where A is amplitude, ω is (angular) frequency, and φ is initial phase, which characterizes the initial position of the point x. Furthermore,

$$\omega = 2\pi/T \quad \text{and} \quad T = 2\pi/\omega \tag{2.8}$$

where T is the period. (In the course of a period T, x sweeps through an angle of 2π radians.) This function is called a **harmonic** of amplitude A, frequency ω, and initial phase φ. If the simplest case is assumed, where $A = 1$, $\omega = 1$, and $\varphi = 0$ (0 phase angle), this function gives a normal sine curve $y = \sin x$, as shown in Figure 2.10A. If in this example a phase angle of 90 degrees ($\pi/2$) is used instead of 0, we get the cosine curve $y = \cos x$. The graph of the cosine function is the same as that of the sine shifted to the left by an amount of 90 degrees ($\pi/2$).

If, instead of plotting the ordinary sine curve $y = \sin x$, we plot $y = \sin 2x$, $y = \sin 3x$, etc., we deform the sine curve by uniform compression along the x axis (Tolstov, 1962). Figure 2.10B shows the harmonic $y = \sin 3x$, of period $T = 2\pi/3$. In addition to uniform compression (or expansion) along the x axis, a shift along the x axis occurs if we change the initial phase (phase angle). Thus, Figure 2.10C represents the harmonic

$$y = \sin (3x + \pi/3) \tag{2.9}$$

FIGURE 2.10
Harmonics of a sine wave. (*After Tolstov, G. P., 1976, Fourier series. Fig. 4, p. 4, reprinted by permission of Dover Publications, Inc., New York.*)

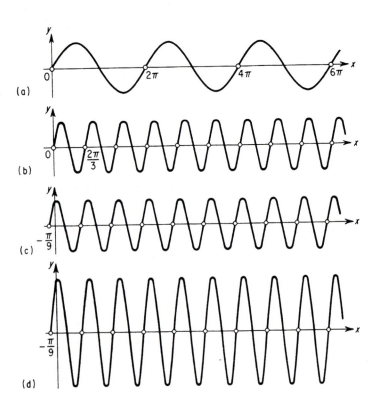

with a period $2\pi/3$ and an initial phase $\pi/3$. Finally, a change in the amplitude of the sine curve can also be effected. Thus, the harmonic

$$y = 2 \sin (3x + \pi/3) \qquad (2.10)$$

has the form shown in Figure 2.10D.

A mathematical theorem owing to Fourier shows that any periodic function with frequency ω can be represented as a superposition of harmonic vibrations of frequencies ω, 2ω, 3ω . . . with varying amplitudes (harmonics 1, 2, 3, . . .). Consider, for example, Figure 2.11. The dashed line in Figure 2.11A has the form of a square "wave." Figure 2.11B shows three harmonics of a sine wave. If these three harmonics are superimposed (added graphically), the result is the heavy, solid line in Figure 2.11A. Note that this solid curve has the approximate shape of the square curve. If additional, higher-order harmonics are superimposed, a more exact approximation of the square curve can be achieved.

This representation of a curve shape by superimposition of successive harmonics of a periodic function is the basis for Fourier analysis of particle shape. The shape of the curve formed by unrolling the outline of a particle, as shown in Figure 2.9, can be synthesized quite faithfully by superimposition of the harmonics of the periodic sine function. As described by Ehrlich and Weinberg (1970), grain shape can be estimated by a closed Fourier series by expansion of the periphery radius of a grain as a function of angle about the grain's center of gravity. To accomplish this, the image of a grain is projected onto a polar grid or a digitizing tablet. The center of the grain (center of gravity)

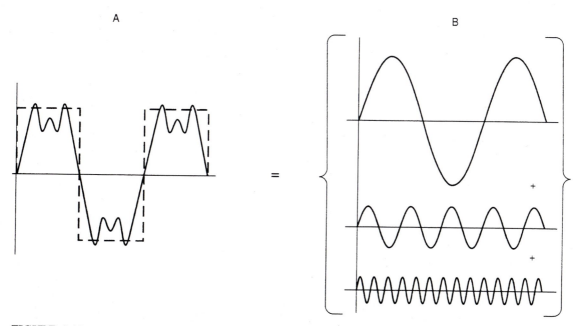

FIGURE 2.11

Fourier synthesis of a square wave by addition of successive harmonics of a sine wave. (*After Borowitz, S., and A. Beiser, 1966, Essentials of physics. Fig. 25-17, p. 432, reprinted by permission of Addison-Wesley Publishing Co., Reading, Mass.*)

is determined by calculation (Ehrlich and Weinberg, 1970). A number of points are then picked along the periphery of the grain and the locations of these points expressed in polar coordinates about the center of gravity.

The radius $R(\theta)$ is determined by the relationship

$$R(\theta) = R_0 + \sum_{N=1}^{\infty} R_n \cos(n\theta - \phi_n) \qquad (2.11)$$

where θ is the polar angle measured from an arbitrary reference line. R_0 is equivalent to the average radius determined from measuring all the radii from the center of gravity to the peripheral points. The remainder of the equation (which is a function of the harmonic amplitude R_n) represents the length to be added to the average radius at each angle n. Thus, R_n is the amplitude at each frequency, n is the harmonic order or number, and ϕ_n is the phase angle.

Each successive harmonic curve represented by this periodic function has a distinctive shape, if you think of the ends of the curves as being joined. The "zeroth" harmonic is simply the R_0 value in Equation 2.11 and is thus a centered circle with an area equal to the total area of the grain (based on the average radius). The curve of the first harmonic has the shape of an offset circle, the second is a figure eight, and the third is a trefoil. The shapes of the first 20 harmonics are illustrated in Figure 2.12. The shape of a particle can be adequately represented by an average radius and 24 harmonic amplitudes representing the relative contribution of each fixed lobate form (harmonic) to the approximation of the outline (Ehrlich et al., 1987). The first few harmonics are sufficient to define the gross shape of the particles (form); successively higher-order harmonics add to the refinement of this shape and allow discrimination of smaller-scale features such as roundness.

Once the harmonics have been calculated, the outline of the grain can be regenerated by superimposing all of the harmonic shapes, as illustrated in the simpler example in

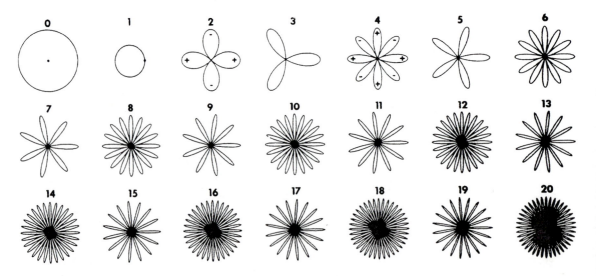

FIGURE 2.12
Graphic representation of the first 20 Fourier harmonics. (*From Eppler, D. T., R. Ehrlich, D. Nummedahl, and P. H. Schulz, 1983, Sources of shape variation in lunar impact craters: Fourier shape analysis: Geol. Soc. America Bull., v. 94, Fig. 1, p. 275.*)

Figure 2.11. Obviously, it would be very difficult to superimpose curves by visual graphic addition; however, this operation is done easily by computer. See Clark (1987) and Telford et al. (1987) for details of these computer applications.

Significance of Fourier Shape.   Because the shapes of particles can be characterized very exactly by Fourier analysis, shape studies based on this technique may provide a more reliable method for studying the source and transport histories of particles than do studies based on roundness and sphericity. An immediate problem is encountered in interpreting Fourier results, however, because the data obtained do not directly represent any physical property of the grains. Fourier data are in the form of amplitudes of the various harmonics. Thus, to compare the shapes of two grains, an investigator compares the amplitudes of each harmonic of the two grains, which in total represent the shape of the grain. These individual amplitudes have no easily visualized relationship to the actual shapes of the grains, and, in fact, interpretation is not related to the physical properties of the grain except in a crude way (Clark, 1981). Nonetheless, this method of comparing amplitudes is the principal technique used in interpretative studies.

Because large numbers of grains are commonly studied in environmental and provenance analysis, the harmonic amplitudes must be presented in some condensed form. This data reduction is commonly done by plotting as a histogram the "mean amplitude spectrum" as a function of number frequency (Fig. 2.13). The mean amplitude spectrum is obtained by averaging the amplitudes of a particular harmonic, say the fifth, as determined from all the grains analyzed from a particular sample. Visual inspection of the histograms makes it possible to distinguish differences owing to gross form (lower-order harmonics) and finer-scale features of shape, such as roundness and surface possibly texture (higher-order harmonics).

In investigations of the shapes of large numbers of grains in numerous samples, the huge amount of data generated in 20 or more harmonics of each grain in each sample makes cumbersome the analysis of each harmonic separately. It becomes desirable to

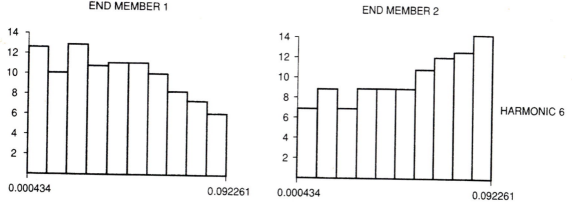

FIGURE 2.13
Mean amplitude spectra for harmonic 6 of two populations of grains. End member 1 represents a population of grains that are on the average smooth and round. Grains in end member 2 are largely angular and irregular in shape. (*After Mazzullo, J., and J. B. Anderson, 1987, Analysis of till and glacial-marine sands, in Marshall, J. R., ed., Clastic particles, Fig. 2, p. 319. Copyright 1987 by Van Nostrand Reinhold. All rights reserved.*)

determine which one or which few harmonics may yield the most useful information about shape for the purposes of a particular study, so that all harmonics need not be studied in equal detail. An investigator can then concentrate on statistical analysis of the data from perhaps four or five harmonics, rather than 20 or more. The problem is to select those four or five harmonics that make the most useful or meaningful contribution to the particle's shape.

Full et al. (1984) suggest a method for selecting the most useful harmonics for more-detailed study by plotting against harmonic number a parameter called **relative entropy**. Entropy, in this context, refers to so-called information entropy and is concerned with the frequency distributions of values. Entropy is defined as

$$E = -\sum_{i=1}^{n} P_i \log_a P_i \tag{2.12}$$

where P_i is the probability of an event occurring in an interval i, a is any base of logarithm (commonly base e), and n is the number of intervals (such as size classes or shape classes) in a frequency distribution (histogram). Maximum entropy is achieved when any single event has an equal probability of occurring in any interval. For example, if we consider the grain size of a sand sample, plotted as a histogram, maximum entropy would be attained if all the grain sizes occurred with equal frequency (same percentage of grains in all size classes). Thus, low entropy values represent frequency plots with large differences between intervals (most values fall in one or two classes, with few values in the other classes), and high entropies characterize plots with relatively little contrast between intervals (Full et al., 1984). Numerically, maximum entropy is equal to the log of the number of intervals. A plot with the distribution divided into five intervals, or frequency classes, for example, would have a maximum entropy of $\log_e 5 = 1.61$. The maximum amount of information is considered to be preserved when the width of class intervals of a frequency distribution is selected in such way as to achieve maximum entropy when calculated according to Equation 2.12. This commonly means selecting class intervals of unequal width.

Relative entropy of an individual sample is the ratio of the calculated sample entropy (Equation 2.12) to the maximum possible entropy ($\log_e$ of the number of class intervals). The relative entropy of an entire data set is the sum of the calculated individual sample entropies divided by the product of the maximum possible entropy and the number of samples.

$$R_e = \Sigma S_e/n \times M_e \tag{2.13}$$

where R_e is relative entropy, S_e is calculated sample entropy, n is the number of samples, and M_e is maximum possible entropy. The smaller the relative entropy of a particular harmonic in a data set, the greater the potential of that harmonic to provide meaningful information about the particle shapes that may have significance with respect to depositional environments or sources of the particles. See Full et al. (1984) for details.

Applications of Fourier Techniques to Provenance Analysis.   Since the introduction to geologists of Fourier shape analysis in the late 1960s and early 1970s, several investigators have applied this technique to the problem of interpreting sediment source. The fundamental assumption in these applications is that the shapes of quartz grains from different sources are sufficiently different and distinctive, as revealed by Fourier shape analysis, that the separate sources can be differentiated and identified. Ehrlich et al. (1980) suggest that both provenance and process history (degree of abrasion) can be evaluated by Fourier techniques. For several examples of provenance applications on the basis of Fourier shape analysis of quartz grains, see Ehrlich and Chin (1980), Mazzullo et al. (1982), Wagoner and Younker (1982), Mazzullo and Withers (1984), Mazzullo et al. (1984), Kennedy and Ehrlich (1985), Prusak and Mazzullo (1987), and Mazzullo and Magenheimer (1987).

Application of Fourier Shape Analysis to Environmental Studies.   Although the techniques of Fourier shape analysis have been used by geologists since the late 1960s, surprisingly few attempts have apparently been made to apply the technique to interpretation of ancient sedimentary environments. At this time, no investigator has published data on the Fourier shape characteristics of quartz grains from a wide range of modern environments, such as beach, shoreline dune, inland dune, fluvial, alluvial fan, continental shelf, delta, etc., that can be used by other investigators as a basis for environmental interpretation of ancient sedimentary rocks. One reason why relatively few "fingerprinting" studies of modern quartz sands have been made may be that the effects of diagenesis (solution, overgrowth cementation, etc.) may make the shape characteristics of quartz sands in modern environments a questionable model for quartz sands in ancient sedimentary rocks. For a few examples of Fourier application to environmental analysis, see Dowdeswell (1982), Mazzullo and Anderson (1987), Mazzullo and Ehrlich (1983), Mazzullo et al. (1986), and Mazzullo (1987).

Fractal Dimension

Fourier analysis does not work for highly irregular or strongly embayed grains, where radii from the grain center may intersect the grain perimeter more than once. For example, many grains that have undergone intensive chemical leaching during diagenesis or weathering would fall into this category of highly irregular particles. Some workers, notably Orford and Whalley (1983, 1987) and Whalley and Orford (1986), have urged the use of the fractal dimension of particles as a method of analyzing the shapes of these irregular particles. The fractal dimension of an object is defined as the degree to which a line fills a two-dimensional space (Whalley and Orford, 1986). The greater the wiggliness of a line, the greater its fractal dimension, which can range in numerical value from 1 for a point or a straight line to values approaching 2 for an extremely wiggly line (Fig. 2.14).

The fractal value D is calculated as follows (Whalley and Orford, 1986). If the length of a grain outline (perimeter) is measured off several times with a pair of dividers, using a different divider step length each time, the length of the perimeter is found to be dependent upon the step length. The smaller the step length, the greater will be the value of the measured perimeter. If a log plot is prepared where the log of the perimeter P is plotted against the log of the steps S for a number of measurements of the perimeter, the plotted values generally show a linear inverse relationship (Fig. 2.15A). If the slope of a line drawn through these points is designated b, then the fractal value D of the grain is the slope term plus 1 (because the slope is negative). Grains having relatively simple outlines or perimeters tend to plot as a single linear element (Fig. 2.15A). More-complex shapes yield plots having two fractal elements, a **textural fractal d** and a **structural fractal f** (Fig. 2.15B).

The value of the fractal is calculated from the relationship

$$D = b + 1$$

(2.14)

A $D \longrightarrow 1$

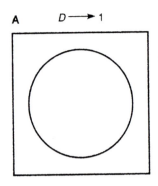

B $D \longrightarrow 2$

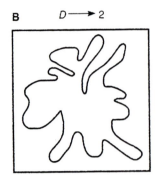

FIGURE 2.14
The smooth, closed loop of a circle (A) gives a fractal dimension of 1. A highly contorted outline such as that shown in (B) has a fractal dimension approaching 2. *(After Whalley, W. B., and J. D. Orford, 1986, Practical methods for analysing and quantifying two-dimensional images, in Sieveking, G. De C., and M. B. Hart, eds., The scientific study of flint and chert. Fig. 26.7, p. 240, reprinted by permission of Cambridge Univ. Press, Cambridge.)*

then

$$\log P \propto (D - 1) \log S \qquad (2.15)$$

so that

$$P \propto S^{D-1} \qquad (2.16)$$

Discrimination between grain shapes is done by comparing log plots of the fractal dimension or by comparing the calculated values of the total fractal element (D_T) and the textural fractal (D_1) and structural fractal (D_2). Fractal analysis works best for grains with highly embayed or lobate outlines. Fourier analysis is regarded to be a better technique for determining the shapes of grains with simpler outlines. The image analysis techniques mentioned above can be applied also to analysis of the fractal dimension (Whalley and Orford, 1986).

2.3.6   SEM Analysis of Grain Surface Texture

Principles and Techniques

Geologists have long recognized through observation with a standard binocular light microscope that the surfaces of some pebbles and mineral grains are either polished or roughened (frosted); however, the details of surface texture could not be recognized at such low magnification. It was not until the early 1960s, when the transmission electron microscope became readily available, that the surface texture of grains could be studied in detail. With the appearance of the scanning electron microscope (SEM) in the late 1960s, the technique for studying surface texture at high magnifications became simpler and more accurate. With the aid of the electron microscope, geologists were able to identify a variety of microrelief markings on grain surfaces, such as V-shaped pits, scratches, tiny sinuous ridges, and conchoidal fractures (Fig. 2.16). The idea of using grain surface textures as a tool for environmental analysis quite naturally followed, and several workers in the early 1960s applied the technique to identification of quartz grains from eolian, beach, and glacial environments.

FIGURE 2.15
A. A plot of log P versus log S to give total fractal dimension (D_T). B. A log P log S plot showing textural (D_1), and structural (D_2) fractal elements. The third fractal element (b''') shown in this plot is probably an artifact of the calculation technique used. Tr refers to the truncation point or break between fractal elements. (*From Whalley, W. B., and J. D. Orford, 1986, Practical methods for analysing and quantifying two-dimensional images, in Sieveking, G. De C., and M. B. Hart, eds., The scientific study of flint and chert. Fig. 26.8, p. 240, reprinted by permission of Cambridge Univ. Press, Cambridge.*)

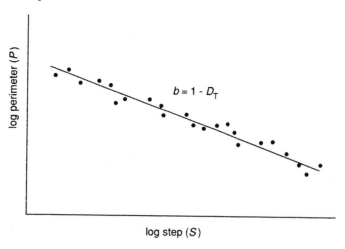

A   Single fractal element

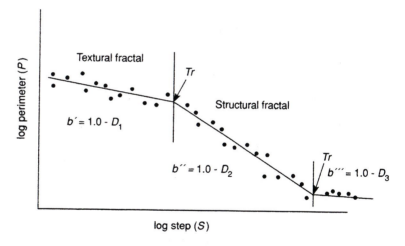

B   Multiple fractal element

The appeal of SEM analysis of grain surface texture as an environmental tool lies in the assumption that surface texture, being easier to modify than either particle form or roundness, is more likely than either form or roundness to preserve the imprint of the last depositional environment. Owing to its superior hardness and chemical stability, quartz is commonly considered to be the most appropriate material to record the surface textural features produced in each environment. The assumption is made that the energy conditions within each environment will tend to produce surface textural features that reflect that environment and that these features will differ from environment to environment. Thus, different textural features are produced by the grinding and sliding motion of glaciers, the collision and abrasion of grains in breaking waves of the beach surf zone, and the traction and saltation movement of grains by wind or river water.

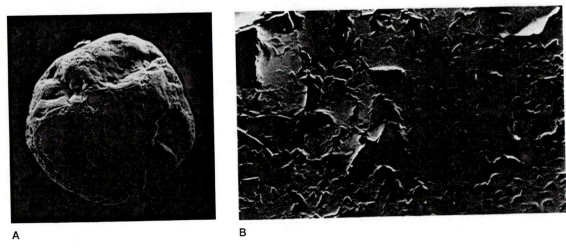

A   B

FIGURE 2.16
A. Electron micrograph of surface markings on a quartz grain from a late Pleistocene high-energy beach deposit near Norwich, Norfolk, England. B. Enlarged portion of the extreme right edge of the grain in A. Note abundant **V**-shaped markings, which are characteristic of quartz grains from high-energy beaches. (*Photographs courtesy David Krinsley, Arizona State University.*)

The surface textures of thousands of quartz grains from known modern environments have now been studied, together with experimentally formed textures produced by grinding or other laboratory techniques to mimic glacial, eolian, or subaqueous transport. On the basis of these studies, more than 30 surface features of quartz grains have now been identified, perhaps as many as 50 or 60 altogether (Bull, 1986). The more common ones are shown in Figure 2.17. Some of these textures are mechanically produced microrelief markings, some are markings produced by chemical leaching and etching, and other features are aspects of form or morphology that presumably are related also to mechanical processes. Unfortunately, few if any of the textural features shown in Figure 2.17 are unique to a particular environment. Likewise, several different kinds of markings can be produced in the same environment. Although the same type of surface features can be produced in more than one environment, some markings tend to be relatively more abundant in some environments than in others. Mechanical **V**-pits, for example, are common on quartz grains from high-energy subaqueous environments but are not common on grains from eolian or glacial environments. Parallel striations and very small conchoidal fractures are common on grains from glacial environments but are less abundant on grains from other environments, and so on. Therefore, environmental analysis of surface features on quartz grains from ancient sedimentary rocks is made on a statistical basis, looking at relative abundances of particular features rather than absolute abundances. Environmental analysis has not yet progressed beyond identifying a few major environments: glacial, high-energy beach (littoral), eolian, fluvial, and possibly the environment of turbidite deposition. Features produced during diagenesis and soil formation (pedogenic sands) can also be identified.

In practice, 30 to 50 grains from a sample of sand from a particular sedimentary rock are selected for analysis. The grains are then treated chemically to remove surface materials such as calcium carbonate, iron oxides, and adhering particles. After cleaning and drying, they are viewed in a scanning electron microscope at an appropriate magnification. The relative abundances of surface textural features are then tabulated on a chart. Comparison of such a chart with a similar chart prepared from analysis of grains from known modern environments and experimental studies (e.g., Fig. 2.17) allows interpretation of the depositional environment. Krinsley and Trusty (1986) suggest several applications of surface textural data in addition to simple environmental interpretation. These applications include identification of multiple environmental episodes, studies of multiple sources, tracing a particular sedimentary environment through a series of progressively younger sedimentary beds, and possibly microcorrelation from one geographic area to another (Table 2.6). SEM studies of the textural features of tephra and other pyroclasts (features such as the fraction of bubbles, shapes of bubbles, the percent of glass coating on crystals, and the shapes of pyroclasts) are now being used also as a tool for investigating such aspects of explosive volcanism as magma viscosities, vesicle growth, interaction of magma with surface water, energy release during eruptions, and the mode

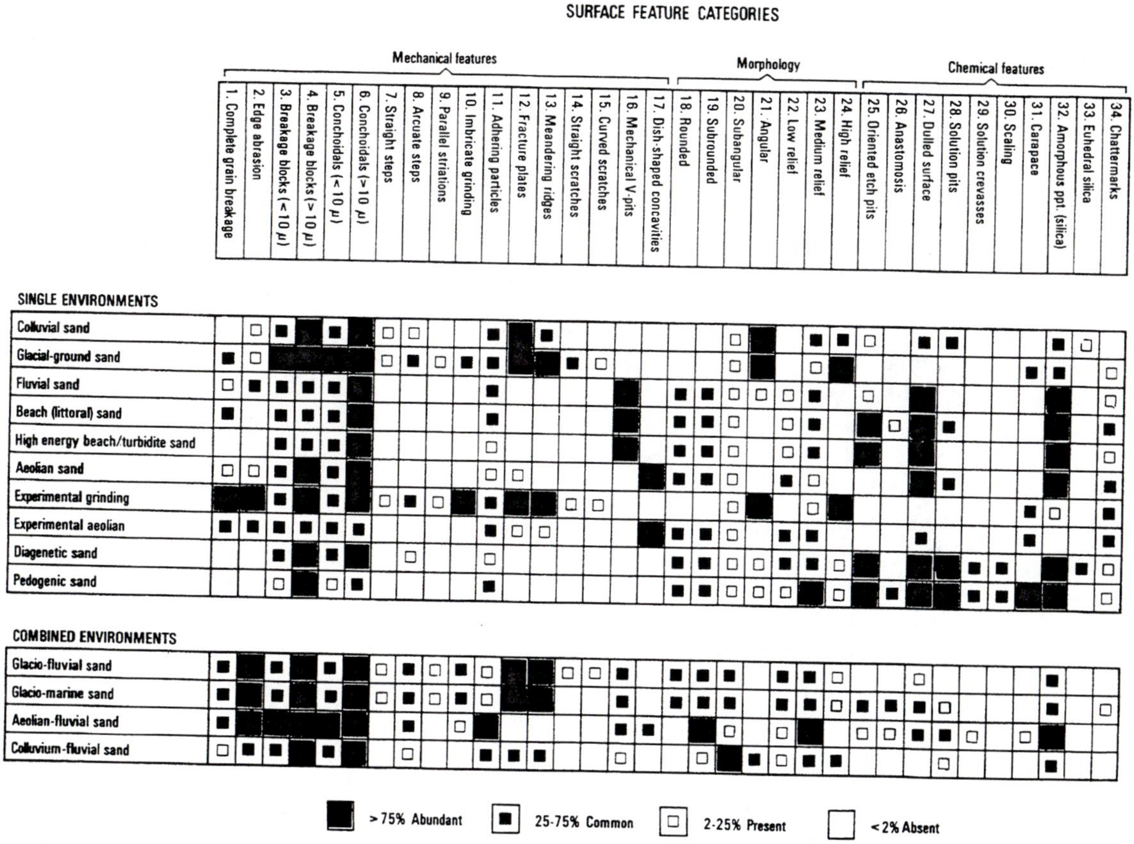

SURFACE FEATURE CATEGORIES

Mechanical features:
1. Complete grain breakage
2. Edge abrasion
3. Breakage blocks (< 10 μ)
4. Breakage blocks (> 10 μ)
5. Conchoidals (< 10 μ)
6. Conchoidals (> 10 μ)
7. Straight steps
8. Arcuate steps
9. Parallel striations
10. Imbricate grinding
11. Adhering particles
12. Fracture plates
13. Meandering ridges
14. Straight scratches
15. Curved scratches
16. Mechanical V-pits
17. Dish-shaped concavities

Morphology:
18. Rounded
19. Subrounded
20. Subangular
21. Angular
22. Low relief
23. Medium relief
24. High relief

Chemical features:
25. Oriented etch pits
26. Anastomosis
27. Dulled surface
28. Solution pits
29. Solution crevasses
30. Scaling
31. Carapace
32. Amorphous ppt (silica)
33. Euhedral silica
34. Chattermarks

SINGLE ENVIRONMENTS:
- Colluvial sand
- Glacial-ground sand
- Fluvial sand
- Beach (littoral) sand
- High energy beach/turbidite sand
- Aeolian sand
- Experimental grinding
- Experimental aeolian
- Diagenetic sand
- Pedogenic sand

COMBINED ENVIRONMENTS:
- Glacio-fluvial sand
- Glacio-marine sand
- Aeolian-fluvial sand
- Colluvium-fluvial sand

Legend: ■ >75% Abundant   ■ 25-75% Common   □ 2-25% Present   □ <2% Absent

FIGURE 2.17

Schematic representation of the relative abundance of surface textural features found on quartz grains from various environments. (*From Bull, P. A., 1986, Procedures in environmental reconstruction by SEM analysis, in Sieveking, G. De C., and M. B. Hart, eds., The scientific study of flint and chert. Fig. 24.1, p. 222, reprinted by permission of Cambridge Univ. Press, Cambridge.*)

TABLE 2.6
Applications of SEM study of quartz surface textures

1. Selection of several environmental episodes in a sample of sand grains and the determination of their respective ages
2. Studies of how diagenesis occurs at the microlevel, e.g., cementation and pressure solution
3. Comparison of grains from multiple sources
4. Tracing of particular environments up the sedimentary column by comparing specific textures in a series of progressively younger samples
5. Determination of environments on land at specific times in Earth history from information located in ocean cores
6. Study of ancient wind velocities
7. Study of eolian and other environmental abrasion mechanisms in the laboratory
8. Studies of the origin of fine silt and clay particles in the geologic column
9. Microcorrelation from one geographic area to another with respect to individual sedimentary layers
10. Relation of quartz sands to various soil horizons and types of soils

Source: Krinsley, D., and P. Trusty, 1986, Sand grains surface textures, *in* Sieveking, G. De C., and M. B. Hart, eds., The scientific study of flint and chert: Cambridge University Press, Cambridge, England, p. 201–207.

of deposition of pyroclastic material (ash fall, pyroclastic surges, etc.) (Heiken, 1987). Application of the SEM to investigation of particle surface textures is also being made in the field of archaeology, where study of microscopic wear traces on artifacts is of considerable interest.

Advantages and Limitations of Surface Textural Analysis

The major advantage of SEM analysis of surface textures is that SEM imagery allows a researcher to study small-scale shape features that cannot be resolved by any other shape-study technique. Even Fourier analysis characterizes only relatively large bumps on a grain periphery—those bumps that subtend angles of 7.5 degrees or more when measured from the grain center (Ehrlich et al., 1987). Furthermore, the SEM permits direct observation of surface features; in contrast, Fourier analysis "sees" grain shapes only indirectly. Another advantage of SEM imagery is that surface texture changes more readily during transport and abrasion than do roundness and sphericity and thus may be a more sensitive indicator of environment. At the same time, quartz is sufficiently durable that a grain that has passed through several environments may retain some diagnostic markings from each environment. Thus, surface texture may reveal multiple episodes in the transport history, making it possible for a researcher to trace its passage through these environments.

On the other hand, the fact that quartz retains its surface markings for some time could lead to incorrect interpretations. For example, quartz sand deposited on the continental shelf from a melting iceberg would bear the textural imprint of glaciation. If such a sand was buried quickly before shelf processes had an opportunity to imprint markings characteristic of a subaqueous environment, the surface texture of the quartz would indicate that the last depositional environment was a glacial setting on land. Another weakness of SEM textural analysis is that the technique is less quantitative than Fourier analysis; it is, in fact, a semiquantitative or qualitative technique in many ways. Because of this fact, operator error and the reproducibility of data can be problems. To take advantage of the best features of both SEM imagery and Fourier analysis, many investigators are now using these two techniques together.

2.3.7   Concept of Textural Maturity

Sedimentologists often use the term textural maturity in reference to the textural characteristics of a particular sediment. Folk (1951) suggested that textural maturity of sandstones encompasses three textural properties: (1) the amount of clay-size sediment in the rock, (2) the sorting of the framework grains, and (3) the rounding of the framework grains. He visualized four stages of textural maturity: immature, submature, mature, and supermature (Fig. 2.18). Any sandstone containing considerable clay, say more than 5 percent, is in the immature stage. Also, the framework grains in immature sediments are poorly sorted and poorly rounded. Presumably, immature sediments have not undergone sufficient sediment transport and reworking to remove fine-size material and produce sorting and rounding of grains. With additional sediment transport and reworking, sediments enter the submature stage, in which the sediments are characterized by low clay content but grains are still not well sorted or well rounded. This stage is followed by the

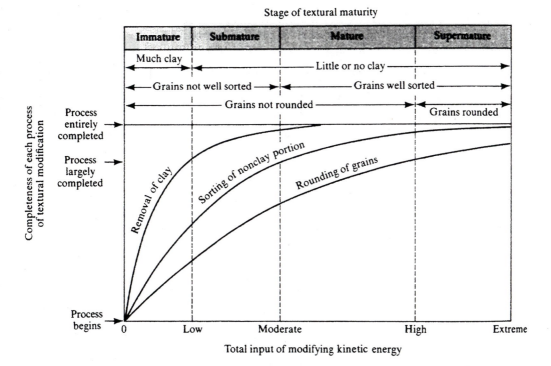

FIGURE 2.18

Textural maturity classification of Folk. Textural maturity of sands is shown as a function of input of kinetic energy. (*From Folk, R. L., 1951, Stages of textural maturity in sedimentary rocks: Jour. Sed. Petrology, v. 21. Fig. 1, p. 128, reprinted by permission of Society of Economic Paleontologists and Mineralogists, Tulsa, Okla.*)

mature stage, in which clay content is low and framework grains become well sorted but are not yet well rounded. Sediments in the supermature stage are essentially clay-free, and framework grains are both well sorted and well rounded.

The textural maturity concept is a useful one for characterizing sediments; however, to attribute progressive increase in textural maturity primarily to increasing total input of modifying kinetic energy is probably overly simplistic. Matlack et al. (1989) have shown, for example, that the vadose infiltration of clays into sandy sediments is influenced by several variables in addition to the hydrologic conditions and depositional processes within a depositional environment. These additional factors may include grain size of the sand, size and shape of clay particles, and concentration of suspended sediment. Another significant difficulty with the concept stems from the likelihood that much of the clay-size material in many sandstones is of diagenetic origin. Therefore, its presence may have little or nothing to do with the spectrum of energy expended during transport and deposition. Also, many mature to supermature sandstones may have been recycled one or more times. Thus, the total expenditure of energy necessary to produce rounding may not have come in a single depositional cycle.

2.3.8   Fabric

General Concept

The properties of grain size and shape have to do with the characteristics of individual grains. Fabric refers to the textural characteristics displayed by aggregates of grains. Fabric encompasses two properties of grain aggregates: grain packing and grain orientation. Grain packing is a function of the size and shape of grains and the postdepositional physical and chemical processes that bring about compaction of sediment. Grain orientation is mainly a function of the physical processes and conditions operating at the time of deposition; however, original grain orientation can be modified after deposition by the activities of organisms (bioturbation) and to some extent by the processes of compaction during diagenesis.

Grain Packing

Interest in the packing of sediment dates back to the late nineteenth century (see summary by Griffith, 1967, p. 166). Griffith points out that most early investigators were primarily concerned with the theoretical and experimental treatment of packing (e.g., the in-depth study of the packing of spheres done by Graton and Fraser, 1935). Serious study of packing in natural sediments did not begin until the late 1930s.

A much quoted definition of packing is that of Graton and Fraser (1935, p. 790), who defined packing as "any manner of arrangement of solid units in which each constituent unit is supported and held in place in the Earth's gravitational field by tangent contact with its neighbors." Kahn (1965, p. 390) defined packing simply as the "mutual spatial relationship among . . . grains." The AGI *Glossary of Geology* (Bates and Jackson, 1980, p. 449) refers to packing as "the manner of arrangement or spacing of the solid particles in a sediment or sedimentary rock . . .; specifically, the arrangement of clastic grains entirely apart from any authigenic grains that may have crystallized between them." Pandalai and Basumallick (1984, p. 87) suggest that packing is "the effective utilization of space by mutual arrangement of the constituent grains of an aggregate."

Whatever its definition, packing is regarded to be a function of several variables or properties, including particle size and sorting, particle shape, and particle orientation or arrangement. Numerous attempts have been made to reduce the concept of packing to an operational level, such that packing can be measured and specified in terms of specific properties of individual grains or sets of grains. Some of these attempts are summarized by Griffith (1967, p. 166), Pandalai and Basumallick (1984, p. 88), and Pettijohn et al. (1987, p. 86). Table 2.7 is a further summary of these operational measures of packing. Packing is of special interest to geologists concerned with the porosity and permeability of sediments and with changes in these parameters as a function of burial depth and sediment compaction.

Particle Orientation

General Statement.   Platy, flaky, or elongated particles in sedimentary rocks commonly display some degree of orientation that reflects the nature of the depositional process. For example, small platy or flaky particles settling from suspension onto a flat bed in the absence of current flow are commonly deposited with their flattened dimensions parallel to bedding surfaces (Fig. 2.19A). Small elongated grains settling under the same conditions also tend to have their long dimensions oriented approximately parallel to the bedding surface; however, the grains may have any orientation (random arrangement) within the bedding plane (Fig. 2.19B). Small or large particles transported and deposited by traction currents or by sediment gravity flows generally display an orientation that reflects the flow direction of the depositing current. Thus, particles may be oriented with long or flattened dimensions parallel to bedding but have orientations within the bedding plane that are either parallel to current flow (Fig. 2.19C) or perpendicular to current flow (Fig. 2.19D), depending upon particle size and the nature of the flow. On the other hand, under some conditions of flow, the particles may lie at an angle to bedding surfaces in a shingled arrangement called imbrication (Fig. 2.19E).

The primary interest in particle orientation arises from its potential usefulness in paleocurrent analysis. In the absence of distinctive directional sedimentary structures in a deposit, particle orientation may provide the only clue to paleocurrent directions. Particle orientation is also believed to have an effect on the permeability of sedimentary rocks to fluid movement and thus is of interest to hydrologists and petroleum geologists. A few investigators have suggested that particle orientation may have significance in paleoenvironmental studies; however, use of particle orientation as an environmental tool is not fully developed. Orientation analysis has been applied to the orientation of clasts and mineral grains in siliciclastic sedimentary rocks, clasts and shells in carbonate rocks, and even to waterlogged wood.

Methods of Studying Orientation.   The orientation of pebble- and cobble-size particles can be measured comparatively easily. If pebbles have a prolate (elongated) shape, the orientation (compass bearing) and dip (plunge) of the long particle axis is commonly measured. The orientation of plate- or disc-shaped particles is generally not expressed in terms of the long axis; instead, the orientation of the L-I plane (the plane containing the long and intermediate axes; often called the ab plane) is specified. The bearing of the strike of the L-I plane and the amount and direction of dip of the plane are all needed to completely describe the orientation of such pebbles. Measuring the orientation of sand-size and smaller particles is much more difficult. Some techniques in use for determining the orientation of grains in fine-size sediments are (1) techniques based on the magnetic susceptibility of sediments (Taira and Lienert, 1979; Rees, 1965), which is related to the shapes and orientations of magnetic particles (as well as some other properties), (2)

TABLE 2.7

Operational measures of grain packing in natural sediment

Types of Grain Contacts and Numbers of Contacts (Taylor, 1950, p. 707).

Recognizes five types of grain contacts: floating grains (no contacts), tangential (point), long, concavo-convex, sutured. Shows that the percentage of grain contact types and the number of contacts per grain changes with increasing depth of burial (compaction).

Floating   Point   Long   Concavo-convex   Sutured

Packing Index (Emery and Griffith, 1954, p. 71). A measure of packing defined as

Number quartz-to-quartz grain contacts per thin-section traverse × average quartz diameter ÷ total length of thin-section traverse

Condensation Index (Allen, 1962, p. 678). Ratio of percent fixed rock fragments to percent free grains. A fixed grain is a grain in which the fixed margin (length of margin touching other grains) exceeds the free margin (not touching other grains); a free grain is the opposite. Applied to lithic sandstones as a measure of compaction.

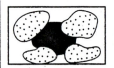

Fixed grain

Rock fragment

Harder grain

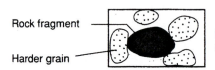

Free grain

Packing Density (Kahn, 1956, p. 390). The sum of all grain diameters intercepted in a thin-section traverse ÷ total length of the traverse × 100.

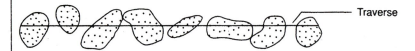

Traverse

Packing Proximity (Kahn, 1956, p. 392). Ratio of grain-to-grain contacts in a thin-section traverse to the total number of contacts (grain to cement, grain to matrrix) × 100 [i.e., the percent grain-to-grain contacts in a traverse of *n* grains].

Horizontal and Vertical Packing Intercepts (Mellon, 1964, p. 799). The average horizontal and vertical distance between framework grains, or grains and patches of matrix (a measure of the size of open pore space, or cement-filled intergranular pore space).

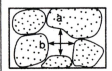

a = vertical packing

b = horizontal packing

Radial Distribution Function (Smalley, 1964a, 1964b). The number of equal-size particles in a spherical shell of thickness dr, at distance r from a randomly chosen particle p, is $4\pi r^2 \, dr \, p_0$ (where p_0 is average density; i.e., number of particles divided by the volume of the system). A plot of this function against r shows the density variations around each particle in the sediment.

Contact Strength (Füchtbauer, 1967, p. 365). A measure of packing utilizing Taylor's (1950) grain-contact types. These contact types are combined linearly in a formula that yields a single number that reflects packing:

$$\text{Contact strength} = \frac{1a + 2b + 3c + 4d}{a + b + c + d}$$

where a = number point contacts, b = number long contacts, c = number concavo-convex contacts, and d = number sutured contacts.

Weighted Contact Packing (Hoholick et al., 1982, p. 79). A modification of Füchtbauer's (1967) contact-strength formula that weights the basic contact types by powers of two and is alleged to be more discriminatory of fabrics in quartz-rich sandstones than is Füchtbauer's formula:

$$\text{WCP} = \frac{1a + 2b + 4c + 8d + 16e}{a + b + c + d + e}$$

where a = number floating grains, b = number point contacts, c = number long contacts, d = number concavo-convex contacts, and e = number sutured contacts.

Grain Volume Fraction (Vinopal and Coogan, 1978, p. 16). A measure of the grain volume (complement of porosity) of carbonate sands of different shapes expressed by the formula:

$$\text{Grain volume} = \frac{\Sigma(\% \text{ of } Xs_i)(\%GVXs_i) + \Sigma(\% \text{ of } Xr_i)(\%GVXr_i)}{\text{number of shape components}} - 7\%GV$$

where GV is grain volume, and Xs and Xr are simple and radical shape components.

Intrinsic Packability Factor (Pandalai and Basumallick, 1984, p. 89). A measure of packing expressed by the formula

$$P_f = \frac{\Sigma V_i}{V_d}$$

where V_i = volume occupied by a packed mass of constituent grains of a particular size, packed in a dense random state, and V_d = the pack volume when all grains are packed together in dense random state.

Tight Packing Index (Wilson and McBride, 1988, p. 664)

$$TPI = \text{average number of long, sutured, and embayed contacts/grain}$$

Applied particularly to studies of porosity change with depth. The *TPI* gradient is 0.89/1000 m for samples with less than 10 percent cement.

photometric or optical method, which involves observations of changes in the interference color of quartz when viewed through a petrographic microscope with crossed nicols and a gypsum plate (these interference colors are related to the orientation of the optic C-axis of quartz, which tends to be parallel to the long crystal dimension of quartz, although its orientation can apparently vary by as much as 40 degrees from parallel) (Sippel, 1971; Taira and Lienert, 1979; Zimmerle and Bonham, 1962), (3) microscopic counting technique, in which the directions of all grains showing definite elongation are determined by point counting in thin section (Dapples and Rominger, 1945), (5) dielectric anisotropy

QUIET WATER

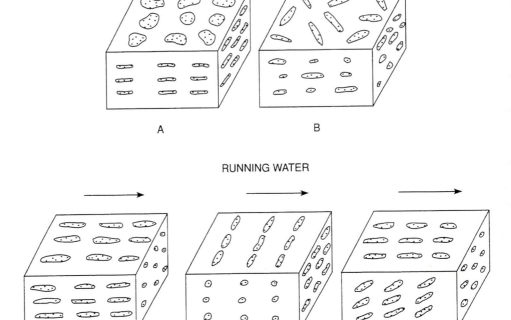

RUNNING WATER

FIGURE 2.19
Hypothetical, schematic arrangement of grains in sediments. A. Platy or flaky grains deposited in quiet water onto a flat bed. B. Elongated grains deposited in a random arrangement in quiet water. C. Elongated grains deposited under current flow with long dimensions parallel to current flow. D. Elongated grains deposited under current flow. E. Elongated grains deposited in an imbricated arrangement under current flow. Arrows indicate the direction of current flow.

(anisotropic electrical conductivity), a technique that involves measuring the maximum electrical capacitance of a sample and in which the direction of maximum capacitance is considered to be the preferred grain orientation (Matalucci et al., 1969; Shelton et al., 1974; Winkelmolen, 1972), (6) image analysis, in which the methods of image analysis are applied to the problem of determining the orientation of elongated grains in thin sections (Schäfer and Teyssen, 1987).

Once the orientation of a statistically significant number of grains (perhaps 100 or more) has been measured, the orientation data are commonly plotted on a fabric diagram (Fig. 2.20). If a particle can be imagined to be located in its properly oriented position in the center of a sphere and the particle axes are elongated to intersect the sphere, each axis would cut the sphere in two places—one in the upper hemisphere and one in the lower. Commonly, only the lower hemisphere is used in such projections, as shown in Figure 2.20A. The points where the axes cut the lower hemisphere are then projected onto the equatorial plane of the hemisphere, as seen from directly above (Fig. 2.20B). The geographic positions of the points within the four quadrants of the equatorial plane indicate

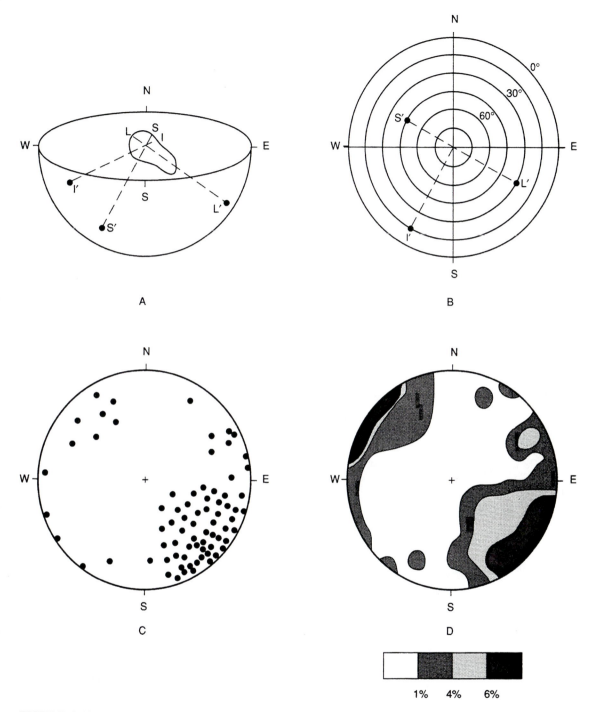

FIGURE 2.20
Method of graphically displaying particle orientations using fabric diagrams. A. Extension of the
long (L), intermediate (I), and short (S) axes of a particle to intersect the lower hemisphere at
points L′, I′, and S′. B. Projection of L′, I′, and S′ onto the equatorial plane of the lower
hemisphere, as seen from directly above. C. A point diagram showing the projections of the
long axes of 75 prolate particles. D. The point diagram contoured to show the percent points per
one percent area of the diagram.

the bearing of the axes (e.g., point L' in Fig. 2.20B indicates that the long particle axis is oriented NW–SE and is dipping SE; point I' indicates that the intermediate axis is oriented NE–SW and is dipping SW). The relative distance of the points from the center of the hemisphere shows the amount of dip of the axes. Points located near the center indicate axes with high dips; those near the edge imply low dips (e.g., point I' has a dip of 15 degrees; point S' has a dip of 45 degrees). We may plot the orientation of one or more axes to specify the orientation of the particle. Commonly, we plot the long axes of prolate particles or the short axes of plate- or disc-shaped particles. Figure 2.20C is a "point" diagram showing the orientations of the long axes of 75 prolate pebbles. Note that the long axes of most of these pebbles are oriented NW–SE, and most dip to the SE at inclinations ranging from horizontal to about 75 degrees. For easier visualization of particle orientation, point diagrams are often contoured as shown in Figure 2.20D. The contours generally represent the percentage of particles that fall in each one percent area of the hemisphere (percent per one percent area). Further details of techniques for contouring point diagrams may be found in Turner and Weiss (1963, p. 61–64).

Interpretation of grain-orientation data involves an understanding of how particles behave under different transport and depositional conditions. Numerous studies of particle orientation, particularly the orientation of pebble-size particles, have been reported since the 1930s. These studies suggest that particle orientation may be a function of several variables, including the nature of flow (e.g., current flow versus sediment gravity flow), bed slopes, flow velocity, particle size, particle shapes, and packing density (the number of particles that touch each other during transport). Unfortunately, the ways in which these different variables interact are not well understood. From the standpoint of paleocurrent analysis, we are particularly interested in (1) the orientation (and dip) of the long axes of prolate (elongated) particles or the L-I plane of less-elongated particles, and (2) the direction and amount (dip) of imbricated particles. Several empirical studies and a limited number of experimental investigations have addressed these problems of interpretation. Table 2.8 summarizes orientation data for various kinds of sedimentary deposits.

I have not attempted to summarize all current ideas and data pertaining to particle orientation. Interested readers may wish to consult Potter and Pettijohn (1977, p. 30–68) for a general synopsis of earlier investigations. Those readers desiring a more rigorous, mathematical treatment of the concepts of particle orientation should consult Allen (1982, p. 179–235).

2.4   Porosity and Permeability

2.4.1   General Statement

Porosity and permeability are important secondary, or derived, properties of sedimentary rocks that are controlled in part by the textural attributes of grain size, shape, packing, and arrangement. Because porosity and permeability are, in turn, controlling parameters in the movement of fluids through rocks and sediment, they are of particular interest to petroleum geologists, petroleum engineers, and hydrologists concerned with groundwater supplies and liquid waste management. Porosity and permeability also play an extremely important role in the diagenesis of sediments by regulating the flow through rocks of fluids that promote dissolution, cementation, and authigenesis of minerals.

TABLE 2.8

Particle imbrication and orientation in various kinds of sedimentary deposits

Type of deposit	Imbrication direction and dip	Orientation of long axes of prolate clasts or L-I planes of blade- or disc-shaped clasts	Remarks	References
Gravels				
Fluvial gravel	Dip upstream 10°–30°	Dominantly transverse to flow; less commonly parallel to flow	Parallel orientation possibly related to high flow velocities and increasing pebble concentration on the bed	Lane and Carlson, 1954; Unrug, 1957; Doeglas, 1962; Sedimentary Petrology Seminar, 1965; Kelling and Williams, 1967; Rust, 1972
Glacial deposits	Poorly imbricated	Strong preferred orientation parallel to direction of ice flow	Parallel orientation weaker in basal lodgment till; water-laid glacial gravels may have random fabric	Lawson, 1979; Dowdeswell et al., 1985; Dowdeswell and Sharp, 1986
Lahars	Weak imbrication with low-angle upflow dips	Variable; both flow-parallel and flow-transverse orientations reported	Clast fabrics complex and variable	Mills, 1984; Major and Voight, 1986
Debris flows	Poorly imbricated	Slight tendency toward flow-parallel orientation	Weak orientation	Lewis et al., 1980
Turbidites	Dip upflow at about 10°	Flow-parallel orientations dominant	Stronger clast orientation than in debris flows	Davies and Walker, 1974; Allen, 1982, p. 219–220
Beach gravel	Dip seaward at angles exceeding foreslope angle	Variable; ranging from parallel to perpendicular to 45° to beach trend	Some clasts may dip landward	Waag and Ogren, 1984; Ogren and Waag, 1986
Sands	Commonly dip upstream 10°–20°; less commonly dip downstream	Dominant flow-parallel orientation, but transverse orientation possible	Sediment size, sorting, transport mode, and flow velocity may all affect orientation	Parkash and Middleton, 1970; Gupta et al., 1987; Rees, 1968, 1979, 1983; Rees and Woodall, 1975

2.4.2 Porosity

Porosity is defined as the ratio of pore space in a sediment or sedimentary rock to the total volume of the rock. It is commonly expressed in percent as

$$\text{porosity } (\%) = V_p/V_b \times 100 \tag{2.17}$$

where V_p is pore volume and V_b is bulk volume. This formula yields the **total** or **absolute porosity**. Petroleum geologists and hydrologists are often more interested in the effective porosity, which is the ratio of the interconnected pore space to the bulk volume of a rock:

$$\text{effective porosity } (\%) = IV_p/V_b \times 100 \tag{2.18}$$

where IV_p is interconnected pore volume. It is the effective porosity, commonly several percent less than total porosity, that controls the movement of fluids through rock.

In terms of origin, porosity may be either **primary** (depositional) or **secondary** (postdepositional). Primary porosity can be of three types: (1) **intergranular** or **interparticle**—pore space that exists between or among framework grains, such as siliciclastic particles and carbonate grains (oolites, fossils, etc.), (2) **intragranular** or **intraparticle**—pore space within particles, such as cavities in fossils and open space in clay minerals, and (3) **intercrystalline**—pore space between chemically formed crystals, as in dolomites. Secondary porosity may include (1) **solution porosity,** caused by dissolution of cements or metastable framework grains (feldspars, rock fragments) in siliciclastic sedimentary rocks or by dissolution of cements, fossils, framework crystals, etc. in carbonate or other chemically formed rocks, (2) **intercrystalline porosity,** arising from pore space in cements or among other authigenic minerals, and (3) **fracture porosity,** owing to fracturing of any type of rock by tectonic forces or by other processes such as compaction and desiccation.

Porosity can be measured by a variety of techniques ranging from the purely qualitative to quantitative. Qualitative ''estimates'' of porosity can be made by scanning a rock specimen with a hand lens or under the microscope. Somewhat more quantitative estimates can be made by point-counting techniques with a petrographic microscope. Before preparing a thin section for porosity analysis, the specimen is impregnated with a colored (commonly blue) epoxy resin. Impregnation prevents the pores from becoming damaged during thin-section grinding, and the colored dye makes the pores easily visible under the microscope (Fig. 2.21). Counts of 500 or more points provide a semiquantitative estimate of porosity. More-accurate measurements of porosity can be made in the laboratory with a variety of instrumental techniques (Monicard, 1980). Most of these techniques determine the effective porosity of a specimen by first measuring the pore volume. Pore volume is calculated by determining the volume of fluid or gas that can be forced into the rock to completely fill the pores. The porosity is then calculated by Equation 2.18, with IV_p considered to be equal to the volume of fluid forced into the pores. An alternative technique is to determine pore volume indirectly by first determining the grain volume of a specimen. The grain volume is determined by dividing the dry weight of the specimen by the rock grain density (about 2.65 in the case of a quartz-rich sandstone). The pore volume is the grain volume subtracted from the bulk volume. This method works reasonably well for clean quartz sandstones and limestones but is less accurate for rocks with highly variable compositions, owing to the difficulty of determining rock grain density.

FIGURE 2.21

Quartz-rich sandstone impregnated with (blue) epoxy resin (arrow) to reveal porosity. Miocene, Louisiana. Field view of 1.3 mm wide. *(Photograph courtesy of D. W. Houseknecht.)*

Petroleum geologists are often required to determine the porosity of particular rock units in well bores after drilling is completed. If the well has not been cored, and thus samples of these rock units are not available, indirect methods for determining porosity must be used. These methods involve semiquantitative measurements of porosity from various types of instrumental well logs or petrophysical logs (electric logs, sonic logs, neutron logs, etc.). Discussion of these techniques is beyond the scope of this book. A brief, elementary discussion of instrumental well logs and porosity measurements from such logs is given by Selley (1985, p. 52–78). Many other more-advanced references on this topic are also available (e.g., Asquith, 1982; Merkel, 1979; Pierson, 1983).

The porosity of sedimentary rocks is affected by numerous variables. Some of these variables are physical characteristics, such as grain size, sorting, shape, packing, and grain arrangement, that are the result, at least in part, of depositional processes. Other factors that affect porosity are caused by postdepositional processes such as compaction (which can rearrange packing), solution, and cementation. Freshly deposited, unconsolidated sediment may have porosities of 40–50 percent, or higher. During diagenesis, this initial porosity can be reduced to essentially zero, depending upon depth of burial, by a combination of compaction and cementation. On the other hand, depending upon geochemical conditions, diagenetic solution of cements and framework grains can generate considerable secondary porosity. The ultimate porosity of a rock thus depends upon both the initial depositional conditions and the diagenetic history of the rock. Additional details of porosity and porosity changes in sedimentary rocks are explored in appropriate sections of subsequent chapters, particularly Chapters 9 and 12.

2.4.3   Permeability

Permeability is commonly defined as the ability of a medium to transmit a fluid. Rock permeability can be thought of more simply as the property of a rock that permits the passage of a fluid through the interconnected pores of the rock. Much original work on fluid flow through porous media was performed by the French scientist Henri Darcy about 1856. Subsequent workers have quantified the passage of fluids through porous substances and formulated an equation for fluid flow that is commonly called Darcy's law. This equation is expressed as

$$Q = K(P_1 - P_2)A/\mu L \tag{2.19}$$

where Q is the flow rate, K is the permeability constant or permeability of the flow medium, $P_1 - P_2$ is the pressure drop across length L of the medium, A is the cross-sectional area of the medium, and μ is the viscosity of the fluid measured in centipoises. From this equation, it can be seen that the rate of flow of a fluid through a porous rock is directly proportional to the rock permeability and indirectly proportional to the fluid viscosity. Strictly speaking, Darcy's law applies when only one fluid is present in the rock and no chemical reaction takes place between the fluid and the rock.

The qualitative permeability of a rock (poor, fair, good) can be guessed by techniques such as putting a drop of water on the rock and noting the rate at which the water is absorbed into the rock. Quantitative permeability is determined in the laboratory with an instrument that can measure the rate of flow of a fluid of known viscosity across a sample of cross section A and length L. These values are then plugged into the above equation to yield the mathematical value of K. Permeability is expressed in darcies or millidarcies (0.001 darcy). A darcy is defined as a unit of permeability equivalent to the passage of one cubic centimeter of fluid of one centipoise viscosity flowing in one second under a pressure differential of one atmosphere through a porous medium having an area of cross section of one square centimeter and a length of one centimeter.

Permeability is a complex function of particle size, sorting, shape, packing, and orientation of sediments. The exact relationship of permeability to each of these variables is still not fully understood, and no attempt is made here to develop a rigorous treatment of these relationships. In a very general way, permeability is believed to decrease with decreasing particle size (owing to the decrease in pore diameters and increase in capillary pressures) and decreasing sorting. It may possibly decrease with increasing sphericity (perhaps owing to tighter packing of spheres) and increasingly tighter or denser packing (which may reduce pore size). Also, permeability is affected by particle orientation; that is, permeability seems generally to be greater parallel to an oriented fabric, e.g., parallel to bedding planes, than perpendicular to an oriented fabric, although this may not be true in every case. Permeability also tends to increase with increasing effective porosity, but it may not be positively correlated with total porosity. For example, very fine-grained sediments may have high total porosity but very low permeability. A more extended discussion of the relationship of permeability to the above variables is given in Pettijohn et al. (1987) and Selley (1985). Interested readers may also wish to consult the voluminous literature on petroleum geology (e.g., publications of the American Association of Petroleum Geologists) that deals with this subject.

▰▰▰▰▰▰ Additional Readings

Carver, R. E., ed., 1971, Procedures in sedimentary petrology: John Wiley & Sons, New York, 653 p.

Folk, R. L., 1951, Stages of textural maturity in sedimentary rocks: Jour. Sed. Petrology, v. 21, p. 127–130.

————1974, Petrology of sedimentary rocks: Hemphill, Austin, Texas, 182 p.

Griffiths, J. C., 1967, Scientific methods in analysis of sediments: McGraw-Hill, New York, 508 p.

Krinsley, D., and J. Dornkamp, 1973, Atlas of quartz sand surface textures: Cambridge Univ. Press, Cambridge, 91 p.

Marshall, J. R., ed., 1987, Clastic particles: Van Nostrand Reinhold, New York, 346 p.

Monicard, R. P., 1980, Properties of reservoir rocks: Core analyses: Gulf Pub. Co., Houston (Editions Technip, Paris), 168 p.

Whalley, W. B., ed., 1978, Scanning electron microscopy in the study of sediments: Geo Abstracts, Norwich, England, 414 p.

Chapter 3
Sedimentary Structures

3.1 Introduction

Study of sedimentary structures has captured the interest of geologists for decades. Some sedimentary structures such as cross-bedding and ripple marks were recognized as early as the late eighteenth century and perhaps well before. Progress in identification, description, classification, and interpretation of sedimentary structures has been especially rapid since the 1950s, and the fundamental origin of most sedimentary structures is now reasonably well understood. Nonetheless, empirical study of modern and ancient sediments and experimental investigation of the mechanisms that form sedimentary structures continue. Geologists are especially interested in understanding how specific sedimentary structures are related to such aspects of ancient depositional environments as relative water energy, water depth, and current flow directions. Investigation of the origin and significance of bedforms such as ripples and dunes has been a particularly active field of research since the 1950s to 1960s.

Many sedimentary structures originate by physical processes involving moving water or wind that operate at the time of deposition. Others are formed by physical processes such as gravity slumping or sediment loading that deform unconsolidated sediment after initial deposition (soft-sediment deformation). Still other structures are of biogenic origin, formed by the burrowing, boring, browsing, or sediment-binding activities of organisms. Some types of bedding, the laminated bedding of evaporites, for example, are generated by primary chemical precipitation processes. A few other structures, such as concretions, form by chemical processes operating within sediment during burial and diagenesis; thus, they are regarded to be secondary in origin. Short discussions of the processes that act to form the major kinds of sedimentary structures are given in appropriate parts of this chapter; however, detailed analysis of the sedimentary processes involved in the formation of sedimentary structures is beyond the intended scope of the book. Other topics covered in this chapter include classification of sedimentary structures and a description of the major kinds of structures.

Readers who wish more information on sedimentary structures may turn to a variety of additional sources. For example, a rigorous treatment of the mechanisms of sediment transport involved in generation of bedforms and other types of structures is given in Allen (1982). Biogenic activities important to the formation of sedimentary structures are discussed in detail in the books of Basan (1978), Crimes and Harper (1970), Curran (1985), Ekdale et al. (1984), and Frey (1975). Collinson and Thompson (1982) provide a general reference work for study and classification of all types of sedimentary structures, as well as an extensive bibliography of pertinent monographs, textbook chapters, and papers published prior to 1982. Recent textbook chapters summarizing sedimentary structures include those of Boggs (1987) and Pettijohn et al. (1987). Other pertinent references are listed under additional readings at the end of this chapter.

3.2   Major Kinds of Sedimentary Structures

Several methods have been proposed for classifying the many kinds of sedimentary structures now recognized (see review by Conybeare and Crook, 1968, p. 6–13; also, Pettijohn and Potter, 1964, p. 5). Structures may be classified purely on the basis of their morphological or descriptive characteristics or on the basis of presumed mode of origin. Neither of these methods is entirely satisfactory. Descriptive classification provides little or no information about the genesis of structures; also, it is somewhat awkward trying to fit all structures into a few descriptive categories. On the other hand, genetic classifications are subjective and can be misleading. Some structures can form by more than one process or by a mixture of processes and hence can be classified under different genetic categories. The classification shown as Table 3.1 attempts a compromise by listing primary sedimentary structures under both morphological and genetic headings. In any case, this table provides a reference point for further discussion of the major kinds of sedimentary structures. The structures are discussed mainly under the descriptive headings shown in Table 3.1; however, some kinds of structures are further subdivided for discussion by genetic category.

3.3   Bedding and Bedforms

3.3.1   Nature of Bedding

All sedimentary rocks occur in beds of some kind. **Beds** are tabular or lenticular layers of sedimentary rock having characteristics that distinguish them from strata above and below. Griffith (1961) suggests that beds are a function of and are distinguished by the composition, size, shape, orientation, and packing of sediment. Beds are separated by **bedding planes** or **bounding planes** into units that may range widely in thickness. Although the term bed is used in an informal sense for any sedimentary layer, most workers formally designate as beds only those layers thicker than 1 cm. Layers thinner than 1 cm are **laminae**. On the other hand, Campbell (1967) suggests that beds have no limiting thickness and can range in thickness from a few millimeters to a few tens of meters. Various schemes have been proposed to describe the thickness of beds in more detail. Figure 3.1 shows one commonly used classification of bed thickness.

TABLE 3.1

Classification of common primary sedimentary structures

MORPHOLOGICAL CLASSIFICATION \ GENETIC CLASSIFICATION	Depositional structures			Erosional structures		Deformation structures						Biogenic structures	
	Suspension-settling and current- and wave-formed structures	Wind-formed structures	Chemically and biochemically precipitated structures	Scour marks	Tool marks	Slump structures	Load and founder structures	Injection (fluidization) structures	Fluid-escape structures	Desiccation structures	Impact structures (rail, hail, spray)	Bioturbation structures	Biostratification structures
STRATIFICATION AND BEDFORMS													
Bedding and lamination													
Laminated bedding	X	X	X										
Graded bedding	X											X	
Massive (structureless) bedding	X											X	
Bedforms													
Ripples	X	X											
Sand waves	X												
Dunes	X	X											
Antidunes	X												
Cross-lamination													
Cross-bedding	X	X											
Ripple cross-lamination	X	X											
Flaser and lenticular bedding	X												
Hummocky cross-bedding	X												
Irregular stratification													
Convolute bedding and lamination							X						
Flame structures							X						
Ball and pillow structures							X						
Synsedimentary folds and faults						X							
Dish and pillar* structures									X				
Channels				X									
Scour-and-fill structures				X									
Mottled bedding												X	
Stromatolites													X
BEDDING-PLANE MARKINGS													
Groove casts; striations; bounce, brush, prod, and roll marks					X								
Flute casts				X									
Parting lineation	X												
Load casts							X						
Tracks, trails, burrows†												X	
Mudcracks and syneresis cracks										X			
Pits and small impressions											X		
Rill and swash marks	X												
OTHER STRUCTURES													
Sedimentary sills and dikes								X					

*Not wholly stratification structures   †Not wholly bedding-plane markings

Source: Boggs, S., 1967, Principles of sedimentology and stratigraphy. Table 6.1, p. 137, reprinted by permission of Merrill Publishing Co., Columbus, OH.

FIGURE 3.1

Terms used for describing the
thickness of beds and laminae.
*(Modified from McKee, E. D., and
G. W. Weir, 1953, Terminology
for stratification and cross-
stratification in sedimentary rocks:
Geol. Soc. America Bull., v. 64,
Table 2, p. 383; and Ingram,
R. L., 1954, Terminology for the
thickness of stratification and
parting units in sedimentary rocks:
Geol. Soc. America Bull., v. 65,
Fig. 1, p. 937.)*

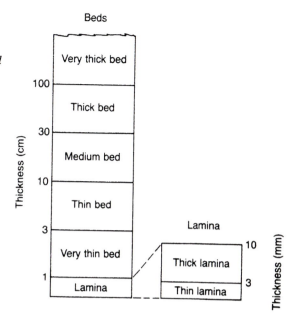

Most bedding planes that separate sedimentary layers represent a plane of nondeposition, an abrupt change in depositional conditions, or an erosion surface (Campbell,
1967), although the processes or events that produce the bedding plane may not affect the
entire bedding plane at one moment. Some bedding planes may be postdepositional
features created by processes such as intense burrowing of some layers by organisms.
Features resembling bedding or bedding planes may also be created by diagenesis or
weathering, although these features are probably not true bedding. The Liesegang banding
described in Section 3.6.2 is an example of such pseudobedding. Bedding surfaces may
have a variety of configurations, as shown in Figure 3.2. Thus, the beds enclosed by these
bounding surfaces may likewise have various geometric forms as seen in cross-section,
forms such as uniform-tabular, lens-shaped, wedge-shaped, or irregular. Beds that contain internal layers that are essentially parallel to the bounding bedding surfaces are said
to be **planar-stratified**. Groups of similar planar beds are called **bedsets** (Fig. 3.3).
Simple bedsets are characterized by similar compositions, textures, and internal structures; **composite bedsets** consist of groups of beds that differ in these characteristics but
that are genetically associated. Beds displaying internal layers deposited at a distinct angle
to the bounding surfaces are **cross-stratified**. A cross-stratified bed is sometimes called a
set of cross-strata, and a succession of such sets is called a **coset**.

3.3.2   Laminated Bedding

Many sandstones and shales, as well as some nonsiliciclastic sedimentary rocks such as
evaporites, display internal laminations that are essentially parallel to bedding surfaces
(Fig. 3.4). Individual laminae in these planar-stratified beds may range in thickness from
a few grain diameters to as much as 1 cm. The laminae are distinguished on the basis of
differences in grain size, clay and organic matter content, mineral composition, and in
rare cases microfossil content of the sediment. Color changes may accentuate the presence
of some laminae.

FIGURE 3.2

Descriptive terms used for the configuration of bedding surfaces. *(From Campbell, C. V., 1967, Lamina, lamina set, bed and bedset: Sedimentology, v. 8. Fig. 2, p. 18, reprinted by permission of Elsevier Science Publishers, Amsterdam.)*

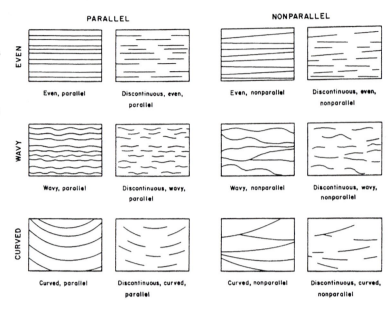

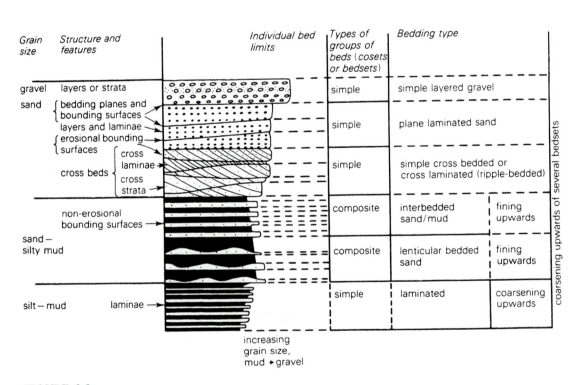

FIGURE 3.3

Diagram illustrating the terminology of bedsets. *(From Collinson, J. D., and D. B. Thompson, 1982, Sedimentary structures. Fig. 2.2, p. 8, reprinted by permission.)*

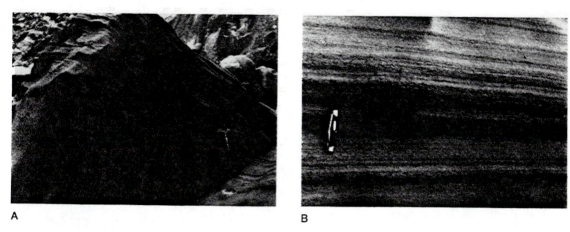

A   B

FIGURE 3.4
A. Laminated sandstone lying above massive sandstone containing rip-up clasts; Elkton Siltstone
(Tertiary), southern Oregon coast. B. Close-up view of laminae in the laminated part of A.

Some laminated bedding forms as a result of suspension settling of fine-size sediment in a variety of depositional environments (lakes, tidal flats, subtidal shelves, deep-sea environments). The laminae in many shales and evaporites, for example, may have formed by such suspension-settling mechanisms. The presence of suspension-deposited laminae in fine-grained sediments suggests slow deposition in quiet-water environments where organic activity (bioturbation) at and below the depositional interface was not intense enough to destroy the lamination.

On the other hand, the laminae in some shales and most laminae in sandstones probably formed by traction transport mechanisms. For example, many laminated deep-sea mud deposits are now being interpreted as the products of dilute, low-velocity turbidity current flows. The laminae are believed to result from the spillover of the dilute upper parts of channelized turbidity currents (Hesse and Chough, 1980; Stow and Bowen, 1980); however, the exact mechanism that produces the lamination is still speculative. Parallel laminae are very common in many sandstones and have been attributed to a variety of causes: (1) swash and backwash on beaches (Clifton, 1969), (2) wind transport (McKee et al., 1971; Hunter, 1977), (3) transport by steady-flow currents in the upper-flow regime (Harms and Fahnestock, 1965; Allen, 1984; Cheel and Middleton, 1986; Cheel, 1990), (4) phases of upper-flow–regime and lower-flow–regime transport during turbidity current flow (Bouma, 1962, p. 97–98, (5) sheet flow—the oscillatory equivalent of plane-bed transport in the upper-flow regime (Clifton, 1976), and (6) transport in the lower-flow regime (McBride et al., 1975; Southard and Boguchwal, 1990).

3.3.3   Graded Bedding

Graded beds are strata characterized by gradual but distinct vertical changes in grain size (Fig. 3.5A). Beds that display gradation from coarser particles at the base to finer particles at the top are said to have **normal grading**. Those that grade from finer particles at the base to coarser at the top have **reverse grading** or **inverse grading**. Grading can occur in beds of almost any thickness, even in laminae, but is most common in beds ranging from a few centimeters to a few meters. Bouma (1962, p. 48–51) describes an ''ideal''

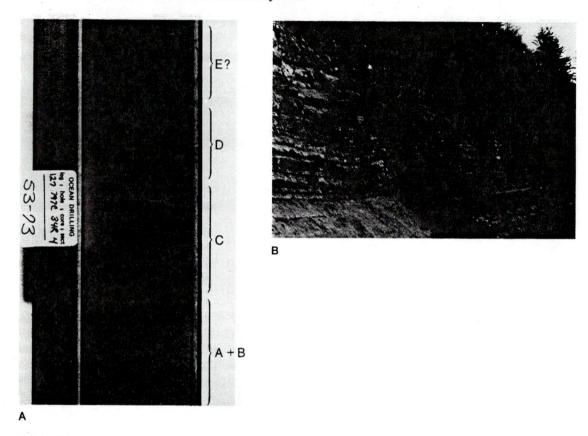

FIGURE 3.5

A. Graded bedding in Miocene deep-sea sandstone (core) from ODP Leg 127, Site 797, Japan Sea. Note the nearly complete Bouma sequence (units A through E) in this core, and compare with Figure 3.6. (*Photograph courtesy of the Ocean Drilling Program, Texas A & M University.*) B. Rhythmically bedded, graded turbidites from the Tyee Formation (Eocene), northern Oregon Coast Range.

graded-bed sequence, in rocks of probable turbidity current origin, that consists of five distinct divisions. This complete sequence of units grades upward from a massive, well-graded basal portion (unit A) through a lower unit characterized by parallel laminae (B), a ripple cross-laminated middle unit (C), an upper unit with parallel laminae (D), and a topmost nearly structureless mud unit (E) (Fig. 3.6, 1). This idealized complete sequence is now commonly referred to as a Bouma sequence; Figure 3.6, 1 shows a nearly complete Bouma sequence. Bouma points out, however, that many graded sequences may be truncated at the top, base, or both and thus do not contain all of the units found in the ideal sequence. In fact, many graded beds display no visible internal structures except size grading.

Hsü (1989, p. 117) claims that Bouma's unit D rarely occurs and that most turbidites can be divided into only two units: a lower horizontally laminated unit (unit A + B, Fig. 3.6, 2) and an upper cross-laminated unit (unit C). Unit E may be pelagic shale and thus may not be part of a turbidite flow unit. The differences between Bouma's model and Hsü's model can be seen by comparing Figures 3.6, 1 and 3.6, 2.

Normal graded bedding can result from any process (e.g., turbidity currents, storm activity on shelves, periodic silting on delta distributaries, deposition in the last phases of a flood, settling of volcanic ash after an eruption) that produces a suspension of sediments of various sizes, which may then settle according to size. Some graded beds may have formed through the bioturbation activities of organisms. Most graded beds in the geologic record, especially graded beds that display complete Bouma sequences (Fig. 3.6, 1), have been attributed to deposition from waning turbidity currents. Nelson (1982) has shown, however, that some shallow-water sediments deposited under the influence of storm-wave surges may also develop graded units that display Bouma sequences. Deep-sea fan deposits are commonly made up of thin, graded sandstone or siltstone beds of turbidite origin with interbeds of pelagic or hemipelagic shale. The graded units repeat one after another, producing what is commonly called rhythmic bedding (Fig. 3.5B).

Inverse grading, although less common than normal grading, occurs in some sediments, particularly in sediment-gravity-flow deposits such as debris flows and possibly some turbidites (e.g., resedimented conglomerates). There is still considerable disagreement about the causes of reverse grading. It has been attributed, among other causes, to (1) dispersive pressures owing to interparticle collisions, (2) kinetic sieving—the process whereby smaller particles fall downward through layers of coarser particles when agitated, and (3) the strength loss that clays undergo on deformation—the lowermost, most strongly sheared layers of debris flows, for example, are weakest and support relatively small clasts compared to the uppermost layers. These and other ideas to explain reverse grading are reviewed by Naylor (1980).

3.3.4 Massive Bedding

The term massive bedding is applied to beds of sedimentary rock that contain few or no visible internal laminae. Truly massively bedded sediments are rare. Many massive-appearing beds have been shown to actually contain internal structures when examined after etching and staining or by X-radiography techniques (Hamblin, 1965). Nonetheless, massive beds do occur, both in graded and nongraded units. They appear to be most common in sandstones (Fig. 3.7).

FIGURE 3.6
Ideal sequence of sedimentary structures in graded-bed units as proposed by Bouma (1) and Hsü (2). Note in Hsü's model that Bouma's units A and B are combined and unit D is omitted. (*After Hsü, K. J., 1989, Physical principles of sedimentology. Fig. 7.8, p. 116, reprinted by permission of Springer-Verlag, Berlin.*)

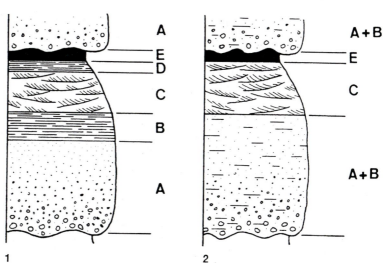

FIGURE 3.7
Massive-bedded sandstone (upper part of photograph) lying above thin, parallel-bedded siltstone and shale. Fluornoy Formation (Eocene), southwestern Oregon.

The origin of massive beds is difficult to explain. Presumably, massive bedding is generated in the absence of fluid-flow traction transport, either by some type of sediment gravity flow or by rapid deposition of material from suspension. For example, turbidites deposited from highly concentrated flows may be massively bedded, particularly at the base. Arnott and Hand (1989) demonstrated experimentally that under upper plane-bed conditions of transport in the presence of a heavy rain of suspended sand (typical of turbidity currents) the formation of laminations is suppressed. This finding suggests that rapid aggradation can account for the massive character of Bouma A divisions of turbidites. The deposits of some grain flows, fluidized or liquefied flows, and debris flows may also appear massive. Nonetheless, extremely thick, massive beds are particularly difficult to explain, inasmuch as the deposits of single sediment gravity flows tend to be much thinner, although some thick, massive beds may be the product of truly catastophic sediment gravity flows. Some very thick units may actually be **amalgamated** units, formed by the "welding" together of the deposits of several successive sediment gravity flows consisting of sediments having about the same grain size and general characteristics. Other massive beds may be of secondary origin, formed either by the homogenizing activities of bioturbating organisms or by postdepositional sediment liquefaction owing to shocking or other mechanisms.

3.3.5   Cross-bedding

Cross-bedding is one of the commoner structures in sedimentary rocks. Although it is most abundant in sandstones, it can occur in any kind of rock made up of grains capable of undergoing traction transport. Thus, cross-bedding has been reported in limestone, salt deposits, ironstones, and phosphorites. Cross-beds, described in the simplest possible terms, are strata in which internal layers, or foresets, dip at a distinct angle to the surfaces that bound the sets of cross-beds. (Cross-bedding is called cross-lamination if thickness of the foresets is less than 1 cm.) The bedding surfaces themselves may be either planar surfaces or surfaces that are curved in some manner. Thus, one common, simple method of classifying cross-bedding is to characterize it as either **tabular cross-bedding,** having bounding surfaces that are planar, or **trough cross-bedding,** having bounding surfaces that are curved (Figs. 3.8, 3.9, 3.10). Bedding that is markedly trough-shaped or scoop-shaped has also been referred to as **festoon bedding**.

Considered in detail, however, cross-bedding of these two basic types can display considerable variability, particularly when observed in three dimensions. Thus, many attempts have been made to formulate more-detailed classifications for cross-bedding (see review by Allen, 1982, p. 346). None of these classifications has been widely accepted. Figure 3.11 summarizes some of the morphological features of cross-strata and gives one set of terminology (Allen, 1982) used to describe these features. For a different, and new, approach to classification of bedforms and cross-bedding, see Rubin (1987, p. 3–6).

Tabular cross-bedding is formed mainly by the migration of large-scale two-dimensional bedforms (dunes). (The geometry of two-dimensional [2-D] bedforms can be described by one transect parallel to flow, whereas three-dimensional [3-D] bedforms must be defined in three dimensions [Ashley, 1990].) Individual beds range in thickness from a few tens of centimeters to a meter or more, but bed thicknesses up to 10 m have been reported. Trough cross-bedding originates by migration of 3-D bedforms, either small current ripples that produce small-scale cross-bed sets or large-scale ripples (dunes) that produce much larger-scale cross-bed sets. Trough cross-bedding formed by migration of large-scale ripples commonly ranges in thickness to a few tens of centimeters and in

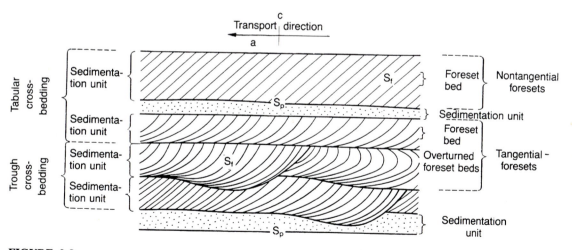

FIGURE 3.8

The terminology and defining characteristics of two fundamental types of cross-bedding. Symbols: (a), direction parallel to the average sediment transport direction; (c), direction perpendicular to (a) and the transport plane (bed) in which (a) lies; (S_p), the principal bedding surface or bedding plane; (S_f), the foreset surface of cross-bedding. (*After Potter, P. E., and F. J. Pettijohn, 1977, Paleocurrents and basin analysis, 2nd ed. Fig. 4.1, p. 91, reprinted by permission of Springer-Verlag, Heidelberg.*)

FIGURE 3.9
Multiple sets of small-scale planar cross-beds (between arrows) with tangential foresets. Coaledo Formation (Eocene), southwestern Oregon.

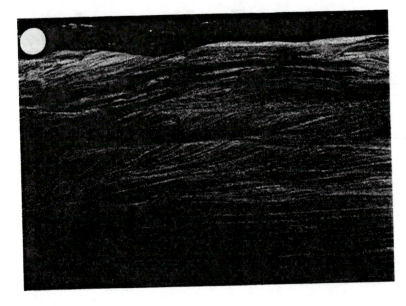

FIGURE 3.10
Three intersecting sets of small-scale trough cross-beds in fine, laminated sandstone. The area marked (B) may be a burrow. Probable hummocky cross-stratification (arrow) is shown in the lower part of the photograph; see also Figure 3.17. Coaledo Formation (Eocene), southwestern Oregon.

width from less than 1 m to more than 4 m. Cross-bedding can also form by filling of scour pits and channels, deposition on the point bars of meandering streams, and deposition on the inclined surfaces of beaches and marine bars. Cross-bedding formed under different environmental conditions (fluvial, eolian, marine) can be very similar in appearance and thus may be difficult to differentiate in ancient deposits.

Some cross-stratified units contain inclined surfaces that separate adjacent foresets, with similar orientations, and truncate the lower foreset laminae (McCabe and Jones, 1977). These surfaces are called **reactivation surfaces** (Collinson, 1970). Reactivation surfaces have been attributed to modification of previously formed ripples (or larger

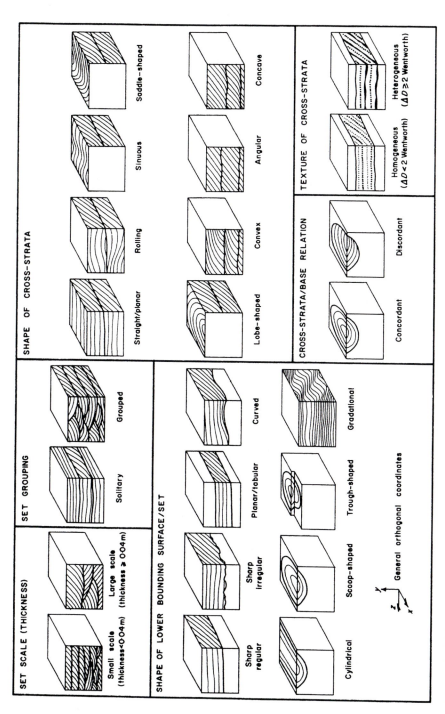

FIGURE 3.11

Morphological features of cross-stratified deposits and terms descriptive of cross-stratification.
The textural term "homogeneous" is used for cross-strata within sets that differ in size grade by
up to two Wentworth size classes; "heterogeneous" is used if the size difference is greater than
two Wentworth classes. *(From Allen, J. R. L., 1982, Sedimentary structures—Their character
and physical basis, v. I, Fig. 9.2, p. 348, reprinted by permission of Elsevier Science
Publishers, Amsterdam.)*

bedforms) by several mechanisms, including (1) erosion during a decrease in water depth owing to wave action or flow around the bedforms (Fig. 3.12A), (2) erosion during a change in current flow direction, as during a tidal reversal (Fig. 3.12B), and (3) modification at constant water depth and flow direction owing either to erosion resulting from random interaction of bedforms (Fig. 3.12C) or to erosion in the lee of an advancing bedform (Fig. 3.12D).

3.3.6   Ripple Cross-lamination

Ripple cross-lamination is a type of cross-stratification that has the general appearance of waves when viewed in outcrop sections cut normal to the wave (ripple) crests (Fig. 3.13). Ripple cross-lamination forms when deposition takes place very rapidly during migration of current or wave ripples (McKee, 1965; Jopling and Walker, 1968). A series of cross-laminae is produced owing to superimposition of one ripple on another as the ripples migrate. The ripples succeed one another upward in such a manner that the crests of vertically succeeding ripples are out of phase and appear to be advancing or climbing in a downcurrent direction; thus, this structure is sometimes called climbing-ripple lamination. In outcrop sections cut at orientations other than normal to ripple crests, the laminae may appear horizontal or trough-shaped, depending upon the orientation and the shape of

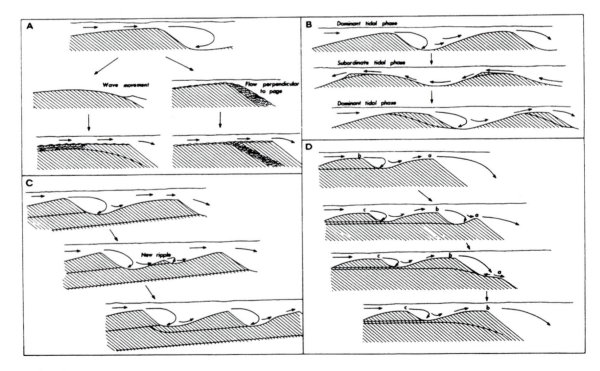

FIGURE 3.12
Methods of forming reactivation surfaces. See text for explanation. (*After McCabe, P. J., and C. M. Jones, 1977, Formation of reactivation surfaces within superimposed deltas and bedforms: Jour. Sed. Petrology, v. 47. Fig. 7, p. 713, reprinted by permission of Society of Economic Paleontologists and Mineralogists, Tulsa, Okla.*)

FIGURE 3.13
Ripple cross-lamination (arrow). Both stoss-side and lee-side ripples are preserved in this example, but lee-side ripples are dominant. Note truncation of underlying parallel-laminated fine sandstone by ripple cross-lamination. Coaledo Formation (Eocene), southwestern Oregon.

the ripples. Two types of ripple cross-lamination are recognized, one in which both the lee side and stoss side of the ripples are preserved and one in which mainly the lee side is preserved.

Ripple cross-lamination forms under conditions where abundant sediment is present, particularly sediment in suspension, which quickly buries and preserves rippled layers. Abundant sediment supply is combined with enough traction transport to produce ripples and to cause the ripples to migrate but not enough to cause complete erosion of laminae from the stoss (upcurrent) side of the ripples. In some deposits, ripple laminae are in phase, indicating that the ripples did not migrate. Ripple cross-lamination of this type forms under conditions where a balance is achieved between traction transport and sediment supply so that ripples do not migrate despite a growing sediment surface. According to Ralph Hunter (personal communication), in-phase laminae form by climbing ripples that climb essentially vertically. Such a process could occur under two possible kinds of conditions: (1) the ripples are oscillating ripples and remain active but nonmigrating during deposition or (2) the ripples are inactive and are being passively draped by suspension settling, in which case the ripple amplitude gradually decreases upward; that is, the ripples are damped. The conditions of rapid sedimentation required to produce ripple cross-lamination occur in a variety of environments, including fluvial floodplains, point bars, river deltas subject to periodic flooding, and environments of turbidite sedimentation.

3.3.7   Flaser and Lenticular Bedding

Flaser bedding is a special type of ripple cross-lamination in which thin streaks of mud occur between sets of ripple laminae (Fig. 3.14). The mud streaks tend to occur in the ripple troughs but may partly or completely cover the crests. Flaser bedding appears to

FIGURE 3.14
Flaser bedding (upper half of
photograph) in Elkton Siltstone
(Eocene), southwestern Oregon.

form under fluctuating depositional conditions marked by periods of current activity,
when traction transport and rippling of fine sands takes place, alternating with periods of
quiescence, when mud is deposited. Repeated episodes of current activity result in erosion
of previously deposited ripple crests, allowing new rippled sands to bury and preserve
rippled beds with mud flasers in the troughs (Reineck and Singh, 1980). The term
lenticular bedding is used instead of flaser bedding for interbedded mud and ripple
cross-laminated sand in which the ripples or sand lenses are discontinuous and isolated in
both vertical and lateral directions (Fig. 3.15). Lenticular bedding appears to form in
environments that favor deposition of mud over sand, whereas flaser bedding forms under
conditions that favor deposition of sand over mud. Both structures are common in the
deposits of tidal flats and some subtidal environments where conditions of current flow or

FIGURE 3.15
Lenitcular bedding. Lenses of
light-colored, fine sandstone
interbedded with dark-colored
mudstone. Elkton Siltstone
(Eocene), southwestern Oregon.

wave activity, which cause sand deposition, alternate with slack-water conditions that favor mud deposition. Flaser and lenticular bedding may also form in marine and lacustrine delta-front environments, where fluctuations in sediment supply and current velocity are common, and possibly on shallow-marine shelves owing to storm-related transport of sand into deeper-water zones of mud deposition.

3.3.8 Hummocky Cross-stratification

Hummocky cross-stratification is a type of cross-stratification originally called "truncated wave-ripple laminae" by Campbell (1966). It was later renamed hummocky cross-stratification by Harms et al. (1975). This structure is characterized by undulating sets of cross-laminae that are both concave-up (swales) and convex-up (hummocks). The cross-bed sets cut into each other with curved erosional surfaces (Figs. 3.16, 3.17). Hummocky cross-stratification commonly occurs in sets 15–20 cm thick. Spacing of hummocks and swales is from a few tens of centimeters to several meters. The lower bounding surface of a hummocky unit is sharp and is commonly an erosional surface. Current-formed sole marks may be present on the base. Hummocky cross-stratification seems to occur typi-

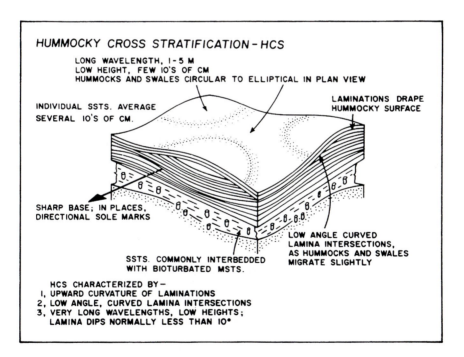

FIGURE 3.16

Schematic diagram of hummocky cross-stratification. (*From Walker, R. G., 1984, Shelf and shallow marine sands, in Walker, R. G., ed., Facies models, 2nd ed.: Geoscience Canada Reprint Ser. 1. Fig. 11, p. 149, reprinted by permission of Geological Association of Canada. Originally after Walker, R. G., 1982, Hummocky and swaley cross stratification, in Walker, R. G., ed., Clastic units of the Front Range between Field, B.C., and Drumheller, Alberta: Internat. Assoc. Sedimentologists, 11th International Congress on Sedimentology [Hamilton, Canada], Guidebook to Excursion 21A, p. 22–30.*)

A   B

FIGURE 3.17
A. Hummocky cross-stratification in Coaledo Formation (Eocene) sandstones, southwest
Oregon. The arrow points to a line marking the erosion surface between underlying, truncated
laminae and overlying draped laminae. B. Detail of the erosional contact between underlying
hummock and overlying draped laminae in a hummocky cross-stratified unit, Coaledo
Formation.

cally in fine sandstone to coarse siltstone that commonly contain abundant mica and fine
carbonaceous plant debris.

Dott and Bourgeois (1982, 1983) propose an ideal hummocky cross-stratification
sequence, somewhat analogous to the Bouma sequence for turbidite units (Fig. 3.18).
Their idealized sequence begins with a sharp base that may or may not have sole marks
and that may or may not contain coarse lag particles immediately above the base. Above
the base is a hummocky zone (H), which gradually changes upward to a zone of flat
laminae (F) that is, in turn, overlain by cross-laminae (X). A burrowed or bioturbated
mudstone (Mb) is commonly present between one bed of hummocky cross-stratified
sandstone and the next. Dott and Bourgeois recognize that considerable variation from
this ideal sequence is possible; some units may be missing, and bioturbation or burrowing
(b) may affect sandstone as well as mudstone units (Fig. 3.16).

Most workers agree that hummocky cross-stratification forms in some manner under
the action of waves and that it appears to be particularly common in ancient sediments
deposited on the shoreface and shelf. The exact process or processes by which hummocky
cross-stratification is formed is, however, still speculative and controversial. A principal
reason for uncertainty about the origin is that few unequivocal examples of this structure
have been observed in modern sediments, although Greenwood and Sherman (1986)
report possible hummocky cross-stratification in modern surf-zone deposits in the Canadian Great Lakes. Harms et al. (1982) suggest that the structure is formed by strong surges
of varied direction (oscillatory flow) that are generated by relatively large storm waves.
Strong-storm wave action first erodes the seabed into low hummocks and swales that lack
any significant orientation. This topography is then mantled by laminae of material swept
over the hummocks and swales. Duke (1985, 1987) argues that most examples of hummocky cross-stratification were formed by tropical hurricanes. Several other workers
(Klein and Marsaglia, 1987; Swift and Nummedal, 1987) take exception to this interpretation, pointing out, among other things, that hummocks are no more likely to form under

hurricanes than under midlatitude storms. Finally, some workers (Kreisa and Nottvedt, 1986) argue that hummocky stratification is simply a type of trough cross-bedding. Southard et al. (1990) report the results of experiments on bed configuration in fine sands under bidirectional oscillatory flow that may have some bearing on the origin of hummocky cross-stratification. Their experimental objective was to simulate the kinds of oscillatory flows with long periods and high velocities that occur during storm-induced deposition in the shallow ocean. Their reported results suggest that some hummocky cross-stratification may be generated during sediment fallout from strong oscillatory flows at moderate to long oscillation periods as large three-dimensional oscillatory-current bedforms develop from a planar bed during strong but waning flow. The question of the origin and significance of hummocky cross-stratification is still open, and it seems likely that hummocky cross-stratification can originate by more than one process. Nonetheless, it is generally regarded to be a fairly reliable indicator of deposition in shelf and shoreface environments.

3.3.9   Ripple Marks

Ripples of various sizes are among the most common sedimentary structures in modern sedimentary environments, where they form in both siliciclastic and carbonate sediments. Experimental and empirical studies have firmly established that ripples occur owing to traction transport of granular materials under either unidirectional current flow or oscil-

FIGURE 3.18

Dott-Bourgeois ideal sequence for hummocky cross-stratification shown with some important variations drawn as spokes from the central-hub ideal sequence. *(From Dott, R. H., and J. Bourgeois, 1983, Hummocky stratification: Significance of its variable bedding sequences: discussion and reply: Geol. Soc. America Bull. v. 94, Fig. 1, p. 1250.)*

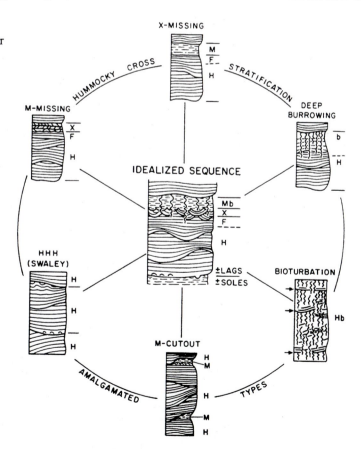

latory flow (wave action). They are most common in sand-size sediment but can occur in finer and coarser sediment.

Experimental work has shown that a progression of bedforms develop in granular materials undergoing traction transport as flow conditions change from lower-flow regime to upper-flow regime (Simons and Richardson, 1961; Southard and Boguchwal, 1990). At low flow velocities, only small ripples (0.05–0.2 m in length and 0.005–0.03 m in height) form. With increase in flow velocity, small ripples are replaced by much larger ripples. Under natural conditions, these larger ripples may reach lengths ranging from 0.5 m to more than 100 m and heights to tens of meters. Earlier workers tended to group these large ripples into two types: sand waves (low, long-wavelength bedforms) and dunes (higher, shorter-wavelength bedforms). More recent workers (e.g., Harms et al., 1982) cast doubt on the reality of making a distinction between these two types of large bedforms. They suggest that both sand waves and dunes be referred to simply as large ripples (some workers use the term megaripples). A symposium convened by the Society of

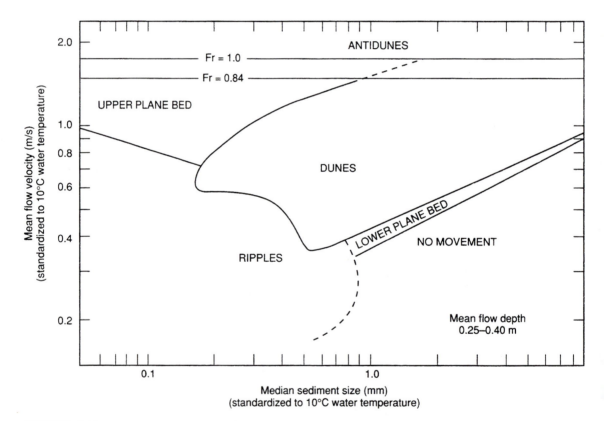

FIGURE 3.19

Plot of mean flow velocity against median sediment size showing the stability fields of bed phases. Note that the recommended terminology for bedforms is (1) lower plane bed, (2) ripples, (3) dunes (all large-scale ripples), (4) upper plane bed, and (5) antidunes. Fr = Froude number. *(After Southard, J. B., and L. A. Boguchwal, 1990, Bed configurations in steady unidirectional water flows. Part 2. Synthesis of flume data: Jour. Sed. Petrology, v. 60. Fig. 3, p. 664, reprinted by permission of Society for Sedimentary Petrology, Tulsa, Okla.)*

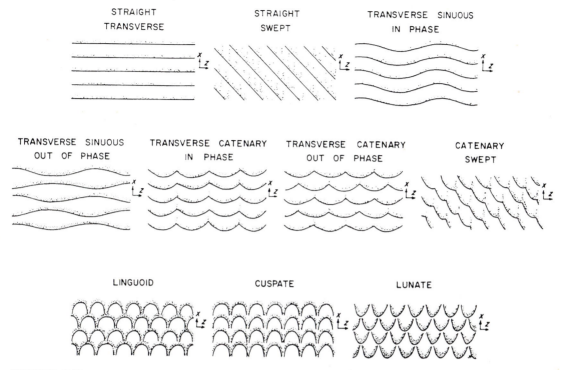

FIGURE 3.20

Idealized classification of current ripples based on plan-view shape. Flow is from bottom to the top in each case. (*After Allen, J. R. L., 1968, Current ripples: Their relation to patterns of water motion: North Holland Pub., Amsterdam. Fig. 4.6, p. 65, reprinted by permission of Elsevier Science Publishers, Amsterdam.*)

Economic Paleontologists and Mineralogists (SEPM) in 1987 examined this problem of bedform nomenclature. Participants in this symposium recommended that large bedforms should have only one name, **dune;** they can be referred to as **subaqueous dunes** when it is important to distinguish them from eolian dunes (Ashley, 1990). At the higher flow velocities that bring flow conditions into the upper-flow regime, dunes are eroded and destroyed and a phase of plane-bed sediment transport occurs. Still-higher flow velocities may produce antidunes, which migrate in an upcurrent direction. Figure 3.19, which shows the kinds of bedforms that develop as a function of mean flow velocity and mean sediment size, summarizes much of the past experimental work on bedform configuration.

When flow velocities eventually diminish, bedforms tend to be destroyed in reverse order to that in which they were produced; therefore, ripples, especially large ripples and antidunes, have low preservation potential. Thus, preserved ripples (ripple marks) are rarely abundant in ancient sedimentary rocks, although small-scale ripple marks are moderately common.

Ripples developed under unidirectional current flow are asymmetrical in cross-sectional shape, with a gently sloping upcurrent stoss side and a more steeply sloping lee side. Ripples of this type are called **current ripples.** In plan view, the crests of current ripples may have a variety of shapes: **straight, sinuous, catenary, linguoid,** and **lunate** (Fig. 3.20). Ripples developed under wind flow are also asymmetrical in cross-sectional

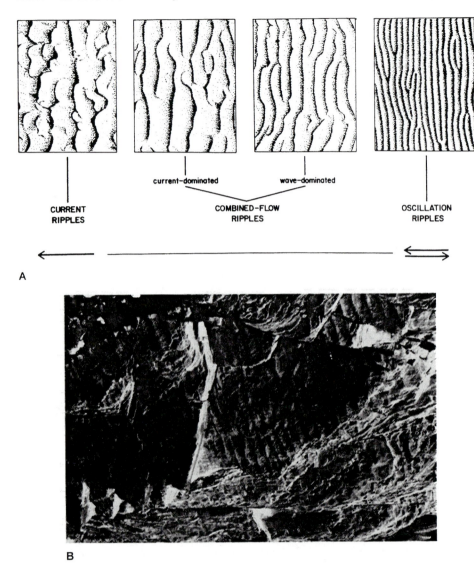

A

B

FIGURE 3.21
A. Plan-view shape of oscillation ripples. The shapes of typical current ripples and combined-flow ripples are shown also for comparison. (*After Harms, J. C., et al., 1982, Structures and sequences in clastic rocks: Soc. Econ. Paleontologists and Mineralogists Short Course 9. Fig. 2–19, p. 2–48, reprinted by permission of SEPM, Tulsa, Okla.*) B. Multiple sets of oscillation ripples on the upper surface of upturned, fine sandstone beds, Elkton Siltstone (Eocene), southwestern Oregon. Note hammer for scale.

shapes, but the crests are predominantly straight. Ripples generated by wave action tend to be symmetrical in cross-sectional shape unless a unidirectional bottom current is superimposed on the oscillatory flow during ripple formation. Wave-formed ripples are called **oscillation ripples** or **wave ripples** (Fig. 13.21A). Figure 3.21B illustrates the typical plan-view shape of oscillation ripples compared to that of current ripples.

Ripple marks in ancient sedimentary rocks furnish extremely useful information about paleoflow conditions and paleocurrent directions. Ripple marks are not, however, indicators of unique depositional environments. Because they can form under unidirectional currents (in both shallow and deep water), by wind transport, and by wave action, great care must be used in interpreting depositional environments on the basis of these bedforms. See Boggs (1987, p. 145–156) for a short synopsis of the origin and significance of bedforms. More-detailed discussion of this topic may be found in Allen (1968; 1982, p. 271–514) and Harms et al. (1982).

3.4 Irregular Stratification

3.4.1 General Statement

Several kinds of sedimentary structures are recognized that display features of bedding or stratification, but that do not show the regular stratification of the structures discussed above. Many of these structures appear to have formed from regular bedding or stratification that was deformed or altered during or after deposition but prior to consolidation (penecontemporaneous deformation). Presumably, deformation occurs by physical processes that involve soft-sediment slumping, loading, squeezing, or partial liquefaction. Structures of this putative origin are commonly referred to as **deformation structures**. Other irregular stratification structures appear to have formed as a result of erosion of unconsolidated beds followed by an episode of sedimentation. Structures of this kind are called simply **erosion structures**. One type of irregular stratification structure, stromatolites, is produced as an original bedding structure through the activities of organisms and is thus a biogenic structure.

3.4.2 Deformation Structures

Convolute Bedding and Lamination

Convolute bedding, or convolute lamination, is the name applied to complexly folded or intricately crumpled beds or laminae that are commonly, although not invariably, confined to a single sedimentation unit. Strata above and below this unit may show little or no evidence of deformation. In cross section, the deformed strata appear as small-scale anticlines and synclines (Fig. 3.22). Axial planes of some folds may lean in the paleocurrent direction, as determined by other structures in the deposit. Convolute bedding is most common in fine sand- to silt-size siliciclastic sediment but can occur also in carbonates. Individual laminae can generally be traced from fold to fold; however, the laminae can be truncated by erosional surfaces, which may also be convoluted. Convolutions tend to increase in complexity and amplitude upward from the base or from undisturbed laminae in the lower part of the unit. They may either die out in the top part of the unit or be truncated by the upper bedding surface (Potter and Pettijohn, 1977). Although commonly confined to beds less than about 25 cm in thickness, convoluted units ranging to several meters thick have been reported in both subaqueous and eolian sediments. The lateral extent of units containing convolute lamination can apparently be

FIGURE 3.22
Convolute laminae in laminated
siltstone overlying a thin mudstone
unit. Elkton Siltstone (Eocene),
southwest Oregon.

considerable. In one example, a convoluted unit only 12 cm thick is reported to occur over
an area of about 750 km^2 (Sutton and Lewis, 1966). Convolute lamination is particularly
common in turbidites but can occur also in a variety of other sediments, including
intertidal-flat, deltaic, river floodplain, point bar, and eolian deposits.

The origin of convolute bedding and lamination is not fully understood. Allen
(1982, v. II, p. 351–352) suggests that convolute lamination may form at three different
stages with respect to the time of sedimentation: **Postdepositional convolute lamination**
arises some time after the start of burial, **metadepositional convolute lamination** arises
just before or immediately after deposition ceases, and **syndepositional convolute lamination** forms episodically to continuously during deposition. The mechanisms or processes that cause convolute lamination to form are, however, poorly understood. Most
workers agree that it is caused in some way by deformation of hydroplastic or liquefied
sediment, but little agreement exists regarding the actual deformation mechanism (although it is probably not slumping in most cases). For additional discussion of the origin
of these structures, see Allen (1982, v. II, p. 351–354) and Potter and Pettijohn (1977,
p. 207–208).

Flame Structures

Flame structures are flame-shaped projections of mud that extend upward from a shale
unit into an overlying bed of different composition, commonly sandstone (Fig. 3.23).
Individual "flames" may range in height from a few millimeters to several centimeters.
In some examples of the structure, the flames extend more or less directly upward into the
overlying layer. In others, the crests of the flames are overturned or bent downward,
commonly all in the same direction. Flame structures are probably caused by squeezing
of low-density, water-saturated muds upward into denser sand layers owing to the weight
of the sand. The oriented, overturned crests of some flames suggest that slight horizontal
(downslope or downcurrent) movement or drag may take place between the mud and sand
layers during the process of loading and squeezing. The orientation direction of over-

FIGURE 3.23
Multiple sets of small-scale flame structures in a thinly bedded and laminated siltstone-mudstone sequence. Elkton Siltstone (Eocene), southwestern Oregon.

turned flame crests is commonly consistent with the paleocurrent direction suggested by other, associated, sedimentary structures.

Ball and Pillow Structures and Pseudonodules

The name ball and pillow structure (Potter and Pettijohn, 1977, p. 201) is applied to the basal portion of sandstone beds, overlying shales, that are broken into masses of various sizes packed vertically and laterally in a mud matrix (Fig. 3.24). These sand masses somewhat resemble the structure of pillow lavas and may have "pillow," "hassock,"

FIGURE 3.24
Ball and pillow structure. The balls are composed of laminated fine sandstone and are enclosed within a thin mudstone layer. Coaledo Formation (Eocene), southwest Oregon.

"kidney," or "ball" shapes. The underlying shale is squeezed between pillows and may extend as tongues into the overlying sandstone. In some deposits, the sand masses may become detached from the overlying sand and be completely surrounded by shale, forming a laterally extensive layer of nearly uniform-size sand balls that may superficially resemble concretions (Allen, 1982, v. II, p. 359). These isolated sand masses are commonly called **pseudonodules**. Individual pillows tend to be curled upward, that is, concave upward and convex downward. If they contain internal laminations, the laminations may be deformed, but they conform roughly to the boundaries of the pillows (Fig. 3.24).

Ball and pillow structures and pseudonodules are believed to be caused by the breakup and foundering of semiconsolidated sand or limy sediment when underlying muds become temporarily liquefied or partially liquefied. As in the formation of flame structures, lower-density mud is pushed upward into higher-density sand. Kuenen (1958) produced structures in the laboratory that closely resemble natural ball and pillow structures by applying a shock to a layer of sand overlying a thixotropic clay, momentarily liquefying the clay. Under such conditions, the overlying sand breaks up and the pieces sink into the partially liquefied clay, becoming somewhat rounded and upward-curved as they sink.

Synsedimentary Folds, Faults, and Rip-up Clasts

Unconsolidated sediment may move downslope under the influence of gravity as slumps, slides, or flows. Potter and Pettijohn (1977) suggest that structures produced by such penecontemporaneous deformation may be caused by two types of movement. The first of these is a décollement type of movement in which the lateral displacement is concentrated along a sole, producing beds that are tightly folded and piled into nappelike structures (Fig. 3.25). Such structures are referred to as **synsedimentary folds**. Slump structures of this type may involve sediments in several beds, and the structures are commonly faulted. Thicknesses of units characterized by slump structures may range from less than 1 m to more than 50 m. Slump units are commonly bounded above and below by strata that show

FIGURE 3.25
Small-scale synsedimentary folds in thick sand laminae (white) in a laminated and thinly bedded mudstone–sandstone sequence. Elkton Siltstone (Eocene), southwestern Oregon.

FIGURE 3.26
Large rip-up clasts in a massive, coarse channel sandstone, Elkton Siltstone (Eocene),
southwestern Oregon. Note well-preserved laminae in some of the clasts.

no evidence of deformation. Such units must be examined with care, however, to be sure
that they are indeed the product of penecontemporaneous deformation and not the result
of deformation of incompetent shale or other beds caught between competent sandstone
or carbonate beds during tectonic folding. Slump structures commonly occur in mudstones and sandy shales and less commonly in sandstones, limestones, and evaporites.
They typically form in sediments deposited in environments where rapid sedimentation
and oversteepened slopes lead to instability: glacial sediments, varved silts and clays of
lacustrine origin, eolian dune sands, turbidites, delta and reef-front sediments, subaqueous dune sediments, and sediments from the heads of submarine canyons, continental
slopes, and the walls of deep-sea trenches.

The second type of penecontemporaneous deformation produced by slumping or
flowing is the product of pervasive movement involving the interior of the transported
mass, producing a chaotic mixture of different types of sediments, such as broken mud
layers embedded in sandy sediment (Potter and Pettijohn, 1977). If muddy and sandy
sediment are both capable of flowing during such transport, the result is a streaked,
''migmatitic'' mixture (Pettijohn et al., 1987, p. 117). If, on the other hand, the clay is
cohesive and resistant to flow, it will break into fragments, which become incorporated
into and surrounded by sand flowing as a slurry (Fig. 3.26). Shale clasts of this type in
sandstone units are sometimes called **rip-up clasts,** although this term is more commonly
used for shale clasts ripped up by currents, e.g., turbidity currents. The clasts may range
from a few millimeters in size to several tens of centimeters. They may be angular or
subrounded and may be bent or curled. Rip-up clasts appear to be particularly common in
turbidites, but they occur also in debris flows and other types of sediment gravity flows
and in fluvial sandstones.

Dish and Pillar Structures

Some sandstone and siltstone beds are characterized in cross-sectional exposures by the presence of thin, dark-colored, subhorizontal, flat to concave-upward clayey laminations (Fig. 3.27). In plan view, these features are polygonal, circular, oval, or elliptical. Because the shapes of these clayey laminations superficially resemble the shapes of saucers or other shallow dishes, they are called **dish structures** (Stauffer, 1967; Lowe and LoPiccolo, 1974; Lowe, 1975; Rautman and Dott, 1977; Pederson and Surlyk, 1977). Dish structures typically occur in laterally extensive, thick beds that may, or may not, be devoid of other structures. Individual dishes range from 1 to 50 cm in width but commonly are only a few millimeters thick. They may also occur in thinner beds (<0.5 m), where they may cut across primary flat lamination or other lamination (e.g., convolute lamination). The laminae-forming dishes are generally darker in color than the surrounding sediment (Fig. 3.27) and commonly contain more clay, silt, or organic matter. Dish structures are particularly common in turbidites and other high-concentration flow deposits, but they have also been reported in deltaic, alluvial, lacustrine, and shallow-marine sediments and in volcanic ash layers.

  Pillar structures commonly occur in association with dishes (Fig. 3.27). They are vertical to near-vertical cross-cutting columns and sheets of structureless or swirled sand that cut through either massive or laminated sands that commonly also contain dish structures and convolute lamination. Pillars range in size from tubes a few millimeters in diameter to large structures greater than 1 m in diameter and several meters in length.

  Dish structures and pillars have generally been considered water-escape structures—formed as a result of rapid deposition, with subsequent escape of water from the

FIGURE 3.27
Dish structures (large arrow) and pillar structures (small arrow) in siliciclastic sediments of the Jackfork Group, southeast Oklahoma. (*From Lowe, D. R., 1975, Water escape structures in coarse-grained sediment: Sedimentology, v. 22. Fig. 8, p. 175, reprinted by permission of Elsevier Science Publishers, Amsterdam.*)

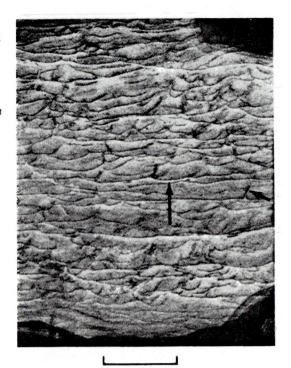

0   4 cm

sediment during compaction and consolidation. Lowe and LoPiccolo (1974) suggest that during gradual compaction and dewatering, semipermeable laminae act as partial barriers to upward-moving water carrying fine sediment. The fine particles are retarded by the laminae and added to them, forming the dishes. Some of the water is forced horizontally beneath the laminae until it finds an easier escape route upward. This forceful upward escape of water forms the pillars. Allen (1982, v. II, p. 373–374) proposes an alternative explanation for formation of dish structures. He suggests that dishes may occur from a kind of stoping process within a water-saturated bed that is slightly cohesive but far from completely consolidated. Shallow, horizontal water-filled cavities must exist in the lower part of a bed (created as a result of sedimentation after liquefaction of zones where sand has remained cohesionless or by forceful injection of external water into a bed). Each such shallow, water-filled cavity is a potentially unstable system under the influence of gravity. Slightly cohesive sand above the cavities will fail and collapse en masse into the cavities, creating subhorizontal failure surfaces, the dishes. As the sand masses are released one after the other from the roofs of the cavities, the cavities will progress upward through the bed, creating a series of dishes. If dish structures form in this manner, as suggested by Allen, they are not, strictly speaking, water-escape structures.

3.4.3   Erosion Structures: Channels and Scour-and-fill Structures

Channels are sediment-filled troughs that show a U- or V-shape in cross section and that cut across previously formed beds or laminae (Fig. 3.28). Channels exposed in outcrops commonly range in width and depth from a few centimeters to a few meters; rarely, they may reach tens of meters across. The long profiles of channels are rarely exposed in outcrops but can, in some cases, be defined by mapping or drilling. Channels are generally filled with sediment that is texturally different, commonly coarser, than that of the beds they truncate. Channels are probably eroded principally by currents, but some may be the result of erosion by sediment gravity flows. They are particularly common in fluvial and tidal sediments but occur also in turbidite sediments, where the long dimensions of the channels tend to be parallel to paleocurrent direction determined by other structures.

Scour-and-fill structures, also called **cut-and-fill structures,** resemble channels but tend to be somewhat smaller, more asymmetrical in cross-sectional shape, and shorter in length. They may be filled with material that is either coarser or finer than the substrate into which they are cut. They are most common in sandy sediment, where they probably form owing to current scour and subsequent backfilling as current velocity decreases. In contrast to channels, several scour-and-fill structures may occur closely spaced in a row. They are particularly prevalent in sediments of fluvial origin and can occur in river, alluvial-fan, or glacial outwash-plain environments. Genetically, scour-and-fill structures are related to flute casts (Section 3.5.1), which occur on the soles of beds.

3.4.4   Biogenic Structures: Stromatolitic Bedding

Stromatolites are laminated structures commonly composed of fine silt- or clay-size, more rarely sand-size, carbonate sediment. They have also been reported in siliciclastic sediment, but such occurrences are rare. Some stromatolites are composed of nearly flat laminae that may be difficult to differentiate from laminae of other origins. Most stromatolites are hemispherical bodies made up of laminae that are curved, crinkled, or

FIGURE 3.28
Channels. A. Shallow channel in
fine sandstone filled with fine sand
and pebbles. B. Small channel
filled with gravel, showing a
lenslike form in cross section. The
sandstones at Floras Lake
(Miocene), southwestern Oregon.

A

B

deformed to various degrees (Fig. 3.29). The term thrombolite was proposed by Aitken
(1967) for structures that resemble stromatolites in external form and size but lack lamination. The laminae of stromatolites are generally less than 1 mm thick and are caused
by variations in the concentrations and properties of fine calcium carbonate minerals, fine
organic matter, and detrital clay and silt.

   Stromatolites are organosedimentary structures formed largely by the trapping and
binding activities of blue-green algae (cyanobacteria). They are known from rocks of
Precambrian age and are forming today in many localities. Modern stromatolites are
confined mainly to the shallow subtidal, intertidal, and supratidal zones of the ocean, but
they have been found also in lacustrine environments. Owing to the requirements of
blue-green algae to carry on photosynthesis, stromatolites are restricted to environments
with adequate light for photosynthesis. The laminated structure forms as a result of
trapping of fine sediment in the very fine filaments of algal mats. Once a thin layer
of sediment covers a mat, the algal filaments grow up and around sediment grains
to form a new mat—a process that is repeated many times to produce the laminated
structure.

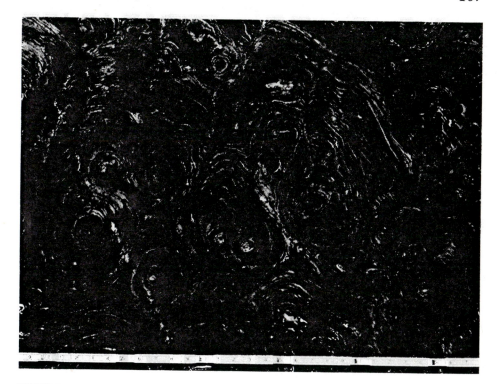

FIGURE 3.29
Well-developed hemispherical stromatolites showing fine-scale lamination. Hearn Formation
(age?), Canada. (Photograph courtesy of Geological Survey of Canada.)

3.5   Bedding-plane Markings

3.5.1   Markings Generated by Erosion and Deposition

General Statement

Bedding-plane markings may occur either on the tops of beds or on the undersides (soles)
of beds. **Sole markings** are particularly common, typically consisting of positive-relief
casts and various kinds of irregular markings, especially on the soles of sandstones and
other coarser-grained sedimentary rocks that overlie shales. Many sole markings are
generated by a two-stage process that involves initial erosion of a cohesive mud substrate
to produce grooves or depressions, followed by an episode of deposition during which the
grooves or depressions are filled by coarser-grained sediment. After burial and lithification, the coarser-grained sediment remains welded or amalgamated to the base of the
overlying bed. After tectonic uplift, weathering processes may remove the softer underlying shale, exposing the sole of the overlying bed and the sole markings (Fig. 3.30).
Erosional markings are particularly common on the soles of turbidite sandstones, but they
can form in any environment where the requisite conditions of an erosive event followed
quickly by a depositional event are met, e.g., fluvial, tidal flat, and shelf environments.

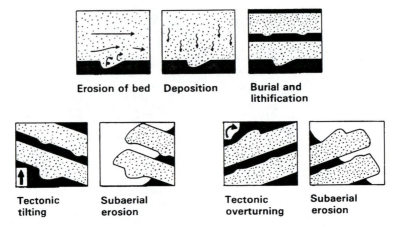

Erosion of bed   Deposition   Burial and lithification

Tectonic tilting   Subaerial erosion   Tectonic overturning   Subaerial erosion

FIGURE 3.30
Suggested stages of development of sole markings owing to erosion of a mud bottom followed
by deposition of coarser sediment. The diagram illustrates also how the sole markings appear as
positive-relief features on the base of the infilling bed after tectonic uplift and subaerial
weathering and suggests how sole markings can be used to tell top and bottom of overturned
beds. (*From Collinson, J. D., and D. B. Thompson, 1982, Sedimentary structures. Fig. 4.1, p.
37, reprinted by permission. After Ricci-Luchi, F., 1970, Sedimentolografia: Sanchelli,
Bologna, Italy.*)

Tool-formed Erosional Structures

General Statement.   The erosional event that initiates the process of forming erosional
sole markings can result from the action of current-transported objects that intermittently
or continuously make contact with the bottom. Such contact may simultaneously deform
(compress) the soft bottom sediment and gouge depressions or grooves in the sediment.
The "tools" can be pieces of wood, the shells of organisms, pebbles or clasts, or any
similar object that can be rolled or dragged along the bottom by normal currents or
turbidity currents. Filling of these tool-formed grooves or depressions by coarser sediment
generates positive-relief casts on the base of the overlying bed.

Groove Casts.   The most common tool-formed structures are probably groove casts.
These structures are elongate, nearly straight ridges (Fig. 3.31) that result from the
infilling of grooves produced by some object dragged over a mud bottom in continuous
contact with the bottom. Typical groove casts are a few millimeters to a few tens of
centimeters wide, with a relief of a few millimeters to a few centimeters. Much larger
groove casts are known. Groove casts are markedly elongated parallel to the paleocurrent
direction; therefore, they can be used to determine the sense of paleoflow, although it is
generally not possible to distinguish the upcurrent from the downcurrent direction. A
special kind of groove cast called chevrons is made up of continuous **V**-shaped crenula-
tions that close in a downstream direction. Thus, this type of groove cast can be used to
determine the true paleoflow direction. Dzulynski and Walton (1965) suggest that chevrons are formed by tools moving just above the sediment surface, but not touching the
surface, causing rucking-up of the sides of the groove.

Bounce, Brush, Prod, Roll, and Skip Marks.   These markings are related in origin to
groove casts, but they are produced by tools that make intermittent contact with the

FIGURE 3.31
Large intersecting groove casts on the base of a graded, turbidite sandstone. The knife near the center of the photograph is about 10 cm long. Fluornoy Formation (Eocene), southwestern Oregon.

bottom rather than continuous contact. Brush and prod marks are positive-relief features produced by the infilling of small gouge marks. They are asymmetrical in cross section, with the deeper, broader part of the mark oriented downcurrent. By contrast, bounce marks are roughly symmetrical. Roll and skip marks are formed either by a saltating tool or by rolling of a tool over the surface, producing a continuous track. The postulated genesis of these structures is illustrated diagrammatically in Figure 3.32.

Current-formed Erosional Structures

General Statement. Under some conditions, currents are capable of eroding depressions in muddy or sandy substrates independently of the action of tools. Current scour owing to eddies and associated velocity fluctuations behind obstacles, or by chance eddy scour, commonly generates asymmetrical, somewhat elongated depressions with the steepest and deepest side of the depression upstream and the more gentle side downstream. Filling of such depressions produces a positive-relief sole marking with an abrupt upstream end and a gradually tapering downstream end. Thus, current-formed structures generally make good paleocurrent indicators because they show the unique direction of current flow.

Flute Casts. Flute casts are elongated welts or ridges that have at one end a bulbous nose that flares out toward the other end and merges gradually with the surface of the bed (Fig. 3.33). Flute casts tend to occur in swarms, with all of the flutes oriented in roughly the same direction, but they can occur singly. All of the flutes on the base of a given bed are generally about the same size, but great variation in size from one bed to another is possible. Flute casts can range in width from a few centimeters to more than 20 cm, in

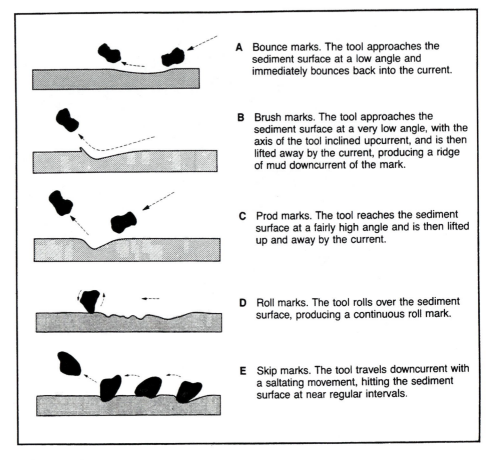

FIGURE 3.32
Postulated development in a cohesive mud bottom of (A) bounce marks, (B) brush marks, (C)
prod marks, (D) roll marks, and (E) skip marks by the action of tools making contact with the
bottom in various ways. These tool-formed depressions are subsequently filled with coarser
sediment to produce positive-relief casts. (*After Reineck, H. E., and I. B. Singh, 1980,
Depositional sedimentary environments, 2nd ed. Figs. 127, 129, 125, 132, 131, p. 82, 83,
reprinted by permission of Springer-Verlag, Heidelberg.*)

height or relief from a few centimeters to more than 10 cm, and in length from a few
centimeters to a meter or more. The plan-view shape of flutes varies from nearly stream-
lined, bilaterally symmetrical forms to more-elongate and irregular forms, some of which
are highly twisted. Flute casts are particularly common on the soles of turbidite sandstones, but they occur also in sediments from shallow-marine and nonmarine environments. Less commonly, they have been reported on the soles of limestone beds.

Current Crescents.   These structures, also called obstacle scours, are common in modern environments, particularly sandy beach environments. They occur as narrow semi-
circular or horseshoe-shaped troughs, which form around small obstacles such as pebbles
or shells owing to current scour (Fig. 3.34). They can form also in muddy sediment. They
are relatively uncommon in ancient sedimentary rocks, where they typically occur as

FIGURE 3.33

Flute casts covering the entire base (sole) of a turbidite sandstone bed. Note also the large groove casts in the lower part of the photograph (running under the hammer). The paleocurrent direction is from the upper right toward the lower left. Hornbrook Formation (Cretaceous), northern California.

FIGURE 3.34

Current crescents formed downflow from pebbles on a modern beach, southern Oregon coast. The knife is about 10 cm long.

positive-relief casts on the soles of sandstone beds. They are most characteristic of ancient fluvial sandstones with shale interbeds, but they have also been reported in turbidite sequences. Similar structures can form in modern eolian sediments as a result of wind transport of sand around obstacles. Wind-produced current crescents, as well as those produced in beach sands, are rarely preserved in ancient sedimentary rocks.

3.5.2   Sole Markings Generated by Deformation: Load Casts

Load casts are rounded knobs or irregular protuberances on the soles of sandstone beds that overlie shales. They differ in appearance from flute casts because they lack the regular form and orientation of flutes, and they differ from ball and pillow structures in that the basal sand layer in which they are formed is never completely pierced by the diapirs of mud rising from the shale layer beneath. Where load casts are present, they tend to cover the entire surface of the sole (Fig. 3.35). They can range in size from a few centimeters to a few tens of centimeters, and the casts on a single sole may display considerable variation in size.

Load casts appear to be most common on the soles of turbidite sandstones, but have been reported in deposits from a variety of environments, including fluvial, lacustrine, deltaic, and shallow-marine. They have been found also in pyroclastic sequences. Load casts originate owing to the gravitational instability arising from the presence of beds of greater density lying above beds of lower density and low strength. Hydroplastic or uncompacted muds with excess pore pressures or muds liquefied by an externally generated shock can be deformed by the weight of the overlying sand, which may sink unequally into the incompetent mud. Such loading allows protrusions of sand to sink down into the mud, creating positive-relief features on the bases of the beds. Load casts are closely related genetically to ball and pillow structures and flame structures. Flute and groove casts, as well as ripples, may be modified by loading, a process that tends to exaggerate their relief and destroy original shapes. The presence of load casts on the bases of some beds and not on others appears to reflect the state of the underlying hydroplastic mud. They apparently will not form on the bases of sandstone beds deposited on muds that have already been compacted or dewatered prior to deposition of the sand.

3.5.3   *Markings Generated by Organisms: Trace Fossils*

Bed-surface markings attributed to the activities of fossil organisms have long been recognized. Such organically produced markings are now commonly called **trace fossils,** but they have also been referred to as **ichnofossils** and **Lebensspuren**. These markings are believed to result from the burrowing, boring, feeding, resting, and locomotion activities of organisms, which can produce a variety of trails, shallow depressions, and open burrows and borings in muds or other semiconsolidated sediments. Filling of trails or other shallow depressions with sediment of different size or packing creates structures that show up as positive-relief features on the soles of overlying beds (Fig. 3.36A). Filled burrows or borings show up as rounded markings on the top surfaces of underlying beds.

FIGURE 3.35
Load casts on the base of a large sandstone boulder. Some of the load casts may be modified organic traces. Unknown formation, southern Oregon coast. The knife is about 10 cm long.

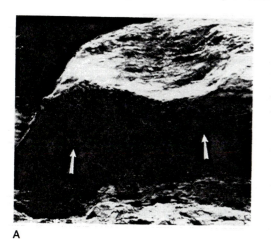

A

B

FIGURE 3.36

Organic markings. A. Copious organic traces (arrows) on the base of a massive sandstone bed. Bateman Formation (Eocene), southwestern Oregon. B. Shallow burrows (arrows) cutting across laminated fine sandstone. Coaledo Formation (Eocene), southwestern Oregon.

Burrows and borings commonly extend down into beds and thus may appear in cross-sectional exposures of beds as filled tubes (Fig. 3.36B). Therefore, filled burrows and borings are not exclusively bedding-plane structures.

Four broad categories of biogenic structures are recognized: (1) **bioturbation structures** (burrows, tracks, trails, root-penetration structures) arising from organic activity that tends to penetrate, mix, or otherwise disturb sediment, (2) **bioerosion structures** (borings, scrapings, bitings), (3) **biostratification structures** (stromatolites, graded bedding of biogenic origin), and (4) **excrement** (coprolites, such as fecal pellets or fecal castings). Not all geologists regard biostratification structures as trace fossils (see Section 3.4.3 for a discussion of stromatolites), and these structures are not commonly included in published discussions of trace fossils. Trace fossils can be classified also in various other ways: on the bases of morphology, presumed behavior of the organisms that produced the structures, and type of preservation (Simpson, 1975; Frey, 1978; Ekdale et al., 1984). On the basis of morphology, they can be grouped into such categories as tracks, trails, burrows, borings, and bioturbate texture (mottled bedding), as shown in Table 3.2. On the basis of behavior (ethological classification), they can be divided into resting traces, crawling traces, grazing traces, feeding traces or structures, and dwelling structures (Fig. 3.37A). Trace fossils can be classified in terms of type of preservation by use of such terms as full relief, semirelief, concave, and convex (Fig. 3.37B).

Geologists have become increasingly interested in trace fossils in the past few decades owing to their potential usefulness in paleoenvironmental and paleoecological interpretation. Several full-length books and extended research papers dealing with trace fossils have now been published (Basan, 1978; Crimes and Harper, 1970, 1977; Curran, 1985; Ekdale et al., 1984, Frey, 1975; Frey and Pemberton, 1984; Häntzschel, 1975; Seilacher, 1964). Trace fossils have special usefulness in paleoenvironmental analysis because they are biogenic features that clearly formed in place. In contrast to body fossils, trace fossils other than excrement cannot be transported from one environment to another and cannot be reworked from older rocks.

TABLE 3.2

Descriptive-genetic classification of trace fossils

A   Tracks and Trails

Track—impression left in underlying sediment by an individual foot or podium

Trackway—succession of tracks reflecting directed locomotion by an animal

Trail—trace produced during directed locomotion and consisting either of a surficial groove made by an animal having part of its body in continuous contact with the substrate surface or of a continuous subsurface structure made by a mobile endobenthic organism

B   Burrows and Borings

Boring—excavation made in consolidated or otherwise firm substrates, such as rock, shell, bone, or wood

Burrow—excavation made in loose, unconsolidated sediments

Burrow or *boring system*—highly ramified and/or interconnected burrows or borings, typically involving shafts and tunnels

Shaft—dominantly vertical burrow or boring or a dominantly vertical component of a burrow or boring system having prominent vertical and horizontal parts

Tunnel (= gallery)—dominantly horizontal burrow or boring or a dominantly horizontal component of a burrow or boring system having prominent vertical and horizontal parts

Burrow lining—thickened burrow wall constructed by organisms as a structural reinforcement; may consist of (1) host sediments retained essentially by mucus impregnation, (2) pelletoidal aggregates of sediment shoved into the wall, like mud-daubed chimneys, (3) detrital particles selected and cemented like masonry, or (4) leathery or felted tubes consisting mostly of chitinophosphatic secretions by organisms; burrow linings of types (3) and (4) commonly called "dwelling tubes"

Burrow cast—sediments infilling a burrow (burrow fill); may be either "active," if done by animals, or "passive," if done by gravity or physical processes; active fill termed "back fill" wherever U-in-U laminae, etc., show that the animal packed sediment behind itself as it moved through the substrate

C   Bioturbation

Bioturbate texture—gross texture or fabric imparted to sediments by extensive bioturbation; typically consists of dense, contorted, truncated, or interpenetrating burrows or other traces, few of which remain distinct morphologically. Where burrows are somewhat less crowded and thus are more distinct individually, the sediment is said to be "burrow mottled"

D   Miscellaneous

Configuration—in ichnology, the spatial relationships of traces, including the disposition of component parts and their orientation with respect to bedding and (or) azimuth

Spreite—bladelike to sinuous, U-shaped, or spiraled structure consisting of sets or cosets of closely juxtaposed, repetitive parallel or concentric feeding or dwelling burrows or grazing traces. Individual burrows or grooves comprising the spreite commonly anastomose into a single trunk or stem (as in *Daedalus*) or are strung between peripheral "support" stems (as in *Rhizocorallium*). "Retrusive" spreiten extended upward or promimal to the initial point of entry by the animal, and "protrusive" spreiten extended downward, or distal to the point of entry.

Source: After Frey, R. W., 1978, Behavioral and ecological implications of trace fossils, in Basan, P. B., ed., Trace fossil concepts: Soc. Econ. Paleontologists and Mineralogists Short Course 5. Table 2, p. 49, reprinted by permission of SEPM, Tulsa, Okla.

FIGURE 3.37A

Classification of trace fossils on the basis of presumed behavior of the organisms producing the structures and the relationship of these traces to body fossils. Note the overlap of some categories of traces. Escape structures overlap several categories of behavioral traces and are not included here. (*From Simpson, S., 1975, Classification of trace fossils, in Frey, R. W., ed., The study of trace fossils. Fig. 3.2, p. 49, reprinted by permission of Springer-Verlag, Heidelberg. As translated from Seilacher, A., 1953, Neues Jahrb. Geologie u. Paläontologie, Abh. 96, p. 421–452.*)

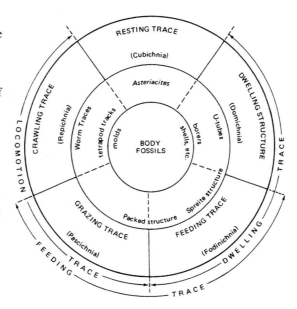

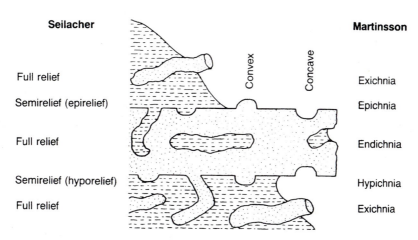

FIGURE 3.37B
Terminology for preservational classification of trace fossils used by Seilacher and Martinsson. (*From Ekdale, A. A., R. G. Bromley, and S. B. Pemberton, 1984, Ichnology: Trace fossils in sedimentology and stratigraphy: Soc. Econ. Paleontologists and Mineralogists Short Course 15. Fig. 2.6, p. 22, reprinted by permission of SEPM, Tulsa, Okla. From Seilacher, A., Sedimentological classification and nomenclature of trace fossils: Sedimentology, v. 3, p. 253–256; Martinsson, A., 1970, Toponomy of trace fossils, p. 109–130, in Crimes, T. P., and J. C. Harper, eds., The study of trace fossils: Springer-Verlag, New York.*)

Although similar trace fossils may be produced by different organisms, certain associations of biogenic structures tend to characterize particular sedimentary facies. These facies, in turn, can be related to depositional environments. The term **ichnofacies** was introduced by Seilacher (1964) for sedimentary facies characterized by a particular association of trace fossils. Trace fossils can form under both subaerial and subaqueous conditions. In subaerial environments, organisms such as insects, spiders, worms, millipedes, snails, and lizards produce various types of burrows and tunnels; vertebrate organisms leave tracks; and plants leave root traces. In subaqueous environments, the principal factors that appear to influence their distribution are water salinity, water depth, oxygenation state of the water, and consistency of the substrate (hard or soft bottom). Numerous organisms such as worms, crustaceans, insects, bivalves, gastropods, fish, birds, amphibians, mammals, and reptiles produce traces in freshwater fluvial or lacustrine environments. Trace fossils of continental deposits are grouped into the *Scoyenia* ichnofacies (Frey et al., 1984), which consists of a rather nondistinctive, low-diversity suite of invertebrate and vertebrate tracks, trails, and burrows.

Marine trace fossils, produced largely by invertebrate organisms and some fish, are grouped into seven marine ichnofacies (Fig. 3.38), each named from a representative trace fossil: *Terodolites, Trypanites, Glossifungites, Skolithos, Cruziana, Zoophycos,* and *Nereites.* The *Teredolites* ichnofacies, not shown in Figure 3.38, occurs only in woody material. The *Trypanites* ichnofacies is characteristic of hard, fully indurated substrates, and the *Glossifungites* ichnofacies typically occurs in firm, but uncemented, substrates. The remaining marine ichnofacies are all soft-sediment ichnofacies whose distribution appears to be controlled mainly by water depth, as illustrated in Figure 3.38.

It is now recognized that trace-fossil depth zones may overlap. For example, some trace fossils originally believed to be exclusively shallow-water forms, such as *Ophiomorpha* (Fig. 3.38), have been found in sediments from greater depths. Also, some supposedly deep-water trace fossils such as *Zoophycos* have turned up in beds that, on the basis of associated coal or algal limestone, are clearly shallow-water deposits (Hallam, 1981). These inconsistencies suggest that factors other than water depth and food supply—oxygen levels, for example—may be involved in controlling the distribution of trace fossils. Also, not all environments at a particular depth may harbor organisms capable of leaving traces. For example, highly saline waters or highly reducing (euxinic) conditions, where low oxygen levels and the production of hydrogen sulfide gas create a toxic environment, may preclude or greatly reduce organic activity.

3.5.4   Bedding-plane Markings of Miscellaneous Origin

A variety of generally small-scale markings of miscellaneous origin can occur on the tops (mainly) of beds; these include mudcracks, syneresis cracks, raindrop and hailstone imprints, bubble imprints, rill marks, swash marks, and parting lineation. **Mudcracks** are common in modern environments and may be preserved on the top or bottom bedding surfaces of ancient sedimentary rocks as positive-relief fillings of the original cracks (Fig. 3.39). They can occur in both siliciclastic and carbonate muds and indicate subaerial exposure and desiccation. They may occur in association with **raindrop imprints** and **hailstone imprints,** which are craterlike pits with slightly raised rims that are commonly less than 1 cm in diameter. These imprints can be confused with **bubble imprints,** caused by bubbles breaking on the surface of sediments. **Syneresis cracks** resemble mudcracks but tend to be discontinuous and vary in shape from polygonal to spindle-shaped or

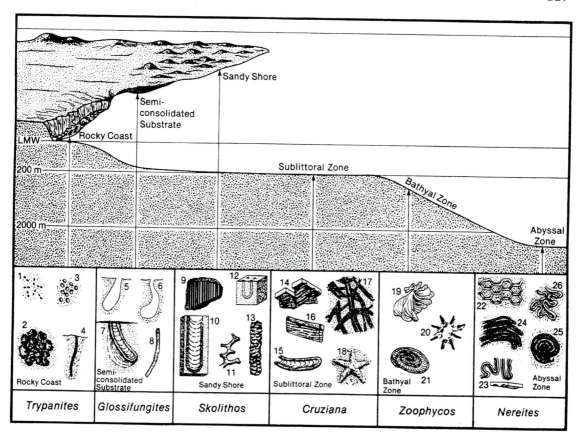

FIGURE 3.38

Schematic representation of the relationship of characteristic trace fossils to sedimentary facies and depth zones in the ocean. Borings of 1, *Polydora;* 2, *Entobia;* 3, echinoid borings; 4, *Trypanites;* 5, 6, pholadid burrows; 7, *Diplocraterion;* 8, unlined crab burrow; 9, *Skolithos;* 10, *Diplocraterion,* 11, *Thalassinoides;* 12, *Arenicolites;* 13, *Ophiomorpha;* 14, *Phycodes;* 15, *Rhizocorallium;* 16, *Teichichnus;* 17, *Crossopodia;* 18, *Asteriacites;* 19, *Zoophycos;* 20, *Lorenzinia;* 21, *Zoophycos;* 22, *Paleodictyon;* 23, *Taphrhelminthopsis;* 24, *Helminthoida;* 25, *Spirorhaphe;* 26, *Cosmorhaphe.* *(From Ekdale, A. A., R. G. Bromley, and S. B. Pemberton, 1984, Ichnology: Trace fossils in sedimentology and stratigraphy: Soc. Econ. Paleontologists and Mineralogists Short Course 15. Fig. 15.2, p. 187, reprinted by permission of SEPM, Tulsa, Okla. Modified from Crimes, T. P., 1975, The stratigraphical significance of trace fossils, in Crimes, T. P., and J. C. Harper, eds., The study of trace fossils. Fig. 7.2, p. 118, reprinted by permission of Springer-Verlag, New York.)*

sinuous (Plummer and Gostin, 1981). They commonly occur in thin mudstones interbedded with sandstones as either positive-relief features on the bases of sandstones or negative-relief features on the tops of the mudstones. Syneresis cracks are believed to be subaqueous shrinkage cracks, formed in clayey sediment by loss of pore water from clays that have flocculated rapidly or that have undergone shrinkage of swelling clay mineral lattices owing to changes in salinity of surrounding water (Burst, 1965). **Rill marks** are small dendritic channels or grooves that form on beaches by discharge of pore waters at

FIGURE 3.39
Mudcracks on the upper surface of a Miocene mudstone bed, Bangladesh. (*Photograph by E. M. Baldwin.*)

low tide by small streams debouching onto a sand or mud flat. They have very low preservation potential and are seldom found in ancient sedimentary rocks. **Swash marks** are thin, arcuate lines or small ridges on a beach formed by concentrations of fine sediment and organic debris owing to wave swash; they mark the farthest advance of wave uprush. They likewise have low preservation potential.

     Parting lineation, also called **current lineation,** forms primarily on the bedding surfaces of parallel-laminated sandstones, although it is reported to form also on the backs

FIGURE 3.40
Parting lineation in sandstone. Current flow parallel to hammer. Haymond Formation, Texas. (*Photograph courtesy of E. F. McBride.*)

of ripples and dunes. It consists of subparallel ridges and grooves a few millimeters wide and many centimeters long (Fig. 3.40); length of the ridges is generally 5–20 times greater than their width (Allen, 1982, v. I, p. 261). Relief on the ridges is commonly on the order of the diameter of the sand grains. Parting lineation appears to be most common in deposits composed of medium sand or coarse silt. Detailed study of parting lineation by several investigators has revealed that the trend of the ridges and grooves is commonly in good agreement with the preferred grain orientation in the deposits, indicating that the linear fabric of current lineation is parallel to current flow direction. Parting lineation occurs in newly deposited sands in modern beach, fluvial, and tidal run-off environments. It is most common in ancient deposits in thin, evenly bedded sandstones, probably of shallow-marine origin. It has been described also in turbidite sandstones. Its origin is obviously related to current flow and grain orientation, probably turbulent flow in the upper-flow–regime plane-bed phase. A more rigorous explanation of the origin of parting lamination is given by Allen (1982, v. I, p. 262–265).

3.6   Other Structures

3.6.1   Sandstone Dikes and Sills

Sandstone dikes are tabular bodies of sandstone that fill fractures in any kind of host rock. They range in thickness from a few centimeters to 10 m or more. Most sandstone dikes lack internal structures except oriented mica flakes and elongated particles that may be aligned parallel to the dike walls. They are apparently formed from liquefied sand forcefully injected upward into fractures, although examples are known where sand appears to have been injected downward into fractures. Sandstone sills are features of similar appearance and origin except that they are sand bodies that have been injected between beds of other rock. **Sandstone sills** may be difficult to differentiate from normally deposited sandstone beds unless they can be traced into sandstone dikes or traced far enough to show a cross-cutting relationship with other beds. For example, Archer (1984) describes sandstone sills in Ordovician deep-sea deposits of western Ireland that are virtually identical to normally deposited beds. She discusses also some of the criteria that can be used to distinguish clastic sills from normal beds.

The sands that form sandstone dikes and sills must have been in a highly water-saturated, liquefied state at the time of injection, and injection seems to have been essentially instantaneous. Although some sandstone dikes and sills may have been injected into host rocks before they were completely consolidated, others could have formed much later—considerably later than the time of lithification of the host. Pettijohn et al. (1987) suggest that host rocks that have been injected early tend to be contorted, whereas those injected after lithification are sharp-edged and straight-walled. Sandstone dikes and sills appear to have relatively little sedimentological significance, although they may be of some value in working out the tectonic history of a region; e.g., sand-filled cracks imply episodes of tensional deformation.

3.6.2   Structures of Secondary Origin

Most structures discussed in preceding sections of this chapter, with the exception of some sandstone dikes and possibly convolute lamination, appear to have formed at or very shortly after deposition of the host sediment. Thus, these structures are commonly called primary sedimentary structures. A few kinds of structures occur in sedimentary rocks that clearly postdate deposition and are thus secondary sedimentary structures. The majority of these secondary structures are of chemical origin, formed by precipitation of mineral substances in the pores of semiconsolidated or consolidated sedimentary rock or by chemical replacement processes. Some secondary structures appear to form through processes involving pressure and solution.

Concretions are perhaps the most common secondary structure. These structures are typically composed of calcite; however, concretions made of other minerals (e.g., dolomite, hematite, siderite, chert, pyrite, gypsum) are known also. Concretions form by precipitation of minerals around a center, building up a globular mass. Some concretions have a distinct nucleus, such as a shell or shell fragment, but many do not. When cut open, concretions may display concentric banding around the center or may display little or no internal structure. They range in shape from nearly spherical bodies to disc-shaped, cone-shaped, and pipe-shaped bodies and in size from less than a centimeter to as much as 3 m. Some concretions may be syndepositional in origin, growing in the sediment as it accumulates; however, most concretions are probably postdepositional, as shown by original bedding structures, such as laminations, that pass through the concretions (Fig. 3.41). Some early workers have suggested that concretions can grow displacively in sediments, pushing aside sedimentary layers as they grow. Most concretions, however, appear to have formed simply by precipitation of minerals in the pore spaces of the sediment. Concretions are especially common in sandstones and shales but can occur in other sedimentary rocks.

An unusual type of concretion was reported by Boggs (1972) that has as its nucleus a large rhombic-shaped prism, with pyramidal terminations, composed of calcite. In some specimens, the nucleus extends through the concretion and projects at either end (Fig. 3.42). Exceptional specimens of these rhombic prisms exceed 4 cm in width and 25 cm in length. Careful study of the nuclear prisms revealed that they are actually pseudomorphs after some earlier mineral. The concretionary calcite that surrounds the rhombic prisms clearly formed later than the prisms, as shown by distinctly different $\delta^{13}C$ values in the concretionary calcite and the nucleus. The identity of the parent mineral was solved in the early 1980s, when a German research vessel reported crystals of **ikaite** in a core from the seabed off Bransfield Strait, Antarctica (Suess et al., 1982). Ikaite is a little-

FIGURE 3.41
Carbonate concretion in laminated sandstone of the Coaledo Formation (Eocene), southern Oregon Coast. Note that laminations pass through the lower part of the concretion, indicating passive precipitation of carbonate in the pores of the sandstone. Laminations above the concretion are domed upward probably owing to differential compaction over the concretion.

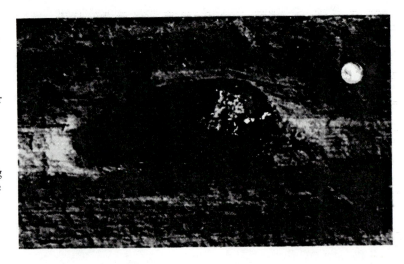

FIGURE 3.42

Ikaite concretions. A. A complete specimen with ikaite nucleus still intact, shown in place within host mudstone. B. Cross-sectional views of three concretions showing typical rhombic cross-sectional shape of the ikaite nuclei. Astoria Formation (Miocene), northwestern Oregon.

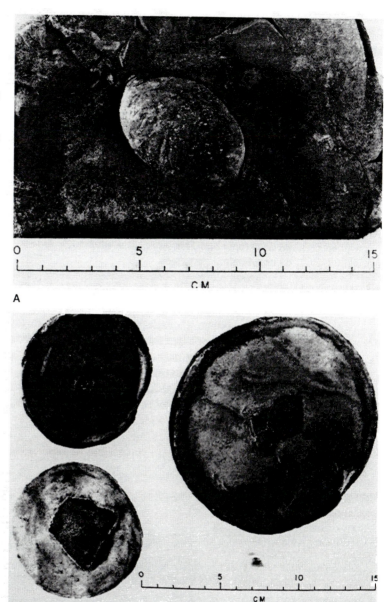

known hydrated calcium carbonate mineral ($CaCO_3 \cdot 6H_2O$) discovered and named in 1963. It is now believed to be the parent mineral of these unusual pseudomorphs (Shearman and Smith, 1985). According to Shearman and Smith, ikaite crystallizes at temperatures near 0°C and breaks down at normal temperatures to form calcium carbonate and water. The calcium carbonate is then redistributed to form the calcite pseudomorphs. Aside from their unusual nature, these ikaite pseudomorphs (formerly called glendonites), not all of which are enclosed in concretionary calcite, may prove to be important as paleothermometers.

Nodules are closely related to concretions. They are small, irregularly rounded bodies that commonly have a warty or knobby surface. They generally have no internal structure except the preserved remnants of original bedding or fossils. Common minerals that make up nodules include chert, apatite (phosphorite), anhydrite, pyrite, and manganese. Some so-called nodules (e.g., manganese and phosphorite nodules) are forming now on the seafloor and are syndepositional in origin (these might better be called concretions). Other nodules (e.g., chert nodules in limestones) are clearly postdepositional. Postdepositional nodules appear to form by partially or completely replacing minerals of the host rock rather than by simple precipitation of mineral into available pore space.

Sand crystals are very large euhedral or subhedral crystals of calcite, barite, or gypsum that are filled with detrital sand inclusions (Fig. 3.43). They appear to form during diagenesis by growth in incompletely cemented sands. Rosettes are radially symmetric, sand-filled crystalline aggregates or clusters of crystals that somewhat resemble the shape of a rose. They are commonly composed of barite, pyrite, or marcasite and form by cementation processes.

Color banding, sometimes referred to as Liesegang banding, is a type of rhythmic layering resulting from the precipitation of iron oxide in fluid-saturated sediments to form thin, closely spaced, commonly curved layers. Layers having various shades of red, yellow, or brown alternate with white or cream layers. Color banding may resemble primary bedding or lamination, but can almost always be distinguished by careful study (e.g., color banding may cut across primary stratification).

Stylolites are suture- or styluslike seams, as seen in cross-section, in generally homogeneous, thick-bedded sedimentary rocks (Fig. 3.44). The seams result from the irregular, interlocking penetration of rock on each side of the suture. They are typically only a few centimeters thick, and they are generally marked by concentrations of difficultly soluble constituents such as clay minerals, iron oxide minerals, and fine organic matter. Stylolites are most common in limestones, but occur also in sandstones, quartz-

FIGURE 3.43
Sand crystals, Miocene sandstone, Badlands, South Dakota. Length of specimen about 15 cm. (*From Pettijohn, F. J., Sedimentary rocks.* © *1975. Fig. 12.3, p. 467, reprinted by permission of Harper & Row, New York.*)

FIGURE 3.44
Sharp-peaked, sutured stylolites in a polished limestone slab. Unknown age and location.
University of Oregon collection.

ites, and cherts. Although many ideas were advanced by early workers to explain stylolites, they are now regarded to form as a result of pressure solution. These structures are
discussed in further detail in Chapter 12.

   Cone-in-cone structure is a type of structure rarely mentioned in recent sedimentological literature. It consists of nested sets of small concentric cones, composed, in most
examples, of calcium carbonate, with individual cones ranging in height mainly from 10
mm to 1 cm (Fig. 3.45). Sides of the cones are slightly ribbed or fluted, and some have
fine striations that resemble slickensides. Cone-in-cone structure generally occurs in thin,
persistent layers of fibrous calcite, commonly in association with concretions. Cone-in-
cone structure in pyrite concretions is also known (Carstens, 1985). The structure is most
common in shales and marly limestones. Considerable difference of opinion about the
origin of this structure emanated from early workers (see review by Franks, 1969). Most
recent workers appear to believe that cone-in-cone is an early diagenetic structure that

FIGURE 3.45
Large specimen showing typical
cone-in-cone structure. Age and
locality of specimen unknown.

forms by growth of fibrous crystals in the enclosing sediment while it is still in a plastic state (Woodland, 1964; Franks, 1969). Alternatively, the structure is attributed to recrystallization of fibrous aragonite to microcrystalline calcite (Gilman and Metzger, 1967).

3.7 Methods for Studying Sedimentary Structures

Sedimentary structures provide important information about depositional conditions and depositional environments. Also, many sedimentary structures have directional significance, making them useful indicators of paleocurrent direction. Analysis of sedimentary structures thus plays a crucial role in basin analysis (Section 1.6.3). Information gained from such evaluation may have economic significance in petroleum and mineral exploration and significance to interpretation of geologic and tectonic history, including plate reconstructions, regional historical geology, and paleogeography. Because study of sedimentary structures constitutes such an important part of environmental interpretation and basin analysis, much work has gone into developing suitable methods for their study. Space limitations do not permit description of those methods here. Good discussions of these techniques may be found in Bouma, 1969; Collinson and Thompson, 1982, Chapter 10, Appendixes A and B; Miall, 1990, p. 315–327; Potter and Pettijohn, 1977, p. 364–397; and Pettijohn et al., 1987, p. 124–132.

Additional Readings

Allen, J. R. L., 1968, Current ripples: Their relation to patterns of water and sediment motion: North Holland Publishing, Amsterdam, 443 p.

———1982, Sedimentary structures: Their character and physical basis: Elsevier, Amsterdam, v. 1, 593 p., v. II, 664 p.

Basan, P. B., ed., 1978, Trace fossil concepts: Soc. Econ. Paleontologists and Mineralogists Short Course 5, 181 p.

Bouma, A. H., 1969, Methods for the study of sedimentary structures: John Wiley & Sons, New York, 457 p.

Collinson, J. D., and D. B. Thompson, 1982, Sedimentary structures: George Allen & Unwin, London, 194 p.

Conybeare, C. E. B., and K. A. W. Crook, 1968, Manual of sedimentary structures: Dept. National Development, Bur. Mineral Resources, Geology and Geophysics, Bull. 102, 327 p.

Crimes, T. P., and J. C. Harper, eds., 1970, Trace fossils: Seel House Press, Liverpool, 547 p.

———1977, Trace fossils 2: Seel House Press, Liverpool, 351 p.

Curran, H. A., ed., 1985, Biogenic structures: Their use in interpreting depositional environments: Soc. Econ. Paleontologists and Mineralogists Spec. Pub. 35, 347 p.

Dzulnyski, S., and E. K. Walton, 1965, Sedimentary features of flysch and greywackes: Developments in sedimentology, v. 7, Elsevier, Amsterdam, 274 p.

Ekdale, A. A., R. G. Bromley, and S. G. Pemberton, 1984, Ichnology, trace fossils in sedimentology and stratigraphy: Soc. Econ. Paleontologists and Mineralogists Short Course 15, 317 p.

Frey, R. W., ed., 1975, The study of trace fossils: Springer-Verlag, New York, 562 p.

Harms, J. C., J. B. Southard, D. R. Spearing, and R. G. Walker, 1975, Depositional environments as interpreted from primary sedimentary structures and stratification sequences: Soc. Econ. Paleontologists and Mineralogists Short Course 2, 161 p.

Harms, J. C., J. B. Southard, and R. G. Walker, 1982, Structures and sequences in clastic rocks: Soc. Econ. Paleontologists and Mineralogists Short Course 9.

Middleton, G. V., ed., 1965, Primary sedimentary structures and their hydrodynamic interpretation: Soc. Econ. Paleontologists and Mineralogists Spec. Pub. 12, 265 p.

Miller, M. F., A. A. Ekdale, and M. D. Picard, eds., 1984, Trace fossils and paleoenvironments: marine carbonate, marginal marine terrigenous and continental terrigenous settings: Jour. Paleontology, v. 58, p. 283–597.

Pettijohn, F. J., and P. E. Potter, 1964, Atlas and

glossary of primary sedimentary structures: Springer-Verlag, New York, 370 p.

Picard, M. D., and L. R. High, Jr., 1973, Sedimentary structures of ephemeral streams: Elsevier, New York, 223 p.

Potter, P. E., and F. J. Pettijohn, 1977, Paleocurrents and basin analysis, 2nd ed.: Springer-Verlag, New York, 460 p.

Reineck, H. E., and I. B. Singh, 1980, Depositional sedimentary environments, 2nd ed.: Springer-Verlag, New York, 439 p.

Rubin, D. M., 1987, Cross-bedding, bedforms, and paleocurrents: Soc. Econ. Paleontologists and Mineralogists, Concepts in Sedimentology and Paleontology, v. 1, 187 p.

Sarjent, W. A. S., ed., 1983, Terrestrial trace fossils: Benchmark Papers in Geology, v. 76, Hutchinson Ross, 415 p.

Chapter 4
Sandstones:
Composition

4.1 Introduction

Sandstones make up nearly one-quarter of the sedimentary rocks in the geologic record. They are common rocks in geologic systems of all ages, and they are distributed throughout the continents of Earth. Sandstones contain many kinds of sedimentary textures and structures that have potential environmental significance, as discussed in the preceding two chapters.

Particle composition is a fundamental physical property of these sandstones and is the chief property used in their classification. Also, particle composition has significant value in interpreting the provenance history of siliciclastic deposits (Chapter 8). Particle composition may also influence the economic importance of sandstones as oil and gas reservoirs because particle composition has an important effect on the course of diagenesis in sandstones (Chapter 9) and thus on the ultimate porosity and permeability of these rocks.

Because of their relatively coarse grain size, the particle composition of sandstones can generally be determined with reasonable accuracy with a standard petrographic microscope. Therefore, petrographic microscopy has remained for many years the primary tool for studying the composition of sandstones. Newer tools for studying particle composition are available also. X-ray diffraction techniques are used to determine the mineralogy of very fine-grained sediments (e.g., Eslinger et al., 1973). The electron microprobe can provide accurate chemical compositions of individual mineral grains, and X-ray fluorescence and ICP (inductively coupled argon plasma emission spectrometry) allow rapid and relatively inexpensive chemical analysis of bulk samples. Cathodoluminescence petrography is being increasingly used to study cements in sandstones and the provenance of quartz and other minerals (Sippel, 1968; Sibley and Blatt, 1976; Owen and Carozzi, 1986). The scanning electron microscope (SEM) allows study of mineral grains at high magnification, and a recent innovation makes possible petrographic examination of sedimentary rocks in the SEM by using back-scattered electron image analysis (Pye and Krinsley, 1984; Diks and Graham, 1985; White et al., 1984; Krinsley and Manley, 1989). Chemical compositions of particular features (and thus mineral identification) can also be determined directly by using an X-ray attachment to the SEM (energy-dispersive x-ray spectrometer) (Minnis, 1984).

Sand- and coarse silt-size particles in sandstones constitute the **framework fraction** of the sandstones. Sandstones may also contain various amounts of **matrix** (material $<\sim0.03$ mm) and **cement,** which are present within interstitial pore space among the framework grains. **Authigenic minerals** that replace some framework grains, matrix, or cement are also present in many sandstones, as well as empty pore space (**porosity**). The composition of the framework fraction is the major emphasis of this chapter; however, discussion of matrix and cement is also included. Thus, in the present chapter, I describe the mineralogy of the important detrital and authigenic constituents of sandstones and briefly consider their chemical composition. The classification of sandstones and the petrographic characteristics of the major groups of sandstones are discussed in Chapter 5.

4.2   Particle Composition

4.2.1   Detrital Constituents

General Statement

Sandstones are composed of a very restricted suite of major detrital minerals and rock fragments, plus a variety of minerals that may be present in accessory amounts (Table 4.1). Detrital constituents are defined as those derived by mechanical-chemical disintegration of a parent rock. Most detrital constituents in sandstones are terrigenous siliciclastic particles that are generated through the process of weathering, explosive volcanism, and sediment transport from parent rocks located outside the depositional basin. Some volcaniclastic particles may, however, originate from volcanic centers located within basins. A few detrital constituents in sandstones may be nonsiliciclastic particles, such as skeletal fragments or carbonate clasts, formed within the depositional basin by mechanical disruption of reef masses or other consolidated or semiconsolidated carbonate bodies. Sandstones may also contain intrabasinal biogenic remains that accumulated at the depositional site as organisms died, along with detrital sediment.

Owing to the wide variety of igneous, metamorphic, and sedimentary rocks that may constitute source materials for detrital sediment, sandstones could theoretically contain an extensive suite of major minerals. The fact that they do not can be attributed to the processes of chemical weathering and physical and chemical attack during transport and deposition that tend to destroy or degrade chemically unstable and mechanically weak sand-size grains. Thus, the framework grains of most sandstones are composed predominantly (commonly >90 percent) of quartz, feldspars, and rock fragments. Clay minerals may be abundant in some sandstones as matrix constituents; however, the detrital origin of such clay minerals is often difficult to establish. Coarse micas, especially muscovite, make up a few percent of the framework grains of many sandstones. Finally, heavy minerals may constitute a small percentage of the detrital constituents of sandstones, particularly the chemically stable heavy minerals such as zircon, tourmaline, and rutile.

Numerous references that describe the silicate minerals are available. One of the earlier English-language descriptions of the physical and optical properties of the detrital

TABLE 4.1
Common minerals and rock fragments in siliciclastic sedimentary rocks

---

MAJOR MINERALS (abundance $> \sim 1-2\%$)

  Stable minerals (greatest resistance to chemical decomposition)

    Quartz—makes up approximately 65% of average sandstone, 30% of average shale; 5% of average carbonate rock

  Less stable minerals

    Feldspars—include K-feldspars (orthoclase, microcline, sanidine, anorthoclase) and plagioclase feldspars (albite, oligoclase, andesine, labradorite, bytownite, anorthite); make up about 10–15% of average sandstone, 5% of average shale, <1% of average carbonate rock

    Clay minerals and fine micas—clay minerals include the kaolinite group, illite group, smectite group (montmorillonite a principal variety), and chlorite group; fine micas are principally muscovite (sericite) and biotite; make up approximately 25–35% of total siliciclastic minerals, but may comprise >60% of the minerals in shales

ACCESSORY MINERALS (abundances $< \sim 1-2\%$)

  Coarse micas—principally muscovite and biotite

  Heavy minerals (specific gravity $> \sim 2.9$)

    Stable nonopaque minerals—zircon, tourmaline, rutile, anatase

    Metastable nonopaque minerals—amphiboles, pyroxenes, chlorite, garnet, apatite, staurolite, epidote, olivine, sphene, zoisite, clinozoisite, topaz, monazite, plus about 100 others of minor importance volumetrically

    Stable opaque minerals—hematite, limonite

    Metastable opaque minerals—magnetite, ilmenite, leucoxene

ROCK FRAGMENTS (make up about 10–15% of the siliciclastic grains in average sandstone and most of the gravel-size particles in conglomerates; shales contain few rock fragments)

  Igneous rock fragments—may include clasts of any igneous rock, but fragments of fine-crystalline volcanic rock and volcanic glass are most common in sandstones

  Metamorphic rock fragments—include metaquartzite, schist, phyllite, slate, argillite, and less commonly gneiss clasts

  Sedimentary rock fragments—any type of sedimentary rock possible in conglomerates; clasts of fine sandstone, siltstone, shale, and chert are most common in sandstones; limestone clasts are comparatively rare in sandstones

CHEMICAL CEMENTS (abundance variable)

  Silicate minerals—predominantly quartz; others may include chalcedony, opal, feldspars, and zeolites

  Carbonate minerals—principally calcite; less commonly aragonite, dolomite, siderite

  Iron oxide minerals—hematite, limonite, goethite

  Sulfate minerals—anhydrite, gypsum, barite

Note: Stability refers to chemical stability.

minerals is that of Milner (1962). Additional references, all of which contain color photomicrographs of detrital minerals and other constituents of sandstones, include Adams et al. (1984), MacKenzie and Guilford (1980), and Scholle (1979). Pettijohn et al. (1987, p. 19–21) provide a bibliography of petrographic manuals and encyclopedias that includes many non-English titles. The works of Deer and others (1963, 1966) on rockforming minerals are excellent general references for the silicate minerals.

Silica Minerals

General Statement.   The silica minerals are the most abundant minerals in most sandstones. Quartz is the dominant silica mineral. Cristobalite and tridymite are high-temperature varieties of quartz that are uncommon in most sandstones. Chalcedony (fibrous quartz) and opal (amorphous and crystalline silica) are present in chert, which is a common rock fragment (described below) in sandstones. The most important characteristics of the silica minerals are summarized in Table 4.2.

Quartz.   The principal crystalline varieties of SiO_2 are quartz, cristobalite, and tridymite. Cristobalite and tridymite are metastable, high-temperature polymorphs of quartz that transform slowly with time to quartz. As mentioned, detrital cristobalite and tridymite rarely occur (or are rarely recognized) in detrital sediments. By contrast, quartz is an extremely stable mineral, and it is a dominant mineral constituent in most sandstones. It

TABLE 4.2
Characteristics of detrital silica (SiO_2) minerals

QUARTZ
   Most abundant detrital SiO_2 mineral; uniaxial ($+$), but can have biaxial ($-$) flash figure when cut nearly parallel to c-axis; low birefringence in sections cut perpendicular to optic axis; white to straw yellow interference colors in parallel sections. Occurs as single grains (monocrystalline quartz) and composite grains (polycrystalline quartz).
      Monocrystalline quartz—consists of single quartz crystals that may be characterized by
         Undulatory extinction—ranging to 30°. Quartz with more than 5° undulatory extinction is called undulatory quartz; quartz with 5° or less undulatory extinction is nonundulatory quartz.
         Inclusions
            Liquid inclusions—liquid- or gas-filled bubbles
               Vacuoles—nonoriented bubbles
               Böhm lamellae—bubbles oriented along strain lines
            Mineral inclusions (microlites)—tourmaline or rutile particularly common
      Polycrystalline quartz—quartz grains composed of aggregates of two or more quartz crystals (composite quartz); some rock fragments (chert, quartzite, quartz-rich sandstone) are also considered to be polycrystalline quartz grains by some authors.

CRISTOBALITE AND TRIDYMITE
   High-temperature varieties of SiO_2 best identified by X-ray methods; metastable and invert slowly with time and temperature to quartz; rare as detrital minerals in most sandstones, but may be present in volcaniclastic sandstones

CHALCEDONY AND MICROCRYSTALLINE QUARTZ
   Chalcedony is composed of sheaflike bundles of radiating, thin fibers of quartz that average ~0.1 mm but range from ~20 μm to 1 mm. Microcrystalline quartz consists of aggregates of nearly equant crystals that are commonly <5 μm but range to ~20 μm (Folk, 1974, p. 80). Both may appear brownish in plain light owing to the presence of minute, liquid-filled bubbles. Chalcedony and microquartz are the principal minerals that make up chert, which may also contain opal.

OPAL
   Opal-A is a hydrous, cryptocrystalline form of cristobalite with submicroscopic pores containing water; isotropic under crossed polarizing prisms; may appear brownish in plain light. Opal-A is metastable and inverts in time to opal-CT and eventually quartz. Rare as a detrital mineral in most sandstones, but may occur as a cement, particularly in volcaniclastic sandstones.

commonly makes up about two-thirds of the average sandstone (Blatt, 1982), although its actual modal abundance may range from less than 5 percent to more than 95 percent. Its relative abundance in most source rocks, coupled with its chemical resistance to weathering and mechanical resistance to abrasion during transport, account for its abundance in sandstones.

Quartz is distinguished optically by its uniaxial positive sign, although quartz that has undergone deformation may yield a biaxial figure with a small 2V (up to 8–10 degrees). It has low birefringence, with gray interference colors in sections cut perpendicular to the optic axis. Sections cut parallel to the optic axis show white to straw yellow (in thicker sections) interference colors. In this orientation, quartz may display a flash figure, which breaks up and quickly leaves the field of view upon slight rotation of the microscope stage. Many quartz grains display undulatory extinction, a pattern of sweeping extinction as the stage is rotated. Undulatory extinction is caused by deformation of quartz after crystallization, which results in displacement of the c crystallographic axis in the optic plane of the grain. That is, the orientation of the c-axis is different in various parts of the crystal, "either as a response to bend gliding or progressive misorientation of various parts of the quartz crystal developed by parallel walls of dislocations" (Young, 1976, p. 597). Extinction angles, measured by rotation of a microscope stage, may be as great as 30 degrees. Quartz having an extinction angle greater than about 5 degrees is called **undulatory quartz;** quartz with extinction angles of 5 degrees or less is **nonundulatory quartz** (Basu et al., 1975). Some workers have suggested that the degree of undulatory extinction may have significance in determining the provenance of quartz (Chapter 8). Intensely strained quartz grains may also be marked by the presence of trains of minute bubbles forming thin lines across the grains. These bubble trains are called **Böhm lamellae**.

Many quartz grains contain inclusions, either bubbles or mineral inclusions. The bubbles, also called **vacuoles,** are filled with liquid, liquid and gas, or gas alone. They are distributed randomly through the quartz, in contrast to the oriented bubble trains of Böhm lamellae, and may be extremely abundant or very sparse. Folk (1974) suggests that quartz crowded with bubbles to the point that it appears milky in reflected light is probably derived from hydrothermal veins. On the other hand, some hydrothermal quartz may have few bubbles, as does quartz from most other sources. Thus, the presence of sparse vacuoles in quartz appears to have no provenance significance. Tiny mineral inclusions are common in quartz and include apatite, rutile, tourmaline, zircon, magnetite, micas, chlorite, and feldspars.

Detrital quartz grains, particularly small grains, tend to be subangular; however, grains that have undergone an episode of intensive eolian transport or polycyclic grains that have undergone several episodes of transport and deposition may be well rounded, with all traces of original crystal faces or original angularity removed. Such rounded quartz grains that undergo secondary silica cementation during diagenesis may assume crystalline outlines owing to precipitation of silica **overgrowths** onto the rounded grains in the presence of open pore space (Fig. 4.1). The original rounded grain outline is commonly revealed by the presence of small specks of hematite, clay, or other material. Not all quartz overgrowths are marked by such inclusions; therefore, it may be necessary to use cathodoluminescence petrography to identify the overgrowths.

Although quartz occurs preferentially in sandstones as individual sand-size crystals (**monocrystalline quartz**), detrital polycrystalline quartz is extremely common in some sediments. **Polycrystalline quartz,** also called composite quartz, is quartz made up of

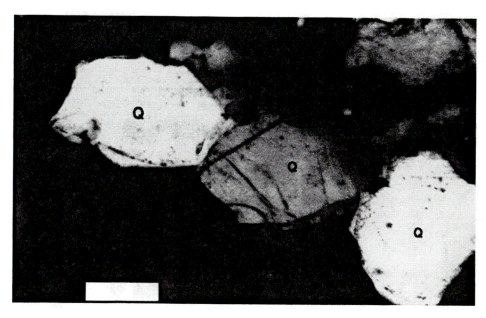

FIGURE 4.1
Detrital quartz grains (Q) that began to take on crystal faces owing to formation of overgrowths. Deadwood Formation (Cambrian-Ordovician), South Dakota. Crossed nicols. Scale bar = 0.1 mm.

aggregates of two or more crystals (Fig. 4.2). The character of polycrystallinity can vary enormously (Fig. 4.3). The individual crystals within a polycrystalline grain may be equant or elongate in shape, may be fine grained or coarse grained, may be all about the same size or variable in size, and may have crystal boundaries that are relatively straight or sutured to various degrees. Young (1976) has shown that polycrystalline quartz can develop from monocrystalline quartz during metamorphism. Under the influence of increasing pressure and temperature, nonundulatory monocrystalline quartz changes progressively to undulatory quartz, polygonized quartz (quartz that shows distinct zones of extinction with sharp boundaries), and finally to polycrystalline quartz. The polycrystalline quartz initially consists of elongated original crystals characterized by sutured boundaries. With increasing metamorphism, the grain recrystallizes to form small new crystals. Some of these new crystals may grow larger at the expense of others until the smaller crystals are replaced by a few large crystals. These large crystals are characterized by polyhedral outlines, lack of undulose extinction, smooth crystal–crystal boundaries, and interfacial angles of 120 degrees at triple junctions.

Polycrystalline quartz grains can also form in plutonic igneous rocks. The exact nature of the polycrystalline grains putatively depends upon the type of igneous or metamorphic rock from which the grains were ultimately derived. Some geologists believe that polycrystallinity can be used as a tool for provenance analysis; however, the provenance significance of polycrystalline quartz is controversial, a subject discussed further in Chapter 8. In addition to polycrystalline quartz grains that originate as single grains in metamorphic or plutonic igneous rocks, clasts of quartzite, chert, and quartz-rich sandstone are considered by some authors (e.g., Pettijohn et al., 1987, p. 30) to be polycrystalline quartz grains.

FIGURE 4.2
Large, well-rounded, polycrystalline quartz grain (arrow) that displays sutured internal boundaries between composite crystals, indicating a probably early stage in development of metamorphic polycrystalline quartz. Crossed nicols. Scale bar = 0.1 mm.

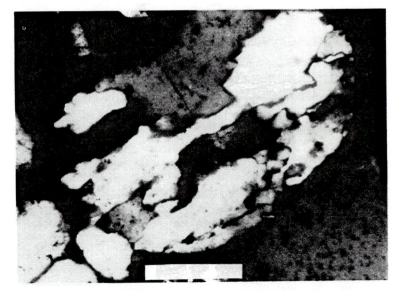

FIGURE 4.3
Typical textural appearance of fine, equigranular, microcrystalline quartz in a chert. Unknown formation. Crossed nicols. Scale bar = 0.1 mm.

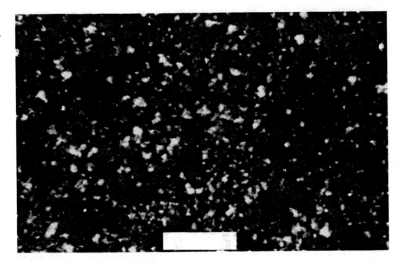

Quartz may be derived from igneous rocks, especially acid plutonic rocks, various kinds of metamorphic rocks, and sedimentary rocks. Owing to the increase in the volume of sedimentary rocks through time (Chapter 1), the principal source for detrital quartz in more-recent geologic time has probably been sedimentary rocks.

Chalcedony.   Chalcedony is a fibrous variety of quartz containing submicroscopic pores. Water may be present in these minerals in the pore spaces or in the form of hydroxyl (Nesse, 1986, p. 254). Some chalcedony may contain cristobalite. Chalcedony typically displays a ''feathery'' texture owing to alignment of quartz fibers in a parallel or spherulitic pattern (Fig. 4.4). The fibers are a few micrometers thick, up to a millimeter long, and average about 0.1 mm long (Folk, 1974, p. 80). Chalcedony commonly contains very tiny bubbles or inclusions that cause it to appear brownish in transmitted light (uncrossed

FIGURE 4.4
Fibrous to feathery texture of
chalcedony. Unknown formation.
Crossed nicols. Scale bar = 0.1
mm.

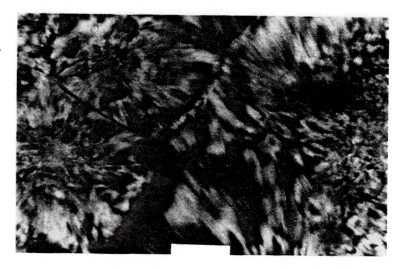

nicols). Although chert (chalcedony) is considered to be a mineral, detrital chert grains are actually rock fragments. They are derived primarily by weathering of bedded chert deposits or nodular cherts in limestones or other rocks. Rock fragments that are similar in composition and texture to sedimentary cherts may also be derived from silicified tuffs.

Opal.   Opal is SiO_2 that contains a few percent water. It consists of mixtures of amorphous and crystalline silica. The amorphous silica commonly consists of very small (0.1–0.5 nm), closely packed spherical masses. The crystalline material consists of extremely small crystals of cristobolite or tridymite. Water is contained in submicroscopic pores. Amorphous opal is metastable and transforms in time to opal-CT (disordered cristobalite/tridymite) and eventually to quartz. Opal is isotropic under crossed polarizing prisms, making it relatively easy to identify in thin section. Opal is rare as a detrital mineral. It occurs mainly in sandstones as an uncommon secondary cement.

Feldspars

General Characteristics and Occurrence.   Feldspars are the most common framework mineral in sands and sandstones after quartz. Data on the feldspar content of Holocene and Pleistocene sands of North America compiled by Pettijohn et al. (1987, p. 35) indicate that feldspars make up about 22 percent of the average river sand and 10 percent of the average beach and dune sand. The overall average content of feldspars in these Holocene and Pleistocene sands is about 15 percent, but their range of abundance extends from less than 1 percent to more than 75 percent. Feldspars are believed to make up about 10–15 percent of the average ancient sandstone, but their reported abundance in sandstones ranges from zero to as much as 60 percent. Sandstones containing more than 25 percent feldspar are considered feldspar-rich.

Feldspars are commonly divided into two main groups: alkali feldspars (potassium feldspars) and plagioclase feldspars (see Deer et al., 1963a, p. 1–178, and J. V. Smith, 1974, for an extended treatment of the feldspars). Both groups are well represented in detrital sediments. Potassium feldspars are generally regarded to be more abundant than plagioclase feldspars in the average sandstone; however, sandstones derived from source areas rich in volcanic rocks may contain more plagioclase than potassium feldspar. Sodic plagioclase tends to be more abundant in sandstones than calcic plagioclase. Some important characteristics of detrital feldspars are summarized in Table 4.3.

Alkali Feldspars.   Alkali feldspars form a group of minerals in which the chemical composition can range through a complete solid-solution series from $K(AlSi_3O_8)$ through $(K,Na)(AlSi_3O_8)$ to $Na(AlSi_3O_8)$. Because potassium-rich feldspars are such common members of this group, geologists often refer to all of these feldspars as potassium feldspars, or K-feldspars. A more accurate name is potassium-sodium feldspars. Orthoclase, sanidine, and microcline all have the chemical formula $KAlSi_3O_8$ but may contain various amounts of sodium. Anorthoclase is a sodium-rich alkali feldspar that can have the end-member formula $NaAlSi_3O_8$; however, it may contain considerable potassium and thus grade to microcline or one of the other potassium feldspars. Orthoclase and microcline appear to be by far the most abundant potassium feldspars in sandstones.

TABLE 4.3
Characteristics of detrital feldspars

ALKALI (POTASSIUM-SODIUM) FELDSPARS
Complete solid-solution series from $K(AlSi_3O_8)$ (orthoclase, sanidine, microcline) to $Na(AlSi_3O_8)$ (anorthoclase); all biaxial $(-)$ with low negative relief
Orthoclase—a common detrital feldspar characterized by birefringence lower than quartz and 2V ranging from $\sim 40°$ to $75°$; extinction angle $\sim 13° - 21°$; twinned (Carlsbad twins most common) or untwinned; may appear cloudy owing to alteration products; distinguished from quartz most easily by staining
Sanidine—a high-temperature feldspar, with similar appearance to orthoclase, derived mainly from volcanic rocks; distinguished from orthoclase by smaller 2V ($0° - 47°$); extinction angle $\sim 5° - 21°$; commonly has fewer alteration products than orthoclase
Microcline—a common detrital feldspar distinguished particularly by distinctive cross-hatch twinning with twin lamellae approximately at right angles; 2V commonly larger than $65°$ (range $50° - 85°$); extinction angle $\sim 5° - 15°$; may be cloudy owing to the presence of alteration products
Anorthoclase—Comparatively rare in sandstones; extinction angle $\sim 5° - 20°$; distinguished from microcline by finer-scale cross-hatch twinning and smaller 2V ($\sim 43° - 54°$)
Perthite—alkali feldspars characterized by platy intergrowths of albite

PLAGIOCLASE FELDSPARS
Complete solid-solution series ranging from $NaAlSi_3O_8$ (albite) to $CaAl_2Si_2O_8$ (anorthite); composition commonly expressed as percent anorthite (An) molecule. Large 2V, which ranges with composition from about $50°$ to $105°$; biaxial $(+)$ or $(-)$; extinction angles vary as a function of composition and temperature of crystallization and may range from $\sim 3°$ to $60°$; indices of refraction also vary with composition. Twinned or untwinned; if albite twinning present, easily distinguished from alkali feldspars. If untwinned, distinguished by large 2V, extinction angles, or by staining. Plagioclase from igneous rocks may display compositional zoning. Very common in sandstones derived from volcanic and metamorphic rocks but may also be derived from plutonic igneous rocks. [Check published references such as Nesse (1986) for details of extinction angles, 2Vs, and indices of refraction.] Members of the series are
   Albite ($An_0 - An_{10}$)
   Oligoclase ($An_{10} - An_{30}$)
   Andesine ($An_{30} - An_{50}$)
   Labradorite ($An_{50} - An_{70}$)
   Bytownite ($An_{70} - An_{90}$)
   Anorthite ($An_{90} - An_{100}$)

Note: Data on extinction angles and 2Vs mainly from Nesse (1986).

Orthoclase has lower birefringence than quartz but can appear very similar in thin section. Thus, it can be confused with quartz and misidentified unless its optic sign is checked (biaxial negative with a large 2V), cleavage faces are apparent, or the orthoclase is clouded with alteration products, particularly if the alteration products are arranged in faint lines parallel to crystal directions. The most reliable technique for distinguishing between orthoclase and quartz is to stain the orthoclase (Houghton, 1980). Determining the chemical composition by use of the electron microprobe is likewise a reliable (but expensive) way of identifying orthoclase. **Sanidine** is a high-temperature feldspar, which occurs in lavas and some intrusive rocks, that closely resembles orthoclase. Orthoclase may be distinguished from sanidine by careful optical examination (orthoclase has a higher 2V, may have a slightly higher extinction angle, and may contain microperthitic textures).

Microcline is generally easy to recognize in thin section owing to the common presence of its distinctive "grid" twinning, with the two sets of twin lamellae approximately at right angles to each other (Fig. 4.5). The lamellae are commonly tapered. Some microcline is untwinned (Nesse, 1986, p. 271). Untwinned microcline may be misidentified as orthoclase or quartz. **Anorthoclase** can have grid twinning resembling that of microcline; however, the twinning is on a much finer scale and anorthoclase has a smaller 2V and extinction angle. **Perthite** is microcline or orthoclase characterized by patchy intergrowths of albite (described below) in the form of small strings, lamellae, blebs, films, or irregular veinlets (Fig. 4.6). It forms by exsolution of sodium-rich albite from the potassium feldspar. The thickness of the perthitic lamellae may have significance in provenance determinations (Chapter 8).

The alkali feldspars are derived particularly from alkali and acid igneous rocks and are especially abundant in syenites, granites, granodiorites, and their volcanic equivalents. They occur also in pegmatites and acid and intermediate-composition metamorphic rocks such as gneisses.

FIGURE 4.5
Microcline grain (center of photograph) with well-developed grid twinning. Fountain Formation (Pennsylvanian), Colorado. Crossed nicols. Scale bar = 0.5 mm.

FIGURE 4.6
Large K-feldspar grain with
perthitic lamellae. Unknown
formation. Crossed nicols. Scale
bar = 0.5 mm.

Plagioclase Feldspars.   The plagioclase feldspars constitute a solid-solution series rang-ing in composition from $NaAlSi_3O_8$ (albite) through $CaAl2Si_2O_8$ (anorthite). The general formula for the series is $(Na,Ca)Al(Al,Si)Si_2O_8$. As mentioned, sodic plagioclases appear to be more abundant in sandstones than calcic plagioclases. Many plagioclase grains are characterized by distinctive albite twinning, with twin lamellae that are straight and parallel (Fig. 4.7). When such twinning is present, plagioclase is easily distinguished from other feldspars and quartz. Unfortunately for the purpose of identification, much plagioclase is untwinned, especially plagioclase from metamorphic source rocks. Some plagioclase displays either oscillatory or progressive zoning (Chapter 9). Untwinned and unzoned plagioclase can be identified by staining (Houghton, 1980) or by electron mi-croprobe analysis. In the past, many sedimentary petrographers have not bothered to identify the individual plagioclase feldspars within the series. The growing usefulness of feldspars in provenance determination has, however, made identification of the individual plagioclases important. Determination of the identity of loose grains of plagioclase can be done by refractive index methods. In thin section, the approximate composition of pla-gioclase can be determined on the basis of extinction angles by the so-called Michel-Lévy method. See, for example, Deer et al. (1966, p. 330–334) or Jones and Bloss (1980, p. 17-3). Very accurate determination of the composition of plagioclase feldspars is now done routinely with the electron microprobe. The principal source for detrital plagioclase is probably basic and intermediate lavas, where it occurs as phenocrysts. It may also be derived from basic intrusive rocks and is a very common constituent of many metamor-phic rocks, where its composition is related to the metamorphic grade of the host rock.

Coarse Micas

The principal coarse micas that occur as detrital grains in sandstones are **muscovite** and **biotite**. These minerals are complex hydrous potassium, aluminum sheet silicates, with biotite, containing in addition iron and magnesium. **Chlorites,** which occur as detrital grains in some cases, are hydrous magnesium, iron, aluminum sheet silicates that resem-

FIGURE 4.7
Large, twinned plagioclase crystal from Miocene deep-sea sandstone, ODP Leg 127, Site 796, Japan Sea, 326 m below seafloor. Crossed nicols. Scale bar = 0.1 mm.

ble the micas in many respects. The micas are distinguished from other minerals by their platy or flaky habit. Also, their layered appearance in sections cut normal to the basal plane somewhat resembles the grain in "birds-eye" maple (Fig. 4.8). Muscovite is colorless in thin section. Biotite is normally yellow, brown, or green, but may be leached pale yellow or almost colorless. Muscovite has very weak or no pleochroism; however, flakes of biotite cut normal to the basal plane commonly are strongly pleochroic and have parallel extinction. Muscovite and biotite have moderately high birefringence in second-order reds, blues, and greens. In addition to its dark color and pleochroism, biotite is distinguished from muscovite by its lower 2V (0–25 vs. 30–47 degrees). Chlorites are generally green and pleochroic. They are commonly distinguished by anomalous interference colors, particularly in blues, but also in gray-blue, russet brown, or khaki yellow.

FIGURE 4.8
Large mica (biotite) grain cut normal to basal section. Belt (Precambrian) facies sandstone, northern Rocky Mountains. Ordinary light. Scale bar = 0.1 mm.

Micas and chlorite are derived primarily from metamorphic rocks, but biotite occurs also in basic intrusive and volcanic rocks and in granites. Muscovite occurs also in granites and pegmatites. Coarse micas rarely form more than about 2 percent of the framework grains of sandstones and commonly much less. Muscovite is chemically more stable than biotite and is much more abundant in sandstones on the average than biotite (Folk, 1974, p. 87, says about four times more abundant). Coarse detrital chlorite is less abundant than biotite, possibly as a result of the greater tendency of chlorite to mechanically degrade to finer-size grains. Detrital micas are rarely rounded, and they are commonly deposited with their flattened dimension parallel to bedding. Owing to their sheet-like shape and consequent low settling velocity, they tend to be hydraulic equivalents of finer grains and thus are commonly deposited in very fine sands and silts rather than in coarser sands (Doyle et al., 1983).

Clay Minerals

Clay minerals are common in sandstones as matrix constituents. They occur also within argillaceous rock fragments. Because of their fine grain size, clay minerals cannot easily be identified under the petrographic microscope. Accurate identification requires X-ray diffraction methods or use of the scanning electron microscope or electron microprobe. Therefore, clay minerals are rarely identified during routine petrographic study of sandstones. In most cases, they are simply lumped together with fine-size (<0.03 mm) quartz, feldspars, and micas as ''matrix.'' Furthermore, there is growing evidence that much of the matrix of sandstones may be authigenic, derived during diagenesis by chemical precipitation of clay minerals into pore space or alteration of framework grains to clays. The clay minerals are not discussed further at this point in the book, but are considered in detail in Chapter 7.

Heavy Minerals

Most sandstones contain small quantities of sand-size accessory minerals. Most of these minerals have specific gravities exceeding 2.85 and are thus called heavy minerals. Because of their generally low abundance in sandstones and siliciclastic sediments, heavy minerals are usually concentrated for study by disaggregating the sandstones and then separating the heavy minerals from the light-mineral fraction by using heavy liquids (Carver, 1971b; Lewis, 1984, p. 145–150; Lindholm, 1987, p. 214–217). After washing and drying, a grain mount of the heavy minerals is prepared for microscopic study (Swift, 1971). The heavy grains are sprinkled on a warmed glass slide covered with a thin layer of resin or epoxy and then covered with a cover slide. Identification of heavy minerals in grain mounts can be challenging for beginning petrographers, and considerable experience is generally required before the grains can be identified rapidly and accurately. A standard reference work for identification of detrital heavy minerals is Milner (1962). See also Tickell (1965).

Heavy minerals are commonly divided into two groups on the basis of optical properties: opaque and nonopaque (Table 4.4). **Opaque heavy minerals** include magnetite, ilmenite, hematite and limonite, pyrite, and leucoxene. Opaque minerals are difficult to identify with an ordinary petrographic microscope, and few petrographers attempt to do so. The **nonopaque heavy minerals** encompass a very large group of more than 100 minerals, of which olivine, clinopyroxenes, orthopyroxenes, amphiboles, garnet, epidote, clinozoisite, zoisite, kyanite, sillimanite, andalusite, staurolite, apatite, monazite, rutile, sphene (titanite), tourmaline, and zircon are particularly common.

TABLE 4.4

Detrital heavy minerals in sandstones

NONOPAQUE HEAVY MINERALS	
Ultrastable	Rutile, tourmaline, zircon, anatase (uncommon)
Stable	Apatite, garnet (iron-poor), staurolite, monazite, biotite
Moderately stable	Epidote, kyanite, garnet (iron-rich), sillimanite, sphene, zoisite
Unstable	Hornblende, actinolite, augite, diopside, hypersthene, andalusite
Very unstable	Olivine
Relative stability not well established	Ankerite, barite, brookite, cassiterite, chloritoid, chrondrite, clinozoisite, corundum, chromite, dumortierite, fluorite, glaucophane, lawsonite, magnesite, monazite, phlogopite, pitotite, pleonaste, pumpellyite, siderite, spinel, spodumene, topaz, vesuvianite, wolframite, xenotime, zoisite (many of the minerals in this group are uncommon as detrital grains in sandstones)
OPAQUE HEAVY MINERALS	
Stable to moderately stable	Magnetite, ilmenite, hematite, limonite, pyrite, leucoxene

Note: Relative stabilities of the nonopaque minerals, excluding those for which relative stability is not well established (after Pettijohn et al., 1987, p. 262). Stability refers to chemical stability.

We frequently see the statement in the literature that the heavy-mineral content of average sandstone is about 1 percent or less. This figure is difficult to verify because the authors of many papers on heavy minerals either do not determine the total heavy-mineral content of the sediments, or they do not report this information. Table 4.5 shows heavy-mineral abundances in selected examples of unconsolidated sands and a few ancient sandstones. Most of these modern/Holocene sands clearly contain more than 1 percent heavy minerals, some considerably more. Heavy minerals can become concentrated under certain conditions, such as on some beaches, resulting in localized placer deposits that may be composed almost entirely of heavy minerals. The lithic sandstones and some of the feldspathic sandstones shown in Table 4.5 also contain more than 1 percent heavy minerals. On the other hand, the quartzose sandstones and some feldspathic sandstones contain less than 1 percent, about the amount suggested for typical sandstones. These data are too few to allow definite conclusions, but they suggest that the heavy-mineral content of modern, unconsolidated sands can vary widely and may be several times greater than that of equivalent ancient sandstones. The data suggest also that lithic sandstones are enriched in heavy minerals compared to feldspathic and quartzose sandstones.

Many factors account for the variation in heavy-mineral abundance in sandstones, including the abundance of heavy minerals in the source rocks, the intensity of chemical weathering processes at the source, the nature and intensity of transport processes, and conditions of the depositional and diagenetic environments. Although the abundance of heavy minerals in source rocks is a primary factor controlling abundance in sediments, the differential stability of heavy minerals to chemical weathering and diagenetic dissolution and their durability to mechanical abrasion may be even more important controlling factors (Chapter 9). Less-stable heavy minerals can be selectively destroyed during weath-

TABLE 4.5
Relative abundance of heavy minerals in selected sands and sandstones (sample localities in the United States unless otherwise noted)

Type of sediment	Range (%)	Average (%)	No. of samples	Reference
MARINE SHELF SANDS				
Taiwan shelf	0.1–20	2.7	77	Boggs (1975)
SE U.S. shelf	?	0.5	78	Pilkey (1963)
FLUVIAL SANDS				
Oregon	2–5	3.0	75	Boggs (1969a)
Midcontinent	0–12	2.8	14	Hunter (1967)
World's major rivers	to 20	?	36	Potter (1978)
BEACH SANDS				
Georgia	0.2–14	2.0	36	Hails and Hoyt (1972)
Oregon (pocket beaches)	0.1–69	8.0	104	Rottman (1970)
Oregon (placers)	1–100	35.0	32	Peterson et al. (1985)
DUNE SANDS				
Beach dunes (Georgia)	0.1–18	4.1	18	Hails and Hoyt (1972)
Fluvial dunes (Georgia)	0.8–34	9.0	9	Hails and Hoyt (1972)
Inland dunes (Midcont., USA)	0–9	2.4	10	Hunter (1967)
FELDSPATHIC SANDSTONES				
Colorado (Penn)	0–3	0.4	68	Boggs (1966)
Colorado (Permo-Penn)	?	0.8	80	Bartleson (1972)
Auvergne, France (Oligocene, Permian)	?	~2.7*	?	Huckenholz (1963)
LITHIC SANDSTONES				
Oregon (Jur, Cret)	0–6	2.3	204	Lent (1969)
Oregon (Tertiary)	0.3–13	3.7	72	Kugler (1979)
QUARTZOSE SANDSTONES				
Midcontinent (Ord)	<0.1	<0.1	64	Thiel (1935)
Colorado (Dev)	0.1–0.9*	?	?	Campbell (1972)

Note: *May not be entirely heavy minerals

ering and transport or can be destroyed by intrastratal solution during burial and diagenesis (Pettijohn, 1941). Thus, older sedimentary rocks tend to contain fewer heavy minerals and a greater percentage of highly stable heavy minerals than do younger sedimentary rocks.

The most stable heavy minerals (**ultrastable minerals**) are zircon, tourmaline, and rutile. Zircon and tourmaline are particularly resistant to both chemical decomposition and mechanical abrasion and, like quartz, can survive multiple recycling. Thus, the presence of abundant rounded zircon and tourmaline in a sandstone that contains few if any other heavy minerals is suggestive of sediment recycling or of an episode of intensive chemical leaching or mechanical abrasion. Hubert (1962) proposed a **zircon-tourmaline-rutile (ZTR) index** as a measure of mineralogic maturity of heavy-mineral assemblages in sandstones. The ZTR index is the percentage of combined zircon, tourmaline, and rutile grains among the transparent, nonmicaceous, detrital heavy minerals. Hubert's data for selected formations suggest the ZTR index is highest for quartzose sandstones and may be least for feldspathic sandstones. Force (1980) points out that rutile grains are rare in most igneous rocks; therefore, use of rutile in the ZTR index is only partially valid owing to its restricted provenance. There is some disagreement about the relative stability of the remaining heavy minerals, although monazite, garnet, apatite, and possibly ilmenite and magnetite appear to have intermediate stability. Other heavy minerals such as hornblende, augite, and olivine have lower stabilities (see Pettijohn, 1975, Tables 13.5 and 13.7; Luepke, 1984).

Rock Fragments

Kinds of Rock Fragments.   Rock fragments are detrital particles made up of two or more mineral grains. Depending upon source rock composition, almost any kind of rock fragment can be present in a sandstone; however, clasts of fine-crystalline or fine-grained parent rocks are generally most abundant. The most common igneous rock fragments are volcanic rock fragments and glass. Metamorphic clasts include schist, phyllite, slate, and quartzite. Common sedimentary rock fragments are shale, fine sandstone and siltstone, and chert. Clasts of limestone and coarse plutonic and metamorphic rock are less common. Figure 4.9 shows examples of some common types of detrital rock fragments.

Clasts of chert (see Section 13.3, Chapter 13) and metaquartzite are rock fragments, but they are also regarded to be a type of polycrystalline quartz grain by many geologists. On the other hand, polycrystalline quartz grains that originate as single grains in plutonic igneous or metamorphic rocks, as discussed by Young (1976), are not actually rock fragments. Conceptually, and for the purpose of provenance interpretation, it is important to differentiate between these kinds of polycrystalline quartz grains and quartzose rock fragments. It may be very difficult in some cases, however, to tell the difference between polycrystalline quartz grains and fragments of metaquartzite or coarse-grained chert.

Identifying Rock Fragments.   Correct identification of individual kinds of rock fragments is important to sandstone classification and especially to provenance determination (Chapter 8). Identification of some types of rock fragments can be extremely difficult, particularly if the fragments become physically deformed or chemically altered during burial diagenesis. Dickinson (1970) recommends texture as the principal operational criterion for recognition because of the difficulty of identifying the very small individual minerals that make up most rock fragments. Composition of minerals within the frag-

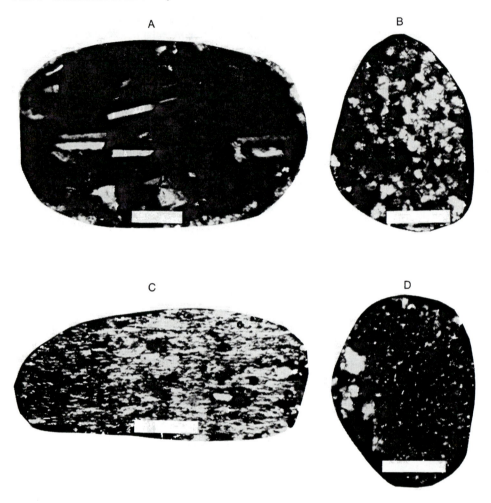

FIGURE 4.9
Examples of several kinds of common detrital rock fragments. A. Volcanic clast with plagioclase phenocrysts, Miocene deep-sea sandstone, ODP Leg 127, Site 796, Japan Sea, 387 m below seafloor. B. Sandstone clast, Taiwan shelf sand. C. Metamorphic clast, Taiwan shelf sand. D. Chert clast, Taiwan shelf sand. Crossed nicols. Scale bar in A = 0.1 mm; scale bars in B, C, and D = 0.5 mm.

ments is a supplementary criterion. Table 4.6 lists the principal types of rock fragments that occur in sandstones and some of the specific criteria that can be used in their identification.

Perhaps the most common problem encountered in rock fragment identification in sandstones is distinguishing between stable chert or fine metaquartzite fragments and some less-stable rock fragments. Gradations exist between these stable grain types and other rock fragments such as felsic volcanic fragments and micaceous phyllites (Dickinson, 1970). Distinguishing between felsic volcanic rock fragments, particularly silicified fragments, and chert is a particularly persistent and difficult problem. Dickinson (1970) suggests the following criteria, in order of increasing difficulty to apply, that can help

TABLE 4.6

Principal kinds of rock fragments in sandstones (identification based on textures, fabrics, and mineralogy of fragments)

IGNEOUS ROCK FRAGMENTS—Igneous textures

VOLCANIC ROCK FRAGMENTS—aphanitic and porphyritic textures

FELSIC GRAINS—characterized by an anhedral, microcrystalline mosaic, either granular or seriate (commonly porphyritic; grain size varies gradually or in a continuous series); composed mainly of quartz and feldspar; derived principally from silicic volcanic rocks, either lavas or tuffs

MICROLITIC GRAINS—contain subhedral to euhedral feldspar crystals in felted or pilotaxitic (ground-mass of holocrystalline rock in which lath-shaped crystals, commonly plagioclase, are arranged in a glass-free mesostasis and generally interwoven in an irregular fashion), trachytic (feldspar crystals in groundmass have a subparallel arrangement corresponding to flow lines), or hyalopilitic (needlelike crystals of the groundmass are set in a glassy mesostasis) patterns of microlites (microscopic crystals); derived from lavas of intermediate composition

LATHWORK GRAINS—characterized by plagioclase laths forming intergranular and insertal (triangular patches of interstitial glass between feldspar laths) textures; derived mainly from basaltic lavas

VITRIC TO VITROPHYRIC GRAINS AND GLASS SHARDS—composed of glass or altered glass; alteration products may be phyllosilicates, zeolites, feldspars, silica minerals, or combinations of these in microcrystalline aggregates

HYPABYSSAL AND PLUTONIC IGNEOUS ROCK FRAGMENTS—hypabyssal types are fine crystalline, subhedral, commonly low in quartz and rich in feldspars; plutonic fragments are fine- to coarse-crystalline, anhedral granular: commonly composed of quartz and feldspar; derived mainly from acid igneous rocks; not common in sandstones owing to coarse crystal size

SEDIMENTARY ROCK FRAGMENTS—fragmental or microgranular textures

EPICLASTIC SANDSTONE-SILTSTONE GRAINS—fragmental textures composed dominantly of silt- to fine-sand size quartz and feldspar; may display cement or interstitial clay (matrix)

VOLCANICLASTIC GRAINS—fragmental textures; zoned plagioclase, embayed quartz, glass shards; may be difficult to differentiate from some epiclastic sandstone grains or some volcanic rock fragments

SHALE CLASTS—fragmental textures; dominantly clay- and silt-size particles

CHERT—mainly microgranular texture; composed entirely of quartz, chalcedony, opal; may be confused with silicified volcanic fragments

CARBONATE GRAINS—microgranular texture; composed of calcite or dolomite; may contain microfossils or fossil fragments

METAMORPHIC ROCK FRAGMENTS—foliated or nonfoliated fabrics

GRAINS WITH TECTONITE FABRIC—grains showing schistose, semischistose, or slaty fabric resulting from preferred orientation of recrystallized mineral grains

METASEDIMENTARY GRAINS—characterized by presence of quartz and mica; include schist, phyllite, slate; slate possibly distinguished from shale by mass-extinction effect

METAIGNEOUS GRAINS—mainly metavolcanic; contain abundant feldspars or mineral assemblages that include chlorite and amphiboles

GRAINS WITH NONFOLIATED FABRIC—mainly microgranular textures; may include hornfelsic grains (containing metamorphic minerals), metaquartzite clasts (composed mainly of quartz with strongly sutured contacts), and marble fragments (coarse-crystalline carbonate)

INDETERMINATE MICROGRANULAR GRAINS—very fine-grained fragments in which individual mineral grains are difficult to distinguish and textures are indeterminate; could be igneous, metamorphic, or sedimentary

INDETERMINATE MICROPHANERITE GRAINS—individual minerals large enough to identify, but identity of clasts (gneiss, granite, etc.) difficult to establish; not common in sandstones

---

Source: Based on Dickinson (1970).

distinguish felsic volcanic fragments from chert grains: (1) under crossed nicols, felsic fragments display subhedral to euhedral microphenocrysts of feldspar set in an anhedral matrix; (2) under crossed nicols, felsic grains may display plagioclase laths or K-feldspar blocks in an anhedral mosaic; (3) in plain light, faint curvilinear outlines of glass shards may be present; (4) in plain light, internal relief, within the grains, between quartz and feldspar—having differences in refringence in excess of the slight difference between the ordinary and extraordinary refractive indices of quartz—distinguishes felsic grains; and (5) the presence of feldspars is revealed by yellow or red stain when thin sections are stained.

Robert Oaks (personal communication, 1990) suggests that the presence of inherited black opaque minerals, as shown in plain light, can be used also to help identify altered felsic volcanic fragments. In contrast to these characteristics of felsic fragments, chert is composed mainly of quartz, although inclusions of precursor minerals such as carbonates, evaporites, or phosphates may be present. In plain light, chert grains may display ovoid or reticulate relics of radiolarians, diatoms, or spicules or even relict limestone textures (ooids, pellets, etc.). Under crossed nicols, networks of planar, crisscrossing veinlets may be visible.

Occurrence of Rock Fragments.   Rock fragments are common constituents of both modern sediments and ancient sandstones and conglomerates. The rock-fragment content of Holocene and Pleistocene sands ranges from a few percent to as much as 90 percent. For example, data compiled by Pettijohn et al. (1987, p. 45) for North American river sands show a range in rock-fragment content from 9 to 69 percent, and the average content in different rivers ranges from 11 to 57 percent. Franzinelli and Potter (1983) determined that the average rock-fragment content of sands from different parts of the Amazon River system ranges from 3 to 36 percent. Hunter (1967) found that the rock-fragment content of midcontinent eolian dune sands ranges from less than 5 percent to more than 20 percent. Potter (1986) reports the average rock-fragment abundance in South American beach sands from different areas to range between 5 and 60 percent. Boggs et al. (1974) determined that the rock-fragment content of Taiwan shelf sands ranges from 11 to 85 percent, with average abundances in the northern, western, southern, and eastern shelves being 28, 45, 65, and 53, respectively. Packer and Ingersoll (1986) report petrographic data for sands cored in Deep Sea Drilling Program (DSDP) holes in the Japan and Mariana forearc and backarc regions. Expressed as percentage of the QFL (quartz, feldspar, rock fragment) fraction of the sands, rock fragments range in abundance from 19 to 91 percent (average 64 percent) in the Japan forearc, 9 to 82 percent (average 47 percent) in the Japan backarc, and 59 to 99 percent (average 82 percent) in the Mariana forearc and backarc. In a similar study of sands from DSDP sites in the North Pacific and the Bering Sea, Gergen and Ingersoll (1986) report 0–95 percent rock fragments (average 40 percent) in the QFL fraction.

Rock fragments probably make up about 10–20 percent of the average sandstone. The range in rock-fragment abundances in ancient sandstones is very broad, however, and extends from less than 1 percent in some quartzose sandstones to well above 50 percent in some lithic sandstones. The abundance of rock fragments in sands and sandstones is a function of several factors. Other factors being equal, coarser sediments tend to contain more rock fragments than finer sediments. Furthermore, coarse-crystalline source rocks such as granites tend to yield fewer sand-size rock fragments than do finer-crystalline source rocks. For example, Boggs (1968) demonstrated experimentally that, when crushed, coarse-crystalline (0.5–1.0 mm) rocks such as gabbro, granite, and diorite yield only about 25 percent rock fragments to the fine-sand-size crushed fraction. By contrast, fine-grained rocks such as basalt, slate, and phyllite yield 90–100 percent rock fragments to this size fraction. The percentage of coarser-grained rock fragments is considerably higher in the coarser fractions of these experimental sands and rises to more than 90 percent in very coarse sands.

Thus, the lithology of the source area exerts control on the abundance of sand-size rock fragments released into the transport mill. The tectonic setting, which ultimately controls source-rock lithology, has an influence also on rock-fragment abundance. Dickinson and Suczek (1979) and Dickinson (1985) suggest in their provenance diagrams that rock fragments are more abundant in sediments derived from magmatic arcs than in sediments derived from recycled orogenic or continental-block provenances. Valloni and Maynard (1981) carry through this same general idea of tectonic influence on petrology. They report that the average rock-fragment content of deep-sea sands, expressed as percentage of the QFL fraction, is 12 percent in basins of trailing-edge continental margins, 31 percent in basins of leading-edge margins, 27 percent in strike-slip basins, 51 percent in backarc basins, and 75 percent in forearc basins (see also Packer and Ingersoll, 1986, and Gergen and Ingersoll, 1986).

The chemical stability, mechanical durability, and the transport and diagenetic history of sediments also influence rock-fragment abundance. Grantham and Velbel (1988) have shown, for example, that the abundance of rock fragments in sand-size sediment derived from source areas is a function of duration and intensity of chemical weathering in the source area, which, in turn, is related to topographic relief and climate. Intensive transport and reworking, or recycling, of sediment through several cycles (Chapter 1) can destroy low-durability clasts by breaking them down into their constituent mineral grains. It is often asserted, for example, that shale clasts cannot survive intensive reworking, such as reworking on beaches (e.g., McBride and Picard, 1987). Cameron and Blatt (1971) suggest that schist fragments are likewise mechanically weak and may be destroyed by as little as 15 miles of stream transport. On the other hand, Mack (1981) has shown that some shale clasts, perhaps well lithified or cemented clasts, may survive 15–25 km of stream transport with little destruction. Intrastratal solution during burial and diagenesis may remove chemically unstable fragments such as limestone clasts.

Thus, on the whole, we expect to find fewer rock fragments in compositionally mature sandstones than in immature sandstones. To my knowledge, relatively little recent work has been done to develop a stability series for sand-size rock fragments such as that developed for minerals. Chert and quartzite clasts are surely the most chemically stable and mechanically durable clasts. Shale and schist fragments may be the least durable mechanically, and limestone fragments are probably the least stable chemically. The mechanical durability and chemical stability of the remaining types of rock fragments are not well demonstrated but appear to depend upon the chemical and mineral composition of the fragments and their degree of cementation and lithification. Thus, fragments of silicic or alkaline igneous rocks are likely to survive better than those of basic rocks, and sedimentary or metamorphic rocks cemented with silica cements will survive better than those cemented with calcite or clay minerals. The relative durability of some gravel-size clasts subjected to abrasion in tumbling mills is reported by Abbott and Peterson (1978).

Skeletal Remains

Skeletal particles, generally in low abundances, are common grains in many sandstones. These skeletal remains may be broken fragments or unbroken remains of larger fossils or may be microfossils. Because of the usefulness of fossils in paleoenvironmental and paleoecological studies, it is often desirable to identify the types of fossils present. Identifying fossils and fossil fragments in thin sections can be an exercise in futility for the uninitiated. Fortunately, several excellent picture-atlas volumes are now available that greatly aid in identification (Johnson, 1971; Majewske, 1971; Horowitz and Potter, 1971; Scholle, 1978; Flügel, 1982). Identification of fossils is particularly important in carbonate rocks (Chapter 10), in which fossils and fossil fragments are commonly abundant.

Organic Matter

Organic matter can occur in sandstones as clay- to silt-size particles or as fragments ranging to several centimeters in size. These organic substances are derived through the breakdown of plant and animal tissue; thus, they can be considered a kind of detrital particle. Some organic material, particularly woody and leafy material and mineral charcoal, originates on land and is subsequently transported into depositional basins. Other particulate organic material originates within depositional basins by disintegration of marine or freshwater organisms.

In plain light in thin sections, organic material appears as black (opaque), brownish translucent, or brownish transparent particles. Darkness of the color is related to increasing organic carbon content (Pettijohn et al., 1987, p. 52). Larger, elongated carbonaceous or woody fragments are commonly oriented parallel or subparallel to bedding, and the fragments may show evidence of compaction and flattening owing to diagenesis. In sandstones that contain abundant silt- to sand-size carbonaceous detritus, the carbonaceous material may be concentrated in layers. These organic-rich layers show up as thin, dark laminae that superficially resemble heavy-mineral laminae.

Sandstones commonly contain less than about 0.1 percent organic matter (Chapter 14). Sandstones deposited in oxygenated environments are particularly depleted in organic matter, which is destroyed by oxidation. Rare sandstones, such as some graywackes, that were deposited under more strongly reducing conditions may contain a few percent organic matter. Mudrocks (shales) commonly contain much higher amounts of organic matter than do sandstones. These higher concentrations probably result both because fine organic matter tends to be deposited preferentially with clay- and silt-size sediment and because the lower permeability of muds, compared to sands, inhibits entry of oxygen-bearing fluids into these sediments. Organic-rich sedimentary rocks that contain extremely high concentrations (>10–20 percent) of organic matter are discussed in detail in Chapter 14.

4.2.2   *Authigenic Minerals*

General Statement

Authigenic minerals are minerals that form in place within sediments either shortly after deposition, while sediment is still in an unconsolidated state, or during burial and diagenesis. They can occur as cements, or crystallize in pore space as new minerals that do not act as cements, or form by replacement of original detrital minerals or rock fragments (see Chapter 9). Authigenic minerals in sandstones can be distinguished from detrital minerals by several criteria: (1) they tend to be smaller than the associated framework grains, although some may grow to sand size, (2) they may display either well-developed euhedral crystal faces or, alternatively, highly irregular, intricate grain outlines that could not have survived transport, (3) they may display definitive replacement textures such as transected grain boundaries or caries texture (bitelike embayments), or (4) they may exhibit cementation textures such as syntaxial rims or drusy (void-filling) texture. Electron-microprobe analysis, to determine chemical composition, and other instrumental techniques such as cathodoluminescence can be used also to identify some authigenic minerals.

Authigenic Silica, Feldspars, Micas, and Clay Minerals

Quartz and Other Silica Minerals.   Authigenic silica in small amounts is common in sandstones. It occurs in several forms, including very small euhedral quartz crystals, rims or syntaxial overgrowths on preexisting quartz grains, and chert (chalcedony) or opal. Larger euhedral quartz crystals can occur as fillings in fractures or cavities, or they can form by replacement of carbonate or evaporite minerals. Quartz overgrowths, which are in optical continuity with the detrital quartz nucleus, are a common and obvious form of authigenic quartz. They are particularly common in quartz-rich Paleozoic sandstones, where they form interlocking frameworks of crystals that firmly cement the sandstones. In many cases, authigenic quartz overgrowths are easily distinguished because the outlines of original detrital quartz grains are picked out by a line of very tiny iron oxide particles, clay particles, or bubbles (Fig. 4.1). Unfortunately, not all overgrowths are marked by such inclusions, making identification of much authigenic quartz with a standard petrographic microscope extremely difficult. Under such circumstances, one can look for euhedral crystal faces in partially filled cavities or irregular overgrowth boundaries in filled cavities. Irregular boundaries can develop also through compaction and pressure solution; therefore, a petrographic microscope equipped for cathodoluminescence, or an electron microprobe, must be used to make unambiguous distinction of authigenic and detrital quartz (Sippel, 1968; Sibley and Blatt, 1976). Detrital quartz luminesces in the visible range (blue and red or orange), whereas authigenic quartz does not.

Austin (1974) reported multiple authigenic overgrowths (up to 9!) on scattered detrital quartz grains in a dolomite formation. Some of these overgrowths are optically continuous syntaxial rims; others are radiating fibrous or microcrystalline forms. Inherited quartz overgrowths are common in many sandstones (e.g., Sanderson, 1984). An inherited origin is suggested by such features as broken or rounded overgrowths. Sanderson suggests that owing to spatial constraints only about half of the quartz grains with inherited overgrowths may actually be observed in normal point counts of thin sections. It is important that these second-cycle overgrowths be recognized and distinguished from authigenic overgrowths because they provide direct evidence of derivation from sedimentary source rocks. Walton (1986) points out that some first-cycle volcanic quartz may have overgrowths that could become rounded during transport, complicating the problem of identifying inherited overgrowths. Volcanic overgrowths can be distinguished by cathode luminescence because they fluoresce the same as the detrital cores.

Some authigenic quartz can apparently grow in sandstones without forming around a detrital core. Authigenic quartz crystals of this type that grow as single crystals within pore spaces are commonly only a few tens of micrometers in size. A scanning electron microscope is required for viewing quartz in this small-size range. Larger euhedral crystals of authigenic quartz, greater than about 20 μm or so, occur in veins and cavities.

Larger crystals form also in carbonate rocks and evaporites by replacement. This type of replacement quartz commonly forms nodular masses of various sizes that may be composed of so-called **flamboyant,** crudely radiating quartz. Friedman and Shukla (1980) report optically length-slow, euhedral quartz that occurs as both isolated grains and fan-shaped clusters in Silurian dolomites of New York. They suggest that this quartz originated by filling cavities created by dissolution of original sulfate crystals in the dolomite.

Some authigenic silica forms in sandstones as a cryptocrystalline mosaic of equidimensional or fibrous grains (chert or chalcedony) that act as a cement for larger framework detrital grains. Chalcedony is a common replacement mineral also in carbonates and evaporites. Most fibrous chalcedony is optically length-fast; that is, the fiber axes are oriented normal to the c-axis. Some fibrous varieties of chalcedony that occur as a replacement of evaporites, or by authigenic precipitation in sulfate-rich, alkaline environments, are unusual in that they are length-slow; that is, they have their c-axis parallel or nearly parallel to the long dimension of the fibers. Two length-slow varieties of chalcedony—**quartzine** (fiber axes parallel to c-axis) and **lutecite** (fiber axis about 30 degrees to c-axis)—have been identified (see Folk and Pittman, 1971). Some workers have used quartzine as an indicator of shallow-water environments where evaporite deposits are forming; however, quartzine has been reported also in deep-sea cherts, where it forms authigenically in sulfate- and magnesium-rich pore waters (Keene, 1983). Another interesting variety of chalcedony is zebraic chalcedony, with fibers helically twisted around the axis of elongation (McBride and Folk, 1977).

Feldspars.   Authigenic feldspars are quantitatively insignificant in most sandstones. They range only from trace amounts to a few percent. Most authigenic feldspars tend to be small (silt size or smaller), with well-developed crystal faces. Many are untwinned. Authigenic feldspars can occur either as overgrowths around detrital feldspar cores or as newly formed crystals without a preexisting core, although the former is apparently most common (Kastner and Siever, 1979). Both authigenic K-feldspars and plagioclase feldspars (mainly albite) occur in sandstones. They are most abundant in feldspathic sandstones and volcaniclastic sandstones, but occur also in some quartz arenites and lithic arenites. Authigenic feldspars can be difficult to identify. Table 4.7 lists analytical methods and criteria for differentiating authigenic feldspars from detrital feldspars. Many authigenic feldspars are albites, and the process of albitization of detrital plagioclase is one of the most important diagenetic changes that occur in feldspathic sandstones and lithic arenites (Boles, 1982; Walker, 1984). Other pertinent references to authigenic feldspars in sandstones include Baskin (1956), Buyce and Friedman (1975), Stablein and Dapples (1977), Waugh (1978), and Ali and Turner (1982).

Micas and Clay Minerals.   Authigenic micas and clay minerals are extremely common products of diagenesis. Detrital clay minerals are altered under the higher temperatures and pressures of burial diagenesis (Chapter 9) to form muscovite, biotite, and chlorite. These authigenic micas are generated principally in clay-rich sandstones and mudstones. In addition to alteration of clay minerals to micas, one type of clay mineral may alter to another, e.g., the conversion of smectite to illite. Clay minerals and fine-size muscovite (sericite) may form also by alteration of feldspars and volcanic rock fragments. Potassium feldspars are commonly altered to sericite and kaolinite, and plagioclase feldspars may be altered to smectite. The formation of authigenic micas and clay minerals is favored at temperatures above those of the depositional environment; however, kaolinite is known to

TABLE 4.7

Methods for analyzing feldspars and criteria for authigenic origin of feldspars in sedimentary rocks, listed in order of decreasing reliability

Analytical method	Criteria for authigenic origin	Remarks
Electron microprobe	High purity of end-member composition	For any size >1 μm high-purity end members may occur in low-grade metamorphic rocks and some pegmatites
Optical petrography	Microscopic textures such as overgrowths and twinning (untwinned crystals or fourling twins) (Kastner, 1971) most common; crystals commonly translucent or transparent; composition as determined by refractive indices and optic axis angle	Good only for grains >50 μm
Cathodoluminescence	No luminescence	Good only for grains >50 μm (cathodoluminescence attachment to microscope; >1 μm (microprobe); nonluminescing grains occur also in low-grade metamorphic rocks
Isotopic analysis	High positive $d^{18}O$ values	Difficult to avoid contamination with detrital feldspars
X-ray diffraction	Purity of composition	Difficult to avoid contamination with detrital feldspars
Bulk chemical analysis	High purity of end-member composition	Difficult to avoid contamination with detrital feldspars and matrix inclusions
Stain tests	Differential staining of detrital and authigenic feldspar	Good only for grains >50 μm; stains insensitive to small differences between detrital cores and authigenic overgrowths

Source: Kastner and Siever (1979)

form also by alteration of feldspars under the low-temperature, low-pressure conditions of near-surface weathering. Authigenic micas and clay minerals form cements and matrix in the interstices of framework sand grains or occur as individual crystals or aggregates of crystals within altered feldspars and rock fragments. They are commonly too small to be identified effectively with a petrographic microscope and must be studied with the scanning electron microscope or electron microprobe or by X-ray diffraction techniques.

Glauconite and Chamosite.   The terms glauconitic and chamositic are applied to groups of chemically complex, green, iron-rich clay minerals containing more than about 15 percent total Fe_2O_3 (Van Houten and Purucker, 1984). Glauconite, or glauconitic mica, is one of the glauconitic minerals (Fig. 4.10). It is an iron- and potassium-rich mica (illite)-type clay mineral. The other glauconitic minerals are glauconitic smectite (McConchie and Lewis, 1980) and green smectite (Odom, 1976, p. 237). Chamosite, one of the two chamositic minerals, is an iron-rich chlorite. Berthierine is a serpentine mineral.

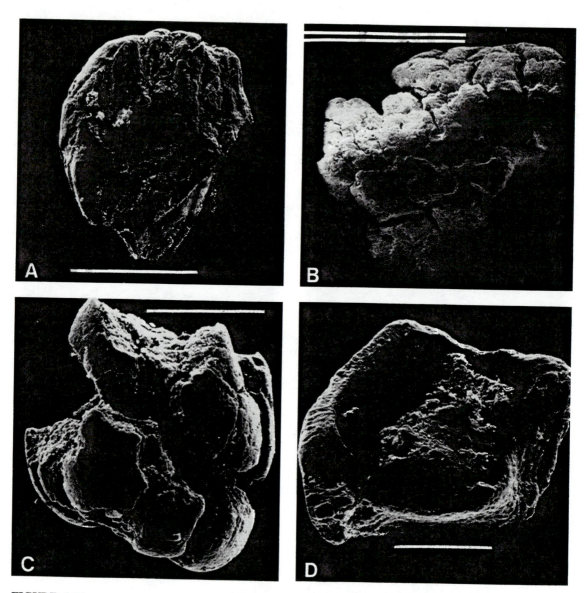

FIGURE 4.10
Modern glauconite grains: (A) glauconized mica grain, (B) pale- green, immature glauconitic grain, (C) very dark green-black, deeply fissured, mature glauconitic grain, (D) glauconitic infilling in a broken test of a benthic foraminifer. Single scale bar = 200 μm; double scale bar = 500 μm. *(From Bornhold, B. D., and P. Giresse, 1985, Glauconitic sediments on the continental shelf off Vancouver Island, British Columbia, Canada: Jour. Sed. Petrology, v. 55, Fig. 5, p. 656, reprinted by permission of Society of Economic Paleontologists and Mineralogists, Tulsa, Okla. Photographs courtesy B. D. Bornhold.)*

Glauconite and chamosite have been reported in rocks ranging in age from Early Proterozoic to Holocene. Chamositic particles are particularly common in Ordovician, Devonian, and Late Mesozoic rocks. Glauconitic grains are especially abundant in Middle Cambrian to Early Ordovician and Middle Cretaceous to Early Cenozoic rocks (Van Houten and Purucker, 1984, Fig. 4). Glauconite peloids are common also in many modern environments—for example, the continental shelf off Vancouver Island, British Columbia, Canada (Bornhold and Giresse, 1985); Queen Charlotte Sound, British Columbia (Murray and Mackintosh, 1968); Monterey Bay, California (Hein et al., 1974); the Atlantic coastal shelf off the United States (Milliman, 1972); and bottom sediments in various parts of the Atlantic, Pacific, and Indian oceans (Logvinenko, 1982). Chamosite pellets are less common in modern environments, but are reported in sediments of the Niger and Orinoco deltas and the shelf off Sarawak (Porrenga, 1965, 1967), the Mahakam Delta in Kalimantan (G. P. Allen et al., 1979), and Loch Etive, Scotland (Rohrlich et al., 1969).

Although glauconite and chamosite can form grain coatings and thin crusts (the nongranular materials in Table 4.8), they typically occur as sand-size particles. In modern environments, the grains are mainly structureless peloids, assumed by many workers to be fecal pellets. Most ancient chamosite grains, and some glauconite particles, have an ooid structure, presumably formed by accretion around some kind of conspicuous or inconspicuous nucleus.

Glauconite peloids can have a variety of shapes, including spheroidal, ovoid, botryoidal, tabular, and vermiform. They occur also as internal molds and casts of microfossils. Modern peloids display both smooth, unfissured grains and deeply fissured or cracked grains (Fig. 4.10). Viewed under a binocular microscope, colors of glauconite grains range from pale light green to dark green-black. In thin section, glauconite appears greenish gray, bluish green, greenish yellow, bright green, or dark green. The grains may be oxidized to a brownish yellow or reddish brown. Internally, glauconite grains typically display a microcrystalline texture, made up of very finely crystalline, overlapping, mi-

TABLE 4.8
Morphology and mineralogy of iron-rich clay minerals

Iron-rich clay minerals		Non-granular		Granular			
				Peloids		Ooids	
		M	A	M	A	M	A
Glauconitic minerals	Glauconite		X	?	XX		
	Glauconitic smectite	x	X	XX	X		
	Green smectite		x	X	X	x	x
Chamositic minerals	Berthierine		x	XX	X	x	XX
	Chamosite		x	x	X		XX

M = modern, A = ancient, XX = abundant, X = common, x = rare. Nongranular refers to minerals that occur as cements, thin crusts, and grain coatings.

Source: After Van Houten, F. B., and M. E. Purucker, 1984, Glauconitic peloids and chamositic ooids—Favorable factors, constraints, and problems: Earth Science Rev., v. 20. Fig. 1, p. 214, reprinted by permission.

caceous plates (Fig. 4.11). Several other kinds of small-scale structures are revealed by the scanning electron microscope (Van Houten and Purucker, 1984, p. 218). Owens and Sohl (1973) report that some glauconite peloids have an accordion-type texture that appears to result from stacked mica plates, and rare grains may display a concentrically layered (ooid) texture.

Chamosite peloids in modern sediments are dark green unless oxidized to yellowish brown goethite. They show no internal structures but may have a thin outer shell of goethite. Some pellets contain inclusions of other minerals (Porrenga, 1965). By contrast, most chamosite grains in ancient sandstones display a concentrically layered texture (Fig. 4.12). The individual layers may be composed entirely of chamosite or may be composed in part of goethite. Nuclei of the grains can include pellets of goethite, quartz grains, ferruginous mudstone pellets (fecal pellets?), and, rarely, pieces of shell or fish teeth (James and Van Houten, 1979). Glauconite and chamosite grains are difficult to identify unequivocally in thin section, and identification may require X-ray verification.

The exact processes by which glauconites and chamosites form are still speculative. Most geologists believe that these minerals are authigenic and that they form primarily under marine conditions, although they have been reported rarely in some saline lakes. They are most common on siliciclastic shelves at water depths to about 200 m, but have been found also on some carbonate banks and shelves. Some glauconites may have formed by alteration or transformation of mixed-layer clays by adsorption of potassium and iron. Most are believed to form by precipitation of dissolved material in the pores of a substrate that is progressively altered and replaced (Odin and Matter, 1981). Precipitation can take place also in the cavities of microfossils to produce internal molds. Chamosite is unstable in the presence of free oxygen and thus is thought to form under anoxic diagenetic conditions from a precursor that originated on an oxygenated seafloor. The literature on glauconites and chamosites is voluminous. Van Houten and Purucker (1984) give an excellent review of this subject, as well as an extensive bibliography of pertinent papers.

FIGURE 4.11
Rounded glauconite grains (G) and quartz. Deadwood Formation (Cambrian-Ordovician), South Dakota. Crossed nicols. Scale bar = 0.1 mm.

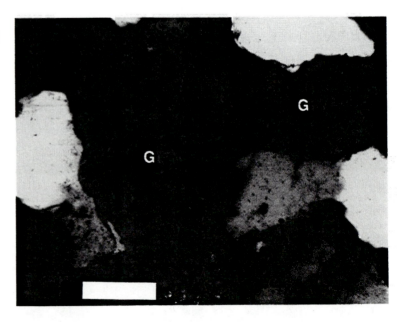

FIGURE 4.12
Concentrically layered ooids in an
ancient chamosite oolite with dark
goethitic cement. Scale bar at
lower left = 1 mm. *(Photograph
from James, H. E., and F. B. Van
Houten, 1979, Miocene goethitic
and chamositic oolites,
northeastern Columbia:
Sedimentology, v. 26, Fig. 3b, p.
130, reprinted by permission of
Elsevier Science Publishers,
Amsterdam.)*

Other Authigenic Minerals.   In addition to the authigenic minerals discussed above,
numerous other minerals occur in sandstones as either authigenic cements or replacement
minerals. These minerals include carbonates (aragonite, calcite, dolomite, and siderite),
iron oxides (limonite, goethite, and hematite), the sulfate minerals (gypsum, anhydrite,
and barite), sulfides (pyrite and marcasite), zeolite minerals (analcime, chabazite, clinoptilolite, heulandite, laumontite, phillipsite, and a few others), and phosphate minerals
(apatites). These minerals and the diagenetic conditions under which they originate are
discussed further in Chapter 9.

4.2.3   Framework Grains vs. Matrix and Cement

Preceding discussion in this chapter suggests that sandstones are composed principally of
three fundamental kinds of constituents: framework grains, matrix, and cement. In addition, replacement minerals and open pore space may be present. **Framework grains** are
coarse silt- and sand-size (0.03–2.0 mm) detrital particles that include quartz, feldspars,
coarse micas, and heavy minerals. **Matrix** is finer-grained material that fills interstitial
spaces among framework grains. The upper size limit of material in sandstones considered
to be matrix is arbitrary and debatable; however, a maximum size of 0.03 mm appears to
be favored by many workers (Dott, 1964). The most common matrix minerals in sandstones are fine silica minerals, feldspars, micas, clay minerals, and chlorite. Matrix may
make up trace amounts to a few tens of percent of the total rock volume. Siliciclastic rocks
that contain more matrix than framework grains are shales or mudrocks. **Cements** are
authigenic minerals that fill interstitial areas that were originally open pore spaces. Cement crystals may be any size up to or larger than the sizes of the individual pores they
fill. A single crystal of calcite, for example, can fill several adjacent pores. Cements

visible under a petrographic microscope rarely make up more than about 30 percent of the total volume of sandstones and commonly are much less abundant.

Earlier workers assumed that most matrix in sandstones was detrital and was sedimented along with framework grains at the time of deposition. That point of view has undergone considerable revision. It now appears that most transport and depositional processes separate clay-size grains from coarser detritus, so that most sand when initially deposited contains very little if any matrix. For example, Walker et al. (1978) examined recent fluvial sands at hundreds of locations in the southwestern United States and found these sands to be essentially matrix-free. Even turbidites, which were once believed to contain large amounts of detrital matrix owing to presumed settling together of fine and coarse particles after turbulence waned, may not contain substantial syndepositional detrital matrix. Current views suggest that much, if not most, matrix in sandstones originates (1) by postdepositional infiltration of clay into interstitial spaces, particularly in fluvial deposits (Walker et al., 1978), or (2) as an authigenic filling owing to diagenetic alteration of unstable rock fragments, feldspars, and ferromagnesian minerals.

With respect to infiltration, Matlack et al. (1989) conclude on the basis of experimental work that vadose infiltration of muddy water through sand is an effective mechanism for emplacing clay into sand. They suggest that this infiltration occurs most effectively in environments characterized by high suspended sediment concentration, fluctuating water levels, and minimum sediment reworking by waves or currents.

Alteration of framework grains during diagenesis may also produce significant amounts of authigenic clay matrix and cement. Alteration takes place mainly by dissolution and replacement of framework grains, and the alteration products are reconstituted as clay minerals, chlorite, micas, and fine quartz and feldspar (Cummins, 1962; Hawkins and Whetten, 1969; Whetten and Hawkins, 1970; Shannon, 1978; Walker et al., 1978; Morad, 1984). In particular, the matrix in so-called graywackes is now believed to be largely authigenic.

Recognition of authigenic clay cements and matrix may be extremely difficult by routine petrographic examination. Coarser-grained clays deposited in the form of radiating crystals that line interstitial voids (drusy texture) (Fig. 4.13) may be identifiable as cements under the petrographic microscope. On the other hand, much authigenic clay is so fine sized that it shows up during petrographic examination as an indistinguishable part of the interstitial matrix. At higher magnification under the scanning electron microscope, authigenic clays may be distinguishable from detrital clays. Walker et al. (1978) show, for example, that authigenic clays in fluvial sandstones have a characteristic boxwork texture (Fig. 4.14), whereas mechanically infiltered clays consist of clay platelets oriented parallel to grain surfaces (Fig. 4.15). Moraes and De Ros (1990) report significant amounts of mechanically infiltrated clays in Jurassic sandstones of northeastern Brazil. According to these authors, mechanically infiltrated clays can be identified in ancient rocks on the basis of several criteria: (1) ridges and bridges between grains, (2) geopetal fabric, (3) loose aggregates, (4) isopachous clay coatings (coatings that do not completely surround grains) or tangentially accreted lamellae, (5) massive aggregates, and (6) shrinkage patterns developed during diagenesis. The matrix of sandstones that have undergone extreme diagenesis or incipient metamorphism may not, however, preserve these distinctions between authigenic clays and infiltered clays. Shannon (1978) reports, for example, that some graywackes may display flattening of detrital framework quartz and feldspar grains and alignment of matrix micas in the plane of S_1 cleavages.

FIGURE 4.13

Thin-section photograph of coarse-clay mineral cement with drusy texture (arrows) filling pore space among framework grains. Cuchara Formation (Eocene), Colorado. Ordinary light. Scale bar = 50 μm. *(From Walker, T. R., et al., 1978, Diagenesis in first-cycle desert alluvium of Cenozoic age, southwestern United States and northwestern Mexico: Geol. Soc. America Bull., v. 89, Fig. 1E and 1F, p. 20. Photographs courtesy T. R. Walker.)*

Philosophically and pragmatically, it is difficult to distinguish between fine-grained silicic cements and authigenic matrix. Dickinson (1970) attacked this problem by grouping the interstitial constituents of sandstones into six categories:

1. **Cement**—easily recognizable authigenic calcite, chalcedony, zeolites, or other minerals not common in the framework
2. **Phyllosilicate cement**—clay-mineral or mica cements displaying growth patterns in open pores. Dickinson suggests five textural characteristics of clay cements that may allow their distinction from inhomogeneous matrix:
   a. clear transparency of clay cements, indicating absence of minute detritus or murky impurities
   b. dominance of a single mineral type
   c. radial arrangement of crystal platelets, projecting inward from surrounding framework grains (drusy texture)
   d. medial sutures within the interstitial clays, indicating lines of junction of radially arranged crystals
   e. concentric color zonation, indicating successive compositional changes in minerals grown in voids from interstitial fluids as coatings built on framework grains
3. **Protomatrix**—unrecrystallized detrital clay in weakly consolidated rocks
4. **Orthomatrix**—recrystallized detrital clay or protomatrix, recognized by relict detrital textures and inhomogeneity, in contrast to the homogeneity of clay cements

FIGURE 4.14
Electron micrograph showing typical boxwork structure of authigenic clay minerals. Gila Group (Pliocene-Pleistocene), Tucson, Arizona. Scale bar = 5 μm. *(From Walker, T. R., et al., 1978, Diagenesis in first-cycle desert alluvium of Cenozoic age, southwestern United States and northwestern Mexico: Geol. Soc. America Bull., v. 89, Fig. 8C, p. 27. Photograph courtesy T. R. Walker.)*

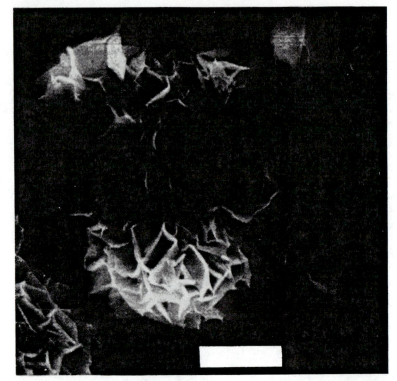

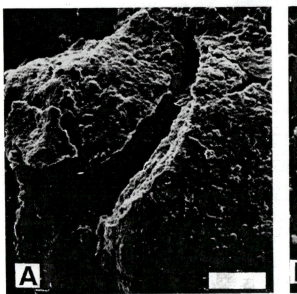

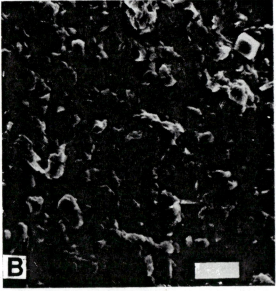

FIGURE 4.15
A. Electron micrograph of infiltered clay particles showing clay platelets oriented parallel to grain surfaces; scale bar = 25 μm. B. Enlargement of the area outlined in A; scale bar = 5 μm. Gila Group (Pliocene-Pleistocene), Red Rock, New Mexico. *(From Walker, T. R., et al., 1978, Diagenesis in first-cycle desert alluvium of Cenozoic age, southwestern United States and northwestern Mexico: Geol. Soc. America Bull., v. 89, Fig. 1E and 1F, p. 20. Photographs courtesy T. R. Walker.)*

5. **Epimatrix**—authigenic interstitial materials grown in originally open interstices but lacking the homogeneity of clay cements
6. **Pseudomatrix**—discontinuous matrixlike material formed by squeezing and flowing of weak detrital grains, such as shale or phyllite fragments, into adjacent pore spaces (see also J. R. L. Allen, 1962; Carrigy and Mellon, 1964). Dickinson suggests that pseudomatrix can be recognized by
   a. Flamelike wisps of mashed rock fragments that extend into pore spaces between rigid framework grains
   b. Deformation of internal fabric of lithic fragments to form concentric drape lines that conform to the margins of confining, rigid grains
   c. Irregular distribution of matrix, forming large matrix-filled patches in the framework (which are separated by areas nearly free of matrix); true matrix should more or less fill all pore spaces

Dickinson's classification of interstitial components is valuable in the sense that it helps us to understand the various possible origins of interstitial materials. The classification is, however, extremely difficult to apply in practice. It is highly doubtful, for example, that even an experienced petrographer using a petrographic microscope can routinely distinguish protomatrix, orthomatrix, and epimatrix, or unambiguously distinguish these matrix types from clay cement. On the other hand, recognition of pseudomatrix can often be made by careful examination. Therefore, this term is a very useful one and has gained general acceptance by geologists.

4.3   Chemical Composition

4.3.1   Significance

Geologists are interested in the chemical composition of sandstones, but they generally regard chemical composition to be less useful than mineralogy as a means of characterizing sandstones and evaluating their provenance and depositional environments. Nonetheless, sandstone geochemistry has a number of important applications. For example, major-element chemistry can provide information about the tectonic setting of sandstone accumulation (Chapter 8), allowing distinction among sandstones derived from oceanic island arc, continental island arc, active continental margin, and passive margin settings (Bhatia, 1983; Roser and Korsch, 1986). Major- and trace-element chemistry have been used to evaluate sedimentation rates and depositional environments in orogenic belts (Sugisaki, 1984). Major-element chemistry has been utilized also to infer the original clastic assemblages in deeply buried and altered sedimentary rocks and to help clarify the processes that produced the sediments (Argast and Donnelly, 1987). Trace elements have value in some kinds of provenance studies. For example, Owen (1987) used the hafnium content of zircons to establish a common provenance for the Jackfork Sandstone of Arkansas and the Parkwood Formation of Alabama. Darby and Tsang (1987) used the Ti, Fe, Mn, Mg, V, Zr, Cr, Ni, and Cu concentrations in detrital ilmenite grains to distinguish sediments in different drainage basins of the Blue Ridge Province, Virginia. Suttner and Leininger (1972) studied the Ti, Mg, and Fe content of plutonic, volcanic, and metamorphic quartz from sources in Montana to provide baseline chemical data for quartz provenance studies. In an electron microprobe study of detrital feldspars, Trevena and Nash (1981) demonstrated that the K, Na, Fe, Ca, and Sr content of feldspars can be used to distinguish feldspars from different sources.

Stable isotopes provide an important tool for studying cements in carbonate rocks and are beginning to be used also in the study of sandstone diagenesis. As an example of this application, Land and Dutton (1978) studied oxygen isotopes in quartz and carbonate cements to unravel the sequence of diagenetic processes that took place during burial and diagenesis of Pennsylvanian deltaic sandstones in Texas. In another study of Texas sandstones, Milliken et al. (1981) used carbon and oxygen isotopic data to deduce the approximate temperature of formation of diagenetic minerals.

As the discussion above suggests, many of the applications of geochemistry require bulk analysis of sedimentary rocks. Such analyses are now fairly routine using techniques such as X-ray fluorescence or ICP. The chemical compositions determined by bulk analyses obviously bear a strong relationship to the mineral composition of sediments; however, it is generally not possible to recreate the mineralogy exactly from the chemical composition owing to variations in the chemical composition of some minerals. Also, bulk chemical analyses do not allow discrimination between original detrital constituents and cements or other authigenic minerals, a fact that limits the usefulness of geochemistry in some applications. Some trace-element and isotope studies require analysis of individual minerals, such as zircons or ilmenites, rather than bulk analyses. Such analyses are commonly performed with an electron microprobe.

4.3.2 Expressing Chemical Composition

It is customary in reporting the results of chemical analyses to express major elements as oxides and report abundances in weight percent of the total oxides. Table 4.9 is an example of the general form used in reporting chemical analyses. Note from this table that trace elements are not reported as oxides; instead, their abundances are given in actual concentration units, in this case parts per million (ppm). The concentrations of elements present in even lower abundances are given in parts per billion (ppb). As suggested by Table 4.9, the major chemical elements in sandstones are silicon, aluminum, iron (expressed either as Fe_2O_3 or FeO), magnesium, calcium, sodium, potassium, titanium, manganese, and phosphorus. Silicon is the most abundant element in sandstones, commonly followed by aluminum and iron. The relative abundances of calcium, magnesium, sodium, and potassium vary considerably in different sandstones, but these elements are commonly much more abundant than manganese, titanium, and phosphorus.

4.3.3 Relation of Chemical Composition to Mineralogy

The **silicon** content of sandstones is a function of all the silicate minerals present, but obviously is most strongly influenced by the presence of quartz. Therefore, SiO_2 values are particularly high in quartz-rich sandstones. **Aluminum** is contained mainly in feldspars, micas, and clay minerals. Because of the abundance of aluminum in fine micas and clay minerals, sandstones containing abundant clay matrix commonly have much higher aluminum contents than those with little or no matrix. **Iron** is a common constituent of many minerals and may be present both in the ferrous (Fe^{2+}) and ferric (Fe^{3+}) states. Ferrous iron is particularly important in chlorites (and to a lesser extent in some other clay minerals such as smectites), biotites, carbonates (siderite and ankerite), and sulfides (pyrite and marcasite). Ferrous iron may occur also in some volcanic rock fragments and in minor quantities in some silicate minerals such as feldspars. Ferric iron is most abundant in the iron-oxide minerals hematite, goethite, and lepidochrocite and in glauconites. Iron is an important constituent also of many heavy minerals such as magnetite; however,

TABLE 4.9

Average chemical composition of sandstones reported by Argast and Donnelly (1987)

n	(1) 11	(2) 23	(3) 30	(4) 16	(5) 18	(6) 12	(7) 119	(8) 12	(9) 59
SiO$_2$ (wt. %)	86.5	67.8	65.6	56.9	56.2	68.4	70.6	37.3	50.3
TiO$_2$	0.53	0.95	0.91	1.42	0.89	0.69	0.64	0.34	0.64
Al$_2$O$_3$	5.71	15.4	15.1	12.3	15.3	13.5	12.6	7.91	14.0
Fe$_2$O$_3$ (t)	2.69	6.46	6.09	6.18	6.48	5.30	4.97	3.18	6.40
MnO	0.02	0.07	0.15	0.11	0.07	0.09	0.08	0.10	0.13
MgO	0.69	1.73	1.82	4.20	2.35	1.68	1.51	1.07	3.25
CaO	0.05	0.42	1.94	5.82	5.74	2.38	1.61	26.0	9.90
Na$_2$O	0.02	1.07	0.87	1.92	1.28	3.15	2.76	0.92	—
K$_2$O	1.55	2.74	3.03	1.90	2.80	2.62	2.20	0.51	2.09
P$_2$O$_5$	0.02	0.16	0.17	0.17	0.17	0.18	0.02	0.10	0.21
V (ppm)	51	123	159	100	126	71	79	103	—
Cr	55	82	88	225	71	55	44	31	—
Ni	19	231	58	130	49	30	8	5	49
Zn	29	52	104	84	114	69	—	66	91
Rb	60	123	133	72	125	93	—	10	79
Sr	29	134	113	233	168	310	110	879	267
Y	17	31	40	21	35	36	37	15	29
Zr	417	238	260	191	187	333	413	58	118

Note: Iron is reported as total Fe$_2$O$_3$; *n* is the number of samples in each average. No reported value is shown with a dash.

Source:

1 Shawangunk Formation, near Ellenville, New York (quartz arenite)
2 Millport Member of the Rhinestreet Formation, Elmira, New York (lithic arenite/wacke)
3 Oneota Formation, Unadilla, New York (lithic arenite/wacke)
4 Cloridorme Formation, St. Yvon and Gros Morne, Quebec (lithic arenite/wacke)
5 Austin Glen Member of Normanskill Formation, Poughkeepsie, New York (lithic arenite/wacke)
6 Rensselaer Member of the Nassau Formation, near Grafton, New York (feldspathic arenite/wacke)
7 Rensselaer Member, averages of analyses from Ondrick and Griffiths (1969) (feldspathic arenite/wacke)
8 Rio Culebrinas Formation, La Tosca, Puerto Rico (fossiliferous volcaniclastics)
9 Turbidites from DSDP site 379A (lithic arenites/wacke)

the content of such minerals is so low in most sandstones that they do not greatly affect the bulk chemical composition of the sandstones. **Sodium** and **potassium** are contained especially in alkali feldspars and muscovite. They are present also in illite and smectite clay minerals and most zeolite minerals. **Calcium** is contributed to sandstones in calcitic plagioclases, calcite cements, and, to a minor extent, smectite clay minerals. Calcite cements probably account for most of the calcium in sandstones that are particularly enriched in calcium. **Magnesium** is contained particularly in chlorite, smectite clay minerals, and dolomite cements. Because dolomite cements in sandstones are much less common on the average than calcite cements, calcium is generally more abundant than magnesium in sandstones. As Table 4.9 indicates, however, magnesium may exceed calcium in an occasional sandstone. **Titanium** and **manganese** each commonly makes up less than 1 percent of the chemical constituents in sandstones. Titanium is contained in some clay minerals and in the heavy minerals ilmenite, rutile, brookite, and anatase. Manganese probably occurs mainly as a substitute for iron in the iron oxide minerals, but minor amounts of manganese oxides occur also. **Phosphorus** is a common element in sandstones but seldom exceeds a few tenths of 1 percent in abundance. Its principal source is detrital and authigenic apatite. Numerous other elements may be present in sandstones in trace amounts, including, as shown in Table 4.9, V, Cr, Ni, Zn, Rb, Sr, Y, and Zr. The **trace-element** composition of sandstones is mainly a function of detrital-mineral composition. Therefore, some trace elements contained in specific detrital minerals are useful in provenance studies, as discussed above. On the other hand, trace elements can be added to sandstones at the depositional site and during diagenesis. Thus, considerable caution is necessary in interpreting provenance on the basis of trace-element abundances in bulk samples.

The preceding discussion clearly suggests that chemical composition of sandstones is strongly correlated with mineral composition, particularly detrital mineral composition. We should expect, therefore, that mineralogically different kinds of sandstones will show considerable variation in chemical composition. This topic is discussed in greater detail in Chapter 5.

4.4   Relationship of Particle and Chemical Composition to Grain Size

Sandstones are composed of mixtures of mineral grains and rock fragments, but mineral grains dominate in most sandstones. Mudrocks (shales) are made up almost entirely of minerals. Most mudrocks, with the exception of some so-called pebbly mudstones, contain few rock fragments. Several studies have shown that particle composition in sediments may vary considerably as a function of particle size. A relationship between particle size and composition of sandstones has long been recognized (e.g., P. Allen, 1945, 1947; Hunter, 1967). In a study of Illinois river sands, for example, Hunter (1967) reports that the percentage of feldspar and heavy minerals increases with decreasing grain size, whereas the amount of quartz, rock fragments, chert, and quartzite decreases. Odom et al. (1976) also report a strong relationship between feldspar content and grain size in quartz-rich sandstones. They found that feldspars are concentrated especially in the <0.125 mm size fraction and that a nearly linear inverse relationship exists between feldspar abundance and grain size in some sandstones. Charles and Blatt (1978) describe a similar concentration of feldspars in finer size fractions and a tendency for chert grains to occur preferentially in coarser fractions. With respect to heavy minerals, Rittenhouse (1943) pointed out that heavy minerals in stream sediments are sorted by density and size. That is, fine-size heavy minerals tend to occur with much coarser, lower-density quartz grains—the concept of a **hydraulic-equivalent size** (the size of a larger or smaller grain that settles with a given mineral grain under the same conditions). In general, the higher the specific gravity of the heavy minerals, the smaller the heavy minerals will be with respect to the size of the quartz grains with which they occur. Rittenhouse proposed use of the **hydraulic ratio** to express the quantity of any given mineral in a sediment. The hydraulic ratio is equal to the weight of a heavy mineral in a given size class divided by the weight of light minerals in the hydraulic-equivalent class, multiplied by 100. The relationship among heavy mineral density, size, and composition is further amplified by van Andel (1959).

These examples all point out a basic sedimentologic tenet as applied to the petrologic study of siliciclastic sediments: comparison of composition between samples of different sediments can be made only if sediments of approximately the same grain size are compared. Ingersoll et al. (1984) suggest that the problems inherent in grain size–composition relationships can be minimized in point counts of sandstones by assigning the sand-size crystals and grains within larger rock fragments to the category of the crystals or grains, rather than to the category of rock fragments. This technique is the so-called **Gazzi-Dickinson** point-counting method. On the other hand, some geologists (e.g., Suttner, 1985) believe that use of the Gazzi-Dickinson method obscures provenance information (see discussion in Chapter 8, Section 8.6.2). Furthermore, it may not adequately address the composition changes when few rock fragments are present and the sizes of particles are very different.

Because the chemical composition of siliciclastic sedimentary rocks is closely related to the mineral composition of these rocks, as discussed, chemical composition varies as a function of grain size along with variations in mineralogy. Pettijohn (1975, p. 270) points out, for example, that SiO_2 abundance decreases progressively from fine sands to fine clays, whereas the Al_2O_3 content systematically increases.

Additional Readings

Adams, A. E., W. S. MacKenzie, and C. Guilford, 1984, Atlas of sedimentary rocks under the microscope: John Wiley & Sons, New York, 104 p.

Carver, R. C., ed., 1971, Procedures in sedimentary petrology: John Wiley & Sons, New York, 653.

Deer, W. A., R. A. Howie, and J. Zussman, 1966, An introduction to the rock-forming minerals: John Wiley & Sons, New York, 528 p.

Lindholm, R. C., 1987, A practical approach to sedimentology: George Allen & Unwin, London, 276 p.

MacKenzie, W. S., and C. Guilford, 1980, Atlas of rock-forming minerals in thin section: John Wiley & Sons, New York, 98 p.

Milner, H. B., 1962, Sedimentary petrography, v. 2, Principles and applications: MacMillan, New York, 715 p.

Scholle, P. A., 1979, A color illustrated guide to constituents, textures, cements, and porosities of sandstones and associated rocks: Am. Assoc. Petroleum Geologists Mem. 28, 201 p.

Tickell, F. G., 1965, The techniques of sedimentary mineralogy: Elsevier, New York, 220 p.

Chapter 5
Sandstones: Classification and Petrography

5.1 Introduction

The principal particle and chemical constituents of sandstones are discussed in Chapter 4. In this chapter, we continue study of these constituents by first taking a look at the way they are used to divide sandstones into identifiable groups. We then examine in greater depth the characteristics of each of these major groups of sandstones. In the broadest sense, sandstones can be separated into two groups: **siliciclastic** (mainly of terrigenous origin) and **nonsiliciclastic** (mainly carbonates and evaporites). Carbonate sands and other carbonate rocks are discussed in Chapter 10. Our concern in this chapter is the siliciclastic deposits, which can be divided further into **epiclastic** and **volcaniclastic** deposits. Epiclastic deposits are formed from fragments of preexisting rocks derived by weathering and erosion. Thus, they are composed mainly of silicate minerals and various kinds of igneous, metamorphic, and sedimentary rock fragments. Volcaniclastic deposits are those especially rich in volcanic debris, including glass. Many volcaniclastic deposits consist principally of pyroclastic materials such as ash or lapilli, derived directly through explosive volcanism. On the other hand, some material in volcaniclastic deposits may be epiclastic debris derived by weathering of older volcanic rock. Epiclastic and volcaniclastic deposits can be further classified on the basis of their composition. Unfortunately, there is little agreement among geologists about sandstone classifications, particularly classifications for epiclastic sandstones.

In the succeeding part of this chapter, we examine the characteristics of the major groups of sandstones. Each group is defined on the basis of outstanding compositional features, and its petrographic characteristics are discussed in detail. Finally, the temporal and spatial distribution, origin, and significance of each group of sandstones are discussed.

5.2   Classification of Epiclastic Sandstone

5.2.1   Parameters for Classification

The framework grains of most sandstones are dominated by quartz, feldspars, and rock fragments. Many other minerals may be present in a given sandstone, but the abundances of these other minerals are so low in most sandstones that they can be ignored for the purpose of sandstone classification. Some sandstones contain **matrix** in addition to sand-size framework grains. As mentioned in Chapter 4, matrix is defined as material less than about 0.03 mm (30 μm) in size. Thus, it is not a framework constituent. Rather, it occupies the interstitial spaces among sand-size grains. The matrix content of sandstones may range from zero to several tens of percent.

Owing to the simple framework composition of sandstones (mainly quartz, feldspars, rock fragments), classification of sandstones ought to be a fairly straightforward process. Practice has proven differently! Sedimentologists have had a difficult time indeed developing a single sandstone classification scheme that is acceptable to most workers in the field. According to Friedman and Sanders (1978, p. 190), more than 50 classifications for sandstones have been published since the late 1940s in ten countries and seven languages. Some of the reasons why geologists have had so much difficulty developing a widely accepted sandstone classification are discussed below.

5.2.2   Problems in Selecting Classification Parameters

First, sedimentologists disagree about the use of genetic vs. descriptive classification. A **genetic classification** is a classification in which the names of the rocks putatively convey some information about the origin of the rocks. By contrast, a **descriptive classification** is based strictly upon observable or measurable properties of a rock without regard to the origin of these properties. Ideally, classification ought to be based upon purely objective criteria that are also genetically significant; however, this concept hasn't always worked out in practice.

A second problem has to do with assigning certain kinds of grains to a particular end-member grain category. For example, assume that quartz (Q), feldspars (F), and rock fragments (R) are chosen as the end members of a classification triangle, the most common practice in sandstone classification. In which of these end-member groups should we include chert and quartzite clasts? Both chert and metaquartzite clasts are rock fragments and thus might logically be included with all other rock fragments. On the other hand, because they are composed almost entirely of quartz, they can just as logically be included as polycrystalline quartz with other polycrystalline and monocrystalline quartz grains at the Q pole of the classification triangle. As another example, if granite or gneiss fragments that contain significant amounts of feldspar grains are present, should these grains be included with the feldspars or the rock fragments?

A third classification problem concerns matrix. Matrix is a textural rather than a compositional attribute of sandstones. In most cases, we do not know what kinds of minerals make up the matrix. We simply know the amount of matrix present. Should matrix be ignored in sandstone classification? If not, how do we deal with it? Some classifiers have indeed omitted matrix as one of the classification variables, presumably because it simply occupies interstitial space and is not part of the framework grain fraction (e.g., McBride, 1963). Other classifiers have attempted to factor in the matrix variable by dividing sandstones into two groups on the basis of matrix content, that is, sandstones without matrix (or having less than a specified amount of matrix, such as 10 or 15 percent) and those containing significant matrix (e.g., Dott, 1964; Okada, 1971; Pettijohn et al., 1987, p. 145). Matrix creates still another problem. Most workers believe that sandstone classification should be based solely on the detrital constituents of sandstones; that is, cements or other authigenic constituents should be ignored. There is a growing belief among sedimentologists, however, that much of the matrix in sandstones is authigenic. Yet it is often extremely difficult in a routine petrographic examination of sandstones to distinguish between authigenic and detrital matrix. For the purpose of classification, should greater efforts be made to distinguish between authigenic and detrital matrix in petrographic analyses? Does it matter?

5.2.3   Classification Choices

The classification problems discussed, plus other factors such as the personal preferences of geologists working with particular kinds of sandstones, are responsible for the proliferation of sandstone classifications. The literature on sandstone classification has been reviewed by Klein (1963) and Okada (1971). A very useful tabulation of selected classifications published since 1933 is given by Pettijohn et al. (1987, Table 5-1, p. 142). These authors provide also a nice summary of major classification trends during that period, as well as a list of all the published names for sandstones. Most sandstone classifications are verbal classification; however, a few workers (e.g., Boggs, 1967b; Dickinson, 1970) have proposed the use of numerical classifications to avoid some of the problems inherent in verbal classifications.

Few sandstone classifications have enjoyed widespread popularity, and many have had a very short "shelf life." It is difficult to say which is the best of the existing classifications because, in the end, what is best often boils down to personal biases and preferences. It appears, however, that most workers in recent years are leaning toward the type of classification first proposed by Gilbert in 1954 (in Williams et al., 1954), later modified by Dott (1964), and still later modified again by Gilbert (Williams et al., 1982, p. 327). A somewhat different version of this classification appears also in Pettijohn et al. (1987, p. 145).

Gilbert's latest version of this classification is included here as Figure 5.1. A principal feature of this classification is its simplicity. On the basis of matrix content, sandstones are divided into two broad groups: **arenites**, containing little (<5 percent) or no matrix, and **wackes**, containing perceptible matrix. Combining the matrix parameter with composition (QFL; see Figure 5.1 for explanation) yields six kinds of sandstones: quartz arenites and wackes, feldspathic arenites and wackes, and lithic arenites and wackes. Figure 5.2 shows two examples of more-complex classifications that divide sandstones into seven or eight compositional types. These classifications do not use matrix as a classification parameter. Note that Figure 5.2B allows even finer subdivision of sandstones by erecting "daughter" compositional triangles on the main compositional diagram.

Two common sandstone names that do not appear in many formal classifications require some additional explanation. The term **arkose** is used in some sandstone classi-

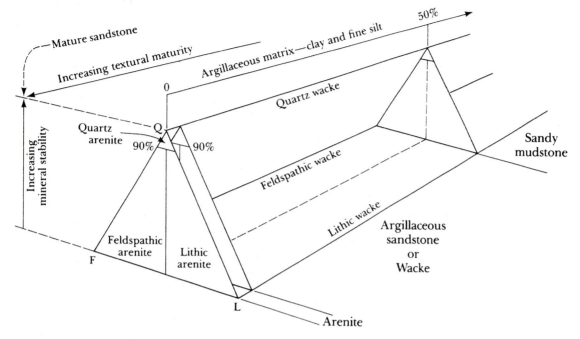

FIGURE 5.1

Classification of sandstones on the basis of three mineral components: Q = quartz, chert, quartzite fragments; F = feldspars; L = unstable, lithic grains (rock fragments). Points within the triangles represent relative proportions of Q, F, R end members. Percentage of argillaceous matrix is represented by a vector extending toward the rear of the diagram. The term arenite is restricted to sandstones essentially free of matrix (<5%); sandstones containing matrix are wackes. *(From Williams, H., F. J. Turner, and C. M. Gilbert, Petrography, an introduction to the study of rocks in thin sections, 2nd ed. W. H. Freeman and Company, San Francisco. © 1982, Fig. 13.1, p. 327. Modified from Dott, R. H., Jr., 1964, Wacke, graywacke, and matrix—what approach to immature sandstone classification?: Jour. Sed. Petrology, v. 34. Fig. 3, p. 629, reprinted by permission of Society of Economic Paleontologists and Mineralogists, Tulsa, Okla.)*

fications and is in general use by many geologists. The exact meaning of this term is elusive because the term has been defined in different ways. It has been applied to any sandstone containing conspicuous amounts of feldspars, as seen in hand specimens, to sandstones containing more than 20 percent, more than 25 percent, or more than 30 percent feldspars and to sandstones derived from granitic source rocks. Probably the most widely accepted definition for an arkose is a feldspathic sandstone containing more than 25 percent feldspars. Presumably, an arkose also has a lower rock-fragment content than feldspar content.

The term **graywacke** is a much-maligned term that is still used by many geologists. Some workers have suggested that the name be abandoned, but it manages to survive. In general, the name graywacke is applied to dark-gray, greenish-gray, or black, matrix-rich, well-indurated sandstones. According to Crook (1970), the origin of the term goes back to Werner in 1787. The name was originally given to rocks on the basis of their field

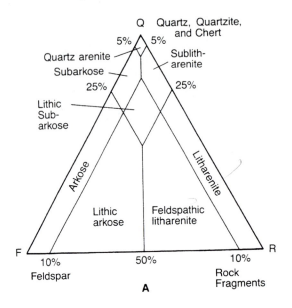

A

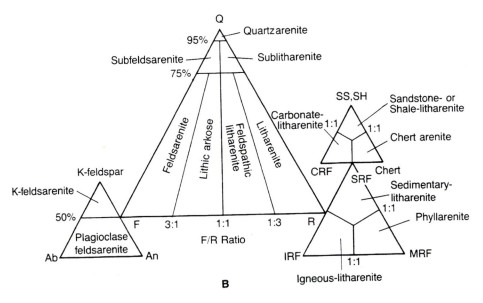

B

FIGURE 5.2

Classification of sandstones according to (A) McBride and (B) Folk. In Folk's classification, chert is included with rock fragments at the R pole, and granite and gneiss fragments are included with feldspars at the F pole. SS = sandstone, SH = shale, CRF = carbonate rock fragments, SRF = sedimentary rock fragments, IRF = igneous rock fragments, MRF = metamorphic rock fragments. *(A, from McBride, E. F., 1963, A classification of common sandstones: Jour. Sed. Petrology, v. 34. Fig. 1, p. 667, reprinted by permission of Society of Economic Paleontologists and Mineralogists, Tulsa, Okla. B, from Folk, R. L., P. B. Andrews, and D. W. Lewis, 1970, Detrital sedimentary rock classification and nomenclature for use in New Zealand: New Zealand Jour. Geology and Geophysics, v. 13. Fig. 8, p. 955, and Fig. 9, p. 959, British Crown copyright, reprinted by permission.)*

characteristics, but the term has subsequently been defined in various other ways. Hence, the suggestion that it be abandoned. The properties of graywackes are further discussed in Section 5.3.4 of this chapter.

5.3   Petrography and Chemistry of Epiclastic Sandstones

5.3.1   Major Petrographic Divisions

The preceding discussion indicates that epiclastic sandstones can be divided by particle composition into three major groups: quartz arenites (and wackes), feldspathic arenites (and wackes), and lithic arenites (and wackes). These major sandstone clans differ in more than just particle composition. They may show also detectable variations in texture, types of cements, matrix content, and bulk chemistry. The important characteristics of each of these major kinds of sandstones are described below.

5.3.2   Quartz Arenites

General Characteristics

By definition (Section 5.2), quartz arenites are composed of more than 90 percent siliceous grains (quartz, chert, quartzose rock fragments). (Some authors, e.g., Folk et al., 1970, and Pettijohn et al., 1987, place the quartz boundary at 95 percent.) Quartz arenites are commonly white or light gray but may be stained red, pink, yellow, or brown by iron oxides. They are generally well lithified and well cemented with silica or carbonate cement; however, some are porous and friable. Quartz arenites typically occur in association with assemblages of rocks deposited in stable cratonic environments such as eolian, beach, and shelf environments. Thus, they tend to be interbedded with shallow-water carbonates and, in some cases, with feldspathic sandstones (e.g., the basal Minturn Formation [Pennsylvanian] of Colorado [Tweto and Lovering, 1977]). Most quartz arenites are texturally mature to supermature according to Folk's (1951) textural maturity classification, but quartz arenites with low maturity exist (Folk, 1974, p. 145). Cross-bedding is particularly characteristic of these rocks, and ripple marks are moderately common. Fossils are rarely abundant in these sandstones, possibly owing to poor preservation or to the eolian origin of some quartz arenites, but both fossils and carbonate grains may be present. Also, trace fossils such as burrows of the *Skolithos* facies may be locally abundant in some shallow-marine quartz arenites (e.g., James and Oaks, 1977). Quartz arenites are common in the geologic record. Pettijohn (1963) estimates that they make up about one-third of all sandstones.

Particle Composition

Quartz arenites have the simplest particle composition of all sandstones. Some representative petrographic data are given in Table 5.1. Siliceous grains in these sandstones range in abundance from 90 percent to more than 99 percent. Quartz dominates the siliceous constituents; however, a few percent chert, metaquartzite clasts, or siliceous sandstones/siltstone clasts may be included. Most quartz in quartz arenites is monocrystalline. Polycrystalline grains commonly make up less than 10 percent of the framework grains, but rare values exceeding 50 percent polycrystalline quartz have been reported (Bond and Devay, 1980). Blatt and Christie (1963) show an average value of only 2 percent poly-

TABLE 5.1

Average particle composition of North American quartz arenites, as determined by petrographic analysis

	A	B	C	D	E	F	G	H	I	J
Quartz (M)	94	89	87?	43	97	86	?	?	?	?
Quartz (P)	6	7	5?	57	1	3	?	?	?	?
Quartz (T)	99	96	91	94	98	89	99	95+	98	~95
Plagioclase	—	Tr	1	5	—	0.6	—	Tr	1–2	~4
K-feldspar	—	3–4	4	Tr	<2	5	—	3–5?	—	—
Chert	—	—	0.7	Tr	—	0.7	Tr	—	—	—
Silicic rock fragments	—	—	Tr	<1	—	0.7	—	—	—	—
Unstable rock frag.	—	—	—	<1	<0.5	Tr	—	—	—	—
Muscovite	Tr	Tr	0.4?	Tr	Tr	0.5	—	—	—	—
Biotite	—	Tr	—	—	?	0.2	—	—	—	—
Heavy minerals	Tr	0.4	0.4	Tr	Tr	0.3	Tr	<0.1	<0.5	<0.5
Total framework grains	68	?	82	84?	89	93.5	?	?	?	95?
Silica cement	30	C	6	?	4	1.4	C	C	?	?
Carbonate cement	Tr	Pr	2	?	3	0.8	R	—	?	?
Hematite cement	—	—	—	—	—	0.3	—	—	—	—
Matrix	2	0?	10	16	4	4	Pr	?	Pr	~5
Number of samples	12	?	494	120	18	46	320	?	?	?

Note: Abundance of framework grains given as percent of total framework grains; cement and matrix are shown as percent of total constituents. M = monocrystalline, P = polycrystalline, T = total, Tr = trace, C = common, Pr = present, R = rare.

Source:
A James and Oaks (1977), Kinnikinic Quartzite (Ord) from 3 localities in Idaho
B Campbell (1972), Parting Formation (Dev) from 11 measured sections in Colorado
C Lobo and Osborne (1976), Precambrian sandstones from 10 localities in southern California
D Bond and Devay (1980), Shoo Fly Formation (Dev) from 4 localities in northern California
E Ojakangas (1963), Lamotte Sandstone (Camb), Missouri
F Schwab (1970), Antietam Formation (Precambrian), Virginia
G Ketner (1966), Ordovician quartz Antietam of California, Nevada, Utah, and Idaho
H Thiel (1935), St. Peter Sandstone (Ord), Minnesota, Iowa, Wisconsin
I Harshbarger et al. (1957), Navajo Sandstone (Jurassic), Colorado Plateau
J Harshbarger et al. (1957), Entrada Formation (Jurassic), Colorado Plateau

crystalline quartz in samples from 20 different quartz-rich sandstones. The low content of polycrystalline grains in quartz arenites has been attributed to size reduction of these grains. Polycrystalline quartz grains apparently break up to form smaller monocrystalline grains owing to sediment transport and other processes involved in grain recycling (Blatt and Christie, 1963; Blatt, 1967a). The high content of polycrystalline quartz reported by Bond and Devay (1980) could be the result of postdepositional metamorphism, although these authors believe that it reflects source-rock types. Blatt and Christie (1963) point out also that quartz arenites tend to be characterized by high percentages of nonundulatory monocrystalline quartz (average about 43 percent in their samples). They suggest that sandstones become relatively enriched in nonundulose quartz over time owing to selective destruction of less-stable undulose quartz by mechanical and chemical processes operating during successive sedimentary cycles.

Feldspars make up trace amounts to as much as 5 percent of the framework constitutents in many quartz arenites. Microcline and orthoclase are probably most common; however, plagioclase is the dominant feldspar in some quartzose sandstones (Table 5.1). Most plagioclase in these sandstones is sodic. For example, Bond and Devay (1980) report the Ab content in 75 plagioclase grains in 10 samples to range between 95 and 100 percent. Some quartz arenites may contain very minor amounts of fine-size authigenic feldspars, particularly in the fine fraction of the sandstones. For example, Odom et al. (1979) report authigenic K-feldspars in several parts of the St. Peter Sandstone. Most of these small, authigenic feldspars appear to represent new crystallization, although some may have formed around detrital feldspar cores.

Chert and silicic rock fragments, both metaquartzite clasts and sandstone/siltstone clasts, occur in some quartz arenites in amounts that rarely exceed 1 percent. Unstable rock fragments are very rare. Micas, commonly muscovite, occur in amounts that generally do not exceed about 0.5 percent. Heavy minerals are common in most quartz arenites but generally make up no more than about 0.5 percent of total framework grains. The most common heavy minerals are the ultrastable types—zircon, tourmaline, and rutile; therefore, the ZTR index of heavy minerals in quartz arenites is commonly very high. The ultrastable heavy minerals are commonly well rounded. Less-common heavy minerals include magnetite, sphene, epidote, monazite, apatite, leucoxene, garnet, hypersthene, and hornblende. Minor amounts of secondary pyrite are present in some quartz arenites.

Cements

Most quartz arenites are well cemented with silica, carbonate, or hematite cement. Table 5.1 provides a few data on cement abundances; however, many investigators do not report the exact amount of cement in their samples. Quartz appears to be by far the most common cement and is typically present as syntaxial overgrowths on detrital quartz cores. When syntaxial overgrowths are picked out by a line of bubbles or impurities (Fig. 4.3), they are easy to see. As discussed in Section 4.2.2, however, overgrowths not marked by impurities are difficult to identify. Application of cathodoluminescence petrography may be necessary to identify such overgrowths. Some quartz grains may appear in thin section to be completely, or almost completely, surrounded by an overgrowth (Fig. 5.3). In extremely well cemented quartz arenites, overgrowths may grow together in pore spaces, in which overgrowths from adjacent grains meet along an irregular boundary. This type of cement forms an interlocking mosaic of grains, essentially eliminating porosity. Quartz arenites that have undergone extensive compaction may also display interlocking fabrics, the result in this case of squeezing and pressure solution (Fig. 5.4). Cathodoluminescence petrography may sometimes be necessary to distinguish between these two types of fabrics; i.e., overgrowths do not luminesce. In less completely cemented quartz arenites, overgrowths may terminate in well-developed crystal faces that incompletely fill pores (Fig. 5.5). Microquartz and opal cements occur in some quartz arenites (Carozzi, 1960, p. 8, 10) but are much rarer than syntaxial overgrowths. Opal is distinguished by its isotropic character. Quartz cement occurs as equidimensional microquartz or fibrous chalcedony. In some pores, microquartz may display well-developed drusy texture.

Carbonate cements are present in some quartz arenites. Calcite is the most common carbonate cement, but dolomite and, less commonly, siderite may occur also. Calcite cement can occur as a mosaic of small crystals within a single pore, commonly with drusy texture. More typically, it forms single, large crystals that completely fill a pore (Fig.

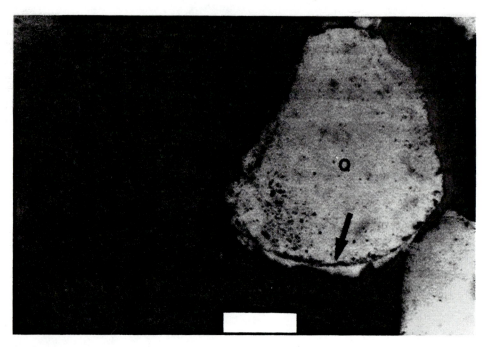

FIGURE 5.3
Photomicrograph of two quartz grains (Q) with overgrowths. Note that original grain outlines are clearly marked by a line of impurities (arrows) and that the grain on the right is almost completely surrounded by the overgrowth. Quartz arenites of the Deadwood Formation (Cambrian-Ordovician), South Dakota. Crossed nicols. Scale bar = 0.1 mm.

FIGURE 5.4
Compaction fabric in a quartz arenite. Note squeezed, elongated grains and highly sutured contacts (arrow) between some grains. Oriskany Formation (Devonian), Pennsylvania. Crossed nicols. Scale bar = 0.1 mm.

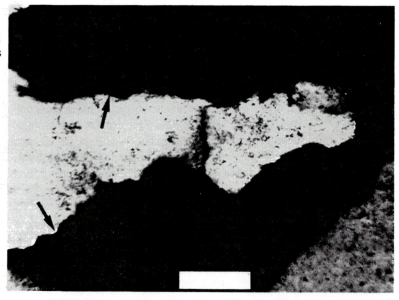

FIGURE 5.5
Quartz overgrowth (arrow)
incompletely filling pore space (P)
in a quartz arenite. Navajo
Sandstone (Jurassic), Utah.
Crossed nicols. Scale bar =
0.1 mm.

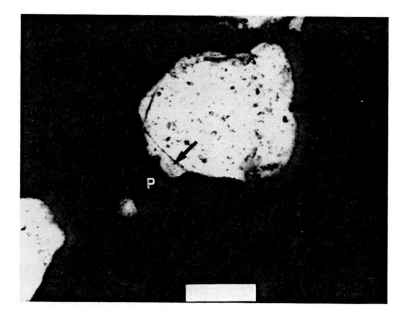

5.6). In fact, in some portions of a rock, a single crystal may fill the pore spaces among several grains, thus enclosing parts or all of several grains to form a poikilitic texture (Fig. 5.7). Some dolomite cements may be present as small, easily recognizable rhombs. Dolomite cements that do not form rhombs may be difficult to distinguish optically from calcite cements.

Minor amounts of hematite occur as cements in some quartz arenites. Authigenic overgrowths on detrital feldspar cores also constitute a minor cement in quartz arenites. Other less common cements that have been reported include anhydrite, barite, celestite,

FIGURE 5.6
A large, twinned calcite crystal (C)
filling pore space among quartz
grains. Quartz-rich example of the
Mauch Chunk Sandstone
(Mississippian), Pennsylvania.
Crossed nicols. Scale bar =
0.1 mm.

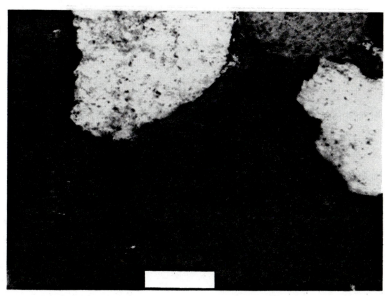

FIGURE 5.7
Poikilitic calcite (C) enclosing quartz (Q). Quartz-rich example of Mauch Chunk Sandstone (Mississippian), Pennsylvania. Crossed nicols. Scale bar = 0.5 mm.

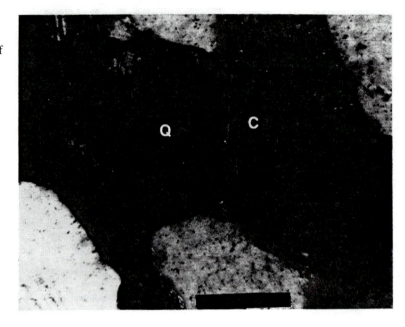

and glauconite. Some quartz arenites are very poorly cemented and have porosities that may exceed 20 percent. Such rocks may be so friable that hand specimens easily crumble.

Matrix

Quartz arenites typically contain very minor amounts of matrix. Many are matrix-free. On the other hand, some quartz arenites have more than 10 percent matrix (Table 5.1). Because quartz arenites appear to form mainly in winnowing environments, it seems likely that little detrital matrix is deposited with the framework grains in these sandstones. The small amount of matrix present in most quartz arenites is probably best explained as the product of diagenesis. The matrix consists mainly of fine-size quartz, feldspars, micas, and clay minerals. It probably forms diagenetically mainly by alteration of detrital feldspars and rock fragments, although alteration of detrital mica is possible also (Bond and Devay, 1980). Matrix in quartz arenites that is very abundant (e.g., sample D, Table 5.1) is probably not the product of diagenesis. It may be the result of some kind of special depositional conditions, such as conditions where quartz sands are blown into a lagoon or other quiet-water environment or are washed over during a storm. Alternatively, abundant matrix might result from bioturbation by organisms that mix mud from an underlying or adjacent layer into a clean quartz sand or by vadose infiltration of clays into sands (see Matlack et al., 1989).

Texture

Quartz arenites tend to be texturally mature to supermature; however, some are submature to immature. They are typically well sorted with low matrix content, as discussed. Subrounded to well-rounded grains predominate in many examples (Fig. 5.8). On the other hand, some quartz arenites are more poorly sorted and may contain high percentages of subangular to angular grains (Fig. 5.9), particularly smaller grains. In poorly sorted sandstones, larger grains tend to be better rounded than smaller grains. Some quartz

FIGURE 5.8
Quartz arenite with subrounded to
well-rounded, well-sorted grains.
Navajo Formation (Jurassic), Utah.
Crossed nicols. Scale bar =
0.5 mm.

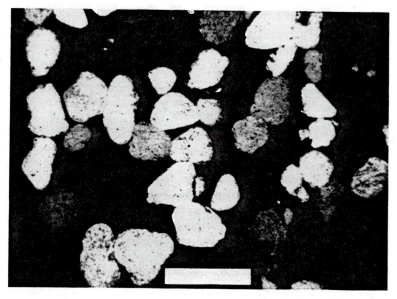

arenites exhibit textural inversions (Folk, 1974, p. 145) such as a combination of poor sorting and high rounding, a lack of correlation between roundness and size, such as small round grains and larger angular grains, or mixtures of rounded and angular grains within the same size fraction. These textural inversions probably result from mixing of grains from different sources, erosion of older sandstones, or environmental variables such as wind transport of rounded grains into a quiet-water environment. Angular grains may result also from diagenetic development of secondary overgrowths.

The bimodal character of some quartz arenites has been attributed by Folk (1968) to selective winnowing by wind action. Wind presumably removes the fine sand fraction

FIGURE 5.9
Angular to subangular quartz in a
quartz arenite. Coconino Formation
(Permian), Grand Canyon area,
Arizona. Crossed nicols. Scale
bar = 0.5 mm.

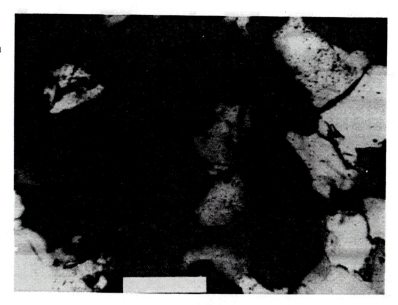

(0.1–0.3 mm) and leaves behind the coarser fraction as a lag deposit with the still-finer fraction trapped within the coarser. Some quartz grains in quartz arenites may display frosted or matte surface textures, particularly grains that are well rounded. Although frosted texture is often attributed to eolian action, frosting may result also from chemical etching.

Chemical Composition

Owing to their rather restricted mineral composition, the bulk chemical composition of quartz arenites varies within very narrow limits. Analyses of 15 representative Paleozoic and Precambrian quartz arenites (Table 5.2) show that SiO_2 makes up about 96–97 percent of these sandstones. On the average, Al_2O_3 accounts for 0.5–1.0 percent of the remaining constituents. Fe_2O_3, FeO, MgO, and CaO together make up about 1.0–1.25 percent of the chemical constituents. Other elements are commonly present in trace amounts. More extreme variations from these averages result mainly from differences in cement and matrix content rather than differences in particle composition. Thus, significant amounts of calcite or dolomite cement are reflected in abnormally high calcium and magnesium abundances, hematite cements increase the iron content of samples, high clay-matrix content is reflected in higher than normal aluminum content, and so on.

TABLE 5.2

Chemical composition of quartz arenites

	A		B	
	Mean	*Standard deviation*	*Mean*	*Standard deviation*
SiO_2	96.14	4.76	97.57	0.84
Al_2O_3	1.12	1.25	0.45	0.25
Fe_2O_3	0.26	0.38	0.22	0.13
FeO	0.08	0.17	0.24	0.27
MgO	0.06	0.07	0.33	0.21
CaO	0.94	2.76	0.22	0.16
Na_2O	0.06	0.05	0.044	0.009
K_2O	0.07	0.09	0.13	0.06
H_2O	0.11	0.18	0.43	0.10
TiO_2	0.04	0.09	0.01	0.01
P_2O_5	0.002	0.006	0.002	0.005
MnO	0.001	0.003	0.006	0.013
ZrO_2	0.007	0.019	0.005[1]	0.004[1]
CO_2	0.69	2.19	0.14	0.17
Number of samples	?		320	

Note: [1]Weight percent Zr

A. Average of values reported by Pettijohn (1963) from the following formations: Mesnard Quartzite (Precambrian), Lorrain Quartzite (Precambrian), Sioux Quartzite (Precambrian), Lauhauori Sandstone (Cambrian?), Tuscarora Quartzite (Silurian), Oriskany Sandstone (Devonian), Berea Sandstone (Mississippian), Mansfield Formation (Pennsylvanian).
B. Average of values reported by Ketner (1966) from the following Ordovician formations: Kinnikinic Quartzite, Swan Peak Quartzite, Eureka Quartzite, Valmy Formation, Petes Summit Formation.

Field Characteristics

Quartz arenites tend to occur in sheetlike units, many of regional extent, that may range in thickness from a few meters to several hundred meters. Thickness of individual units can vary considerably from place to place. The Jurassic Navajo Formation of the U.S. Colorado Plateau, for example, reaches a thickness of 425 m in some areas, but thins to about 5 m in other localities (Harshbarger et al., 1957). The Ordovician St. Peter Sandstone has an average thickness of about 30 m over much of northern Arkansas, Missouri, Illinois, Iowa, and southern Wisconsin and Minnesota in the U.S. northern midcontinent region, but actual thicknesses range from less than 1 m to about 150 m (Thiel, 1935). Quartz arenites may occur in dominantly nonmarine stratigraphic sections or dominantly marine sections. In marine sections, they are typically present in association with shallow-water limestones, dolomites, and, less commonly, feldspathic sandstones. On the other hand, some quartz arenites are associated with siliceous shales and cherts, suggesting a deeper-water origin (Bond and Devay, 1980). In nonmarine sections, they may occur in association with fluvial, lithic, or feldspathic sandstones, siltstones, conglomerates, and evaporites.

Cross-bedding is a prominent feature of some quartz arenites; some units display giant-scale sets of cross-beds several meters thick that have foreset dips of 20 to 35 degrees near the top of each bed. Other quartz arenites consist of thick-bedded, massive-appearing units that lack cross-beds or that display only indistinct cross-beds with low-angle foresets. Parallel-laminated quartz arenites are common also. Ripple marks occur in some quartz arenites. Body fossils are present in some, the Devonian Oriskany Sandstone, for example (Seilacher, 1968), but are rare in most. On the other hand, bioturbation features and trace fossils may be locally abundant (e.g., James and Oaks, 1977; Lobo and Osborne, 1976). Both fining-upward and coarsening-upward grain-size trends have been reported in quartz arenites (Lobo and Osborne, 1976).

Origin

The extreme compositional maturity of quartz arenites requires that these sandstones originate under rather specialized conditions. If quartz arenites are first-cycle deposits, they must form under weathering, transport, and depositional conditions so vigorous that most grains chemically and mechanically less stable than quartz are eliminated (see, for example, Johnsson et al., 1988). Conceivably, extreme chemical leaching under hot, humid, low-relief weathering conditions, prolonged transport by wind, intensive reworking in the surf zone or tidal zone (reworking by reversing tides), or a combination of these factors might be adequate to generate a first-cycle quartz arenite. Stream transport is known to produce little rounding of sand-size quartz grains, whereas wind transport is an effective rounding agent (Kuenen, 1959, 1960). Therefore, the well-rounded and well-sorted character of many quartz arenites, plus the common presence of large-scale sets of cross-beds with high-angle foresets, suggests that these quartz arenites may be eolian deposits. On the other hand, many quartz arenites are clearly of marine origin, as shown by their association with carbonates or other marine deposits. These marine quartz arenites may have been deposited in beach or barrier-island settings; however, the effectiveness of surf action in rounding of sand-size quartz is still not well known. Ferree et al. (1988) conclude that both modern beach sand and ancient beach sandstones have frameworks that are mineralogically more mature than their fluvial counterparts. They suggest that breakdown of rock fragments on beaches where wave power is high tends to enrich the sand in quartz and is an important factor in increasing compositional maturity of the sands.

Suttner et al. (1981) believe that first-cycle quartz arenites cannot be produced under "average" conditions. A unique combination of extreme climatic conditions, transportation, and low sedimentation rates would be required to produce such a sandstone. These authors conclude that most quartz arenites in the geologic record are multicycle deposits. Less-intensive weathering/transport processes operating through several sedimentation cycles (Chapter 1) may be able to selectively remove less-stable grains to produce a quartz arenite. To account for the extreme grain rounding, many quartz arenites may have been subjected to an episode of wind transport during at least one of these cycles, although not necessarily the last cycle. On the other hand, Johnsson et al. (1988) report that definite first-cycle quartz arenites are forming in the Orinoco River basin in Venezuela and Colombia. They cite two conditions that are necessary to produce first-cycle quartz arenites: an environment of intense chemical weathering and a mechanism to provide extended time over which weathering can operate (e.g., temporary storage on extensive alluvial plains).

Quartz arenites can conceivably be derived from any parent rock containing quartz; however, the quartz in these sandstones is most likely derived from quartz-rich sandstones or silicic plutonic or metamorphic rocks. Dickinson and Suczek (1979) and Dickinson (1984) suggest that three fundamental types of tectonic provenances exist: continental block (continental interiors), recycled orogen (uplifted zones along the suture belts of collided crustal plates), and magmatic (volcanic) arc (see Fig. 1.4). Except for those dissected arcs that expose deep-seated plutonic rocks, magmatic arcs supply abundant volcanic material but relatively little quartz. Considerable quartz may be derived from collision orogens or foreland uplift areas where sedimentary or metasedimentary sequences are exposed along suture belts. Most quartz arenites are probably derived from continental-block provenances, where the quartz is released by intense weathering of crystalline basement rocks or of sedimentary or metasedimentary cover. Quartz arenites may be deposited within platform or interior basins, or they may be carried off the continent to be deposited in stable, marginal oceanic settings. Rarely, they may be transported into deeper-water basins by sediment-gravity-flow processes such as grain flow. The high preponderance of North American quartz arenites in rocks of Precambrian and early Paleozoic age, discussed below, probably reflects derivation of the quartz from the stable, low-lying interior of the Canadian Shield.

Occurrence and Examples

Quartz arenites are present in geologic systems ranging in age from Precambrian to Tertiary. Even some Holocene eolian, coastal plain, and fluvial sands are quartz-rich (e.g., see Potter, 1978). At least 60 formations composed totally or in part of quartz arenites are known from North America alone, and many more occur in other parts of the world. Distribution of North American quartz arenites by age is skewed toward those of Precambrian, Cambrian, and Ordovician ages. Some well-known examples include the Precambrian Baraboo Quartzite of Wisconsin and the Sioux Quartzite of Minnesota, the Cambrian Franconia Sandstone of the upper Mississippi Valley and the Potsdam Sandstone of New York, the Ordovician St. Peter Sandstone of the upper Mississippi Valley, and the Eureka Quartzite of Utah, Nevada, and California. Other well-known quartz arenites include the Silurian Tuscarora Quartzite of Pennsylvania and New Jersey, the

Devonian Oriskany Sandstone of Pennsylvania, the Pennsylvanian Tensleep Sandstone of Wyoming, the Permian Coconino Sandstone of the Colorado Plateau, the Triassic Wingate Sandstone of the Colorado Plateau, the Jurassic Entrada (in part) and Navajo sandstones of the Colorado Plateau, and parts of the Cretaceous Dakota Formation of the Colorado Plateau and the Great Plains. See Pettijohn et al. (1987, p. 179–184) for additional examples of quartz arenites from North America, Europe, and other parts of the world. For a well-described case history of a fairly typical quartz arenite, the Ordovician Kinnikinic Quartzite of Idaho, see James and Oaks (1977).

5.3.3   Feldspathic Arenites

General Characteristics

Feldspathic arenites contain less than 90 percent quartz grains, more feldspar than unstable rock fragments, and minor amounts of other minerals such as micas and heavy minerals. They may contain as little as 10 percent feldspar grains, but most feldspathic arenites show greater feldspar enrichment. In fact, rare feldspar values exceeding 80 percent have been reported (e.g., Crook, 1960). As discussed in Section 5.2, sandstones that contain more than about 25 percent feldspar are commonly called arkoses. Some feldspathic arenites are colored pink or red owing to the presence of K-feldspars or iron oxides; others are light gray to white. They are typically medium to coarse grained and may contain high percentages of subangular to angular grains. Matrix content may range from trace amounts to more than 15 percent, and sorting of framework grains can range from moderately well sorted to poorly sorted. Thus, feldspathic sandstones are commonly texturally immature or submature. Feldspathic arenites are not especially characterized by any particular kinds of sedimentary structures. Bedding may range from essentially structureless to parallel-laminated or cross-laminated. Fossils may be present in marine examples. Feldspathic arenites typically occur in cratonic or stable shelf settings, where they are associated with conglomerates, shallow-water quartz arenites or lithic arenites, carbonate rocks, and evaporites. Less typically, they occur in sedimentary sequences that were deposited in unstable basins or other deeper-water, mobile-belt settings. Feldspathic arenites of the latter type, which are matrix-rich and well indurated owing to deep burial, are often called feldspathic graywackes. Graywackes are further described in Section 5.3.4. The abundance of feldspathic arenites in the geologic record is not well established. Pettijohn (1963) estimates that arkoses make up about 15 percent of all sandstones. Feldspathic arenites in total are probably more abundant than 15 percent, especially if feldspathic graywackes are included.

Particle Composition

The particle composition of some representative feldspathic arenites is given in Table 5.3. Quartz is the predominant particle constituent in most feldspathic arenites and typically makes up about 50–60 percent of the framework grains. Reported quartz abundances range from about 10 percent to more than 75 percent. Note from Table 5.3 that quartz tends to be less abundant in plagioclase-rich feldspathic arenites than in K-feldspar–rich feldspathic arenites. Most quartz is monocrystalline. Polycrystalline grains generally make up less than 6 or 7 percent of the framework constituents. Blatt and Christie (1963) report an average value of 2.6 percent polycrystalline quartz in the feldspathic arenites they examined. Blatt and Christie's data on monocrystalline quartz in feldspathic arenites

TABLE 5.3

Average particle composition of some feldspathic arenites, as determined by petrographic analysis

	K-feldspar–rich arenites						Plagioclase-rich arenites						Tertiary/Quaternary sediments		
	A	B	C	D	E	F	G	H	I	J	K	L	M[4]	N	O
Total quartz	50.5	52.5	39.9	75.6	72.0	65.5	65.2	24.0	44.2	45.9	8.2	42.8	29.2	58.0	50.0
Monocrystalline			33.3	72.9			56.6			43.9			25.2		43.0
Polycrystalline			6.6	2.6			8.6			2.0			4.0		7.0
Total feldspars	34.3	35.2	35.6	14.9	23.0	21.9	18.0	34.2	43.2	26.3	75.6	52.4	36.9	17.0	29.0
Plagioclase	18.6	15.5	0.2	0.1	2.0	8.7	4.4	25.6	23.6	23.5	75.6	51.6	28.1	2.0	12.0
Orthoclase	14.6	19.7[1]	2.0	2.0	8.0	13.2[1]	11.3	8.6[1]	19.6[1]	2.8		0.8	8.8	15.0	17.0[1]
Microcline	1.1		33.4	12.8	13.0		2.3								
Chert	—	—	—	—	—	1.4	8.3	—	3.1	7.5	—	—	9.4	?	2.0
Silicic rock frag.	5.5	1.4	5.2	4.3	—	1.0	5.2[4]	3.6	—	—	—	—	—	9.0	1.0
Unstable rock frag.	0.8	3.5	12.4	4.8	5.0	7.3		32.2	4.0	13.6	4.7	—	—	2.0	14.0
Muscovite	2.6	2.7	1.8	0.3	<1.0	1.7	1.1	0.3	0.7	Tr	—	4.8[2]	24.6	Tr	—
Biotite	3.8	3.9	3.5	0.4	<1.0	—	0.8	2.0	1.4	1.7	—	—	—	Tr	—
Heavy minerals	0.6	1.0	1.2	0.3	3.0	0.1	1.5	2.9	2.1	1.7	1.5	—	?	8.0	1.0
Total framework grains	85.9	80.2	82.4	79.9	?	70.3	79.3	83.7	91.5	82.7	80.5	88.0	?	?	?
Silica cement	0.2	Tr	0.1	0.1	?	0.2	8.4	—	—	—	—	—			
Carbonate cement	11.4	19.8	2.6	6.0	?	20.3	1.9	—	2.9	Tr	14.9	—			
Other cement	—	—	0.6	3.0	?	—	3.8	5.6[3]	—	—	—	—			
Matrix	2.5	Tr	14.3	11.0	?	9.2	6.6	10.7	5.6	17.3	4.6	12.0			
Number of samples	68	70	49	32	42	131	46	10	26	38	3	?	23	10	26

Note: Composition of framework grains given as percent of total framework grains; cement and matrix shown as percent of total constituents. [1]may include microcline [2]may include biotite [3]type of cement not specified [4]0.074–0.25 mm size fraction

Source:

A. Boggs (1966), Minturn Formation (Pennsylvanian), Colorado (See text for explanation of feldspar ratios)
B. Bartleson (1968), Gothic Formation (Pennsylvanian), Colorado
C. Hubert (1960), Fountain Formation (Permo-Pennsylvanian), Colorado
D. Hubert (1960), Lyons Formation (Permian), Colorado
E. Higgs (1978), Mesozoic and Cenozoic sediments, Labrador and western Greenland continental margin
F. Cadigan (1971), Moenkopi sandstones (Triassic), Colorado Plateau
G. Phillips (1986), Arumbera sandstones (Precambrian), central Australia
H. Dickinson (1971), Permian and Jurassic sandstones, Eastern Axial Facies, New Zealand
I. Van de Kamp et al. (1976), Eocene and Oligocence arkoses, California
J. Lent (1969), feldspathic sandstones of Otter Point Formation (Jurassic), Oregon
K. Crook (1960), plagioclase-rich sandstones (Carboniferous), Australia
L. Huckenholz (1963), Permian arkoses, Auvergne, France
M. Stewart (1976), Paleogene turbidites, Aleutian abyssal plain
N. Hubert and Neal (1967), Holocene-Pleistocene deep-sea sands, western North Atlantic, Eastern Arkosic Subprovince
O. Combellick and Osborne (1977), Holocene beach sands, Monterey Bay (southern beaches), California

show an average of 73.8 undulose grains, as compared to 55.1 percent undulose quartz in quartz arenites. This difference suggests less recycling, or at least less selective destruction, of the undulose quartz grains that make up feldspathic arenites.

Total feldspar content of feldspathic arenites can range from as little as 10 percent to more than 75 percent of grains; a more typical range is 20 to 40 percent. Available data indicate that feldspathic arenites in which K-feldspars exceed plagioclase feldspars are most common; however, plagioclase is the predominant feldspar in many sandstones, particularly those from mobile-belt settings (Table 5.3). Pettijohn et al. (1987, p. 150) suggest that microcline is the most common K-feldspar in feldspathic arenites; however, K-feldspar identified as orthoclase equals or exceeds microcline in many feldspathic sandstones (Table 5.3). Plagioclase is typically sodic, in the range of An_5 to An_{50}, with oligoclase and andesine most common. Calcic plagioclases have been reported (e.g., Stewart, 1976). Both twinned and untwinned, and zoned and unzoned, plagioclases occur. The feldspars in feldspathic arenites may range from fresh, unaltered grains to those showing various degrees of alteration to sericite or kaolinite. Feldspar grains may also be stained with hematite and partially replaced by calcite or other minerals. Figure 5.10 shows a typical K-feldspar–rich arkose; Figure 5.11 is a plagioclase-rich feldspathic arenite.

Coarse muscovite and biotite are common minor constituents of feldspathic arenites. Muscovite is generally unaltered, but may be stained with hematite or limonite. Biotite grains may show alteration to hematite or chlorite. Heavy minerals commonly make up only 1 or 2 percent of the framework grains. In contrast to the quartz arenites, the heavy minerals in feldspathic arenites may include unstable or metastable types such as olivine, pyroxenes, amphiboles, magnetite, epidote, garnet, and kyanite, as well as ultrastable rutile, tourmaline, and zircon. Therefore, the ZTR index of heavy minerals in many of these sandstones is typically lower than that in quartz arenites. On the other hand, some feldspathic arenites contain very restricted suites of heavy minerals (e.g., Boggs, 1966), possibly owing to destruction of less-stable heavy minerals by intrastratal solution

FIGURE 5.10
Large microcline grain (grid twinning) in a K-feldspar–rich arkose. Fountain Formation (Pennsylvanian), Colorado. Crossed nicols. Scale bar = 0.5 mm.

FIGURE 5.11
Plagioclase grains (P) in a plagioclase-rich feldspathic arenite. Bushnell Rock Member of Lookingglass Formation (Eocene), southwest Oregon. Note alteration of the large plagioclase grain to sericite and clay minerals. Crossed nicols. Scale bar = 0.1 mm. *(Sample courtesy of R. L. Kugler.)*

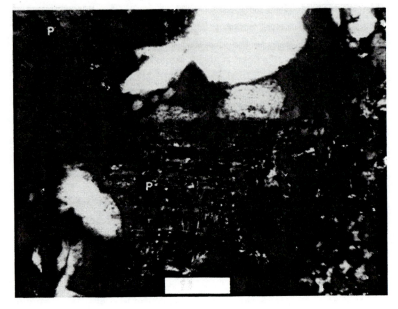

during diagenesis (see T. R. Walker, 1967). Rock fragments are common but are much less abundant than feldspars. These fragments may include quartzose clasts (mainly metaquartzite) and various kinds of less-stable plutonic igneous, volcanic, metamorphic, and sedimentary clasts. Rock fragments, especially volcanic rock fragments, appear to be most abundant in plagioclase-rich feldspathic arenites. Feldspathic arenites can grade compositionally into lithic arenites.

Cements

Carbonate cements appear to be the most common cements in feldspathic arenites and range from trace amounts to more than 20 percent of total constituents (Table 5.3). Carbonate cements are typically calcite but may include dolomite. The cement may occur as a mosaic of smaller crystals (Fig. 5.12) but more commonly occurs as large, single crystals that entirely fill pores or even surround grains. Other cements, which commonly occur in minor amounts, may include silica (quartz overgrowths), authigenic feldspars (overgrowths), hematite, and sulfate minerals such as barite, pyrite, and clay minerals.

Matrix

The matrix content of feldspathic arenites may range from trace amounts to more than 15 percent of total constituents. Many of these sandstones contain very little matrix and have correspondingly higher amounts of cement, commonly carbonate cement. The matrix in many feldspathic arenites appears to consist mainly of sericite and kaolinite, commonly stained with hematite. It was probably derived by alteration of biotite or other iron-bearing minerals. In others, the matrix materials may include also chlorite, fine quartz, feldspar, and organic matter. Although some matrix in feldspathic arenites may be detrital, or may have been introduced by vadose infiltration, much of it is probably derived by diagenetic alteration of feldspars, micas, and ferromagnesian minerals. Therefore, the matrix content of these rocks is probably not a reliable indicator of depositional conditions.

FIGURE 5.12

Mosaic calcite (center of photograph) filling pore space among grains of microcline, plagioclase, and quartz in an arkose. Fountain Formation (Pennsylvanian), Colorado. Crossed nicols. Scale bar = 0.1 mm.

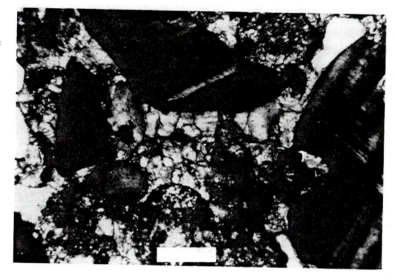

Texture

Feldspathic arenites are typically coarse grained, but grain size can range from coarse silt or very fine sand to very coarse sand. A relationship appears to exist between grain size and feldspar content. For example, Odom (1975) demonstrated a nearly linear relationship between the grain size and feldspar content of some Cambrian arenites, with feldspar content increasing with decreasing grain size. Cadigan (1967) reported a similar trend of increasing feldspar content with decreasing grain size in feldspathic arenites of the Jurassic Morrison Formation. On the other hand, Cadigan reported an opposite trend in one unit of the Morrison Formation. Also, Okada (1966) reported decreasing feldspar content with finer grain size in Silurian and Ordovician turbidites. The inverse correlation of increasing feldspar abundance with decreasing grain size in some sandstones may reflect the greater tendency of feldspars to break into smaller fragments (owing to the presence of twin planes, etc.) during sediment transport.

Sorting of framework constituents in feldspathic arenites can range from moderately well sorted to poorly sorted (Fig. 5.13), with moderate sorting perhaps most common. Both quartz and feldspars in feldspathic arenites are typically angular to subangular (Fig. 5.14). Subrounded to well-rounded quartz occurs in some feldspathic arenites, but much less commonly. Owing to the presence of matrix in many feldspathic arenites and generally poor rounding of these grains, most feldspathic arenites appear to be texturally immature to submature.

Chemical Composition

Relatively few recent chemical analyses of feldspathic arenites have been published. Table 5.4 summarizes some available data for both K-feldspar–rich arenites and plagioclase-rich arenites. Comparison of this table with Table 5.2 shows that feldspathic arenites contain markedly less silica and considerably more aluminum, sodium, and potassium than do quartz arenites. Silica content is strongly affected by quartz abundance; however, Table 5.4 shows a rather poor correlation overall between quartz content and total SiO_2 concentration. This poor correlation indicates that rock fragments, feldspars, and other silicate minerals also influence total silica content. Aluminum content is influ-

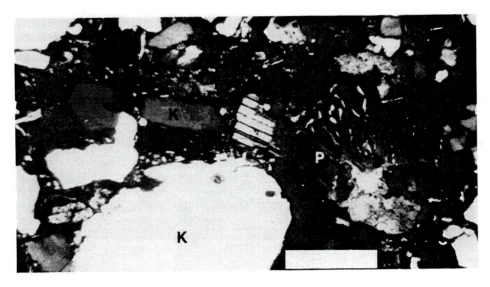

FIGURE 5.13
Very poorly sorted feldspathic arenite containing both K-feldspar (K) and plagioclase feldspar (P). Unknown formation. Crossed nicols. Scale bar = 0.5 mm.

enced particularly by feldspars, micas, and clay minerals, all of which are commonly more abundant in feldspathic arenites than in quartz arenites. Iron in both the oxidized (ferric) and reduced (ferrous) states is common in feldspathic arenites. The limited data in Table 5.4 suggest that Fe_2O_3 exceeds FeO in K-feldspar arenites, but FeO is more abundant in plagioclase arenites. This difference may reflect the greater tendency for K-feldspar arenites to form under subaerial conditions, as suggested by Pettijohn et al. (1987, p. 150). The high content of sodium and potassium in feldspathic arenites compared to that in quartz arenites is a function mainly of the greater content of sodium and

FIGURE 5.14
Poorly rounded feldspars (M = microcline, O = orthoclase, P = plagioclase) and quartz (Q) in a feldspathic arenite. Belt (Precambrian) facies, Montana. Crossed nicols. Scale bar = 0.5 mm.

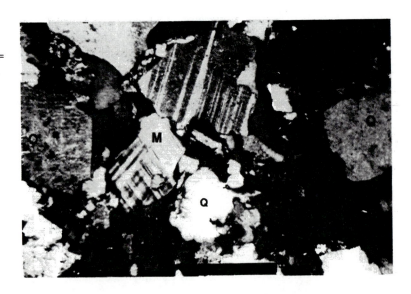

TABLE 5.4

Chemical composition of some feldspathic arenites; framework constituents, cements, and matrix shown also for reference

	K-feldspar-rich arenites					Plagioclase-rich arenites			
	A	*B*	*C*	*D*	*E*	*F*	*G*	*H*	*I*
Chemical constituents									
SiO_2	76.6	92.6	75.8	85.7	66.2	69.2	69.6	79.3	59.1
Al_2O_3	12.4	3.5	11.7	6.8	10.2	14.4	12.8	12.8	23.8
Fe_2O_3	0.7	0.4[1]	0.6	0.8[1]	7.0[1]	1.0	1.8	0.7	0.7
FeO	0.2		1.3			2.6	1.7	0.2	0.7
MgO	0.3	0.4	0.5	1.1	4.5	1.4	1.7	0.3	2.7
CaO	0.4	0.06	1.4	0.5	2.0	1.4	1.7	0.4	1.2
Na_2O	0.3	2.9[2]	2.4	1.2	1.8	1.9	2.7	4.8	4.7
K_2O	3.8		4.5	2.2	1.6	3.6	3.1	0.5	6.9
H_2O	2.7	0.2	1.0	—	—	2.2	1.8	1.0	—
TiO_2	0.6	0.2	0.2	0.4[3]	0.5	0.5	0.4	0.5	0.5
P_2O_5	0.2	0.02	0.6	0.01	—	0.2	0.2	0.1	0.1
MnO	—	—	0.06	—	—	0.08	0.06	—	0.4
CO_2	—	0.06	—	—	—	—	1.1	—	—
Other	1.7	—	—	—	6.2?	0.1	—	—	—
Framework constituents[4]									
Quartz	65	68	?	?	72	24	44	43	11
Feldspar	24	29	?	?	15	34	43	52	83
Rock frag.	0	?	?	?	6	36	6	0	3
Other grains	11	3	?	?	7	6	6	5	3
Cements[5]	2	?	?	?	8	6	3	—	18
Matrix[5]	18	?	?	?	38	11	6	12	—
No. of samples	?	5	3?	?	21	10	26	?	?

Note: [1]total iron [2]$Na_2O + K_2O$ [3]contains ZrO_2 [4]percent of framework constituents [5]percent of total constituents

Source:
A. Huckenholz (1963), Oligocene arkoses, Auvergne, France
B. Wiesnet (1961), Potsdam Sandstone (Cambrian), New York
C. Kennedy (1951), Torridonian (Precambrian), Kinlock, Skye
D. Swineford (1955), Whitehorse Group (Permian), Kansas
E. Condie et al. (1970), Fig Tree Group (Precambrian), South Africa
F. Dickson (1971), Permian-Jurassic sandstones of Eastern Asian Facies, New Zealand
G. Van de Kamp et al. (1976), Eocene and Oligocene arkoses, California
H. Huckenholz (1963), Permian arkoses, Auvergne, France
I. Crook (1974), Carboniferous sandstones of Australia

potassium feldspars in these rocks. Calcium and magnesium abundances are affected particularly by the content of carbonate cements.

Field Characteristics

As defined in this book, feldspathic arenites include all sandstones that contain less than 90 percent quartz and more feldspars than rock fragments. Thus, they encompass both classic arkoses (25 percent or more feldspars) and sandstones that are less feldspar-rich, some of which may be considered graywackes by some workers. Feldspathic arenites may be deposited under a wide range of conditions, discussed below, ranging from nonmarine to deep marine (turbidite). Therefore, feldspathic arenites may display considerable variation in their field characteristics.

Many feldspathic arenites appear to have been deposited in fluvial, lacustrine, or transitional-marine environments. These arkoses are typically red or pink and tend to form thick (to 2000 m or more), wedge-shaped units adjacent to ancient uplifts. These so-called aprons, or fans, commonly become thinner and finer grained basinward, where the arkoses may interfinger with finer-grained lacustrine or marine deposits. Arkoses may be interbedded with a variety of nonmarine to marine deposits, including conglomerates, shales, limestones, and evaporites. Individual arkose beds may range in thickness from a few centimeters to several meters and tend to be poorly sorted, irregularly bedded, and laterally discontinuous. Trough cross-bedding is common, and some cross-bed units display normal grading (Hubert, 1960). Other evidence of fluvial deposition such as cut-and-fill structures and plant fossils may be present. Feldspathic arenites deposited in marine environments tend to have less feldspar than nonmarine arkoses and are better sorted, more evenly bedded, and laterally continuous. Also, they may contain marine fossils. Feldspathic arenites deposited by turbidity currents may display graded bedding and sole markings of various types.

Origin

Feldspathic arenites originate mainly by weathering of feldspar-rich crystalline rocks, either plutonic igneous rocks or feldspar-rich metamorphic rocks. Therefore, most feldspathic arenites are probably first-cycle deposits. Most reported feldspathic arenites contain considerably more K-feldspar than plagioclase; however, several plagioclase feldspathic arenites are known. These plagioclase-rich feldspathic arenites are derived mainly from volcanic sources. Most plagioclase in these sandstones is sodic, commonly oligoclase and andesine. The feldspar composition of feldspathic arenites thus suggests that most of these sandstones were derived from acid (felsic) to intermediate crystalline rocks. Rarely, plagioclase arenites contain calcic plagioclase, indicating derivation from more-basic igneous rocks. The preservation of large quantities of feldspars during the process of weathering appears to require that feldspathic arenites originate either (1) under very cold or very arid climatic conditions, where chemical weathering processes are inhibited, or (2) in warmer, more-humid climates where marked relief of local uplifts allows rapid erosion of feldspars before they can be decomposed. Dickinson and Suczek (1979) and Dickinson (1984) suggest that feldspar-rich sandstones are derived especially from fault-bounded, uplifted basement areas in continental-block provenances, where high relief and rapid erosion of uplifted sources gives rise to quartzo-feldspathic sands of classic arkosic character. These sandstones may accumulate in basins related to transform ruptures of continental blocks, incipient rift blocks, or zones of wrench tectonism within continental interiors. Some feldspathic arenites, particularly plagioclase-rich arenites, could conceivably be derived from dissected magmatic-arc settings, where they might be deposited in forearc or backarc basins.

As discussed above, most feldspathic arenites, particularly arkoses, appear to have been deposited as clastic wedges or fans very close to their sources. Pettijohn et al. (1987, p. 152) suggest that some arkoses are *in situ,* or residual, deposits that formed essentially in place by disintegration of coarse crystalline rocks. These residual deposits tend to have very poor sorting, high detrital matrix content, and very angular grains. Residual arkosic debris shifted downslope short distances by mass-transport processes probably has much the same characteristics as residual arkoses. On the other hand, arkosic sediment transported greater distances by traction currents is "cleaned up." Thus, transported arkoses display better sorting and various degrees of grain rounding. Less detrital matrix is likely present in these transported arkoses than in residual arkoses, but the transported arkoses may contain considerable diagenetic matrix. Arkosic material transported into marine environments probably undergoes additional sorting and perhaps mixing of sediment from other sources, which may dilute the feldspar content below that typical of residual arkoses. Feldspathic sediments may be retransported by turbidity currents into deep water. Well-indurated feldspathic turbidites are sometimes referred to as graywackes, but they are nonetheless feldspathic arenites.

Occurrence and Examples

Feldspathic arenites are present in geologic formations ranging in age from Precambrian to Holocene. They do not appear to be especially characteristic of any particular geologic periods. In contrast to the quartz arenites, they tend to be of local rather than regional extent. They apparently formed whenever and wherever tectonic processes generated marked uplifts of crystalline basement rocks, allowing rapid erosional stripping of these tectonic blocks and accumulation of thick, localized deposits of arkosic debris in adjacent, subsiding basins. Some formations contain feldspathic arenites only at their bases. These basal feldspar-rich units grade upward into either lithic arenites (e.g., the Cretaceous Hornbrook Formation in southern Oregon) or quartz arenites (e.g., the Cambrian Lamotte Sandstone of Missouri; Ojakangas, 1963). Some well-known U.S. examples of feldspathic arenites include the Precambrian Torridon sandstones of Scotland, the Silurian Clinton Formation of Pennsylvania, the Devonian Old Red Sandstone of England, and the Triassic Newark Group of Connecticut. See Boggs (1966) and Tweto and Lovering (1977) for case histories of a typical K-feldspar-rich feldspathic arenite, the Pennsylvanian Minturn Formation of Colorado. A good example of a plagioclase-rich arkose is the Paleocene Swauk Formation of Washington (Foster, 1960).

5.3.4   Lithic Arenites

General Characteristics

Lithic arenites are an extremely diverse group of rocks that are characterized by generally high content of unstable rock fragments. Classified according to Figure 5.1, any sandstone that contains less than 90 percent quartz (plus chert and quartzite) and unstable rock fragments in excess of feldspars is a lithic arenite. Colors may range from light gray, "salt and pepper," to uniform medium to dark gray. Many lithic arenites are poorly sorted; however, sorting ranges from well sorted to very poorly sorted. Quartz and many other framework grains are generally poorly rounded. Lithic arenites tend to contain substantial amounts of matrix, most of which may be of secondary origin. Lithic arenites may range from irregularly bedded, laterally restricted, cross-stratified fluvial units to evenly bedded, laterally extensive, graded, marine turbidite units. They occur in association with fluvial conglomerates and other fluvial deposits and in association with generally deeper-water, marine conglomerates, pelagic shales, cherts, and submarine basalts.

Classified as mentioned, lithic arenites include many sandstones that are called **graywackes**. Some authors (e.g., Pettijohn et al., 1987) consider graywackes to be distinctly different from lithic arenites (or feldspathic arenites) and treat them as different groups of rocks. The origin, meaning, and usage of the term graywacke remain controversial. Even the spelling of the term varies: greywacke vs. graywacke. The "graywacke problem" has been discussed by several workers, including Cummins (1962), Dott (1964), Crook (1970), and Pettijohn et al. (1987, p. 163). According to Crook (1970), the name seems to have been mentioned first by Werner about 1787; however, the "type" graywacke is generally regarded to be the Upper Devonian–Lower Carboniferous Kulm strata of the Harz Mountains in Germany, studied by Lasius in 1789 (see Dott, 1964). Although consistent usage of the term graywacke is still lacking, it appears to be used mainly for sandstones that are dark gray or dark green, well indurated or lithified, and matrix rich. The matrix, often referred to as a chloritic "paste," tends to pervade the rock and obscure the boundaries of rock fragments and other grains, making identification of grains difficult. Most graywackes appear to be turbidites and thus are associated with deep-water deposits of various types.

As indicated, graywackes are defined largely on the basis of field characteristics and matrix content, rather than on particle composition, and they all look very much alike in overall field appearance. When examined petrographically, they turn out to be mainly lithic wackes or feldspathic wackes. Graywackes that classify as quartz wackes are rare. In this book, I do not treat graywackes as a separate group of sandstones. Rather, I regard them to be special types of either lithic or feldspathic arenites and include them in discussion of these sandstone types. One can, of course, attempt to classify them petrographically as feldspathic, lithic, or quartzose graywackes, as suggested by Gilbert (Williams et al., 1982, p. 329), provided that the difference between wackes and graywackes is a recognizable one. Some workers have suggested that we abandon any precise definition of graywacke and simply use these terms for imprecise field descriptions (e.g., Dickinson, 1970). Pettijohn (1963) estimates that lithic arenites and graywackes together make up nearly one-half of all sandstones.

Particle Composition

Sandstones that classify as lithic arenites may have highly variable compositions, as indicated in Table 5.5. Additional petrographic data are given in Pettijohn et al. (1987, p. 157, 165). The quartz-grain content of lithic arenites can range from less than 1 percent to more than 70 percent, and no particular range of quartz values seems typical. Available data suggest that the percentage of polycrystalline quartz is much higher than that in either feldspathic arenites or quartz arenites. Blatt and Christie (1963) give an average value of 9.7 percent polycrystalline quartz in the graywackes (lithic arenites) they studied. Feldspar content may range from less than 1 percent to more than 25 percent of grains. Plagioclase predominates over K-feldspars in many lithic arenites. Plagioclase is typically sodic, ranging in composition from about An_{15} to An_{35} (albite to andesine). Among the K-feldspars, orthoclase appears to be more common than microcline.

Rock fragments are clearly the most distinctive feature of lithic arenites. Both stable, e.g., chert, and unstable fragments may be present. Although detrital chert is included with quartz in some sandstone classifications (e.g., Fig. 5.1; Fig. 5.2A), chert is nonetheless a rock fragment. The chert content of some lithic arenites is shown in Table 5.5. Many lithic arenites contain only minor amounts of chert; however, chert concentrations ranging to 75 percent or more have been reported. Examples of chert-rich sandstones include some sandstones of the Jurassic Otter Point Formation in Oregon (Lent, 1969), the Jurassic Morrison Formation of Montana (Suttner, 1969), and the Cretaceous

Cut Bank Sandstone of Montana (Sloss and Feray, 1948). Sandstones that have a very high content of chert grains create an awkward classification problem when one uses a classification such as Gilbert's (Fig. 5.1), which includes chert with quartz grains at the Q pole of the classification diagram. Such sandstones could conceivably be classified as either a quartz arenite or feldspathic arenite, depending upon the relative abundance of quartz and feldspar grains in the sandstones, even though rock fragments (chert) may make up the bulk of the framework constituents. Figure 5.15 shows chert clasts in a chert-rich sandstone from the Otter Point Formation in southwestern Oregon.

It is, of course, the presence of abundant unstable, fine-grained rock fragments that particularly characterize lithic arenites. The abundance of these rock fragments may range from less than 10 percent to as much as 80 percent (Table 5.5), but abundances of 20–40 percent are more typical. Volcanic rock fragments (Fig. 5.16) are especially common in lithic arenites derived from magmatic-arc settings. Granite and other coarse plutonic igneous rock fragments are reported in some lithic arenites (Table 5.5), but generally in very minor amounts. Common metamorphic rock fragments include slate, phyllite, and schist (Fig. 5.17). Coarser crystalline gneiss fragments are much less common, as are serpentinite fragments. Sedimentary rock fragments include fine-grained sandstone (Fig. 5.18), siltstone, shale (Fig. 5.19), and, less commonly, carbonates. Owing to the large variety of rock fragments that may be present in lithic sandstones, some workers have suggested subdividing these sandstones on the basis of rock-fragment type (e.g., Folk, 1974, p. 129). This procedure yields such special names as volcanic arenite, phyllarenite, chert arenite, and calcilithite (composed predominantly of carbonate clasts), depending upon rock-fragment composition.

Micas are absent or very scarce in some lithic arenites; however, they are very abundant in others, particularly some fluvial sandstones. Rarely, mica values exceeding 10 percent have been reported (Table 5.5). Muscovite appears to be somewhat more common than biotite. Heavy minerals are present in most lithic arenites in amounts ranging from traces to more than 10 percent. Unusually high heavy-mineral contents ranging to 25 percent are reported by Lerbekmo (1961) in some Tertiary sandstones in central California. Although some lithic arenites contain mainly ultrastable zircon, tourmaline, and rutile, most contain a variety of unstable heavy minerals. These less-stable heavy minerals may include epidote, sphene, garnet, hornblende, glaucophane, clinopyroxenes, orthopyroxenes, apatite, fluorite, pumpellyite, clinozoisite, laumontite, lawsonite, cordierite, magnetite, pyrite, leucoxene, chromite, and corundum. Thus, the ZTR index of these sandstones is generally low compared to that of quartz arenites. Many lithic arenites of Devonian or younger age contain charcoal fragments or opaque organic material that is probably mostly macerated plant debris. In some lithic arenites, fine organic matter is concentrated in thin laminae that give the sandstones a laminated appearance.

Cements and Matrix

The average content of cement and matrix in some representative lithic arenites is given in Table 5.5. Some lithic arenites contain no visible cements, whereas others contain cements in amounts ranging to more than 30 percent of total constituents. Carbonate cements appear to be most common and may include calcite, dolomite, and siderite. Silica cement (quartz overgrowths) is moderately abundant in some lithic arenites. Other common cements are chlorite, clay minerals, iron oxides, and pyrite. The matrix content of lithic arenites ranges from trace amounts to as much as 40 percent, although most contain less than 20 percent matrix.

TABLE 5.5
Average particle composition of lithic arenites

	A	B	C	D	E	F	G	H	I	J	K	L	M	N	O	P	Q	R
Total quartz	36.8	67.0	9.3	0.5	56.4	72.5	11	8.5	47.6	26.3	62.1	39.6	46.0	41.7	41.1	46.0	50.8	2.4
Monocrystalline					30.9	52.6	5	6.8		21.7				28.9		32.8		
Polycrystalline					25.5	19.9	6	1.7		4.6				12.8		13.2		
Total feldspar	13.9	12.0	15.2	15.9	5.2	7.2[6]	13	25.7	0.3	13.9	6.4	4.0	12.6	17.2	13.4	25.4		23.7
Plagioclase	5.1	4.4	9.5	15.9	1.1		12	25.6	0.3	11.8	6.4	4.0	12.6	14.2	6.3	4.3		23.7
Orthoclase	8.8[1]	5.4	5.7[1]		4.1[1]		1[1]	0.1[1]		2.1[1]		?	?	3.0[1]	7.1[1]			
Microcline		2.2														21.1		
Chert		Trace?	19.3		1.2	1.0	38	5.2	21.7	14.8	6.6	15.8	1.9	6.4	19.3		24.5	
Silicic rock frag.			6.7												0.8			
Unstable rock frag.	35.9[2]	19.7[3]	44.8	79.8	37.3	8.6[7]	37	64.9	28.9	36.0	11.2[11]	27.6	28.0	30.1	24.2	26.5[16]	22.4	69.9
Volcanic			17.9[2]	79.8			27	52.3	4.8	19.1				20.3	12.1			66.6
Metamorphic			15.1		30.5		10	0.3	5.7	11.8		6.7	27.9	6.4			4.7	
Sedimentary			11.8		6.8			12.3	18.4	5.1		20.9	0.1	3.4	12.1		17.7	3.3
Muscovite		0.7[4]				5.6[4]		Trace		Trace		1.4	10.6	0.5	0.7[4]			
Biotite								Trace		0.7				0.5				
Heavy minerals	12.6	Pres.			Pres.	Pres.		1.0	1.5	2.8		2.9	0.3	3.7	0.7			4.1

	A	B	C	D	E	F	G	H	I	J	K	L	M	N	O	P	Q	R
Other grains	0.7	0.6	4.7	3.8		5.1[8]	1			5.5[10]	13.7[12]	8.9[15]	0.6			2.2[17]	2.3	
Total framework gr.	67.9	68.1	82.5	78.7	84.1	89.5	93.0	76.7	67.9	78.7	80.7	54.9	76.5	76.1	75.3	69.3	85.4	79.2
Silica cement	0.8				3.4[3]	2.4			6.0		7.8[13]					0.59	5.3	2.7
Carbonate cement		15.3		~1.0		4.4		6.5[9]	12.0			35.3	4.3	1.0	19.2	1.26	6.5[18]	1.1
Other cement									1.2								0.5	8.9[19]
Matrix	31.3	16.6	17.5	20.3	12.5	3.7	7	16.8	12.9	21.3	11.5	9.8	19.2	22.9	5.5	28.85	2.3	8.1
Number of samples	119	19	13	31	10	28	11	63	47	100	225	?	?	72	22	20	63	9

Note: Composition of framework grains shown as percent of framework constituents; cements and matrix shown as percent of total constituents.

A. Ondrick and Griffiths (1969), Rensselaer Graywacke (Cambrian/Ordovician), New York
B. Hiscott (1978), Tourelle Formation (Ord.), Quebec, Canada
C. Condie and Snansieng (1971), Gazelle Formation (Silurian), California
D. Chappell (1968), Baldwin Formation (Devonian), New South Wales, Australia
E. Jones (1972), lithic arenites of the Trenchard Group (Carboniferous), England (fluvial)
F. Hoque (1968), Mauch Chunk sandstones (Miss.), Pennsylvania (fluvial)
G. Dickinson et al. (1979), Vesper Formation (Upper Triassic), Oregon
H. Goodfellow (1987), Otter Point Formation (Jurassic), Oregon.
I. Rapson (1965), Jurassic-Cretaceous rocks, southern Rocky Mountains, Canada

J. Lent (1969), Rocky Point Formation (Cretaceous), Oregon
K. Ranganathan and Tye (1984), Shannon Sandstone (Cret.), Wyoming
L. Sestini (1970), Ostia Fm. (Cretaceous), N. Apennines, Italy
M. Sestini (1970), Monte Senario Fm. (Oligocene), N. Apennines, Italy
N. Kugler (1979), Lookingglass Fm. (Eocene), Oregon.
O. Nanz (1954), Oligocene "Frio" Sandstone, Texas
P. Morad (1984), Visingsö Group, Upper Proterozoic, Sweden
Q. Morris (1987), Ivishak Formation (Triassic), Alaska
R. Boles (1974), North Range Group (north peak of Glenore Formation) (Triassic), New Zealand

[1] may include some micro line
[2] includes granite and gneiss fragments
[3] includes sediments, volcanic, and minor plutonic igneous clasts
[4] may include biotite
[5] undifferentiated cement
[6] includes both plagioclase and K-feldspars
[7] includes sedimentary and metamorphic rock fragments
[8] includes some heavy minerals
[9] includes some replacement calcite
[10] includes muscovite, opaque organic material, and calcite
[11] includes granite, gneiss, and shale
[12] includes glauconite, siderite, and grains replaced by calcite
[13] includes calcite, siderite, quartz, chlorite
[14] mainly carbonate clasts
[15] includes detrital calcite and fossils
[16] argillaceous and granite clasts
[17] includes heavy minerals and micas
[18] mainly siderite
[19] mainly chlorite

FIGURE 5.15
Detrital chert grains (C) in a chert-rich lithic arenite. Otter Point Formation (Jurassic), southwest Oregon. Crossed nicols. Scale bar = 0.5 mm. *(Sample courtesy of R. L. Lent.)*

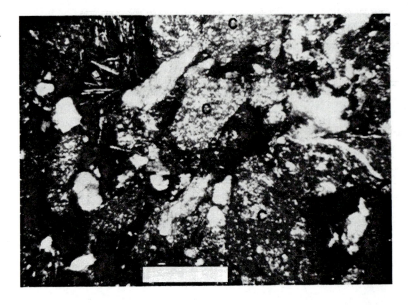

FIGURE 5.16
Large volcanic clast showing plagioclase laths set in a glassy groundmass. Miocene volcaniclastic sandstones, ODP Leg 127, Site 796, Japan Sea (depth 283 m below seafloor). Crossed nicols. Scale bar = 0.1 mm.

As discussed in Section 4.2.3, clay matrix and clay cements are difficult to distinguish under a petrographic microscope, unless the clays are fairly coarse grained and exhibit distinctive cement textures such as the drusy texture shown in Figure 4.14. This problem of identification is compounded owing to the authigenic origin of much matrix. So-called authigenic matrix that fills original pore space or pore space created by dissolution of cement or framework grains is, in fact, cement, although it may not be recognizable as such under a petrographic microscope. Other authigenic matrix forms by replacing framework grains or cements. Finally, some authigenic matrix is simply original detrital or infiltrated matrix that has undergone recrystallization. One can attempt to use

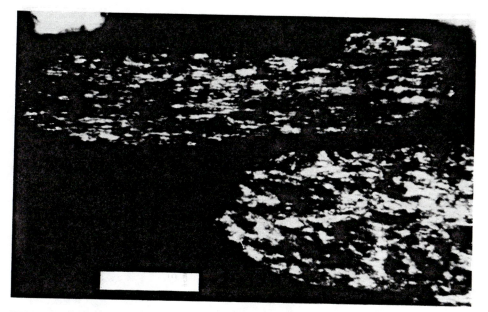

FIGURE 5.17
Large metamorphic fragments (fine schist/phyllite), modern shelf sediments, Taiwan Strait. Crossed nicols. Scale bar = 0.1 mm.

FIGURE 5.18
Large, well-rounded sandstone clast. Modern shelf sediment, Taiwan Strait. Crossed nicols. Scale bar = 0.5 mm.

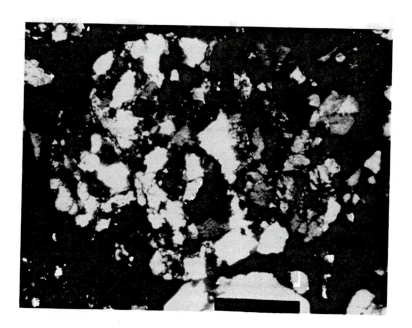

FIGURE 5.19
Silty shale (mudstone) clast,
modern shelf sediment, Taiwan
Strait. Crossed nicols. Scale bar =
0.5 mm.

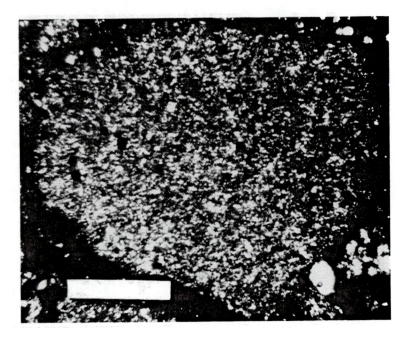

Dickinson's (1970) criteria (see Section 4.2.3) to distinguish between clay cements and
clay matrix; however, even with these criteria it is commonly difficult to discriminate
between clay cements and clay matrix. Also, the boundaries between lithic fragments and
matrix in lithic arenites that have undergone considerable diagenetic modification may be
very fuzzy (Fig. 5.20), making it difficult to differentiate matrix from some very fine-
grained lithic fragments. Squeezing of soft rock fragments to produce pseudomatrix (Fig.
5.21) creates another identification problem. Owing to these analytical difficulties, it
seems likely that the matrix content of many lithic arenites, particularly the lithic graywackes, is overestimated by petrographic methods.

The data in Table 5.5 suggest that lithic arenites deposited in fluvial or nearshore
marine environments may have less matrix on the average than deeper-water, graywacke-
type lithic arenites. This difference may simply reflect different diagenetic histories rather
than significant differences in original, detrital matrix content. Therefore, we should be
extremely careful, as Walker et al. (1978) point out, about making environmental interpretation on the basis of matrix content of sandstones.

Texture
Many lithic sandstones are coarse grained; however, the grain size of these rocks may
range from very fine sand to very coarse sand or granules. Classified according to Folk's
(1951) textural maturity classification, most lithic arenites are texturally immature owing
to their high matrix content. If we ignore matrix as being largely authigenic and classify
arenites on the basis of the sorting and rounding characteristics of the framework grains,
they are mainly submature to mature. Sorting of framework grains may range from well
sorted to very poorly sorted. There appears to be some tendency for increasing grain size
to correlate with poorer sorting (e.g., Jones, 1972). Although lithic arenites derived from
sedimentary source rocks may contain moderately well rounded quartz, or quartz with
rounded outlines beneath inherited overgrowths, the roundness of quartz in most lithic

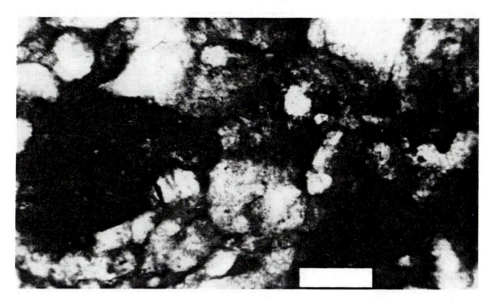

FIGURE 5.20
Lithic arenite with indistinct boundaries between matrix and many of the lithic fragments.
Dothan Formation (Jurassic), southwest Oregon. Crossed nicols. Scale bar = 0.1 mm.

FIGURE 5.21
Pseudomatrix in a lithic arenite.
Note how the pseudomatrix (black)
is squeezed among the more rigid
grains owing to plastic deformation
during compaction. Moenave
Formation (Triassic), Colorado
Plateau. Ordinary light. Scale bar
= 0.1 mm.

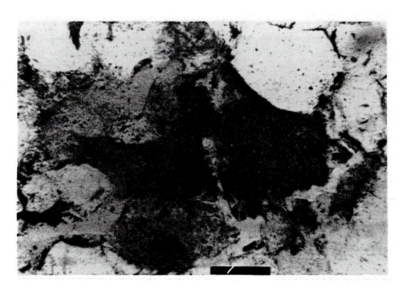

arenites tends to range between angular to subrounded. Subangular quartz is probably
most common. Lithic fragments and other framework grains in lithic arenites are also
commonly subangular to subrounded. Few lithic arenites are composed predominantly of
well-rounded grains.

Chemical Composition

The chemical composition of lithic arenites is quite variable (Table 5.6), and the composition of some lithic arenites is very similar to that of some feldspathic arenites (Table

5.4). SiO_2 values are moderate and commonly range between 50 and 70 percent, whereas Al_2O_3 values tend to be high and commonly exceed 10 percent. Iron, potassium, and magnesium values are moderately high also. The high Al, Fe, K, and Mg concentrations may reflect the generally high clay content of lithic arenites. Argast and Donnelly (1987) suggest that Al_2O_3, K_2O, Fe_2O_3, and MgO all tend to be enriched in the fine-grained, phyllosilicate-rich fraction of sandstones, whereas SiO_2 and Na_2O concentrations are related to the coarser-grained, tectosilicate fraction. The Na_2O content is probably strongly correlated to the albitic composition of sodium-rich plagioclase feldspars. Magnesium content can be affected by the presence of detrital dolomite clasts and dolomite cements, and calcium abundance is affected by detrital limestone clasts and calcite cements.

Origin and Field Characteristics

The high content of unstable rock fragments and the moderately high feldspar content of lithic arenites suggest that they are derived from rugged, high-relief source areas. Detritus is stripped rapidly from these elevated areas before weathering processes can destroy unstable clasts and other framework grains. Furthermore, most lithic arenites contain fine-grained clasts derived from source regions composed mostly of fine-grained rocks, that is, volcanic rocks, schists, phyllites, slates, fine-grained sandstones, shales, and limestones. Source areas with these characteristics occur primarily in orogenic belts located along the suture zones of collision plates and in magmatic arcs (Dickinson and Suczek, 1979; Dickinson, 1984).

Lithic arenites are probably derived only rarely from continental-block provenances, which typically yield quartz arenites and feldspathic arenites. Lithic arenites derived from recycled orogen provenances that formed by collision of continental blocks may be deposited in proximate alluvial fans or other fluvial environments. They are light gray, ''salt and pepper,'' or brown sandstones that display bedding characteristics ranging from evenly bedded to irregular and laterally discontinuous. Both tabular and trough stratification may be prominent. Scour-and-fill structures, current lineation, and ripple marks are common sedimentary structures. Many beds may be highly micaceous. Fluvial lithic arenites are commonly associated with thin shales and conglomerates and may be locally very coarse grained.

In some settings, deposition can take place in marine foreland basins adjacent to fold-thrust belts. Alternatively, lithic detritus may be transported by large rivers off the continent into deltaic or shallow, pericontinental shelf environments. Lithic arenites deposited in deltaic environments can have many characteristics in common with alluvial deposits, but the deltaic sands tend to be somewhat better sorted and more evenly bedded. They may be interbedded with shales containing marine fossils or with tidal flat or marsh deposits. Lithic arenites deposited in shallow-marine environments are commonly more evenly bedded and laterally persistent than fluvial and deltaic sandstones. Horizontal lamination and both tabular and trough cross-bedding may be common. Some units may display hummocky cross-stratification. Ripple marks are common, especially oscillation ripples. Marine fossils may be present, and bioturbation structures can be scarce to abundant. Macroscopic plant fragments may be locally abundant in marine lithic arenites of Devonian age or younger. Shallow-marine shelf arenites may be interbedded with shelf muds that are commonly bioturbated also.

Lithic arenites derived from magmatic-arc settings are typically enriched in volcanic rock fragments and plagioclase feldspars. Lithic detritus may be deposited in nonmarine settings within intra-arc basins, but most is probably carried by rivers to coastal areas. There, much of it is retransported into deeper water by turbidity currents or by other sediment-gravity-flow mechanisms. Ultimately, this detritus is deposited in forearc basins, backarc basins, or subduction-zone trenches. Lithic arenites deposited in these settings are particularly likely to undergo deep burial and incipient metamorphism, leading to development of characteristics generally ascribed to graywackes. They are typically dark gray, dark green, or black, well indurated and well bedded. The bedding may be repetitious or rhythmic (Fig. 5.22) and laterally extensive. Individual sandstone beds may display distinct vertical size grading and Bouma sequences (Fig. 3.5). Lithic arenites of turbidite origin are typically interbedded with pelagic clays and may also be associated with resedimented conglomerates, bedded cherts, and submarine basalts.

Examples and Case Histories

Introduction.   Lithic arenites are very abundant in the geologic record. Little purpose is served by a simple listing of the names of the many formations composed of lithic arenites; however, the examples shown in Tables 5.5 and 5.6 provide some insight into the variability of lithic arenites deposited in different environmental settings. Fluvial lithic arenites shown in these tables include the Mississippian Mauch Chunk Formation of Pennsylvania, sandstones of the Carboniferous Trenchard Group in England, some Cretaceous and Jurassic sandstones of the southern Rocky Mountains of Canada, and some

FIGURE 5.22
Rhythmic bedding in thin, graded lithic arenites of the Roseburg Formation (Eocene), southern Oregon Coast Range. The beds are tilted at an angle of about 70 degrees. Note the hammer for scale.

TABLE 5.6

Average chemical composition of some lithic arenites; the average percentages of framework grains, cements, and matrix given also for reference

	A	B	C	D	E	F	G	H	I
SiO_2	64.6	71.9	70.6	63.8	57.6	55.8	84.0	65.0	59.9
TiO_2	0.7	0.5	0.6	0.8	1.2	1.6	<0.1		0.9
Al_2O_3	13.5	12.4	12.6	12.9	16.5	15.6	2.6	9.6	17.1
Fe_2O_3	8.4[1]	6.4[1]	5.0[1]	8.4[1]	9.3[1]	3.4	0.2	1.6	2.4
FeO						5.0	0.3	1.1	4.3
MgO	2.5	1.7	1.5		3.9	3.06	0.7	<0.1	1.9
Cao	1.8	0.1	1.6	2.9	5.3	4.07	5.4	10.1	1.6
Na_2O	1.9	0.1	2.8	2.6	5.0	4.2	0.2	2.1	1.4
K_2O	2.0	4.2	2.2	1.1	0.7	1.24	0.9	1.4	4.0
H_2O						4.99	0.7	1.1	6.0
P_2O_5	0.1	0.1				0.23	<0.1		
MnO	0.1	0.1	0.1	0.1		0.12	<0.1		0.05
CO_2						0.36	4.7	6.9	
Other			0.5			0.04	Tr		
Totals reported	95.6	97.5	93.0	92.6	99.5	98.8	99.7	98.9	99.55
Quartz + chert[2]	?	?	36.8	28.6	0.5	2.4	86	60.4	46.0
Feldspars[2]	?	?	13.9	15.2	15.9	23.7		13.3	25.4
Rock fragments[2]	?	?	35.9	51.5	79.8	69.9	14	25.0	26.5
Other grains[2]	?	?	13.4	4.7	3.8	4.1		1.4	2.2
Cement[6]	?	?				12.7[5]	20.4[3]	19.2[4]	1.8[5]
Matrix[6]	>15	>15	31.3	17.5	20.3	8.7	6.8	5.5	28.9
Number of samples	46	22	119	13	10	10	96	10	6

Note: [1] total iron
[2] normalized to 100% framework grains
[3] includes silica and carbonate cement
[4] carbonate cement
[5] undifferentiated cement
[6] percent of total constituents

Source:
A. Argast and Donnelly (1986), Dhwar Supergroup (Precambrian), India, high-sodium facies (may include some feldspathic graywacke)
B. Argast and Donnelly (1986), Dhwar Supergroup (Precambrian), India, low-sodium facies (may include some feldspathic graywacke)
C. Ondrick and Griffiths (1969), Rensselaer Graywacke (Cambrian/Ord.), New York
D. Condie and Snansieng (1971), Gazelle Formation (Silurian), California
E. Chappell (1968), Baldwin Formation (Dev.), Australia
F. Boles (1974), North Range Group (Malakovian or older) (Triassic), New Zealand
G. Reported in Pettijohn (1963, Tables 3 and 4), Saltwash Member of Morrison Formation (Jurassic), Colorado Plateau
H. Nanz (1954), Frio Fomation (Oligocene), Zone 19B sands, Texas
I. Morad (1984), Vishingö Group (Upper Proterozoic), Sweden

sandstones of the Jurassic Morrison Formation in the Colorado Plateau. Probable deltaic lithic arenites in these tables include the Oligocene Frio Formation of Texas and the Triassic Ivishak Formation in Alaska. Shallow-marine lithic arenites include the Cretaceous Shannon Sandstone of Wyoming, the Eocene Bushnell Rock Member of the Lookingglass Formation in Oregon (fluvial to shallow marine), and the North Peak and Glenure formations of New Zealand. Most of the remaining lithic arenites shown in Tables 5.5 and 5.6 are deeper-water turbidites. Overall, turbidites are probably the most abundant type of lithic sandstones.

5.4   Volcaniclastic Sandstones

5.4.1   Introduction

The term volcaniclastic is applied to all siliciclastic rocks enriched in volcanic fragments regardless of the mechanism that produced the fragments. Volcaniclastic deposits can be emplaced or deposited in any environment—on land, under water, or under ice—and may be mixed in any significant proportion with any nonvolcanic fragment types (Fisher, 1961, 1966). As thus defined, volcaniclastic deposits include both epiclastic sedimentary rocks, made up of products generated by fragmentation of preexisting volcanic rocks owing to weathering and erosion, and rocks formed by primary volcanic processes. Primary volcanic processes include pyroclastic eruptions and autoclastic processes.

Volcaniclastic rocks composed of abundant sand-size, epiclastic volcanic clasts are simply volcanic-derived epiclastic sandstones. These sandstones are studied and classified as any other epiclastic sandstones and differ from other sandstones primarily in their high content of volcanic-derived particles. They classify as either lithic arenites or feldspathic arenites, depending upon feldspar/rock fragment ratios. Sand-size volcaniclastic rocks that form by primary volcanic processes are called tuffs and are commonly regarded to be igneous rocks rather than sedimentary rocks. Such rocks are not, however, wholly igneous. They originate by igneous processes (e.g., magmatic explosions), but they are deposited by sedimentary processes (e.g., airfall). Owing to their mixed origin, the classification and genetic affinities of these rocks are intriguing problems.

Because both sedimentary petrologists and igneous petrologists are interested in volcaniclastic rocks, considerable information is now available regarding their origin and characteristics. Publications dealing with nomenclature and classification include those of Fisher (1961, 1966), Schmid (1981), G. P. L. Walker (1973), Wentworth and Williams (1932), and Wright et al. (1980). Important books that treat volcaniclastic sediments include the works of Cas and Wright (1987), Fisher and Schmincke (1984), and Rittman (1962). Useful short summaries of volcaniclastic rocks are those of Lajoie (1984) and Pettijohn et al. (1987, Ch. 6). Pyroclastic deposits are discussed by Williams and McBirney (1979, Ch. 6, 7) and Williams et al. (1982, Ch. 9).

5.4.2   Processes That Form Volcaniclastic Rocks

Primary Volcanic Processes

Pyroclastic Eruptions.   As mentioned, the processes that bring about fragmentation of volcanic rocks to form volcaniclastic deposits include both primary volcanic processes and the secondary surficial processes of weathering and erosion. Primary volcanic processes include pyroclastic eruptions and autoclastic processes (Table 5.7). Pyroclastic eruptions are explosive eruptions, either magmatic explosions or phreatic (steam) explo-

TABLE 5.7

Processes that bring about fragmentation of volcanic rocks

Magmatic explosions Phreatic or steam explosions Phreatomagmatic explosions	PYROCLASTIC ERUPTIONS	PRIMARY VOLCANIC PROCESSES
Quench- or chill-shatter fragmentation Flow fragmentation (autobrecciation)	AUTOCLASTIC PROCESSES	
Epiclastic fragmentation		SECONDARY SURFACE PROCESSES

Source: After Cas, R. A. F., and J. V. Wright, 1987, Volcanic successions: modern and ancient, p. 34 reprinted by permission.

sions. Both generate fragmented products that are deposited contemporaneously with the eruptions. Magmatic explosions are created owing to rapid exsolution from magmas of dissolved water and carbon dioxide, creating enormous increase in pressure within a magma chamber. Eruption can occur either in sealed, near-surface magma chambers or in erupting magmas. Phreatic explosions result from interaction of magmas with groundwater in subsurface rocks, surface bodies of water, or water-saturated near-surface sediment. Pyroclastic flows moving into a body of water or over water-saturated sediment can create phreatic explosions also. Fragmentation probably results both from quenching and from explosive activity (Cas and Wright, 1987, p. 42). Fisher and Schmincke (1984, p. 74) use the terms **hydroclastic eruptive processes** to describe explosive eruptions that involve magma and external water. For simplicity, eruptions of this kind are considered in this book to be a type of pyroclastic eruption. The clasts produced by pyroclastic eruptions are called **pyroclasts**. The term **tephra** is commonly used as a collective term for all pyroclastic deposits. The explosive activity of pyroclastic eruptions generates fragments that may be partially crystallized or uncrystallized pieces of the erupting magma; free crystals, or pieces of crystals, of feldspar or other minerals; fragments of the country rock explosively ejected during eruption; or lithic fragments picked up locally by pyroclastic flows.

Autoclastic Processes.   In contrast to explosive pyroclastic eruptions, autoclastic processes are nonexplosive processes. They generate fragmental material owing to fracturing caused by sudden quenching or chilling of magmas or by flow fragmentation (autobrecciation). Quenching occurs when magmas flow into water, beneath ice or snow, or over water-saturated sediments or where magma is intruded into water-saturated sediments. The sudden contact between the hot, coherent magma body and cold water or sediment sets up tensile stresses that result in shattering of the hot rock into glassy fragments. Rittman (1962, p. 72–73) introduced the term **hyaloclastites** for such quench-fragmented volcanic debris. Autobrecciation occurs owing to continued movement of viscous, partially congealed lava. Internal stresses are generated that cause the partially congealed lava to stretch plastically and break into slabs or blocks (Cas and Wright, 1987, p. 56).

Emplacement of Pyroclastic Debris.   Pyroclastic materials are emplaced by two primary sedimentation processes: airfall and flows. **Airfall** involves the free fall of material ejected explosively from the vent. Material of all sizes from large blocks to fine ash may be ejected. The distribution of this material outward from the vent depends upon the height of the eruption column, velocity and direction of atmospheric winds, and clast size. Larger particles and the greatest amount of material fall closer to the vent. Therefore, particle size and debris thickness commonly decrease abruptly away from the vent. Very fine ash that is carried rapidly into the upper atmosphere may be transported by winds hundreds or even thousands of kilometers from the vent. Wright et al. (1980) suggest two kinds of pyroclastic flow mechanisms. **Flows** are high–particle-concentration, gas–solid dispersions that are gravity driven and that move predominantly by laminar transport. They tend to travel along the surface, generating deposits that are topographically controlled and that fill valleys and depressions. **Surges** are similar mechanisms, but differ in that they consist of low–particle-concentration, gas–solid dispersions that undergo turbulent flow, somewhat like turbidity currents under water. They transport particles along the surface and generate deposits that tend to mantle topography, although they are also topographically controlled (Cas and Wright, 1987, p. 98). There is apparently some ambiguity regarding these processes, as both pyroclastic flows and pyroclastic surges are considered to be types of flows (see discussion by Lajoie, 1984, p. 43–44).

Epiclastic Processes and Redeposition of Pyroclastic Deposits

The term epiclastic processes is used here to refer to processes that bring about fragmentation and transport of volcanic debris but that are not directly vent-related. Chemical and physical weathering operate to break down ancient lava flows and consolidated pyroclastic deposits. The resulting volcaniclastic debris is subsequently eroded; transported by wind, water (suspension and traction), ice, or sediment-gravity-flow processes (debris flows, mud flows, grain flows); and deposited. These processes that form volcaniclastic sandstones differ from processes that form nonvolcaniclastic sandstones only in that they involve transport predominantly of volcanic debris.

Recent, unconsolidated pyroclastic deposits may be eroded and the pyroclastic detritus retransported by the same kinds of processes that transport epiclastic detritus. Pyroclastic debris may thus be removed from original source areas and redeposited in a variety of settings, such as fluvial and deltaic environments, on beaches or shallow-marine shelves, or in the deeper-ocean basins.

5.4.3   Volcaniclastic Deposits

Classification

Pyroclastic Deposits.   Modern pyroclastic deposits are classified genetically as **pyroclastic fall deposits, pyroclastic flow deposits,** and **pyroclastic surge deposits.** Pyroclastic flow and surge deposits that consist largely of pumiceous materials are commonly called **ignimbrites.** More-detailed genetic classification of these three basic types of pyroclastic deposits is possible, as discussed by Wright et al. (1980) and Cas and Wright (1987, p. 351–353). Genetic classification of ancient pyroclastic deposits can be exceedingly difficult, however, and can generally be accomplished only after extensive field investigation and laboratory analyses. Descriptive classification depends upon grain size of the deposits and the nature of the particulate constituents. Pyroclastic eruptions produce fragmental material in sizes ranging from cobbles to silt. The most commonly used size classification of pyroclastic fragments is that of Fisher (1961), as modified from Wentworth and Williams (1932) (Table 5.8). Table 5.9 summarizes the principal components in pyroclastic deposits.

Our concern here is primarily with the sand-size fraction of pyroclastic deposits. Note that these materials are referred to in Table 5.8 as **ash** when unconsolidated and as **tuff** when consolidated into volcaniclastic sandstone. The term tuff should be used only when it can be confidently established that the rocks are of pyroclastic rather than epiclastic origin. Tuffs can vary considerably in their content of glass, crystals, and lithic fragments. The terms crystal tuff, lithic tuff, and vitric tuff can be used to express the relative proportions of these components (Fig. 5.23). **Vitric tuffs** have a predominant composition of mainly uncrystallized, sand-size or smaller glassy fragments. **Crystal tuffs** are enriched in crystal grains over either glassy fragments or lithic fragments, and **lithic tuffs** are predominated by lithic fragments, which may be cognate, accessory, or accidental lithics (see Table 5.9 for an explanation of these terms). If a chemical analysis of the components is available, compositional terms can be added to this classification; e.g., andesitic crystal tuff, basaltic lithic tuff.

Hyaloclastites.   As mentioned, hyaloclastites are deposits formed by flowing or intrusion of lava or magma into water, ice, or wet sediments, leading to granulation or shattering of the lava into small, angular fragments. Many ancient hyaloclastites are breccias, composed of angular fragments coarser than 2 mm; however, finer-grained hyaloclastites are common also. Considerable amounts of sand-size fragments have been observed to form during historic eruptions of lava that flowed into water, e.g., some Hawaiian eruptions (Moore et al., 1973). The particles that make up hyaloclastites are mainly angular lithic fragments or glass shards, all essentially of the same composition. They are commonly closely associated with pillow basalts or other lava flows. When the quench origin can definitely be established, the genetic terms **hyaloclastite breccia** or **hyaloclastite sandstone** can be used. The terms **peperite** and **peperitic hyaloclastite** are nongenetic terms used for mixtures of hyaloclastite and sediment. Peperites are believed to form when magma is intruded through, or lava flows over, wet unlithified sediment (Cas and Wright, 1987, p. 361).

TABLE 5.8

Grain-size limits of pyroclastic fragments and aggregates

Grain size phi (φ) mm	Pyroclastic fragments		Name of unconsolidated aggregate	Lithified equivalent (pyroclastic rock)
−8	Coarse 256------------Bombs (rounded) Fine	Blocks (angular)	Agglomerate (bombs) or pyroclastic breccia	Agglomerate (bombs) or pyroclastic breccia
−6	64-------------------------------------			
	Lapilli		Lapilli deposit	Lapillistone
−1	2-------------------------------------			
+4	Coarse 0.0625------- Ash Fine		Ash deposit	Tuff

Source: After Fisher (1961, 1966).

FIGURE 5.23
Subdivision of tuffs and ashes on the basis of relative abundance of crystals, lithic fragments, and glass. *(After Schmid, R., 1981, Descriptive nomenclature and classification of pyroclastic deposits and fragments: Recommendations of the IUGS Subcommission on the systematics of igneous rocks: Geology, v. 9, Fig. 1, p. 42. Published by Geological Society of America, Boulder, Co.)*

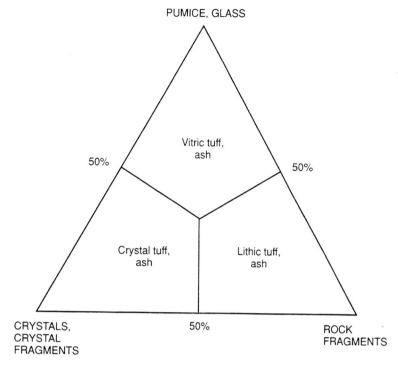

Epiclastic and Redeposited Pyroclastic Deposits. Epiclastic volcaniclastic deposits form by erosion of volcanic sources and transport and deposition of the eroded products by sedimentary processes, as noted. Also as noted, recent, unconsolidated pyroclastic deposits such as airfall ash can be eroded, transported, and redeposited by these same kinds of processes. For example, debris flows or mudflows, called **lahars,** commonly transport unconsolidated pyroclastic materials down the flanks of volcanoes and redeposit the material at lower elevations. Subsequently, this detritus may be moved by streams to an ocean basin. There, turbidity currents may transport the material to deeper water and redeposit it as turbidites. When dealing with ancient volcaniclastic deposits, it can be a challenging and subjective task to determine whether the particles in the deposit originated as epiclastic or pyroclastic detritus. Furthermore, under some conditions, epiclastic volcanic or nonvolcanic sediment and pyroclastic sediment can be deposited together to form mixed epiclastic–redeposited pyroclastic deposits. To emphasize the pyroclastic contribution to these deposits, these mixed sandstones are often referred to as **tuffaceous sandstones.**

Thus, both epiclastic volcanic detritus and redeposited pyroclastic detritus (as well as nonreworked pyroclastic detritus) form volcaniclastic deposits. When volcaniclastic sandstones are classified on the basis of particle composition, they classify as lithic arenites (wackes) or feldspathic arenites (wackes). Typically, volcanic clasts exceed feldspars; however, plagioclase feldspars may be more abundant than volcanic clasts in some volcaniclastic sandstones. Owing to their high content of unstable volcanic rock fragments that can alter diagenetically to form matrix, most epiclastic volcaniclastic sandstones are wackes rather than arenites. Some volcaniclastic sandstones, such as

TABLE 5.9

Components of pyroclastic deposits

Pyroclastic flows and surges			
	Essential components		
Type of flow or surge	Vesicular	Nonvesicular	Other components
Pumice flow and surge	pumice	crystals	accessory and accidental lithics
Scoria flow and surge	scoria	crystals	cognate, accessory, and accidental lithics
Block and ash flow and surge (nuée ardente)	poor to moderately vesicular juvenile clasts	cognate lithics and crystals	accidental lithics

Pyroclastic falls				
		Essential components[1]		
Predominant grain size	Type of fall	Vesicular	Nonvesicular	Other components
>64 mm (−6φ)	agglomerate	pumice or scoria		cognate and accessory lithics
	breccia	pumice or scoria	cognate or accessory lithics or both	
<2 mm (−1φ)	lapilli deposit	pumice or scoria	cognate or accessory lithics or both	crystals
<2 mm	ash deposit	pumice or scoria	crystals and/or cognate and/or accessory lithics	

Note: [1]Depending upon the type of deposit
Cognate lithics = nonvesicular, juvenile magmatic fragments
Accessory lithics = fragments of country rocks that have been explosively ejected during eruption
Accidental lithics = clasts picked up locally by pyroclastic flows and surges
Source: After Fisher (1961) and Cas and Wright (1987, p. 356).

sandstones deposited as lahars, may also contain detrital matrix. The characteristics of lithic and feldspathic arenites are discussed in preceding sections.

Undifferentiated Volcaniclastic Sediments.   In some cases, it may be extremely difficult, or impossible, to determine if an ancient volcaniclastic deposit is a pyroclastic, autoclastic, epiclastic, or redeposited pyroclastic deposit. For such sediments, Fisher (1961) suggests a nongenetic classification (Table 5.10) whereby these sediments are called simply volcanic breccia, volcanic conglomerate, volcanic sandstone, or volcanic mudstone, as appropriate.

Petrographic Characteristics of Volcaniclastic Sandstones

Pyroclasts.   The compositional feature that particularly distinguishes volcaniclastic sediments, whether pyroclastic or epiclastic, is their content of particles derived from volcanic sources. Particles generated as a direct result of volcanic action are called pyroclasts. Three principal kinds of pyroclasts are recognized: glassy fragments, crystals, and lithic fragments.

TABLE 5.10

Nongenetic classification for volcaniclastic rocks that cannot be determined to have a definite pyroclastic or epiclastic origin

```
Volcanic breccia
   Clast-supported (matrix-poor)
   Matrix-supported (matrix-rich)
      Noncohesive, granular matrix
      Cohesive, mud-sized matrix
Volcanic conglomerate
   Clast-supported (matrix-poor)
   Matrix-supported (matrix-rich)
      Noncohesive, granular matrix
      Cohesive, mud-sized matrix
2 mm (−1φ) ---------------------------------------------------------------------
   Volcanic sandstone
0.0625 mm (+4φ)----------------------------------------------------------------
   Volcanic mudstone
      Volcanic siltstone ____
                              |  If sufficiently well sorted and
                              |  volcanic origin is clear
      Volcanic claystone ____|
```

Source: After Fisher (1961) and Cas and Wright (1987).

1. **Glassy fragments** are fragments consisting of partially recrystallized to unre-crystallized, consolidated pieces of erupted magma. They may be either magmatic or phreatomagmatic (quench-chill) in origin. The phreatomagmatic products of basaltic magmas are commonly light- to medium-brown, glassy fragments called **sideromelane** (Fig. 5.24). These glass fragments are transparent in thin section in ordinary light and isotropic under crossed nicols. Glasses may have other colors such as red, yellow, and black depending upon impurities and the oxidation state of iron in the glasses. More-viscous

FIGURE 5.24

Sideromelane glass fragment (G) in Miocene deep-sea sandstone, ODP Leg 127, Site 796, Japan Sea (depth 368 m below seafloor). Ordinary light. Scale bar = 0.1 mm.

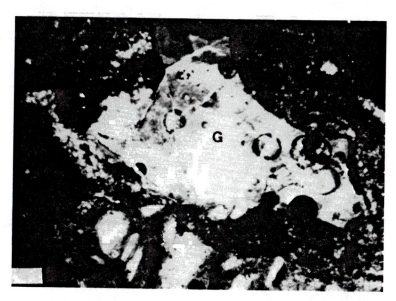

silicic to intermediate lavas give rise to highly vesicular magmas that during explosive eruptions yield pumice fragments of various sizes. Sand-size pumice fragments (Fig. 5.25) are called **pumice ash**. Magmatic explosions cause fragmentation of pumice vesicle walls or bubble walls, producing angular, ash-size glass particles called **shards**. Magmatically produced glass shards tend to have **Y** and cuspate shapes (Fig. 5.26) and may become plastically deformed and welded if they are still hot enough when deposited. The shapes of these shards contrast with those of phreatomagmatic glasses, which are generally more blocky and less vesicular (Heiken, 1972). Wohletz (1983) has shown, however, that a variety of shapes can occur in phreatomagmatic glasses, some of which, particularly the shapes of very small fragments, may closely resemble those of magmatic glasses. In addition to sand- or silt-size glass fragments, glass may be present also as a matrix that helps to cement particles together. Thin sections containing glass should be examined under both ordinary light and under crossed nicols to make sure that glass is not missed.

2. **Crystals,** both entire crystals and angular fragments of crystals, occur in many pyroclastic deposits. Broken crystals may be particularly abundant. Crystals of plagioclase feldspar are most common and may display zoning (Fig. 5.27). Also, pyroxene and amphibole crystals are common in some tuffs. If quartz crystals are present they generally show the slight rounding and embayment typical of volcanic quartz. Fragments of glassy material may remain attached to the crystals as blebs or crusts. Most such crystals are phenocrysts derived by magmatic or phreatomagmatic explosive disruption of porphyritic magmas. Less commonly, crystals or crystal fragments may be derived by fragmentation of accessory or accidental lithic fragments.

3. **Lithic fragments** are the denser, generally nonvesicular, nonglassy volcanic fragments in pyroclastic deposits. Some glass may be present within the fragments as interstitial material between small crystals. These lithic fragments may include fragments

FIGURE 5.25
Large pumice fragment (arrows) surrounded by small glass shards. Colestin Formation (Eocene-Oligocene), southern Oregon. Ordinary light. Scale bar = 0.5 mm. *(Sample courtesy of E. A. Bestland.)*

FIGURE 5.26
Glass shards in a volcaniclastic deposit. Note large Y-shaped clast near center of photograph, surrounded by smaller shards with curved and straight shapes. Colestin Formation (Eocene-Oligocene), southern Oregon. Ordinary light. Scale bar = 0.5 mm. *(Sample courtesy of E. A. Bestland.)*

FIGURE 5.27
Compositional (oscillatory) zoning in a plagioclase crystal. Colestin Formation (Eocene-Oligocene), southern Oregon. Crossed nicols. Scale bar = 0.1 mm. *(Sample courtesy of E. A. Bestland.)*

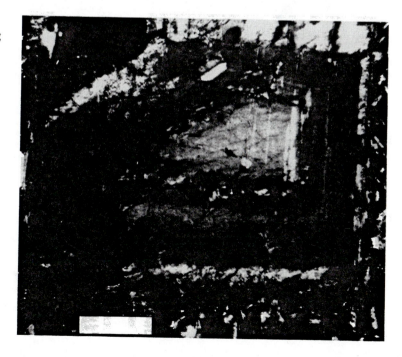

formed from erupting magma (cognate lithics), pieces of country rock blown out by magmatic or phreatic eruptions (accessory lithics), and older fragments picked up locally during pyroclastic flows or surges (accidental lithics) (Cas and Wright, 1987, p. 54). A variety of rock-fragment textures may be present, including fragments with small oriented laths (Fig. 5.28A), small unoriented laths (Fig. 5.28B), large phenocrysts set in a fine-grained matrix (Fig. 5.28C), and both large phenocrysts and small laths in a fine ground-mass (Fig. 5.28D).

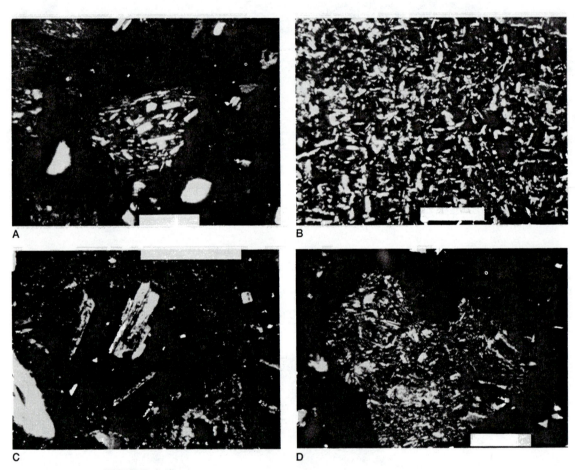

FIGURE 5.28
Typical volcanic rock fragments in volcaniclastic sediments. A. Small lithic fragment (center of photograph) containing very small, mainly oriented plagioclase laths; Miocene deep-sea sandstone, ODP Leg 127, Site 796, Japan Sea (depth 388 m below seafloor). B. Large lithic fragment containing numerous small, unoriented plagioclase laths; Colestin Formation (Eocene-Oligocene), southern Oregon. Crossed nicols. *(Sample courtesy of E. A. Bestland.)* C. Lithic fragment (center of photograph) containing large phenocrysts in a glassy groundmass. D. Lithic fragment containing both large phenocrysts and small, unoriented phenocrysts. Miocene deep-sea sandstone, ODP Leg 127, Site 796, Japan Sea (depth 388 m below seafloor). Crossed nicols. Scale bar = 0.5 mm for all photographs.

The scanning electron microscope (SEM) is now being used extensively to study pyroclasts. For example, see several papers on this subject, complete with numerous electron micrographs, that appear in the volume *Clastic Particles* edited by Marshall (1987). SEM study allows characterization of pyroclasts in terms of properties such as shape, rounding, surface textural features (grooves, scratches, fractures, **V**-shape depressions, microcrystalline encrustation, etc.), fraction of bubbles, and shapes of bubbles. From these features, investigators can interpret details of eruptive activity such as the relative importance of magmatic vesiculation and hydrovolcanic water interaction in the production of pyroclasts, transport of pyroclastic particles, and diagenetic alteration.

Epiclasts.   Epiclasts are crystals, crystal fragments, glass fragments, and rock fragments that have been released from preexisting volcanic rock by weathering or erosion and transported from their place of origin by gravity, air, water, or ice (Schmid, 1981). An interesting nomenclature problem arises with regard to the particles in retransported, unconsolidated (recent) pyroclastic debris. Are such particles pyroclasts or epiclasts? Fisher and Schmincke (1984, p. 89) suggest that reworking or recycling of unconsolidated pyroclastic debris by water or wind does not transform pyroclasts into epiclasts. According to these authors, such fragments are simply called reworked pyroclasts. Not all geologists may agree with this interpretation. In any case, distinguishing between epiclasts and pyroclasts in ancient volcaniclastic deposits can be an extremely difficult pragmatic problem. If we accept Fisher and Schmincke's interpretation, even the roundness of grains may not be a reliable criterion for distinction. Some epiclasts can be very angular, and some redeposited pyroclasts may be rounded by transport processes. Pettijohn et al. (1987, p. 192) suggest that the following features constitute criteria that can be used to identify pyroclasts.

1. Euhedral feldspars, many of which are broken; commonly zoned and generally oscillatory
2. Volcanic quartz, generally rounded or embayed owing to magmatic resorption
3. Minerals such as olivine and pyroxenes not commonly abundant in epiclastic sandstones
4. Glassy fragments
5. Low quartz content and high (euhedral) feldspar to quartz ratio

Alteration of Volcaniclastic Sandstones

Volcaniclastic sandstones contain a variety of unstable constituents (fine-grained lithic fragments, glassy fragments, mafic minerals, plagioclase feldspars) that alter readily during diagenesis (Fisher and Schmincke, 1984, Ch. 12). Glass is particularly susceptible to alteration. It typically alters to smectite clay minerals, zeolite minerals, or silica minerals. Devitrification of silicic glass can generate microcrystalline textures that resemble sedimentary chert. Such fragments may be extremely difficult to distinguish from chert unless distinctive euhedral crystals, such as feldspar laths, are preserved in the fragments. Also, glass can be replaced by minerals such as zeolites or calcite. Alteration of glass, feldspars, and lithic fragments can produce large amounts of authigenic clay matrix/cement. Silica cement may be present also as chalcedony or opal. Finally, squeezing of soft volcanic grains can produce pseudomatrix. These diagenetic changes in volcaniclastic sandstones make identification of original constituents difficult and differentiation of volcanic clasts from matrix a challenging task.

Chemical Composition

The chemical composition of some epiclastic(?) volcaniclastic sandstones is given in Table 5.6 (examples E and F). The SiO_2 content of the sandstones in these examples is moderately low (55–58 percent). By contrast, Al_2O_3 content (15–16 percent), total iron (8–9 percent), MgO (3–4 percent), CaO (4–5 percent), and Na_2O (4–5 percent) are all moderately high. These are only a few examples, and their chemistries may not be typical of all epiclastic volcaniclastic sandstones.

The chemistry of pyroclastic deposits depends upon the chemistry of the parent magma and the diagenetic history of the deposits. Thus, the SiO_2 content of tuffs could conceivably range from that of basic igneous rocks (45–52 percent), through that of intermediate rocks (52–66 percent), to that of felsic (acid) rocks (>66 percent). On the other hand, pyroclastic eruptions appear to result more commonly from viscous, silicic magmas than from basaltic magmas, which are commonly more fluid. Therefore, the SiO_2 content and overall chemistry of pyroclastic deposits is commonly in the range of intermediate to acid magmas. Interested readers may consult Fisher and Schmincke (1984, Table 2.2) for an extensive list of chemical compositions of pyroclastic deposits.

Occurrence and Field Characteristics

Volcaniclastic sandstones are common throughout the geologic record. Accurate measurements of the overall abundance of volcaniclastic sediments are not available; however, the volume of such ancient sediments is believed to be large (Fisher and Schmincke, 1984, p. 3). They occur preferentially in rock sequences deposited in convergent-margin, magmatic-arc settings, including trenches, forearc basins, backarc basins, and intra-arc basins. Epiclastic and reworked volcaniclastic sediments can be deposited by various fluid-flow and sediment-gravity-flow processes in both continental and marine environments. Many ancient epiclastic or redeposited pyroclastic volcaniclastic sandstones appear to be turbidites that are associated with deeper-water deposits such as pelagic shales and cherts.

Pyroclastic detritus can also be deposited in both subaerial and subaqueous environments. Deposition of pyroclastic deposits is probably most common on land, including deposition in lakes. Ancient pyroclastic deposits are commonly closely associated with lava flows. Airfall, flows, and surges may generate well-bedded sequences of deposits (Fig. 5.29) that can appear very similar to epiclastic deposits. See Pettijohn et al. (1987, Table 6.4, p. 224, and Table 6.6, p. 236) for some distinguishing characteristics of pyroclastic deposits and redeposited volcaniclastic deposits. Pyroclastic deposits may contain graded beds (normal, reverse, complex grading), cross-beds, massive beds, and thin, parallel beds. Other sedimentary structures include antidunes, chute-and-pool structures, convolute bedding, load casts, and mudcracks. Readers are referred to Fisher and Schmincke (1984) and Cas and Wright (1987) for more detailed treatment of volcaniclastic deposits.

5.5   Miscellaneous Sandstones

As indicated in preceding sections, the framework grains in epiclastic and volcaniclastic sandstones are dominantly quartz, feldspars, and rock fragments. Rock fragment composition can be quite varied. Most clasts are pieces of fine-grained igneous rocks, silicic metamorphic rocks, or siliciclastic sedimentary rocks; however, fragments of limestone, dolomite, and chert are common also. Occasionally, we find sandstones that contain

FIGURE 5.29
Well-bedded Quaternary andesitic airfall pyroclastic deposits. The uppermost layers are draped over previously formed deposits along an irregular erosional surface. The beds are not folded; the inclined layers accumulated at the attitudes shown in the photograph. Oshima Island, Japan. *(Photograph courtesy of A. R. McBirney.)*

significant amounts of framework constituents other than quartz, feldspars, and common rock fragments. These constituents may include heavy minerals such as magnetite, ilmenite, zircon, and rutile that were probably concentrated by placer-forming processes, as on beaches. Rarely, unusual rock fragments such as serpentinite fragments may make up a substantial fraction of the framework grains. Sand-size fragments of intrabasinal precipitates such as carbonate grains (ooids, pellets, intraclasts, fossils) can be abundant in some sandstones. Pettijohn (1975, p. 272) suggests the name **calcarenaceous sandstones** for sandstones enriched in carbonate grains. **Greensands,** sandstones containing significant amounts of glauconite, are moderately common in the geologic record, especially along unconformity surfaces. **Phosphatic sandstones** containing appreciable amounts of phosphatic skeletal fragments (Chapter 13), ooids, etc. occur also but are much less common.

Additional Readings

Carozzi, A. V., 1960, Microscopic sedimentary petrography: John Wiley & Sons, New York, 485 p.

Cas, R. A. F., and J. V. Wright, 1987, Volcanic successions: modern and ancient: George Allen & Unwin, London, 528 p.

Fisher, R. V., and H.-U. Schmincke, 1984, Pyroclastic rocks: Springer-Verlag, Berlin, 472 p.

Pettijohn, F. J., P. E. Potter, and R. Siever, 1987, Sand and sandstone, 2nd ed.: Springer-Verlag, New York, chapters 5 and 6.

Chapter 6
Conglomerates

■■■■

6.1 Introduction

Siliciclastic sedimentary rocks that consist dominantly of gravel-size (>2 mm) clasts are called conglomerates. The Latin-derived term **rudite** is also sometimes used for these rocks. Conglomerates are common rocks in stratigraphic sequences of all ages but make up less than about 1 percent by weight of the total sedimentary rock mass (Garrels and Mackenzie, 1971, p. 40). During the 1980s and the preceding few decades, geologists focused a level of attention on conglomerates, judged by the number of published papers, quite out of proportion to their relative abundance. This focus stems from their usefulness in tectonic and provenance analysis (Chapter 8) and the growing interest of sedimentologists in the rather specialized depositional environments of conglomerates. Also, some conglomerates serve as reservoir rocks for oil and gas.

The framework grains of conglomerates are composed mainly of rock fragments (clasts) rather than individual mineral grains. These clasts may consist of any kind of rock. Some conglomerates are composed almost entirely of highly durable clasts of quartzite, chert, or vein quartz. Others are composed of a variety of clasts, some of which, limestone and shale clasts, for example, may be unstable or weakly durable. Conglomerates may contain various amounts of matrix, which commonly consists of clay- or sand-size particles or a mixture of clay and sand.

Owing to their coarse grain size, conglomerates do not lend themselves readily to study in the laboratory. They are studied primarily in the field, by a variety of techniques. Clast counts, which consist of identifying several hundred randomly chosen clasts from a given outcrop, are used to determine clast composition. Size and sorting of clasts can be determined by measuring the dimensions of individual pebbles with a caliper. Some shape parameters such as sphericity can be derived also from these size measurements. Roundness is commonly determined by visual estimates. Conglomerate beds exposed in three-dimensional outcrops can be studied to determine aspects of fabric such as fabric support (clast-supported vs. matrix-supported), the long-axis orientation of clasts, and the imbrication dips of clasts. All of these textural elements are useful in environmental interpretation.

In this chapter, we examine the compositions, textures, and structures of conglomerates. Further, we see how some of these properties can be used as a basis for conglomerate classification and as tools for environmental analysis. Finally, some generalized depositional models for conglomerates are presented and discussed.

6.2   Definition of Conglomerates, Breccias, and Intraformational Conglomerates

In contrast to sandstones, conglomerates contain a substantial fraction of gravel-size (>2 mm) particles. The percentage of gravel-size particles required to distinguish a conglomerate from a sandstone or shale (mudstone) is arguable. Folk (1974, p. 28) sets the boundary between gravel and gravelly mud or gravelly sand at 30 percent gravel. That is, he considers a deposit with as little as 30 percent gravel-size fragments to be a gravel. On the other hand, Gilbert (Williams, Turner, and Gilbert, 1982, p. 330) indicates that a sedimentary rock must contain more than 50 percent gravel-size fragments to be called a conglomerate. Siliciclastic sedimentary rocks that contain fewer than 50 percent gravel-size clasts (possibly fewer than 30 percent according to Folk's usage) are conglomeratic sandstones or conglomeratic mudstones. Some geologists use the terms pebbly sandstone or pebbly mudstone to describe such sandstones or mudstones, although Gilbert reserves these terms for rocks with less than about 25 percent gravel-size clasts. Crowell (1957) suggested the term **pebbly mudstone** for any poorly sorted sedimentary rock composed of dispersed pebbles in an abundant mudstone matrix. The term "pebbly" is not an appropriate adjective for all such rocks with sparse clasts, however, because many of these rocks contain larger clasts (cobbles or boulders). Because of this somewhat confusing terminology for conglomeratic rocks that contain considerable mud or sand matrix, Flint et al. (1960) proposed the term **diamictite** for nonsorted to poorly sorted siliciclastic sedimentary rocks that contain larger particles of any size in a muddy matrix.

In this book, I accept diamictite, the term of Flint et al. (1960) for coarse-grained sedimentary rocks (rudites) that have too much matrix to classify as conglomerates. Thus, there are two kinds of rudites: (1) conglomerates, with low to moderate amounts of matrix, and (2) diamictites, with abundant matrix. Although the term diamictite is often applied to poorly sorted glacial deposits, it may be used for any sedimentary rock having the characteristics described. Flint (1971, p. 154) indicates that the term diamictite can be used "for nonsorted terrigenous sediments and rocks containing a wide range of particle sizes, regardless of genesis."

Unfortunately, we don't know exactly what "abundant matrix" means when applied to diamictites. Flint et al. (1960) did not specify the minimum amount of matrix that characterizes a diamictite. If we accept Gilbert's definition of a conglomerate as a sedimentary rock that contains >50 percent gravel-size fragments, does a diamictite then contain <50 percent gravel-size fragments (>50 percent matrix)? Alternatively, if we accept Folk's 30 percent gravel boundary between gravel and gravelly mud or sand, does a diamictite contain fewer than 30 percent gravel-size fragments (>70 percent matrix)? Rather than trying to distinguish between conglomerates and diamictites on the basis of a fixed matrix content, perhaps a better way to make this distinction is on the basis of fabric support. If mud or sand matrix is so abundant that the clasts in a rudite or gravelly sediment do not form a supporting framework, the fabric is commonly referred to as **matrix-supported** (Fig. 6.1). Rudites or gravelly sediments that contain so little matrix that the gravel-size framework grains touch and thus form a supporting framework are

FIGURE 6.1
Matrix-supported fabric in the
Bushnell Rock Member of the
Lookingglass Formation (Eocene),
southern Oregon Coast Range.
Note that clasts appear to "float"
in the mud-sand matrix.

called **clast-supported** (Fig. 6.2). I suggest that the definition of diamictites be modified to include the concept of matrix support. Thus, diamictites are matrix-supported rocks that contain larger clasts in a mud or sand matrix, whereas conglomerates are grain-supported rocks. Loosely, diamictites are conglomeratic sandstones or mudstones.

Because fabric support depends upon clast shape as well as upon the relative abundance of framework clasts and matrix, there is no fixed percentage of matrix that characterizes a clast-supported fabric vs. a matrix-supported fabric. To determine the kind of fabric support in a gravel deposit or in a consolidated rock, one must commonly examine the deposit in a three-dimensional outcrop. Clasts that do not appear to touch in a two-dimensional outcrop may actually touch in three dimensions. Furthermore, clast-supported fabrics may grade to matrix-supported fabrics within the same depositional

FIGURE 6.2
Clast-supported fabric in terrace
gravels of the Umpqua River,
southwest Oregon. Clasts are in
contact and thus form a supporting
framework.

unit. Careful, detailed field examination of three-dimensional outcrops of gravel deposits or ancient conglomerates and diamictites is the only way to determine fabric support.

The gravel-size material in conglomerates (and diamictites) consists mainly of rounded to subrounded rock fragments (clasts). By contrast, **breccias** are aggregates of angular, gravel-size fragments. The particles in breccias are distinguished from those in conglomerates by their sharp edges and unworn corners, although no specified roundness limit for breccias is in common use. Many breccias, such as volcanic and tectonic breccias, are nonsedimentary in origin. Table 6.1 lists the major kinds of conglomerates and breccias, classified on the basis of origin.

The most common kinds of rudites are **extraformational, epiclastic conglomerates** and **breccias**. These rocks are called extraformational because they are composed of clasts that originated outside the formation itself. They are **epiclastic** because they are generated by breakdown of older rocks through the processes of weathering and erosion. Thus, they are formed by the same kinds of processes that create epiclastic sandstones.

Intraformational conglomerates and **breccias** are deposits that formed by pene-contemporaneous fragmentation of weakly consolidated beds and subsequent redeposition

TABLE 6.1

Fundamental genetic types of conglomerates and breccias

Major types	*Subtypes*	*Origin of clasts*
Epiclastic conglomerate and breccia	Extraformational con-glomerate and breccia	Breakdown of older rocks of any kind through the processes of weathering and erosion; deposition by fluid flows (water, ice) and sediment gravity flows
	Intraformational con-glomerate and breccia	Penecontemporaneous fragmentation of weakly consolidated sedimentary beds; deposition by fluid flows and sediment gravity flows
Volcanic breccia	Pyroclastic breccia	Explosive volcanic eruptions, either magmatic or phreatic (steam) eruptions; deposited by air-falls or pyroclastic flows
	Autobreccia	Breakup of viscous, partially congealed lava owing to continued movement of the lava
	Hyaloclastic breccia	Shattering of hot, coherent magma into glassy frag-ments owing to contact with water, snow, or water-saturated sediment (quench fragmentation)
Cataclastic breccia	Landslide and slump breccia	Breakup of rock owing to tensile stresses and im-pact during sliding and slumping of rock masses
	Tectonic breccia: fault, fold, crush breccia	Breakage of brittle rock as a result of crustal move-ments
	Collapse breccia	Breakage of brittle rock owing to collapse into an opening created by solution or other processes
Solution breccia		Insoluble fragments that remain after solution of more soluble material; e.g., chert clasts concen-trated by solution of limestone
Meteorite impact breccia		Shattering of rock owing to meteorite impact

Source: Modified from Pettijohn, F. J., *Sedimentary rocks*, 3rd ed., 1975, p. 165.

of the resulting fragments within the same general depositional unit. Sedimentary processes, such as storm waves or mass flows, that bring about fragmentation and reposition of clasts to create intraformational conglomerates and breccias are probably very short-term events, possibly requiring only a few hours or days. These intraformational deposits commonly occur as thin units that are generally localized in extent. The fragments in these deposits may be well rounded or angular depending upon the amount of transport and reworking. Very commonly, they are flat or disc-shaped, giving rise to the term **flat-pebble conglomerate**. Flat-pebble conglomerates characterized by flattened pebbles stacked virtually on edge, owing to strong current activity, are called **edgewise conglomerates**. The most common intraformational conglomerates and breccias are those formed of (1) limestone or dolomite clasts in a limestone or sandy limestone matrix and (2) shale (mudstone) clasts in a sandy matrix (Pettijohn, 1975, p. 184). In this book, epiclastic sedimentary breccias are considered to be a kind of angular conglomerate and are not further differentiated from conglomerates.

Many breccias are generated by nonsedimentary processes such as volcanism. **Volcanic breccias** are formed by primary volcanic processes that may include explosive volcanism, autobrecciation of partially congealed lavas, or quench fragmentation of hot magmas that come into contact with water, snow, or water-saturated sediment. (Note: Volcanic conglomerates made up of clasts formed by weathering and erosion of older volcanic rocks are epiclastic conglomerates.) Less-common breccias are those that form through the processes of cataclasis or collapse (**cataclastic breccias**) and solution of soluble rocks such as limestone or salt, leaving insoluble gravel-size residues (**solution breccias**). **Meteorite impact breccias** are even less common.

Although nonsedimentary breccias and intraformational conglomerates are interesting, our concern in this chapter is primarily with the more common and abundant extraformational, epiclastic conglomerates. Therefore, most of the remaining part of this chapter deals with these conglomerates. A very short discussion of volcanic breccias and agglomerates is included near the end of the chapter.

6.3   Composition of Epiclastic Conglomerates and Diamictites

6.3.1   Composition of Framework Clasts

Gravel-size particles are regarded to be the framework grains of conglomerates and diamictites, even though diamictites are matrix-supported and thus commonly contain greater amounts of sand or mud than gravel. Conglomerates may contain gravel-size pieces of individual minerals such as vein quartz; however, the framework fraction of most conglomerates consists of rock fragments (clasts). Virtually any kind of igneous, metamorphic, or sedimentary clast may be present in a conglomerate, depending upon source rocks and depositional conditions. Some conglomerates are made up of only the most stable and durable types of clasts, that is, quartzite, chert, or vein-quartz clasts. The term **oligomict conglomerate** is often applied to stable conglomerates composed mainly of a single clast type, as opposed to **polymict conglomerates,** which contain an assortment of clasts. Polymict conglomerates made up of a mixture of largely unstable or metastable clasts such as basalt, limestone, shale, and phyllite are commonly called **petromict conglomerates**.

Some conglomerates that are enriched in quartzose clasts may be first-cycle deposits (i.e., not recycled from an older generation of conglomerates) that formed by erosion of a quartzite, quartz arenite, or chert-nodule limestone source. Others were probably derived from mixed parent-rock sources that included less-stable rock types. Continued recycling of mixed ultrastable and unstable clasts through several generations of conglomerates leads ultimately to selective destruction of the less-stable clasts and concentration of stable, quartzose clasts. Conglomerates made up of weakly durable clasts such as limestone and basalt, are more likely than conglomerates enriched in quartzose clasts to be first-cycle deposits. The clast composition of first-cycle conglomerates depends upon both the composition of the source rocks and the nature and intensity of the transport and depositional processes. These processes may destroy some very weakly durable clasts even in a single depositional cycle.

In addition to these factors, the clast composition of conglomerate deposits may be a function also of sorting by clast size (Boggs, 1969b). As a result of weathering, some parent rocks typically yield large clasts, whereas others break down to yield smaller clasts. For example, metaquartzites and dense, volcanic-flow rocks such as rhyolite tend to yield large fragments whose sizes are determined by the thickness of bedding and spacing of joints, whereas shales and argillites yield clasts in the finer pebble sizes owing to their fissile nature and closely spaced joint patterns in outcrop (Blatt, 1982, p. 144). Furthermore, less-durable fragments such as shale clasts tend to break into still finer-sized clasts during transport, whereas metaquartzite and rhyolite are more durable. Thus, the clasts of a given size in a conglomerate may be biased toward a particular rock type. Meaningful comparison of conglomerate composition from one unit to another requires that comparison be made between units of comparable clast size.

Other than the distinction made here between oligomict and petromict conglomerates, few generalities can be stated about clast composition. Almost any combination of clast types is possible in conglomerates, as shown in Table 6.2. The examples in Table 6.2 are not suggested to be representative of conglomerates in general. They are provided to illustrate the wide range of conglomerate compositions that is possible. Note from this table that conglomerates may be made up of various mixtures of igneous, sedimentary, and metamorphic clasts, or they may be composed dominantly of a single clast type. For example, some conglomerates are composed dominantly of volcanic clasts, others of metamorphic quartzite clasts, and still others of sedimentary chert, limestone, or dolomite clasts. For additional examples of conglomerate compositions, see Seiders and Blome (1988).

6.3.2 Composition of Matrix and Cements

The matrix of conglomerates (and diamictites) is composed mainly of clay- and sand-size particles. In contrast to the upper size limit of about 30 μm set for the matrix in sandstones, no grain-size limit has been established for the matrix of conglomerates. Conglomerate matrix is simply the finer material that fills the interstitial spaces among gravel-size clasts. Any kind of mineral or small rock fragment, including glassy fragments, can be present as matrix. Thus, the matrix may consist of various kinds of clay minerals and fine micas and/or silt- or sand-size quartz, feldspars, rock fragments, heavy minerals, and so on. The matrix itself may be cemented with quartz, calcite, hematite, clay, or other cements. Together, these cements and matrix materials bind the framework grains of the conglomerates and diamictites.

TABLE 6.2
Average clast composition of some North American conglomerate units

Clast type	A	B	C	D	E	F	G	H	I	J	K	L	M	N	O	P	Q
Quartzite	0.6	2.4	5.3	16.6	5.8	—	—	1.2	26.3	89.7	—	46[1]	—	—	10[2]	95+	1.0
Chert	14.8	1.2	—	17.8	—	13	1.9	—	8.4	0.3	95+	—	—	4	—	—	83.1
Vein quartz	4.5	—	—	1.9	—	20	0.7	—	—	—	—	18[3]	—	—	—	—	0.6
Basalt/andesite	7.8	50.6	1.5	26.4	91.5[3]	—	2.4[3]	42.2	—	—	—	—	—	—	—	—	0.1
Rhyolite/dacite	—	0.9	77.7	4.6	—	—	—	54.5	—	—	—	—	—	—	—	—	2.6
Granite/diorite	8.6	10.8	5.6	6.2	3.5	—	15.8	—	3.8	2.3	—	12	—	—	—	—	0.3
Metavolcanics	15.6	15.1	—	—	—	8	3.7	—	55.5	7.0	—	—	—	—	—	—	—
Metasediments (undiff.)	3.3	9.4	—	2.3	—	—	—	—	5.4	—	—	—	—	—	—	—	—
Schist/argillite	1.7	—	—	—	—	29	22.6	—	—	—	P	6	10[4]	48	Tr	—	1.6
Sandstone/siltstone	40.3	6.9	—	19.8	—	22	48.4	2.2	0.3	0.7	P?	7?	—	48	Tr	—	8.4
Shale	—	—	—	5.6	—	6	5.5	—	—	—	—	—	—	—	—	—	—
Conglomerate	1.3	3.8	—	—	—	—	—	—	—	—	—	—	—	—	90	—	—
Limestone/dolomite	1.1	—	—	—	—	2[5]	—	—	—	—	P	11[6]	90	—	—	?	2.3
Other	—	—	9.9	—	—	2[5]	—	—	—	—	P	—	—	—	—	—	—

Note: [1] Includes some quartz; [2] includes chert; [3] undifferentiated volcanic clasts; [4] undifferentiated sedimentary clasts; [5] serpentinite; [6] gneiss; P = present; Tr = trace.

Source:

A Kugler (1979), Lookingglass Formation (Eocene), eastern Coast Range, southwestern Oregon
B Ahmad (1981), Lookingglass Formation (Eocene), western Coast Range, southwestern Oregon
C Bellemin and Merriam (1958), Poway Conglomerate (Eocene), San Diego County, California
D Popeno (1941), Trabuco and Baker conglomerates, Santa Ana Mountains, California
E Howell and Link (1979), Eocene conglomerates of San Diego area and southern California borderland
F Brownfield (1972), Pleistocene conglomerates of Floras Creek, Oregon
G Farooqui (1969), Port Orford Conglomerate (Pliocene), Cape Blanco, Oregon
H Wolff and Huber (1973), Copper Harbor Conglomerate (Precambrian), Michigan
I Barats et al. (1984), nonquartzose conglomerate of Hornbrook Formation (Cretaceous), Oregon/California
J Barats et al. (1984), quartzose conglomerate of Hornbrook Formation (Cretaceous), Oregon/California
K Armin (1987), Earp Formation (Lower Permian), Pedrogosa Basin, Arizona, New Mexico, northern Mexico
L Rust (1966), conglomerates of Wheeler Gorge (Late Cretaceous), California
M Lindholm et al. (1979), Leesburg Conglomerate Member of Bull Run Formation (Jurassic), Virginia
N Lindholm et al. (1979), Barboursville Conglomerate Member of Bull Run Formation (Jurassic), Virginia
O, P Wilson (1970), Beaverhead and Monida formations (Upper Cretaceous), Wyoming
Q Seiders and Blome (1988), Galice Formation (Jurassic), Klamath Mountains, Oregon/California

6.4   Texture

6.4.1   Matrix Content and Fabric Support

High-energy processes, such as fluvial and beach processes, that transport and deposit gravels may remove most fine-size detritus and deposit gravels with little sand or mud matrix. On the other hand, gravels transported by glaciers and sediment-gravity-flow processes such as debris flows may contain abundant matrix. In fact, many gravel deposits that originate by these processes may contain more muddy matrix than framework clasts. Thus, as mentioned, some gravel fabrics are matrix-supported, and ancient rocks with these fabrics are called diamictites.

Conglomerates that contain essentially no matrix (voids among pebbles unfilled) are called **openwork** conglomerates (Pettijohn, 1975, p. 157). Openwork conglomerates are uncommon, in contrast to many sandstones that are deposited with open pores. That is, the pores of most sandstones are filled at the time of deposition only by fluids and not by matrix. Thus, in contrast to sandstones, conglomerates tend to have two size modes, one in the gravel-size range (the framework grains) and one in the sand- to mud-size range (the matrix grains).

6.4.2   Clast Shape and Orientation

The shape and orientation of sedimentary particles are discussed in Chapter 2. The discussion in Chapter 2 applies to the particles in conglomerates as well as those in sandstones; however, the shapes and orientations of particles in conglomerates may differ in some respects from those of associated sandstones. For example, gravel-size particles can become moderately rounded with comparatively short distances of stream transport, whereas sand-size particles undergo very little rounding. Therefore, clasts in a fluvial conglomerate may be well rounded, whereas grains in associated sandstones may be subangular to angular (Pettijohn, 1975, p. 163). The form (sphericity) of conglomerate clasts tends to be related to the shapes of the initial rock fragments released from the parent rocks. Parent rocks with a schistose or fissile fabric are likely to release tabular or disk-shaped fragments, whereas more-massive rocks such as metaquartzite tend to release more-equant-shaped clasts. Clast shape may, of course, be subsequently modified to some extent during transport owing to abrasion and clast breakage.

Because many conglomerate clasts do have an elongated or tabular shape, these clasts may assume a preferred orientation during transport and deposition. For example, elongated, gravel-size clasts tend to become oriented transverse to current flow (Fig. 2.19D) during stream transport, whereas sand-size grains are more likely to become oriented parallel to current flow. Tabular and elongated clasts also tend to develop an imbricated or shingled fabric under strong unidirectional currents (Fig. 2.19E). For more details on particle orientation, see Section 2.3.7.

6.5   Structures in Conglomerates

Many of the sedimentary structures described in Chapter 3 do not occur in conglomerates, which tend to have restricted suites of structures. Many conglomerates are massive (structureless); however, crude to well-developed planar horizontal or inclined stratification is moderately common. Tabular and trough cross-bedding is also present in some conglomerates, such as fluvial conglomerates; however, cross-bedding is much less common in conglomerates than in sandstones. Conglomerates may be nongraded, or they may display normal, inverse, or normal-to-inverse size grading. Other structures include gravel-filled scours and channels, and gravel lenses. Conglomerates may be associated with sandstones that display a much greater variety of sedimentary structures.

6.6   Descriptive Classification of Conglomerates and Diamictites

6.6.1   General Statement

In spite of the strong interest that geologists have in coarse-grained sedimentary rocks, we lack an adequate system for formally classifying conglomerates. Instead of formal names for conglomerates, geologists commonly use a variety of informal names. Some of these names are based on composition of the clasts (e.g., quartzite conglomerates), others on grain size (e.g., pebble conglomerates), and still others on presumed depositional environment or depositional process (e.g., fluvial conglomerates, debrisflow conglomerates). Pettijohn's (1975) classification is probably the best known of the few published conglomerate classifications, but it appears to be little used.

The paucity of formal conglomerate classifications stands in sharp contrast to the more than 50 classifications for sandstones that have been proposed (Friedman and Sanders, 1978, p. 190). Although sandstones may be overclassified, there appears to be a need for an adequate conglomerate classification. One reason that few conglomerate classifications have been proposed may be because conglomerates are not as easily classified as sandstones. For example, the many kinds of framework clasts that can be present in conglomerates make it difficult to reduce clast lithologies to three principal kinds. Thus, in contrast to classification of sandstones, it is not easy to plot the clast composition of conglomerates on a ternary classification diagram.

6.6.2   Classification by Relative Clast Stability

One way to deal with the composition of conglomerates for the purpose of classification is to place clasts into two groups on the basis of relative clast stability: (1) **ultrastable clasts** (quartzite, chert, vein quartz) and (2) **metastable** and **unstable clasts** (all other clasts). Thus, on the basis of clast stability, we recognize two kinds of conglomerates. Conglomerates made up of framework grains that consist dominantly of ultrastable clasts (>90 percent) are **quartzose conglomerates**. Conglomerates that contain less than 90 percent ultrastable clasts are **petromict conglomerates** (Table 6.3). As indicated in Section 6.3.1, the term petromict conglomerate is already in common use for conglomerates containing abundant unstable or metastable clasts. Diamictites can be classified also

TABLE 6.3

Classification of conglomerates and diamictites on the basis of clast stability and fabric support

Percentage of ultrastable clasts	Type of fabric support	
	Clast-supported	*Matrix-supported*
>90	Quartzose conglomerate	Quartzose diamictite
<90	Petromict conglomerate	Petromict diamictite

on the basis of clast stability in the same way as conglomerates, that is, as quartzose diamictite or petromict diamictites (Table 6.3). The compositional boundary between quartzose and petromict conglomerates is placed at 90 percent stable clasts. This 90 percent boundary is arbitrary; however, it does agree with the boundary between stable grains and unstable grains set in some sandstone classifications. Also, it coincides with the boundary suggested by Pettijohn (1975, p. 165) in his conglomerate classification.

Classifying conglomerates on the basis of clast stability has some important genetic significance. For example, some quartzose conglomerates probably originate as a result of intense chemical weathering of source rocks such as chert-nodule limestones. Others form as a result of prolonged transport or multiple recycling of clasts—processes that mechanically eliminate less-durable clasts. By contrast, petromict conglomerates are more likely to be first-cycle deposits that originate under less-intensive weathering conditions or that undergo less-prolonged transport and abrasion. Thus, the proposed stability classification has important interpretative value.

6.6.3   Classification by Clast Lithology

Although classification of conglomerates on the basis of clast stability is useful, as discussed, it reveals little about the actual clast lithology of conglomerates. Because conglomerates can contain a wide variety of clast types (Table 6.2), it is often desirable in their study to focus on clast composition rather than on clast stability, particularly in source-rock studies.

For classification purposes, conglomerate clasts can be grouped into three fundamental kinds: igneous, metamorphic, and sedimentary. The relative abundance of the different kinds of clasts (e.g., basalt, schist, sandstone) in a conglomerate is established in the field by clast counts. Once composition has been determined, clasts are grouped by genetic type (igneous, metamorphic, sedimentary) as end members of a classification triangle. Thus, all clasts, including ultrastable clasts, are normalized in terms of these three fundamental end members.

Figure 6.3 shows how conglomerates can be classified on the basis of these end-member clast types. The classification triangle in Figure 6.3 is divided into four fields to yield four kinds of conglomerates: **metamorphic-clast conglomerates, igneous-clast conglomerates, sedimentary-clast conglomerates,** and **polymict conglomerates.** The term polymict can be applied informally to all conglomerates containing clasts of mixed lithology; however, it is used in a formal sense in this classification to denote conglomerates made up of roughly subequal amounts of metamorphic, igneous, and sedimentary clasts. Although this classification is largely descriptive, these conglomerate names obviously have some provenance significance because the terms identify major genetic categories of source rocks from which conglomerates are derived.

To arrive at the actual name of a conglomerate on the basis of dominant clast lithology requires that daughter triangles be erected on the basic classification diagram, as shown in Figure 6.3. For conglomerates that fall into the metamorphic-clast, igneous-clast, or sedimentary-clast fields, the clasts are renormalized in the daughter triangle to 100 percent metamorphic, igneous, or sedimentary clasts, as appropriate. Clast composition is then plotted on the corresponding daughter triangle. Thus, sedimentary conglomerates, for example, could be classified as limestone or dolomite conglomerates, sandstone conglomerates, chert conglomerates, and so on, depending upon the relative percentages of various kinds of sedimentary clasts. Metamorphic conglomerates include

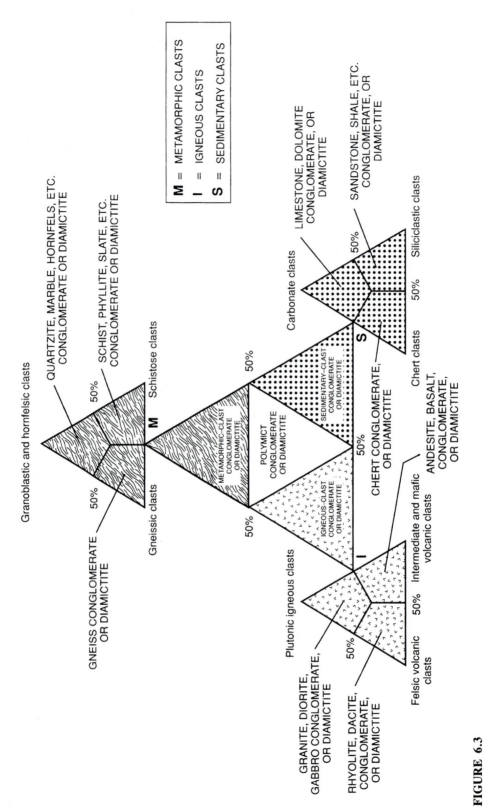

FIGURE 6.3

Classification of conglomerates on the basis of clast lithology and type of fabric support.

schist conglomerates, gneiss conglomerates, quartzite conglomerates, and so on. Igneous conglomerates include granite conglomerates, basalt conglomerates, and so on. Informal names such as limestone conglomerate have been in use for years. This recommended classification simply formalizes the classification procedure and places some definite compositional limits on each kind of conglomerate on the basis of clast types. Note that polymict conglomerates, which contain roughly equal amounts of metamorphic, igneous, and sedimentary clasts, are not further classified by specific lithologic types.

This compositional classification scheme can be used also for diamictites. If one substitutes the word diamictite for conglomerate (Fig. 6.3), this procedure yields names for diamictites such as basalt diamictite, sandstone diamictite, and so on.

In verbal descriptions of conglomerates, the stability class name from Table 6.3 can be combined, if desired, with the appropriate lithologic name derived from the classification diagram to provide additional clarification of the composition. Thus, for example, a chert conglomerate might be called a quartzose chert conglomerate if it contains more than 90 percent stable clasts or a petromict chert conglomerate if it contains less than 90 percent stable clasts and chert is the dominant clast type.

6.6.4   Classification by Clast Size

Finally, terms for the relative sizes of clasts in a conglomerate can be used as adjectives that are added to appropriate compositional terms, if desired. Thus, we could have a cobble-rich quartzose rudite, a boulder-rich petromict diamictite, and so on.

6.7   Occurrence of Quartzose and Petromict Conglomerates (and Diamictites)

6.7.1   General Statement

Conglomerates occur in sedimentary sequences of all ages from Precambrian to Holocene and on all continents of the world. The clasts that make up conglomerates are derived from many kinds of igneous, metamorphic, and sedimentary source rocks. These clasts can be transported and deposited by a variety of fluid-flow and sediment-gravity-flow processes (Section 6.9). Space does not permit detailed discussion of conglomerate occurrence; however, a brief discussion of quartzose and petromict conglomerates and some generalities about their sources are given below.

6.7.2   Quartzose Conglomerates

Quartzose conglomerates, which consist dominantly of metaquartzite, vein-quartz, or chert clasts, are derived from metasedimentary, sedimentary, and some igneous rocks. As suggested by Pettijohn (1975, p. 166), these clasts are a residuum concentrated by destruction of a much larger volume of rock. Quartzite clasts are derived from metasedimentary sequences containing quartzite beds. Less stable metasedimentary clasts such as argillite, slate, and schist, originally present in such metasedimentary sequences, must have been destroyed by weathering, erosion, and sediment transport. Quartz-filled veins occur broadly scattered through mainly igneous and metamorphic rocks. Concentrations of vein-quartz clasts in quartzose conglomerates implies destruction of large bodies of such primary igneous or metamorphic rock. Likewise, the concentration of chert clasts in a conglomerate implies destruction of large volumes of chert-nodule limestone to yield a chert-clast concentrate. Some chert clasts may, of course, be derived also by erosion of bedded chert deposits.

Such wholesale destruction of rock masses to yield relatively small concentrates of quartzose clasts implies either extremely intensive chemical weathering or vigorous transport that mechanically destroyed less-durable clasts. More than one cycle of weathering and transport could be involved. The source rocks that yield quartzose clasts are most likely to occur in recycled orogen or continental block provenances (Chapter 1). Because quartzose clasts represent only a small fraction of a much larger original body of rock, the total volume of quartzose conglomerates is commonly small. They tend to occur as scattered pebbles, thin pebbly layers, or lenses of pebbles in dominantly sandstone units (Pettijohn, 1975, p. 166). They appear to be largely of fluvial, particularly braided-stream, origin, but marine, wave-worked quartzose conglomerates also exist.

Examples of quartzose conglomerates are known from throughout the world in sedimentary units ranging in age from Precambrian to Tertiary. Precambrian examples include the jasper-bearing Lorrain quartzites (Huronian) of Ontario, and Paleozoic examples include parts of the Silurian Tuscarora Quartzite and the Devonian Chemung and Mississippian Pocono formations of the U.S. central Appalachians. Mesozoic and Tertiary examples include quartzose conglomerates in the Cretaceous Hornbrook Formation and Mississippian Bragdon Formation of the northern Klamath Mountains of northern California, the Tertiary Brandywine upland gravels of Maryland, and the Lafayette upland gravels of western Kentucky. Many quartzose conglomerates also occur in Tertiary sequences in the U.S. Rocky Mountains (e.g., Kraus, 1984). The stratigraphic characteristics of one of these conglomerates are illustrated in Figure 6.4. Additional examples of quartzose conglomerates are discussed by Pettijohn (1975, p. 166), and still others are shown in Table 6.2. This tabulation of examples is by no means comprehensive.

6.7.3   Petromict Conglomerates

As discussed, petromict conglomerates contain significant amounts of metastable rock fragments. Most are polymictic conglomerates, made up of a variety of metastable clasts. They may be derived from many different types of plutonic igneous, volcanic, metamorphic, or sedimentary rock. As shown in Table 6.2, the clasts in a given conglomerate unit may be dominantly volcanic, dominantly metamorphic, or dominantly sedimentary, depending upon the source-rock lithology. Conglomerates composed dominantly of plutonic igneous clasts appear to be uncommon, probably because plutonic rocks such as granites tend to disintegrate into sand-size fragments rather than forming larger blocks. Among petromict conglomerates containing significant amounts of sedimentary clasts, clasts of siliciclastic sedimentary rock are generally more common than clasts of carbonate rocks. On the other hand, some conglomerates are composed mainly of carbonate clasts (Table 6.2). Owing to their lesser stability compared to quartzose clasts, the clasts of petromict conglomerates are more likely than quartzose clasts to be of first-cycle origin. Nonetheless, the fact that some petromict conglomerates contain clasts of a previous generation of petromict conglomerates (Table 6.2) shows that some metastable clasts can survive recycling.

The volume of ancient petromict conglomerates is far greater than that of quartzose conglomerates. The petromict conglomerates form the truly great conglomerate bodies of the geologic record, and they may reach thicknesses of thousands of meters. Some conglomerates composed mainly of volcanic clasts form especially thick sequences. The

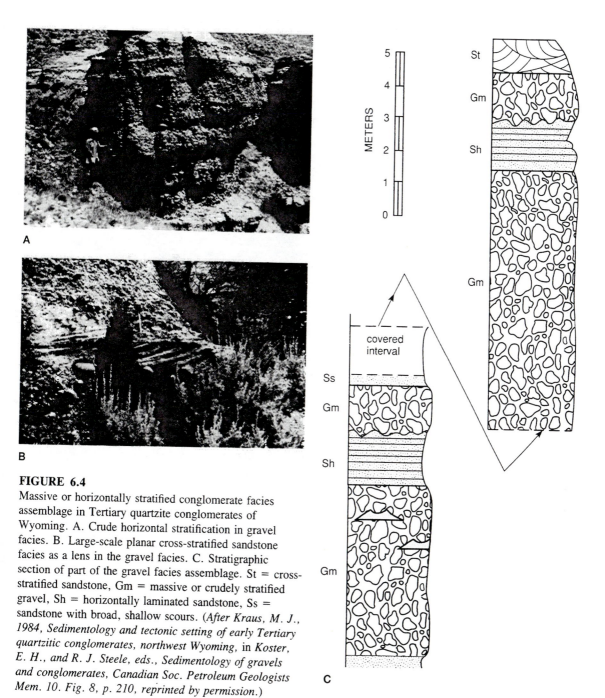

FIGURE 6.4

Massive or horizontally stratified conglomerate facies assemblage in Tertiary quartzite conglomerates of Wyoming. A. Crude horizontal stratification in gravel facies. B. Large-scale planar cross-stratified sandstone facies as a lens in the gravel facies. C. Stratigraphic section of part of the gravel facies assemblage. St = cross-stratified sandstone, Gm = massive or crudely stratified gravel, Sh = horizontally laminated sandstone, Ss = sandstone with broad, shallow scours. (*After Kraus, M. J., 1984, Sedimentology and tectonic setting of early Tertiary quartzitic conglomerates, northwest Wyoming, in Koster, E. H., and R. J. Steele, eds., Sedimentology of gravels and conglomerates, Canadian Soc. Petroleum Geologists Mem. 10. Fig. 8, p. 210, reprinted by permission.*)

FIGURE 6.5
Polymict conglomerates of the
Miocene Kotanbetsu Formation of
Hokkaido, Japan. This facies
displays inverse-to-normal grading
and overlies finer-grained turbidite
units. The vertical scale bar is 100
cm long. (*From Okada, H., and
S. K. Tandon, 1984, Resedimented
conglomerates in a Miocene
collision suture, Hokkaido, Japan,
in Koster, E. H., and R. J. Steel,
eds., Sedimentology of gravels and
conglomerates: Canadian Soc.
Petroleum Geologists Mem. 10.
Fig. 9, p. 419, reprinted by
permission. Photograph courtesy
H. Okada.*)

preservation and accumulation of such thick sequences of metastable clasts, particularly
clasts of highly soluble limestone or dolomite, imply rapid erosion of sharply elevated
highlands (or areas of active volcanism in the case of volcanic conglomerates). Alternatively, some petromict conglomerates, such as some limestone conglomerates, may have
accumulated at lower elevations, but under very cold conditions, where glacial activity
provided the erosion mechanism. In any case, the metastable clasts of petromict conglomerates must have been stripped from source areas before chemical weathering processes could bring about solution or promote disintegration to sand-size particles. Furthermore, they must have been transported only short distances from the source, or they
were transported by processes that did not mechanically destroy the clasts. Petromict
conglomerates can accumulate in any tectonic provenance (continental block, recycled
orogen, or magmatic arc) where the requisite conditions that allow their preservation are
met. They are deposited in environments ranging from fluvial through shallow-marine to
deep-marine, although the bulk of the truly thick petromict conglomerate bodies are
probably nonmarine.

Examples of petromict conglomerates in the geologic record are so numerous and
so varied that a listing of occurrences is essentially pointless. Examples could be cited
from sedimentary sequences of all ages and from all continents of the world. Several are
listed in Table 6.2, and others are described by Seiders and Blome (1988). Space does not
permit discussion of specific examples, but one example of a mixed-clast petromict
conglomerate from the Miocene Kotanbetsu Formation of Japan is shown in Figure 6.5.
Readers who wish additional information about petromict conglomerates may consult
some of the references given in Table 6.2 and in the additional readings at the end of this
chapter.

6.8   Depositional Environments and Characteristic Properties of Conglomerates (and Diamictites)

Descriptive classification of conglomerates is useful as a means of identifying the compositional differences among conglomerates and in relating composition to sediment

sources. On the other hand, many conglomerate workers focus on study of depositional environments and are thus more interested in the properties of conglomerates that relate to depositional origin. They are commonly more concerned with the textural and structural properties of conglomerates than with clast composition. Instead of classifying conglomerates on the basis of clast type, they are more likely to classify them on the basis of presumed depositional origin. Thus, they may refer to a particular conglomerate as a wave-worked conglomerate, a resedimented conglomerate, and so on.

The textural and structural characteristics of conglomerates are generated owing to deposition under a particular set of environmental conditions. A goal of many conglomerate studies is to identify the paleoenvironments of conglomerates on the basis of these characteristic properties. Gravels are deposited in modern environments ranging from fluvial to deep marine by a variety of fluid-flow, ice-flow and rafting, and sediment-gravity-flow processes. We assume that ancient conglomerates were deposited in similar environments and that the characteristics of ancient conglomerates and modern gravels deposited in similar environments are also similar. The scope of this book does not allow detailed discussion of depositional environments. Therefore, readers who lack background knowledge of depositional environments may wish to consult appropriate references on this subject before continuing this chapter. Some suggested references are Blatt et al. (1980, Ch. 19), Boggs (1987, Ch. 10–13), Reading (1986), and Selley (1978).

Table 6.4 provides a listing of the major environments of modern gravel deposition and the putative mechanisms that operate within these environments to deposit gravels. This table shows that a particular type of transport and depositional mechanism may operate in more than one environment and thus produce similar types of gravel deposits in these different environments. Sheetflood gravels, for example, can be deposited on alluvial fans, outwash plains, and possibly under glaciers by streams flowing beneath glacial ice. Resedimented gravels, reentrained from previously deposited sediment, may be redeposited by subaqueous debris flows or related processes in glaciolacustrine or glaciomarine environments, on lacustrine or marine deltas and fan-deltas, or in deep-marine environments, and so on. To more clearly show this relationship between process and environment, Table 6.5 groups gravels according to probable depositional process and indicates the principal environments in which modern gravel deposition occurs.

For many years, geologists have informally classified ancient conglomerates on the basis of presumed mode of origin or depositional environment. Although classifications ought to be as genetically significant as possible, the subjectivity inherent in genetic classifications makes them difficult to use. What, for example, characterizes fluvial conglomerates? What are the physical differences between streamflow and debrisflow fluvial conglomerates? What distinguishes fluvial conglomerates from glacial conglomerates? Can we unambiguously distinguish fluvial conglomerates from deep-water, resedimented marine conglomerates? These questions suggest that although genetic classification is highly desirable, genetic classification of ancient conglomerates is likely to be difficult. We begin study of such conglomerates knowing neither their depositional process nor their depositional environment—both must be inferred from the characteristics of the conglomerates. To construct a practical genetic classification for conglomerates, we must formulate the classification in such a way that we have a realistic expectation of using it in the field. Thus, the preceding questions about our ability to discriminate conglomerates deposited in different environments or by different processes become highly pertinent.

TABLE 6.4

Environmental setting of modern gravel deposits and the major transport and depositional processes that generate gravel deposits

Major environmental setting	Subenvironment	Transport and depositional processes	Process-related gravel name
Fluvial	Alluvial fans	Sheetflood from a network of braided, shallow distributary channels	Sheetflood gravel
		Sheetflood with rapid infiltration of water (sieveflood), producing gravel lobes	Sieveflood gravel
	Fan and nonfan stream channels	Debris flow (unchanneled, sheet mass flow)	Debrisflow gravel
		Channelized, flood-stage, streamflow traction transport in channels	Streamflow gravel
Glacial and proglacial	Grounded glacier meltout zone	Transport by ice and deposition on land as meltout or lodgment deposits	Meltout-lodgment gravel
	Glacier margin	Mass flow processes (mainly debris flows) near glacier margin	Glacial debrisflow gravel
	Subglacial zones	Sheetflood flow or channelized stream flow beneath grounded ice	Subglacial sheetflood gravel or subglacial streamflow gravel
	Proglacial outwash fans and plains (glaciofluvial)	Transport and deposition of gravel beyond ice margin by meltwater, mainly in a network of shallow, braided channels	Subaerial outwash gravel
	Proglacial lakes (glaciolacustrine)	Melting of gravel-charged ice within lakes (ice-contact lakes), discharge of meltwater streams into lakes (distal, glacier-fed lakes)	Sublacustrine outwash gravel

Environment	Process	Gravel type
Proglacial marine zones (glaciomarine)	Retransport of gravels within lakes by massflow processes (subaqueous debris flow, grain flow, turbidity currents)	Resedimented sublacustrine outwash gravel
	Melting of gravel-charged ice in coastal zones; discharge of meltwater into the ocean, melting of floating ice in deeper water	Submarine outwash gravel
	Retransport of subaqueous outwash gravels by massflow processes	Resedimented submarine outwash gravel
Lacustrine		
Lake shoreface/beachface	River input of gravel, reworking by wave swash, storm surges	Wave-worked lacustrine gravel
Deltas, fan-deltas, deep lake basin	Retransport of river-input gravels by massflow processes	Resedimented lacustrine gravel
Transitional marine		
Marine beachface, shoreface, inner shelf	River input of gravel, reworking by wave swash, storm surge, longshore currents, rip currents	Wave-, storm-, and current-worked marine gravel
Nearshore tidal zone	Tidal current flow, particularly in tidal channels	Tide-worked gravel
Deltas, fan-deltas	Retransport of river-input gravels by subaqueous massflow processes	Resedimented transitional-marine gravel
Marine		
Shallow-marine shelf	Gravel deposited by fluvial or beach processes during low sea level, stranded in deeper water, with little reworking, by rising sea level	Relict gravel
	Relict gravel reworked by marine processes during rising sea level	Palimpsest gravel
Deep-marine slope, basin (submarine fans, canyons)	Retransport of nearshore gravels into deeper water by turbidity currents or other subaqueous massflow processes	Resedimented marine gravel

TABLE 6.5
Major transport/depositional processes and depositional environments of conglomerates

Transport mechanism	Type of flow or depositional process	Suggested conglomerate name based on process	Depositional environment	Environmental name
	Sheetflood flow in shallow braided channels	Sheetflood conglomerate	Alluvial fans, proglacial outwash fans, subglacial zones, fan-deltas	Braided-stream conglomerate
	Channelized flow in deeper fluvial channels	Streamflow conglomerate	Entrenched channels in alluvial and outwash fans, subglacial channels, channels of nonfan streams, subaerial delta/fan-delta distributaries	Alluvial-fan conglomerate, fluvial-channel conglomerate, deltaic conglomerate, fan-delta conglomerate, glaciofluvial conglomerate
Fluid (water) flow	Wave upswash and backwash	Wave-worked conglomerate	Marine beachface, possibly lacustrine beachface	Beachface conglomerate
	Shoreward movement of wave bore; flow of longshore and rip currents, storm surges	Wave-, storm-, and current-worked conglomerate	Marine shoreface, possibly lacustrine shoreface	Shoreface conglomerate, shelf conglomerate
	Tidal-current flow, largely in channels	Tide-worked conglomerate	Marine nearshore environments, especially in tidal channels	Tidal conglomerate
Ice flow	Subaerial ice melt and glacial overriding	Meltout/lodgment conglomerate	Meltout zones of grounded glaciers	Glacial conglomerate
	Subaqueous ice meltout, meltwater (traction) underflow, ice rafting	Subaqueous meltout conglomerate	Proglacial lakes, proglacial marine environments, including deltaic environments	Glaciomarine conglomerate, glaciolacustrine conglomerate
Sediment gravity flow	Subaerial debris flow	Subaerial debrisflow conglomerate	Alluvial fans, proglacial outwash fans, glacier margins	Alluvial-fan conglomerate, proglacial conglomerate
	Subaqueous debris flow	Subaqueous debris flow conglomerate	Subaqueous outwash plains, marine and lacustrine deltas and fan-deltas, submarine channels and fans	Deltaic conglomerate, fan-delta conglomerate, submarine-fan conglomerate
	Density-modified grain flow	Subaqueous grain flow conglomerate		
	High-density turbidity-current flow	Turbidite conglomerate		

Resedimented conglomerate (bracket spanning the subaqueous debris flow, subaqueous grain flow, and turbidite conglomerate rows under the Depositional environment column)

They indicate that an extremely detailed genetic classification for ancient conglomerates, with numerous categories and subcategories, is unlikely to be workable.

Inspection of Tables 6.4 and 6.5 suggests that genetic typing of modern gravels could be made on the basis of either depositional environment or depositional process. On the basis of depositional environment, gravels could be referred to by names such as fluvial gravel, glacial gravel (till), or marine gravel. More specific environmental names might be applied, such as fluvial, braided-channel gravel; glaciofluvial gravel; or submarine-fan gravel. Alternatively, modern gravel could be called by names that reflect depositional process, such as sheetflood gravel, streamflow gravel, or wave-worked gravel. Conglomerate workers at this time appear to favor process-related names rather than names based on depositional environment. Table 6.5 shows suggested process-related names for the principal kinds of ancient conglomerates. Some of these names are already in common use, e.g., sheetflood conglomerate, resedimented conglomerates. Others, such as meltout/lodgment conglomerate and subaqueous meltout conglomerate, are introduced here. The process classification in Table 6.5 divides conglomerates into nine fundamental types: sheetflood conglomerate, streamflow conglomerate, wave-worked conglomerate, wave-, storm-, and current-worked conglomerate, tide-worked conglomerate, meltout/lodgment conglomerate, subaqueous meltout conglomerate, sub-aerial debris-flow conglomerate, and resedimented conglomerate. Resedimented conglomerates are further divided into subaqueous debrisflow conglomerates, subaqueous grainflow conglomerates, and turbidite conglomerates.

Note that Table 6.5 also lists suggested names for conglomerates on the basis of depositional environment. Thus, ancient conglomerates can be referred to as braided-stream conglomerates, fan-delta conglomerates, beachface conglomerates, and so on whenever we are sufficiently confident that we can identify the depositional environment. Because we can generally interpret the depositional process (sheetflood flow, stream flow, debris flow, etc.) with greater confidence than we can interpret the depositional environment, conglomerates are categorized in this book primarily on the basis of depositional process.

6.9   Distinguishing Characteristics of Conglomerates and Diamictites Classified by Depositional Process

6.9.1   General Statement

To make practical use of the process-related classification in Table 6.5 for study and classification of ancient conglomerates requires the availability of reasonable models for each major genetic type of conglomerate. That is, we must have a good understanding of the physical attributes that characterize each kind of conglomerate. Such models are constructed whenever possible by observing the characteristics of conglomerates in modern environments. Models of this kind are commonly referred to as **actualistic models**. In the case of some conglomerates, such as turbidite conglomerates, we do not have modern analogs that are readily available for observation. Therefore, we must turn to the ancient record to understand the characteristics of these conglomerates. Clearly, models of such conglomerates are less firmly based than those generated by study of modern gravel deposits. Very simple models summarizing the principal physical characteristics of the major kinds of conglomerates are discussed in the succeeding sections.

6.9.2   Conglomerates Deposited by Fluid Flow

Sheetflood (Braided-stream) Conglomerates

Gravels transported by braided streams are derived principally by reworking of mass-transported deposits or glacial deposits. Braided-stream conglomerates are particularly common in alluvial-fan settings, including settings where fans build into standing water, creating **fan deltas** (Ethridge and Wescott, 1984; Holmes, 1965; McGowen, 1970). They are also deposited on glacial-outwash fans where large volumes of meltwater and gravel are supplied by melting glaciers (Boothroyd and Ashley, 1975; Bull, 1972; Hein and Walker, 1977; Jopling and McDonald, 1975; Miall, 1977, 1978; N. D. Smith, 1974). A principal fluid-flow process that operates in braided-stream settings to transport gravel is sheetflood. (Hogg, 1982, defines sheetflood as a sheet of unconfined flood water moving down a slope.) Sheets of sediment are deposited by surges of sediment-laden water that spread out from the ends of the stream channels on a fan (Bull, 1972). Deposition is caused by widening of the flow into shallow bands or sheets and concurrent decrease in depth and velocity of flow. Depths of water are generally less than about 0.3 m. The shallow distributary channels fill rapidly with sediment, then shift a short distance to another location. The resulting deposit is sheetlike, and it may be traversed by shallow channels that repeatedly divide and rejoin. In short, sheetflood gravels are deposited by traction processes in a network of shallow, braided distributary channels. Under conditions where little sand, silt, or gravel is present in the source, sheetflood waters may infiltrate completely before reaching the toe of the fan, leaving **sieve** deposits.

The distinguishing characteristics of sheetflood gravels are summarized in Table 6.6. Sheetflood gravels are dominantly clast-supported (Nemec and Steel, 1984; Wells, 1984); matrix is principally silt or sand. According to Nemec and Steel (1984), sheetflood gravels deposited from ephemeral (flash) floods tend to have the least textural maturity. More texturally mature gravels probably result from stronger channelized transport, more-continuous runoff, and effective contemporaneous reworking. Some texturally mature gravels may initially have an openwork fabric (no matrix) or may contain a matrix of

TABLE 6.6

Characteristics of sheetflood (braided-stream) conglomerates

Characteristic	*Description*
Fabric support	Dominantly clast-supported
Texture	Clasts moderately to poorly sorted; commonly contain a silt or sand matrix; clasts subangular to well rounded
Vertical size grading	Dominantly nongraded, but normal and inverse grading common
Clast orientation	Commonly imbricated with *ab* plane dipping upstream; long axes typically oriented transverse to flow direction
Stratification	Stratification common; crude to well-developed planar horizontal or inclined stratification; both planar and trough cross-bedding common, and giant-scale (to 25 m thick) cross-strata can occur; may be associated with well-laminated sandstone
Vertical facies sequences	Fining-upward sequences most typical, but coarsening-upward sequences also occur

sand entrained from the stream bed during gravel transport. Clasts may be well rounded to subrounded (Ramos and Sopeña, 1983); however, the deposits of flash floods may contain abundant angular clasts. Size sorting of clasts is variable, probably ranging mainly from moderate to poor, although Bull (1972) suggests that many sheetflood gravels are well sorted. Imbrication of clasts may be well developed. The ab planes of the clasts commonly dip upstream, and the long axes of prolate pebbles are oriented transverse to stream flow (Rust, 1978). Sheetflood gravels may be nongraded (Fig. 6.6), normally graded, or inversely graded, although nongraded gravels appear to be most common. Fining-upward sequences are particularly common in sheetflood gravel deposits (Rust, 1978; Nemec and Steel, 1984), but coarsening-upward sequences occur (Middleton and Trujillo, 1984). Miall (1977, 1978) presents models for several types of braided-river depositional profiles, all of which are characterized by fining-upward cycles.

Stratification in sheetflood gravels is common, ranging from crudely developed to well developed. Stratification may be either planar horizontal (Fig. 6.6) or inclined. Cross-bedded units can range from planar cross-bedded to trough cross-bedded, and foreset dips can range from low- to high-angle. Gravel beds may have lensoid geometry and abundant curved erosion surfaces (Nemec and Steel, 1984). Sheetflood gravel units are commonly associated with well-laminated sandstones. A schematic representation of stratification types in braided-stream conglomerates and associated sandstones by Rust (1978) is shown in Figure 6.7. Miall (1977) recognized three main types of gravel facies in braided stream deposits, which he labeled Gm, Gt, and Gp. The **Gm** facies consists of massive or crudely bedded gravel with minor sand, silt, or clay lenses (longitudinal-bar or channel-lag deposits), the **Gt** facies consists of stratified gravel with broad, shallow

FIGURE 6.6
Poorly graded, massive to crudely stratified sheetflood conglomerate, Port Orford Formation (Early Pleistocene), southern Oregon coast.

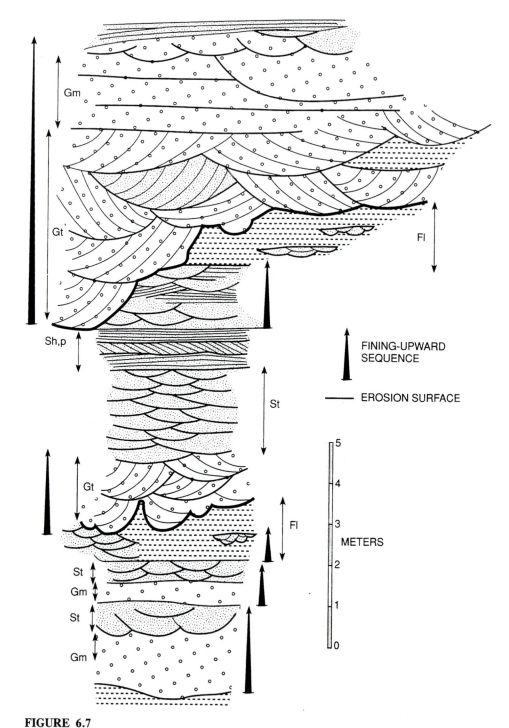

FIGURE 6.7

Schematic representation of typical stratification and structures in braided-stream (sheetflood) conglomerates. Gm = massive, horizontally bedded gravel; Gt = trough cross-bedded gravel; Sh = horizontally stratified sand; Sp = planar cross-stratified sand; St = trough cross-stratified sand; Fl = laminated or cross-laminated fine sand, silt, or mud. (*After Rust, B. R., 1978, Depositional models for braided alluvium, in Miall, A. D., ed., Fluvial sedimentology: Canadian Soc. Petroleum Geologists Mem. 5. Fig. 9, p. 618, reprinted by permission.*)

trough cross-beds and gravel imbrication (minor channel fills), and the **Gp** facies is stratified gravel with planar crossbeds (deposits of linguoid bars or deltaic growths from older bar remnants.

Sheetflood conglomerates can apparently contain clasts of almost any lithology, ranging from compositionally immature (weakly durable) to compositionally mature. The deposits of texturally immature flash floods probably contain higher percentages of weakly durable clasts than those in more texturally mature gravels. Ethridge et al. (1984) report clasts in Precambrian braided-stream conglomerates that include schist, phyllite, gneiss, amphibolite, greenstone, vein quartz, quartzite, jasper, chert, and clasts of banded iron.

Features of sheetflood braided-stream conglomerates that appear particularly characteristic include (1) predominance of clast-supported fabrics, (2) well-developed, upstream-dipping clast imbrication and transverse long-axis orientation, (3) common occurrence of fining-upward sequences, (4) common presence of planar stratification and/or cross-bedding, and (5) lensoid geometry of gravel units.

Streamflow Conglomerates

Streamflow conglomerates are deposited by normal traction-transport, streamflow processes in single channels of relatively straight rivers, or high-gradient nonbraided rivers. Even some moderately meandering rivers can transport gravels. Gravelly, nonbraided streams have been described by Jackson (1978) and Forbes (1983), among others. Gravel transport in these streams occurs during seasonal extreme flooding or during high-magnitude flood events that happen only rarely (Baker, 1984; Gupta, 1983). These flood events result in the deposition of coarse gravel in stream channels as crudely horizontal sheets or as flow-transverse gravel waves, in chute and point bars, and on terraces, floodplains, and levees (Baker, 1977; Gupta, 1983). Transport of gravel within single channels can occur also on alluvial fans or outwash fans where channels are entrenched temporarily into the fans.

Distinguishing characteristics of streamflow conglomerates are summarized in Table 6.7. Many of the characteristics of ancient conglomerates presumed to be of channel-fill origin appear very similar to those of braided stream conglomerates (compare Tables 6.6 and 6.7). Streamflow conglomerates are dominantly clast-supported, and matrix is commonly sand or silt. Some streamflow conglomerates were deposited with an openwork fabric and contain only small amounts of sand/clay matrix that subsequently washed into interstitial spaces. The clasts of streamflow conglomerates are generally moderately well sorted to moderately sorted (Gustavson, 1978); however, high-magnitude floods can deposit extremely large clasts, generating deposits with a wide range of gravel sizes (Baker, 1984). Clast roundness ranges from subangular to well rounded, although well-rounded clasts are probably most common. Clasts tend to be well imbricated (Fig. 6.8), with **ab** planes dipping upstream, but variations from upstream dips have been reported (e.g., Folk and Ward, 1957). The long axes of prolate pebbles are commonly oriented transverse to current flow; however, a dominant mode of long-axis orientation parallel to current flow has been reported by a few investigators (e.g., Schlee, 1957). Most streamflow conglomerates are probably nongraded, but both normal and reverse grading can occur (Gustavson, 1978). They may grade upward into overbank sand or mud deposits (e.g., Arche, 1983).

Streamflow gravels may be deposited as transverse bars or in gravel sheets. Structures within bars are mainly large-scale planar cross-beds and horizontal beds but may

TABLE 6.7

Characteristics of streamflow conglomerates

Characteristic	Description
Fabric support	Dominantly clast-supported
Texture	Matrix commonly sand or silt, and may be rare to abundant; sorting of clasts may range from moderate to very poor; maximum clast size can range to several meters; clasts sub-angular to well rounded
Vertical size grading	Dominantly nongraded, but may display normal or inverse grading
Clast orientation	Commonly imbricated with *ab* plane dipping upstream; long axes typically oriented transverse to flow direction
Stratification	Transverse bar deposits characterized by large-scale planar cross-beds and horizontal beds, some trough-stratified units; gravel sheet deposits commonly massive to horizontally stratified; intercalated sand/silt lenses may be present
Vertical facies sequences	May grade upward into sandstone or shale (mudstone)

include some trough cross-stratified units and shallow scours with gravel lags. Giant-scale cross-bed sets to 25 m thick and with high-angle foresets have been reported from some conglomerates (see Nemec and Steel, 1984). Channel-floor gravels are commonly massive or horizontally stratified (Forbes, 1983). Gravel sheets can display crude horizontal layering that may be inversely graded. As mentioned, streamflow conglomerates may pass upward or laterally into sandy gravels and sandstones or silty sandstones (Fig. 6.9), and fining-upward sequences are common. Fining-upward sequences are typically only a few meters thick; however, vertical stacking of gravel units, owing to channel migration, can produce very thick conglomerate bodies. Any kind of clasts may be present in streamflow conglomerates, depending upon the source. Multicycled conglomerates may be enriched in highly durable clasts such as quartzite. S. A. Smith (1990) provides a detailed description of an ancient gravel-bed stream deposit from the Lower Triassic of southwest

FIGURE 6.8

Well-imbricated gravels in terrace deposits of the Umpqua River, southwest Oregon. The arrow shows direction of stream flow.

FIGURE 6.9
Clast-supported streamflow gravels
grading upward into planar-
laminated sand. Terrace deposits of
the Umpqua River, southwest
Oregon.

England. This sequence of deposits ranges from 20 m to 30 m thick and consists dominantly of conglomerate with subordinate pebbly sandstone (Fig. 6.10). The conglomerate
is typically poorly sorted, clast-supported, and poorly imbricated. Most conglomerate
units have closed-framework fabrics; however, a few have open-framework fabrics.
Smith recognizes three depositional styles in the conglomerates: (1) planar cross-bedded
conglomerate (sets 1–3 m thick) interbedded with horizontally bedded conglomerate and
sandstone (Fig. 6.10B1), which occurs mainly in the lower portion of the sequence and
which reflects deposition from linguoid-shaped macro-bedforms, (2) couplets of trough
cross-bedded sandstone and horizontally bedded conglomerate (Fig. 6.10B2), which occur in the upper part of the sequence and which reflect deposition in less-channelized
fluvial systems by migration of relatively low-relief macro-bedforms and large sand
ripples, and (3) localized large-scale trough cross-bedded conglomerate wedges (Fig.
6.10B3), which formed by erosion and subsequent fill of scours cut into thick, sandy
substrate.

   The outstanding features of streamflow conglomerates appear to be that they (1) are
dominantly clast-supported, in some cases with openwork fabrics, (2) display unimodal
clast orientation with dominant upstream imbrication dips and transverse long-axis orientation, (3) have fair sorting of clasts and generally are well rounded, (4) exhibit abundant cross-stratification, and (5) may contain very large clasts—probably larger than
those common to braided-stream conglomerates. Clearly, the characteristics of channelized streamflow conglomerates and sheetflood conglomerates are generally very similar.
Differentiating between these conglomerates in ancient deposits may require study of
stratification, sedimentary structures, vertical and lateral facies sequences, and so on in
associated finer-grained deposits.

Wave-worked (Beachface) Conglomerates
Wave-worked conglomerates occur in nearshore environments where wave energy is
sufficient to transport and rework gravels supplied by fluvial input or coastal erosion.
Reworking takes place on the beachface in the active swash zone, particularly on wave-
dominated coasts where wave energy overwhelms tidal energy. Wave-worked conglom-

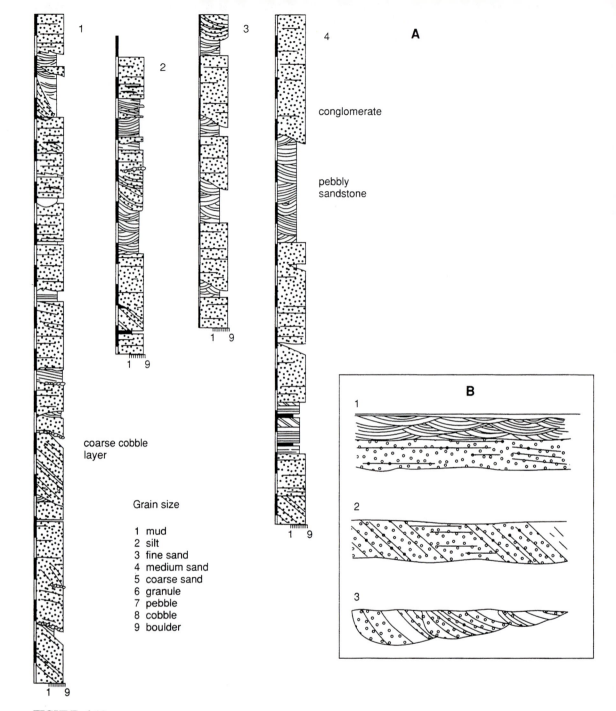

FIGURE 6.10

Ancient streamflow conglomerates of the Lower Triassic Budleigh Salterton Pebble Beds of east Devon, England. A. Stratigraphic sections at (1) Budleigh Salterton, (2) Blackhill Quarries, (3) Foxenholes and Beggar's Roost, and (4) Hillhead Quarry; scale is in meters. B. Schematic sketches of depositional styles in the conglomerates, not to scale: (1) couplets of trough cross-bedded sandstone and horizontally bedded conglomerate, (2) planar cross-bedded conglomerate interbedded with horizontally bedded conglomerate and sandstone, (3) large-scale trough cross-bedded conglomerate wedges. (*After Smith, S. A., 1990, The sedimentary and accretionary styles of an ancient gravel-bed stream: the Budleigh Salterton Pebble Beds (Lower Triassic), southwest England: Sed. Geology, v. 67. Fig. 2, p. 201, and Fig. 12, p. 214, reprinted by permission of Elsevier Science Publishers, Amsterdam.*)

erates originate in both lacustrine and marine environments; however, little information is available about the characteristics of lacustrine wave-worked conglomerates. Therefore, the models for wave-worked conglomerates described below are based mainly on marine examples. Presumably, the characteristics of wave-worked conglomerates deposited in large lakes would be similar.

Wave-worked (beachface) conglomerates are deposited in the active swash zone extending from mean high tide to mean low tide. Transport of clasts is dominated by wave-driven upswash and gravity-driven backswash (sheetflow). During flood tide, water is lost by percolation into the beach and the backswash is weaker than the shoreward swash. During ebb tide, the backswash may be stronger than the upswash owing to water added from the beach water table (Komar, 1976). Gravels can be transported both landward and seaward on beaches, although net landward movement during normal sea conditions is probably small (Bluck, 1967; Carter and Orford, 1984). During storm events, particles of many different sizes, including extremely large clasts, are probably moved up the beachface. If a berm is present, water may be carried over the berm and down the backbeach (Dupré et al., 1980; Kirk, 1980).

The constant reworking of clasts in the surf zone tends to produce well-sorted, well-rounded, clast-supported gravel deposits (Fig. 6.11); however, the characteristics of wave-worked conglomerates actually vary considerably, as indicated in Table 6.8. Gravelly beaches may range from those characterized dominantly by gravel to beaches composed of mixed sand and gravel or of sand with only scattered pebbles. Beachface conglomerates are typically clast-supported with a sand matrix; however, on dominantly sand beaches, pebbles may be segregated into thin layers interspersed in a dominantly sand section (Clifton, 1973). Also, isolated pebbles can occur scattered through the sand. Some thicker units may be sand-matrix–supported and tend to have a well-developed, subhorizontal, thin stratification (Kleinspehn et al., 1984). Framework clasts are com-

FIGURE 6.11
Moderately well-sorted, well-rounded, clast-supported modern beach gravels, southern Oregon coast. Note also the imbrication of the clasts.

TABLE 6.8

Characteristics of wave-worked (beachface) conglomerates

Characteristic	Description
Fabric support	Typically clast-supported with a sand matrix; on dominantly sand beaches, clasts may occur in thin gravel layers in a dominantly sand section
Texture	Commonly well sorted, but bimodal or polymodal size textures may occur; clasts tend to be well rounded, and disc-shaped clasts are particularly characteristic; fossils rare
Vertical size grading	Dominantly nongraded, but normal grading can occur
Clast orientation	Tend to be well imbricated with *ab* planes dipping in a seaward direction; long axes of clasts may be oriented at various angles to the strike or trend of the beach
Stratification	Well stratified; most strata dip seaward, but backbeach units may have a landward dip; associated sands may contain planar laminations rich in heavy minerals
Vertical facies sequences	Progradational sequences coarsen upward; locally, fining-upward sequences occur in which deposits grade from gravels to sand

monly well sorted in individual beds, but successive beds may vary considerably in coarseness (Nemec and Steel, 1984). Sand or small pebbles may fill interstitial spaces among clasts, producing bimodal or polymodal size textures (Nemec and Steel, 1984; Bluck, 1967). Beach clasts are well rounded, and disc-shaped clasts are especially characteristic (Fig. 6.12). On some beaches, disc-shaped clasts tend to occur preferentially on the upper beachface, whereas spherical and rod-shaped clasts are more characteristic of the lower beachface (Bluck, 1967; Bourgeois and Leithold, 1984).

Gravels on beaches tend to be well imbricated, with **ab** planes dipping in a seaward direction (Fig. 6.13), and long axes may be oriented at various angles to the strike or trend of the beach. Work by Ogren and Waag (1986) and Waag and Ogren (1984) on pocket beaches in Michigan indicates that most clasts dip seaward at angles exceeding the foreshore slope, although a few clasts may dip landward. On the landward side of the berm, some clasts may dip landward also (Bluck, 1967; Dupré et al., 1980; Maejima, 1982). Beachface conglomerates are commonly nongraded, but normal grading may be present. Little information has been reported about vertical facies sequences in beachface conglomerates. Progradational sequences presumably coarsen upward on a gross scale; however, Hunter (1980) describes fining-upward gravelly Pleistocene beach deposits that grade from gravels and pebbly sand at the base through a middle unit of fine to coarse sand to an upper unit of interbedded mud, muddy sand, and sand.

Conglomerates deposited on the beachface are commonly well stratified (Fig. 6.14). Most strata dip gently seaward, but strata may dip landward on backbeach sequences (Clifton, 1973; Maejima, 1982). Beds tend to be thin and laterally persistent, or they may consist of discontinuous horizons of isolated pebbles. Associated sands may contain planar laminations rich in heavy minerals (Clifton, 1969). Fossils are uncommon in beachface conglomerates. If present, they consist of thick, robust shells of organisms thrown onshore during storms and sorted by waves and backswash (Bourgeois and Leithold, 1984; Leithold and Bourgeois, 1984). Biogenic structures are also rare in the

FIGURE 6.12
Well-rounded, disc-shaped clasts on a modern beach in Tahiti. (*Photograph courtesy R. L. Folk.*)

FIGURE 6.13
Shingled beach gravels on the southern Oregon coast. Note the strong seaward dip of the imbricated clasts. See also Figure 6.11.

FIGURE 6.14
Stratification in Pleistocene beachface gravels, Cape Blanco, southern Oregon coast.

conglomerates, but may be fairly common in associated sandstones (Clifton and Thompson, 1978). If biogenic structures are present in conglomerates, they are likely to be vertical or U-shaped escape structures (Bourgeois and Leithold, 1984). Beachface conglomerates may include any kind of durable clast, including limestone clasts. Weakly durable clasts such as shale clasts probably do not survive in the beach environment.

Features of beachface conglomerates that are particularly characteristic include (1) an abundance of disc-shaped clasts, (2) generally good sorting of clasts, (3) clast-supported fabrics, (4) well-developed, seaward-dipping imbrication, (5) gentle, seaward-dipping stratification, (6) heavy mineral laminae in associated sandstones.

Wave-, Storm-, and Current-worked (Shoreface and Shelf) Conglomerates

The dominantly wave-worked beachface grades seaward into the shoreface. The shoreface, which extends from mean low tide level seaward to a point where waves begin to break, is the zone of breaking waves, longshore currents, and rip currents and is also affected by storms. Therefore, conglomerates deposited on the shoreface are reworked by both waves and currents. Water depth is generally less than 10 m, and bars and troughs are common features of this zone (Hunter et al., 1979). As shoaling waves become oversteepened and break, the waves are transformed into a bore that moves through the surf zone. The breaking waves and bores move water shoreward and pile it against the shoreline, thereby creating a hydraulic head that generates longshore and rip currents. According to Bourgeois and Leithold (1984), shoreward migration of bars and megaripples, as the wave bore moves through the surf zone, causes deposition of shoreward-dipping cross-bedded conglomerates and pebbly sandstones on the lower shoreface. Offshore movement of gravelly sediment can occur also owing to storm transport (Hart and Plint, 1989). The upper shoreface is dominated by longshore- and offshore-dipping cross-bedded pebbly sandstones and gravel-filled scours generated by longshore currents and rip currents.

Seaward of the shoreface, at water depths greater than about 10 m, bottom sediment is mainly below normal wave base. Such sediment may, however, be resuspended and reworked by storm waves and currents during storm activity. Therefore, gravelly sediment may also accumulate on the shelf seaward of the shoreface.

Most shoreface deposits that have been reported in modern settings (e.g., Hunter et al., 1979; Shipp, 1984) are sand-dominated systems. Many ancient systems (e.g., Clifton, 1981a; Leckie and Walker, 1982; Cant, 1984) are also sand-dominated systems, although some sand-dominated systems (e.g., Leithold and Bourgeois, 1984; DeCelles, 1987; Leckie, 1988) also contain numerous conglomerate lenses and beds. Gravel-dominated shoreface sequences appear to be much rarer than sand-dominated systems. Hart and Plint (1989) report a modern gravel-dominated shoreface at Chesil Beach, England, and also describe an ancient gravel-dominated shoreface sequence from the Cardium Formation (Cretaceous) in Alberta, Canada.

General characteristics of shoreface conglomerates are summarized in Table 6.9. The characteristics of some conglomerates in sand-dominated systems are illustrated diagrammatically in Figure 6.15, and the characteristics of a gravel-dominated shoreface sequence are shown in Figure 6.16. Shoreface conglomerates may be clast-supported or sand-matrix–supported (diamictites) (Table 6.9). Sand-matrix–supported conglomerates or pebbly sandstones typically consist of thin, laterally discontinuous beds that exhibit low-angle cross-stratification (Fig. 6.17A). Clast-supported conglomerates tend to occur in laterally discontinuous, channel-form lenses (Fig. 6.17B; Bourgeois and Leithold, 1984). Few data are available on the size sorting of shoreface conglomerates, although they may be more poorly sorted in general than beachface conglomerates. For example, DeCelles (1987) reports poorly sorted conglomerates in Tertiary shoreface conglomerates of California. Probably, fewer disc-shaped pebbles occur in shoreface conglomerates than in beachface conglomerates. Imbrication in shoreface conglomerates may be bimodal or

TABLE 6.9

Characteristics of wave-, storm-, and current-worked (shoreface and shelf) conglomerates and diamictites

Characteristic	Description
Fabric support	Clast-supported or sand-matrix–supported
Texture	Moderately to poorly sorted(?), generally well rounded, fewer disc-shaped clasts than in beachface conglomerates; fossils common, bioturbation in sands
Vertical size grading	Nongraded or graded; crude normal grading common in upper-shoreface conglomerates; lower-shoreface conglomerates may or may not show normal grading; storm-deposited shelf sediments may display normal grading
Clast orientation	Not well established; possibly bimodal or polymodal imbrication dips
Stratification	Occur in lenticular or sheetlike units or in units dominated by low-angle shoreward-dipping or offshore-dipping cross-strata; longshore- and offshore-dipping, cross-bedded pebbly sandstones and gravel-filled scours also common; commonly associated with hummocky cross-strata
Vertical facies sequences	Progradational sequences coarsen upward on a gross scale; individual bar sequences may display fining-upward trends

FIGURE 6.15

Schematic representation of wave- and current-dominated (shoreface) conglomerates in the sandstones of Floras Lake (Miocene), southwestern Oregon. (*After Leithold, E. L., and J. Bourgeois, 1984, Characteristics of coarse-grained sequences deposited in nearshore, wave-dominated environments—examples from the Miocene of southwest Oregon: Sedimentology, v. 31. Fig. 3, p. 572, reprinted by permission of Elsevier Science Publishers, Amsterdam.*)

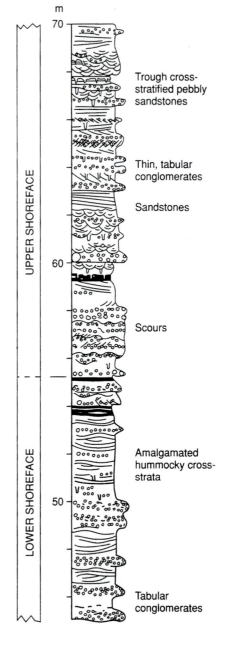

Trough cross-stratified pebbly sandstones

Thin, tabular conglomerates

Sandstones

Scours

Amalgamated hummocky cross-strata

Tabular conglomerates

EXPLANATION

∪∪ BURROWS

⊃⊂ SHELLS

☛ ORGANIC DEBRIS

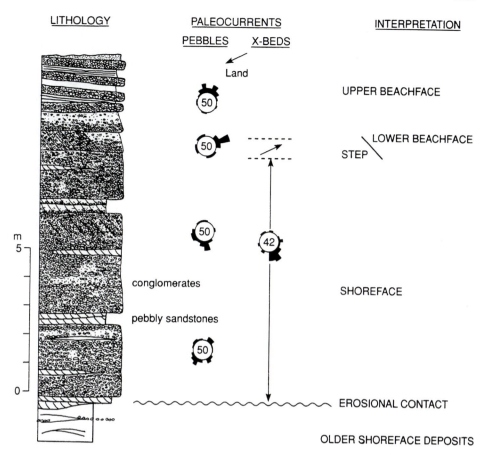

FIGURE 6.16

Characteristics of a gravel-dominated shoreface sequence in the Baytree Member of the Cardium Formation (Cretaceous), Alberta, Canada. Numbers inside the paleocurrent rose diagrams refer to number of measurements. (*After Hart, B. S., and A. G. Plint, 1989, Gravelly shoreface deposits: a comparison of modern and ancient facies sequences: Sedimentology, v. 36. Fig. 4, p. 556, reprinted by permission of Elsevier Science Publishers, Amsterdam.*)

polymodal owing to different depositional processes (waves, longshore currents, rip currents, storm surges), but few data are available to confirm this possibility. Bourgeois and Leithold (1984) suggest that crude normal grading is common in upper-shoreface conglomerates. Dupré (1984) reports upper-shoreface gravels that range from graded to massive, or cross-stratified. Lower-shoreface conglomerates may or may not show normal grading. Progradation probably produces coarsening-upward sequences on a gross scale, but individual bar sequences can display a fining-upward trend (J. Bourgeois, personal communication). Locally, shoreface conglomerates may be characterized by alternating, thin, gravelly lenses or layers and thicker sandstone layers.

Lower-shoreface conglomerates occur in lenticular or sheetlike units that represent storm-wave–reworked lag pavements or in units dominated by low-angle, shoreward-

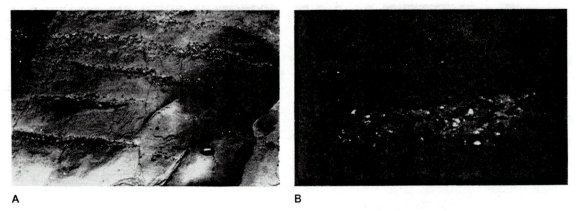

A   B

FIGURE 6.17
Shoreface conglomerates in sandstones of Floras Lake (Miocene), southwest Oregon. A. Low-angle cross-stratificatied, thin conglomerates in a sand-dominated unit. B. Channel-form lenses of clast-supported gravel in shoreface sandstones.

dipping cross-strata. On the other hand, seaward-dipping strata have been reported from modern shoreface deposits at Chesil Beach, England (Hart and Plint, 1989). The cross-bedded units typically become sandy in the upper half of individual sets, and the conglomerates are commonly associated with **hummocky cross-strata** (Leckie, 1988). Sheetlike lag pavements can occur also in upper-shoreface sequences, along with longshore- and offshore-dipping trough cross-bedded pebbly sandstones and high-angle gravel-filled scours (Bourgeois and Leithold, 1984; Leithold and Bourgeois, 1984; Nemec and Steel, 1984). Fossils, especially bivalves and gastropods, are more common in shoreface conglomerates than in beachface conglomerates. Trace fossils may be locally abundant in associated sandstones. Clasts of virtually any lithology can occur, with the possible exception of weakly durable clasts.

Features especially characteristic of shoreface conglomerates include (1) the common presence of low-angle cross-stratification, (2) polymodal dips of cross-bed sets, (3) presence of associated hummocky cross-strata, and (4) moderately abundant fossils and bioturbation features in associated sandstones.

As mentioned, conglomerates can also occur in shelf environments seaward of the shoreface at water depths below normal wave base (Leckie and Walker, 1982; Leckie, 1988). Presumably, the gravels are transported offshore into this environment by storm rip currents, where they accumulate in graded gravel bars or large ripples (Fig. 6.18). Gravel ripples are reported by Leckie (1988) to occur in modern marine environments at depths ranging from 3 to 160 m. The ripples have crest spacings of 0.25–3.0 m and amplitudes of 5–35 cm. Leckie also cites several examples of ancient coarse-grained ripples that formed in shelf environments.

Ancient storm-wave–formed shelf conglomerates appear to have the following characteristics: (1) poor to moderate sorting, (2) fabrics are clast-supported or matrix-supported, (3) beds are sharp-based and range to 1 m or more in thickness, (4) individual beds may be cross-bedded, structureless, graded (commonly normally), or imbricated in lower portions, (5) clasts commonly show bimodal dip directions (on opposing sides of ripple crests), (6) commonly occur in close juxtaposition, both vertically and laterally, with hummocky cross-stratified sandstones.

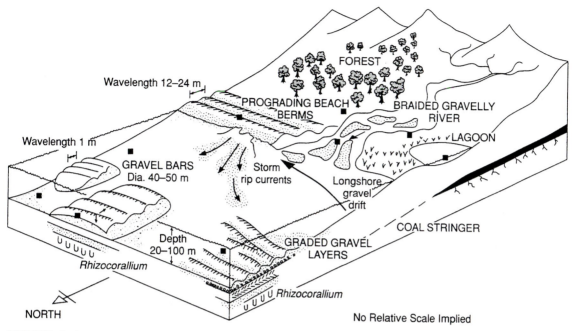

FIGURE 6.18

Schematic representation of storm-deposited offshore gravel bars and graded gravel layers in the Cretaceous Gates Formation, western Canada. Note that gravels were likely transported offshore from the beachface or shoreface by storm rip currents. (*After Leckie, D. A., and R. G. Walker, 1982, Storm- and tide-dominated shorelines in Cretaceous Moosebar–Lower Gates Interval —Outcrop equivalents of deep basin gas trap in western Canada: Am. Assoc. Petroleum Geologists Bull., v. 66. Fig. 22, p. 155, reprinted by permission of AAPG, Tulsa, Okla.*)

Tide-worked Conglomerates and Diamictites

Tidal currents are generated by movement of a progressive tidal wave, characterized by a tidal ellipse with nearly horizontal motion. This motion reverses as the crest of the tidal wave passes. Tides are amplified as they move into shallower water and narrower channels (Howarth, 1982), reaching surface velocities that may exceed 2 m/s and bottom velocities of 1 m/s or more, particularly in narrow channels. The reversing tidal currents generated as the tidal stage passes from flood to ebb commonly display velocity asymmetry (Boggs and Jones, 1976; Howarth, 1982; Klein, 1977); therefore, some net transport of sediment can occur under tidal conditions. Tidal currents on macrotidal and mesotidal coasts may thus be effective agents for reworking and transporting gravel.

Tide-worked surface gravels are known to occur in modern marine-shelf environments (Belderson and Stride, 1966; review by Phillips, 1984); however, relatively little appears to be known about the characteristics of these gravels. Likewise, relatively little is apparently known about the characteristics of ancient tide-worked conglomerates, as few examples of such conglomerates have been described in the literature. One well-studied example of a possible tide-worked conglomerate is the basal Santa Margarita Sandstone in central California (Phillips, 1983, 1984). Characteristics of this conglomerate are summarized in Table 6.10, and some bedding characteristics of the conglomerate are shown in Figure 6.19.

TABLE 6.10

Characteristics of tide-worked conglomerates and diamictites

Characteristic	Description
Fabric support	Clast-supported or matrix-supported; matrix-supported conglomerates (diamictites) probably most common
Texture	Moderately to poorly sorted, commonly bimodal; sand infill; mainly rounded to well rounded clasts; fossils and trace fossils abundant in associated sandstones
Vertical size grading	Nongraded, or normally or inversely graded; predominance of a particular type of grading not well established
Clast orientation	Preferred orientation typically poorly developed; some clasts may be imbricated
Stratification	Tabular conglomerate units, trough cross-beds, and tabular cross-beds all common; directions of foreset dips may reverse from one layer to another; may overlie reactivation surfaces
Vertical facies sequences	Display a general fining-upward trend, with clasts diminishing upward in both size and abundance

Source: Mainly Phillips, R. L., 1984, Depositional features of Late Miocene, marine cross-bedded conglomerates, *in* Koster, E. H., and R. J. Steel, eds., Sedimentology of gravels and conglomerates: Canadian Soc. Petroleum Geologists Mem. 10, p. 345–358.

A   B

FIGURE 6.19
A. Large-scale (>5m thick) trough cross-bed in conglomerates of the basal Santa Margarita Sandstone (miocene), Santa Cruz Mountains, California. Ruler scale = 20 cm. B. Conglomerate and sand cross-beds in the Santa Margarita Sandstone. Scale bar = 50 cm. *(Photographs courtesy of R. L. Phillips.)*

6.9.3 *Conglomerates and Diamictites Deposited by Ice Flow*

Meltout/Lodgment Conglomerates and Diamictites (Tills)

Rock materials dropped by melting of grounded glaciers form poorly sorted, matrix-rich gravelly deposits that are commonly called **till** or **glacial diamictite**. Glaciers acquire their gravel load by plucking and abrasion along the walls and floors of the glaciers and, to some extent, by rockfall from above. The source of the gravel may be either crystalline, metasedimentary, or sedimentary bedrock or previously transported glacial sediments such as glacial outwash gravels or stream gravels that are reentrained. Gravels are transported within and along the sides, base, or top of the glacier. Glacial transport results in grinding of the rock load against the walls and floor of the glacial valley, generating abundant fine silt-size and clay-size material. Melting of grounded glacial ice thus deposits poorly sorted, poorly stratified glacial debris characterized by a large range of particle sizes. The term **lodgment till** has been applied to glacial materials deposited from the sliding base of a dynamically active glacier by pressure melting and/or other mechanical processes. **Meltout till** is formed by the slow release of debris from glacial ice that is not sliding or deforming internally (Ashley et al., 1985; Edwards, 1986). Although gravel deposits formed by meltout and lodgment have somewhat different characteristics, they are considered together here as a single, closely related type of grounded ice deposit. **Meltout/lodgment conglomerate (diamictite)** is suggested as a process-related name for ancient coarse-grained grounded-ice deposits.

   Some characteristics of gravelly grounded ice deposits are summarized in Table 6.11. Owing to the abundance of fine-size detritus carried by glaciers, meltout/lodgment gravels are dominantly matrix supported. They are commonly very poorly sorted, with bimodal or polymodal grain-size distributions (Karrow, 1976), and they may contain clasts of widely ranging sizes—from granules to boulders. Matrix includes sand, silt, and clay. Clasts are rounded to various degrees, and some may be faceted, striated, polished, or fractured. Well rounded gravels may have been incorporated into ice riding over stream gravels (Easterbrook, 1981). The long-axis orientation of prolate pebbles and the ab planes of blade- and plate-shaped clasts in glacial deposits have long been regarded to be

TABLE 6.11

Characteristics of meltout/lodgment conglomerates and diamictites

Characteristic	*Description*
Fabric support	Dominantly matrix-supported
Texture	Very poorly sorted, bimodal or polymodal; clast size can range to meter-size boulders; matrix may be sand, silt, or clay; clasts angular to well rounded; some clasts may be faceted or striated
Vertical size grading	Nongraded
Clast orientation	Long dimension of clasts dominantly parallel to ice-flow direction
Stratification	Mostly massive, but may contain lenses and beds of stratified, better-sorted sediment
Vertical facies sequences	May be succeeded upward by stratified, finer-grained glaciofluvial or glaciolacustrine deposits

parallel to the direction of ice flow, and these features are regarded to be generally reliable guides to flow direction (Lawson, 1979; Dowdeswell et al., 1985b; Dowdeswell and Sharp, 1986).

Dragging of clasts across underlying deformable material can erode grooves that are subsequently filled by till. Thus, some meltout/lodgment tills are characterized by large **groove casts** on their bases (Ashley et al., 1985, p. 34). Meltout/lodgment gravels are nongraded and commonly show no definite pattern of vertical grain-size change; however, massive, nongraded meltout/lodgment gravels can be succeeded upward by finer-grained, stratified glaciofluvial or glaciolacustrine deposits.

Meltout/lodgment conglomerates (diamictites) tend to be massive and unstratified in general (Fig. 6.20); however, they may contain lenses and beds of stratified, better-sorted deposits and may show crude banding, particularly in the basal parts of thick units (Edwards, 1986). Nonetheless, they clearly rank among the most poorly stratified of all gravel deposits. Clasts in meltout/lodgment tills may be of any lithology, depending upon the lithology of the bedrock over which they move or the composition of previously deposited gravels that may be reentrained. Clasts may include exotic or erratic lithologies that were transported considerable distances.

The most important characteristics of meltout/lodgment conglomerates and diamictites include (1) dominant matrix support, (2) extremely poor sorting, (3) generally massive, unstratified character of deposits, (4) strongly developed preferred long-axis orientation of clasts in an ice-flow parallel direction, but little or no imbrication, and (5) possible presence of faceted, striated, and polished clasts. Although not abundant in the geologic record, meltout/lodgment conglomerates and diamictites have been reported particularly in rocks of Precambrian, Ordovician, Carboniferous, Permian, and Pleistocene ages. On the other hand, some deposits originally believed to be tills have been reinterpreted as massflow deposits.

Subaqueous Meltout Conglomerates and Diamictites (Aquatillite)

The somewhat cumbersome term subaqueous meltout conglomerate (diamictite) is suggested for conglomerates that form in either lacustrine or marine environments owing to

FIGURE 6.20
Erosional pinnacles formed of massive, unstratified glacial meltout/lodgment conglomerate. Valley of Herron, Switzerland. (*Photograph by E. M. Baldwin.*)

direct deposition by melting of gravel-charged ice. Other names that have been applied to the general category of glacial materials deposited in water include paratillite (Harland et al., 1966), aquatillite (Schermerhorn, 1974), or simply glacial marine, glaciomarine, glacial lacustrine, or glaciolacustrine deposits, as appropriate. Two principal depositional processes may be involved in direct deposition of subaqueous ice-meltout conglomerates and diamictites: (1) deposition as a traction carpet by dense underflows of meltwater, probably issuing from subglacial tunnels at the front of retreating ice (Rust and Romanelli, 1975; Rust, 1977) (this process leads to sedimentation of coarse debris close to the ice front), and (2) deposition of ice-rafted material, melted out of floating ice. Most such ice-rafted "dropstones" and associated meltout debris are deposited relatively close to the ice front, but some may be transported to far-distant areas. Glacial deposits formed in water by these mechanisms constitute a type of outwash, which Rust and Romanelli (1975) termed **subaqueous outwash**. As suggested by Edwards (1986), such outwash could be called **sublacustrine outwash** if deposited in a lake. Following that line of reasoning, it could be called **submarine outwash** if deposited in a marine environment. It is necessary to make a distinction between subaqueous outwash deposited directly from melting ice, as described above, and resedimented subaqueous outwash formed by re-mobilization of subaqueous outwash by subsequent massflow processes such as subaqueous debris flows.

Characteristics of subaqueous meltout conglomerates are summarized in Table 6.12. Conglomerates in ice-rafted deposits are mainly matrix supported (diamictites); however, clast-supported fabrics can occur, especially in underflow traction deposits. Sorting of clasts in these conglomerates and diamictites is typically poor, but it is generally better, particularly in underflow deposits, than that in subaerial ice-meltout/ lodgment conglomerates and diamictites. Very large dropstones can occur in finer sediment at any distance from the shoreline. Powell (1983) suggests that the sorting of

TABLE 6.12

Characteristics of subaqueous meltout conglomerates and diamictites

Characteristic	*Description*
Fabric support	Dominantly matrix-supported; rarely clast-supported
Texture	Sorting generally poor, but traction-underflow deposits may be better sorted than subaerial meltout/lodgment conglomerates; clasts angular to well rounded, depending upon source, possibly striated, polished, faceted
Vertical size grading	Nongraded, or very poorly developed grading
Clast orientation	Orientation of dropstones commonly random unless reworked by bottom currents; deposits of traction-underflows may show parallel-to-flow long-axis trends, with upstream imbrication dips
Stratification	Dropstone deposits generally massive; underflow deposits may show crude horizontal stratification; may be associated with better-stratified pebbly sands and silts that display cross-bedding and ripple cross-lamination
Vertical facies sequences	May show no pronounced vertical trends or may pass upward into finer-grained sediments that formed as an ice front retreated

ice-rafted meltout debris may depend upon the rate of melting. If melting is episodic, fall sorting occurs as the sediment falls to the basin floor, producing sorted and graded deposits. If melting is continuous, coarser and finer debris fall together, producing poorly sorted deposits. Coarser debris may occur also as lenses in finer sediments. Powell believes that these lenses form when surface ablation on an iceberg causes debris to accumulate on the ice surface. Faster subaqueous melting of the iceberg eventually creates a mass imbalance, causing the iceberg to roll, dumping the debris as a pile on the ocean floor. Clast orientation in dropstone deposits tends to be random; however, there may be a slight preference toward vertical orientation of elongated dropstone clasts. Also, reworking of dropstone debris by bottom currents may concentrate clasts and produce better-oriented fabrics (Anderson, 1983). Furthermore, the deposits of dense underflows of meltwater may show parallel-to-flow long-axis trends with upstream imbrication dips (Rust, 1977; Mustard and Donaldson, 1987). Clasts may be rounded and may display faceting, striations, or polish. Subaqueous ice-meltout conglomerates and diamictites commonly have poorly developed grading patterns, further supporting the suggestion of rapid deposition from meltwater streams (Mustard and Donaldson, 1987). These deposits may pass upward into finer-grained sediments that formed as an ice front retreated.

Ice-rafted, dropstone deposits are generally massive, but dense underflow deposits may show crude horizontal stratification. Conglomerates and diamictites may be associated with better-stratified pebbly sands and silts that can display sedimentary structures such as cross-bedding and ripple cross-lamination (Rust and Romanelli, 1975). Marine fossils may occur in submarine outwash deposits, but they are rarely abundant.

Subaqueous ice-meltout conglomerates and diamictites may be difficult to differentiate from subaerial ice-meltout conglomerates and diamictites. Possible distinguishing characteristics include (1) clast-supported fabric in some subaqueous conglomerate units, (2) better sorting and better-developed stratification, especially in underflow deposits, (3) association with fine-grained varved sediments (in lacustrine deposits), and (4) the presence of fossils (in marine deposits).

6.9.4 Conglomerates and Diamictites Deposited by Sediment Gravity Flows

Subaerial Debrisflow Conglomerates and Diamictites

Subaerial debris flows constitute extremely important transport mechanisms for gravel on alluvial fans and proglacial outwash fans (Bull, 1972; Harvey, 1984; Larsen and Steel, 1978; Lawson, 1982; Wells, 1984). In these environments, debrisflow gravels may be interbedded with sheetflood or streamflow deposits. Subaerial debris flows are sediment gravity flows composed of gravel-size particles and characterized by the presence of a cohesive matrix of clay particles and fine sand. The matrix has sufficient cohesive strength and buoyancy to support large cobbles and boulders. Debris flows may occur under any conditions where large amounts of unconsolidated sediment or colluvium can become mobilized (slope failure) by heavy rainfall or melting snow. Once in motion, the flows can move down relatively gentle slopes. When the shear stress owing to gravity no longer exceeds the yield strength at the base of the flow, the flow "freezes" and stops moving. Once the debris flow is deposited, the cohesive strength of the matrix keeps clasts of all sizes suspended within the flow, so they do not settle to the base. Consequently, debrisflow deposits tend to be poorly sorted, poorly graded, and poorly stratified. Not all

subaerial debris flows are highly viscous. They can range from slow-moving, high-strength, high-viscosity flows to markedly fluidal, turbulent(?) flows such as those described by Lawson (1982). The classification and characteristics of these sediment gravity flows are considered in detail by Lowe (1979) and Nemec and Steel (1984).

Some typical characteristics of subaerial debrisflow conglomerates and diamictites are summarized in Table 6.13. The gravel deposits of debris flows can range from clast-supported to very mud-rich matrix-supported (Nemec and Steel, 1984; Schultz, 1984; Wells, 1984), although matrix-supported fabrics appear to be much more common (Fig. 6.22). Debrisflow conglomerates are poorly sorted and tend to be bimodal or polymodal (Lawson and Steel, 1978). Clast size can be highly variable, ranging from pebbles to boulders. A rough correlation exists between bed thickness and clast size, with thicker units having larger clasts (Gloppen and Steel, 1981). Debrisflow conglomerates are commonly nongraded, but some units may show crude to well developed inverse grading (owing to dispersive pressures?) or normal grading (deposition from more watery and possibly turbulent flows?) (Nemec and Steel, 1984; Schultz, 1984). Inverse-to-normal grading has been reported also (Larsen and Steel, 1978). Shape of clasts can range from well-rounded to angular, depending upon the source (reentrained gravels vs. weathered bedrock clasts). Clasts commonly show no preferred orientation; however, the deposits of more-fluidal flows may display crude upflow imbrication, with the long axes of prolate pebbles oriented either parallel or transverse to flow (Lawson, 1982). Apparently, flow-parallel orientation is somewhat more common than flow-transverse. The deposits typically show no internal stratification, but they may be crudely layered owing to a succession of deposits (Fig. 6.21). Deposits of more-fluidal flows can show some internal layering, including thick, irregular cross-strata. Debrisflow deposits may be capped by a thin layer of tightly packed gravel, overlain by stratified (stream reworked?) sandstone

TABLE 6.13

Characteristics of subaerial debrisflow conglomerates and diamictites

Characteristic	*Description*
Fabric support	Mud-rich, matrix-supported to clast-supported; matrix-supported conglomerates most common
Texture	Poorly sorted, bimodal to polymodal; clast size highly variable; thicker flow units tend to have larger clasts; clasts commonly poorly rounded, but may be well rounded depending upon source
Vertical size grading	Commonly nongraded, but may display crude to well-developed inverse or normal grading
Clast orientation	Commonly show no preferred orientation, but deposits of more-fluidal flows may display crude upflow imbrication with long axes of prolate pebbles either parallel or transverse to flow; flow-parallel orientation most common
Stratification	Typically show no internal stratification, but may be crudely layered owing to succession of deposits; deposits of more-fluidal flows may display crude internal layering, including thick, irregular cross-strata
Vertical facies sequences	May be capped by a thin layer of tightly packed gravel, overlain by stratified, stream-reworked(?) sandstone

FIGURE 6.21
Massive to crudely stratified
subaerial debrisflow conglomerate.
Clarno Formation (Eocene–
Oligocene), eastern Oregon.
(*Photograph courtesy G. J.
Retallack.*)

(Nemec and Steel, 1984; Wells, 1984). Some of the typical characteristics of subaerial debrisflow deposits are illustrated in Figure 6.22.

The principal distinguishing characteristics of subaerial debrisflow conglomerates and diamictites include (1) very poor sorting, (2) a rough, positive correlation between bed thickness and clast size, (3) mud-rich, matrix-supported ranging to clast-supported, (4) common presence of nongraded units with poor to absent internal stratification; (graded units are much less common), and (5) weak to absent clast orientation. Debrisflow conglomerates deposited in proglacial environments may possibly be distinguished from those in nonglacial environments by the presence of faceted, striated, or polished pebbles. They may be difficult to distinguish from meltout/lodgment conglomerates and diamictites, although meltout/lodgment conglomerates and diamictites should have much better-developed preferred clast orientation.

Resedimented Conglomerates and Diamictites

General Statement.   Resedimented conglomerates and diamictites form by remobilization of sedimentary material that has a previous history of deposition, in a fluvial, lacustrine, coastal, or shelf setting, and retransport into generally deeper water (Howell and Normark, 1982; Walker, 1975a). Reentrainment and retransport of sediment is particularly common in nearshore environments where rates of sedimentation are very high, as on subaqueous outwash plains, deltas, and fan-deltas. Thus, for example, fan-deltas can grade seaward into deeper-water slope and submarine fan deposits (Wescott and Ethridge, 1983) owing to retransport processes. Although no specific process is implied by the name, geologists appear to use the term "resedimented" mainly to describe retransport of sediment by subaqueous massflow processes. Subaqueous massflow processes capable of transporting gravel are probably confined mainly to subaqueous debris flows, density-modified grain flows, and high-density turbidity currents (Lowe, 1982).

The classification of gravelly mass flows is discussed by Lowe (1979, 1982) and Nemec and Steel (1984), among others. Subaqueous debris flows are distinguished from grain flows in that gravel-size particles are supported by the cohesiveness of a sediment-water matrix rather than by dispersive pressures among grains. Subaqueous debris flows

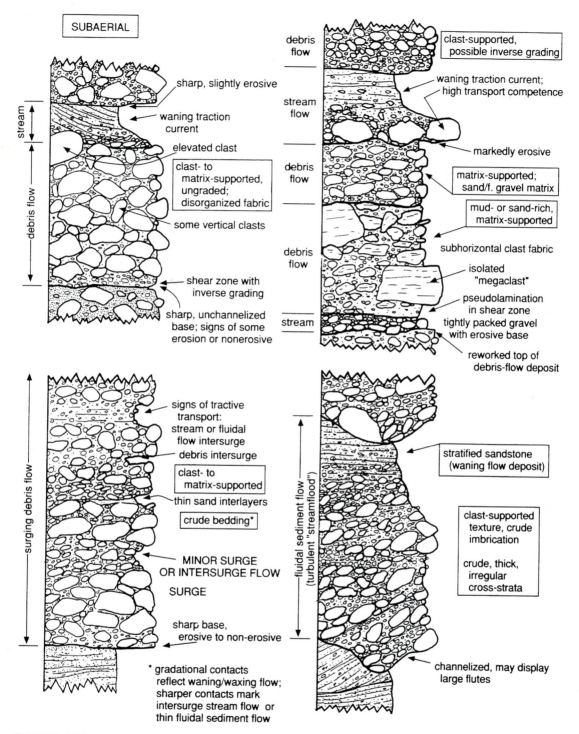

FIGURE 6.22

Schematic sections illustrating typical features of subaerial debrisflow deposits. (*From Nemec, W., and R. J. Steel, 1984, Alluvial and coastal conglomerates: their significant features and some comments on gravelly massflow deposits, in Koster, E. H., and R. J. Steel, eds., Sedimentology of gravels and conglomerates: Canadian Soc. Petroleum Geologists Mem. 10. Fig. 15, p. 15, reprinted by permission.*)

may originate from subaerial debris flows that enter water, or they may originate sub-aqueously when excess pore pressures induce failure. When mixed with sufficient water, these flows can apparently become swift, fully turbulent, fluidal flows (Lawson, 1982). Subaqueous debris flows have also been called slurry flows (Carter, 1975).

Lowe (1982, 1976) used the term ''density-modified grain flow'' for grain flows in which the excess mass of larger particles is supported by the buoyant lift of a dense, cohesionless sediment-water matrix, either silt-sand or clay-silt-sand suspensions between gravel-sized clasts. Density-modified grain flows thus resemble subaqueous debris flows to the extent that larger particles are supported in part by a buoyant matrix, but they differ in that the matrix is cohesionless rather than cohesive.

High-density turbidity currents contain sufficiently high sediment concentrations that support of larger grains comes in part from hindered settling of grains or by dispersive pressures and matrix buoyant lift, as well as from turbulence. Lowe (1982) divides high-density turbidity currents into (1) sandy flows, with grain support provided mainly by turbulence and hindered settling, and (2) gravelly flows, with grain support provided in large part by dispersive pressures and matrix buoyant lift.

A close genetic relationship obviously exists between subaqueous debris flows, density-modified grain flows, and high-density turbidity-current flows. These flow types probably form a continuum. Both subaqueous debris flows and grain flows can likely evolve into fully turbulent turbidity currents with downslope mixing and dilution. Because of this close relationship in depositional process, it can be difficult to differentiate the deposits of these three types of flows—the deposits probably also form a continuum. Subaqueous mass flows can occur both in marine settings and in lakes, including glacially influenced marine settings and lakes. Marine subaqueous massflow deposits may closely resemble the massflow deposits of large lakes, but possibly may be differentiated by features in associated beds such as marine or nonmarine fossils or lake varves.

Walker (1975b) proposed three models for resedimented conglomerates, which he classified mainly on the basis of grading and stratification: (1) inverse–to–normally graded model, (2) graded-stratified model, and (3) disorganized-bed model. Subsequently, Walker (1977, 1984) added a fourth type, the graded-bed model, which he considered to be transitional between graded-stratified and inverse-to-normally graded. Walker identifies these models as resedimented, deep-water turbidite-association conglomerates, but does not specifically link a particular model to a particular massflow process. Presumably, they can all be deposited by turbidity-current flow (R. G. Walker, 1984). Lowe (1982) suggests that at least some of Walker's (1975a) conglomerates are density-modified grainflow deposits. Lowe (Fig. 12, 1982) further suggests that disorganized beds are characteristic of cohesive flows (debris flows), inversely graded beds are especially characteristic of grain flows, inverse-to-normal grading may indicate flows transitional from density-modified grain flows to high-density turbidity currents, and normally graded conglomerates are particularly characteristic of high-density turbidity currents. Unfortunately, the deposits of a particular type of flow can exhibit apparent characteristics of more than one of these idealized models (discussed below). Thus, interpretation of ancient massflow deposits on the basis of grading characteristics can be equivocal. Some of the more important characteristics of subaqueous massflow gravelly deposits are listed in Table 6.14 and are further discussed below.

Subaqueous Debrisflow Conglomerates and Diamictites.   Subaqueous debrisflow conglomerates may range from fully clast-supported to mud/sand matrix-supported (diamictites) (Nemec and Steel, 1984; Kessler and Moorhouse, 1984). Within a given conglom-

TABLE 6.14

Characteristics of resedimented conglomerates and diamictites

	Subaqueous debrisflow conglomerates	Subaqueous grainflow conglomerates	Turbidite conglomerates
Fabric support	Clast-supported to matrix-supported; matrix may increase upward within beds	Clast-supported with sand, silt, or clay matrix	Clast-supported to matrix-supported; matrix may increase upward within beds
Texture	Sorting generally poor; pebble roundness variable depending upon source; poor correlation between clast size and bed thickness	Sorting moderate to poor; clast roundness variable	Sorting poor to moderate; variable clast roundness; fossils in associated shales (mudstones)
Vertical size grading	Typically nongraded, but may display inverse, inverse-to-normal, or normal grading	Inverse grading especially common, but may be nongraded or normally graded	Normal grading particularly common; inverse, complex, or nongraded less common
Clast orientation	Poorly developed, but better than in subaerial debrisflow conglomerates	Imbrication common; long axes of clasts dominantly oriented in flow-parallel direction	Upflow imbrication common; long axes dominantly oriented in flow-parallel direction
Stratification	Poor internal stratification; may overlie stratified granule sandstones or contain thin mud/silt interbeds, possibly with wave-generated structures	Crude stratification on a large scale; may be separated by sandstone interbeds; internally, conglomerate units lack stratification; associated sandstones planar-laminated or cross-laminated	Horizontally stratified to trough cross-bedded, more rarely massive; channel deposits more massive than nonchannel deposits
Vertical facies sequences	Sequences that fine upward to sandy capping beds common; common tendency toward upward increase in matrix content	May become finer grained upward, grading to sand cappings	Fining- and thinning-upward sequences most common; rare coarsening-upward sequences; may pass upward into a sand capping

erate unit, the amount of matrix may increase upward in the bed; thus, beds may grade from clast-supported in the base to mud-supported in the top (Nemec and Steel, 1984). Extremely mud-rich, pebbly mudstones occur also. Pebble roundness can be quite variable, depending upon the roundness of clasts in the original deposits. Sorting is poor, but it is probably better than that in subaerial debrisflow conglomerates. Also, preferred clast alignment and imbrication are better developed than in subaerial debrisflow deposits. Internal stratification is poor, but massive subaqueous debrisflow deposits may overlie thinly bedded, stratified sandstones or may contain thin interbeds of muds or silt. Depending upon water depth, interbeds may contain wave-generated structures. Poorly stratified subaqueous debrisflow conglomerates may be associated with better-stratified turbidites. Nongraded beds are probably most typical of subaqueous debrisflow conglomerates; however, these conglomerates can display moderately well-developed inverse, inverse-to-normal, or normal grading (Nemec et al., 1980; Nemec and Steel, 1984; Kessler and Moorhouse, 1984). A pattern of fining-upward to sandy capping beds is common. In contrast to subaerial debrisflow deposits, subaqueous deposits commonly show a poor correlation between maximum size of clasts in individual conglomerate units and thickness of the units (Kessler and Moorhouse, 1984). In marine units, fossils may be present in mud/silt interlayers. Bioturbation may be common. Typical characteristics of subaqueous debrisflow deposits are illustrated diagrammatically in Figure 6.23.

Criteria that distinguish subaqueous debrisflow conglomerates and diamictites from subaerial debrisflow deposits include (1) association with turbidites, (2) presence of fossiliferous mud/silt interlayers, (3) bioturbation, (4) possible presence of wave-generated structures in sandy interlayers, (5) better clast alignment, imbrication, and grading than that in subaerial deposits, (6) marked upward increase in matrix content, (7) presence of sandy cappings, and (8) poor correlation between maximum clast size and flow thickness. These criteria may not be sufficient in all cases to distinguish subaqueous debrisflow deposits from other subaqueous massflow deposits. Read on.

Subaqueous Grainflow Conglomerates.   Lowe (1982) suggests that density-modified grainflow conglomerates consist of clast-supported pebbles and cobbles set in a poorly sorted sand, silt, and clay matrix. Inverse grading is common in cobble beds, reflecting high dispersive pressures between large clasts. On the other hand, pebbly beds tend to be ungraded or to show poorly developed inverse grading, indicating low dispersive pressures and little size-sorting capability. Some normal grading might be present if the flows became turbulent—tending toward high-density turbidity currents. Clifton (1981b, 1984) describes dominantly clast-supported, inversely graded conglomerates at Point Lobos, California, that he interprets as grainflow deposits (Fig. 6.24). Pebble imbrication is prominent in many of the conglomerate beds, and long axes of pebbles are oriented parallel to flow direction. The conglomerates are stratified on a large scale, and they may be separated by sandstone interbeds. Internally, the conglomerates lack stratification, but associated sandstone beds are planar-laminated or, less commonly, show high-angle foresets. As with some subaqueous debrisflow conglomerates described above, Clifton reports that some Point Lobos conglomerate units become finer grained upward and grade to sand cappings.

Eyles (1987) describes glacially influenced conglomerate deposits from Middleton Island, Alaska, that she interprets also as grainflow deposits. These deposits occur as thin (20 cm to 1 m) beds or thicker (102 m) channel fills. The conglomerates are inversely graded, and the lower, finer portions of the beds are better sorted than the upper, coarser

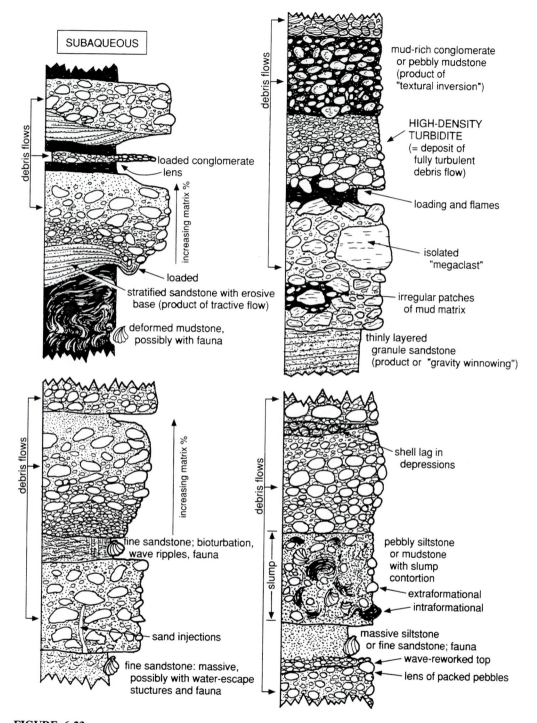

FIGURE 6.23

Schematic sections illustrating typical features of subaqueous debrisflow conglomerates. (*From Nemec, W., and R. J. Steel, 1984, Alluvial and coastal conglomerates: their significant features and some comments on gravelly massflow deposits, in Koster, E. H., and R. J. Steel, eds., Sedimentology of gravels and conglomerates: Canadian Soc. Petroleum Geologists Mem. 10. Fig. 16, p. 17, reprinted by permission.*)

FIGURE 6.24
Inversely graded conglomerate in the Carmelo Formation (Tertiary), Point Lobos, California. Scale about 7 cm long. (*Photograph courtesy H. E. Clifton.*)

parts. Imbrication is poor. Eyles' illustrations suggest that the thick conglomerate units pass upward into sand cappings; however, her thin units are succeeded upward by either massive or normally graded conglomerates. The thick, channel-fill units pass laterally into normally graded conglomerates.

Available data thus suggest that the most characteristic property of grainflow conglomerates may be well-developed inverse grading. They are mainly clast-supported, and pebble imbrication can apparently range from good to poor. Some, at least, grade upward into sand cappings.

Turbidite Conglomerates and Diamictites. Turbidite conglomerates may range from clast-supported to matrix-supported (diamictites). Clast-supported types appear most common and have been reported from many areas. In a given unit, conglomerates with clast-supported framework may pass upward into conglomerates with a matrix-supported fabric (diamictites). Sorting of clasts is generally poor to moderate. The long axes of clasts commonly show a flow-parallel orientation, with upflow imbrication dips (R. G. Walker, 1984). Downflow imbrication dips can apparently occur also (Hein, 1984). Normal size-grading is very common (Fig. 6.25). Nongraded fabrics are less common than normally graded fabrics but are more common than those that display inverse or complex grading. R. G. Walker (1984) suggests a downcurrent trend from inverse−to−normally graded conglomerates to graded-stratified conglomerates. That is, graded-stratified conglomerates tend to be deposited in a more-downcurrent position on a fan than do graded-bed conglomerates, and graded-bed conglomerates in a more-downcurrent position than inverse-to-normally graded conglomerates. Fining- and thinning-upward sequences are common, and conglomerates may pass upward into sandstone cappings. Coarsening-upward sequences occur but are less common. Stratification can range from massive, unstratified to horizontal stratified or trough cross-bedded. Channel deposits tend to be more massive and structureless than nonchannel deposits. In marine deposits, turbidite conglomerates may be associated with shales containing marine fossils.

FIGURE 6.25
Normally graded resedimented conglomerate, Late Cretaceous, Hungary. The scale is marked in
centimeters. (*Photograph courtesy H. E. Clifton.*)

Inasmuch as subaqueous debris flows or density-modified grain flows may evolve
into turbidity currents, the characteristics of turbidite conglomerates may resemble those
of other subaqueous sediment gravity flows. Some characteristics that possibly distinguish
them are (1) better-developed clast alignment and imbrication, (2) better-developed stratification (in some turbidites), (3) greater preponderance of normal grading, and (4) a
common tendency toward fining- and thinning-upward sequences.

6.10   Volcaniclastic Conglomerates and Breccias

The composition, classification, and origin of volcaniclastic sediments, with particular
emphasis on volcaniclastic sandstones, are discussed in Section 5.4. Volcaniclastic deposits composed mainly of gravel-size volcanic clasts of epiclastic origin are best regarded
as volcanic-derived, epiclastic conglomerates or breccias. They are classified and interpreted as any other epiclastic conglomerate. Coarse volcaniclastic deposits of primary
volcanic origin are classified differently. Pyroclastic particles ranging in size from 2 to 64
mm are called **lapilli**. Particles coarser than 64 mm are called **blocks** if angular and
bombs if round and fluidally shaped. The lithified deposits formed from these coarse
pyroclastic particles are called **lapillistone, pyroclastic breccia,** and **agglomerate
(bombs)** (Table 5. 8).

As discussed in Section 5.4, coarse pyroclasts are produced by explosive eruptions
and may be deposited by pyroclastic fall, pyroclastic flow, or pyroclastic surge. Intrusion
of lava or magma into water, ice, or wet sediment can lead to granulation or shattering of
the lava into angular fragments, producing so-called **hyaloclastites**. Debris flows or mud

flows may retransport pyroclastic materials down the slopes of volcanoes to produce lahar deposits.

If neither the epiclastic nor the primary volcanic origin of coarse volcaniclastic deposits can be established, these coarse materials are referred to simply as volcanic conglomerates or breccias, as appropriate (Table 5.10). Extended discussion of the characteristics of coarse pyroclastic deposits is beyond the scope of this book. Readers are referred to Cas and Wright (1987) and Fisher and Schmincke (1984) for further details of these deposits.

▰▰▰▰▰ *Additional Readings*

Ashley, G. M., J. Shaw, and N. D. Smith, 1985, Glacial sedimentary environments: Soc. Econ. Paleontologists and Mineralogists Short Course 16, 246 p.

Collinson, J. D., and J. Lewin, eds., Modern and ancient fluvial systems: Internat. Assoc. of Sedimentologists Spec. Pub. 6, 575 p.

Koster, E. H., and R. H. Steel, eds., 1984, Sedimentology of gravels and conglomerates: Canadian Soc. of Petroleum Geologists Mem. 10, 441 p.

Miall, A. D., ed., 1978, Fluvial sedimentology: Canadian Soc. Petroleum Geologists Mem. 5, 859 p.

Chapter 7
Shales

7.1 Introduction

Fine-grained, siliciclastic sedimentary rocks, composed mainly of particles smaller than about 63 μm (coarse silt and finer), make up approximately 50 percent of all sedimentary rocks in the stratigraphic record (Chapter 1). Thus, they are about twice as abundant as sandstones and conglomerates combined. These fine-grained rocks are known by a variety of names, including lutites, siltstones, mudstones, mudrocks, claystones, and shales. Tourtelot (1960) reviews in detail the history of fine-sediment terminology. He points out (p. 242) that historically the term shale has been used in two ways: (1) in a restricted sense to mean a laminated clayey rock and (2) as a broad, group name for all fine-grained siliciclastic rocks. He concludes that it is acceptable practice to include both these meanings for shale, and, therefore, that shale is the appropriate class name for fine-grained rocks, of equal standing with sandstones and limestones as group names.

> Certainly our comprehension is broad enough to include two meanings of the word "shale": First, the reasonably precise meaning of "laminated clayey rock" to which the origin of the word entitles it, and second, the meaning of the "general class of fine-grained rocks," which our historical use of the word bequeaths to it.

Although some geologists (e.g., Potter et al., 1980, p. 15) agree with Tourtelot's conclusion that shale is an acceptable class name for all fine-grained rocks, others (e.g., Lundegard and Samuels, 1980; Spears, 1980; Stow and Piper, 1984) consider the dual use of the term in this way to be confusing. They favor using the term mudrock as a group name for fine-grained sedimentary rocks and prefer to reserve the term shale for use in its more restricted sense to mean a laminated or fissile fine-grained rock. Inasmuch as the name for a particular rock cannot be mandated, it is likely that both shale and mudrock, or mudstone, will continue to be used as group names for fine-grained siliciclastic rocks, depending upon individual preferences. I have chosen in this book to follow Tourtelot's usage of shale as a group name. In any case, we shall subsequently see that shales (mudrocks) can be subdivided on the basis of textural characteristics and composition into subtypes, just as sandstones are subdivided by mineral composition into different types.

Geologists in the past have focused less attention overall on shales than on sandstones. Probably this is so because shales are more difficult to study owing to their fine grain size and because they contain fewer interesting textures and structures. For example, petrographic analysis of all but the very coarsest-grained shales is quite difficult. The mineralogy of finer-grained shales must be determined by X-ray diffractometer techniques, the scanning electron microscope, or other techniques that are commonly time consuming and often quite expensive. Furthermore, many shales occur as thick, repetitive sequences of thin- to medium-bedded rocks that may contain few visible sedimentary structures or textural features of interest. They can be pretty monotonous to work with in the field. Also, many shale units are poorly exposed and difficult to observe.

The study of fine-grained siliciclastic sedimentary rocks has thus lagged behind that of coarser-grained rocks. Nonetheless, readers should not conclude from these remarks that geologists have no interest in shales. In fact, considerable research has been done on shales, and the literature on these fine-grained rocks is extensive. See, for example, the bibliography of shales provided by Potter et al. (1980). Information about fine-grained sediments, including recent, unconsolidated sediments, has been accumulating particularly rapidly since the late 1960s, owing in part to research involved in the deep-sea drilling program (DSDP) and to the interest of petroleum geologists in shales as possible source rocks for petroleum. Geologists are interested in such aspects of fine-grained sediments as their source (provenance), modes of transport (e.g., surface currents, bottom currents, nepheloid plumes, turbidity currents, and other mass movements), depositional environments, and organic content (petroleum source potential). To evaluate the genetic significance of shales requires that we have a knowledge of their textures, structures, mineralogy, chemistry, and fossil content. In this chapter, we examine some of the characteristic properties of shales, discuss their classification, and briefly describe their origins and occurrences.

7.2   Methods of Study

Owing to their fine grain size, shales require somewhat different methods of study than do sandstones and conglomerates. Detailed discussion of these techniques is outside the scope of this book. A few of the more important techniques are mentioned below, together with some references that readers may wish to consult. X-ray powder diffraction methods have for many years been used as a standard technique for determining clay mineralogy and the bulk mineral composition of fine-grained sediments (e.g., Wilson, 1987b). Thermal analysis or thermoanalytical methods (Paterson and Swaffield, 1987; Mackenzie, 1982) are used particularly to study differences in characteristics of clay minerals in shales. Infrared methods (Russell, 1987; Fripiat, 1982b) are used to study the structure and composition of clay minerals. The scanning electron microscope (Eberhart, 1982; McHardy and Birnie, 1987) is especially useful for studying the microfabric and texture of shales and the growth forms of clay minerals and other minerals. It can be useful also in mineral identification, particularly when used with back-scattered electron imaging. Chemical analysis of shales is possible by a variety of wet chemical and instrumental techniques, such as X-ray fluorescence spectroscopy, the electron microprobe, and atomic emission and absorption (Bain and Smith, 1987). X-radiography of thin slabs (Hamblin, 1971) is a useful technique for studying microlaminations and other sedimentary structures that are not visible to the eye. Instrumental well logs provide an important tool for studying stratigraphic relationships and establishing correlation of subsurface shale units.

See Boggs (1987, p. 554) for a brief review of instrumental well logs. For additional information on study of shales, see Carver (1971a), Fripiat (1982a), O'Brien and Slatt (1990), Potter et al. (1980, ch. 2), Tucker (1988), and Wilson (1987a).

7.3   Physical Characteristics of Shales

7.3.1   Texture

Grain Size

Shales differ from sandstones and conglomerates particularly in their finer grain size. Owing to their fine particle size, the grain size of shales cannot be determined by sieving methods. The particle sizes of shales that can be disaggregated are measured by methods based on particle settling velocity, or they are measured by electronic particle-sizing techniques (Chapter 2). The accuracy of settling-tube methods is influenced by particle shape. Many particles in shales are platy or flaky and thus settle more slowly than spheres. This hampered settling leads to discrepancies when calculations of grain size are based on settling velocities of spheres, although the discrepancies are commonly not serious. See Singer et al. (1988) and Jones et al. (1988) for an evaluation of analytical techniques for size analysis of fine-grained sediment. Also, some shales are so firmly cemented and so well indurated that they cannot be disaggregated into individual particles, making accurate size determination impossible. Even if complete disaggregation of samples is possible, there is no guarantee that sedimentation took place by settling of individual particles—a problem that complicates environmental interpretation on the basis of particle size. Under some conditions, clay-size particles flocculate and settle as aggregates rather than as individual particles (Kranck, 1975, 1984). In fact, both flocculated particles and individual particles may be present in the same sample. Also, diagenesis (Chapter 9) may affect the size of particles as a result of partial dissolution of grains or by addition of cement coatings. Owing to these various factors, grain-size analyses may not yield data that can be easily related to depositional conditions. Thus, fine-sediment grain-size data must be interpreted with considerable caution.

The grain size and sorting of muddy sediments and sedimentary rocks are highly variable. Data compiled by Picard (1971) suggest that modern muds contain on the average about 45 percent silt, 40 percent clay, and 15 percent sand. (Texturally, clay is defined as all material finer than 4 μm; silt ranges in size from 4 to 63 μm; sand ranges from 63 μm to 2 mm.) Picard's compiled data for a small number of analyses of ancient shales indicate that these samples contain about 80 percent silt, 17 percent clay, and 3 percent sand. Because of the difficulty of disaggregating and measuring the grain size of these fine-grained sediments and the relatively few analyses of ancient shales that are available, these data may not accurately reflect the true average grain size of ancient shales. Pettijohn (1975, p. 262) suggests that the average shale contains about two parts silt and one part clay. Comparatively little is yet known about the distribution of grain sizes in ancient shales or the environmental significance of grain size as a property of fine-grained sediments. Nonetheless, grain size is one of the principal parameters used in classification of these rocks.

Particle Shape

The shapes of the small particles that make up shales, unlike the shapes of sand-size and larger particles, are little modified by sediment erosion and transport. For example, Kuenen (1959, 1960) demonstrated that very small quartz particles ($<\sim 0.1$ mm) do not become rounded very effectively by any type of eolian or stream transport. Therefore, the shapes of fine silt- and clay-size particles in shales reflect mainly the original shapes of the detrital particles, largely unmodified by transport abrasion, or they reflect the shapes of minerals generated during diagenesis. Thus, most particles in shales are very angular. Many particles, especially clay minerals and fine micas, have very low sphericities. Electron microscopy (e.g., Sudo et al., 1981) reveals that most clay minerals have platy, flaky, or acicular shapes (Fig. 7.1). Some investigators (e.g., Ehrlich and Chin, 1980; Mazzullo et al., 1982) have used Fourier shape analysis (Chapter 2) to study the shapes of quartz silt as a clue to provenance. Otherwise, few efforts appear to have been made to relate the shapes of fine particles to the genesis of shales.

Fabric

Fissility.   Because shales or mudrocks contain high concentrations of clay minerals that have platy or flaky shapes, as discussed, these rocks may exhibit microfabrics resulting from the preferred orientation of flaky clay minerals. Well-developed preferred orientation of clay minerals is often alleged to be the cause of fissility in shales; however, not all workers agree with this conclusion. Also, the term fissility itself appears to be used in more than one way.

Fissility is defined as the property possessed by some rocks of splitting easily into thin layers along closely spaced, roughly planar, and approximately parallel surfaces (Bates and Jackson, 1980). Alling (1945) suggests that the ease of splitting of shales ranges from **very hard** [to split] (massive shales) to **very easy** [to split] (fissile shales). The term **fissile** is used in two ways to mean (1) capable of being easily split along closely spaced planes, i.e., possessing fissility, and (2) frequency of splitting, i.e., the thickness of layers between fissile planes or planes of parting.

Potter et al. (1980) use the term fissile to indicate a class of parting, with parting defined as the tendency of a rock to split along lamination or bedding — a tendency greatly

FIGURE 7.1
Electron micrograph of kaolinite clay minerals magnified 4700×. Note the platy or flaky appearance of the pseudohexagonal kaolinite crystals, which are arranged in this specimen in distinct "books."

enhanced by weathering. The relationship of parting to bedding and lamination is shown in Table 7.1, as modified by Potter et al. from earlier schemes of Alling (1945), Ingram (1954), and McKee and Weir (1953). Table 7.1 shows that fissile is a subdivision of parting and that a fissile parting is thinner than a lamina (laminae are strata less than 10 mm thick); that is, fissile partings range in thickness from 0.5 mm to 1.0 mm. Parting thinner than 0.5 mm is referred to as **papery,** and parting thicker than 1 mm is **platy, flaggy,** or **slabby,** depending upon thickness.

Ingram (1953) proposed three kinds of breaking characteristics for shales: massive, flaggy-fissile, and flaky-fissile. **Massive shales** have no preferred direction of cleavage or splitting and thus upon breaking yield fragments that are blocky. **Flaggy shales** split into fragments that have two flat sides that are approximately parallel and that have width and length dimensions many times greater than the thickness. **Flaky shales** split along irregular surfaces parallel to the bedding and yield uneven flakes, thin chips, and wedgelike fragments. Ingram (1953) does not state any thickness limits for flaggy or flaky units. Other workers (Payne, 1942; McKee and Weir, 1953) have proposed different thickness limits for the papery, fissile, platy, flaggy, etc. classes of parting. For example, Payne (1942), as well as Spears (1980), proposes for fissile parting an upper thickness limit of 2 mm rather than the 1 mm upper limit shown in Table 7.1.

Thus, in summary, a fissile shale is a shale that tends to split relatively easily into thin, approximately parallel layers that range in thickness from about 0.5 mm to 1.0 mm.

TABLE 7.1

Stratification and parting in shales

Thickness	Stratification		Parting	Composition
30 cm	Thin	Bedding	Slabby	Clay and organic content ———— Sand, silt, and carbonate content ————
3 cm	Very thin			
10 mm	Thick	Lamination	Flaggy	
5 mm	Medium		Platy	
1 mm	Thin		Fissile	
0.5 mm	Very thin		Papery	

Source: Potter, P. E., et al., 1980, Sedimentology of shale. Table 1.3, p. 16, reprinted by permission of Springer-Verlag, Berlin.

Shales that split into thinner layers are called papery shales, and those that split into thicker units are platy, flaggy, or slabby shales depending upon thickness of the parting, as shown in Table 7.1. Potter et al. (1980) suggest that the clay and organic contents of shales decrease and the sand, silt, and carbonate contents increase as thickness of parting units increases (Table 7.1).

Let's return now to the subject of shale fabric. The exact nature of shale microfabrics and the reasons why some mudrocks are fissile and some are not is still not well understood in spite of considerable research. A useful review of the history of clay microfabric research is given by Moon and Hurst (1984). Geologists have long believed (e.g., Ingram, 1953) that fissility of shales is related to the orientation of clay minerals. Moreover, some clay minerals were thought to settle as individual particles, whereas others flocculate and settle as aggregate particles. Settling of individual clay minerals was believed to produce oriented fabrics, whereas settling of flocculated aggregates was thought to produce random fabrics. Is there actually a difference between the fabric of fine sediment deposited by settling of individual particles and that produced by settling of flocs? Let's see.

Orientation of Clay Particles during Suspension Settling.   Van Olphen (1977) theorized that three different modes of particle association may occur when a suspension of platelike clay particles flocculates: face-to-face (F-F), edge-to-face (E-F), and edge-to-edge (E-E). The various possible ways that clay particles can be associated in a clay suspension are illustrated in Figure 7.2. Figure 7.2a shows dispersed particles that have not become flocculated. Figure 7.2b illustrates clay particles in F-F association. Strictly speaking, clay particles in this association do not constitute flocs. Van Olphen uses the term **aggregation** to refer to this type of association. Figures 7.2c through 7.2g show flocs with various kinds of E-E and E-F arrangements. E-F–type arrangements such as Figure 7.2e are frequently referred to as **cardhouse-type** structures. Experimental confirmation of these particle arrangements is difficult to obtain because the arrangements of particles in suspensions are difficult to observe directly. Moon and Hurst (1984) indicate that available, but limited, evidence suggests that single-plate, cardhouse-type structures are present in sedimenting clays under normal conditions of pH and salinity. These structures appear to consist of about a dozen individual particles arranged in a simple E-F array.

The factors that control the behavior of clay particles settling in a suspension are complex and are apparently still poorly understood. Some factors that may be important include pH, Eh, the presence or absence of surface-active ions (electrolytes), mineralogy, particle size, particle concentration, organic compounds, and water energy or turbulence.

Clay particles are negatively charged. (See van Olphen, 1977, p. 17, for a discussion of the origin of the electric charge on particles.) In fresh water, both interparticle attraction and repulsion operate simultaneously but the repulsive forces between clay particles of like charge are greater than the van der Waals attraction forces. Therefore, the individual particles remain dispersed. In saltwater, in the presence of ionized salts, repulsion forces are reduced to the point where attraction dominates, causing the clay particles to flocculate and settle as flocs (van Olphen, 1977, p. 11). Thus, increase in salinity promotes flocculation. Gibbs (1983) suggests that flocculation of clays can begin at salinities as low as 0.5–1‰, or about 1/70 to 1/35 the average salinity of seawater. The pH of a suspension appears also to have an influence on the flocculation behavior of a suspension as well as the type of clay particle arrangements that occurs in the flocs. At lower pH values, flocculation occurs and E-F arrangements may predominate in the flocs

FIGURE 7.2

Kinds and terminology of particle associations in clay suspensions: (a) dispersed and deflocculated, (b) aggregated but deflocculated (face-to-face association, or parallel or oriented aggregation), (c) edge-to-face flocculated but dispersed, (d) edge-to-edge flocculated but dispersed, (e) edge-to-face flocculated and aggregated, (f) edge-to-edge flocculated and aggregated, (g) edge-to-face and edge-to-edge flocculated and aggregated. *(From van Olphen, H., An introduction to clay colloid chemistry, 2nd ed., Fig. 23, p. 97, © 1977, John Wiley & Sons, Inc. Reprinted by permission of John Wiley & Sons, Inc., New York.)*

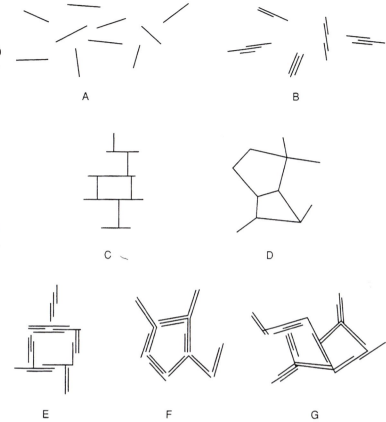

(Fig. 7.3). At higher pHs (above ?), E-E arrangements may be more common (Flegmann et al., 1969; Rand and Melton, 1977). At some higher(?) pH value, deflocculation occurs. Note in Figure 7.3 that increasing pH correlates with decreasing Bingham yield strength, which means that increasing destruction or dispersion of flocs occurs at higher pH values, and results in greater fluidity of the suspension.

Some investigators have suggested that clay mineralogy may influence the tendency of fine particles to flocculate; however, Kranck (1980) believes that mineralogy has only a minor effect on flocculation, whereas particle size and concentration are important. The tendency to flocculate appears to increase with increasing sediment concentration, which increases the likelihood of particle collisions, and decreasing sediment size. Kranck suggests that very fine sediment rarely, even in fresh water, reaches the bottom unflocculated. Fine sediment that does settle as individual particles (Stokes settling) settles at a rate dependent upon particle size, producing a graded deposit with good sorting within individual layers of the deposit. When flocs form early during the settling of a suspension, they are made up of all particle sizes present in the suspension. Thus, the sediment that results from sedimentation of flocs tends to be poorly sorted. Kranck (1984) points out that both individually settled particles and flocs can be present in the same sediment. Turbulence tends to promote flocculation because it brings about increased interparticle collision. At the same time, turbulence limits flocculation by disrupting flocs and limiting floc size (Kranck, 1980).

FIGURE 7.3

Effects of pH on the Bingham yield stress of kaolinite suspensions with respect to the mode of particle interaction. (*After Rand, B., and I. E. Melton, 1977, Particle interactions in aqueous kaolinite suspensions I. Effect of pH and electrolyte upon the mode of particle interaction in homoionic sodium kaolinite suspensions: Jour. Colloid and Interface Science, v. 60, Fig. 1, p. 310, as modified slightly by Bennett, R. H., and M. H. Hulbert, 1986, Clay microstructure, D. Reidel Publishing Co., Dordrecht. Fig. 3.27, p. 89, reprinted by permission.*)

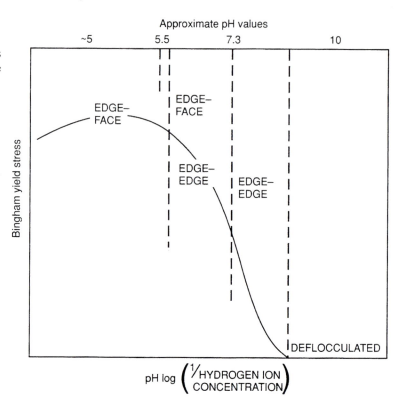

The presence of organic substances that can influence the Eh of a suspension may also have an effect on clay particle settling behavior. Negative Eh (reducing) settling environments, characterized by low oxygen content and abundant organic matter, tend to produce dispersion of clay particles and settling of particles as individual flakes. Various kinds of organic substances such as amino acids and humic acids are known to be capable of dispersing, or peptizing, clay minerals (van Olphen, 1977, Ch. 11). These organic substances are present in a number of depositional environments. Clay particles settling in waters that are completely anoxic (reducing) probably settle in a dispersed, grain-by-grain state. Clays settling in environments characterized by oxic (oxygenated) nearsurface waters but anoxic bottom waters might form flocs in the oxic water, but the flocs would likely become dispersed upon reaching the anoxic bottom water.

Fabric Characteristics of Freshly Sedimented Clays. Freshly deposited clays are those that have not been subjected to any significant degree of compaction by superjacent sediment. The structure of freshly deposited clays is difficult to study because the processes involved in sample collection and preparation tend to alter the clay structure. Nonetheless, investigators working with only slightly compacted sediments have, by use of the transmission electron microscope and the scanning electron microscope, gained some insight into the possible structure of uncompacted clays. A significant development in understanding the structure of clayey sediments was introduction of the concept of **domains**. Aylmore and Quirk (1960) envisaged domains as microscopic or submicroscopic regions within which the clay particles are in parallel array. These groups of crystals (domains) within which the clay particles are in parallel orientation are randomly oriented with respect to each other throughout the clay matrix. They were referred to by Aylmore and Quirk as **turbostratic groups** because the domains are in turbulent array. Subsequent workers (Sloane and Kell, 1966; Smalley and Cabrera, 1969) identified two major types of domains: **bookhouse** and **stepped face-to-face** (Fig. 7.4A). Bennett et al. (1991) define a domain as ''a multiplate particle composed of parallel or nearly parallel plates that may be stacked either as sheets in a book or with an offset or stair-step arrangement.'' Figure 7.4B illustrates random arrangement of domains within a clayey sediment.

Domains should not be equated to flocs, although some domains may have formed through the process of flocculation. They may form also as a result of consolidation, or by authigenic processes, or they may have developed as domains from a parent mineral prior to its erosion and subsequent deposition as a detrital particle (Bennett et al., 1981). Domains have been identified in unconsolidated clays, along with single-plate particles. On the basis of a study of Mississippi Delta and DSDP submarine clays, Bennett et al.

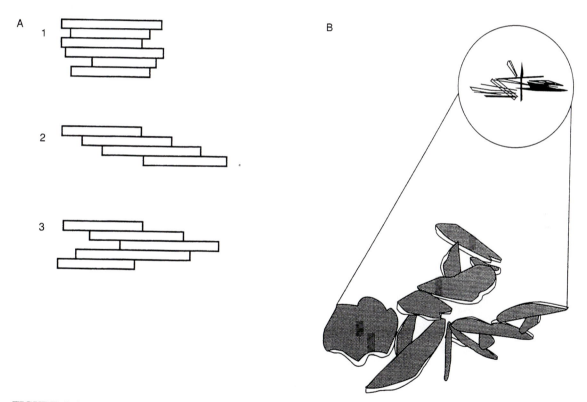

FIGURE 7.4

A. Domain structure: (1) bookhouse, (2) and (3) variants of stepped face-to-face structures. B. Plane and cross-sectional views of the microfabric of a hypothetical sediment constructed of numerous domains in a random arrangement. (*A after Moon, C. F., and C. W. Hurst, 1984, Fabric of muds and shales: an overview, in Stow, D. A. V., and D. J. W. Piper, eds., Fine-grained sediments: Deep-water processes and facies, Blackwell, Fig. 1, p. 581. B from Bennett, R. H., et al., 1991, Determinants of shale microfabric signatures: Processes and mechanisms, in Bennett, R. H., et al., eds., Microstructures of fine-grained sediments. Fig. 2.3, p. 9, reprinted by permission of Springer-Verlag, New York.*)

(1981, p. 228) conclude that (1) the basic "building blocks" of clay fabric are domains and/or "single plate-like" particles, (2) the basic particles (domains or "single plates") can form flocs or chains, (3) the framework of clay fabric can be developed by domains, "single plate-like" particles, flocs, chains, or a combination of these basic fabric entities. Bennett et al. found that DSDP clay fabric samples were characterized by flocs and linking chains, whereas domains and "turbostratic-type" clay fabrics are characteristic of high-porosity (high-void-ratio) Mississippi Delta sediments. Their proposed model for clay fabrics is shown in Figure 7.5. Moon and Hurst (1984, p. 582) conclude that domain structures are commonly present in freshly deposited, uncompacted clay sediments and that "there must be a transition from single-plate to domain structures at, or very near to, the depositional interface." Reasons for this transition are not as yet well understood.

Fabric Characteristics of Compacted Clay Sediments.   As discussed, both single-plate and domain structures may be present in freshly deposited clays; however, these clayey sediments do not display fissility. Domains are present in consolidated shales also, although the domains may be forced so closely together by compaction that individual domains are difficult to differentiate. A factor of significant interest to geologists is the reason why some clayey sediments develop fissility (Fig. 7.6) upon consolidation whereas

FIGURE 7.5
Fabric model for smectite- and illite-rich submarine sediments. *(After Bennett, R. H., et al., 1981, Clay fabric of selected submarine sediments: Fundamental properties and models: Jour. Sed. Petrology, v. 51, Fig. 6, p. 229, reprinted by permission of Society of Economic Paleontologists and Mineralogists, Tulsa, Okla.)*

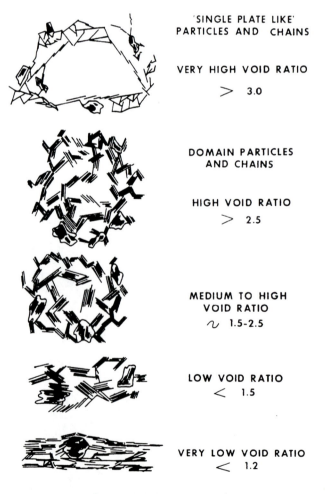

'SINGLE PLATE LIKE' PARTICLES AND CHAINS

VERY HIGH VOID RATIO

> 3.0

DOMAIN PARTICLES AND CHAINS

HIGH VOID RATIO

> 2.5

MEDIUM TO HIGH VOID RATIO

~ 1.5-2.5

LOW VOID RATIO

< 1.5

VERY LOW VOID RATIO

< 1.2

FIGURE 7.6

Electron micrograph of fissile shale showing distinct orientation of shale fabric. Scale bar = 20 μm. *(From Moon, C. F., and C. W. Hurst, 1984, Fabric of muds and shales: an overview, in Stow, D. A. V., and D. J. W. Piper, eds., Fine-grained sediments: Deep-water processes and facies. Fig. 2b, p. 584, reprinted by permission of Blackwell Scientific Publications Limited, Oxford.)*

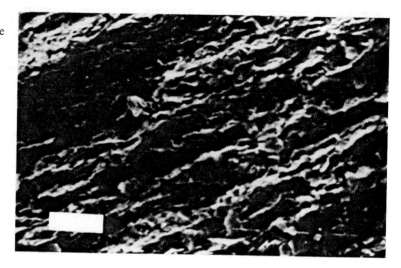

others do not (Fig. 7.7). Overburden pressure alone cannot be the complete reason for development of fissility because nonfissile shales may occur in a stratigraphic section below fissile shales. Some workers (e.g., Byers, 1974) have attributed the lack of fissility in shales to bioturbation, which destroys microlaminations. Byers suggests that fissility of shales is a primary characteristic of the depositional fabric, but is maintained only when burrowing infauna are absent in the original deposit. Others (e.g., Spears, 1976) have suggested that parallel orientation of clay minerals is not responsible for fissility and that fissility is primarily due to weathering of organic-rich laminae. Most workers, however, apparently believe that fissility in shales is related to orientation of clay minerals; that is, fissile shales have a parallel fabric (parallel-oriented domains or single plates), whereas a random domain structure is present in nonfissile shales. If this is so, how is the orientation produced, and are there factors other than bioturbation that may account for nonfissile shales?

FIGURE 7.7

Electron micrograph of nonfissile shale showing random orientation of clay platelets. Scale bar = 20 μm. *(From Moon, C. F., and C. W. Hurst, 1984, Fabric of muds and shales: an overview, in Stow, D. A. V., and D. H. W. Piper, eds., Fine-grained sediments: Deep-water processes and facies. Fig. 2e, p. 584, reprinted by permission of Blackwell Scientific Publications Limited, Oxford.)*

Moon and Hurst (1984) suggest that the geochemistry of the environment may be the crucial factor in producing fissility of shales. In particular, the presence or absence of peptizing or dispersing agents in the water determines if clay sediments reach the bottom as dispersed single plates or as flocs. Organic substances are suggested by Moon and Hurst to be the main peptizing agents. In anoxic environments where organic compounds are abundant, the peptizing qualities of these organic substances cause dispersion of flocs or prevent flocs from forming in the first place. Thus, clay settles as single plates. In oxic environments where peptizing agents are absent, clay settles as flocs (Fig. 7.8). Domains form (exactly how is not yet clear) in the freshly deposited sediment. These domains are randomly oriented in oxic sediments and more or less parallel-oriented in anoxic sediments. Lithification causes additional alignment or parallel orientation of the domains in the muds deposited in anoxic environments, producing fissility. Lithification does not destroy the random orientation of domains in muds of oxic environments, although the domains become more closely packed. Thus, muds deposited in oxic environments form nonfissile shales.

FIGURE 7.8
Schematic representation of major microstructural changes that occur in clay sediments during deposition and lithification. (*From Moon, C. F., and C. W. Hurst, 1984, Fabric of muds and shales: an overview, in Stow, D. A. V., and D. J. W. Piper, eds., Fine-grained sediments: Deep-water processes and facies. Fig. 5, p. 588, reprinted by permission of Blackwell Scientific Publications Limited, Oxford.*)

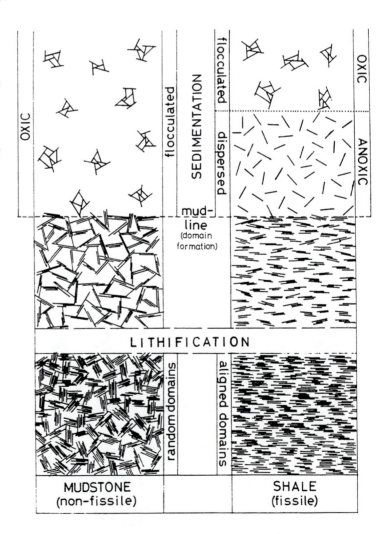

As Byers (1974) suggested, oriented fabrics produced in anoxic environments would likely be preserved owing to limited bioturbation in such toxic environments. O'Brien (1987) reports that black, organic-rich shales from anoxic environments are typically unbioturbated and are characterized by fine lamination and a microfabric of parallel clay flakes. He recognizes two types of fabrics in gray shales that presumably formed in oxic environments. Highly bioturbated gray shales, as seen in a scanning electron microscope, have fabrics characterized by a random microfabric, that is, randomly oriented individual particles. According to O'Brien, this fabric is very similar to that of mudstones formed from lithified flocculated clay, except that it does not possess the stepped domains or stepped cardhouse fabric of flocculated clays. The second type of oxic gray clay is characterized by indistinct bedding. It possesses some remnant original lamination and preferred parallel particle orientation indicative of less intense bioturbation.

O'Brien appears to suggest from his observations that parallel orientation of clay particles can develop in clayey sediments deposited in oxic environments and that this fabric can be preserved, at least in part, if the sediment is not too intensively bioturbated. He does not explain how the parallel orientation develops, but refers to a 1981 paper (O'Brien, 1981) in which he discusses the possibility that parallel-oriented fabric is a secondary feature, which develops after deposition as a result of the mechanical rearrangement during the early stages of consolidation of flocculated clay. Curtis et al. (1980) propose also that preferred orientation in clay-rich sediments results mainly from compaction strain. They suggest that fissility is not due to clay-mineral orientation but instead is related to fine-scale lamination. They suggest further that the degree to which clay mineral orientation can be attained is limited by the presence of nonplaty minerals such as quartz, which prevent planar fabric development in their immediate vicinity. The problem of ascribing clay-mineral orientation to postdepositional compaction, as indicated near the beginning of the preceding section, is that shales that do not display oriented fabrics can occur stratigraphically below other shale units that do display oriented fabrics.

Processes That Affect Clay and Shale Microfabrics.   It appears likely that several kinds of depositional and diagenetic processes and mechanisms operate to produce clay and shale microfabrics (Bennett et al., 1991). These processes and mechanisms are summarized in Table 7.2. **Physicochemical processes** take place by three mechanisms: **electrochemical** (forces that hold particles together internally and that bind particles), **thermochemical** (forces arising from temperature and temperature differences, e.g., Brownian motion, freezing of water), and **interface dynamics** (differential motion of settling particles under the influence of gravity, differential flow of water masses of differing density, impact of particles on sediment interface, flow at the interface, and microroughness of the interface). **Bioorganic processes** represent the effects of living organisms on sediment properties and are brought about by three mechanisms: **biomechanical** (bioturbation), **biophysical** (aggregation or agglutination of particles by organic processes), and **biochemical** (chemical production and destruction of chemical entities, e.g., production of gases by organisms). **Burial diagenesis** processes that can affect shale microfacies take place by **mass gravity mechanisms** (mechanical rearrangement of particles owing to overburden stresses) and **cementation** and other diagenetic phenomena. As shown in Table 7.2, each of these processes produces a characteristic fabric signature

TABLE 7.2
Processes and mechanisms that determine clay and shale microfabrics; effects of the mechanisms (fabric signatures) and the physical and temporal scales associated with the mechanisms also shown

Processes	Mechanisms	Fabric signatures (predominant)*	Scales Physical	Time	Remarks
Physiochemical	Electrochemical	E-F	Atomic and molecular to ~4 μm	μs to ms	Two particles may rotate F-F
	Thermomechanical	F-F (some E-F)	Molecular to ≡0.2 mm	ms to min	Initial contacts E-F, then rotation to F-F; common in selective environments
	Interface dynamics	F-F and E-F	μm to ~ ≡0.5 mm	s	Some large compound particles may be possible at high concentrations
Bioorganic	Biomechanical	E-F	~0.5 mm to >2.0 mm	s to min	Some F-F possible during bioturbation
	Biophysical	E-E and F-F	μm to mm	s to min	Some very large clay organic complexes possible
	Biochemical	Nonunique (unknown)	μ to mm	hr to yr	New chemicals formed, some altered
Burial diagenesis	Mass gravity	F-F localized swirl	cm to km	≡yr	Can operate over large physical scales
	Diagenesis-cementation	Nonunique (unknown)	molecular	≡yr	New minerals formed, some altered, changes in morphology

*E-F, edge-to-face; E-E, edge-to-edge; F-F, face-to-face.

Source: Bennett, R. H., et al., 1991, Determinants of clay and shale microfabric signatures: Processes and mechanisms, *in* Bennett, R. H., W. R. Bryant, and M. H. Hulbert, Microstructure of fine-grained sediments: Springer-Verlag, New York.

in terms of clay-particle arrangement. Table 7.2 also shows the physical scale at which these mechanisms operate, as well as the approximate time involved.

Fabric Study by Petrographic Microscopy.   Detailed study of fabric, as discussed in the preceding section, requires the use of a scanning electron microscope. The presence or absence of oriented fabrics can be detected in thin sections by observing extinction patterns of platy minerals in thin sections cut perpendicular to bedding. Clay minerals and micas have nearly perfect platy cleavage <001>. Although individual clay minerals are too small to be seen effectively with the microscope, the parallel extinction of large numbers of clay minerals all oriented in approximately the same direction gives a mass-extinction phenomenon that makes the thin sections look as if they were cut from a single crystal (Pettijohn, 1975, p. 263). Thin sections of shales characterized by random orientation of clay minerals do not display a mass-extinction effect.

Kuehl et al. (1988) report two kinds of mass-extinction patterns that occur in fine-grained sediments of the Amazon delta. The first fabric displays a "layered" extinction pattern that consists of thin (~0.01 mm) layers of oriented clays that have a lateral continuity of about 1 mm. These layers of oriented clay are vertically stacked at ~0.1 mm intervals and are separated by layers of randomly arranged platy minerals that display no obvious extinction orientation. The second extinction pattern results from essentially uniform distribution of platy minerals that are oriented parallel to bedding and that display no apparent vertical change in orientation on scales less than 1 mm. Both types of extinction patterns can occur in the same thin section.

7.3.2   Sedimentary Structures

The characteristics and origins of most common structures that occur in sedimentary rocks are described in Chapter 3. Many of these structures are present in shales, including parallel stratification, massive bedding, graded bedding, flaser bedding, ripple marks, convolute lamination, trace fossils and bioturbation structures, mud cracks, concretions, cone-in-cone structures, and color banding. Because of the prevalence of parallel laminated stratification in many shales, the relationship of laminae to fissile partings, and the use of lamination as a criterion in the classification of shales, some additional discussion of lamination is given here.

As discussed in Chapter 3, **laminae** are defined as strata having a thickness less than 10 mm. Laminae may or may not display fissility. They are produced by short-lived fluctuations in depositional conditions that cause variations in grain size, content of clay and organic material, mineral composition, or microfossil or fecal pellet content of sediment. Laminae produced by alternating layers of finer- and coarser-grained sediment (Fig. 7.9) are probably most abundant overall in sedimentary rocks; however, laminae that contain concentrations of organic matter (Fig. 7.10) are common also in some shales.

Parallel laminae are known to form both by deposition from suspension and by traction currents. Most laminae in shales are probably deposited in some manner from suspension. Possible suspension mechanisms may include (1) slow suspension settling in lakes, where levels of organic activity are commonly low—the formation of varves, for example, (2) sedimentation on some parts of deltas, where abundant fine sediment that is periodically supplied by distributaries leads to rapid deposition, (3) deposition on tidal flats in response to fluctuations in energy levels and sediment supply during tidal cycles, (4) deposition in subtidal shelf areas, where thin sand layers that accumulate owing to storm activity may alternate with thin mud laminae formed during periods of slower

FIGURE 7.9
Fine laminations in Devonian black shales, New York. Light layers are silt (large arrows), darker layers are dominantly clay, with organics and scattered quartz (small arrow). Scale bar = 1 mm. (*After O'Brien, N., 1989, Origin of lamination in Middle and Upper Devonian black shale, New York State: Northeastern Geology, v. 11, Fig. 2A, p. 161. Photograph courtesy of N. O'Brien.*)

FIGURE 7.10
Fine laminations in Devonian black shales, New York. Fine organic material is concentrated in the thicker, dark layers. Thin silt laminae (arrows) separate the organic-rich laminae. Scale bar = 1 mm. (*After O'Brien, N., 1989, Origin of lamination in Middle and Upper Devonian black shale, New York State: Northeastern Geology, v. 11, Fig. 2C, p. 161. Photograph courtesy of N. O'Brien.*)

accumulation, (5) surface-wave activity that causes resuspension of shelf muds, and (6) slow sedimentation in deep-sea pelagic or hemipelagic environments by deposition from nepheloid layers.

Turbidity currents provide an additional mechanism for formation of mud laminae in deep water that results from a combination of suspension settling and near-bottom current shear (Stow and Bowen, 1978, 1980; Kranck, 1984). As sediment settles out of suspension from a turbidity current, individual silt grains and mud flocs settle together. As sediment settles toward the bottom, increasing shear in the near-bottom boundary layers initially breaks up the flocs. The mud is temporarily dispersed and remains suspended, whereas the larger silt grains fall through the viscous sublayer to form a silt lamina. With continued sedimentation and mud concentration, reflocculation of the clay eventually occurs. At some critical mud concentration, the clays are able to form flocs large enough to overcome shear breakup and are deposited rapidly through the viscous sublayer as a mud lamina over the coarser silt lamina (Stow and Bowen, 1978). Thus, shear sorting in the near-bottom boundary layer produces alternating silt and clay laminae.

Hesse and Chough (1980) also relate the origin of laminated turbidite muds to shear sorting in the boundary layer; however, they visualize a different mechanism for producing the laminae. They attribute laminae formation to "burst and sweep" events in the turbulent boundary layer. **Bursts** refer to low-velocity streaks in the boundary layer that periodically lift up from the sublayer, enlarge, start to oscillate and then break up into chaotic vortices within the very near-bottom region. Bursts alternate with **sweeps** or inrushes of higher-velocity water from the upper part of the boundary layer; these sweeps may be the triggering mechanism for the bursts. Individual silt laminae of laminated turbidite muds putatively form owing to shear sorting during burst-and-sweep events, which have a winnowing effect on the clay fraction.

O'Brien (1989) describes laminations in Devonian black shales of New York that consist of alternating fine, organic-rich clay layers and coarser, silty clayey layers. He suggests that the silty laminae were deposited from low-density turbidity currents, which periodically interrupted the background sedimentation of hemipelagic organic-rich clay. Repeated turbidity-current flows thus produced a succession of clay-silt couplets.

Some silty laminae in shales can form by traction current mechanisms. These mechanisms may involve migration of low-relief ripples, as suggested by McBride et al. (1975), or they may be related to bedload sorting and winnowing processes. Kuehl et al. (1988) postulate that the sharp basal contacts of most silt laminae in Amazon Delta muds indicate that lamina formation is associated with conditions of increased bottom shear stress and consequent erosion of the seabed. They suggest that the silt particles that form the laminae probably are supplied both by sorting processes operating at the seabed and by changes in the nature of sediment supplied by suspension. Laminae in some siltstones may be the result of tidal activity. Kvale et al. (1989) report laminated siltstones in the Pennsylvanian Mansfield Formation in Indiana (USA) in which they recognize daily, monthly, and yearly tidal cycles.

O'Brien (1990) reports three styles of lamination in Lower Jurassic shales from Yorkshire, Great Britain, which he relates to changing deposition during marine transgression. He characterizes these lamination types as fine, thick, and wavy laminated. The **finely laminated shales** display (in X-radiographs and thin sections) alternating light and dark laminae that are commonly less than 1 mm thick and that have smooth, parallel contacts (Fig. 7.11A). The laminae are composed of fine, parallel-oriented carbonaceous flakes mixed with platy clay minerals. **Thick lamination** is also characterized by alternating light and dark layers, but individual layers are greater than 1 mm thick. Contacts between layers are parallel, but may display microscours (Fig. 7.11B). Microcrosslamination may also be present (Fig. 7.11C). The alternating layers are distinguished by different grain sizes. Darker, organic clay-rich layers are finer grained than adjacent lighter, coarser, silty layers. **Wavy lamination** is similar in layer thickness to fine lamination, but it displays wavy or undulating contacts (Fig. 7.11D, E, F). The wavy laminae are composed mainly of fine-grained organic and clayey material, but a few silt grains may be present. The shales display a progressive vertical upward change in lamination type from thick lamination to wavy lamination to fine lamination.

O'Brien suggests that the different styles of lamination in these shales are the result of deposition under changing conditions of water depth and oxygen state imposed during

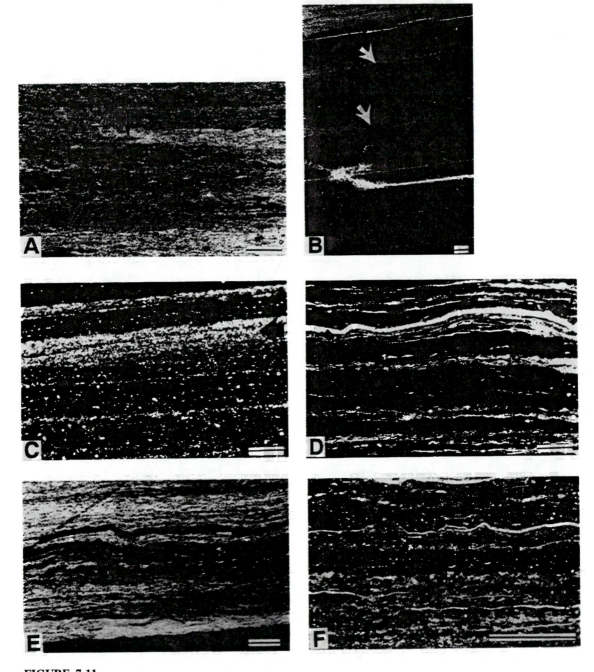

FIGURE 7.11
Thin-section photomicrographs of lamination types in Lower Jurassic shales from Yorkshire, Great Britain. A. Fine lamination with alternating light and dark layers; arrows point to carbonaceous flakes. B. Thick lamination; arrows point to scour features; coarser-grained components are silt. C. Thick lamination; arrow points to current lamination feature; coarser-grained components are silt. D, E, F. Wavy lamination. Scale bar = 1 mm except for E = 1 cm. (*From O'Brien, N. R., 1990, Significance of lamination in Toarcian (Lower Jurassic) shales from Yorkshire, Great Britain: Sed. Geology, v. 67. Fig. 3, p. 28, reprinted by permission of Elsevier Science Publishers, Amsterdam.*)

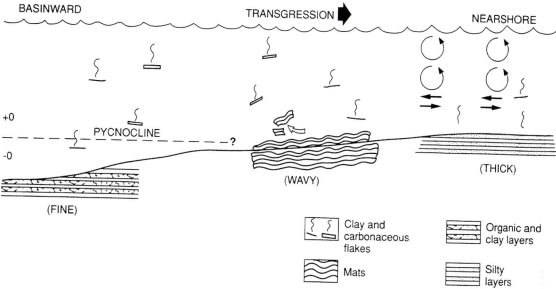

FIGURE 7.12

Schematic interpretation of the depositional setting of Lower Jurassic shales from Yorkshire, Great Britain. The pycnocline refers to the zone of abrupt density change in the water column. +O signifies oxic conditions, −O indicates low-oxygen conditions. (*From O'Brien, N. R., 1990, Significance of lamination in Toarcian (Lower Jurassic) shales from Yorkshire, Great Britain: Sed. Geology, v. 67. Fig. 5, p. 30, reprinted by permission of Elsevier Science Publishers, Amsterdam.*)

marine transgression (Fig. 7.12). He interprets the progressive vertical change in lamination as follows. Thick lamination indicates initial dominance of bottom-current activity at moderately shallow depths. The wavy laminae suggest subsequent colonization of the seafloor by cyanobacteria, which formed microbial mats that trapped fine sediment. With continuing transgression, the fine-laminated facies was deposited in deeper water under more-anaerobic conditions by suspension settling of fine inorganic particles and organic detritus. Vertical stacking of the three laminated facies is the result of shifting environmental conditions during transgression (i.e., the principle of Walther's Law).

7.4 Composition

7.4.1 Mineralogy

Bulk Mineral Composition

Clay minerals, fine-size micas, quartz, and feldspars are the most abundant minerals in shales. A variety of other minerals may occur in shales in minor quantities, including zeolites, iron oxides, heavy minerals, carbonates, sulfates, and sulfides, as well as fine-size organic matter (Table 7.3). Because of the difficulties involved in petrographic analysis of fine sediments, most investigators have concentrated on the clay mineralogy of shales, which can be determined fairly easily by X-ray diffraction methods. Quantitative to semiquantitative determination of fine quartz, feldspars, and other minerals in shales can be made also by X-ray diffraction methods (e.g., Shaw and Weaver, 1965; Till

TABLE 7.3

Principal constituents in shales

Constituents	Remarks
Silicate minerals	
Quartz	Makes up 20 to 30 percent of the average shale; probably mostly detrital; chalcedony and biogenic opal-A and opal-CT may also be present
Feldspars	Commonly less abundant than quartz; plagioclase generally more abundant than alkali feldspars
Zeolites	Commonly present as an alteration product of volcanic glass; phillipsite and clinoptilolite are common zeolites in modern marine sediments
Clay minerals	
Kaolinite (7 Å)	Forms under strongly leaching conditions: abundant rainfall, good drainage, acid waters; in marine basin tends to be concentrated nearshore
Smectite-illite-muscovite (10 Å and greater)	Smectite is a hydrated, expandable clay; common in soils and as an alteration product of volcanic glass; alters to illite during burial; illite is the most abundant clay mineral in shales; derived mainly from preexisting shales; alters to muscovite during diagenesis; muscovite may also be detrital
Chlorite, corrensite, and vermiculite	Chlorite forms particularly during burial diagenesis; second in abundance only to illite in Paleozoic and older shales; during burial, vermiculite may convert to corrensite and finally to chlorite
Sepiolite and attapulgite	Mg-rich clays that form under special conditions where pore waters are rich in Mg, e.g., saline lakes
Oxides and hydroxides	
Iron oxides	Hematite most common in shales, but goethite or limonite may be more common in modern muds; commonly present as coatings on clay minerals; may be converted to pyrite or siderite in reducing environments
Gibbsite	Consists of $Al(OH)_3$; may be associated with kaolinite in marine shales derived from the weathering of tropical landmasses
Carbonates	
Calcite	More common in marine than nonmarine shales
Dolomite	An important cementing agent in some shales
Siderite and ankerite	Occur in shales most commonly as concretions
Sulfur minerals	
Sulfates: gypsum, anhydrite, and barite	Occur as concretions in shales and may indicate the presence of hypersaline conditions during or after deposition
Sulfides	Mainly the iron sulfides pyrite and marcasite; these sulfides are most abundant in marine shales and indicate reducing conditions either at the time of deposition or during diagensis
Other constituents	
Apatite	Occurs particularly as nodules in marine shales that accumulated slowly in areas of high organic productivity (see Chapter 13)
Volcanic glass	Common in modern continental and marine muds in areas of volcanic activity; converts to zeolites and smectites during burial diagenesis
Heavy minerals	Occur in shales, but little is known about patterns of occurrence and relative abundance
Organic substances	
Discrete and structured organic particles	Mostly palynomorphs or small coaly fragments (vitrinite) (see Chapter 14)
Kerogen	Occurs in all shales except red ones; see Chapter 14 for the characteristics of kerogen

Source: Drawn mainly from Potter, P. E., et al., 1980, p. 47–49.

and Spears, 1969; Griffin, 1971), with an electron microscope equipped with an energy-dispersive X-ray unit (e.g., Bryant and Williams, 1982), or with an electron microprobe (e.g., Siever and Kastner, 1972). Also, quartz, feldspars, and other nonclay minerals can be separated from clay minerals in shales by chemical techniques (Blatt et al., 1982), allowing these nonclay minerals to be studied more effectively by petrographic methods. See also Wiegmann et al. (1982) for a discussion of methods of determining the complete mineral composition of shales.

In spite of the availability of these analytical techniques, few quantitative analyses of the bulk composition of shales have been reported. Some available data are summarized in Table 7.4. Note that the average content of quartz in the ancient shales reported in this table ranges from about 15 to 35 percent. Charles and Blatt (1978) report a substantially higher average value, about 50 percent, for the quartz content of some modern fluvial muds. Reported average values of feldspar content in shales range from trace amounts to more than 30 percent. Reported average clay- mineral abundances range between about 25 and 65 percent. Readers should keep in mind that the variations in mineral abundances shown in Table 7.4 may be due to several factors. First, different analytical techniques were used to obtain the data and some of the data are quite old. For example, some investigators used X-ray diffraction methods to obtain mineral composition, whereas others calculated mineral composition from chemical data. Also, mineral composition is known to vary markedly with grain size. Quartz tends to be more abundant in coarser-grained shales, whereas clay minerals are more abundant in finer-grained shales. A considerable amount of the variation shown in Table 7.4 could be due to this factor alone. Mineral composition may vary also owing to tectonic setting or depositional environment of the shales. For example, Bhatia (1985a) reports that quartz ranges from a low of 17 percent in passive-margin shales to as much as 46 percent in shales deposited in oceanic island arcs, whereas clay-mineral abundance ranges from 20 percent in oceanic island arcs to more than 75 percent in passive-margin shales. Blatt and Totten (1981) report significant variation in the quartz content of marine shales as a function of distance from shoreline: 47 percent quartz at a distance of 60 km from shore vs. 11 percent quartz at a distance of 270 km. Because so many different factors may affect the mineral composition of shales, it is difficult to generalize about the composition of average shales. Many additional data are needed to better define their bulk mineralogy. The values shown in Table 7.4 suggest, however, that the average quartz content of shales may be about 30 percent, average feldspar content about 10 percent, and average clay mineral content approximately 50 percent. Iron oxides, carbonate minerals, and various other minerals (Table 7.3) may make up about 10 percent of the average shale.

Clay Mineral Composition

Silicon-oxygen Tetrahedra.   Because clay minerals form such a significant fraction of most shales, some additional discussion of the clay minerals is necessary. Clay minerals belong to the group of silicate minerals known as **phyllosilicates**. They are characterized particularly by SiO_4^{4-} ionic groups in combination with metallic cations. The SiO_4^{4-} groups consist of a silicon atom surrounded by four oxygen atoms in a tetrahedral configuration (Fig. 7.13a). Therefore, they are called **silica tetrahedra** or **silicon-oxygen tetrahedra**. As discussed below, silica tetrahedra can be linked together to form indefinitely extending tetrahedral sheets. Phyllosilicates also contain groups of OH^- ions joined with cations (aluminum, magnesium, iron) in a sixfold coordination (Fig. 7.13d),

TABLE 7.4

Average mineral composition of selected shales reported in the literature

	1	2	3	4	5	6	7	8	9	10	11
Quartz	22.3	32.0	29.8	31.0	22.0	51.0	27.6	24.5	32.2	16.4	28.9
Feldspars	30.0	18.0	4.4	25.0	8.8	8.7	7.0	Trace	3.2	2.1	10.7
Clay minerals	25.0	34.0	59.0	42.0	55.0	40.3	?	62.3	63.4	63.9	50.79
Iron oxides	5.6	5.0	<0.5		3.3					—	1.4
Carbonates	5.7	8.0	3.5	2.0	7.7			13.2		13.5	5.4
Other minerals	11.4	1.0	1.9		3.3			Trace	1.1	4.3	2.3
Organic matter	—	1.0	1.0								0.2

Source:
1 Clark (1924)
2 Leith and Mead (1915)
3 Shaw and Weaver (1965)
4 Bhatia (1985b)
5 Yaalon (1962)
6 Charles and Blatt (1978)
7 Blatt and Schultz (1976)
8 Scotford (1965)
9 Evans and Adams (1975)
10 Raup (1966)
11 Average of values reported in columns 1 through 10

FIGURE 7.13

Diagrammatic sketches of clay-mineral lattice components: (a) a single silica tetrahedron; (b) silica tetrahedrons arranged in a hexagonal network; (c) three hexagons joined into a double chain; this chain, extended in all directions in this plane, would form a silicon-oxygen sheet; the silicons and the apical oxygens are projected onto the plane of the base of the tetrahedrons in this view; (d) a single octahedral unit; and (e) several octahedral units joined into the octahedral sheet structure. *(From Grim, R. E., Clay mineralogy. © 1968. Figs. 4.2, 4.3, 4.1, p. 52 and 55, reprinted by permission of McGraw-Hill, New York.)*

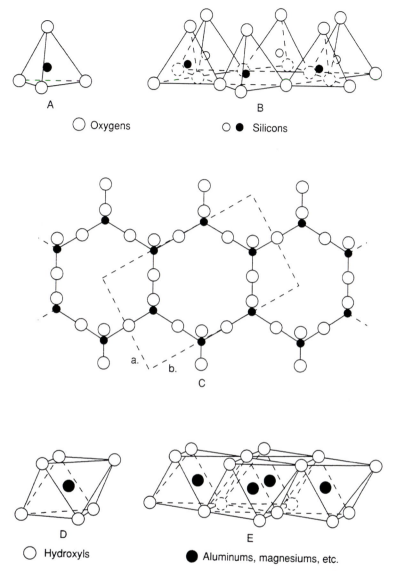

leading to an octahedral configuration. These octahedra can be linked also to form octahedral sheets. All the members of the phyllosilicates can be formed by linking together of tetrahedral and octahedral sheets in various ways.

Tetrahedral Sheets.   Silica tetrahedra can join together by sharing of apical oxygens (Fig. 7.13b) to form sheets. The smallest structural unit of a sheet consists of six tetrahedra, each of which shares three of the four oxygens with adjacent tetrahedra, forming hexagonal rings (Fig. 7.13c). The unshared oxygens in each sheet all point in the same direction (Fig. 7.13b). The sharing of three of the four oxygens in each SiO_4 tetrahedron with neighboring tetrahedra yields a ratio of two silicons to five oxygens. Thus, the formula for a tetrahedral sheet is $Si_2O_5^{2-}$ (or $Si_4O_{10}^{4-}$); however, aluminum ions may replace up to half of the silicon ions, yielding sheets such as $AlSi_3O_{10}^{5-}$ and $Al_2Si_2O_{10}^{6-}$.

Because of the negative charge, tetrahedral sheets cannot exist alone. They must be joined with cations, commonly silicon or aluminum, and additional oxygens to form mineral structures.

Octahedral Sheets.   The basic structural unit of an octahedral sheet is an eight-sided solid called an octahedron, which consists of cations, generally aluminum, magnesium, or iron, surrounded by six hydroxyl (OH^-) ions in an octahedral configuration (Fig. 7.13d). Linking together of octahedra through sharing of hydroxyl ions creates an octahedral sheet (Fig. 7.13e). Hydroxyl ions are shared between adjacent octahedra, so there can be three divalent cations or two trivalent cations for each octahedron of hydroxyl ions. For example, in the mineral brucite, Mg_3OH_6, there are three Mg^{2+} ions for each ring of six hydroxyl ions. The sheet formed by this configuration is called a trioctahedral sheet. In the mineral gibbsite, $Al_2(OH)_6$, there are only two Al^{3+} ions for each ring of six hydroxyl ions because of the higher charge of aluminum. That is, aluminum ions can enter only two-thirds of the possible cation sites; otherwise a charge imbalance would result. An octahedral sheet formed by a configuration of two cations for each octahedron of hydroxyls is called a dioctahedral sheet. Because octahedral sheets carry no charge, they can exist alone.

Basic Clay Mineral Structures

Tetrahedral sheets are often referred to as T layers, and octahedral sheets are called O layers. All phyllosilicate minerals are formed by various combinations of T layers and O layers. Remember that octahedral sheets include both trioctahedral (brucite) layers and dioctahedral (gibbsite) layers. Tetrahedral layers and octahedral layers are joined or linked by sharing in common the oxygen and hydroxyl ions of the individual sheets. A single T layer can be linked to a single O layer to form a one-layer, or T-O, structure. In this process, the apical oxygens of an Si_2O_5 sheet are substituted for two of the six hydroxyl ions of a brucite or gibbsite sheet. If a T layer is linked to one side of a brucite or gibbsite layer and another T layer is linked to the other side, a two-layer, or T-O-T, structure results. Figure 7.14 diagrammatically illustrates the structure of kaolinite, which has a T-O dioctahedral structure. Figure 7.15 shows the dioctahedral T-O-T structure of muscovite. Figure 7.16 further illustrates schematically the manner by which brucite sheets and gibbsite sheets are linked in various ways with tetrahedral sheets to form clay minerals and related phyllosilicates. Remember that T sheets are negatively charged and O sheets are electrically neutral. When T and O sheets are linked, two OH^- ions are lost for each Si_2O_5 tetrahedral unit gained; therefore, T-O and T-O-T structures are electrically neutral units.

Classification and Identification of Clay Minerals

The clay minerals and related phyllosilicates can be divided into a few distinct groups on the basis of layer type and cation content of the octahedral sheet (dioctahedral or trioctahedral) (Table 7.5). Subdivision of these basic groups into subgroups and species is made on the basis of layer charge, type of interlayer material, type of layer stacking, chemical composition, and type of component layers and nature of stacking (ordered, random) for mixed-layer clays (Eslinger and Pevear, 1988, p. 2–10). The clay minerals are a complex group of minerals, and detailed discussion of these minerals and their distinguishing characteristics is beyond the scope of this book. Readers are referred to specialized works on clay minerals such as Brindley and Brown (1980), Eslinger and Pevear (1988), and Velde (1985) for additional information.

FIGURE 7.14

Sketch of kaolinite, 1:1 dioctahedral phyllosilicate. (*From Grim, R. E., Clay mineralogy. © 1968. Fig. 4.4, p. 58, reprinted by permission of McGraw-Hill, New York.*)

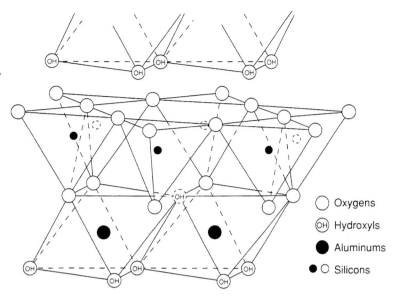

Oxygens
Hydroxyls
Aluminums
Silicons

FIGURE 7.15

Sketch of muscovite, a 2:1 dioctahedral phyllosilicate. (*From Grim, R. E., Clay mineralogy. © 1968. Fig. 4.16, p. 93, reprinted by permission of McGraw-Hill, New York.*)

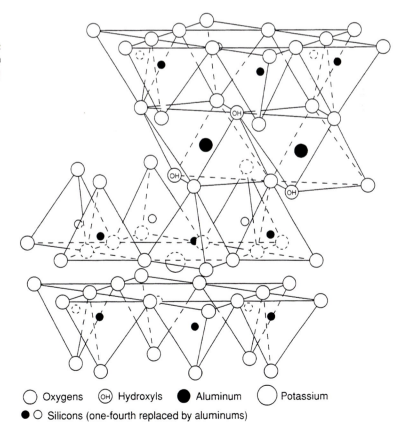

Oxygens   Hydroxyls   Aluminum   Potassium
Silicons (one-fourth replaced by aluminums)

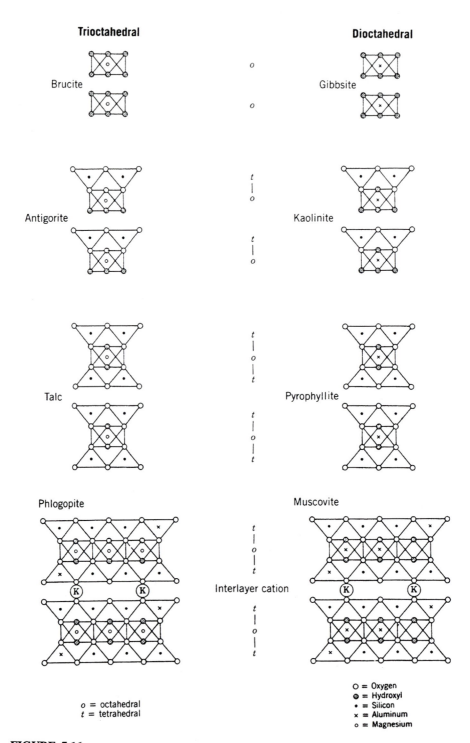

FIGURE 7.16

Schematic representation of the linking of T and O layers to form clay minerals and micas. (*From Klein, C., and C. S. Hurlbut, Jr., Manual of mineralogy, 20th ed. Fig. 11.69, p. 421,* © *1985, John Wiley & Sons, Inc. Reprinted by permission of John Wiley & Sons, Inc., New York.*)

TABLE 7.5
Classification of clay minerals

Layer type	Interlayer material	Group	Subgroup	Species (examples)
1:1	None or H$_2$O only	Serpentine-kaolin (x ~ 0)	Serpentine	Chrysotile, lizardite, berthierine
			Kaolin	Kaolinite, dickite, nacrite, halloysite
2:1	None	Talc-pyrophyllite (x ~ 0)	Talc	Talc, willemseite
			Pyrophyllite	Pyrophyllite
	Hydrated exchangeable cations	Smectite (x ~ 0.2–0.6)	Saponite	Saponite, hectorite, sauconite, stevensite, etc.
			Montmorillonite	Montmorillonite, beidellite, nontronite
	Hydrated exchangeable cations	Vermiculite (x ~ 0.6–0.9)	Trioctahedral vermiculite	Trioctahedral vermiculite
			Dioctahedral vermiculite	Dioctahedral vermiculite
	Nonhydrated cations	True mica (x ~ 0.5–1.0)	Trioctahedral true mica	Phlogopite, biotite, lepidolite, annite
			Dioctahedral true mica	Muscovite, illite, glauconite, paragonite, celadonite
	Nonhydrated cations	Brittle mica (x ~ 2.0)	Trioctahedral brittle mica	Clintonite
			Dioctahedral brittle mica	Margarite
	Hydroxide	Chlorite (x variable)	Trioctahedral chlorite	Clinochlore, chamosite, nimite, pennantite
			Dioctahedral chlorite	Donbassite
2:1 mixed-layer (regular)	Variable	None	None	Hydrobiotite, rectorite, corrensite, aliettite, tosudite, kulkeite
Modulated 1:1 layer	None	No group name (x ~ 0)	No subgroup name	Antigorite, greenalite
Modulated 2:1 layer	Hydrated exchangeable cations	Sepiolite-palygorskite (x variable)	Sepiolite	Sepiolite, loughlinite
			Palygorskite	Palygorskite
	Variable	No group name (x variable)	No subgroup name	Minnesotaite, stilpnomelane, zussmanite

x = layer charge/O$_{10}$ (OH)$_2$

Source: Eslinger, E., and D. Pevear, 1988, Clay minerals: Soc. Econ. Paleontologists and Mineralogists Short Course Notes 22. Table 2.1, p. 2–12, reprinted by permission of SEPM, Tulsa, Okla.

Our primary concern in this book is with the more common clay mineral groups: kaolin or kaolinite, smectite (particularly montmorillonite), the true micas (especially muscovite, biotite, illite), and chlorite. Mixed-layer clay minerals, chlorite/smectite or illite/smectite, are common also. The most common way of identifying clay minerals is by X-ray diffraction methods, which are based on measuring the spacing between successive identical layers in a crystal. This spacing is called the d-spacing and is measured in angstrom units. See Eslinger and Pevear (1988) for a review of X-ray diffraction techniques.

Relative Abundance of Major Clay Minerals

The relative abundance of clay minerals in shales may vary as a function of provenance, depositional environment, and age and diagenetic history. With respect to provenance, different kinds of source rocks furnish different kinds of clay minerals. For example, chlorite and micas are common in low- and medium-rank metamorphic rocks; kaolinite, illite, smectite, chlorite, mixed-layer clays, and micas may be present in sedimentary rocks; and few clay minerals of any kind occur in igneous rocks. Further, climate affects the kinds of clay minerals that form in source areas owing to weathering processes. For example, kaolinite tends to form under warm, humid conditions where chemical leaching processes are intense, whereas smectites and illites are more likely to form under drier conditions.

The effects of the depositional environment on the clay-mineral composition of fine-grained sediments is controversial. The most controversial aspect of environment is the possible role of the depositional environment in alteration of clay minerals. Although it has often been suggested that one kind of clay minerals can transform to another within the depositional environment, many investigators now suggest that there is little evidence to indicate that equilibrium of detrital clay-mineral assemblages is achieved in most depositional environments (e.g., Eslinger and Pevear, 1988, p. 4–8; Hathaway, 1979; Velde, 1985, p. 33). In fact, many researchers believe that the clay mineralogy of modern oceanic sediments more commonly reflects the mineralogy of the source areas from which the clay minerals were derived than the depositional environment (e.g., Griggs and Hein, 1980). A notable exception is the formation of smectite clays by alteration of silicic glasses in volcanic ash beds or by alteration of seafloor basalts. The distribution of clay minerals in the modern ocean may also be a reflection in part of the differential transportability of clay minerals. Coarser-grained kaolinite tends to settle faster, and be transported less far, than finer-grained smectite. Thus, kaolinite settles first, and smectite is enriched in sediments in a seaward direction (Porrenga, 1966).

Finally, diagenesis (Chapter 9) can bring about significant changes in clay mineral composition. Smectite clays tend to alter to illite at burial temperatures ranging from about 70°C to 150°C; smectites may alter also to chlorite. Kaolinite tends to alter to illite at about the same temperatures. Therefore, with deeper burial (and increasing age), illite and chlorite tend to increase at the expense of smectite and kaolinite.

Owing to the many factors that can affect clay-mineral composition, only two generalities are offered. First, sedimentary rocks older than Mesozoic commonly contain more illite and chlorite and less smectite and kaolinite than younger rocks. Second, marine rocks tend to contain more illite and chlorite than nonmarine rocks, which are more likely to be enriched in kaolinite and smectite. Many exceptions to these generalities are known. Table 7.6 provides a few examples of the clay-mineral composition of some ancient shales and modern oceanic sediments.

TABLE 7.6
Average clay-mineral composition of some shales and muds reported in the literature

	1	2	3	4	5	6	7	8	9	10
Kaolinite	5.7			65.4	26.5	14.1	6.2		12.1	14.4
Smectite		2.0	13.0			2.5	37.6	40.8		10.7
Illite	20.0	37.0	27.0	21.5	32.0	62.0	47.5	26.8	34.3	34.2
Chlorite	Trace			5.7	26.5	21.5	8.9		11.1	8.2
Mixed-layer illite/smectite	74.4								42.5	13.0
Kaolinite/chlorite		61.0	60.0					32.2		17.0
Montmorillonite/ mixed-layer clays				7.1	13.5					2.3

Source:
1 Johnsson and Reynolds, 1986 (Cretaceous–Eocene shale)
2 Molnia and Hein, 1982 (shelf sediments, NE Gulf of Alaska)
3 Molnia and Hein, 1982 (Miocene-Quaternary shale)
4 Lonnie, 1982 (Cretaceous shale)
5 Darby, 1975 (Arctic Ocean sediments)
6 Naidu et al., 1971 (Beauford Sea sediments)
7 Stoffers and Müller, 1972 (Black Sea sediments)
8 Griggs and Hein, 1980 (shelf sediments, California continental shelf)
9 Raup, 1966 (Pennsylvanian shales)
10 Average of values reported in columns 1 through 9

7.4.2   Chemical Composition

The principal minerals that occur in shales are indicated in Table 7.3. The relative abundance of these various minerals, in turn, determines the chemical composition of shales. Table 7.7 gives the chemical composition of some average shales from North America and Russia. The SiO_2 content of average shales ranges from about 57 to 68 percent. Compare this value with the average SiO_2 content of sandstones, which is greater than 90 percent for quartz arenites (Table 5.2), 60–90 percent for feldspathic arenites (Table 5.4), and 55–85 percent for lithic arenites (Table 5.6). Thus, most shales contain distinctly lower concentrations of SiO_2 than most sandstones, but the SiO_2 content of shales is comparable to that of some matrix-rich sandstones. The SiO_2 content of shales is influenced by all the silicate minerals present but it is particularly affected by quartz, which makes up almost 30 percent of the average shale, as mentioned. Most quartz in shales is commonly assumed to be detrital. Actually, very little is known about the abundance of authigenic quartz in shales or the ratio of detrital to authigenic quartz. Some shales, the so-called siliceous shales, have abnormally high silica content, ranging to more than 90 percent SiO_2 (Baltuck, 1982), and correspondingly low Al_2O_3 content. Some siliceous shales owe their high SiO_2 content to the presence of biogenic remains, chiefly diatoms and radiolarians, that were initially precipitated as opal. Siliceous shales of the Miocene Monterey Formation of California (Bramlette, 1946) are a classic example. In fact, some diatom-rich siliceous shales grade to diatomites. Diagenetic alteration of opaline skeletal remains can cause remobilization of the silica and deposition in pores of the shales as opal or quartz cement. Abnormally high concentrations of SiO_2 may be related also to high concentrations of volcanic ash in shales.

Aluminum is the second most abundant chemical constituent in shales. The Al_2O_3 content of average shales shown in Table 7.7 ranges from about 16 to 19 percent. Compare this narrow range to the much broader range of values that occurs in sandstones, from less than 1 percent (some quartz arenites) to more than 15 percent (some feldspathic and lithic arenites). The composition of aluminum in shales is a function of both feldspar content and clay-mineral content. Owing to the generally high abundance of clay minerals in shales, these minerals have a particularly strong influence on shale chemistry. Table 7.7 provides only a general indication of clay-mineral chemistry. Readers should consult Newman and Brown (1987) for a detailed discussion of the chemistry of clay minerals and variations in chemical composition among the clay-mineral species of each clay-mineral group. The aluminum content of kaolinites is particularly high; therefore, shales that contain exceptionally high aluminum (20 percent Al_2O_3), called **high-alumina shales,** are likely to contain high percentages of clay minerals, especially kaolinite.

K_2O makes up about 2.5–5.0 percent and MgO about 2.5–4.5 percent of average shale. Potassium and magnesium abundance in shales is related also to clay-mineral abundance, although some magnesium may be supplied by dolomite, and potassium is contained in potassium feldspars. Shales containing more than 5 percent K_2O are comparatively rare. These **potassic shales** have been suggested to owe their high potassium content to the presence of authigenic K-feldspars. Na_2O makes up about 0.75–2.8 percent of average shale. Sodium abundance is related both to clay minerals (e.g., smectites) and to the content of sodium plagioclase. The average abundance of iron oxides (Fe_2O_3 + FeO) in shales ranges between about 4.0 and 7.7 percent. Iron in shales is supplied by iron oxide minerals (hematite, limonite, goethite), some fine micas and clay minerals, e.g., biotite, smectites, and chlorite, and the carbonate minerals siderite and ankerite. Also,

TABLE 7.7

Average chemical composition of selected shales reported in the literature

	1	2	3	4	5	6	7	8	9	10	11	12	13
SiO_2	60.65	64.80	59.75	56.78	67.78	64.09	66.90	63.04	62.13	65.47	64.21	64.10	63.31
Al_2O_3	17.53	16.90	17.79	16.89	16.59	16.65	16.67	18.63	18.11	16.11	17.02	17.70	17.22
Fe_2O_3	7.11											2.70	0.82?
FeO		5.66	5.59	6.56	4.11	6.03	5.87	7.66	7.33	5.85	6.71	4.05	5.45
MgO	2.04	2.86	4.02	4.56	3.38	2.54	2.59	2.60	3.57	2.50	2.70	2.65	3.00
CaO	0.52	3.63	6.10	8.91	3.91	5.65	0.53	1.31	2.22	4.10	3.44	1.88	3.52
Na_2O	1.47	1.14	0.72	0.77	0.98	1.27	1.50	1.02	2.68	2.80	1.44	1.91	1.48
K_2O	3.28	3.97	4.82	4.38	2.44	2.73	4.97	4.57	2.92	2.37	3.58	3.60	3.64
TiO_2	0.97	0.70	0.98	0.92	0.70	0.82	0.78	0.94	0.78	0.49	0.72	0.86	0.81
P_2O_5	0.13	0.13	0.12	0.13	0.10	0.12	0.14	0.10	0.17			—	0.10
MnO	0.10	0.06	—	0.08	—	0.07	0.06	0.12	1.10	0.07	0.05	—	0.06

Source:

1 Moore, 1978 (Pennsylvanian shale, Illinois Basin)
2 Gromet et al., 1984 (North American shale composite)
3 Ronov and Migdisov, 1971 (average North American Paleozoic shale)
4 Ronov and Migdisov, 1971 (average Russian Paleozoic shale)
5 Ronov and Migdisov, 1971 (average North American Mesozoic shale)
6 Ronov and Migdisov, 1971 (average Russian Mesozoic shale)
7 Cameron and Garrels, 1980 (average Canadian Proterozoic shale)
8 Ronov and Migdisov, 1971 (average Russian Proterozoic shale)
9 Cameron and Garrels, 1980 (average Canadian Archean shale)
10 Ronov and Migdisov, 1971 (average Archean shale)
11 Clark, 1924 (average shale)
12 Shaw, 1956 (compilation of 155 analyses of shale)
13 Average of values in columns 1 through 12

substantial amounts of iron in some organic-rich shales are contained in sulfide minerals (pyrite, marcasite). Shales containing more than about 15 percent iron oxides are called **ferruginous shales** or **ferriferous shales**. Discussion of iron-rich sedimentary rocks is covered in Chapter 13.

Table 7.7 shows that the CaO content of "average" shales in North America and Russia ranges from about 0.5 to 8.9 percent. Calcium in shales is derived from calcium-rich plagioclase and from carbonate minerals, particularly calcite and dolomite. Calcium is contained also in some of the clay minerals and in gypsum and anhydrite. Pettijohn (1975, p. 285) suggests that the average shale contains about 2.6 percent CO_2, equivalent to about 6 percent calcite. The carbonate in shales occurs both as fossil remains and as cements, much of which is probably derived by dissolution of fossils. Shales particularly enriched in carbonates are called **calcareous shales**.

Oxides of titanium and phosphorus make up less than 1 percent each of average shales. Although the average P_2O_5 content of shales ranges from about 0.1 to 0.17 percent, some so-called **phosphatic shales** contain substantially greater abundances. Phosphatic shales that contain more than about 20 percent P_2O_5 are called **phosphorites**. The origin of phosphorites is discussed in Chapter 13. In addition to major elements, shales contain a variety of trace elements: B, Ba, Ga, Cr, V, Li, Ni, Co, Cu, Sc, Zr, Sr, Pb, and a host of others. The trace-element abundance of shales is a complex function of provenance, depositional environments, and diagenesis.

Because chemical composition is a direct function of mineralogy, and mineral composition varies with grain size, the major-element chemical composition of shales is related also to grain size. Coarser-grained shales contain more quartz than finer-grained shales and thus tend to have a higher SiO_2 content. Finer-grained shales contain higher percentages of clay minerals, resulting particularly in aluminum enrichment and lower SiO_2 concentrations. Calcium, magnesium, and potassium tend also to be concentrated in the finer fraction of shales; however, calcium and magnesium content can be strongly influenced by secondary carbonate cements that may be particularly abundant in coarser-grained shales.

The chemical data presented in Table 7.7 are recast in Figure 7.17 in the form of graphs in which relative abundance of a major element is plotted against geologic age. Note from Figure 7.17 that silica abundance shows a rather erratic trend with respect to geologic age, neither increasing nor decreasing steadily as a function of increasing age of samples. The other major elements, with the possible exception of iron and sodium, likewise show somewhat erratic abundance trends. Total iron shows a slight increase in abundance with increasing age, especially in North American shales, and sodium increases in abundance in Precambrian shales over Paleozoic shales. Although these data are few, they suggest that diagenesis may not play a significant role in determining the major-element geochemistry of shales. On the other hand, Figure 7.17 shows a remarkably similarity between the abundance trends of major elements in North American and Russian shales. Does this similarity indicate some kind of global tectonic or sedimentological control on the mineralogical and chemical composition of shales? This question has been considered by Ronov (1983), who shows several figures in which major-element abundances and elemental ratios are plotted as a function of age of shales. He explains variations in major-element abundances, particularly abundances of sodium and potassium, with time as a complex function of geologic events that include closure of Precambrian ocean basins and expansion of continental platforms, decrease in extrusive volcanism from Precambrian time, initial formation of granitic rocks and subsequent

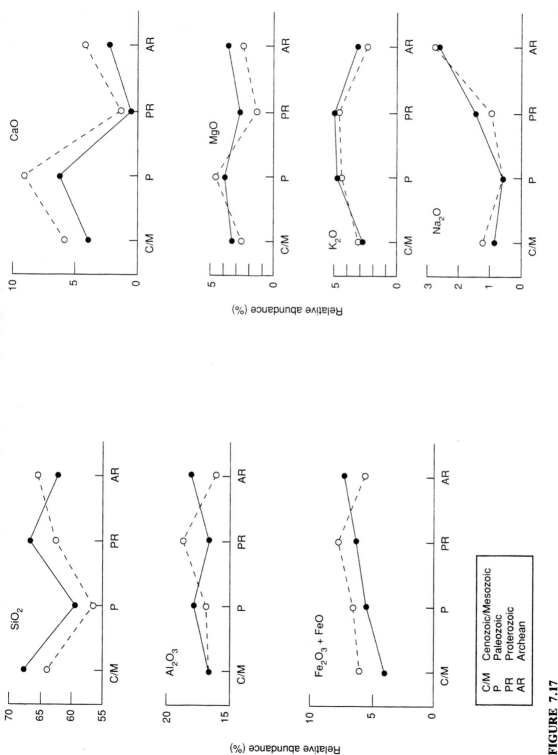

FIGURE 7.17

Major-element geochemistry of North American and Russian average shale plotted as a function of geologic age. Filled circles are North American samples; open circles are Russian samples. Data from Table 7.7.

erosion of these granitic materials, deposition of arkosic associations in platform sand-stones, incorporation of potassium by clay minerals, and repeated cycles of weathering and sedimentation. Ronov concludes that the role of ancient sedimentary rocks as a source for material in younger sediments has increased with time, whereas that of volcanic and granitic rocks has become less significant during the Phanerozoic. See also Ronov and Migdisov (1971).

7.4.3   Organic Content

Kinds of Organic Matter

Most of the organic matter in shales is sapropel, which consists largely of the remains of phytoplankton, zooplankton, spores, pollen, and the macerated fragments of higher plants. During burial and diagenesis, this organic matter goes through a series of complex changes, brought about by combined biochemical and chemical attack. These diagenetic processes destroy much of the organic matter and convert the remainder to an insoluble substance called kerogen. Most of the organic matter in shales consists of kerogen. Additional details about the origin of kerogen, as well as discussion of the chemical and isotope composition of organic matter, are given in Chapter 14.

Abundance of Organic Matter

Most sedimentary rocks deposited under aqueous conditions contain some organic mate-rial. Degens (1965) suggests that the average sedimentary rock contains about 1.5 percent organic matter by weight. Organic matter contains 50–60 percent carbon; therefore, according to Degens, the average sedimentary rock contains somewhat less than 1 percent organic carbon. Shales commonly contain more organic matter than either sandstones or carbonate rocks. Data compiled by Tissot and Welte (1984, p. 97) show that average relative abundance of organic carbon in shales, reported by various workers, ranges between about 1.0 and 2.2 percent. Some black shales contain considerably greater amounts than this average, ranging to 10 percent or more. Sedimentary rocks that are exceptionally rich in organic carbon are discussed in Chapter 14. The abundance of organic matter in sedimentary rocks depends upon a number of factors, including grain size of the sediment, depositional environment, and sediment mineralogy. These factors are examined below.

Relationship of Organic-matter Abundance to Sediment Grain Size

Studies of modern oceanic sediments demonstrate that in a given part of the ocean, clay-sized sediments contain more organic matter than silt-sized sediments and that silts contain more organic matter than sands. For example, Yemel'yanov (1975) reports in one study of shelf sediments that shelf sands contain an average of 0.73 percent organic carbon, silts contain 1.35 percent, and clays 2.86 percent. A similar trend of increasing organic content with decreasing grain size has been observed in some ancient shales. For example, Hunt (1979, p. 272) reports that a sample of Viking shale from Alberta, Canada, contains 1.79 percent organic matter in the silt-size fraction, 2.08 percent in the 2–4 μm clay-size fraction, and 6.50 percent in the <2 μm clay-size fraction. Thus, in general, shales and fine-grained carbonate rocks contain more organic matter than sandstones, and very fine-grained shales contain more organic matter than coarser-grained shales. Barker (1979, p. 27) suggests that three main processes operate to create the observed relation-ship between sediment grain size and organic matter content: (1) wave and current action winnows clay-size sediment and fine organic particles from coarser sediment and deposits the low-density organic material together with fine sediment in quieter water, (2) sandy sediment tends to be deposited in higher-energy environments characterized by oxidizing conditions; organic matter is destroyed in these oxidizing environments because the coarse sediment allows ample diffusion of oxygen through large-size pores, and (3) owing to their large surface areas, settling clay particles may absorb some types of organic materials from solution and transfer them into the sediments.

Depositional Environment of Organic Matter

Significant amounts of organic matter can accumulate in nonmarine, swampy environments to form peats and coals. Abundant fine organic matter may also be deposited with fine sediment in some lacustrine (lake) and marine environments. For example, some organic-rich oil shales (Chapter 14) are lacustrine deposits. Owing to the disproportionally large volume of marine sediments relative to other sediments, geologists and geochemists are particularly interested in the origin and preservation of organic matter in marine sediments. Several processes control the distribution of organic matter in marine sediments. Organic matter is transferred from the land to the ocean by river runoff and winds. Dissolved organic carbon also enters the ocean by groundwater input. Within the ocean, enormous quantities of organic matter are generated by marine organisms, particularly phytoplankton and zooplankton. Upon the death of these organisms, their remains rain to the ocean floor, where they are incorporated into accumulating sediment. Redistribution of organic matter takes place within the ocean by suspension transport, bottom currents, turbidity currents, and sediment slumping and sliding. Much of the organic matter is destroyed by oxidation to CO_2 and by bacterial decomposition while in the zone of oxidation. Much of the surviving organic matter eventually finds its way into deeper water, where it is deposited with fine sediments below wave base and below the zone of well-oxygenated water.

Important depositional factors that influence the amount of organic material in shales include the organic productivity of waters overlying the ocean floor, the sedimentation rate of detrital material, and the oxidation state of the depositional environment. In the modern ocean, zones of greatest organic productivity are concentrated in areas where an abundance of nutrients, especially phosphorus and nitrogen, are constantly available in surface and near-surface waters. These zones occur principally in high latitudes, where ocean waters are essentially isothermal from surface to depth and the pycnocline is absent, and in zones of upwelling along continental margins and some parts of the equator. In much of the deeper waters of the ocean, organic productivity is very low (these areas are sometimes called biologic deserts), and the amount of organic material that reaches the ocean floor is extremely small. In some nearshore areas of the ocean, particularly near the mouths of large rivers, organic matter transported to the oceans from the land may be quantitatively important (Barker, 1979, p. 27). In a few areas of modern sedimentation, pollen, which is transported by wind or aqueous processes, makes up a substantial fraction of the organic material in bottom sediments (Roman, 1974; Heusser, 1988).

High rates of organic productivity in a particular part of the ocean do not necessarily, however, translate to high organic content in the sediment. If rates of clastic sedimentation are high, the clastic sediments "dilute" the organic matter, resulting in low relative organic carbon abundances. This masking of organic matter owing to high sedimentation rates of clastic detritus has been invoked by Shimkus and Trimonis (1974), for example, to explain why the areas of highest organic carbon concentration (>3 percent) in the modern Black Sea do not lie directly beneath the areas of highest primary organic productivity. The Black Sea areas of high organic productivity are apparently also areas of rapid sedimentation. On the other hand, rapid sedimentation leads to rapid burial of organic matter, promoting preservation, as discussed below. Thus, an intermediate rate of sedimentation is probably most favorable overall to generation of high organic content in sediments.

The preservation potential of the environment is another important factor that can affect the ultimate organic carbon content of shales. Organic matter is destroyed fairly quickly in well oxygenated environments by oxidation to CO_2 and by the activities of aerobic bacteria. Thus, little organic matter is preserved in nonmarine sediments except that deposited in some aqueous reducing environments such as swamps or lakes. In the marine realm, a wide variety of environments can exist, ranging from those characterized by high-energy, well-oxygenated waters to those marked by low-energy waters and reducing conditions. In oxic environments, destruction of organic matter by oxidation and aerobic bacteria is aided by the activities of burrowing, benthonic organisms. These organisms directly consume some of the organic matter. In addition, they rework and stir up the bottom sediment, allowing oxygen to more easily come in contact with organic matter in the sediment. Therefore, relatively little organic matter may be preserved in these environments.

Anoxic environments are commonly characterized by both reducing conditions and the presence of toxic hydrogen sulfide (H_2S). Under these conditions, few benthonic organisms are present to stir up the sediment, little molecular oxygen is present, and bacterial decomposition is limited to that accomplished by anaerobic bacteria. As mentioned in Section 7.3.1, sediments deposited under these conditions tend to be well laminated. Thus, considerable organic matter may be preserved in the sediments of anoxic environments. Anoxic conditions can result from poor circulation of water, as in silled or enclosed basins, because oxygen is not replenished in the water as it is used up by organic respiration and chemical oxidation. Anoxia can result also from high organic productivity. In areas of high organic productivity, decay of organic material and its oxidative conversion to CO_2 use up oxygen, creating reducing conditions. This process may be accompanied by production of H_2S owing to the activities of sulfate-reducing bacteria. Deposition of organic-rich, black shales was particularly prevalent during certain geologic periods such as the Devonian and Cretaceous. Some workers have suggested that during such periods anoxia may have been global or oceanwide, owing perhaps to lack of strong oceanic mixing because of stable density stratification and possible low oceanic thermal gradients (e.g., Arthur and Schlanger, 1979). On the other hand, Waples (1983) suggests, for the Early to Middle Cretaceous at least, that anoxia was controlled mainly by local conditions in deep, restricted basins. At times, sluggish ocean circulation may have permitted expansion of the oxygen-minima layer, permitting deposition of anoxic sediments in nonbasinal settings. Waples suggests, however, that development of more-widespread anoxia was a self-damping process owing to the concomitant rise of the calcium carbonate compensation depth, promoting oxidation of organic carbon and causing contraction of the oxygen-minimum layer.

In contrast to the commonly held view that black shales accumulate during periods of anoxia, Pederson and Calvert (1990) propose that deposition under conditions of anoxia has little to do with the origin of organic-rich sediments. They suggest instead that sporadic temporal and spatial increases in primary productivity, which reflect changes in the behavior and/or state of the ocean–atmosphere system, provide a more tenable explanation for the occurrence of organic-rich sediments and black shales. Thus, according to Pederson and Calvert, organic-rich facies do not represent deposition under euxinic conditions and, further, do not indicate stagnant oceans or basins. We will have to wait to see how this interesting idea is received by other workers who are researching black shales.

An additional factor that affects preservation of organic matter is the rate of deposition of detrital sediment. High rates of detrital sedimentation promote rapid burial of organic matter and thus enhance preservation. On the other hand, as mentioned, high sedimentation rates lead to dilution of organic matter.

In summary, the organic carbon content of shales varies rather widely from less than 1 percent to more than 10 percent of total constituents. High organic content of shales appears to be favored by (1) fine grain size of sediment and (2) deposition under conditions where surface waters have high organic productivity, bottom waters are anoxic, and sedimentation rates are intermediate.

7.5 Color of Shales

Shales may have a variety of colors ranging through red, brown, yellow, green, light gray, and dark gray to black. An excellent discussion of the causes of colors in shales is given by Potter et al. (1980, p. 53). The colors of shales appear to be a function mainly of the carbon content and the oxidation state of iron in the shales. The progression of colors from light gray through dark gray to black correlates with increasing carbon content of shales, whereas variations in color from red through purple to greenish gray correlate with decreasing ratios of Fe^{3+}/Fe^{2+}. The oxidation state of iron appears to be much more important in determining the color of sediments (Fe^{3+} iron gives red colors, Fe^{2+} iron gives green colors) than the total amount of iron present in the rocks. On the other hand, a few studies have shown that there is less iron in some greenish or gray sediments than in associated red sediments. The oxidation state of iron is controlled by the amount of organic matter in sediments, which can furnish electrons to drive iron into the reduced state. Therefore, green or greenish gray shales tend to have a higher organic matter content (less organic matter is destroyed by oxidation) than red or yellow shales.

Color is associated in a very general way with depositional conditions. That is, black shales tend to form in relatively deep, restricted basins where reducing conditions prevail and abundant organic matter is preserved. Alternatively, they form in some shallow-water tidal-flat and estuarine environments where organic matter is abundant. Red shales are characteristic of oxidizing environments. Such environments occur particularly in continental settings. Red muds accumulate also in deep-sea basins where sedimentation rates are low, and the muds are thus in contact with oxidizing bottom waters for long periods of time. The colors of muds that prevail at the time of deposition can be changed during burial diagenesis and uplift, if the sediments are brought into a different chemical environment. Thus, the ferric iron (Fe^{3+}) that characterizes red shales at the time of deposition may subsequently be reduced to ferrous iron (Fe^{2+}) during diagenesis to yield green shales. Likewise, green shales can be changed to red shales after deposition if initial reducing conditions are subsequently followed by oxidizing conditions. Because color can be affected so markedly by diagenesis, it is not a reliable indicator of depositional conditions.

7.6   Classification of Shales

7.6.1   Introduction

Classification of sandstones and conglomerates is based primarily on the mineralogy or particle composition of these rocks, although texture (matrix abundance) is used as a secondary classification parameter in some classifications. By contrast, classification of shales primarily on the basis of mineralogy has generally been regarded to be impractical because of the difficulty in obtaining quantitative data on shale composition owing to their fine grain size. Although semiquantitative to quantitative mineralogical data can be obtained by X-ray diffraction analysis of shales, and perhaps some other techniques, routine analysis of shales by these techniques has lagged behind petrographic analysis of sandstones. Consequently, relatively few workers have attempted to classify shales on the basis of mineralogy. In fact, only a few classifications of any kind for shales have been proposed, and most of these proposed classifications are based mainly on textures and structures of shales. Many geologists are content to use informal names for shales that reflect properties such as color (red shales, green shales, black shales), organic content (carbonaceous shales), or relative abundance of particular kinds of cementing materials or chemical constituents (calcareous shales, ferriferous shales, high-alumina shales, phosphatic shales, siliceous shales). Nonetheless, a few formal classifications for shales do exist, as discussed below.

7.6.2   Shale Classification Mainly on the Basis of Texture and Structure

Most classifications of shales are based upon the relative abundance of silt-size and clay-size particles in the shales and upon the presence or absence of lamination or fissility (Blatt et al., 1980, p. 382; Lundegard and Samuels, 1980; Pettijohn, 1975, p. 262; Potter et al., 1980, p. 14). The estimated abundance of silt-size particles in shales is commonly used to divide the particles in shales into three broad groups: **silts** ($>2/3$ silt), **muds** ($>1/3$, $<2/3$ silt), and **clays** ($>2/3$ silt). Some authors (e.g., Blatt et al., 1980, p. 382) further divide shales into two groups on the basis of fissility: fissile mudrocks and non-fissile mudrocks. Other authors (e.g., Lundegard and Samuels, 1980) regard fissility as an inappropriate property for classification because, among other things, fissility is related to the degree of weathering of rocks. These authors use lamination rather than fissility to subdivide shales into laminated and nonlaminated types. Thus, if laminated, shales are called laminated siltstones (or silt-shales), mudshales, or clayshales, depending upon grain size. If nonlaminated, they are called siltstones, mudstones, or claystones, depending upon grain size. The classification of Potter et al. (1980, p. 14), reproduced here as Table 7.8, is another example of a classification that uses lamination as a classification parameter. Potter et al. (1980, p. 15) suggest that other properties of shales such as mineralogy, color, fossil content, and types of organic constituents can be used as adjectival modifiers for the basic textural names shown in Table 7.8.

For an additional viewpoint on classification of shales on the basis of fabric, see Nuhfer (1981). See also the discussion of fabric classification by several authors in the *Journal of Sedimentary Petrology*, v. 51, p. 1027–1033 (1981).

7.6.3   Shale Classification Based in Part on Mineralogy

Picard (1971) proposed a classification for fine-grained sedimentary rocks based on both texture and particle composition. On the basis of grain size, he divided the fine-grained

TABLE 7.8
Textural classification of shales

			Percentage clay-size constituents	0–32	33–65	66–100
			Field Adjective	Gritty	Loamy	Fat or Slick
NONINDURATED	Beds	Greater than 10 mm		BEDDED SILT	BEDDED MUD	BEDDED CLAYMUD
	Laminae	Less than 10 mm		LAMINATED SILT	LAMINATED MUD	LAMINATED CLAYMUD
INDURATED	Beds	Greater than 10 mm		BEDDED SILTSTONE	MUDSTONE	CLAYSTONE
	Laminae	Less than 10 mm		LAMINATED SILTSTONE	MUDSHALE	CLAYSHALE
METAMORPHOSED	Degree of metamorphism LOW → HIGH			QUARTZ ARGILLITE	ARGILLITE	
				QUARTZ SLATE	SLATE	
				PHYLLITE AND/OR MICA SCHIST		

Source: Potter, P. E. et al., 1980, Sedimentology of shale. Table 1.2, p. 16, Reprinted by permission of Springer-Verlag, New York.

rocks into nine textural types and sandstones into three textural types (Fig. 7.18). Picard indicates that the mineral composition of medium to coarse siltstones can be determined reliably with a petrographic microscope. For these rocks, he suggests use of existing terminology for sandstones. Considering the wide range of choices for existing sandstone classification, that may not be much help. He further recommends that the composition of the clay-size fraction be determined by X-ray diffraction methods. The name of the dominant clay mineral can then be combined with the textural classification and sandstone nomenclature to yield names for fine-grained rocks such as **clayey, siltstone, illite-**

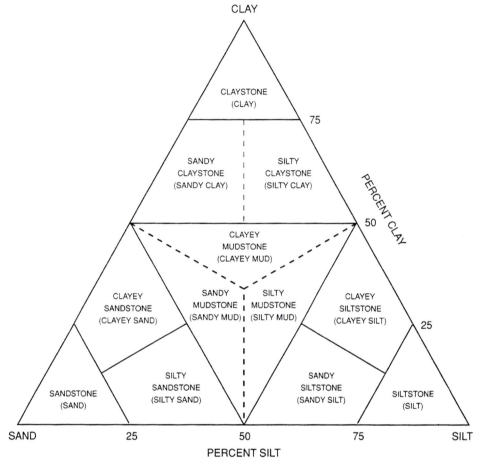

FIGURE 7.18
Textural classification of fine-grained rocks and sediments. (*After Picard, M. D., 1971, Classification of fine-grained sedimentary rocks: Jour. Sed. Petrology, v. 41. Fig. 3, p. 185, reprinted by permission of Society of Economic Paleontologists and Mineralogists, Tulsa, Okla.*)

sublitharenite; and **siltstone, kaolinite-subarkose.** See Lewen (1978) and Spears (1980) for additional examples of shale classifications based in part on mineralogy.

7.7   *Distribution and Significance of Shales*

7.7.1   *Occurrences*

As mentioned in the chapter introduction, shales make up about 50 percent of all sedimentary rocks. Most of these shales are of marine origin. Shales occur in rocks of all ages and on all continents of the world. Ronov (1983) indicates that shales have been the dominant sedimentary rock deposited throughout most of geologic time (Fig. 1.3). Their overall abundance in the geologic record reflects the high average abundance of fine-grained material generated by the processes of weathering and erosion (which are related to elevated land masses and high rainfall), the general availability through time of oceanic and interior basins to serve as catchment areas for fine sediment, and the relative efficiency of transporting agencies such as bottom currents, turbidity currents, suspension transport mechanisms, nepheloid transport mechanisms, and wind for movement of fine sediment to more distal parts of depositional basins.

The wide distribution of shales in space and time suggests that the environments that favor shale deposition are common and that they have recurred throughout time. Shales are not especially characteristic of any particular geologic time. They are abundant in stratigraphic sequences of all ages; however, their geographic distribution varies as a function of age. Thus, their relative abundances reflect changing patterns of land and seas through time. The relative abundances of shales in stratigraphic sequences of a particular age are related to depositional setting. The thickest shale sections tend to occur in sedimentary sequences that were deposited in mobile, marine basins. Nonetheless, moderate thicknesses of nonmarine shales (some oil shales, for example) can occur in stratigraphic sequences deposited in interior basins, either in lacustrine or alluvial basins.

Potter et al. (1980, p. 75) conclude that ancient shales are particularly prevalent in marine deltaic sequences. They are abundant both in sequences that formed by progradation seaward onto oceanic crust and in those that formed by progradation onto cratons or that developed within cratons. They suggest that other important settings for shale deposition include basins lying seaward from and adjacent to rimmed carbonate platforms, rift zones ranging in size from a few hundred kilometers to almost continental or oceanic dimensions, basins associated with volcanic island arcs, and the basinal areas of deep oceans.

7.7.2   Environments of Shale Deposition

Fine-grained sediment may be deposited on land by wind to form loess or other eolian deposits; however, most fine, siliciclastic sediment is deposited in water. Deposition takes place mainly in quiet water, below wave base, at water depths ranging from tens of meters to thousands of meters. Such low-energy environments can be present in a variety of depositional settings on continents (i.e., lakes, alluvial piedmonts or plains, river floodplains), in transitional marine environments (deltas, estuaries, lagoons, tidal flats), on the shallow-marine shelf and the continental slope, and in deeper-marine basins. In continental settings, fine sediment is transported in suspension by wind and by stream flow. It may also be transported, together with sand and gravel, by glacial ice and by mass-transport processes (mud flows, debris flows, turbidity currents in lakes). Fine sediment that is delivered to the ocean by any of these processes can be retransported farther seaward by several mechanisms that may include storm-generated waves and currents, suspension transport in near-surface plumes, near-bottom nepheloid transport, turbidity currents, and bottom currents, as mentioned. Overall, turbidity currents are probably the most important mechanism for moving large quantities of fine sediment to distal parts of deep basins. Wind may also transport fine sediment from land to distant regions of the ocean and may be the principal mechanism for transport of pelagic sediment. See Boggs (1987, p. 458, 498) for a review of these transport processes. Within the depositional environment, deposition of fine sediment takes place by slow gravity settling of dispersed single particles or by more-rapid settling of clay flocs, as discussed in Section 7.3.2. Organisms may aid in deposition of fine sediment by pelletizing muds and by creating baffles or other features (e.g., algal mats) that trap and hold fine sediment.

Shales deposited in a given environment acquire characteristic properties (geometry, lithology, facies associations, textures, structures, fossil content) that may allow them to be distinguished from shales of other environments. Thus, shales may be important paleoenvironmental indicators. See Potter et al. (1980, Table 1.9) for a useful summary of the characteristic properties of shales deposited in major environments. Potter et al. also give a brief description of the depositional processes that prevail in each of these environments.

7.7.3   Significance of Shale Occurrence

The presence of thick units of shales in ancient stratigraphic sequences implies derivation of fine-grained sediment from large-volume land masses that were weathered and eroded under generally humid, high-rainfall conditions. Furthermore, the presence of shales in a stratigraphic section suggests deposition in a quiet-water paleoenvironment. Most thick shale units were deposited in marine basins and tend to be laterally extensive. Individual marine shale beds may be relatively thin, but they can commonly be traced laterally for considerable distances. By contrast, associated sandstone, conglomerate, or limestone facies are generally more restricted in their lateral extent. Therefore, many shale beds make excellent stratigraphic markers. In comparison to marine shales, shale sequences deposited in nonmarine settings tend to be thinner overall and laterally less extensive.

As discussed, shales may display lithologic characteristics, sedimentary structures, fossil content, organic content, geochemical characteristics, and facies associations that have environmental significance. Furthermore, shales, or facies closely associated with shales, may contain directional sedimentary structures that are useful as paleocurrent indicators. Clay minerals and feldspars in shales may have some importance as provenance indicators, although shales are generally less useful than sandstones in provenance studies (Chapter 8). Finally, shales have considerable economic significance as source beds for petroleum and natural gas. Careful geochemical evaluation of the amount and types of kerogen in shales (Chapter 14) is now routine procedure in petroleum exploration programs.

In summary, shales have paleoclimatic, paleoenvironmental, provenance, and economic significance. They deserve greater attention in future sedimentological studies than they have received in the past.

7.8   Examples

Because shales are so common in sedimentary sequences of all ages and on all continents, examples of shale are numerous. Geologists from any continent can readily compile an extensive list of shale formation names. Some shales are internationally "famous" owing to their lithologic makeup, thickness and geographic extent, fossil content, or tectonic and stratigraphic significance or to their economic potential as oil source beds or sites of copper mineralization or other mineralization. Examples of such shales include the Precambrian Figtree shales of South Africa, known for their early fossils; the Middle Cambrian Burgess Shale of British Columbia, famous for its preserved remains of soft-bodied organisms; the Silurian Gothlandian shales that extend throughout western Europe and much of northern Africa and the Persian Gulf; the organic-rich Devonian to Mississippian black shales (Ohio Shale, New Albany Shale, Chattanooga Shale, etc.) that cover much of the central and eastern parts of North America; the Permian Kupferschiefer of western

Europe, known for its mineralization of copper, lead, and zinc; the Cretaceous Mowery Shale of the western interior of the United States, a thick, siliceous shale that is believed to be a major source bed for oil; and the Eocene Green River Oil Shale of Utah, Colorado, and Wyoming (USA), a well-known oil shale that contains high kerogen content with significant recoverable oil potential. These, and some other world-famous shales, are pointed out by Potter et al. (1980, p. 83).

Additional Readings

Arthur, M. A., Organizer, 1983, Stable isotopes in sedimentary geology: Soc. Econ. Paleontologists and Mineralogists Short Course 10.

Bennett, R. H., W. R. Bryant, and M. H. Hulbert, 1991, Microstructures of fine-grained sediments: Springer-Verlag, New York, 582 p.

Bennett, R. H., and M. H. Hulbert, 1986, Clay microstructure: D. Reidel, Dordrecht, 161 p.

Brindley, G. W., and G. Brown, eds., 1980, Crystal structures of clay minerals and their X-ray identification: Mineralog. Soc. Mon. 5, Mineralogical Society, London, 495 p.

Chamley, H., 1989, Clay sedimentology: Springer-Verlag, Berlin, 623 p.

Durand, B., ed., 1980, Kerogen: Éditions Technip, Paris, 519 p.

Eslinger, E., and D. Pevear, 1988, Clay minerals for petroleum geologists and engineers: Soc. Econ. Paleontologists and Mineralogists Short Course Notes 22.

Fripiat, J. J., 1982, Advanced techniques for clay mineral analysis: Elsevier, Amsterdam, 235 p.

O'Brien, N. R., and R. M. Slatt, 1990, Argillaceous rock atlas: Springer-Verlag, New York, 141 p.

Potter, P. E., J. B. Maynard, and W. A. Pryor, 1980, Sedimentology of shale: Springer-Verlag, New York, 306 p.

Stow, D. A. V., and D. J. W. Piper, eds., 1984, Fine-grained sediments: Deep-water processes and facies: Geol. Soc. Spec. Pub. 15, Blackwell, Oxford, 659 p.

Velde, B., 1985, Clay minerals: A physico-chemical explanation of their occurrence: Elsevier, Amsterdam, 427 p.

Chapter 8
Provenance of Siliciclastic Sedimentary Rocks

8.1 Introduction

In the preceding chapters, we characterized sedimentary rocks in terms of their physical and chemical properties. Such characterization is not, however, the principal reason that we normally undertake research on sedimentary rocks. Determining the physical and chemical properties of these rocks is simply a means to a more important end, which is to reconstruct the history of the rocks. Our ultimate aim in studying siliciclastic sedimentary rocks is to develop a fuller understanding of (1) the source(s) of the particles that make up the rocks, (2) the erosion and transport mechanisms that moved the particles from source areas to depositional sites, (3) the depositional setting and depositional processes responsible for sedimentation of the particles (the depositional environment), and (4) the physical and chemical conditions of the burial environment and the diagenetic changes that occur in siliciclastic sediment during burial and uplift. These objectives, in turn, are important to the broader goal of developing reliable paleogeographic models of Earth for particular times in the past.

In this chapter, we deal with one of these important objectives of geologic research, understanding the sources of siliciclastic sediment. We commonly refer to sediment source as provenance. The term **provenance** is derived from the French *provenir,* meaning to originate or come forth (Pettijohn et al., 1987, p. 254). The term is also spelled provenience. Terms such as source area and sourceland are sometimes used as synonyms for provenance. As the word provenance is commonly used by sedimentologists today, however, it has a broader meaning than just source area. The meaning of provenance has been extended to encompass the location of the source area (how far away was it and in what direction?), its size or volume, the lithology of the parent source rocks, the tectonic setting of the source area, and the climate and relief of the source area. Provenance studies are especially important to our understanding of paleogeography. When coupled with studies of depositional environments, they help us interpret the relative positions of ancient oceans and highlands at given times in the geologic past. From such studies, we are able to reconstruct the location, size, and lithologic composition of mountain systems that have long since vanished. We may even be able to make intelligent guesses about the climate and relief of these highlands, as well as the tectonic setting in which the source area lay.

Most early studies of provenance focused on determining the lithology of parent source rocks, as interpreted from the particulate components of sandstones and conglomerates. Many investigators also sought to identify the locations of source areas on the basis of paleocurrent analysis of directional sedimentary structures and by mapping grain-size or grain-shape trends. Beginning in the 1970s, emphasis shifted to interpretation of tectonic setting in terms of plate-tectonic concepts, that is, characterization of source areas as magmatic arcs, collision orogens, or continental blocks. Only a few studies have focused on interpreting climate and relief of source area, possibly owing to the fact that such interpretations are difficult to make and their reliability is somewhat tenuous. Most provenance studies have involved analysis of sandstones or conglomerates; relatively few studies of shale provenance have been attempted. In this chapter, we examine some of the tools and techniques that sedimentologists use to interpret provenance.

8.2   Tools for Provenance Analysis

8.2.1   General Statement

The most important features of siliciclastic sedimentary rocks upon which provenance interpretations are based are (1) the modal and chemical composition of the detrital components in the rocks, from which we read source-rock lithology and tectonic setting, and (2) the presence of directional features in the rocks (sedimentary structures and size and shape gradients) that allow interpretation of paleocurrent patterns. Other characteristics of siliciclastic deposits useful in provenance analysis include the paleomagnetic characteristics of the rocks, which help establish the paleolatitude of the source area; vertical and lateral facies relationships of stratigraphic units, which are related to sediment transport directions; and the overall thickness and volume of siliciclastic units, which reflect to some degree the size of source area (Fig. 8.1).

8.2.2   Composition of Detrital Constituents

The mineralogy of the detrital particles in siliciclastic sedimentary rocks furnishes the primary evidence for the lithology of the parent rocks in the source area. Our knowledge of the lithology of vanished ancient mountain systems rests mainly on analysis of detrital framework modes of siliciclastic deposits. Mineralogy also provides our most useful evidence for interpreting tectonic setting because source-rock lithology is linked fundamentally to tectonic setting. Detrital mineralogy may also provide some insight into climatic conditions on the basis of the assumption that mineralogic maturity of sediments is determined in part by selective destruction of minerals at weathering sites. Weathering under very cold or very dry conditions presumably allows preservation of unstable min-

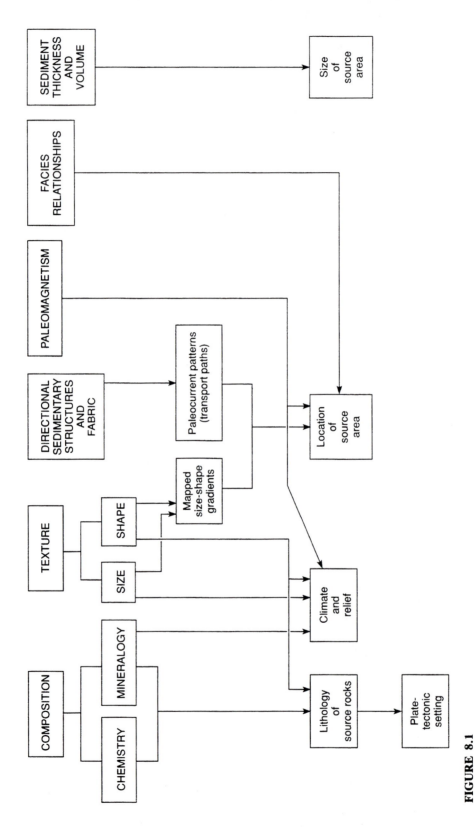

FIGURE 8.1
Properties of siliciclastic sedimentary rocks and their relationships to provenance.

erals, whereas intensive weathering under hot, humid conditions does not. Thus, detrital minerals provide some clues to climate. Chemical composition of some minerals, particularly their trace-element chemistry, may aid also in identifying the lithology of source rocks. The luminescence colors of minerals such as quartz are likewise believed to have significance in source-rock interpretation.

8.2.3 Textures and Structures

The textures and structures of siliciclastic sedimentary rocks provide the most direct evidence for location of source areas. Mapping grain-size trends and grain-shape trends, together with evidence derived from directional sedimentary structures such as flute casts and cross-beds, allows interpretation of paleocurrent directions. Thus, textural and structural data help fix the general direction in which the source area was located. Grain-size and -shape data may also provide some insight into the climate and relief of source areas, under the assumption that the nature and intensity of weathering can affect the sizes of particles released from source rocks. High-relief source areas tend to promote more-rapid erosion, and thus derivation of coarser particles, than low-relief source areas, where slower erosion allows more time for size reduction by weathering processes. The grain size of sediments is a function of numerous complex variables, however, as shown in Table 8.1, and is not easily related to climate and relief. In terms of grain shape, intense rounding of grains, for example, could point to eolian activity and desert conditions in source areas. The application of Fourier analysis to quantification of grain shape (Section 2.3.5) has increased the importance of grain shape as a provenance tool.

8.2.4 Thickness and Volume of Siliciclastic Units

The existence of vanished mountain systems and other highlands that acted as sediment sources in the past is known only indirectly from the preserved record of sediments

TABLE 8.1
Processes that generate clastic particles and the relationship of particle size to the grain size of parent rocks

---

EXPLOSIVE VOLCANISM—generates fragments ranging in size from ash to boulders. Initial size, shape, and sorting of particles depends upon the nature of primary volcanic processes. Size, shape, and sorting of particles can be modified by subsequent retransport.

EPICLASTIC PROCESSES—generate particles ranging in size from clay to boulders. Processes that generate particles and that affect their sizes include

   Glacial grinding—produces silt-size grains, particularly silt-size quartz

   Weathering and soil building—grain sizes of fragments released during weathering is a function of

     Sizes of grains in parent rock

     Spacing of microsheet fractures in parent rocks

     Planes of weakness in individual grains; e.g., microfractures and dislocation structures in quartz grains; cleavages in micas and feldspars

     Chemical environment—Clay minerals and fine-size oxides and quartz may be formed at the weathering site by chemical recombination and crystallization

   Selective size (and shape) sorting during transport—Transport modifies the original size, shape, and sorting characteristics of particles imparted to them by weathering processes

   Grain breakage and pebble abrasion during transport—Grains may break along microfractures or other planes of weakness; pebble abrasion produces mainly silt-size particles

   Minor breakage during burial diagenesis—owing to compaction and displacive growth of cements

stripped from these highlands and deposited in nearby or distant basins. The overall size of these vanished highlands can be estimated roughly from the volume of siliciclastic sediments preserved in the basins. Such estimates are indeed rough because the source area may have included considerable volumes of soluble rocks such as limestones. Also, much mud-size sediment derived from the source area may have been transported to and dispersed within distal areas where it cannot readily be tied to a particular source area.

8.2.5   Paleomagnetism

The magnetic inclination preserved in iron-bearing minerals in rocks of any kind is now used routinely as a tool for interpreting the paleolatitude at which rocks formed. See Wyllie (1976, p. 161) for a simple explanation of how this technique works. When applied to siliciclastic sedimentary rocks, a knowledge of paleolatitude allows geologists to determine if the rocks were deposited in the same latitude in which they are now located. If they were not, they are part of a displaced or exotic terrane. In such a case, the sedimentary rocks may (or may not) have been rifted away from their source area, which could now exist at some completely different location with respect to the depositional basin. A knowledge of the paleolatitude at which the sediments were deposited can be useful also in interpreting the climate of the source area or at least in constraining interpretations of climate on the basis of other factors.

8.2.6   Facies Relationships

When displayed in maps and cross sections, vertical and lateral facies relationships allow interpretation of source directions. Facies relationships maps can be related to directional features such as cross-bedding and thus to paleoslope. Clastic ratio maps or sand-clay ratio maps, for example, give a clear indication of the general source direction of clastic detritus. See Potter and Pryor (1961, Fig. 9) for an example of such facies maps. Examination of a series of stratigraphic units of different ages can reveal the persistence, or lack of persistence, of a particular source area or can show geographic shifts in major sediment dispersal centers as a function of time.

8.3   Locating Source Areas

8.3.1   Paleocurrent Analysis

Methods for determining the relative positions of ancient source areas and depositional basins are based primarily on paleocurrent analysis. The properties of sediments used in paleocurrent analysis are listed in Table 8.2, after Potter and Pettijohn (1977, p. 3). Paleocurrent indicators help establish paleoslope and paleotransport directions, which, in turn, reveal the direction in which the source area lay with respect to the depositional basin. Overall, the most useful paleocurrent indicators are probably the various kinds of directional sedimentary structures such as flute casts, ripple marks, and cross-beds. Methods of measuring the orientation of directional structures and presentation of directional data are discussed in Section 3.7.3. Fabric elements of sediments such as long-axis orientation and imbrication of pebbles also have directional significance (Section 2.3.7). When paleocurrent data are plotted on maps such as that shown in Figure 8.2, the relative position of the source area with respect to the depositional basin may be clearly indicated.

TABLE 8.2

Properties of sediments and sedimentary rocks used in paleocurrent analysis

Property	*Definition*	*Geologic example*	*Remarks*
Attribute	Presence or absence	Trace mineral, pebble or boulder in till, gravel or sand; multicomponent mineral assemblage	Directional significance only when mapped
Scalar	Magnitude	Thickness, lithologic and mineral proportion, grain size, sorting, roundness, etc.	Directional significance only when mapped
Directional	Specified by azimuth; may indicate either a line or direction of movement	All sedimentary structures with directional significance such as sand fabrics, fossil orientation, flute marks, etc.	A vector property, if magnitude could be specified
Tensor	Directions and lengths of principal axes of ellipsoid	Fluid permeability, dielectric constant, sonic transmissibility, etc.	An anisotropic property that is fabric-dependent.

Source: Potter, P.E., and F. J. Pettijohn, 1977, Paleocurrents and basin analysis. Table 1–1, p. 3, reprinted by permission of Springer-Verlag, Berlin.

8.3.2   Mapping Attribute or Scalar Properties

In the absence of directional sedimentary structures or fabric elements, mapping the presence of distinctive minerals or pebbles of a particular lithology may point toward the source area. Such maps may reveal dispersal patterns and thus sediment source directions. Mapping the grain-size distributions within sandstone or conglomeratic deposits may likewise have directional significance. In conglomeratic rocks, grain-size distribution maps are typically based on maximum clast size, which is commonly given as the mean intermediate clast diameter of the ten largest clasts measured at each sample point. In sandy sediments, maps of either mean or maximum grain size may be prepared. If one works under the general assumption that sediments become finer grained away from the source, grain-size distribution maps may reveal distinctive grain-size trends that point toward the sediment source area. Figure 8.2 shows contours of modal grain size as well as paleocurrent patterns. Note that grain-size change and paleocurrent vectors indicate that the source area lay to the west of the map area. Attempts have been made to quantify the rate of change of sediment grain sizes and thus to relate rate of change to distance of transport from the source. For example, in a study of a modern stream system in Italy, McBride and Picard (1987) conclude that downstream decrease in gravel size follows an exponential formula that can be expressed in linear form using phi size: size $(\phi) = 0.1167 \times$ distance (km) $- 10.29$. Owing to the complexity of sediment transport processes, however, efforts to relate grain size of ancient sediments to distance of transport have not been very successful.

Mapping grain shape (roundness, sphericity, form) to establish grain-shape trends is another possible technique to identify sediment-dispersal directions. When such techniques are applied to pebbles, we might expect as a gross approximation that pebble

FIGURE 8.2

Example of the use of paleocurrent data to locate source areas. Brandywine gravel of Maryland. (*From Potter, P. E., and F. J. Pettijohn, 1977, Paleocurrents and basin analysis. Fig. 8–9, p. 282, reprinted by permission of Springer-Verlag, Berlin.*)

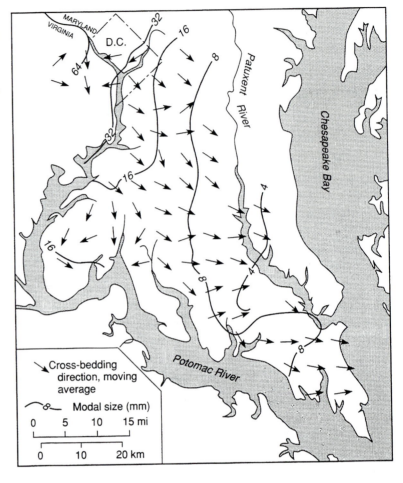

roundness would increase in the direction of sediment transport. Further, particle form and sphericity might change also as a result of shape sorting and possibly abrasion during sediment transport. For example, roller-shaped and equant-shaped pebbles might be expected to outrun disc- and blade-shaped pebbles. Problems arising from factors such as pebble recycling, pebble breakage, and addition in fluvial systems of sediment from downstream tributaries and from uncertainties about the relative effects of current shape sorting and abrasion on the shapes of pebbles all weaken the reliability of interpretations based on such techniques. These common shape techniques do not work well with sand-size particles because gross properties such as roundness and sphericity of sand-size quartz are not greatly affected by stream transport. More-sophisticated techniques for determining the shapes of quartz sand and silt by using Fourier shape analysis (Chapter 2) appear to hold greater promise for detecting meaningful downcurrent changes in particle shape (e.g., Kennedy and Ehrlich, 1985). Studies of this type are, however, still in a formative stage.

Facies relationships that can be expressed in terms of thickness or lithologic character of various lithofacies can be mapped also. Such maps may be used as further evidence for sediment dispersal patterns and thus source-area location. Lithofacies thickness (isopach) maps, maps that show the ratio of siliciclastic constituents to nonsiliciclastic constituents (clastic ratio maps), sand-shale ratio maps, facies maps that show the geographic distribution of major lithofacies types, and so on all provide evidence of sediment-dispersal patterns. The construction and use of stratigraphic maps is discussed in detail by Krumbein and Sloss (1963, Ch. 12). Application of these maps to interpretation of sediment-dispersal patterns is discussed by Potter and Pettijohn (1977, Ch. 8), who also give an in-depth discussion of sediment-dispersal systems.

8.3.3   Studies of Paleomagnetism

As mentioned, the paleomagnetic characteristics of sedimentary rocks can be used to determine the paleolatitude of the depositional basin. Assuming that source areas are located within a few degrees of latitude of depositional basins, paleomagnetic evidence can be used to at least approximately locate the paleolatitude of source areas. Such evidence is particularly important in working with displaced terranes that may have been transported thousands of kilometers from their original locations and may even have been rifted away from their source areas.

8.4   Factors That Affect the Composition of Siliciclastic Sedimentary Rocks and Provenance Interpretation

8.4.1   General Statement

Research on provenance analysis has focused especially on the particle framework composition of siliciclastic sedimentary rocks as a tool for interpreting lithology and tectonic setting of source areas. It has long been recognized, however, that the composition of these rocks is controlled by more than the composition of the source rocks alone. Source-rock composition is certainly the first-order control on sediment composition, but the original composition may be severely modified by several factors, particularly (1) climate and relief of the source region, which control weathering and erosion, (2) the nature of the sediment-transport process, which can affect composition by selective destruction of less-stable grains and selective sorting on the basis of size, shape, and specific gravity, (3) the depositional environment, where further selective grain destruction and sorting may take place, as well as mixing of sediment from different sources, and finally (4) diagenetic processes that may result in partial or complete dissolution of less-stable grains or in their replacement by other minerals.

8.4.2   Effects of Climate and Relief

Each of these factors has a complex relationship to the ultimate composition of sandstones and cannot be treated in detail here. Perhaps the least studied and most poorly understood is the influence of climate and relief. We recognize in a general way that warm, humid climates bring about, in a given period of time, greater destruction of parent-rock minerals than very cold or very dry climates. Also, it is commonly accepted that rugged relief and steep slopes promote more-rapid erosion than low relief and more-gentle slopes. Mechanical and chemical weathering processes can bring about size reduction of stable grains such as polycrystalline quartz, and chemical and biochemical weathering can cause alteration or even complete solution and destruction of less-stable minerals. Relatively few studies have explored these relationships rigorously or in depth. See Basu (1985b) for a short discussion of the problem. Some of these problems become more apparent also in subsequent discussion in Section 8.4.4 of the bases for interpreting the climate of source areas. Owing to the potentially great influence that weathering has on the composition of siliciclastic detritus fed into the transport mill, the topic of climate and relief of source areas deserves greater research attention than geologists have given it in the past.

8.4.3   Effect of Sediment Transport

Sediment transport is the crucial link between source and depositional areas. Transport affects the composition of sediment in a variety of ways depending upon the mode of transport. Transport of sediment by current flow is a selective process. Particles are separated into a suspension load and a bedload, and within the bedload, gravels tend to be separated from sand. These processes thus bring about sorting of particles by size, shape, and density. Obviously, such sorting changes the mix of particle compositions that was initially fed into the transport system and may thus affect provenance interpretation. For example, current transport causes most clay minerals and other very fine minerals to become separated from coarser-grained quartz, feldspars, and rock fragments. Furthermore, gravel, which is composed mainly of rock fragments, tends to become separated from sand-size fragments. Additional sorting occurs within the sand fraction owing to shape and density differences of grains. To illustrate, heavy minerals become sorted by density into different size groups, resulting in fine-size, dense, heavy minerals being transported along with hydraulically equivalent, coarser-grained quartz and feldspar. Even the composition of pebbles may be affected by size sorting during transport. For example, the clast composition of coarse gravels on river point bars can differ significantly from the clast composition of fine gravels on the same bars (Boggs, 1969b).

Current transport also causes significant abrasion of gravel-size detritus, as well as mechanical shattering and breaking owing to impact. Abrasion and impact shattering thus tend to selectively destroy mechanically weak or soft grains. Abrasion of sand-size grains during current transport is not extremely significant, apparently owing to the cushioning effect of water. Transport of sediment under waves, particularly in the littoral zone, may be even more effective than current flow in selective sorting of sediments because prolonged reworking by waves provides a greater opportunity for sorting than does the unidirectional movement of currents. Transport of sediment by ice or sediment gravity flows, such as turbidity currents, commonly produces less selective sorting by size and density and possibly less mechanical destruction than does current transport. Nonetheless, these transport processes do have some effect on particle composition, e.g., abrasion by ice transport and sorting by size in the graded beds of turbidites.

8.4.4   Effects of the Depositional Environment

Depending upon the depositional setting, sediments may or may not undergo additional modification within the depositional environment. Sediments deposited rapidly by fluvial processes or mass flows—on alluvial fans, for example—may undergo little further change before burial. By contrast, sediments brought into the littoral zone may undergo significant change within the beach environment before final deposition takes place. The littoral zone acts as a filter, holding back sands and gravels and allowing muds to pass outward onto the shelf. If sands and gravels do not bypass the shelf by resedimentation transport (mass transport) down submarine canyons that head up close to shore, they may be held within the littoral zone for long periods of time, where they are subject to intensive reworking by wave swash and longshore currents. Such intensive reworking could ultimately erase initial compositional differences in sands, leading to high compositional maturity (quartz arenites) and effectively removing much of the compositional evidence of provenance. On the other hand, under conditions where several streams carrying sediments of diverse composition empty onto a shelf, these sediments may become mixed by littoral transport processes or storm surges. Such mixing creates deposits with compositions that are extremely difficult to link to a single source area. Bioturbation within the depositional environment may also mix sediments of diverse origins, further complicating provenance analysis. Sediment transported off the shelf into deeper water probably does not undergo very much additional change, except perhaps some bioturbation, once it reaches its final depositional site.

8.4.5   Effects of Diagenesis

Finally, sediments become subjected to changed geochemical conditions within the burial environment that can produce profound postdepositional modifications in composition. Metastable grains may be destroyed completely by dissolution, or they may be replaced by another mineral (e.g., the replacement of quartz by calcite) or altered to another mineral (e.g., the alteration of feldspars to clay minerals or micas). See McBride (1985) for a more extended discussion of the diagenetic processes that affect provenance determinations in sandstones. These changes are also discussed in Chapter 9. They can result in selective destruction of less-stable constituents and thus significant loss in provenance information.

Collectively, the changes brought about in the composition of sediment by weathering and erosion, transport, reworking at the depositional site, and diagenesis can be significant. Obviously, provenance analysis requires that we be alert to the possibility of such changes and that we eschew negative evidence. That is, we cannot use the absence of particular constituents as a guide to provenance interpretation; we can use only their presence. The fact that feldspars and heavy minerals may be absent or scarce in a sandstone, for example, does not mean that they were necessarily absent or scarce in the source rocks. In practice, we may tend to forget this principle, but we do it at our peril.

8.5   Provenance of Sandstones

8.5.1   Interpreting Source-rock Lithology

Introduction

Most research on provenance of siliciclastic sedimentary rocks has focused on sandstones. Much of this research has been concerned with attempts to interpret source-rock lithology from detrital mineralogy. Petrographic analysis of quartz, feldspars, micas, heavy minerals, and rock fragments forms the basis for most such studies; however, investigators are now turning also to supplementary techniques such as cathodoluminescence microscopy and geochemical studies of trace elements and isotopes in detrital minerals. Do the major kinds of igneous, metamorphic, and sedimentary rocks contain uniquely different suites of minerals that provide a distinctive provenance signature? Are these characteristics preserved in sediments? Let's examine what is known about this interesting topic.

Reading Provenance from the Properties of Quartz

General Statement.   Because quartz is the most abundant constituent in most sandstones, and essentially the only constituent of some sandstones, geologists have been intrigued by the potential provenance significance of quartz since the early years of petrographic study (e.g., Sorby, 1880; Mackie, 1896). These early studies focused particularly on the provenance significance of inclusions in quartz, although the possible provenance significance of undulatory extinction was also recognized quite early (Gilligan, 1919). In the United States, Krynine, in 1940 and in a series of later papers, appears to be the first worker to make in-depth studies of the properties of quartz (inclusions, undulatory extinction, polycrystallinity, nature of subgrain contacts, grain shape) as provenance indicators. Krynine proposed a genetic classification of quartz based on these properties that is still used by some workers (e.g., Folk, 1974, p. 70). Several subsequent workers have attempted to refine the use of these properties as provenance indicators, including Bokman (1952), Blatt and Christie (1963), Conolly (1965), Blatt (1967a, b), Basu et al. (1975), and Young (1976). The properties of quartz that have been considered to have possible provenance significance include inclusions, undulatory extinction, polycrystallinity, nature of subgrain contacts, grain shape or crystal shape, trace-element composition, and cathodoluminescence colors.

Quartz Inclusions.   Inclusions in quartz include both fluid and mineral inclusions (Section 4.2.1). Krynine (1940), drawing on earlier studies, apparently had considerable confidence in his ability to trace quartz grains to their source on the basis of inclusions and patterns of undulatory extinction. Since Krynine's time, little additional research has been focused on quartz inclusions. R. L. Folk, in the various editions of his well-known petrology textbook (e.g., Folk, 1974), suggests that quartz grains so crowded with bubbles (vacuoles) as to appear milky in reflected light come from hydrothermal veins. Otherwise, liquid inclusions do not appear to have much provenance significance. Volcanic quartz is believed to be nearly water-clear and contains almost no inclusions of any type. Quartz of other origins can contain varying amounts of mineral inclusions and bubbles. The relative abundance of mineral inclusions in these grains does not appear to be diagnostic. Only if the inclusions consist of a mineral that is in itself provenance-diagnostic (such as kyanite or sillimanite, which form only in metamorphic rocks) do mineral inclusions have provenance value. Altogether, given what seems to be our present state of knowledge, quartz inclusions appear to have very limited value in provenance interpretation.

Undulatory Extinction and Polycrystallinity.   Undulatory extinction and the polycrystalline nature of some quartz grains (Section 4.2.1) have received more attention as potential provenance indicators than any other properties of quartz. They are also the most controversial properties. Krynine (1940) and Folk (1974, and earlier editions of his textbook) used degree of undulatory extinction and the nature of polycrystalline grains to assign quartz to various presumed igneous and metamorphic parent rocks. These authors believed that most highly undulose quartz (undulose extinction $> \sim 5$ degrees) is diagnostic of metamorphic rocks and nonundulose quartz is diagnostic of igneous rocks. Polycrystalline quartz was believed to be most indicative of metamorphic origin. Blatt and Christie (1963) and Blatt (1967a) took an opposing view. They suggested that undulatory extinction and polycrystallinity are of little value in provenance determination. They also concluded that the percentage of nonundulatory quartz in mineralogically mature sandstones (quartz arenites) may be higher than that of their source, owing presumably to selective destruction of less-stable undulatory quartz by weathering and transport processes during recycling. Basu et al. (1975) reexamined the validity of Blatt and Christie's conclusions by analyzing a large number of samples from plutonic igneous and metamorphic rocks. They concluded that undulatory extinction and polycrystallinity can be used to distinguish plutonic quartz from low-rank metamorphic quartz in first-cycle sandstones. Their findings are summarized in Figure 8.3. In this figure, the percentages of nonundulatory, undulatory, and polycrystalline quartz in rock samples from known source rocks are plotted on a double-triangle diagram. If more than 75 percent of the polycrystalline quartz grains are composed of either two or three crystal units, the values are plotted on the upper triangle. If more than 25 percent of the polycrystalline grains consist of three or more crystal units, they are plotted on the lower triangle. See Basu et al. (1975) for details of this method.

FIGURE 8.3

A four-variable plot of the nature of quartz populations in Holocene sands derived from source areas indicated by symbols. (*After Basu, A., et al., 1975, Reevaluation of the use of undulatory extinction and polycrystallinity in detrital quartz for provenance interpretation: Jour. Sed. Petrology, v. 45. Fig. 6, reprinted by permission of Society of Economic Paleontologists and Mineralogists, Tulsa, Okla.*)

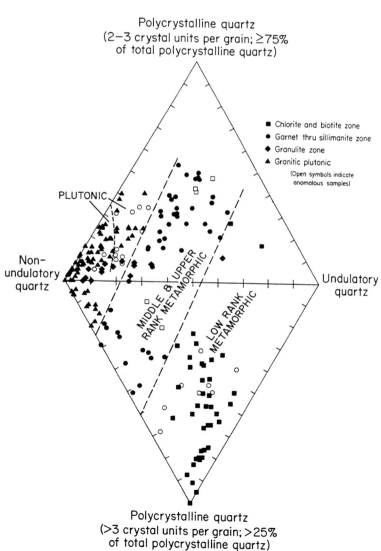

Young (1976) suggests that the internal textures of detrital polycrystalline quartz may have provenance significance. He proposes that during metamorphism quartz grains undergo deformation, which with increasing pressure and temperature proceeds through a series of stages that constitute a continuum. Monocrystalline nonundulose quartz is first altered to monocrystalline undulose quartz, then to polygonized quartz (quartz with distinct zones of extinction with sharp boundaries, i.e., segmented undulosity). The next stage is formation of a polycrystalline aggregate with elongated original crystals, followed by primary recrystallization that produces, first, a polycrystalline aggregate with small nonundulose crystals and, finally, secondary recrystallization that results in large nonundulose polygonal crystals (Fig. 8.4). Young suggests that polycrystalline grains produced by secondary recrystallization occur in medium- to high-grade metamorphic rocks and intrusive igneous rocks and that other types of polycrystalline grains are indicative of low-grade metamorphic rocks. He cautions, however, that polycrystalline quartz grains derived from recrystallized chert may be indistinguishable from those derived from medium- to high-rank metamorphic rocks.

Quartz Grain Shape.   Many early petrographers (e.g., Bokman, 1952) believed that elongated quartz grains denoted derivation from metamorphic rocks. Blatt (1967a) pointed out, however, that although extreme grain elongation does appear to be charac-

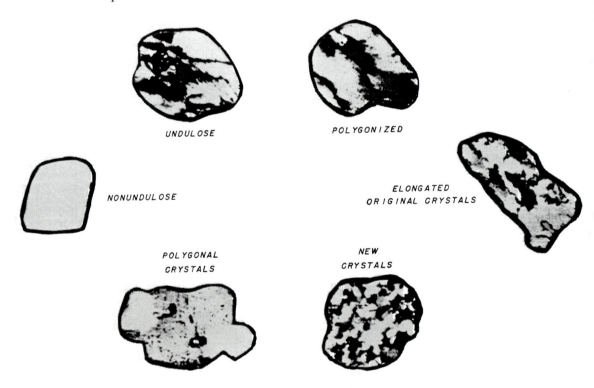

FIGURE 8.4
Schematic diagram illustrating postulated process of forming undulose and polycrystalline quartz owing to increasing pressure and temperature during diagenesis. (*After Young, S. W., 1976, Petrographic textures of detrital polycrystalline quartz as an aid to interpreting crystalline source rocks: Jour. Sed. Petrology, v. 46. Fig. 2, p. 598, reprinted by permission of Society of Economic Paleontologists and Mineralogists, Tulsa, Okla.*)

teristic of metamorphic rocks, moderate elongation is not a reliable criterion of metamorphic origin. Within a given size range of grains, elongation of quartz grains may be a function in part of postdepositional changes. Also, weathering and transport processes (Blatt, 1967a; Cleary and Conolly, 1971, 1972) may bring about breakage of original grains or selective destruction that obscures original characteristics of the quartz grains.

Part of the problem of using quartz grain shape as a criterion for provenance determinations is the difficulty of expressing quartz shape precisely by conventional methods (e.g., sphericity or Zingg shape). The problem of a more exact method of quantifying grain shape may now have been solved by application of Fourier analysis to shape studies. The Fourier method is discussed in detail in Section 2.3.5. Numerous papers have appeared since the early 1970s describing specific applications of Fourier shape analysis to provenance interpretation (e.g., Ehrlich and Chin, 1980; Kennedy and Ehrlich, 1985; Mazzullo and Magenheimer, 1987; Prusak and Mazzullo, 1987; Haines and Mazzullo, 1988; Mazzullo et al., 1988). As promising as these early results are, however, this more exact means of characterizing shape does not remove the problems caused by changes in original grain properties owing to selective destruction of quartz grains during weathering, transport, and diagenesis. Also, the method is apparently not fine-tuned enough as yet to permit discrimination between igneous and metamorphic rocks or among the various grades of metamorphic rocks.

Trace-element and Isotope Composition of Quartz.   If it could be shown that each kind of potential source rock contains quartz with a unique trace-element or isotopic signature, chemical analysis of quartz grains could be a powerful provenance tool. Few elements readily substitute for silicon in the quartz structure. Dennen (1966) has shown that the dominant contaminating elements in quartz are those elements that dominate in the physicochemical environment of quartz genesis and not necessarily those that substitute most easily for silicon. Therefore, trace-element composition of quartz may reflect quartz provenance. A potential problem in applying this concept is that chemical analyses of bulk populations of quartz grains may be misleading or difficult to interpret owing to the possibility of mixing of quartz from multiple sources. If, however, suitable techniques are employed for preparing samples for analysis, and problems of diagenetic contamination are avoided, analysis of individual quartz grains by using an electron probe microanalyzer or other suitable equipment holds the potential to assign individual quartz grains to a specific parent rock (Basu, 1985b).

Unfortunately, reference standards for the trace-element and isotopic composition of quartz from a variety of parent igneous, metamorphic, and sedimentary rocks of various ages and geographic distributions do not appear to be available. Or, if available in scattered publications, the information has not been compiled. Only a few attempts to use the chemical composition of quartz as a provenance tool have been reported. Dennen (1967) reported success in correlating the composition of Al, Ca, Fe, Li, and Ti in quartz from a sandstone with quartz from its granitic source. Suttner and Leininger (1972) determined that the Ti, Mg, and Fe composition of quartz from a granitic batholith is different from that of quartz derived from volcanic rocks that covered the batholith. They found measurable differences also in some trace elements between the quartz of two different batholiths and between quartz derived from the batholiths and that derived from metamorphic rocks. There has apparently been little subsequent effort to utilize the trace-element composition of quartz as a provenance indicator. I am not aware of any studies that have utilized the oxygen-isotope composition of quartz for provenance interpretation. What is needed at this point is an intensive study of quartz in a variety of rock types to determine if the trace-element or oxygen-isotope composition of quartz is truly source-rock–sensitive. Only then will we know if chemical analysis of quartz is likely to become a useful tool in provenance studies. This approach may be a fruitful field for future research.

Cathodoluminescence of Quartz.   The principles of cathodoluminescence are reviewed by Matter and Ramseyer (1985). Thin sections are bombarded by electrons that have been accelerated to between -8 and -50 kV. Luminescence is caused by the emission of photons in the ultraviolet, visible, and infrared range of the electromagnetic spectrum. In the visible range, luminescence occurs in a variety of colors (white, yellow, orange, red, violet, blue, green, brown), depending upon the material bombarded. Luminescence is examined with an electron-probe microanalyzer, a scanning electron microscope equipped with a cathodoluminescence detector, or a special cathodoluminescence microscope. Cathodoluminescence techniques have been applied to study of a variety of materials, including carbonates, feldspars, and quartz. In the case of quartz, luminescence colors appear to be related to a number of factors, including the degree of lattice order (a function of temperature), the stress rate, the Ti/Fe ratio, the Al concentration, and the occurrence of trace amounts of positively charged ions with small ionic radii (Matter and Ramseyer, 1985). Because these factors may vary considerably in quartz of different origins, cathodoluminescence has considerable potential for distinguishing the provenance of quartz.

Research on cathodoluminescence in quartz, summarized by Matter and Ramseyer (1985), suggests that quartz from various source rocks has the following luminescence colors: (1) plutonic rocks and phenocrysts from volcanic rocks display a range of colors from blue through mauve to violet, (2) volcanic quartz shows a zonation or irregular distribution of luminescence colors that commonly allows distinction from plutonic quartz, and red luminescing quartz is also of volcanic origin, (3) strongly deformed quartz (strong undulatory extinction) luminesces bluish-black, (4) quartz in regionally metamorphosed rocks has brown luminescence, and (5) metamorphic quartz that recrystallized at high temperatures reverts to blue luminescence colors comparable to plutonic quartz. As discussed in Section 4.2.2, cathodoluminescence microscopy also allows distinction between detrital quartz and authigenic quartz (overgrowths, etc.) because authigenic quartz does not luminesce in the visible range. As an example of a provenance study utilizing cathodoluminescence of quartz, Owen and Carozzi (1986) used the ratio of brown-luminescing to blue-luminescing quartz to identify metamorphic rocks as the source of quartz in the Carboniferous Jackfork and Parkwood formations of the south-central United States.

Provenance Significance of Feldspars

Introduction.   Feldspars are generally very scarce or absent in quartz arenites; however, they are common constituents of most other sandstones. Several properties of feldspars make them useful provenance indicators. First, because feldspars are chemically and mechanically less stable than quartz, they are less likely to be recycled. Some feldspars may indeed be recycled, depending upon conditions; however, the presence of moderately abundant feldspars in a sandstone suggests derivation from crystalline source rocks. Thus, feldspars are more likely than quartz to provide information about first-generation source rocks. Both the mineralogy and the chemistry of feldspars may have provenance significance. For example, microcline tends to be derived from felsic igneous or metamorphic rocks and calcic plagioclase from basic igneous or metamorphic rocks. Also, the chemical composition of feldspars is known to be a function of source rocks. The presence of zoning and twinning in plagioclase feldspars is likewise related to source rock type.

Feldspar Mineralogy.   A particular type of feldspar may be derived from more than one kind of igneous or metamorphic rock. Therefore, feldspar mineralogy provides only a crude guide to provenance. Nonetheless, it is useful to see how feldspar mineralogy is related to source conditions.

The **potassium-sodium feldspars** are essential constituents of felsic igneous rocks, pegmatites, and many felsic and intermediate gneisses (Deer et al., 1963a, p. 66). The feldspar in plutonic felsic rocks and high-grade metamorphic rocks is commonly ortho-clase and microcline, including perthitic orthoclase and microcline. In volcanic rocks it is sanidine and anorthoclase, including cryptoperthitic sanidine and anorthoclase. The feld-spars occur as phenocrysts and as part of the groundmass. Tuttle (1952) suggests that the potassium-sodium feldspars can be grouped into seven classes on the basis of the size of the exsolved perthitic lamellae. These classes are related to temperature of formation and increasing time subsequent to crystallization of a homogeneous feldspar. In order of decreasing temperature, they are as follows.

Sanidine or anorthoclase	Homogeneous crystals
Sub–X-ray perthite	<15 Å
X-ray perthite	<1 μm
Cryptoperthite	$1-5$ μm
Microperthite	$5-100$ μm
Perthite	$100-1000$ μm (1 mm)
Orthoclase or microcline and albite	

Sanidine or anorthoclase, sub–X-ray perthite, and X-ray perthite are characteristic of volcanic rocks, cryptoperthites are presumably formed in small hypabyssal dikes and sills and chilled borders of larger intrusives, microperthite and perthite should occur only in small plutons, and orthoclase and albite should occur in large plutons of granite or syenite.

Plagioclase feldspars are particularly common in volcanic rocks, where they occur as both phenocrysts and in the groundmass. They may be abundant also in some plutonic igneous and metamorphic rocks. The plagioclase phenocrysts in volcanic rocks have an average composition of about An_{70} in basalts and An_{27} in siliceous rhyolites; plagioclase phenocrysts richer in sodium than An_{20} are uncommon (Deer et al., 1963a, p. 144). Albite is the most distinctive mineral of spilitic basic lavas. The composition of plagioclase in calcalkaline plutonic rocks can include almost the entire range of plagioclase composi-tions. In basic plutonic rocks, the plagioclase tends to be calcium-rich; however, the cores of zoned plagioclases in some gabbroic rocks have been reported to be as low in calcium as An_{44}. The plagioclase in pegmatites is generally albite or oligoclase, and that in anorthosites is commonly in the andesine–labradorite range. Plagioclase occurs both as porphyroblasts and as small crystals in the finer-grained areas of metamorphic rocks (Deer et al., 1963a, p. 148). Albite occurs preferentially in low-grade metamorphic rocks, that is, in greenschist-, blueschist-, prehnite-pumpellyite-, and zeolite-facies rocks. In higher-grade metamorphic rocks, plagioclase may have a wide range of compositions, from about An_{22} to very calcium-rich varieties, apparently reflecting, at least in part, the feldspar composition of the precursor rock. For example, high-grade metamorphism of a basic igneous rock containing calcium-rich plagioclase would yield a metamorphic rock with calcium-rich plagioclase. Unfortunately, plagioclase can become albitized during diagenesis (Chapter 9), which effectively erases the parent-rock composition signature.

Zoning of Feldspars.   Although zoning can be present in some plagioclase porphyroblasts in metamorphic rocks (Deer et al., 1963a, p. 149), most zoned feldspars occur in igneous rocks. Thus, the presence of zoning in feldspars can generally be regarded to signify igneous origin. Both plagioclase and potassium-sodium feldspars can be zoned (Smith, 1974, p. 212); however, zoning in plagioclase appears to be more common than that in the alkali feldspars. Only zoned plagioclase has received much attention as a potential provenance indicator. Zoned crystals commonly have calcium-rich cores and more-sodic margins. In some zoned plagioclase crystals, the cores may be as calcic as An_{80} and margins as sodic as An_{45}. Other zoned plagioclase crystals may display only minor compositional changes between cores and margins. Two general styles of zoning are recognized. Oscillatory zoning appears under crossed polarizing prisms as successive thin bands of alternating extinction (Fig. 8.5). Progressive zoning is more poorly defined and appears as a broad wave of extinction with or without sharp lines of demarcation (Pittman, 1963). Thus, oscillatory zoning is fine-scale zoning, and progressive zoning is characterized by coarse, indefinite texture. Chemical zoning in feldspars can be studied by electron microprobe (Trevena and Nash, 1981) as well as by optical methods.

The most comprehensive application of feldspar zoning to provenance interpretation is that of Pittman (1963). Pittman examined specimens of volcanic, plutonic igneous, hypabyssal, and metamorphic rocks to establish the proportion of zoned to nonzoned plagioclase in these rocks (Fig. 8.6). He also examined zoning in plagioclase grains in several sandstones whose provenance had been established by other means. He concludes that (1) the presence of any kind of zoning in a plagioclase grain is strongly indicative of igneous origin, (2) oscillatory zoning in detrital plagioclase grains is indicative of a

FIGURE 8.5
Photomicrograph of a large, broken plagioclase crystal that displays well-developed oscillatory zoning. Miocene deep-sea sandstone, Japan Sea. ODP Leg 127, site 796, 243 m below seafloor. Crossed nicols. Scale bar = 0.1 mm.

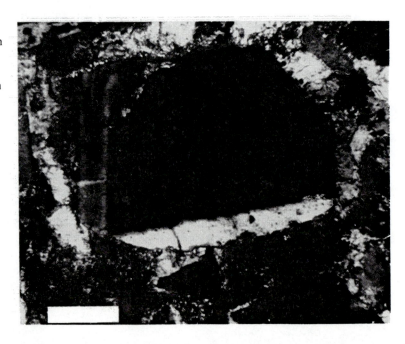

FIGURE 8.6

Proportion of progressively zoned, oscillatory-zoned, and unzoned plagioclase in volcanic and hypabyssal rocks (dots), plutonic rocks (squares), and metamorphic rocks (triangles). *(After Pittman, E. D., 1963, Use of zoned plagioclase as an indicator of provenance: Jour. Sed. Petrology, v. 33. Fig. 4, p. 384, and Fig. 5, p. 385, reprinted by permission of Society of Economic Paleontologists and Mineralogists, Tulsa, Okla.)*

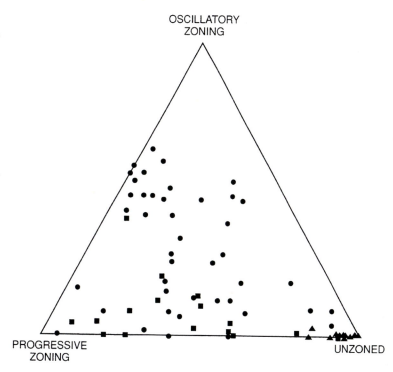

volcanic or hypabyssal source, (3) progressive zoning has no value in distinguishing between a volcanic-hypabyssal and a plutonic igneous source, (4) metamorphic plagioclases tend to be unzoned; however, the absence of zoning does not necessarily indicate metamorphic origin because much volcanic-hypabyssal and plutonic igneous plagioclase is unzoned also.

Trevena and Nash (1981) examined the chemical composition of zoned feldspars by using an electron microprobe. They found that strongly zoned and moderately zoned plagioclase grains are mainly andesine and that most of these grains fall on the high-K (i.e., igneous) trend. They report that oscillatory zoning occurs only among andesine grains of the high-K trend and conclude that this finding is consistent with Pittman's (1963) conclusion of volcanic-hypabyssal origin. Most plagioclase of the low-K metamorphic-plutonic trend is weakly zoned or unzoned.

Helmold (1985) reemphasizes Pittman's caution about using unzoned plagioclase in provenance interpretation. Not only are many igneous plagioclase grains unzoned, but breakage may create small feldspar fragments that do not show zoning. That is, a large, coarsely zoned igneous plagioclase grain may, upon transportation and impact, break down into several smaller pieces that may not display zoning. Therefore, it is not safe to draw any provenance inferences from unzoned plagioclase.

Feldspar Twinning. The relative abundance and types of twinning in feldspars have been suggested by several investigators to have provenance significance (Gorai, 1951; Turner, 1951; Toby, 1962; Pittman, 1970). Most provenance studies based on twinning have considered only the plagioclase feldspars; however, Plymate and Suttner (1983) utilized twinning in K-feldspars for provenance determination. Much of the definitive work on plagioclase twins goes back to the 1950s and 1960s. Gorai (1951) suggests that twinned plagioclase crystals can be grouped into two basic types: A-twins and C-twins. **A-twins** are twinned according to the albite, pericline, and acline laws. **C-twins** include crystals twinned according to essentially all the other twin laws (Manebach, Baveno, parallel, complex). Readers who may be a bit rusty with respect to twins and twin laws may wish to consult a standard mineralogy text, such as Klein and Hurlbut (1985, p. 100), for a review of twin laws.

   Figure 8.7, after Gorai (1951), shows the appearance of four basic types of twin plagioclase crystals, as seen on a flat stage. Type 1 twins are A-twins, and Types 3 and 4 are C-twins. Type 2 twins may include both A-twins and C-twins. To distinguish between Type 2 twins, Gorai (p. 888) suggests the following procedure. Place the twinning line in or near the 45 degree position with reference to the microscope cross hairs. Observe the retardation. If the two different individual twin crystals show different retardation, then the twin is a C-twin. If they show nearly equal retardation, insert a gypsum plate and again observe the retardation. If the retardation is different with the gypsum plate in, it is a C-twin. If the retardation is equal with the gypsum plate in, the twin type is indeterminate; however, indeterminate types are rare.

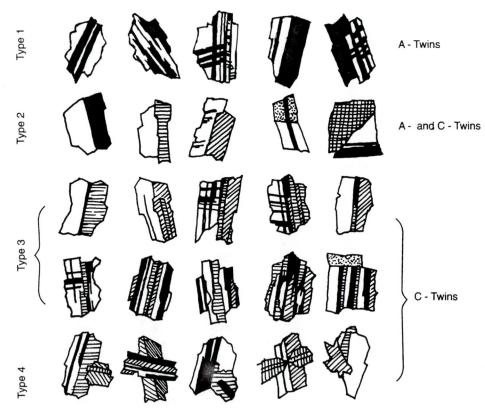

FIGURE 8.7
Sketches illustrating four principal kinds of twinned plagioclase: Type 1—polysynthetic twins and their modifications, Type 2—simple twins and their modifications, Type 3—complex twins and their modifications, Type 4—penetration twins. (*After Gorai, M., 1951, Petrological studies on plagioclase twins: Am. Mineralogist, v. 36. Fig. 1, p. 889, reprinted by permission.*)

Gorai concludes from his study that C-twins are characteristic of volcanic and plutonic igneous rocks, although they may occur in very minor amounts in hornfels. A-twins occur in both igneous and metamorphic rocks but are relatively less abundant in hornfelsic rocks. Volcanic rocks may contain C-twins, A-twins, and untwinned plagioclase. Metamorphic rocks contain mainly untwinned plagioclase and A-twins, and the abundance of twinned plagioclase appears to increase with increasing grain size. Gorai also noted some correspondence between twin types and plagioclase composition in volcanic and plutonic rocks. C-twins tend to be more abundant in calcium-rich plagioclase, especially plagioclase richer than An_{50}, whereas both A-twins and untwinned crystals are more abundant in sodic plagioclase. Apparently, these relationships between twin type and plagioclase composition do not exist in metamorphic rocks. As in the case of unzoned feldspars, considerable caution should be exercised in interpreting the significance of untwinned plagioclase. Volcanic and plutonic igneous rocks, as well as metamorphic rocks, may contain untwinned plagioclase. Therefore, untwinned plagioclase does not necessarily indicate a metamorphic origin. For a specific application of Gorai's conclusions to provenance analysis of sandstones, see Pittman (1970).

Twinning in K-feldspars has been used relatively little in provenance interpretation, but does appear to have some potential. The most distinctive twins in microcline are the cross-hatch twins produced by combined albite and pericline twinning, visible in sections cut parallel to both composition planes (Fig. 4.5). Carlsbad twins are most common in orthoclase and sanidine; less commonly, Baveno and Manebach twins occur (Deer et al., 1963a, p. 29, 31). Plymate and Suttner (1983) made a petrographic study of K-feldspar twins in Holocene stream alluvium derived directly from a granodiorite pluton. By plotting the percentage of cross-hatch microcline in the detrital K-feldspar, they were able to demonstrate that the distribution of microcline in the alluvial sediments very closely reflects the distribution of microcline-rich rocks in the source area.

Structural State of Feldspars.   The structural state of feldspars is determined by the degree of Al-Si ordering in tetrahedral sites. Because the degree of ordering in feldspars is related to genetic factors, their structural state is of interest as a possible provenance indicator. Data on the structural state of plagioclase feldspars can be obtained by high-resolution electron microscopy and careful X-ray diffraction analysis (Ribbe, 1983); however, the plagioclases are complex. Apparently no studies have been made of plagioclase structural state as a provenance indicator (Helmold, 1985). On the other hand, the alkali feldspars are less complex — data on their structural state can be readily obtained by X-ray diffraction methods (Wright and Stewart, 1968; Suttner and Basu, 1977; Plymate and Suttner, 1983). Three general structural states of alkali feldspars are recognized: (1) totally disordered — Al is randomly distributed among the four possible tetrahedral sites (e.g., sanidine and high-albite), (2) totally ordered — most of the Al is concentrated in $T_1(o)$ sites (microcline and low-albite), and (3) intermediately or moderately ordered — Al is only partially concentrated in the $T_1(o)$ sites (orthoclase) (Barth, 1969).

Temperature apparently is a first-order control on ordering of alkali feldspars. Feldspars that form at high equilibration temperatures tend to have poorly ordered structures. Those that form at lower temperatures have more-ordered structures because the slow cooling allows more time for Al-Si ordering (Martin, 1974). Martin indicates also that ordering is enhanced by an adequate supply of water. Suttner and Basu (1977) suggest that additional genetic factors may affect ordering, including deformation (shear stress), grain size, and the total chemistry of the environment of feldspar genesis.

The main applications of feldspar structural state to provenance interpretation have apparently been made by Suttner and Basu (1977) and Plymate and Suttner (1983). Suttner and Basu report that detrital K-feldspar from volcanic source rocks has a disordered structure similar to that of high sanidine, whereas K-feldspar derived from plutonic rocks ranges from moderately ordered to well ordered. Feldspars from the more slowly cooled interiors of plutons are well ordered. Those from the more rapidly chilled, younger borders of plutons are moderately ordered. Detrital K-feldspars from metamorphic rocks are mainly well-ordered microcline. Because the various kinds of K-feldspars (sanidine, orthoclase, microcline) are reflections of structural state of the feldspars, structural state can be estimated, using petrographic microscopy, by determining the percentage of cross-hatch twinning in the K-feldspar. Plymate and Suttner (1983) conclude that the optical method is more efficient for estimating structural state than are X-ray methods because a larger number of grains can be analyzed in a given time. On the other hand, X-ray diffraction methods have greater resolving power, and X-ray analysis reveals more about the true structural state and mode of origin than the mere presence or absence of cross-hatch twinning.

Feldspar Chemistry.   The use of the electron microprobe to study the chemical composition of individual feldspar grains, and other minerals, is steadily growing. Probably the most comprehensive studies of detrital feldspar composition are those of Trevena and Nash (1979, 1981). These authors compiled, from numerous sources, 2939 electron microprobe analyses of plagioclase feldspars and 1476 analyses of alkali feldspars. These analyses were performed on a variety of volcanic, plutonic igneous, and metamorphic rocks. The results of their compilation are shown in Figure 8.8. In this figure, chemical composition is plotted on triangular composition diagrams in terms of An (Ca), Ab (Na), Or (K) end members. Diagram A shows the composition of volcanic feldspars, B the composition of plutonic feldspars, and C the composition of metamorphic feldspars. Diagram D summarizes the ranges of all the provenance groups.

Figure 8.8 shows that feldspars from volcanic rocks have a wide range of compositions, whereas the composition of feldspars from plutonic and metamorphic rocks is somewhat more restricted. Trevena and Nash conclude that detrital alkali feldspars more sodic than 50 weight percent albite molecule are derived almost exclusively from volcanic sources. Alkali feldspars more potassic than 87 percent orthoclase molecule are derived from metamorphic and plutonic rocks. The potassium content of volcanic plagioclase increases significantly as sodium content increases. For the more-sodic detrital plagioclase, this relationship allows discrimination between volcanic plagioclase and plagioclase derived from plutonic and metamorphic rocks. In an earlier, but similar, microprobe study of feldspars, Sibley and Pentony (1978) used electron microprobe analyses of feldspars to identify different source regions for sediments from DSDP drilling sites in the Sea of Japan.

Trevena and Nash (1981) suggest that other chemical elements in feldspars may have some provenance value. For example, detrital plagioclase of volcanic origin generally contains more Fe and Sr than nonvolcanic plagioclase. On the whole, however, the trace-element composition of feldspars does not appear to have been investigated very thoroughly as a possible tool for provenance analysis.

Discussion of Feldspar as a Provenance Indicator.   Feldspars are one of the most useful detrital minerals as indicators of source-rock lithology. Source-rock interpretation

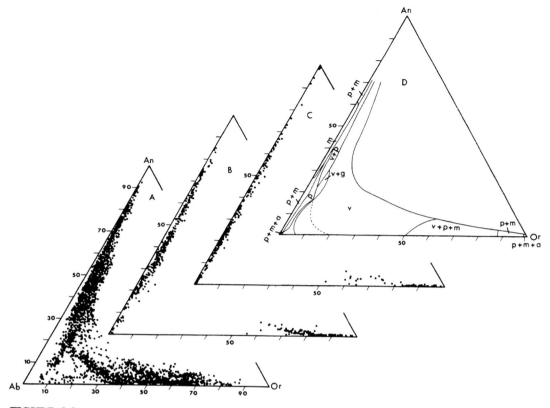

FIGURE 8.8

Chemical composition of feldspars as determined by approximately 5300 electron microprobe analyses. A. Composition of feldspars from volcanic rocks. B. Composition of plutonic feldspars. C. Composition of feldspars from metamorphic rocks. D. Composition range of eight provenance groups of feldspar. v = volcanic; p = plutonic; m = metamorphic, v + g = volcanic or granophyre, v + p = volcanic or plutonic, p + m = plutonic, or metamorphic, v + p + m = volcanic, plutonic or metamorphic, p + m + a = plutonic, metamorphic, or authigenic. (*After Trevena, A. S., and W. P. Nash, 1981, An electron microprobe study of detrital feldspars: Jour. Sed. Petrology, v. 36. Fig. 1, p. 138, reprinted by permission of Society of Economic Paleontologists and Mineralogists, Tulsa, Okla.*)

is made on the basis of feldspar mineralogy, zoning, twinning, structural ordering, and chemical composition. Figure 8.9 summarizes the relationship between some of these important properties of feldspars and their source rocks. The reliability of provenance interpretation is enhanced if interpretation is based on as many properties of feldspars as possible and practical.

Workers should be aware, however, that the reliability of provenance interpretation can be seriously affected by changes in the original feldspar composition of source rocks. These changes may be brought about by the processes of weathering (James et al., 1981), transportation (abrasion and impact) (Pittman, 1969), and diagenesis (Milliken, 1988). See Helmold (1985) for further discussion of these problems.

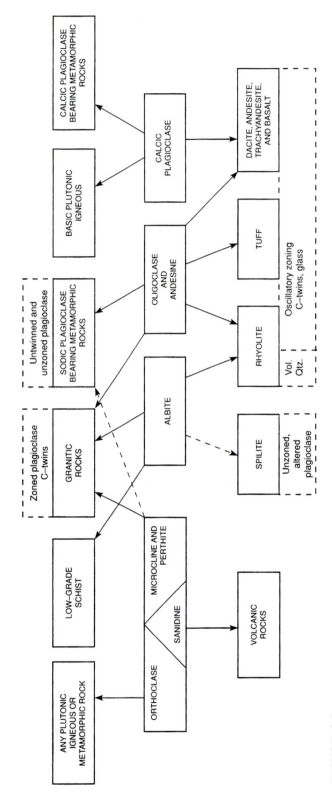

FIGURE 8.9

Summary of provenance-significant properties of plagioclase as determined with a standard petrographic microscope. Unbroken lines indicate more-probable sources of minerals. (*From Pittman, E. D., 1970, Plagioclase feldspars as an indicator of provenance in sedimentary rocks: Jour. Sed. Petrology, v. 40. Fig. 5, p. 596, reprinted by permission of Society of Economic Paleontologists and Mineralogists, Tulsa, Okla.*)

Heavy Minerals

General Statement.   The use of heavy minerals as a tool for provenance analysis is one of the most venerable techniques in sedimentary petrology, and the literature on heavy minerals is extensive. Discussions of heavy minerals appear in several early textbooks (Milner, 1926, p. 101; Boswell, 1933, p. 47; Krumbein and Pettijohn, 1938, p. 463) as well as in numerous articles. Early studies focused on petrographic analysis of heavy-mineral separates and interpretation of provenance on the basis of both heavy-mineral assemblages and individual species of heavy minerals. Petrographic analysis still forms the basis for many heavy-mineral provenance studies; however, more-sophisticated methods are now in use also, including cathodoluminescence, trace-element analysis by electron microprobe and fission-track techniques, and radioactive dating. A short summary of the use of heavy minerals for provenance interpretation is given by Morton (1985a).

Provenance Interpretation from Heavy-Mineral Petrography.   Section 4.2.1 briefly describes the procedures used in petrographic analysis of heavy minerals. Two general approaches are used in provenance interpretation of heavy-mineral petrographic data. The first approach is to interpret source-rock lithology on the basis of assemblages of heavy minerals, under the assumption that each major kind of source rock yields a distinctive suite of heavy minerals (Table 8.3). To use this approach, it must be further assumed that the heavy-mineral assemblages that occur in detrital sedimentary rocks accurately reflect those originally present in the parent rocks. This assumption may be far from valid. Because heavy minerals have a wide range of chemical stabilities (Table 8.4), they are subject to selective chemical destruction during weathering and diagenesis. See discussion in Pettijohn (1975, Ch. 13), the various papers in Luepke (1984), and Morton (1985a). Heavy minerals also have differential stabilities with respect to mechanical destruction during transport (Table 8.5), and they are subject to selective sorting by size and shape during transport (Rittenhouse, 1943). Morton (1985a) describes some techniques to minimize the effects of such hydraulic sorting in provenance studies.

Assume for the moment that variations in heavy-mineral abundances owing to selective transport are insignificant and that one is dealing with an extensive suite of heavy

TABLE 8.3
Heavy-mineral suites characteristic of principal kinds of source rocks

Association	Source
Apatite, biotite, brookite, hornblende, monazite, muscovite, rutile, titanite, tourmaline (pink variety), zircon	Acid igneous rocks
Cassiterite, dumortierite, fluorite, garnet, monazite, muscovite, topaz, tourmaline (blue variety), wolframite, xenotime	Granite pegmatites
Augite, chromite, diopside, hypersthene, ilmenite, magnetite, olivine, picotite, pleonaste	Basic igneous rocks
Andalusite, chondrodite, corundum, garnet, phlogopite, staurolite, topaz, vesuvianite, wollastonite, zoisite	Contact metamorphic rocks
Andalusite, chloritoid, epidote, garnet, glaucophane, kyanite, sillimanite, staurolite, titanite, zoisite-clinozoisite	Dynamothermal metamorphic rocks
Barite, iron ores, leucoxene, rutile, tourmaline (rounded grains), zircon (rounded grains)	Reworked sediments

Source: After Feo-Codecido, G., 1956, Heavy mineral techniques and their application to Venezuelan stratigraphy: Am. Assoc. Petroleum Geologists Bull., v. 40, p. 979, as modified by Pettijohn et al., 1987, p. 261, reprinted by permission of Springer-Verlag, New York.

TABLE 8.4
Relative stabilities of heavy minerals under conditions of weathering and intrastratal solution during diagenesis; stability increases downward in the columns

| | Intrastratal solution | |
Surface weathering (A)	Epidiagenesis (meteoric zone) (B)	Anadiagenesis (deep burial) (C)
Olivine	Olivine, pyroxene	Olivine, pyroxene
Apatite	Amphibole	Andalusite, sillimanite
Pyroxene	Sphene	Amphibole
Garnet	Apatite	Epidote
Amphibole	Epidote, garnet	Sphene
Kyanite	Chloritoid, spinel	Kyanite
Staurolite	Staurolite	Staurolite
Monazite	Kyanite	Garnet
Tourmaline	Andalusite, sillimanite, tourmaline	Apatite, chloritoid, spinel
Rutile	Rutile, zircon	Rutile, tourmaline, zircon
Zircon		

Source: Column (A) based on Goldich (1938), Dryden and Dryden (1946), and Sindowski (1949). Information in columns (B) and (C) from Morton (1985a).

minerals that has not been severely affected by weathering or diagenesis. That is, the heavy-mineral suite contains abundant unstable heavy minerals, showing that the effects of weathering and diagenesis must be minimal. How can provenance of sediment be determined on the basis of such a complex suite of minerals? The current trend in heavy-mineral analysis is to utilize some kind of statistical factor analysis, such as Q-mode factor analysis, to provide a reduction of variables and an objective classification of samples. For some recent examples of studies that make use of this statistical approach, see Stattegger (1987) and Mezzadri and Saccani (1989).

The second method of using heavy minerals in provenance studies, which avoids the problems imposed by hydraulic fractionation and stability relationships, is to work with

TABLE 8.5
Experimentally derived order of relative mechanical stability of heavy mineral grains; listed from top to bottom in order of increasing stability

Enstatite
Kyanite
Augite
Apatite
Hypersthene
Rutile
Hornblende
Zircon
Epidote
Garnet
Sphene
Staurolite
Tourmaline

Source: Relative stabilities from Thiel, G. A., 1940, Jour. Sed. Petrology, v. 10, p. 103–124, and Thiel, G. A., 1945, Geol. Soc. America Bull. v. 56, p. 1207.

a single heavy mineral or heavy-mineral group. Such studies are called varietal studies. Different varieties of a given heavy mineral are usually distinguished on the basis of color, shape, and inclusions. For example, Krynine (1946) divided tourmalines into four main provenance types: (1) granitic tourmaline—dark brown, green, pink; small- or medium-size idiomorphic crystals, commonly full of bubbles and cavities; (2) pegmatitic or vein tourmaline—blue, with pleochroism in shades of mauve and lavender; large crystals; inclusions rare; (3) metamorphic tourmaline—colors variable—brown, pink, green, colorless; generally small crystals; may contain black, carbonaceous inclusions; and (4) sedimentary authigenic tourmaline—shows authigenic overgrowths that are typically colorless to very pale blue.

The petrographic characteristics of some varieties of amphiboles may likewise have provenance significance. Oxyhornblende, or basaltic hornblende, appears dark reddish brown (foxy red) in thin section and displays strong pleochroism, birefringence, and a small extinction angle. Blue-green hornblende is commonly considered to be indicative of metamorphic origin; however, Deer et al. (1963b, p. 307) suggest that color of metamorphic hornblende varies with metamorphic grade. Hornblende ranges from colorless through light blue-green, blue-green, dark green, and brown-green. Glaucophane, which is distinguished by its pale blue color and moderate pleochroism, is derived from blueschist-facies metamorphic rocks. Actinolite and tremolite are also essentially metamorphic minerals. Van Andel and Poole (1960) used hornblende colors to map heavy-mineral provinces in the western Gulf of Mexico.

Garnets come in a variety of colors, and the different varieties of garnets can be related to specific source rocks (Deer et al., 1966, p. 21). The most common varieties are pyrope (pinkish red), almandine (deep red to brownish black), spessartine (black, red-brown, yellowish-orange), and grossular (cinnamon, yellow-brown, green). Pyrope occurs in ultrabasic rocks; almandine is most typical of metamorphic rocks, but may occur in some igneous rocks; spessertine occurs in granite pegmatites and some metamorphic rocks; and grossular is especially characteristic of metamorphosed, impure calcareous rocks. Ratios of colored garnets have been used, among other purposes, to map the source of glacial sediments (e.g., Connally, 1964).

Only a few studies of rutile have been made; however, Force (1980) suggests that rutile occurs chiefly in metamorphic rocks. The elongation or elongation ratio (mean length divided by mean width) and rounding of zircons have long been regarded to have provenance significance (e.g., Wyatt, 1954; Heimlich et al., 1975). Fourier shape analysis has also been applied to zircon provenance studies as a more sensitive method of determining shape. Byerly et al. (1975) report that zircons from three different parts of the Sierra Nevada Batholith, western United States, can be distinguished by Fourier shape analysis.

Although heavy-mineral varietal studies minimize problems imposed by stability variations and density differences among heavy minerals, they have some important drawbacks. First, there is considerable subjectivity in judging properties such as colors of minerals. Second, some overlap of properties may occur between heavy-mineral varieties derived from different kinds of parent rocks. Quite probably owing to this subjectivity, optical methods of studying varietal heavy minerals are rarely used any more. Instead, these methods have been largely replaced by chemical analysis of individual heavy minerals by electron microprobe or other suitable methods. These techniques are discussed below. Fourier shape analysis of resistant heavy minerals such as zircon, rutile, and tourmaline appears to have provenance potential but has not yet been widely applied.

Geochemical Analysis of Heavy Minerals.   Trace-element analysis of individual heavy-mineral species can be useful in provenance interpretation and can be accomplished by suitable techniques such as electron microprobe or atomic absorption analysis. These techniques are commonly applied to the more-stable heavy minerals such as zircon, garnet, ilmenite, and magnetite. Examples of studies that make use of this technique include those of Morton (1985b), who used an electron microprobe with an energy-dispersive X-ray analyzer to determine the chemical composition, and thus the identity, of detrital garnets in Jurassic sandstones in the North Sea, and Owen (1987), who used the hafnium content of detrital zircons, determined by electron microprobe, to establish a common provenance for the Upper Jackford Sandstone of Arkansas and the Parkwood Formation of Alabama. Darby (1984) and Darby and Tsang (1987) determined the elemental composition of Ti, Fe, Mn, Mg, V, Zn, Cr, Ni, and Cu in detrital ilmenites from coastal sands and river sands in Virginia and North Carolina by atomic absorption techniques. They were able to distinguish sands from different provenances and demonstrate compositional differences in ilmenites from granitic, mafic igneous, metamorphic, and sedimentary sources. Luepke (1980) utilized atomic absorption techniques to determine Ti/Cr ratios in detrital magnetites from beach sands in southwestern Oregon. She was able to determine that magnetite from different pocket beaches had distinctive Ti/Cr ratios that were source-related. See also the studies by Basu and Molinaroli (1989) and Grigsby (1990).

Ages of Heavy Minerals.   A different provenance approach is to determine the ages of detrital heavy-mineral species by radiometric techniques. These ages can then be correlated to the ages of known crystalline rocks, thereby establishing the ages of the ultimate, although not necessarily the last, source of the sediments. Such data place age constraints upon the choice of possible source rocks for a given sedimentary deposit. See, for example, Drewery et al. (1987), who used U-Pb dating methods to determine the ages of detrital Carboniferous zircons of northern Britain, and who traced their source to Precambrian crystalline rocks from Greenland and/or Fennoscandinavia. Froude et al. (1983) determined the ages of detrital zircons from quartzites in western Australia by using an ion microprobe and established their ages to be older than any known terrestrial rocks. Ages of alluvial zircons from the Blue Tier region of northeast Tasmania were determined by fission-track techniques by Yim et al. (1985). Their results showed two sources for the zircons: ~367 Ma Upper Devonian granites and 46.7 Ma Blue Tier basalts. Robb et al. (1990) applied this geochemical technique to study of U-Pb ages of detrital zircon grains from the Witwatersrand Basin, South Africa, and determined that the zircons were derived from granites that formed virtually continuously in the Witwatersrand source area from about 3300 to 2900 Ma.

Micas as Source-rock Indicators

Micas are common constituents of sandstones, although they generally occur in accessory amounts. Overall, muscovite appears to be more abundant in sandstones than is biotite. Muscovite is abundant in many metamorphic rocks in all but the highest-grade rocks. It is less common than biotite in acid igneous rocks, but does occur in some granites, aplites, and rhyolite porphyries. Biotite is common in both metamorphic and igneous rocks, including volcanic rocks. Muscovite's greater abundance in sandstones, compared to biotite, probably reflects both its superior chemical stability and its abundance in metamorphic rocks. Abundant micas in a sandstone suggest, but do not prove, a metamorphic source.

Heller et al. (1985) did an interesting provenance study of muscovites in Eocene sedimentary rocks of the Oregon Coast Range of the United States by utilizing geochemical techniques. On the basis of $\delta^{18}O$ isotope values and K-Ar ages of muscovite, they were able to determine that the micas did not come from the nearby Klamath Mountains but were derived instead from the more distant Idaho Batholith. More recently, Renne et al. (1990) used $^{40}Ar/^{39}Ar$ laser-probe dating of detrital micas from the Eocene Montgomery Creek Formation of northern California to study the provenance of this formation. Analyses of muscovite yielded ages ranging mainly from 80.2 to 47.3 Ma, indicating to Renne et al. that the muscovite was also derived from the Idaho Batholith. Biotites in the lower part of the Montgomery Creek Formation yielded ages >95 Ma. These biotites were probably derived from granitic bodies in the Blue Mountains of eastern Oregon.

Provenance Significance of Rock Fragments

Detrital rock fragments in sandstones provide the most-unequivocal evidence of source-rock lithology. There can be little doubt that a volcanic rock fragment indicates ultimate derivation from a volcanic source, a phyllite or schist clast must have come from a metamorphic parent rock, a limestone fragment indicates a limestone source rock, and so on. Therefore, rock fragments are valuable source-rock indicators in lithic arenites and many feldspathic arenites. With the exception of chert grains, rock fragments are scarce in most quartz arenites. Petrologic studies that include plotting the relative proportion of various kinds of rock fragments, such as the study by Graham et al. (1976) of lithic sandstones in the Ouachita Mountains and Black Warrier Basin (United States), can be extremely valuable adjuncts to other provenance tools in source-rock or source-terrane analysis.

Although rock fragments are excellent source-rock indicators, several problems may arise in their use. The first of these is the problem of rock-fragment identification (see Section 4.2.1 and Table 4.3). Differentiating between chert and some felsic rock fragments or silicified volcanic rock fragments can be especially difficult. A second problem in interpretation arises from differences in chemical and mechanical stability of rock fragments that can result in selective destruction of less-stable fragments during weathering, transport, and diagenesis. As discussed in Section 4.2.1, a stability ranking for the various kinds of rock fragments has not yet been established by well-controlled research. Nonetheless, we know that certain kinds of rock fragments are less stable than others. For example, carbonate fragments are easily susceptible to solution in acidic surface waters or formation waters, and even some siliciclastic rock fragments such as volcanic clasts can be replaced or otherwise destroyed during diagenesis. Further, some rock fragments such as shale, slate, and phyllite fragments may be mechanically destroyed by abrasion and impact during transport and deposition unless they are well cemented. Selective destruction of rock fragments, by whatever means, obviously biases the final assemblage of rock fragments in sandstones and weakens source-rock interpretation. To put it another way, we cannot be sure that chemically unstable or weakly durable source rocks were absent in a source area just because clasts of these rocks are absent in a sandstone. Conversely, because stable rock fragments such as chert and metaquartzite can be recycled, their presence in a sandstone may not indicate that quartzite and chert were present in the proximate source area.

The final problem has to do with the inherent relationship of parent-rock grain size and size of rock fragments. Only fine-size parent rocks yield substantial quantities of rock fragments of sand size. Boggs (1968) demonstrated experimentally that coarse-grained rocks (0.5–1.0 mm), when crushed, yield only about 25 percent or less rock fragments to the fine-sand-size crushed fraction. Therefore, coarse-grained parent rocks are poorly represented by rock fragments in sandstones. Establishing the relative abundance of various kinds of rock fragments in a sandstone does not necessarily establish the relative importance of equivalent source rocks in the source area.

Whole-rock Geochemistry as a Source-rock Indicator

General Statement.   An increasing number of studies are now appearing in the geological literature that utilize whole-rock geochemistry in provenance analysis. Provenance interpretations are based on two general approaches: (1) analysis of certain major elements or trace elements and (2) estimation of sediment ages on the basis of isotope values, often combined with chemical analyses.

Major- and Trace-element Analysis.   These studies attempt to correlate the chemical composition of sandstones, expressed as abundances of certain major or minor elements or ratios of elemental abundances, to the chemical composition of appropriate source rocks. Such studies appear to allow identification of source rocks only in terms of rather broad compositional types. They may not have the resolving power of detrital mineralogy to identify specific source-rock types. In particular, they may not be able to determine in a given sandstone unit that mixing of sediment from two or more distinct sources has occurred. Also, chemical composition is affected by grain size of the sediments. Some examples illustrate the method.

Roser and Korsch (1988) used discriminant-function analysis of major-element data to interpret the provenance of New Zealand graywackes. On the basis of petrologic characteristics, they identified four distinct groups of sandstones: (1) mafic sandstones, composed of first-cycle basaltic and lesser andesitic detritus, (2) sandstones of intermediate composition, made up dominantly of andesitic detritus, (3) felsic sandstones, composed of acid plutonic and volcanic detritus, and (4) recycled sandstones, composed of mature polycyclic quartzose detritus. They report that SiO_2/Al_2O_3 and K_2O/Na_2O ratios increase and that total iron (expressed as Fe_2O_3t) + MgO decreases from group 1 to group 4. Bulk compositional variations owing to variations in grain size were minimized by discriminant-function analysis using Al_2O_3, TiO_2, Fe_2O_3, MgO, CaO, Na_2O, and K_2O.

Bhatia and Taylor (1981) examined abundances of La, Th, U, and Hf in graywackes of the Tasman Geosyncline, Australia. They report an increase in La, Th, U, and Hf abundances in graywackes as the parent material varies from andesite, through dacite, to granites and sedimentary rocks. These variations in parent material reflect changes of tectonic setting of sedimentary basins from magmatic arc, through interarc, to rifted continental margin–marginal basin. They suggest that high La/Th and low Th/U ratios in sandstones indicate a substantial contribution of material from volcanic provenances. In a similar study of Australian Paleozoic graywackes, Bhatia (1985b) reports an increase in the total abundance of rare earth elements (REEs), a light to heavy REE ratio, and a decrease in the chondrite-normalized Eu anomaly with increase in SiO_2/Al_2O_3 and K_2O/Na_2O ratios. Bhatia attributes these chemical trends to change in the dominant source rocks from andesite to dacite to granite-gneiss and sedimentary rocks.

Argast and Donnelly (1986) report that Precambrian graywackes from the Upper Dharwar Supergroup of south India contain slightly more Fe_2O_3, MgO, Cr, and Na_2O and slightly less K_2O than Paleozoic graywackes from northeastern North America. They suggest that the slightly higher average Na and ferromagnesian elements indicate existence of more-tonalitic and mafic-rich source rocks. They conclude that neither intermediate nor felsic volcanic rocks nor an extensive metasedimentary terrane were prominent in the source area.

In a provenance study of lithic-volcanic sandstones in the Permanente Terrane of the Franciscal Complex in California, Larue and Sampayo (1990) analyzed the major- and trace-element composition of Franciscan sandstones and Permanente greenstones. Geochemical data were displayed on Harker variation diagrams, with major-element (Fe, Mn, Na, Ti, Ca, Mg, Al) oxides plotted against SiO_2, and on variation diagrams of immobile elements (Nb, La, Ce, V). These plots show that the geochemical compositions of the sandstones occupy fields very similar to those of the greenstones in the Permanente terrane. Thus, in this case, sandstones on a continental margin appear on the basis of geochemical data to have been derived from oceanic crustal sequences.

Isotopic Studies.   Isotopic data derived from whole-rock analyses and analyses of mineral separates in sandstones can be used to estimate ages of detrital minerals. These ages allow detrital materials to be traced to surviving source regions that have comparable radiometric ages. Together with other chemical data, isotope data can help to establish if a particular potential source area is likely to be the actual source area. For example, Heller et al. (1985) used whole-rock Nd-Sm and Rb-Sr values, together with other chemical and petrographic data, to determine that sandstones of the Eocene Tyee Formation in the Oregon Coast Range included two-mica granites that formed in Late Jurassic time from a source area having an old (~700 Ma) crustal component. Ka-Ar dating of white micas shows that minerals in the granite underwent thermotectonic age resetting in Late Cretaceous time. Heller et al. report that rocks in the Klamath Mountains and the northern Sierra Nevada, previously considered the most likely source for the Tyee Formation, do not possess these features. They conclude that the Tyee Formation most likely was derived from the (now) more distant Idaho Batholith, which does possess these features.

Peterman et al. (1981) studied the chemical composition of Eocene graywackes near Agness in southwestern Oregon. They correlated Rb/Sr and $^{87}Sr/^{86}Sr$ ratios in the sandstones for the time of deposition to produce a pseudoisochron of 110 m.y. Correlations of major and trace elements with initial Sr-isotope ratios were interpreted by Peterman et al. to indicate mixing of detritus from two dominant sources, one mafic (broadly basaltic) in composition and the other felsic. They suggest that older Mesozoic graywacke formations could have contributed a major part of the felsic component to the Eocene graywackes and that lowermost Eocene basalts and older units associated with the Mesozoic graywackes may have been sources for the mafic constituents.

In a more recent paper, Nelson and DePaolo (1988) utilized Sm-Nd and Rb-Sr isotopic data together with petrographic data to interpret the provenance of Tertiary continental sediments in New Mexico (USA). The isotopic and petrographic data indicate that the sediments were derived from varying proportions of Precambrian basement and Oligocene volcanic rocks. Nelson and DePaolo found good correspondence between isotopic and petrographic data as provenance indicators. They suggest that the isotopic tracers are more sensitive to minor detrital input from a second source than is framework petrology.

Isotopes, as a tool for provenance interpretation, appear to be most useful for relating detrital sediments to particular source regions rather than for identifying specific types of source rocks. Readers who are interested in this approach should consult the papers cited here for more details. Also, an extensive bibliography of additional papers on this subject is provided by Nelson and DePaolo (1988).

8.5.2   *Interpreting Tectonic Provenance*

Introduction

Section 8.5.1 describes the methods and problems involved in interpreting source-rock lithology on the basis of detrital mineralogy and geochemistry. Interpreting source-rock lithology is, however, only part of the process of provenance analysis. To develop a fuller understanding of provenance and paleogeography requires that we understand also the relationship between source area, depositional basin, and the regional tectonic framework. Geologists have long had an interest in this relationship between tectonics and sedimentation. In North America, the role of tectonics in sedimentation was championed particularly by Krynine in the 1940s (e.g., Krynine, 1943). Interest was rekindled in the early 1960s with the development of the concept of plate tectonics. Research on the relationship of sediment composition and tectonics has continued to be extremely active since that time.

The basis for interpreting tectonic setting is the assumption that detrital mineralogy and geochemistry reflect not only source-rock lithology but also the general plate-tectonic setting. Plate-tectonic setting encompasses two features: (1) **major provenance terranes** (cratonic blocks, volcanic arc systems, collision belts) and (2) **types of plate boundaries** (passive or rifted continental margins, active or orogenic continental margins, transform-fault margins). Presumably, each major plate-tectonic setting generates distinctive suites of source rocks. Erosion of these source rocks furnishes sediment to a variety of depositional basins located in different positions with respect to plate boundaries, e.g., foreland basins, backarc basins, forearc basins, intraarc basins, and trenches (Chapter 1). The modal composition of quartz, feldspars, and rock fragments deposited in these basins, and possibly the geochemistry of the sediments as well, are putatively accurate reflectors of provenance terranes that vary with plate-tectonic setting. This idea was pursued in the early 1970s by a small number of investigators (Dickinson and Rich, 1972; Crook, 1974; Schwab, 1975). It was developed most fully and perhaps stated most elegantly by Dickinson and Suczek (1979). Since that time, the concept has been reevaluated and applied by numerous workers such as Dickinson and Valloni (1980), Valloni and Maynard (1981), Dickinson et al. (1983), Ingersoll (1983), Mack (1984), Maynard (1984), Dickinson (1985, 1988), Valloni (1985), and Ingersoll (1990), to mention but a few.

Interpreting Provenance Terranes from Sandstone Detrital Modes

General Concept.   Dickinson and Suczek (1979), Dickinson et al. (1983), and Dickinson (1985, 1988) suggest that all tectonic provenances can be grouped under three main types: continental blocks, magmatic arcs, and recycled orogens (see Fig. 1.4). Each of these provenance settings includes distinctive groupings of source rocks that shed sediments into associated basins.

Continental Block Setting.   Included in this category of provenances are major shields and platforms, as well as locally upfaulted basement blocks. Major shields, or **craton interior provenances,** are composed dominantly of basement rocks consisting of largely felsic plutonic igneous and metamorphic rock. Associated platform successions may include abundant sedimentary rocks. Sands derived from craton interiors are typically quartzose sands containing minor feldspars, reflecting multiple recycling and perhaps intense weathering and long distances of transport on cratons of low relief. K-feldspar–to–plagioclase ratios tend to be high. Thus, craton interior provenances yield largely quartz arenites, although some lithic arenites could be derived from positive areas located marginally to continental blocks. Sediments derived from cratons are deposited in local basins within the craton, in foreland basins, or along rifted continental margins in shelf, slope, or deeper-water environments (Fig. 1.4A). Fault-bounded, **uplifted basement blocks** typically consist of granitic rocks and gneisses. High relief on these blocks results in rapid erosion, which generates relatively coarse feldspathic arenites and arkoses. Some blocks may have an initial cover of sedimentary or metamorphic rocks that can yield lithic arenites. Sediments from uplifted basement blocks are commonly deposited without much transport in nearby interior basins.

Magmatic Arc Settings.   Magmatic arcs (Fig. 1.4C) consist of volcanic highlands located along active island arcs or on some continental margins. Some magmatic arcs, such as the Japan arc, may not have a continuous volcanic cover and may be associated with igneous, metamorphic, and sedimentary rocks. Magmatic arcs along continental margins that become deeply eroded or dissected may also expose deep-seated plutonic rocks. Young, **undissected arcs** tend to have a nearly continuous cover of volcanic rocks. Therefore, largely volcaniclastic debris is shed from undissected arcs (Dickinson and Suczek, 1979). This debris consists largely of plagioclase feldspars and volcanic lithic fragments, many of which contain plagioclase phenocrysts. If quartz is present, it is volcanic quartz. Thus, sandstones derived from undissected magmatic arcs are almost exclusively volcanic lithic arenites. Sediment may be deposited in backarc basins, forearc basins, intraarc basins, or trenches (Fig. 1.4C). **Dissected arcs** that expose deep-seated plutonic rocks shed a mixture of volcanic and plutonic detritus and under some conditions may even shed metamorphic or sedimentary detritus. Thus, K-feldspars and plutonic quartz may be present in this detritus along with volcaniclastic material. Sandstones derived from dissected arcs are thus less lithic-rich than those from undissected arcs.

Recycled Orogen Settings.   Recycled orogens are source regions created by upfolding or upfaulting of sedimentary or metasedimentary terranes, allowing detritus from these rocks to be recycled to associated basins (Fig. 1.4B). Many recycled orogens were formed by collision of terranes that were once separate continental blocks. This process creates uplift and welds the terranes together along a suture zone (collision orogen provenances). **Collision orogens** are composed dominantly of nappes and thrust sheets of sedimentary and metasedimentary rocks but may include subordinate amounts of plutonic or volcanic rocks, or even ophiolitic mélanges. Therefore, complex suites of sediments can be derived from such orogens. Dickinson and Suczek (1979) suggest that typical sandstones are composed of recycled sedimentary materials, have intermediate quartz contents, and contain an abundance of sedimentary-metasedimentary lithic fragments. Less-typical sandstones derived from collision orogens are quartz arenites, feldspathic arenites, and chert-rich sandstones. Sediments shed from collision orogens may be shed into foreland basins or may be transported longitudinally into adjacent ocean basins. Some recycled orogens are foreland uplifts associated with foreland fold-thrust belts (Fig. 1.4B). These **foreland uplift provenances** may contain a complex variety of source rocks, including siliciclastic sediments, carbonate rocks, metasediments, plutonic rocks in exposed basement blocks, and volcanic rocks. Thus, a variety of sandstone types can be derived from foreland uplifts, some of which may be indistinguishable from sandstones derived from continental blocks, collision orogens, or even subduction complexes (below). Dickinson and Suczek (1979) suggest that the most characteristic sandstones couple moderately high quartz contents with strikingly high feldspar content.

Finally, orogens consisting of uplifted subduction complexes composed of oceanic sediments and lavas are called subduction-complex provenances. Source rocks in these orogens may include deformed ophiolitic materials, greenstones, chert, argillite, gray-wackes, and limestones, which are exposed as constituents of mélanges, thrust sheets, and isoclines formed by deformation within the subduction zone (Dickinson and Suczek, 1979). Sediments may be shed away from the subduction complex into forearc basins or into the adjacent trench (Fig. 1.4B). According to Dickinson and Suczek, the key signal for recognizing sandstones derived from subduction complexes is an abundance of chert, which, in their samples, greatly exceeded combined quartz and feldspar. They caution, however, that sandstones derived from subduction complexes containing abundant sandy source rocks may have a much weaker chert signal. Also, mixing of detritus from subduction-complex orogens, magmatic arcs, and collision orogens is possible.

Discussion.   The principal petrographic characteristics of sediments derived from these major types of tectonic provenances are summarized in Table 8.6. These are generalized characteristics, and readers should be aware that some overlap occurs in the characteristics of sediments derived from different provenances.

Knowing the major kinds of sediments derived from each major tectonic setting, it should be possible to interpret from the mineral composition of sandstones their probable tectonic provenance. Dickinson and Suczek tested this idea by plotting on triangular composition diagrams the framework modal composition of a large number of sandstones whose provenance was reasonably well known. Plots were made using various combinations of total quartz, feldspar, and unstable rock fragments and of monocrystalline quartz, polycrystalline quartz, volcanic rock fragments, sedimentary rock fragments, plagioclase, and K-feldspar. These plots demonstrated that most sands from each major provenance setting formed well-defined clusters on the plots, although some individual samples plotted outside the clusters. The fact that gradational and overlapping field boundaries occur is not surprising, given the overlapping types of source rocks that occur in the major tectonic settings. On the basis of these sample studies, Dickinson et al. (1983) and Dickinson (1985) subsequently constructed generalized provenance plots for framework modes of detrital sandstones showing provisional subdivisions according to inferred tectonic provenance type (Fig. 8.10).

Ingersoll (1990) expands upon this concept of actualistic sandstone petrofacies models—that is, models based upon petrographic data obtained from locally derived sands of known provenance and used to statistically discriminate compositions according to their source rocks. Ancient petrofacies are then compared to these ''actualistic'' petrofacies to better constrain provenance and paleotectonic reconstructions. Ingersoll suggests that this approach can be applied on a first-order scale (specific source rocks), second-order scale (source regions within a given tectonic setting), and third-order scale (continents and ocean basins).

Published applications of these tectonic provenance concepts are too numerous to discuss here. For some recent applications, readers may wish to see Girty and Armitage (1989), Cavazza (1989), and Critelli et al. (1990).

Plate Boundaries and Detrital Modes of Sandstones

Can the nature of plate boundaries and the major kinds of depositional basins be inter-preted on the basis of sandstone compositional modes? Dickinson and Valloni (1980) and Maynard et al. (1982) analyzed turbidite sands collected from a wide variety of sites on the modern ocean floor to examine the relationship of sandstone composition to known plate settings. The geographic position of the samples allowed them to be related to specific continental margins or oceanic island chains. The mean detrital framework modes of the sands were plotted on QFL and related diagrams to see if each major type of plate boundary shed sands with recognizable and distinctive provenance signature. Dickinson and Valloni's (1980) results are shown in Table 8.7. Note that sands derived from rifted continental margins have the highest quartz content and lowest percentage of lithic frag-ments, whereas sands derived from island chains have the lowest quartz content and the highest content of lithic fragments.

Maynard et al. (1982) conclude that sands derived from rifted, or trailing-edge, margins typically have more than 40 percent quartz, whereas those from active-margin settings (backarc, forearc, continental-margin arc, strike-slip) have less than 40 percent (Fig. 8.11). Among active-margin settings, sands from intraocean forearc basins are composed dominantly of volcanic rock fragments, whereas those from strike-slip–related settings have a significantly lower proportion of volcanic grains in the rock fragments. Sands from other arc-related settings overlap considerably in composition. All share the characteristics of low quartz and high proportions of volcanic rock fragments; however, basins lying on the oceanic side of continental-margin arcs and backarc basins of island arcs contain sands of almost identical composition.

Reliability of Framework-mode Provenance Models.   How reliable are the models described above for provenance analysis? There is always a danger with any fairly simple and elegant model that provisional or tentative field boundaries or conclusions become fixed boundaries or conclusions once the model is published. Subsequent workers tend to ascribe to the model a degree of rigor that was never actually intended by the authors. Therefore, it is necessary to be aware that in the application of these provenance models, as with any model, exceptions can occur.

Interpretations based on modal composition may not always agree with interpreta-tions made on the basis of stratigraphic and structural relationships. For example, Mack (1984) points out four categories of sandstones that may plot in error on provenance framework diagrams: (1) sandstones deposited during the transition between tectonic regimes may be derived, in part, from relict source rocks, (2) sandstones enriched in detrital quartz owing to weathering and/or depositional reworking may lead to inaccurate interpretation of tectonic setting from compositional data, (3) sandstones deposited in tectonic settings as yet unrepresented on the provenance diagram may plot between the provenance fields or overlap existing fields, and (4) sandstones containing abundant detrital carbonate rock fragments may affect the location of data points on provenance diagrams. Schwab (1981) applied Dickinson and Valloni's (1980) model based on study of sands from the modern ocean to evaluation of ancient sandstones in the French-Italian Alps. He found general support for the model, but underscored the necessity for defining the relationship between sandstone mineralogy and plate-tectonic setting more specifi-cally. Girty et al. (1988) also sound a cautionary note about use of the model when sand or sandstone samples are from a limited geographic area and when metavolcanics are

TABLE 8.6
Framework composition of sandstones as related to tectonic provenance

Tectonic provenance	Source rocks	Derived sediment		Type of depositional basin	Influence of climate and transport
		Sand	Gravel		
Continental block					
A. Craton interior	Granitic and gneissic basement; subordinate sedimentary and metasedimentary rock from marginal belts	Quartz arenites and minor arkoses; high ratio of K-feldspar to plagioclase; minor lithic arenites	Minor quartzite(?); most clasts probably do not survive transport(?)	Platform settings, interior basins, foreland basins, passive continental margins and bordering oceans	Severe under humid conditions and long transport
B. Uplifted basement blocks	Granitic and gneissic basement plus sedimentary or metasedimentary cover; possible volcanic rocks	Feldspathic arenites and arkoses; minor sedimentary/ metasedimentary or volcanic lithic arenites	Granite and gneiss clasts; minor sedimentary/metasedimentary, or clasts	Fault-bounded interior basins formed by incipient rifting or wrench faulting	Probably minimal owing to rapid erosion and short transport distance
Magmatic arc					
A. Undissected	Mainly andesitic to basaltic volcanic rocks	Lithic arenites composed of volcanic rock fragments and plagioclase grains; minor volcanic quartz	Andesite or basalt clasts	Forearc, backarc, and intraarc basins; trenches; possibly abyssal-plain basins	Probably minimal owing to rapid erosion and short transport distance
B. Dissected	Andesitic to basaltic volcanic rocks; plutonic igneous, metaigneous(?)	Mixtures of volcanic-derived rock fragments and plagioclase plus K-feldspar and quartz from plutonic sources	Andesite, basalt, plutonic igneous, or metaigneous clasts	Same as undissected arcs	Moderate effect of climate(?); minimal effect of transport

Recycled orogen					
A. Subduction complexes	Ophiolite sequences (ultramafic rocks, volcanic rocks, chert); greenstones; argillites, graywackes; limestones; blueschists	Chert a key component (may exceed combined quartz and feldspar); may include sedimentary, ultramafic, volcanic rock fragments	Chert, greenstone, argillite, sandstone, limestone, serpentinite	Forearc basins, trenches; possibly abyssal-plain basins	Probably minimal owing to rapid erosion and short transport distance
B. Collision orogens	Mainly sedimentary and metasedimentary rocks; subordinate ophiolite sequences, plutonic basement rocks, volcanic rocks	Intermediate quartz content; high quartz/feldspar ratio; abundant sedimentary and metasedimentary clasts, which may include chert from mélange terranes or nodular limestones	Sedimentary and metasedimentary clasts, minor plutonic igneous, volcanic clasts, chert	Remnant ocean basins, foreland basins, basins developed along suture belts	Probably moderate to minimal
C. Foreland uplifts	Mainly sedimentary successions within fold-thrust belts; minor plutonic igneous, and metamorphic(?) rocks	Most diagnostic is high quartz with low feldspar content, but variable association of quartz, feldspar, chert	Sedimentary clasts, chert, minor plutonic igneous or metamorphic clasts	Mainly in foreland basins	Probably moderate to minimal

Source: Adapted from Dickinson, W. R., and C. Suczek, 1979, Plate tectonics and sandstone composition: Am. Assoc. Petroleum Geologists Bull., v. 63, p. 2164–2182.

FIGURE 8.10

Relationship between framework composition of sandstones and tectonic setting as indicated in Table 8.6. (*After Dickinson, W. R., et al., 1983, Provenance of North American Phanerozoic sandstones in relation to tectonic setting: Geol. Soc. America Bull., v. 94. Fig. 1, p. 223.*)

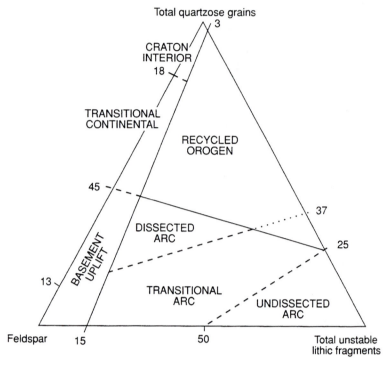

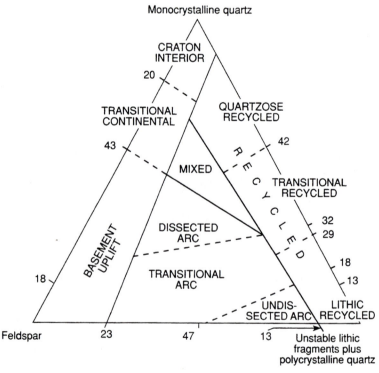

TABLE 8.7

Composition of sandstone detrital constituents as a function of major types of plate boundaries

	Rifted margin	*Active margin*	*Island chain*
Quartz content	highest	intermediate	lowest
Polycrystalline/mono-crystalline quartz ratio	lowest	highest	little quartz
Plagioclase content	lowest	high	high
K-feldspar content	highest(?)	intermediate	little or none
Lithic-fragment content	lowest	intermediate	highest(?)

Source: Dickinson, W. R., and R. Valloni, 1980, Plate settings and provenance of sands in modern ocean basins: Geology, v. 8, p. 82–86.

suspected in the provenance. Also, there is some disagreement among provenance workers about the use of the so-called Gazzi-Dickinson point-counting method vs. the conventional point-counting method in provenance studies (Ingersoll et al., 1984; Suttner and Basu, 1985). (In the Gazzi-Dickinson method, large recognizable crystals within a rock fragment are identified and counted as that crystal—e.g., plagioclase—rather than the grain being counted as a rock fragment.) Even Dickinson (1988, p. 3) cautions against overextension of these models when he points out that petrofacies of mixed provenance are common because dispersal paths connecting sediment sources to basins of deposition may be complex. "Consequently, the geodynamic relations of different types of sedimentary basins as revealed by their overall morphology, structural relations, and depositional systems do not predict reliably the nature of the petrofacies that some basins contain." For further insight into potential problems involved in use of tectonic-

FIGURE 8.11

Framework composition of modern deep-sea sands from various tectonic settings. (*From Maynard, J. B., et al., 1982, Composition of modern deep-sea sands from arc-related basins, in Leggett, J. K., ed., Sedimentation and tectonics on modern and ancient active plate margins. Fig. 2, p. 553, reprinted by permission of Blackwell Scientific Publications Limited, Oxford.*)

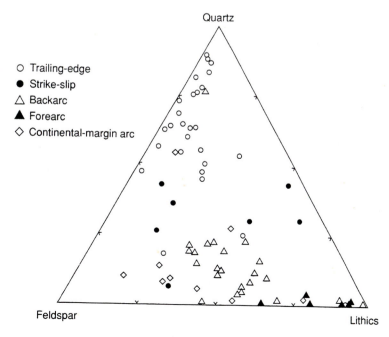

○ Trailing-edge
● Strike-slip
△ Backarc
▲ Forearc
◇ Continental-margin arc

provenance plots, see the discussions by Dickinson and Ingersoll (1990) and Johnsson and Stallard (1990).

Interpreting Tectonic Setting from Sandstone Geochemistry

Although most studies of tectonic setting have relied on interpretations based on sandstone mineralogy, certain aspects of sandstone geochemistry are believed also to reflect tectonic setting (Bhatia, 1983; Cawood, 1990; Crook, 1974; Maynard, 1984; Maynard et al., 1982; Potter, 1978; Roser and Korsch, 1986; Schwab, 1975). The SiO_2 content and K_2O/Na_2O ratios appear to be particularly sensitive indicators of geotectonic setting. Crook (1974) suggests that these chemical parameters plus quartz content can be used to distinguish modern deep-sea sands and ancient graywackes from the three major types of continental margins: (1) sediments that accumulate along passive or trailing-edge (Atlantic-type) margins of continental blocks have greater than 65 percent quartz, an average of 70 percent SiO_2, and K_2O/Na_2O ratios equal to or greater than 1, (2) sediments deposited along active margins where subduction of an oceanic plate occurs adjacent to and beneath a continent (Andean-type margins) have quartz contents ranging from 15 to 65 percent, an average of 68–74 percent SiO_2, and K_2O/Na_2O ratios less than 1, and (3) sediments that occur in trenches located adjacent to volcanic island-arc systems along western-Pacific–type margins have less than 15 percent quartz, an average of 58 percent SiO_2, and K_2O/Na_2O ratios less than 1. Schwab (1975) suggests that Crook's premise is valid and can possibly be extended to include sandstone varieties other than graywacke, but that more data from modern sands are needed to test the concept. Maynard et al. (1982) also used K_2O/Na_2O ratios and SiO_2 content to discriminate between passive-margin and active-margin sandstones. They suggest that passive-margin sandstones have K_2O/Na_2O ratios greater than 1 and active-margin sandstones have ratios less than 1. Also, passive-margin sandstones are enriched in SiO_2 compared to active-margin sandstones.

Bhatia (1983) applied this premise to study of Paleozoic turbidites of eastern Australia. He concluded that sandstones from sedimentary basins adjacent to oceanic island arcs (e.g., Marianas- and Aleutians-type arcs) are characterized by high abundances of Fe_2O_3 + MgO (8–14 percent) and TiO_2 (0.8–1.4 percent) and low Al_2O_3/SiO_2 (0.24–0.33) and K_2O/Na_2O (0.2–0.4) ratios. Sandstones from basins adjacent to continental island arcs (e.g., Cascades-type arc, western USA) are distinguished from oceanic island-arc types by lower Fe_2O_3 + MgO (5–8 percent) and TiO_2 (0.5–0.7 percent) and higher Al_2O_3/SiO_2 (0.15–0.22) and K_2O/Na_2O (0.4–0.8) ratios. Sandstones from basins on active continental margins (Andean-type) have very low Fe_2O_3 + MgO (2–5 percent) and TiO_2 (0.25–0.45 percent) and K_2O/Na_2O ratios of approximately 1. Passive margin (Atlantic-type) sandstones are enriched in SiO_2 and depleted in Na_2O, CaO, and TiO_2, reflecting their recycled and mature nature. Roser and Korsch (1985) took exception to some of Bhatia's conclusions, alleging that some of his discriminant-function scores are not correct and that he did not take into account the effect of grain size on chemical composition. Roser and Korsch (1986) present their own chemical model, based on K_2O/Na_2O ratios and SiO_2 content, for discriminating tectonic setting (Fig. 8.12). Using these chemical parameters, they were able to discriminate among samples from three major tectonic settings: passive margin (PM), active continental margin (ACM), and oceanic island-arc margin (ARC). Some overlaps occur between the composition fields shown in Figure 8.12, but overall the discriminating power of the technique appears to be reasonably good.

FIGURE 8.12

Chemical model for discriminating tectonic setting of sandstones and argillites on the basis of K_2O/Na_2O ratios and SiO_2 content. ARC = oceanic island-arc margin, ACM = active continental margin, PM = passive margin. (*After Roser, B. P., and R. J. Korsch, 1986, Determination of tectonic setting of sandstone-mudstone suites using SiO_2 content and K_2O/Na_2O ratios: Jour. Geology, v. 94. Fig. 2, p. 638, reprinted by permission of University of Chicago Press.*)

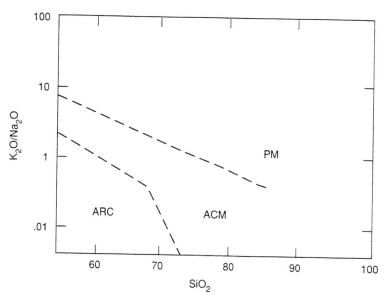

The composition of plagioclase feldspars appears also to have tectonic significance. Maynard (1984) analyzed plagioclases from different settings with an electron microprobe. He determined that the average An content of plagioclase feldspars in sands was, for trailing-edge margins, 19 percent; for strike-slip settings, 26 percent; for forearc basins on continental-margin magmatic arcs, 29 percent; for backarc basins, 38 percent; and for forearc basins of intraoceanic arcs, 56 percent. Maynard concludes that island arcs, having mostly volcanic rocks at the surface, produce sediments with calcium-rich plagioclase, whereas passive continental margins, having extensive sedimentary covers, produce sands richer in pure albite.

Finally, rare-earth–element (REE) geochemistry appears to have some discriminating power in provenance studies. Bhatia (1985b) suggests that oceanic island-arc–type graywackes are characterized by their lower total rare-earth element abundance, only slight enrichment of light rare-earth elements (LREEs) over heavy rare-earth elements (HREEs), and the absence of a negative Eu anomaly on chondrite-normalized plots. Continental island-arc–type graywackes have higher REE abundance and La/Yb ratio and a small negative Eu anomaly on the chondrite-normalized plot. Sediments from Andean-type active continental margins, passive margins, platform, and cratonic basins are all characterized by high enrichment of LREE over HREE and the presence of a pronounced negative Eu anomaly on the chondrite-normalized plots; they cannot be discriminated from each other by REE patterns alone.

8.5.3   Interpreting Climate and Relief of Source Areas

Interpreting Climate

Climate, through its influence on weathering processes, can have a significant effect on the ultimate composition of sandstones and thus on provenance interpretation. Hot, humid climates promote alteration and destruction of less-stable minerals and rock fragments, whereas very cold or very dry climates favor preservation of these less-stable constituents. Furthermore, conditions of low relief and gentle slopes enhance chemical weathering because particles are eroded from such areas slowly. By contrast, high relief and steep slopes promote rapid erosion of detritus before it is significantly weathered. Are the effects of climate sufficiently strong that they leave a climatic "signature" on detrital mineral assemblages that we can read? That is, can we determine from the petrographic characteristics of sandstones what the climate and relief of the source area might have been?

Folk (1974, p. 85) suggests that climate and relief of ancient source areas can be evaluated on the basis of the relative size and degree of rounding of quartz and feldspar and the degree of weathering of feldspar. For example, well-rounded quartz indicates prolonged reworking in source areas of low relief, and heavily weathered feldspars suggest weathering in hot, humid climates. The simplicity of Folk's concept is appealing; however, it is probably oversimplified, and it is certainly difficult to apply in practice. For example, grain roundness may be inherited from a previous sedimentary cycle and feldspars may become altered during diagenesis as well as during weathering.

Is it possible on the basis of sandstone mineralogy to develop reliable criteria for interpreting climatic conditions that overcome these difficulties? Several investigators have attempted to clarify the relationship between climate and sandstone mineralogy by studying the composition of Holocene sands released from similar kinds of source rocks undergoing weathering under different climatic conditions (Young et al., 1975; Basu, 1976; Mack and Suttner, 1977; Suttner et al., 1981; Grantham and Velbel, 1988; Mack and Jerzykiewicz, 1989). A second approach is to study the composition of ancient sandstones that formed under climatic conditions that are believed to be known on the basis of independent paleoclimatic evidence (e.g., Suttner and Dutta, 1986).

Investigators using the first approach attempt to set up a sampling program by which source-rock lithology, relief, and sediment transport are as invariant as possible, with climate being the only major variable. Variations in the composition of first-cycle Holocene sands released from similar source rocks under these conditions should thus reflect differences in climate. For example, Young et al. (1975) studied composition vs. size of first-cycle sands derived from metamorphic and plutonic rocks in semiarid and humid climates. They conclude that regardless of crystalline source-rock type, weathering in semiarid climates generates greater amounts of rock fragments, feldspars, and accessory minerals. Weathering in humid climates produces relatively more polycrystalline and monocrystalline quartz because less-stable minerals and rock fragments are more easily destroyed by vigorous chemical weathering under humid conditions. For additional examples of this approach to paleoclimate interpretation, see Basu (1976) and Suttner et al. (1981).

Suttner and Dutta (1986) extrapolated these techniques developed with Holocene sands to study of ancient sandstones. These authors suggest that systematic variations in compositional maturity can be related to changing climatic conditions during deposition. For example, compositional maturity of Permian-Triassic Gondwana sandstones of India, from base to top of the group, increases progressively from immature to mature to submature to supermature. Suttner and Dutta link this change to the changing paleoclimate of India (glacial, arid → temperate, humid → warm, semiarid → warm, humid) associated with the Permian to Triassic drift of India through different latitudinal zones and the overall change in global climate. As another example of this approach, Mack and Jerzykiewicz (1989) suggest that the ratio of volcanic rock fragments to plagioclase plus volcanic rock fragments (RFP index) and the ratio of volcanic rock fragments to accessory minerals plus volcanic rock fragments (FRA index) can be used as paleoclimate indicators. They present data that show that the RFP and RFA indices are statistically lower in sandstones derived from humid regions than in those derived from regions with semiarid climates. This difference reflects a greater degree of chemical breakdown of the parent rock and gravel- and sand-size rock fragments, and release of monocrystalline grains, in humid climates.

There are many factors that make paleoclimate interpretation from sandstone composition very tenuous. The techniques discussed above appear to work only for first-cycle sediments derived from crystalline plutonic or metamorphic rocks that have been transported relatively short distances, deposited in nonmarine environments without significant reworking, and undergone relatively little diagenetic alteration. Sediment recycling, extensive reworking in a marine environment, and extensive diagenetic alteration can all apparently destroy the climate signature.

Interpreting Relief and Slope

Estimating the relief of source areas is even more tenuous than estimating climate. Although geologists have generally accepted the dictum that rugged relief results in generation of coarser-grained, fresher detritus than that formed under conditions of low relief, that belief appears to be based largely on intuitive reasoning rather than actual research. Little quantitative work has apparently been done to evaluate the effects of relief on sandstone composition. Potter (1978) concludes that relief and rainfall affect the relative volumes of sediment of different composition produced during weathering (Table 8.8); however, Potter (1986) implies that when relief is extreme, a wet climate may simply enhance volume of sediment rather than alter its composition. Basu (1985a) suggests that it is actually slope angle, not relief, that controls the residence time of sediment in soil horizons. If slopes are greater than the angle of repose of sediment, the effects of such steep slopes on rapid removal of soils could presumably mask all effects of climate even under the most humid, hot climatic conditions.

A recent study by Grantham and Velbel (1988) provides some additional insight into this problem. These authors suggest that rock fragments are the most sensitive indicators of cumulative weathering effects. They maintain that abundance of rock fragments in detritus entering a fluvial transport system does not correlate directly with climate but, instead, correlates with total or cumulative chemical weathering in the source area. Total chemical weathering is, in turn, a function in part of duration of weathering. Duration of weathering is measured as the relief ratio (maximum relief divided by maximum length of

TABLE 8.8

Inferred relationship of sand production to climate and relief

		Relief	
		High	Low
Rainfall	High	Large volume of lithic arenite	Small volume of quartz arenite
	Low	Small volume of lithic arenite	Small volume of variable composition (quartz, lithic, or feldspathic arenite)

Source: Potter, P. E., 1978, Petrology and chemistry of modern big river sands: Jour. Geology, v. 86. Fig. 14, p. 444, reprinted by permission.

each watershed) and is related to the total time weathering fluids and weatherable minerals and rock fragments are in contact. Because of higher rates of erosion, high slopes have shorter mineral–fluid times and overall shorter durations of weathering. Chemical weathering is a function also of intensity of weathering, measured as effective precipitation (stream discharge per watershed unit area). Grantham and Velbel express the relationship between total chemical weathering and these duration and intensity factors as

$$\text{Cumulative chemical weathering index} = \text{Effective precipitation} \times \frac{1}{\text{Relief ratio}}$$

They conclude that soils in watersheds with low relief ratios and high discharge per unit area experience the most extensive chemical weathering; therefore, sediments derived from these source areas contain the lowest percentages of rock fragments. Thus, both climate and topographic slope affect sediment composition, and it may be difficult to separate the effects of these two variables.

We can perhaps conclude from this discussion that high relief, or more accurately steep slopes, promotes preservation of rock fragments, which appear to be the most sensitive indicators of weathering conditions. Rock fragments are preserved relatively less well on low slopes that experience high rainfall. If slopes are very high, exceeding the angle of repose, then, presumably, abundant rock fragments will be preserved even under high-rainfall conditions. The relative abundance of rock fragments in a first-cycle sandstone is thus a rough indicator of slope conditions. These relationships have not been quantified, however, and provide only the roughest sort of guide to topographic relief and slope of source areas. We are still a long way from developing a full understanding of the relationship between climate, slope, and sandstone composition.

8.6   Provenance of Conglomerates

8.6.1   General Statement

Most published provenance studies deal with sandstones, probably because sandstones are much more abundant than conglomerates and are easier to study than shales. When conglomerates are present in a stratigraphic section, however, they commonly provide more-reliable provenance interpretation than do sandstones. Conglomerates are composed mainly of rock fragments, which typically preserve textures and structures that make identification of the parent rock quite easy. Thus, the coarse clasts in conglomerates can be readily traced to specific kinds of plutonic igneous, volcanic, metamorphic, or sedimentary source rocks. This virtually unequivocal interpretation of source-rock lithology from conglomerate clasts stands in sharp contrast to provenance interpretation from sandstones, in which minerals such as quartz, feldspars, and micas may be derived from a variety of source rocks.

Interpretation of provenance from conglomerates is not, however, without problems, because depositional assemblages of clasts may not faithfully represent the types and proportions of parent rocks in the source area. A variety of factors may account for these differences. Blatt (1982, p. 144) points out, for example, that the initial size of rock fragments differs with different lithologies owing to factors such as different petrophysical properties (degree of cementation, fissility, foliation) of source rocks, thickness of bedding, and spacing of joints. Thus, for example, well-cemented quartzite tends to weather into larger blocks than does granite or shale. As pointed out in Section 8.4, a strong relationship exists between clast composition and clast size; gravels may become sorted by size and composition during transport. Boggs (1969b) reports that analysis of coarse and fine gravel populations from the same point bar in the Sixes River, southwest Oregon, shows that sandstone, siltstone, and argillite fragments occur preferentially in the fine population, whereas conglomerate, volcanic, schist, phyllite, and greenstone clasts occur preferentially in the coarser population. It is extremely important to keep this size-composition relationship in mind during provenance analysis because fine gravels derived from a particular source region may have a distinctly different composition than coarse gravels derived from the same region. The problems caused by differential susceptibility of different kinds of rock fragments to destruction by chemical weathering and mechanical destruction during transport cause additional difficulties in interpretation. Blatt (1982, Fig. 5-1) shows, for example, that detectable destruction of soft gravel clasts such as sandstone, limestone, and schist can occur during fluvial transport in distances of as little as 15 km.

Abbott and Peterson (1978) performed tumbling experiments on various kinds of clasts to determine their relative durability to abrasion during transport (Fig. 8.13). Their experiments showed that when a population of mixed clasts is tumbled for a short time (equivalent to a short distance of transport), the final population of clasts differed little from the original assemblage. When the same population of mixed clasts was tumbled for a long time (equivalent to a long distance of transport), ultradurable clasts, which made up only a small proportion of the original population, predominated in the final population—demonstrating that less-durable clasts are eliminated during transport. Finally, Paola (1988) suggests that small basins with rapid proximal subsidence tend to sort clasts by size, whereas large, slowly subsiding basins tend to sort clasts by durability.

Thus, relative clast stability, and the susceptibility of clasts to destruction during weathering and transport, are important factors that must be considered in interpretation of provenance from conglomerates. Nonetheless, because the particles in conglomerates are typically deposited closer to their source areas than are sand-size particles, they may survive better than sand-size particles that commonly undergo much longer distances of transport. Thus, on the whole, conglomerates make more reliable provenance indicators than sandstones. Clearly, however, it is necessary to be cautious about interpreting prov-

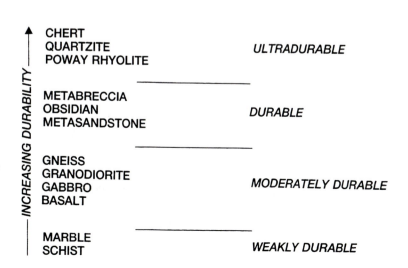

FIGURE 8.13
Abrasional durability scale for clasts. (*From Abbott, P. L., and G. L. Peterson, 1978, Polycrystallinity: effect on the durability of detrital quartz: Jour. Sed. Petrology, v. 48. Fig. 6, p. 36, reprinted by permission of Society of Economic Paleontologists and Mineralogists, Tulsa, Okla.*)

enance from compositionally mature conglomerates made up mainly of ultradurable clasts. Such conglomerates may have been recycled from an earlier generation of conglomerates, or they may have undergone sufficient transport to selectively remove less-durable clasts. Conglomerates made up of mixed populations of clasts that include moderately or weakly durable types are more likely to be first-cycle deposits and can generally be more safely used for provenance determination.

8.6.2   *Interpreting Source-rock Lithology from Conglomerates*

Basic Procedure

Interpreting source-rock lithology from conglomerates is a comparatively straightforward process. First, due regard must be given to the problems imposed by clast size and selective destruction of clasts, as discussed above. The relative abundance of various kinds of clasts in a conglomerate is established by means of random pebble counts or clast counts (Section 6.3), which is the rough equivalent of determining sandstone composition by thin-section modal point counts. Statistical treatment of these clast composition data then allows interpretation of the approximate relative abundance or importance of various kinds of source rocks.

Lithologic Provenance Modeling

Lithologic provenance modeling is a relatively new deductive technique that has been used to identify matches between actual compositions of conglomerates and hypothetical compositions modeled from preserved source sections (Graham et al., 1986; Decelles, 1988). This technique can be used only where some part of all the source rocks remains intact; that is, the full range of source units is at least partially preserved. It cannot be used if the source region has been so severely eroded that some of the potential source rocks are missing and thus not available for examination.

The steps involved in provenance modeling, illustrated in Figure 8.14, are as follows (Graham et al., 1986):

1. Measure and describe the preserved stratigraphic units in the source region. From these data, construct a stratigraphic section for the source region.

2. Measure and tabulate the thickness of erosion-resistant lithologies in each formation (designated Rtn) in the stratigraphic section; that is, determine the relative proportion of rock material that each formation in the source terrane is likely to contribute as gravel to a depositional basin when the source region is eroded. Group these resistant lithologies into units that can be reliably recognized in clast counts of conglomerate units in the depositional basin.

3. Analyze the composition of conglomerates in the depositional basin by making clast counts. These clast counts show which stratigraphic units were exposed at specific times in the source region and subsequently were eroded to form conglomerates. As a source region is slowly unroofed by erosion, progressively older source rocks would be exposed. Clasts of the youngest source rocks would be present in the lower part of the stratigraphic section, and clasts of progressively older formations would appear higher in the section. By dividing the conglomerate into a number of sampled units, clast counts of each sampled unit show which resistant stratigraphic units in the source terrane are represented in that particular conglomerate sample. Thus, for each sample, a ''window,''

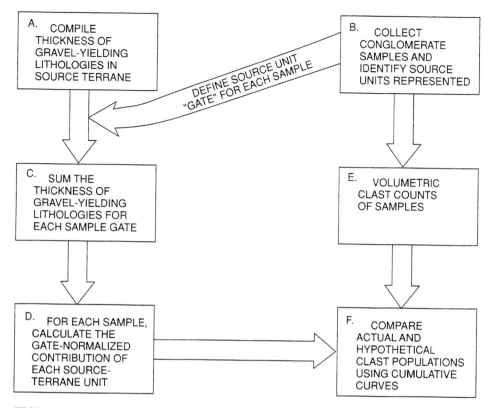

FIGURE 8.14
Schematic representation of the successive steps involved in provenance modeling of conglomerates. (*From Graham, S. A., et al., 1986, Provenance modelling as a technique for analysing source terrane evolution and controls on foreland sedimentation, in Allen, P. A., and P. Homewood, eds., Foreland basins. Fig. 2, p. 427, reprinted by permission of Blackwell Scientific Publications, Oxford.*)

or "gate," of stratigraphic units of the source terrane represented in the clast count can be tabulated. For example, a particular sample might include pebbles from five different formations ranging in age from Cretaceous to Pennsylvanian. Thus the source gate for that sample, using appropriate symbols for each formation, might be K_L, J_E, T_D, P_P, and MP_U.

4. Having established the gate of source units for a particular sample, the thickness of resistant lithologies in the source unit gate (designated *RtG*) is obtained by summing the thickness of resistant lithologies in each formation (n_1 to n_G) in the gate:

$$RtG = \sum_{n_1}^{n_G} Rtn \qquad (8.1)$$

This thickness is the total thickness of resistant lithologies contributing clasts to that particular sample.

5. We now want to calculate the relative abundance of clasts that each resistant lithology in the source gate contributed to the conglomerate sample. This relative, gate-normalized contribution (C) is calculated for each formation in the gate by dividing the thickness of resistant lithologies in that formation by the thickness of the total resistant lithologies in the gate:

$$C = \frac{Rtn}{RtG} \tag{8.2}$$

Graham et al. (1986) suggest that a more sophisticated approach to determining the relative clast contribution from each source formation might be to assess planimetric as well as vertical distribution of resistant source lithologies. Also, it might be possible to include some kind of weighting factor to account for differential transport durability of the various kinds of clasts.

6. The final step in the modeling process is to compare the relative abundances of clasts calculated in step 5 with the actual relative abundances obtained by clast counts (step 3). This comparison can be made by plotting both the calculated relative clast abundance and the measured relative clast abundance on the same cumulative-curve diagram or histogram (Fig. 8.15).

If step 6 demonstrates a close correlation between hypothetical relative abundance curves and actual curves, then we can assume with reasonable confidence that the relative abundance of the various kinds of clasts in the conglomerate adequately represents the relative abundance of these particular lithologies in the source region and that the provenance of the clasts has been correctly identified. If the relative abundance curves do not closely coincide, then some other explanation is required. This explanation might be that some lithologies present in the source area were selectively destroyed by weathering or transport processes. Alternatively, the conglomerates may not have been derived from the presumed source region or not entirely from that source region. Thus, the technique of lithologic provenance modeling, when circumstances permit its application, should provide a more reliable means of relating conglomerate clast composition to a particular provenance than simple observation of clast-count data.

8.6.3 Interpreting Tectonic Setting from Conglomerates

The process of interpreting tectonic setting from conglomerate compositional data is much the same as that used in interpreting tectonic setting from sandstone compositional data. That is, if we know the types of source rocks that are unique to each major tectonic setting, we should be able from clast-count data to interpret, at least approximately, the tectonic setting. Table 8.6 lists the major tectonic provenances, the major kinds of source rocks that occur in each provenance, the principal kinds of sand-size and gravel-size sediment that are derived from these source rocks, and the kinds of depositional basins in which these materials are deposited. No one has put together for conglomerates the kinds of provenance diagrams used by Dickinson and Suczek (1979) to study the tectonic setting of sandstones; however, Table 8.6 provides a rough guide to tectonic setting. For example, a predominance of volcanic clasts in a conglomerate suggests a magmatic-arc provenance, an abundance of sedimentary or metasedimentary clasts indicates a probable recycled-orogen provenance, and abundant granitic and gneissic clasts are most likely derived from uplifted basement blocks. For an application of this principle, see Herbig and Stattegger (1989).

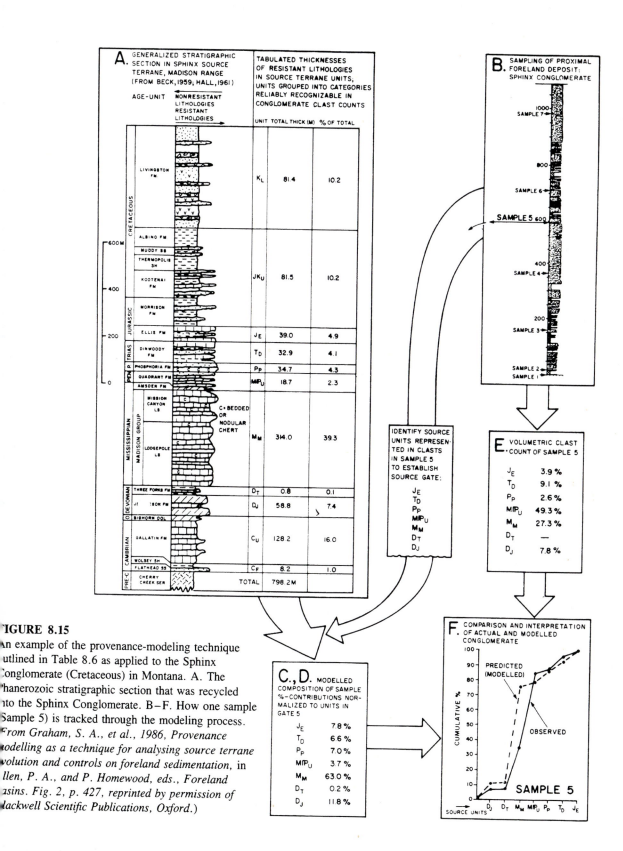

FIGURE 8.15

An example of the provenance-modeling technique outlined in Table 8.6 as applied to the Sphinx Conglomerate (Cretaceous) in Montana. A. The Phanerozoic stratigraphic section that was recycled into the Sphinx Conglomerate. B–F. How one sample (Sample 5) is tracked through the modeling process. *From Graham, S. A., et al., 1986, Provenance modelling as a technique for analysing source terrane evolution and controls on foreland sedimentation, in Allen, P. A., and P. Homewood, eds., Foreland Basins. Fig. 2, p. 427, reprinted by permission of Blackwell Scientific Publications, Oxford.)*

8.7   *Provenance of Shales*

In spite of the overall abundance of shales in the stratigraphic record, few provenance studies of shales have been reported, presumably because the fine grain size of shales makes them difficult to analyze petrographically. Blatt and Caprara (1985) and Jones and Blatt (1984) studied the feldspar content of shales and report that feldspars in shales can be used for provenance interpretation in much the same way as feldspars in sandstones. Nonetheless, few investigators appear to have made such provenance studies of shales (see discussion by Blatt, 1985). The chemical composition of shales has also been used for provenance interpretation, again in a manner similar to that described for sandstones (e.g., Bhatia, 1985a,b; Roser and Korsch, 1986, 1988). Blatt (1987) used the oxygen isotope composition of quartz in shales to infer origin of the quartz. On the basis of oxygen isotope data, he concludes that schists and phyllites are the most important sources of quartz in mudrocks; however, it does not appear possible at this time to use oxygen isotopes as a routine tool for provenance analysis.

Perhaps the greatest interest in shale provenance has been in the study of clay minerals as possible climate indicators. To be useful as indicators of source-area climates, several assumptions must be made about the origin of clay minerals: (1) the kinds of clay minerals formed during weathering are directly related to climatic factors, (2) once formed in the weathering regime, clay minerals are stable as long as the climate remains unchanged, (3) clay minerals remain stable (are not destroyed or altered to other clay minerals) when transported into a depositional environment where conditions are different from the weathering environment, and (4) clay minerals remain stable during burial diagenesis. We know that not all of these assumptions are correct. Singer (1980, 1984) reviews in detail the problems involved in paleoclimate interpretation from clay minerals. He suggests that (1) factors other than climate (e.g., topography, source-rock lithology, time) can affect the formation of clay minerals, (2) clay minerals may undergo differential transport owing to size sorting or flocculation that will alter their composition at the final depositional site, and (3) postdepositional diagenetic changes are quite common. His overall conclusion is that the study of clay minerals in sediments can yield no more than broad groupings of paleoclimates at best. Therefore, it appears, with our present state of knowledge, that the validity of using clay minerals as paleoclimate indicators is tenuous at best.

Additional Readings

Allen, P. A., and P. Homewood, eds., 1986, Foreland basins: Internat. Assoc. Sedimentologists Spec. Pub. No. 8, Blackwell, Oxford, 453 p.

Kleinspehn, K. L., and C. Paola, eds., 1988, New perspectives in basin analysis: Springer-Verlag, New York, 453 p.

Zuffa, G. G., ed., 1984, Provenance of arenites: D. Reidel, Dordrecht, 408 p.

Chapter 9
Diagenesis of Sandstones and Shales

9.1 Introduction

Petrologic study of siliciclastic sedimentary rocks commonly focuses on interpretation of sediment provenance and depositional paleoenvironments. It may aim also at evaluating the economic significance of these sediments as source rocks or reservoir rocks for petroleum or as hosts for ore deposits. Discussion in previous chapters suggests that genetic interpretations can be compromised by postdepositional changes in composition or texture brought about by diagenetic processes. Provenance interpretation, in particular, can be severely affected by diagenetic processes that selectively remove feldspars, heavy minerals, or rock fragments by intrastratal solution. Diagenesis may also play an extremely important role in postdepositional modification of porosity, causing either decrease in porosity as a result of compaction and cementation or increase in porosity owing to solution processes. Thus, the economic importance of a particular sandstone body as a reservoir rock for oil, for example, may very well depend as much upon the diagenetic history of the sandstone as upon its original depositional characteristics. Therefore, to avoid erroneous interpretations of petrogenesis or costly errors in economic evaluations, it is necessary to understand as much as possible about the diagenetic processes that can affect siliciclastic sedimentary rocks. We are concerned not only with the nature of diagenetic processes but also with the types of changes that are produced in the rocks as a result of these processes. That is, we strive to recognize the diagenetic characteristics of siliciclastic rocks and to differentiate these diagenetic features from original depositional characteristics.

Diagenesis in the broadest sense encompasses all of the processes that act to modify sediment after deposition. The net effect of these diagenetic modifications is to move initially incompatible assemblages of siliciclastic minerals toward conditions of chemical equilibrium that are more in adjustment with the diagenetic environment. Our ability to extract genetic information from ancient siliciclastic sedimentary rocks is strongly dependent upon our ability to understand these diagenetic changes and to reconstruct the postdepositional histories of these rocks.

Early studies of diagenesis focused on petrographic examination of thin sections and the evidence of compaction, cementation, dissolution, and so on that could be read from petrographic study. Petrographic examination remains the mainstay of diagenetic study; however, it has been increasingly supplemented since the 1960s by use of the electron microscope, cathodoluminescence petrography, X-ray diffraction analysis, electron microprobe analysis, and various techniques for geochemical evaluation, including isotope studies. In this chapter, we explore some of the more important and interesting aspects of siliciclastic sediment diagenesis.

9.2   Diagenetic Stages and Regimes

Diagenesis takes place during burial at depths ranging from the depositional interface to perhaps 10 km or more. Thus, the pressure–temperature conditions under which diagenesis occurs essentially extend from those that characterize weathering to those that characterize metamorphism. There are no clear-cut boundaries on either end of this scale, although diagenesis is commonly regarded to take place at temperatures below about 300°C and at pressures below 1–2 kb (Fig. 9.1).

Several investigators have suggested that diagenesis proceeds through a series of recognizable stages (Table 9.1). For example, Choquette and Pray (1970) used the terms **eogenetic, mesogenetic,** and **telogenetic** to indicate the times of early burial, deeper burial, and late-stage erosion. Schmidt and McDonald (1979a) suggested the terms **eodiagenesis, mesodiagenesis,** and **telodiagenesis** to refer to diagenesis during these times. Alternatively, Burley et al. (1985) use the terms eogenesis, mesogenesis, and telogenesis to mean the same thing. According to Burley et al., the **eogenetic regime** is the depositional environment. Nonmarine environmental conditions range from arid-oxidizing to wet-reducing pore waters, and marine environments may be characterized by either oxidizing or reducing pore waters. Significant early diagenetic changes may occur during

FIGURE 9.1

Generalized pressure-temperature conditions that control diagenesis and metamorphism. The pressure-temperature region below a geothermal gradient of 10°C/km is not realized in nature. (*After Winkler, H. G. F., 1967, Petrogenesis of metamorphic rocks, 2nd ed. Fig. 1, p. 4, reprinted by permission of Springer-Verlag, New York.*)

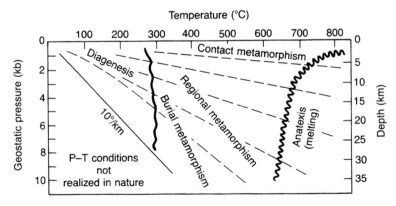

TABLE 9.1

Stages of diagenesis according to various authors

Source	Stage	Description
Choquette and Pray (1970)	Eogenetic	Time of early burial
	Mesogenetic	Time of deeper burial
	Telogenetic	Time of late-stage erosion
Dunoyer de Segonzac (1970)	Early diagenesis	Shallow-burial stage
	Middle diagenesis	Deep-burial stage
	Late diagenesis	Temperature >100°C
	Anchizone diagenesis	Transitional phase to metamorphism
Schmidt and McDonald (1979a)	Eodiagenesis	Diagenesis in the depositional environment
	Mesodiagenesis	Deep-burial phase of diagenesis
	Telodiagenesis	Late-stage diagenesis following uplift
Dapples (1979)	Redoxomorphic	Early-burial stage
	Locomorphic	Deeper-burial state
	Phyllomorphic	Late-burial stage
Fairbridge (1983)	Syndiagenesis	Early diagenesis synchronous with deposition
	Anadiagenesis	Deep-burial phase of diagenesis
	Epidiagenesis	Late-stage diagenesis following uplift
Singer and Müller (1983)	Preburial stage	Occurs in presence of oxygen
	Shallow-burial stage	Conversion of mud to mudstone or shale
	Deep-burial stage	Conversion of mudstone to argillite
Burley et al. (1985)	Eogenesis	Diagenesis in the depositional environment
	Mesogenesis	Deep-burial phase of diagenesis
	Telogenesis	Late-stage diagenesis following uplift
Pettijohn et al. (1987, p. 426)	Stage 1	Diagenesis in the depositional environment
	Stage 2	Burial to a few 10s of m
	Stage 3	Burial to moderate depths of ~1000 m
	Stage 4	Burial to thousands of meters, perhaps accompanied by folding
	Stage 5	Incipient metamorphism, or ''high-grade diagenesis''
	Stage 6	Diagenesis during and after uplift and erosion

this stage, with the course of diagenesis determined by the specific Eh, pH, and chemical composition of the pore waters. The **mesogenetic regime** is the environment of deeper burial. Little further change occurs in the sediment after the eogenetic stage until it is brought into an environment of increased temperature and changed pore-water composition. Burley et al. suggest that elevated temperatures add energy to the reacting system, lowering reaction barriers and increasing reaction rates. Large-scale, deep circulation of pore fluids having different compositions and Eh, pH characteristics than those of the depositional environment causes transport of large quantities of dissolved constituents, bringing about major regional diagenetic changes. Uplift and exposure of the sediment pile bring sediments into the **telogenetic regime,** characterized by lowered pressures and temperatures and by generally oxidizing, meteoric pore waters. Mineral assemblages formed under high temperatures and pressures during mesogenesis tend to become unstable under the changed conditions of telogenesis.

A variety of different names has been suggested by other investigators for the diagenetic stages. Some investigators have also suggested more than three stages. Table 9.1 summarizes the various terminology that has been applied to diagenetic stages.

Whatever the terminology used to describe different diagenetic stages, it is clear that siliciclastic sediments undergo sequential changes with burial that occur as a result of biogenic activity, higher pressures and temperatures, and changed pore-water compositions. These diagenetic changes include postdepositional mixing of sediment owing to bioturbation; rearrangement of grain packing and loss of porosity as a result of compaction from sediment loading; loss of porosity through cementation; partial or complete destruction of some framework grains owing to pressure solution; destruction of some framework grains, cements, or matrix by dissolution, creating secondary porosity; replacement of some minerals by others; and clay-mineral authigenesis. These processes are discussed in the remaining part of this chapter.

9.3   Biologic and Physical Diagenesis

9.3.1   Bioturbation

Organisms rework sediment at or near the depositional interface through various crawling, boring, burrowing, and sediment-ingesting activities. These activities may extend from the depositional interface to depths of several tens of centimeters, resulting in churning or mixing sediment to various degrees. Bioturbation causes extensive destruction of primary sedimentary structures and textural patterns, thereby bringing about loss of important environmental information. Only sediments deposited in toxic or anoxic environments, or those deposited so rapidly that organisms do not have time to rework them, appear to escape bioturbation. As a result of their activities, organisms may destroy primary depositional features and create in their place a variety of traces such as mottled bedding, burrows, tracks, trails, and so on (Fig. 9.2). These features are referred to as trace fossils.

Although bioturbation can destroy primary depositional features such as lamination, it probably has little effect on the composition of siliciclastic sediment, except possibly under those conditions where sediment in one layer is mixed by bioturbation with sediment of a different composition in another layer. Such mixing may also affect textural patterns if sediments of different grain size or shape are mixed. Organisms in deep burrows may bring about interchange of pore waters with overlying seawater; however, the effect of this process on composition of the sediment is probably inconsequential. Also, bioturbation may act locally to alter the porosity and permeability of sediments, but subsequent changes in porosity owing to compaction during burial overshadow the effects of bioturbation. Except for destruction of sedimentary structures, which can be substantial, and minor changes in texture owing to mixing of sediments of different sizes or shapes, bioturbation probably does not greatly modify the characteristics of sediments. For a mathematical treatment of bioturbation effects, see Berner (1980, p. 42).

9.3.2   Compaction

Causes of Compaction

Compaction is the lessening of sediment volume, and concomitant decrease in porosity, brought about by grain rearrangement and other processes that result from sediment

FIGURE 9.2
Mottled bedding in a core of
highly bioturbated, deep-sea mud
(Quaternary), Japan Sea. Ocean
Drilling Program Leg 127, Site
795, 74 m below seafloor.
*(Photograph courtesy Ocean
Drilling Program, Texas A & M
University.)*

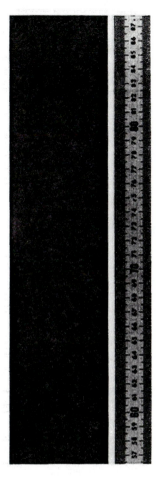

loading and tectonic forces. In regions of folding and thrust faulting, the compactive
stresses resulting from tectonic forces may be greater than those resulting from sediment
loading, thereby causing overprinting of burial compaction. In particular, tectonic stresses
may result in fracturing of grains. On the whole, however, the overburden weight creates
the major stress that causes compaction of sediments. The overburden weight results
mainly from the weight of the sediment or rock mass. The average change in geostatic
pressure with depth—the geostatic gradient—is about 244 bars/km (Fig. 9.3). Thus, at
a depth of 10 km, for example, the geostatic or rock pressure is roughly 2.5 kb. The
average change in fluid pressure with depth—the hydrostatic gradient—is about 104
bars/km. At 10 km depth, the hydrostatic pressure is thus about 1.0 kb. Under most
conditions, however, the weight of the fluid column does not add to the overburden
weight (Chilingarian, 1981). That is, the overburden pressure that causes compaction is
the result primarily of the rock weight alone. Temperature also influences compaction by
promoting pressure solution of grains (Houseknecht, 1984) and by decreasing resistance
of grains to deforming stresses. The maximum effects of temperature on porosity reduction are discussed by Stephenson (1977).

FIGURE 9.3
Average geostatic pressure
gradient, hydrostatic pressure
gradient, and geothermal gradient
in sedimentary basins. *(From
Boggs, S., Jr., 1987, Principles of
sedimentology and stratigraphy.
Fig. 9.1, p. 285, reprinted by
permission of Merrill Publishing
Co., Columbus, Ohio.)*

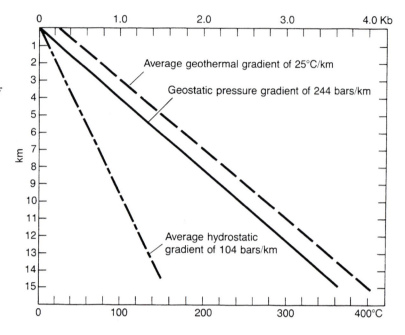

Compaction of Sands

Newly deposited, unconsolidated sand has high porosity and a loosely packed fabric. Under sediment loading, the fabric becomes more tightly packed and porosity is reduced. Changes in packing and reduction of porosity with burial depth have long been subjects of considerable interest, particularly to petroleum geologists (see Chilingarian and Wolf, 1975, 1976). We now recognize that compaction of sands can involve (1) mechanical rearrangement of grains into tighter packing owing to slippage of grains past each other at points of contact, (2) bending of flexible grains such as micas, (3) ductile and plastic deformation, particularly of malleable grains such as rock fragments, (4) brittle fracture, particularly of carbonate shell material but possibly of silicate grains also, and (5) pressure solution of quartz and other minerals (Füchtbauer, 1967; Wilson and McBride, 1988). The first four of these processes produce purely mechanical changes in packing. The fifth, pressure solution, is often referred to as chemical compaction because it also involves solution at grain contacts.

The compactability of sands is a complex function of numerous variables, the most important of which are grain size and sorting, grain shape, grain orientation, composition, matrix content, and cements (Wolf and Chilingarian, 1976). Theoretical considerations show that even-sized spheres packed into the loosest arrangement (cubic packing) have a porosity of 47.6 percent. Simple rearrangement of the spheres to yield the tightest packing (rhombohedral packing) reduces the porosity to 26 percent (Graton and Fraser, 1935). The grains in natural sands are neither spheres nor even-sized grains. Therefore, neither the initial porosity of noncompacted sands nor the porosity loss that these sands undergo owing to tighter packing can be predicted theoretically. Empirical studies show that porosity of loosely consolidated sands may range from about 30 to 50 percent (Singer and Müller, 1983). Chilingarian (1981) suggests that initial porosity ranges from about 37 percent for well-sorted, well-rounded, medium- to coarse-grained sands to more than 50 percent for poorly sorted, fine-grained sands with irregularly shaped grains. Houseknecht

(1987) suggests that a natural, well-sorted sand composed of nonductile grains and having an initial porosity of 40 percent can potentially be reduced in porosity by mechanical compaction (reorientation, repacking, fracturing of grains) to about 30 percent. Further reduction in porosity occurs only as a result of chemical compaction. Sands with ductile grains and more poorly sorted sands may undergo a greater percentage change in porosity owing to mechanical compaction.

Porosity is progressively reduced with burial owing to factors cited in preceding paragraphs. According to Schmidt and McDonald (1979a), mechanical compaction dominates to depths of about 0.6 km (~2000 ft) to 1.5 km (~5000 ft), depending upon the type of sandstone (Fig. 9.4). Dominantly mechanical effects extend to greater depths for quartz arenites than for sandstones containing abundant feldspars or lithic fragments. Below these depths, pressure solution becomes important in reducing primary porosity. Chemical compaction continues to depths of as much as 4.5 km (~15,000 ft) to 8 km (~26,000 ft), where primary porosity reaches near-zero levels.

Figure 9.5 shows the relationship between primary porosity and burial depths for some California feldspathic arenites containing an average of about 7 percent largely carbonate and clay-mineral cement. Porosity–depth relationships for distinctly different kinds of sands—quartz arenites vs. lithic arenites, for example—may show even greater variations. Therefore, it is probably not prudent to overgeneralize about the rate of change of primary porosity in sandstones as a function of burial depth. Furthermore, the rate of change of primary porosity with depth may not be an entirely linear function. For other examples of porosity–depth curves, see the idealized compaction curves of Damanti and Jordan (1989), which were constructed for Tertiary foreland-basin sandstones of Huaco, Argentina.

Recognizing Compaction Effects in Sandstones

Principal Methods.   Recognition of compaction effects and estimates of burial depth in sandstones are based upon analysis of primary porosity, measures of packing such as the numbers of contacts between grains and types of grain contacts, and visible grain deformation such as bent mica flakes and fractured grains. The fundamental applications of these methods are discussed briefly below.

Analysis of Porosity.   Estimates of compaction on the basis of porosity analysis requires first that primary porosity, as measured in thin sections, be differentiated from secondary porosity created by dissolution of framework grains or cements. Criteria for differentiating secondary (solution) porosity from primary porosity include partial dissolution of grains, the presence of molds, inhomogeneous packing, oversize pores, elongate pores, corroded grains, pores within grains, and fractured grains. Schmidt and McDonald (1979b) and Shanmugam (1985) provide full details of these criteria. The amount of primary porosity remaining in a sample, assuming only minor cement in the rock, provides a rough qualitative estimate of the severity of compaction; that is, the lower the porosity, the more severe the effects of compaction. The approximate depth of burial can be estimated from porosity–depth curves such as that shown in Figure 9.5, assuming that appropriate curves for a particular sandstone type (e.g., quartz arenites) are available. Alternatively, burial depth can be estimated from porosity–depth plots based on total porosity, such as those shown in Sclater and Christie (1980) and Baldwin and Butler (1985). Obviously, porosity is decreased by precipitation of cements in pore spaces as well as by compaction. Furthermore, early cementation tends to inhibit compaction.

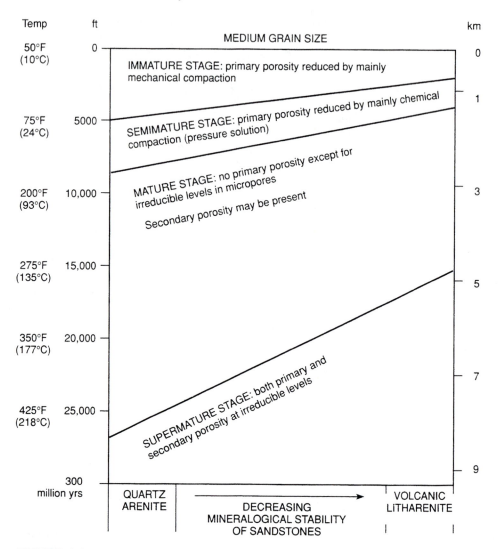

FIGURE 9.4

Reduction in porosity with increasing burial depth owing to mechanical compaction and pressure solution at grain contacts. The assumed sedimentation rate is 30.5 m (100 ft)/million years, and the geothermal gradient is 2.7°C/100 m(1.5°F/100 ft). Note that porosity loss occurs at a shallower depth for mineralogically immature sandstones than for quartz arenites. *(Modified from Schmidt, V., and D. A. McDonald, 1979, Aspects of diagenesis: Soc. Econ. Paleontologists and Mineralogists Spec. Pub. 26. Fig. 10, p. 187, reprinted by permission of SEPM, Tulsa, Okla.)*

Therefore, sandstones containing abundant cement are not very amenable to this type of porosity analysis. Feldspathic or lithic sandstones containing abundant matrix may likewise yield poor results in estimating burial depth from porosity values.

Smosna (1989) suggests that the effects of burial compaction on porosity can be evaluated in terms of depth of burial (d), the percent ductile grains such as phyllitic rock fragments (P), and matrix content (M). On the basis of empirical studies of Cretaceous sandstones on Alaska's North Slope, he derives from these parameters what he calls the

FIGURE 9.5

Changes in primary porosity with depth of burial in Pliocene sandstones, Ventura Basin, California. 1 = empirical best-fit line through most porous samples; 2 = regression line for samples with less than 10 percent cement and omitting samples from Dos Cuadras Field. *(After Wilson, J. C., and E. F. McBride, 1988, Compaction and porosity evolution of Pliocene sandstones, Ventura Basin, California: Am. Assoc. Petroleum Geologists Bull., v. 72. Fig. 4, p. 669, reprinted by permission of AAPG, Tulsa, Okla.)*

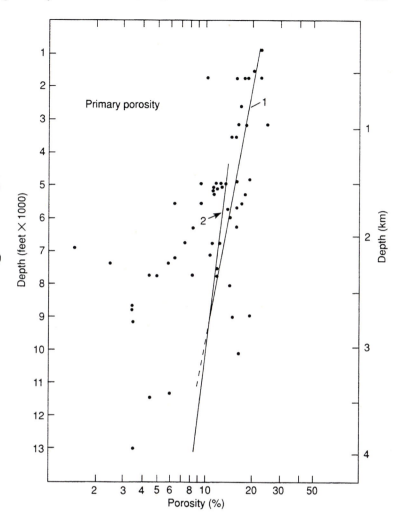

compaction law. According to this law the grain fraction (*GF*), which is the grain volume as a percent of total rock volume (i.e., the inverse of porosity plus matrix and cement), is given by the relationship

$$GF = 90 + 0.23P - 0.72M + 0.0018d \qquad (9.1)$$

Note from this relationship that compaction is enhanced by an increased percentage of ductile grains and increased burial depth but is inhibited by increased matrix content. Whether or not this compaction law holds for other sandstones remains to be tested.

Measurement of Packing.   Taylor (1950) identified five principal types of contacts between grains viewed in thin sections: floating (no contacts), tangential (point contacts), long, concavo-convex (embayed), and sutured (serrated). These grain types are illustrated in Table 2.8, Chapter 2. Long contacts and concavo-convex contacts develop during deposition, depending upon initial grain shapes, and also during diagenesis as a result of grain deformation and pressure solution. Sutured contacts are the result of pressure solution. Taylor (1950) suggested that the relative abundance of grain-contact types in a sandstone provides a measure of burial compaction. Floating and tangential (point) contacts occur in unconsolidated sands and disappear with increasing burial depth. Long and concavo-convex contacts tend to increase with burial depth as floating and tangential contacts disappear. With greater burial depth, sutured contacts appear and increase at the expense of long and concavo-convex contacts. Taylor also points out that the number of contacts per grain increases progressively with burial depth.

Füchtbauer (1967) and Hoholick et al. (1982) used Taylor's concept to create indices of packing, which Füchtbauer calls **contact strength** and Hoholick et al. call **weighted contact packing** (see Table 2.7). Large packing values calculated by the formulas of Füchtbauer and Hoholick et al. indicate close packing and severe compaction and vice versa. Wilson and McBride (1988) use the term **tight packing index** for the average abundance per grain of long, concavo-convex, and sutured contacts. Pettijohn et al. (1987, p. 86) use the term **contact index** for the mean number of contacts per grain. In any case, calculated values for packing indices can be used to estimate depth of burial provided that appropriate empirically developed models are available to which calculated results can be compared. Figure 9.6 shows the relationship between contact index, tight packing index, and burial depth for several sandstones. It is clear from Figure 9.6 that the relationship between packing indices and burial depth varies widely in different sandstones.

FIGURE 9.6

Contact index (CI = mean number of contacts per grain) and tight packing index (TPI = average abundance per grain of long, concavo-convex, and sutured contacts) of selected sandstones plotted as a function of burial depth. A (all samples), B (samples with less than 10 percent cement), and E (samples with less than 10 percent ductile grains) are CI regression lines for Pliocene sandstones, Ventura Basin, California. A′ (all samples) and B′ (samples with less than 10 percent cement) are TPI regression lines for Ventura Basin Pliocene sandstones. C is the CI regression line and C′ the TPI regression line for Eocene Wilcox sandstones of the Texas Gulf Coast. D is the CI regression line for Cretaceous and Jurassic sandstones of Wyoming, and F is the CI regression line for Pennsylvanian (Minnelusa) sandstones of Wyoming. *(From Wilson, J. C., and E. F. McBride, 1988, Compaction and porosity evolution of Pliocene sandstones, Ventura Basin, California: Am. Assoc. Petroleum Geologists Bull., v. 72. Fig. 8, p. 677, reprinted by permission of AAPG, Tulsa, Okla.)*

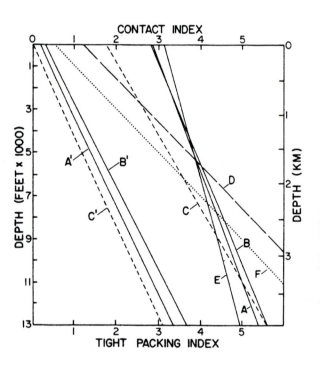

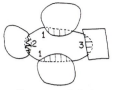

FIGURE 9.7
Schematic representation of textural criteria used to estimate volume loss in sandstones owing to compaction. The hachured areas indicate rock volume lost by grain deformation and pressure solution. *(From Wilson, J. C., and E. F. McBride, 1988, Compaction and porosity evolution of Pliocene sandstones, Ventura Basin, California: Am. Assoc. Petroleum Geologists Bull., v. 72. Fig. 10. p. 679, reprinted by permission of AAPG, Tulsa, Okla.)*

Plastic and Ductile
Grain Deformation

Flexible Grain
Deformation

Pressure Solution
1 Concavo-Convex Contact
2 Sutured Contact
3 Long Contact

Grain Deformation.   Qualitative visual evidence of compaction is provided by bent flexible grains such as mica flakes, deformation of ductile grains, and fractured grains. Figure 9.7 illustrates compaction effects in terms of volume loss owing to ductile and flexible grain deformation, as well as to pressure solution. One should remember that deformation of ductile grains can create pseudomatrix, as discussed in Section 4.2.3. We commonly assume that broken grains in a sandstone are likewise the result of compaction. Braithwaite (1989) and Buczynski and Chafetz (1987) suggest that quartz grains can also be broken during some conditions of early diagenesis (e.g., diagenesis in calcretes) owing to displacive growth of calcite crystals.

Compaction of Muds
Owing to the dominance of platy, flaky, or acicular clay minerals in muds, the initial porosity of muddy sediment is considerably higher than that of sands. Recent clay muds from the ocean floor and lake bottoms may have porosities as high as 70–90 percent, corresponding to a water content of 50–80 percent (Singer and Müller, 1983). The structure or fabric of clayey sediment is discussed in detail in Chapter 7. An important element of mud fabrics is the presence of domains, which are microscopic or submicroscopic regions within which the clay particles are in parallel array. Clay fabrics are illustrated in Figure 7.5.

Compaction during burial sharply reduces the porosity of muds and may aid also in producing fissility in shales, although compaction is probably not the most important factor that causes fissility (Chapter 7). Porosity of muddy sediments may be reduced by compaction to values less than 20 percent at burial depths ranging from a few thousand feet (~1 km) to 10,000 ft (3.3 km) (Fig. 9.8). Loss of porosity by compaction is accompanied by expulsion of pore water and thinning of beds. Dzevanshir et al. (1986) suggest that the porosity of muddy sediments is a function of depth of burial, geologic age, and the ratio of shale thickness to total thickness of terrigenous deposits. They combine these variables into a single equation

$$\phi = \phi_0 \exp \left[-0.014(13.3 \log A) - 83.25 (\log R + 2.79) \times 10^{-3} D \right] \quad (9.2)$$

where ϕ is porosity at a given depth, ϕ_0 is initial porosity of the sediment, A is geologic age in millions of years, R is the ratio of thickness of muddy sediment to total thickness of the terrigenous complex, and D is burial depth. Additional information on compaction of shales is given by Reicke and Chilingarian (1974).

FIGURE 9.8
Relationship between porosity and depth of burial in shales and argillaceous sediments (see Dzevanshir et al., 1986, for reference to numbers). *(After Dzevanshir, R. D., et al., 1986, Simple quantitative evaluation of porosity of argillaceous sediments at various depths of burial: Sed. Geology, v. 46. Fig. 1, p. 170, reprinted by permission of Elsevier Science Publishers, Amsterdam.)*

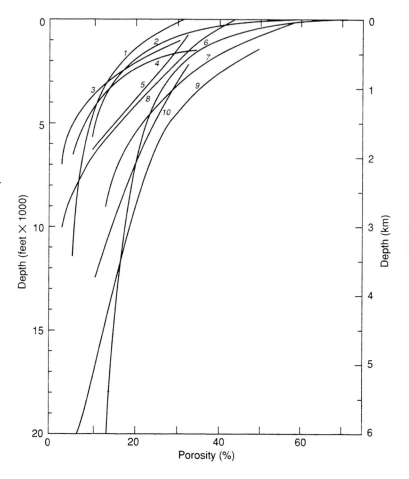

9.4 Biochemical and Chemical Diagenesis

9.4.1   General Statement

Most siliciclastic sediments, other than some continental sediments such as dune sands and subaerial debris flows, are deposited in a subaqueous environment. Even these continental sediments become saturated with water as they are buried below the water table. Therefore, as siliciclastic sediments undergo burial during diagenesis, the mineral grains are in constant contact with water that fills the pore spaces of the sediment. The initial composition of this water depends upon the environment in which the sediments are deposited. As shown in Table 9.2, the salinity of water in different environments can range from fresh to hypersaline, pH can range from acid to alkaline, and redox conditions can range from oxic to anoxic. All natural waters contain the same few major chemical elements; however, the relative abundance of these elements varies considerably between fresh and saline waters.

In addition, natural waters contain dissolved organic carbon (DOC) that can range from a few tens of mg/L in surface and shallow ground waters to several thousand mg/L in some subsurface waters associated with petroleum. Also, most sediments contain small amounts of particulate organic carbon (Table 9.2). Sandstones commonly contain less than about 0.5 percent organic carbon, and the average organic carbon content of green shales (0.5 percent) and red shales (<0.1 percent) is also low; however, black shales commonly contain between 2 and 37 percent organic carbon (Gautier et al., 1985). Thus, during burial, sediments are in contact with waters having particular chemical characteristics that depend initially largely upon the depositional environment. They are also in contact with various amounts of dissolved or particulate organic carbon. The initial pore-water composition and organic content of sediments exert important controls on the early diagenesis of sediments. The diagenesis of shales, in particular, appears to be strongly influenced by the amount of organic matter present during burial (Curtis, 1977). With deeper burial, organic matter undergoes diagenetic modifications, and pore-water compositions change. These changes significantly affect the subsequent course of diagenesis.

9.4.2   Diagenetic Reactions in the Eogenetic Zone

Marine Sediments

According to Garrels and Christ (1965, p. 364), there is little potential initially for chemical reaction between normal marine pore waters and most detrital silicate minerals, and many detrital minerals are metastable in the marine environment. The most reactive constituents of marine sediments are organic compounds and soil-derived ferric oxides. The reactions of these constituents can, however, generate reaction products, particularly hydrogen and bicarbonate ions, that change initial pore-water compositions, rendering other mineral components unstable (Curtis, 1980). Bacterial activity plays a primary role in diagenetic reactions at shallow depth, especially in shales (Table 9.3). Bacterial oxidation of organic matter in oxic sediments at very shallow burial depths generates bicarbonate ions, together with some ammonia and phosphates. At slightly greater depths, oxygen is depleted and reducing conditions prevail. In this zone, bacterial sulfate reduction is a dominant process. This process leads to formation of bicarbonate and hydrogen sulfide, with consequent reduction in pH. Some of the products of bacterial oxidation and sulfate reduction diffuse into the sea through the overlying sediment; however, part is fixed within the shales or is expelled into associated sandstones (Burley et al., 1985). Adequate seawater sulfate to sustain sulfate-reducing bacteria apparently does not diffuse downward below a depth of about 10 m. Therefore, below that depth, sulfate reduction gives way to fermentation. Fermentation produces methane, bicarbonate ions, and hydrogen ions (Table 9.3).

Owing to the processes described above, eogenetic reactions in marine sediments are dominated by dissolution of unstable, fine-grained components and formation of characteristic new minerals. The formation of pyrite in the zone of sulfate reduction (owing to the reaction of H_2S with iron-bearing minerals such as iron oxides, chlorite, and biotite) is particularly characteristic. Other important reactions include formation of authigenic iron-rich chlorites or chamosite, glauconite, illite/smectite clay minerals, and precipitation of K-feldspar overgrowths, quartz overgrowths, and carbonate cements. The commonly reported order of formation of these minerals is precipitation of pyrite, chlorite (in anoxic pore waters), illite/smectite (in oxygenated pore waters), followed by precipitation of quartz and feldspar overgrowths and, finally, precipitation of carbonate cements (Burley et al., 1985). The chemical diagenesis of aluminosilicate minerals in marine sediments during eogenesis appears to be arrested by carbonate cementation. In marine sediments deposited in evaporite basins, the increased ionic concentration of the pore

TABLE 9.2

Chemical characteristics of natural waters in major depositional settings and selected sediments

Environment	Salinity	pH	Eh	Major chemical constituents (in order of relative abundance)	Dissolved organic carbon (DOC) (mg/L) or total particulate organic carbon (%)
NONMARINE					
Rivers	Fresh	~5–9	Oxic; ~+0.5 to +0.3 V	HCO_3^- + CO_3^{2-}, Ca^{2+}, H_4SiO_4, SO_4^{2-}, Cl^-, Na^+, Mg^{2+}, K^+	<1–>30; ave. ~6.0 (DOC)
Lakes	Fresh to hypersaline	~5–9	Oxic to anoxic; ~+0.5 to −0.1 V	Freshwater lakes: similar to rivers. Saline lakes: Cl^-, Na^+, SO_4^{2-}, HCO_3^- + CO_3^{2-}, Mg^{2+}, K^+, Ca^{2+}, H_4SiO_4	<1–>100; ave. ~6.0(?) (DOC)
NORMAL MARINE					
Surface and shallow water	Saline (~35‰)	~7.5–8.4 (ave. 8.0)	Mainly oxic; dominantly ~+0.4 to +0.2 V	Cl^-, Na^+, SO_4^{2-}, Mg^{2+}, Ca^{2+}, K^+, HCO_3^- + CO_3^{2-}	<0.1–~3; ave. ~1.0 (DOC)
Deeper water	Saline	~7.5–8.4	Oxic to anoxic; values?	Same as surface ocean water	~0.5 (DOC)
RESTRICTED MARINE					
Evaporative	Saline to hypersaline	~7.5–9.0	Oxic to anoxic; values?	Mg^{2+} increases relative to most other ions owing to precipitation of evaporite minerals	?
Nonevaporative	Saline to brackish	~7.5–8.4	Oxic to anoxic; values?	Same as surface ocean water	?

Environment	Salinity	pH	Eh	Chemistry	Organic carbon
SOILS AND SEDIMENTS					
Soils	Fresh	~3.5–8.5	Oxic to anoxic (waterlogged); ~+0.75 to −0.35 V	Variable, depending upon soil composition	~1–10% particulate organic carbon
Shallow, fresh-water sediments	Fresh	~5–8	Oxic to anoxic; ~+0.5 to −0.15 V	See subsurface waters below	<~0.25% particulate organic carbon; mainly in shales
Marginal marine sediments	Brackish to saline or hypersaline	~5.5–9.5	Oxic to anoxic: ~+0.5 to −0.3 V	See subsurface waters below	~1.0% particulate organic carbon; mainly in shales
Normal marine sediments	Saline	~7.0–8.5	Oxic to anoxic; ~+0.6 to −0.35 V	See subsurface waters below	~1.0% particulate organic carbon, but some black shales may contain >10%
Evaporites	Hypersaline	~6.0–9.5	Oxic to anoxic; ~+0.55 to −0.45 V	See subsurface waters below	?
SUBSURFACE WATERS					
Shallow ground waters	Fresh to brackish	~6.0–8.5	Oxic to anoxic; ~+0.45 to −0.10 V	Variable depending upon bedrock; commonly HCO_3^- + CO_3^{2-}, SO_4^{2-}, Ca^{2+}, Na^+, Cl^-, Mg^{2+}, Si^{4+}, K^+	<1–~20; ave. ~2.0 (DOC)
Connate waters	Brackish to hypersaline	~6.0–8.0	Oxic to anoxic; ~+0.4 to −0.25 V	Dominantly HCO_3^- or SO_4^{2-} rich at shallow to intermediate depth; also Na^+, K^+, Ca^{2+}, Mg^{2+}; dominantly Cl^- rich at greater depth + Na^+, K^+, Ca^{2+}, Mg^{2+}	<10–>4500 (DOC)

Source: Data derived from Baas Becking et al. (1960), Boggs et al. (1985), Drever (1982), Faust and Aly (1981), Nadler and Magaritz (1980).

TABLE 9.3

Bacterial processes that influence diagenesis at shallow burial depths in organic-rich marine shales

Zone (depth)	Organic matter degradation reactions	Selected mineralogical changes
Depositional waters	Bacterial processes as per Zones 1 and 2, depending on oxygen status Extent of reaction dependent on water depth	Clay exchange reactions
Zone 1 Aerobic oxidation (mm–cm)	Bacterial aerobic respiration; dissolved oxygen depleted $CH_2O + O_2 \rightarrow H^+ + HCO_3^-$ Ammonia $\}$ released in small quantities Phosphate HCO_3^-, HPO_4^{2-}, NH_4^+ lost to depositional waters Residual organic matter more refractory	Clay exchange reactions Phosphate fixation by ferric hydroxides
Zone 2 Sulfate reduction (1–10 m)	Bacterial sulfate reduction; dissolved sulfate depleted $2CH_2O + SO_4^{2-} \rightarrow 2HCO_3^- + HS^- + H^+$ HCO_3^-, HPO_4^{2-}, NH_4^+ mostly lost to depositional waters Residual organic matter more refractory still	Detrital hydrated iron oxides reduced $CH_2O + 4\,FeO \cdot OH \rightarrow HCO_3^- + 4Fe^{2+} + 7OH^-$ Iron monosulfides precipitated $Fe^{2+} + HS^- \rightarrow FeS + H^+$ Sulfur addition $FeS + S^0 \rightarrow FeS_2$ Iron-poor carbonates (calcite, dolomite) may precipitate *locally* as concretions
Zone 3 Fermentation (below sulfate reduction to uncertain depths)	Bacterial degradation via microbiological pathways Net reaction corresponds to fermentation $2CH_2O \rightarrow CH_4 + HCO_3^- + H^+$	If detrital iron hydroxides reduced as per Zone 2, saturation with respect to iron-rich carbonates likely: Siderite $FeCO_3$ Ankerite $Ca(MgFe)(CO_3)_2$ Ferroan calcites
Deeper zones	Thermally induced organic degradation in at least two stages: (*a*) dehydration, decarboxylation (*b*) cracking to produce hydrocarbons	Decarboxylation reactions most likely to lead to aggressive porewaters; mineral dissolution likely; secondary porosity develops where pore solutions migrate (norm)

Source: Curtis, C. D., 1980, Diagenetic alteration of black shales: Jour. Geol. Soc. London, v. 137. Table 1, p. 191, reprinted by permission.

waters may lead to precipitation of sulfates, zeolites, and possibly smectite clays in addition to carbonates (Bjørlykke, 1983).

The organic processes described above apply particularly to organic-rich shales and are less important in organic-poor sandstones. As indicated, however, the reaction products of shale diagenesis (bicarbonates, hydrogen, hydrogen sulfide) can be expelled from shales into associated sandstones and thereby affect the course of sandstone diagenesis.

Nonmarine Sediments

Initially, siliciclastic sediments deposited in nonmarine settings have dominantly fresh pore waters, acid or alkaline pHs, and oxic to slightly anoxic redox conditions. Bicarbonate may be abundant, but sulfate content of pore waters is generally much lower than that of marine sediments. Nonmarine sediments deposited in areas characterized by warm, humid climates and intense chemical weathering in source areas tend to have acidic pore waters with relatively high concentrations of dissolved chemical species (Burley et al., 1985), although concentrations are much lower than those in pore waters of marine sediments. Fine organic matter tends to be preserved in the anoxic sediments in this environment and thus affects diagenetic reactions. Under these conditions, bacterial degradation of organic matter in muddy sediment is similar to that shown in Table 9.3 except that the low concentration of sulfate ions in solution precludes extensive bacterial sulfate reduction. Instead, ferric iron oxides are reduced, releasing hydroxyl ions and increasing pH (Curtis, 1977).

Owing to the lower salinities of pore waters in general, low concentration of SO_4^{2-}, and the general prevalence of oxidizing conditions, early diagenetic reactions in nonmarine sediments differ somewhat from those in marine sediments. For example, pyrite formation is greatly inhibited, although small amounts of pyrite may form in anoxic muds. In nonmarine sediments of warm, humid climates, early diagenetic reactions include precipitation of chlorite (in anoxic pore waters) or kaolinite (in oxygenated pore waters), followed by precipitation of quartz overgrowths and eventually of siderite (anoxic) or calcite (oxic). In acid or neutral pore waters with low concentrations of alkalis and silica, dissolution of feldspars and some other silicate minerals by hydrolysis may also occur (Bjørlykke, 1983). Such dissolution is important, however, only under conditions where pore waters are constantly being renewed by water flow through the sediment. Otherwise, saturation of water in individual pores, with respect to feldspar, is quickly achieved with a very small amount of dissolution. Nonmarine sediments deposited under conditions of low precipitation and relatively little source-area chemical weathering (semiarid to arid environments) tend to have dilute and slightly alkaline pore waters (Burley et al., 1985). In these environments, or other environments of very slow burial where generally oxidizing conditions prevail, organic matter is rapidly oxidized near the sediment–water interface and is effectively removed from the reacting system. Therefore, modification of initial pore-water compositions by processes that involve bacterial decomposition of organic matter is much less important.

Under these conditions of oxic, dilute pore waters, a number of diagenetic reactions occur that can include partial to complete dissolution of unstable heavy minerals, feldspars, and rock fragments, partial replacement of rock fragments or silicate minerals by illitic or montmorillonitic clays, and precipitation of authigenic feldspars, quartz, zeolites, clay minerals (mainly montmorillonite or mixed-layer illite-montmorillonite), iron oxides, and calcite (Walker et al., 1978). Much of the hematite that occurs in redbeds may be generated authigenically under these kinds of conditions. Mechanical infiltration of clays into sandy sediments may also occur during the early stages of diagenesis.

9.4.3   Diagenetic Reactions in the Mesogenetic Zone

General Statement

Diagenetic processes operating in the eogenetic regime may thus result in the dissolution or replacement of many unstable constituents of sandstones as well as precipitation of authigenic pyrite, quartz, feldspars, clay minerals, zeolites, and carbonates. Once these diagenetic reactions are complete and sediments have achieved a state of quasi-equilibrium with their pore waters, little further reaction may occur until sediments are subjected to changed conditions in the mesogenetic regime. These changed conditions may include higher temperatures and geostatic pressures accompanying deeper burial and changes in pore-water composition and pH. Although the boundary between the eogenetic and mesogenetic regimes is difficult to define exactly, mesodiagenesis begins when sediments are buried to depths below the influence of processes directly operating from or closely related to the surface (Choquette and Pray, 1970). Mesodiagenesis continues until the onset of metamorphism or until uplift and reexposure of sediments to surface or near-surface conditions.

Effects of Increased Temperature

Increase in temperature with deeper burial has several important effects on diagenesis. First, it causes an increase in the rate at which chemical reactions occur. An increase in temperature of 10°C can double or perhaps triple the reaction rate (Hunt, 1979, p. 127). Thus, mineral phases that were stable or metastable under eogenetic temperature conditions may become unstable with greater burial depth. Among other things, increasing temperature favors formation of denser, less-hydrous minerals. Increase in temperature also causes an increase in the solubility of most common minerals except the carbonate minerals. Therefore, pore waters are capable of dissolving more silica at higher temperatures. On the other hand, increasing temperature causes a decrease in the solubility of carbonate minerals, which are more likely to precipitate at greater burial depths and higher temperatures unless the pH decreases.

Some cations, such as Mg^{2+} and Fe^{2+}, are strongly hydrated at surface temperatures. Therefore, these cations do not readily enter the carbonate mineral lattice at low temperatures. With increase in burial temperature to 60°C or more, these ions become less hydrated and thus can form iron and magnesium carbonates in pore waters with relatively low Mg/Ca ratios. Thus, carbonate cements precipitated during mesodiagenesis commonly include iron- and magnesium-rich varieties (dolomite, ankerite, siderite).

Effects of Increased Pressure

Increasing geostatic pressure with deeper burial is important because the solubility of minerals increases with increasing stress at grain contacts (following Reicke's principle). In sandstones, this increase in solubility leads to pressure solution of grains at contact points. Therefore, pressure solution can bring about significant increase in silica concentration in pore waters. The silica dissolved by pressure solution may quickly reprecipitate locally on grain surfaces that face open pores, or it may migrate to more-distant areas before reprecipitation occurs. In addition to furnishing a source of silica for cementation, pressure solution reduces porosity and brings about some overall thinning of beds.

Effects of Changed Pore-water Composition

Changes in pore-water composition during mesodiagenesis may strongly influence dissolution and precipitation reactions. Pore-water composition can change as a result of chemical reaction of pore waters with clay minerals or other minerals, large-scale circulation of formation waters, and interaction between organic matter and mineral phases.

Chemical reactions between pore waters and minerals include ionic exchange, breakdown of minerals in contact with pore waters, such as the breakdown of smectite or kaolinite to illite and the alteration of opal to chert; and solution of soluble minerals such as evaporites. The more soluble the minerals, the greater the effect on pore-water composition. These reactions can result in release of silica, sodium, calcium, iron, magnesium, and other ions into solution in pore waters. The reactions that involve dehydration of minerals also add water to the pore-water system.

The diagenetic significance of the smectite-to-illite transition has been of particular interest to researchers, especially since the late 1950s (see review by Freed and Peacor, 1989). The transition begins at a temperature of about 55°C by development of randomly interstratified illite/smectite, with smectite dominating. At higher temperatures, an ordered interstratification occurs in which illite dominates. The dehydration of smectite thus apparently occurs over a range of temperatures from about 55°C to as much as 200°C. Whether or not dehydration occurs more or less continuously over this temperature range or in two or more distinct stages is yet to be firmly established. Large volumes of structured water held in smectite are released during the transition process. The water thus released is of special interest to petroleum geologists seeking an explanation for the primary migration of petroleum out of shale. It also has special significance for the diagenesis of associated sandstones into which it may migrate when expelled from shales. The smectite-to-illite conversion can conceivably occur by two reactions (Boles and Franks, 1979):

$$\text{smectite} + 4.5\ K^+ + 8\ Al^{3+} \rightarrow \text{illite} + Na^+ +$$
$$2\ Ca^{2+} + 2.5\ Fe^{3+} + 2\ Mg^{2+} + 3Si^{4+} + 10\ H_2O \tag{9.3}$$

or

$$1.57\ \text{smectite} + 3.93\ K^+ \rightarrow \text{illite} + 1.57\ Na^+ +$$
$$3.14\ Ca^{2+} + 4.28\ Mg^{2+} + 4.78\ Fe^{3+} + \tag{9.4}$$
$$24.66\ Si^{4+} + 57\ O^{2-} + 11.40\ OH^- = 15.7\ H_2O$$

The smectite-to-illite conversion process requires a source of K, presumably from K-feldspars and micas. A source of Al^{3+} is also required according to reaction 9.3 but not according to reaction 9.4. During the conversion, Si^{4+}, Fe^{2+}, Mg^{2+}, Ca^{2+}, and Na^+ are released from the clay minerals into solution. These ions may migrate into adjacent sandstones, where the silica may be precipitated as quartz overgrowths or may, in shallower sandstones, form authigenic kaolinite. Calcium may combine with available carbonate ions to form calcite or (with Mg) ankerite cements, and Fe and Mg may be taken up by late-stage chlorite or ankerite (Fig. 9.9).

The alteration of kaolinite to illite also releases some water (Bjørlykke, 1983) according to the reaction

$$\text{kaolinite} \rightarrow \text{illite} + 2\ H^+ + 3H_2O \tag{9.5}$$

FIGURE 9.9

Postulated influence on diagenesis of the Eocene Wilcox Sandstone of smectite-to-illite transition in associated shales. Vertical arrows indicate ion transfer between illite/smectite reactions and diagenetic phases in the sandstone. *(From Boles, J. R., and S. G. Franks, 1979, Clay diagenesis in the Wilcox sandstones of southeast Texas: Implications of smectite diagenesis on sandstone cementation: Jour. Sed. Petrology, v. 49. Fig. 9, p. 68, reprinted by permission of Society of Economic Paleontologists and Mineralogists, Tulsa, Okla.)*

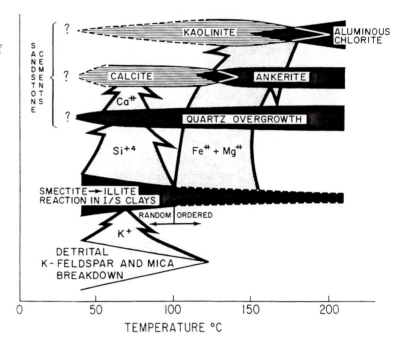

Note that H^+ ions are released during reaction 9.5. Bjorlykke suggests that these may be neutralized by dissolution of feldspars.

Pore-water composition may change also by downward migration of formation waters under a regional hydrodynamic gradient or by upward migration of pore waters expelled by compaction. Downdip migration of fresh waters occurs where meteoric waters enter formations along uplifted basin margins. Theoretically, fresh water can migrate downward into sediment containing pore waters of normal marine salinity to a depth equivalent to about 40 times the height of the water table above sea level (Ghyben-Herzberg principle). Commonly, fresh water does not migrate this deeply, either because it must displace formation waters of greater density or because it encounters higher pore pressures related to restricted basin circulation. Nonetheless, fresh water is known to fill pore spaces to several hundred meters depth in some basins. As meteoric water moves downward, it can cause leaching of carbonates and feldspars or alteration of feldspars to kaolinite. Mesogenetic changes owing to flushing of formations by fresh water are probably not extremely important except in intracratonic basins and some deltaic sequences; however, meteoric waters play a significant diagenetic role during telogenesis. As meteoric waters migrate downward into a basin, the existing formation waters are flushed from pores ahead of the fresh water and are forced to move deeper into the basin.

This process promotes circulation of water through more deeply buried formations. It is well to remember that, owing to the very small amount of water held in each sediment pore, dissolution or precipitation reactions can proceed only if the pore water is constantly renewed by flow of water through the sediment. Thermal convection owing to contrast in geothermal gradients (Leder and Park, 1986) has also been suggested to create basin-wide circulation of fluids. Compaction of sediments expels pore waters, which move mainly upward. Because compaction effects are important primarily at shallow depths, vertical migration of water owing to compaction is confined essentially to the early stages of diagenesis, leading to precipitation of minerals near the sediment–water interface.

Geologists have long recognized that the salinity of subsurface waters increases with increasing depth in many depositional basins. Numerous explanations have been offered for this observed salinity trend, including cation exchange on clay minerals and solution of evaporites such as gypsum and halite. One of the most interesting ideas advanced to explain the trend involves circulating waters and a kind of reverse osmosis process called **salt sieving** (White, 1965). As pore waters migrate through fine sediments containing clay minerals such as smectite and illite, the fine sediments act as semipermeable membranes. They allow uncharged water particles to pass through unimpeded but filter out and retain dissolved ions. The water on the upflow sides of the membranes becomes increasingly saline with time, whereas the water on the downflow sides (the water passing through the membranes) contains relatively few ions. Thus, most shales tend to have lower salinities than associated sandstones. This postulated process may also account for differences in relative abundance of some ions in sandstones and shales under the assumption that the shale membranes are not ideal semipermeable membranes and will allow partial fraction-ation and passage of some ions. The ions that get through the membranes are most likely those with the smallest ionic size and electrical charge. Ions with larger ionic size and charge are preferentially retained in the increasingly saline water behind the mem-brane.

Recently, isotope studies of paragenetic sequences of diagenetic minerals have been used to evaluate changes in pore-water composition during diagenesis. For example, Ayalon and Longstaffe (1988) evaluated changes in diagenetic pore waters in some western Canada sedimentary basins by using oxygen isotope ratios in diagenetic calcite, quartz overgrowths, chlorite, and other clay minerals. On the basis of isotope values, they were able to identify diagenetic minerals that formed during (1) shallow diagenesis in fresh or brackish pore waters, (2) deeper burial after pore waters had changed owing to water/rock interactions, and (3) a late stage of diagenesis after pore water had been replaced by meteoric water during tectonic uplift.

Effects of Organic Matter

Diagenesis in the mesogenetic zone cannot be fully evaluated without considering the role of organic matter. As discussed in Section 9.4.2, bacterial oxidation of organic matter and bacterial sulfate reduction in the presence of organic matter are dominant processes at very shallow burial depths. At greater depths, sulfate reduction gives way to fermentation, which produces methane, bicarbonate ions, and hydrogen ions (Table 9.3). Bacterial fermentation may extend to depths corresponding to a maximum temperature of about 75°–80°C. Commonly, this depth is less than 1 km, but the depth can be greater where the geothermal gradient is low (Tissot and Welte, 1984, p. 202). During burial, organic matter is initially altered mainly by microbial activity, but it subsequently undergoes additional chemical changes owing to processes such as polymerization, polycondensa-tion (formation of humic and fulvic acids by combination of organic molecules), and insolublization (conversion of humic and fulvic acids to insoluble humin). These pro-cesses convert the organic matter into a highly complex geopolymer called **kerogen** (Tissot and Welte, 1984, p. 90). As part of its complex organic structure, kerogen contains carboxylic and phenolic functional groups. These functional groups may undergo thermal and oxidative cracking at temperatures below about 80°C to form soluble organic acids (e.g., carboxylic acids and phenols) in shale source rocks (Carothers and Kharaka,

1979). Presumably, these organic acids are expelled from the source rocks during clay mineral dewatering (smectite/illite transition) and make their way into associated sandstones. As temperatures increase into the range of 120°–200°C with further burial, carboxylic acid anions are destroyed by thermal decarboxylation, as in the reaction

$$CH_3COOH \rightarrow CH_4 + CO_2 \qquad (9.6)$$

or

$$CH_3COOH + H_2O \rightarrow CH_4 + HCO_3^- \qquad (9.7)$$

Thus, at temperatures ranging between about 80°C and 120°C, organic acids predominate in the pore waters of sandstones, whereas between temperatures of about 120°C and 160°C decarboxylation gradually destroys most of the carboxylic acids, releasing CO_2 in the process and increasing the partial pressure of CO_2.

The results of these organic reactions strongly affect the solubility of both carbonates and aluminosilicates. At temperatures below about 80°C, the stability of carbonate minerals in pore waters is a function mainly of pH and the aqueous species Ca^{2+}, H_2CO_3, HCO_3^-, CO_3^{2-}, and CO_2, where $\Sigma CO_2 = H_2CO_3 + HCO_3^- + CO_3^{2-}$. That is, the carbonate system acts as an internal buffer, and carbonate stability is controlled by the partial pressure of CO_2 (P_{CO_2}). Carbonate solubility increases as P_{CO_2} increases; however, P_{CO_2} is commonly low in this temperature range and precipitation of carbonate cements is more common than carbonate dissolution. At temperatures between about 80° and 120°C, the high concentration of carboxylic acid anions acts as an external buffer for the pore-water system (Carothers and Kharaka, 1978; Surdam, Crossey et al., 1989). Although P_{CO_2} is still low, carbonate solubility may be high owing to the high concentration of carboxylic acid. Thus, carbonate dissolution (or lack of carbonate precipitation) will occur. Carboxylic acids also have the effect of significantly enhancing aluminum solubility (Surdam et al., 1984) owing to the ability of these organic substances to form chelates with aluminum (see Boggs et al., 1985, for a discussion of metal chelates). Plagioclase, for example, can be destabilized in the presence of carboxylic acids. Depending upon pH, the aluminum may either be mobilized and removed from solution, creating secondary porosity, or (at higher pH values) plagioclase will alter *in situ* to kaolinite or other alteration phases (MacGowan and Surdam, 1988). Thus, both carbonate minerals and feldspars may undergo dissolution in the temperature range of about 80°–120°C.

With increasing temperature, P_{CO_2} begins to increase as decarboxylation becomes important; however, the system will still be externally buffered and will therefore remain at a constant pH until much of the carboxylic acid is destroyed. Owing to increase in ΣCO_2 (see definition of ΣCO_2 above), the concentration of carbonate species in solution will increase during this process and carbonate minerals will become more stable. Therefore, carbonate precipitation may occur. At some temperature between 120° and 160°C, decarboxylation will have depleted carboxylic acid to the point where the carbonate system becomes internally buffered and the system is no longer confined to a constant pH (Surdam et al., 1989a). When this happens, increasing P_{CO_2} will move the system toward lower pH, thereby initiating another episode of carbonate dissolution.

Thus, as temperature increases progressively during mesodiagenesis, two episodes of carbonate dissolution and one episode of feldspar dissolution may occur in response to organic-inorganic interactions. These and other mesodiagenetic changes are illustrated in

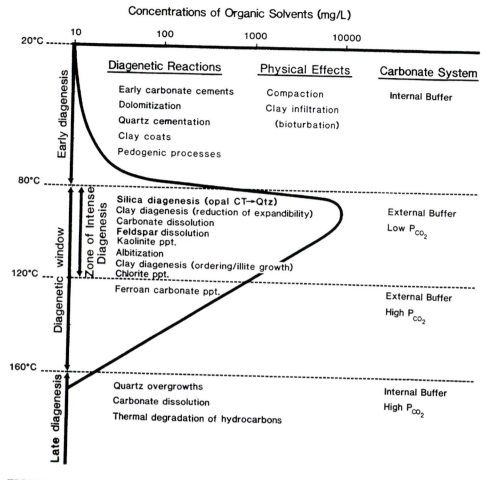

FIGURE 9.10

Mesodiagenetic reactions superimposed on concentration vs. temperature curve for organic acids in formation waters. *(From Surdam, R. C., et al., 1989, Organic-inorganic interactions and sandstone diagenesis: Am. Assoc. Petroleum Geologists Bull., v. 73. Fig. 16, p. 13, reprinted by permission of AAPG, Tulsa, Okla. Modified from Crossey, L. J., 1985, The origin and role of water-soluble organic compounds in clastic diagenetic systems: Ph.D. dissertation, University of Wyoming.)*

Figures 9.10 and 9.11. This postulated sequence of events is based on diagenesis associated with oil-field waters and is somewhat idealized. A more generalized depiction of mesogenetic changes in shales and associated sandstones is shown in Figure 9.12. Note from Figure 9.12 that organic matter as well as the inorganic components undergoes change as mesogenesis progresses toward higher temperatures. At temperatures of about 80°C, thermal or thermocatalytic cracking of kerogen begins to distill oil from the kerogen, and ultimately, at higher temperatures, gas. The presence of oil in sediment pores is important because it may affect diagenetic processes by, e.g., inhibiting cementation.

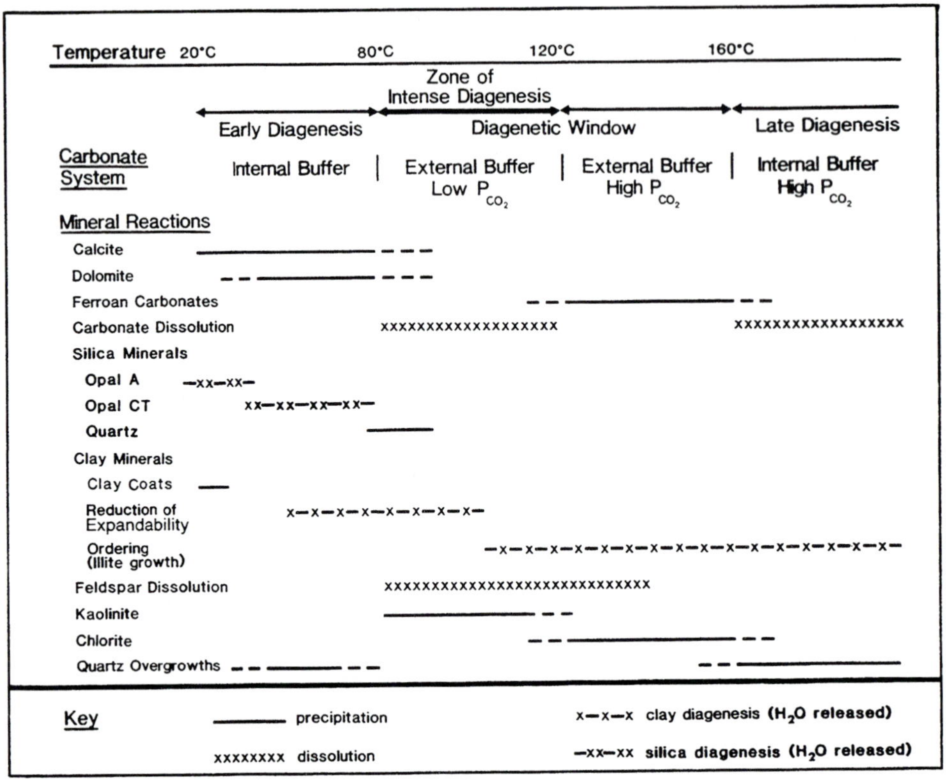

FIGURE 9.11

Paragenetic sequence postulated on the basis of commonly observed diagenetic reactions within the idealized diagenetic zones defined in Figure 9.10. *(From Surdam, R. C., et al., 1989, Organic-inorganic interactions and sandstone diagenesis: Am. Assoc. Petroleum Geologists Bull., v. 73. Fig. 17, p. 14, reprinted by permission of AAPG, Tulsa, Okla. Modified from Crossey, L. J., 1985, The origin and role of water-soluble organic compounds in clastic diagenetic systems: Ph.D. dissertation, University of Wyoming.)*

9.4.4   Diagenetic Reactions in the Telogenetic Zone

If a sediment pile is uplifted and unroofed by erosion, mineral assemblages formed during mesodiagenesis are brought into an environment of lower temperature and pressure. Mesogenetic pore waters may be flushed and replaced by meteoric waters of low salinity that are commonly oxidizing and acidic. Diagenetic mineral assemblages formed during mesogenesis may become unstable under these changed conditions and may undergo dissolution or alteration. Also, original detrital mineral assemblages can undergo further diagenesis. Telogenetic modifications of mineral assemblages may occur at some depth below Earth's surface, but within the zone of meteoric water circulation, or they may occur at the surface by weathering processes. They may include dissolution of previously formed cements or framework grains (creating porosity) or *in situ* alteration of framework grains to clay minerals (occluding porosity). Alternatively, depending upon the nature of pore waters, silica or carbonate cements can be precipitated. For example, carbonate

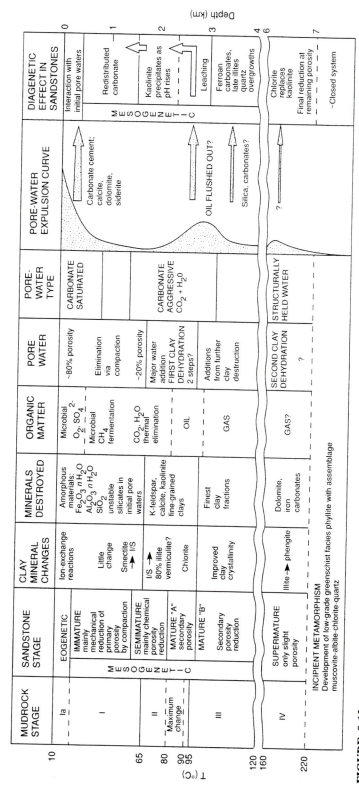

FIGURE 9.12

Depth-related diagenetic changes in shales and the effects of these changes on diagenesis of associated sandstones. (*After Burley, S. D., et al., 1985, Clastic diagenesis, in Brenchley, P. J., and B. P. J. Williams, eds., Sedimentology: Recent developments and applied aspects, reprinted by permission of Blackwell Scientific Publications. As modified from Curtis, C. D., Fig. 17, p. 214, 1983, Geochemical studies on development and destruction of secondary porosity; Geol. Soc. London Spec. Pub. 12, p. 112–125. Sandstone stage is from Schmidt and McDonald, 1979a.*)

cements can be precipitated in the vadose zone by evaporative pumping under climatic conditions that favor evaporation (Surdam, Dunn, et al., 1989b).

Observed diagenetic changes that have been attributed to telogenetic modification include oxidation and destruction of organic matter, oxidation of iron carbonates to form iron oxides and eventually hematite, oxidation of sulfides to form sulfates such as gypsum, dissolution of pyroxenes, amphiboles, and other heavy minerals during weathering, and alteration of detrital feldspars to kaolinite. Detrital feldspars are also reported to alter to illite and authigenic K-feldspars. Other telogenetic changes may include alteration of chlorite to vermiculite and dissolution of iron oxides and carbonates. Telogenesis may, therefore, modify previously formed diagenetic mineral assemblages, or it may produce changes in original detrital mineral assemblages that escaped modification during eogenesis and mesogenesis. Thus, telogenesis may cause additional loss of information about detrital mineral composition and loss of information about eogenetic and mesogenetic authigenic mineral assemblages. Furthermore, it may either enhance or reduce porosity, depending upon the specific conditions of the telogenetic environment.

9.5   Major Chemical Diagenetic Processes

9.5.1   Cementation and Porosity Reduction

Introduction

As discussed in Section 9.4, precipitation reactions can begin during eogenesis, resulting in formation of authigenic clay minerals, K-feldspar overgrowths, quartz overgrowths, and carbonate cements at shallow burial depths. Cementation continues during various stages of mesogenesis and may even occur under some conditions of telogenesis. Cementation plays a major role in reducing porosity in sandstones, and it can also affect compaction of sediments. For example, early-formed cements that fill the pore spaces of a sandstone or shale before significant compaction tend to inhibit further compaction of the sediment. Also, early cementation can restrict subsequent movement of fluids through sediments. Identification of cements and interpretation of the paragenetic sequence of cementation is, therefore, an essential step in diagenetic study. Petroleum geologists concerned with the reservoir properties of sandstones are particularly interested in cementation phenomena because of the adverse effect of cementation on porosity of sandstones. Most of the primary porosity of sandstones is believed to be destroyed at burial depths of about 3 km by a combination of cementation and compaction (Schmidt and McDonald, 1979a).

Minerals that commonly occur as cements in siliciclastic rocks include quartz, chalcedony, opal, K-feldspars and albite, calcite, aragonite, dolomite, siderite, ankerite, hematite, goethite (limonite), pyrite, gypsum, anhydrite, barite, chlorite and other clay minerals, zeolites, tourmaline, and zircon. Of these, carbonate, silica, and clay minerals are by far the most important and abundant cements, although zeolites may be abundant in some volcaniclastic sediments.

Carbonate Cementation

Occurrence of Carbonate Cements.   Carbonate cements occur in all types of sandstones as well as in many shales. They appear to be most typical of quartz arenites, but they are also quite common in many feldspathic arenites (e.g., Morad et al., 1990). Paleozoic and Precambrian lithic arenites (graywackes) generally contain little or no carbonate cement;

however, carbonate cements are abundant in many Cenozoic and Mesozoic graywackes (Pettijohn et al., 1987, p. 448). Older graywackes tend to have pore spaces filled with clay-mineral alteration products rather than with carbonate cements. Carbonate cements may occur as uniform pore fillings over large areas of a rock unit. Alternatively, their distribution can be very patchy. Well-cemented zones can grade to poorly cemented zones over distances of a few meters. Carbonate cement can occur also as lenses or stringers or as concretions. Carbonate concretions are a particularly common feature of some shales. The patchy distribution of carbonate cements may reflect initially patchy carbonate precipitation, or it may be due to subsequent partial removal of more evenly distributed cement owing to dissolution during burial or outcrop exposure.

Calcite is the principal carbonate cement in siliciclastic sediments. Dolomite, ankerite (ferroan dolomite), siderite, and aragonite occur also; however, aragonite exists primarily in very young rocks.

Cementation Process.   As discussed, both precipitation and dissolution of carbonate minerals can take place during diagenesis. Carbonate equilibrium is expressed by the relationships

$$H_2O + CO_2 \longleftrightarrow H_2CO_3 \longleftrightarrow H^+ + HCO_3^- \longleftrightarrow H^+ + CO_3^{2-} \qquad (9.8)$$

$$H_2O + CO_2 + CaCO_3 \longleftrightarrow Ca^{2+} + 2HCO_3^{2-} \qquad (9.9)$$

Precipitation occurs when the solubility of calcium carbonate in pore waters decreases in a carbonate-saturated system. Solubility may decrease as a result of increase in the activity (concentration) of calcium or bicarbonate ions, increase in temperature, or decrease in ΣCO_2. In a pore-water system that is not externally buffered, solubility of calcium carbonate minerals is dependent upon the CO_2 pressure, and pH depends upon ΣCO_2 ($H_2CO_3 + HCO_3^- + CO_3^{2-}$). If ΣCO_2 decreases, the pH increases, causing carbonate precipitation, and vice versa. If the system is externally buffered, such as by organic acids, then the pH does not depend upon ΣCO_2. Under these conditions, carbonate precipitation may occur as a result of increasing ΣCO_2 while the externally buffered system is at a constant pH (see discussion under "Effects of Organic Matter" in Section 9.4.3).

In arid or semiarid nonmarine environments, very early formation of calcite may occur in surface or near-surface sediments owing to caliche precipitation. Early, preburial carbonate cementation has been suggested also for some marine sandstones deposited in seawater supersaturated with carbonate. Putatively, aragonite or calcite may be deposited in a limestone-forming environment simultaneously with deposition of siliciclastic grains as a kind of micrite matrix. Although some carbonate cements likely formed in this fashion, such a penecontemporaneous origin for carbonate cements is difficult to prove conclusively. Most carbonate cement probably precipitated after burial.

Curtis and Coleman (1986) suggest that important volumes of carbonate cement in fine-grained rocks are precipitated in the zone of bacterially driven sulfate reduction and the zone of fermentation, or methanogenesis. During the early stages of eodiagenesis, organic reactions such as oxidation and sulfate reduction (Table 9.3) tend to increase ΣCO_2. Because CO_2 pressure is increased, these reactions do not result in precipitation of carbonates. If the reaction products are expelled into associated sandstones, they cause dissolution rather than precipitation. Thus, the early stages of eogenesis are commonly dominated by dissolution processes (Section 9.4.2). Carbonate precipitation can occur only if during oxidation or sulfate reduction a parallel reaction occurs that consumes H^+

ions or produces proton-consuming species (Surdam, Dunn, et al., 1989). Organic-matter oxidation of Fe or Mn are examples of such oxidation reactions (Curtis, 1987):

$$CH_2O + 2Fe_2O_3 + 3H_2O \rightarrow 4Fe^{2+} + HCO_3^- + 7OH^- \tag{9.10}$$

or

$$CH_2O + 2MnO_2 + H_2O \rightarrow 2Mn^{2+} + HCO_3^- + 3OH^- \tag{9.11}$$

In the sulfate reduction zone, for example, the reduction reaction (Curtis and Coleman, 1986) is

$$15CH_2O + 2Fe_2O_3 + 8SO_4^{2-} \rightarrow 4FeS_2 + 7H_2O + 15HCO_3^- + OH^- \tag{9.12}$$

This reaction leads to precipitation of pyrite and concentration of bicarbonate in pore waters. Because Fe is tied up in pyrite, low-iron calcite or dolomite cement will precipitate from the pore waters, depending upon the concentration of Ca^{2+}, Mg^{2+}, and SO_4^{2-} in the waters: calcite precipitates from marine pore waters rich in dissolved sulfate, and dolomite precipitates from waters depleted in sulfate and enriched in magnesium. Early precipitation of iron-rich carbonates (siderite) appears to occur only in sediments deposited in bicarbonate-rich, freshwater environments with low sulfate content and available Fe^{2+}. It may be possible to identify the origin of carbonate cements formed by these postulated processes by using carbon isotope analysis. Carbonate cements produced through processes involving bacterially driven organic matter oxidation or sulfate reduction should have $\delta^{13}C$ values as low as about $-25‰$ compared to values of about $0‰$ for normal seawater carbonate (Curtis and Coleman, 1986). As an example of such negative $\delta^{13}C$ values, Morad et al. (1990) report $\delta^{13}C$ values ranging from ~ -2 to $-10‰$ in Permo-Triassic sandstones of the Iberian Range, Spain. They attribute these low values to derivation of carbon from oxidation of organic matter in early-formed carbonates.

The reactions described above apply particularly to diagenesis of organic-rich shales. Because many sedimentary sequences consist of interbedded shales and sandstones, it is probably safe to assume that these reactions will also affect diagenesis of associated sandstones. Compaction of shales during early diagenesis may expel shale pore waters into overlying or laterally adjacent sandstones, thereby affecting diagenesis of the sandstones.

Organic-matter reactions may also affect carbonate cementation during mesogenesis. When burial reaches the zone of methanogenesis, microbial fermentation and Fe reduction can combine to cause precipitation of iron-rich carbonates. According to Curtis (1987) this precipitation process is represented by the reaction

$$7CH_2O + 2Fe_2O_3 \rightarrow C_3H_4 + 4FeCO_3 + H_2O \tag{9.13}$$

Iron carbonate (siderite) produced by this process may have a $\delta_{13}C$ signature ranging from -3 to $+10‰$ (Curtis and Coleman, 1986). As discussed in Section 9.4.3, organic acids formed by breakdown of kerogen may predominate in pore waters between burial temperatures of about 80°C and perhaps 160°C. These organic acids act as an external buffer and control the pH of the pore waters. Under these conditions of externally buffered pH, increase in ΣCO_2 owing to decarboxylation reactions increases the activity of carbonate species in the pore waters. Thus, at temperatures of about 120°C–160°C, while decarboxylation continues to destroy the organic acids but while the pH is still externally buffered, precipitation of iron carbonates can occur. Carbonates formed as a result of

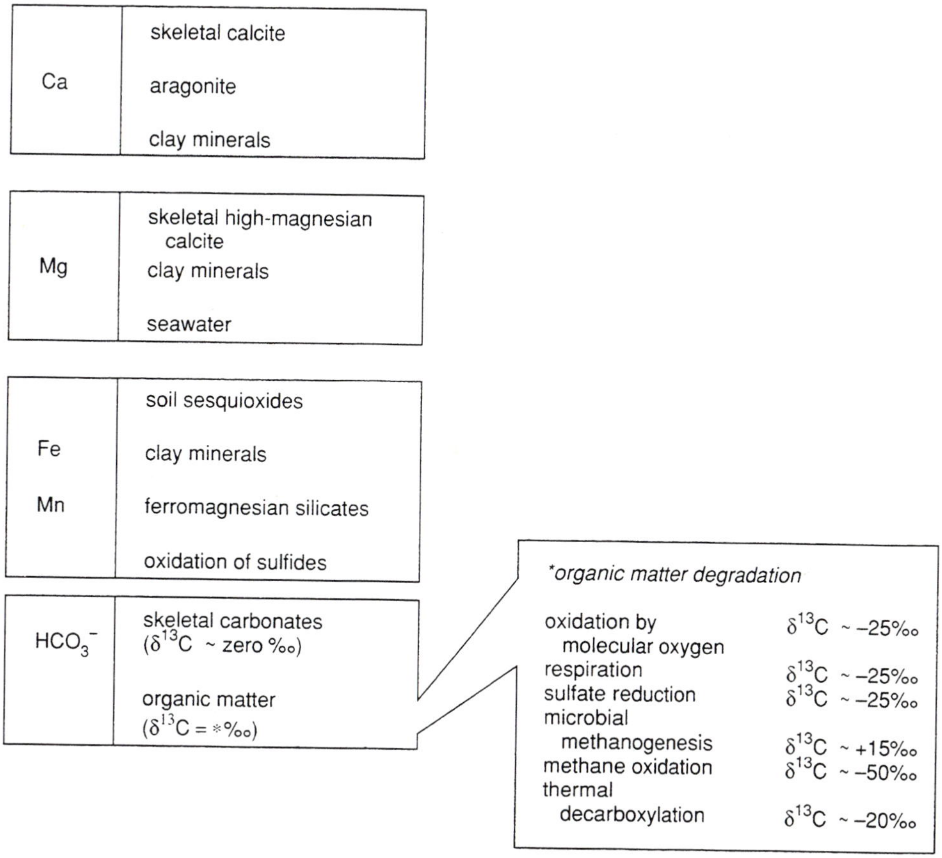

Ca	skeletal calcite
	aragonite
	clay minerals

Mg	skeletal high-magnesian calcite
	clay minerals
	seawater

Fe	soil sesquioxides
Mn	clay minerals
	ferromagnesian silicates
	oxidation of sulfides

| HCO_3^- | skeletal carbonates ($\delta^{13}C$ ~ zero ‰) |
| | organic matter ($\delta^{13}C$ = *‰) |

*organic matter degradation

oxidation by molecular oxygen	$\delta^{13}C$ ~ −25‰
respiration	$\delta^{13}C$ ~ −25‰
sulfate reduction	$\delta^{13}C$ ~ −25‰
microbial methanogenesis	$\delta^{13}C$ ~ +15‰
methane oxidation	$\delta^{13}C$ ~ −50‰
thermal decarboxylation	$\delta^{13}C$ ~ −20‰

FIGURE 9.13

Principle sources of bicarbonate ions, Ca, Mg, Fe, and Mn for diagenetic carbonate precipitation. *(From Curtis, C. D., and M. L. Coleman, 1986, Controls on the precipitation of early diagenetic calcite, dolomite and siderite concretions in complex depositional sequences, in Gautier, D. L., ed., Roles of organic matter in sediment diagenesis; Soc. Econ. Paleontologists and Mineralogists Spec. Pub. 38. Fig. 1, p. 24, reprinted by permission of SEPM, Tulsa, Okla.)*

decarboxylation processes tend to have low $\delta^{13}C$ values (Fig. 9.13; see also Morad et al., 1990). Thus, an episode of siderite or ankerite precipitation may occur within this temperature range before decarboxylation destroys all of the organic acids and the system again becomes internally buffered (Fig. 9.10).

Source of Carbonate.   In the cementation processes described above, bicarbonate ions are supplied by organic reactions of some type. To form calcite cements, a source of calcium must also be available. Dolomite precipitation further requires a source of magnesium, and ankerite and siderite precipitation requires a source of iron. The content of these cations in original seawater is inadequate to supply all the Ca, Mg, and Fe needed for cementation during diagenesis. As many workers have pointed out, precipitation of carbonate cement from static pore waters can supply only a minute amount of cement. A steady contribution of both cations and anions must be renewed in pore water for cementation to proceed and fill a pore space. Berner (1980, p. 124) suggests that 300,000 pore volumes of water are needed to fill one volume of pore space completely. In a sandstone that contains fossils, dissolution of skeletal grains composed of aragonite, calcite, or high-magnesian calcite can furnish an important source of bicarbonate, calcium, and magnesium. Bicarbonate supplied by this source has a higher $\delta^{13}C$ carbon isotope signature ($\sim 0\text{‰}$) than that supplied by sulfate reduction and a lower carbon isotope value than that furnished by methanogenesis (Fig. 9.13). Therefore, mixing within pore waters of bicarbonate derived from skeletal dissolution and that derived by degradation of organic matter can confuse the isotope signature.

Bicarbonate ions and cations can be supplied also by dissolution of limestones or dolomites in stratigraphic sections containing interbedded siliciclastic and carbonate rocks. In this regard, export of calcium carbonate from limestones undergoing pressure solution may be an important process (Blatt, 1979). Additional Ca, Mg, and Fe are supplied by reactions with silicate minerals such as conversion of smectite to illite and dissolution of calcium feldspars and ferromagnesian minerals (Fig. 9.13). Under geologic conditions where downward circulating groundwaters move through mafic volcanic rocks, significant amounts of Ca, Mg, and Fe can be supplied by alteration of these rocks. Finally, carbonate cements precipitated in one part of a sediment pile may be dissolved and redistributed by advective transfer to other parts of the pile.

Carbonate Cement Textures.   Carbonate cements are commonly easy to recognize with a petrographic microscope. They may fill pore spaces among detrital grains with a mosaic of fine crystals, or a pore may be filled with a single large crystal (Fig. 5.6). In fact, a single crystal may grow large enough to surround several detrital grains to produce a poikilotopic texture (Friedman, 1965), as shown in Figure 5.7. Locally, where carbonate cement is abundant, displacive crystallization of the cement (Folk, 1965) may occur, forcing the detrital grains apart into a loosely packed fabric. Although some skepticism has been expressed about the occurrence of displacive calcite, displacive calcite in caliche profiles in sandstones has been reported recently by Saigal and Walton (1988) and Braithwaite (1989). Partial replacement of silicate grains by carbonate cement can further enlarge the area of the cement. Where crystallization occurs within pores containing interstitial clays, the clays may be engulfed by and disbursed within the cement (Dapples, 1979). Dolomite and iron carbonates may occur as distinct rhombs, allowing them to be distinguished from calcite. If they do not exhibit rhombic shape, it is very difficult to differentiate optically among the carbonate cements unless thin sections are stained for carbonates.

Silica Cementation

Occurrence of Silica Cements.   Silica cements are common in both sandstones and shales. In sandstones, they typically occur in arenites with little or no matrix, but silica cement can occur in rocks containing matrix. Dutton and Diggs (1990) show that the volume of quartz cement in sandstones tends to decrease exponentially with increasing matrix content of the sandstones (Fig. 9.14). Silica cement most commonly takes the form of optically continuous overgrowths on quartz grains (see Section 4.2.2 and Fig. 5.3). Such overgrowths are especially typical of silica-cemented quartz arenites, but they can be present also in quartz-rich feldspathic and lithic arenites. They rarely occur on the quartz grains of wackes such as turbidite sandstones. Quartz overgrowths appear to be uncommon in modern sands (Pettijohn et al., 1987, p. 452), although Müller (1967)

FIGURE 9.14

Exponential decrease in quartz cement as a function of increasing matrix content in Travis Peak (Cretaceous) sandstones, east Texas. *(From Dutton, S. P., and T. N. Diggs, 1990. History of quartz cementation in lower Cretaceous Travis Peak Formation, East Texas: Jour. Sed. Petrology, v. 60. Fig. 5, p. 195, reprinted by permission of Society of Economic Paleontologists and Mineralogists, Tulsa, Okla.)*

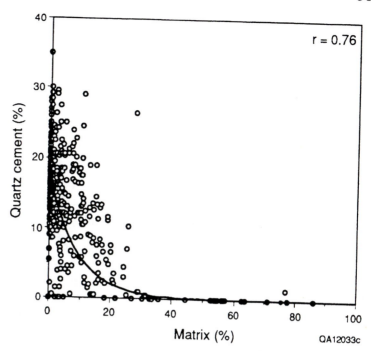

reports authigenic quartz in Recent fine-grained sediments of the Gulf of Naples. Also, quartz overgrowths can be precipitated experimentally in the laboratory (e.g., McKenzie and Gees, 1971). Opal and chalcedony or microquartz (chert) are less-common silica cements. Opal occurs primarily in Tertiary and younger sediments. It is most common as a cement in volcaniclastic sequences but has been reported also from feldspathic sandstones (e.g., the Ogallala Formation of Kansas, Franks and Swineford, 1959). Chalcedony or microquartz cement generally does not occur in quartz arenites but may be common in sandstones with subwacke texture (Dapples, 1979). According to Dapples, microquartz (chert) can occur as either (a) an epitaxial cement on quartz grains that show interpenetration boundaries or (b) as intergrowths in small masses of interstitial or matrix clay. As the sandstone framework approaches that of a wacke, the latter type of cement is likely to be the kind of quartz cement present. Some silica cement may consist of opal-CT (cristobalite-tridymite), which is an intermediate diagenetic stage between opal-A and quartz (chert).

Silica Solubility.   Precipitation of silica cements requires that the pore waters of sediments be supersaturated with respect to quartz or opal. The precipitation of enough cement to fill pore spaces further requires that large volumes of water circulate through the sediments. The excess silica in the pore water of a single pore can precipitate only a minute amount of silica. Once precipitation has occurred, no further cementation can take place until the water in the pore is replaced by new water supersaturated with silica. Thus, the process of filling a pore with cement demands that silica-supersaturated water circulate through the pore many times, a process that takes a very long period of time.

The solubility of silica minerals is a function of the pH, hydrostatic pressure, temperature, and crystallinity of the mineral. At pH values above about 9.0–9.5, the solubility of silica increases sharply; however, at lower values, pH has relatively little effect on silica solubility. Most pore waters are buffered at a pH of around 8.0. Values greater than 9.0 occur only in some groundwaters associated with altering volcanic sequences, alkaline lakes in arid regions, and sabkha environments. Therefore, pH probably does not exert an important control on silica solubility under most diagenetic conditions. Solubility increases with increasing hydrostatic pressure. In seawater, this increase is about 35 percent at 1 kb (Willey, 1974, p. 242).

The equilibrium solubility of quartz at 25°C has been determined by several investigators to be about 6 ppm (e.g., Fournier, 1983). Amorphous silica (opal-A) has a much higher solubility of about 120–140 ppm (Siever, 1957), and opal-CT has a solubility of about 70 ppm (Fournier and Rowe, 1962). Silica solubility increases markedly with increasing temperature, as shown in Figure 9.15. For example, the solubility of quartz at 150°C is more than 20 times that at 25°C. Thus, temperature appears to exert the major control on the solubility of the SiO_2 minerals. Given these characteristics of silica behavior, we would logically expect that quartz, with its lower solubility, will precipitate in preference to opal unless concentrations of silica in pore waters exceed about 120 ppm. On the other hand, experimental work shows that quartz precipitates very slowly; therefore, pore waters supersaturated with respect to quartz can persist for long periods of time before precipitation occurs. Pore waters in shallow sediments buried to a few hundred meters depth have been reported to have concentrations of dissolved SiO_2 as high as 80 ppm, well above the equilibrium solubility of quartz (e.g., Gieskes et al., 1983).

Timing of Quartz Cementation.   Factors of interest in evaluating silica precipitation during diagenesis include the timing of cementation, mechanisms for precipitating silica, and sources of silica. Discussion in preceding sections suggests that most silica cementation occurs during eodiagenesis and early mesodiagenesis. Furthermore, empirical evidence of early cementation has been reported by numerous workers (see review by Blatt,

FIGURE 9.15

Solubility of silica as a function of temperature. The single points showing solubility values for the maximum and mean of silicate rocks and for quartz were obtained by crushing specimens in distilled water. The values for natural waters are from groundwaters in silicate rocks. *(From Dapples, E. C., 1979, Silica as an agent in diagenesis, in Larsen, G., and G. V. Chilingar, eds., Diagenesis in sediments and sedimentary rocks. Fig. 3-1, p. 104, reprinted by permission of Elsevier Science Publishers, Amsterdam.)*

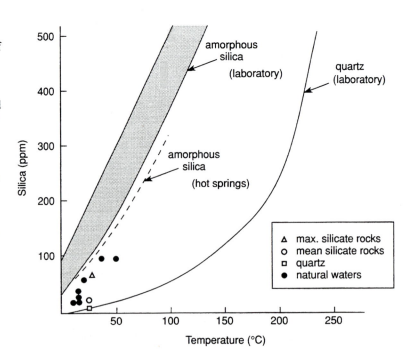

1979). On the basis of time constraints arising from presumed typical silica concentrations in pore waters and average flow rates of fluids through sediments undergoing diagenesis, Blatt (1979) suggests that most silica cementation must occur at very shallow depths. Recently, Dutton and Diggs (1990) used oxygen isotope data to interpret that quartz cements were precipitated in the Cretaceous Travis Peak Formation of Texas (USA) at burial depths of 1 to 1.5 km and temperatures of 55°–75°C.

Although most silica cementation may occur at moderately shallow burial depths and low temperatures, some workers report formation of quartz overgrowths at much greater depths and higher temperatures. Surdam, Crossey, et al. (1989) suggest, for example, that quartz overgrowths can be precipitated at temperatures ranging to about 80°C during early stages of mesogenesis and also at temperatures between 160° and 200°C during late mesogenesis (see Fig. 9.11). On the basis of fluid inclusion studies, Girard et al. (1989) report that quartz overgrowths found in Lower Cretaceous arkoses of the Angora margin formed at temperatures ranging between 128° and 158°C. Also, Walderhaug (1990) interpreted the temperature of formation of quartz overgrowths from subsurface sandstones from offshore Norway to lie between 90° and 118°C. Boles and Franks (1979) imply that precipitation of quartz overgrowths occurs over the temperature range of about 50°C–200°C (see Fig. 9.9).

With deeper burial and progressively increasing temperature, concomitant increase in solubility of silica requires that pore waters contain much higher saturations of silica than that in shallow pore waters before quartz precipitation can occur. Presumably, precipitation of quartz at temperatures of 80°C would require that pore-water silica saturation exceed about 30 ppm, and precipitation at 160°C would require a saturation in excess of 140 ppm (Fig. 9.16). Silica concentrations in deep subsurface waters are reported by White et al. (1963) to be as high as 100 ppm; however, values as high as 140 ppm are reported in some oil-field waters (Fig. 9.16). Land (1984) suggests that most of

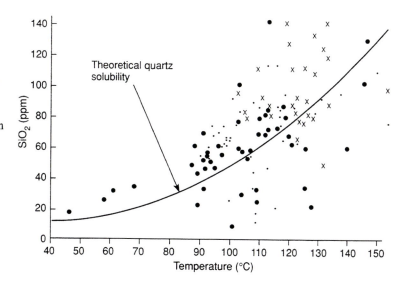

FIGURE 9.16
Concentration of SiO_2 in pore waters of the Frio Formation, Texas Gulf coast, plotted as a function of temperature. Note that the SiO_2 concentration of formation waters at some temperatures is as much as 50 ppm above the theoretical quartz solubility curve; X symbols are waters having *in situ* density less than 0.990; heavy dots are waters having density greater than 1.010; small dots are waters of intermediate density. (*After Land, L. S., 1984, Frio Sandstone diagenesis, Texas Gulf Coast: a regional isotope study,* in McDonald, D. A., and R. C. Surdam, eds., Clastic diagenesis: Am. Assoc. Petroleum Geologists Mem. 37. Fig. 6, p. 52, reprinted by permission of AAPG, Tulsa, Okla.)

the pore waters in the Frio Formation shown in Figure 9.16 are oversaturated with quartz by about 50 ppm, suggesting that formation waters, at least in this formation, must be oversaturated with silica by about 50 ppm before precipitation of quartz occurs.

The timing of cementation in sandstones is commonly depicted in burial-history curves, which show the interpreted time of cementation as a function of burial depth and paleotemperature. Figure 9.17 is an example of this kind of plot. Figure 9.17, constructed for the Cretaceous Travis Peak Formation in east Texas (USA), shows that the major phase of quartz cementation in this formation occurred at a burial depth between about 0.75 and 1.5 km and at temperatures of about 50°–75°C. Minor quartz cementation also took place at depths ranging to about 2 km. Several kinds of data are used to construct burial-depth curves of this type. Burial depths are estimated by using present-day thicknesses of stratigraphic units, which are corrected for compaction. The thicknesses of eroded strata are estimated from vitrinite reflectance measurements. (For an explanation of vitrinite reflectance, see Boggs, 1987, p. 286.) Paleotemperatures are estimated from oxygen isotope data, fluid-inclusion data, and/or vitrinite reflectance data, or from heat-flow extension models. Ages are determined by radiocarbon dating methods, for example, K/Ar ages of diagenetic illite (e.g., Lee et al., 1989).

To provide an adequate explanation for precipitation of silica overgrowths at high burial temperatures requires that we identify both a viable source for the silica and a mechanism for precipitation. These topics are discussed below.

Mechanisms of Silica Precipitation.   Mechanisms that have been suggested for quartz precipitation in sandstones include (1) cooling a hot, saturated solution, (2) lowering pore pressure, (3) mixing silica-saturated fluid with more-saline formation waters at a given

FIGURE 9.17
Burial-history curves for the top and base of the Travis Peak Formation (Cretaceous), east Texas. Interpreted times at which quartz precipitation and other major diagenetic events occurred are shown as a function of burial depth and paleotemperature. *(From Dutton, S. P., and T. N. Diggs, 1990, History of quartz cementation in lower Cretaceous Travis Peak Formation, East Texas: Jour. Sed. Petrology, v. 60. Fig. 3, p. 194, reprinted by permission of Society of Economic Paleontologists and Mineralogists, Tulsa, Okla.)*

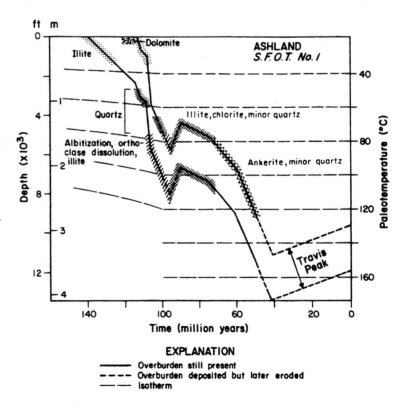

temperature and pressure, (4) lowering the pH of the saturated solution, (5) evaporating an undersaturated solution, and (6) semimembrane osmosis through a clay-rich shale layer (Leder and Park, 1986). As mentioned, pH probably has little effect on precipitation under normal diagenetic conditions; however, any mechanism that increases silica concentration in pore waters above the equilibrium concentration can presumably cause precipitation.

Many workers have suggested that cooling of saturated solutions may be a major mechanism of quartz precipitation. For example, Leder and Park (1986) propose a model for quartz cementation that requires basinwide circulation of silica-saturated fluids driven by thermal convection or hydrostatic head. Silica is dissolved in downward-migrating waters as they encounter progressively higher temperatures. As the continuously circulating fluids that dissolved silica on their downward migration move upward toward the basin edges, they cool and precipitate quartz in pore spaces. Diffusion from deeper regions of warmer temperature to shallower formations at cooler temperature can also transfer silica upward, although this process is probably less effective than convection. Cooling of pore waters offers a reasonable explanation for late-stage quartz precipitation at moderate or shallow burial depths; however, it appears difficult to apply this mechanism to precipitation of quartz at temperatures of 160°C or more. Presumably, some local source would have to be present that could supply enough silica to drive pore-water saturations above the critical level needed for quartz precipitation.

Silica Sources.   The source of silica for quartz cementation during diagenesis has been the subject of considerable discussion and surmise. All of the following have been mentioned as possible mechanisms to supply the needed silica: (1) dissolution of silica by meteoric waters during subaerial weathering, (2) dissolution of the opaline skeletons of organisms such as diatoms and radiolarians, (3) dissolution of silica from quartz grains by circulating pore waters flowing over quartz grains in sandstones, (4) pressure solution of quartz grains, and (5) release of silica by mineral reactions, including clay-mineral reactions, dissolution of feldspars, and alteration of volcanic fragments, glass, and mafic minerals. These mechanisms are summarized in Table 9.4 along with some comments about their probable relative importance.

At the low temperatures typical of shallow burial, precipitation of quartz could occur in pore waters containing as little as about 13 ppm dissolved silica (average concentrations in nonmarine surface waters). Thus, at very shallow burial depths in nonmarine environments or where telogenetic uplift brings sediments near the surface, adequate silica for quartz precipitation may be present in meteoric waters, although large volumes of such dilute water would have to pass through the sediment pores to precipitate a significant amount of quartz cement. On the other hand, marine waters contain an average silica concentration of only about 1 ppm, well below the solubility of quartz. Therefore, silica would have to be supplied from a source other than the original seawater in order for quartz cementation to take place during early burial in marine environments.

Pressure solution of quartz grains at points of contact and various mineral reactions that release silica appear to be the most important sources of silica for quartz cementation at intermediate and deeper depths. There is disagreement among workers, however, about the relative importance of pressure solution and mineral reactions in supplying silica. It is difficult to make quantitative mass-balance calculations between dissolution and precipitation reactions because we commonly do not know if systems undergoing diagenesis are closed, isochemical systems or if they are open systems. Is much of the silica released

TABLE 9.4

Possible sources of silica cement

Source mechanism	Importance for quartz cementation
Dissolution of silica in meteoric waters owing to weathering of silicates on land	Provides a source of silica for precipitation of quartz overgrowths at shallow depths, but low average concentration ($\sim$13 ppm) requires circulation of large volumes of water to fill pores with cement
Dissolution of opaline tests of silica-secreting organisms such as radiolarians and diatoms	May furnish a minor source of silica in marine sands; however, most siliceous skeletons occur in fine-grained, deep-sea sediments where siliceous oozes are converted from opal-A to opal-CT to quartz with burial, forming cherty layers; not an important source of cement in quartz arenites
Dissolution of silica from quartz grains by circulating pore fluids	Most effective at high temperatures; quantitative importance unknown, but probably small
Pressure solution of quartz grains at points of contact at moderate to deep burial depths	Regarded to be a major source of silica; it is commonly assumed that dissolved silica is reprecipitated locally in adjacent pores; however, a mass balance calculation cannot be made for many sandstones; in some sandstones, volume loss by pressure solution is matched by cement precipitation; in others, less cement is present than would be predicted from pressure solution, and in still others, more cement is present than can be accounted for by pressure solution
Liberation of silica by mineral reactions: K-feldspar dissolution with kaolinite formation $2KAlSi_3O_8 + 2H^+ + 9H_2O \rightarrow$ kaolinite $+ 49H_4iO_4$ (aq) $+ 2K^+$	Conversion of all K-feldspar in a sandstone containing 10% K-spar would produce enough silica cement to fill 4% of original pore space (Leder and Park, 1986)
Dissolution of plagioclase by complexing with organic acids (Surdam, Dunn, et al., 1989) $4H^+ + 2H_2C_3H_2O_4 + CaAl_2Si_2O_8 \rightarrow$ $2AlC_3H_2O_4 + 2H_4S{:}O_4 + Ca^{2+}$	Probably a minor source of silica for quartz overgrowths—silica may recombine with other elements to form kaolinite
Conversion of smectite to illite Clay transformation (Hower et al., 1976) $4.5K^+ + 8Al^{3+} +$ smectite $\rightarrow$ illite $+ 3\ Si^{4+} + H_2O +$ cations Smectite cannibalization (Boles and Franks, 1979) $2.5K^+ +$ smectite $\rightarrow 0.64$ illite $+ 15.7\ Si^{4+}$ $+ H_2O +$ cations Conversion of smectite to kaolinite (Siever, 1962) Smectite $+ 3.2H^+ \rightarrow$ kaolinite $+ 5.6\ SiO_2$ (aq) $+ H_2O +$ cations	Clay mineral reactions occur at temperatures of $\sim$50°C to 200°C; commonly regarded to be a major source of silica in shales; assuming that silica can be exported from shales to sandstones, clay mineral reactions may account for much of the silica cement in some sandstones (Boles and Franks, 1979; Leder and Park, 1986)
Alteration of volcanic rock fragments, glass, plagioclase, and mafic minerals to zeolites and authigenic clays (Surdam and Boles, 1979)	An important source of silica in volcaniclastic sequences, but not in quartz arenites; probable source of opal in volcanogenic sediments

during clay-mineral reactions eventually expelled into associated sandstones? Is all of the silica produced by pressure solution precipitated locally in unfilled pores, or is much of it exported to other parts of a sedimentary pile?

The above questions do not have easy answers, and it is commonly difficult to specifically identify the source of silica that forms cements in a particular formation. A possible technique that makes use of oxygen isotope analysis is receiving increasing attention for identifying the source of silica cements. The oxygen isotope composition of quartz overgrowths, for example, is a function of the kind of waters in which the overgrowths form and the temperature of the water. Meteoric waters commonly have negative $\delta^{18}O$ values, but the values can range from about $-55\%_o$ to $+0\%_o$ Ocean water has $\delta^{18}O$ values of about $0\%_o$. The oxygen isotope composition of formation waters is complex, and interpretation of the isotopes is poorly understood. Observations indicate that formation-water samples collected at the highest temperatures (and greatest depths), and with the highest salinities, are generally enriched in $\delta^{18}O$ relative to cooler and less-saline waters (Longstaffe, 1989). Thus, with increasing depth, the $\delta^{18}O$ values in formation waters of a given basin may range from negative values approaching those of meteoric waters to positive values of $+10$ % or more. Suggested causes of isotopic variations in formation waters include mixing of meteoric waters and connate evaporite brines, isotopic exchange with minerals such as calcite, membrane filtration (salt sieving), and clay dehydration (Longstaffe, 1989).

Silica cements that precipitate out of formation waters will thus acquire oxygen isotope signatures that depend first upon the isotope composition of the water. Additional oxygen isotope fractionation takes place, however, as a function of temperature. That is, $\delta^{18}O$ values in minerals decrease with increasing temperature of formation. This principle constitutes the basis for oxygen isotope geothermometry, using coexisting mineral pairs such as quartz and illite (Eslinger et al., 1979). Although problems exist in use of oxygen isotope geothermometry, the technique has been successfully used in shales to estimate the temperature of formation of authigenic minerals (e.g., Eslinger and Savin, 1973; Eslinger and Yeh, 1986). It is apparently less useful in sandstones, although successful applications have been reported (Milliken et al., 1981; Land, 1984; Longstaffe, 1989; Dutton and Diggs, 1990). Quartz geothermometry places constraints on the temperature and depth of formation of silica cements but does not identify the actual source of cement. Nonetheless, we may be able to tell from the isotope data if precipitation occurred at low temperatures within the range of cementation by meteoric waters, at somewhat higher temperatures but below the temperature of smectite-to-illite conversion, or at higher temperatures within the range of most mineral reactions. A major problem in isotope studies of quartz overgrowths is isolation of the overgrowths for analysis. Techniques for overgrowth isolation are discussed by Lee and Savin (1985). As mentioned, study of fluid inclusions in quartz overgrowths is another technique for determining the temperature of formation of the overgrowths (Roedder, 1979; Walderhaug, 1990).

Opal and Microquartz (Chert) Cement.   Because of its high solubility (120–140 ppm SiO_2 for opal-A), opal is a comparatively rare cement in sandstones except in some volcaniclastic sequences where the opal is derived from alteration of volcanic glass. Microquartz (chert) cements are also much less common than quartz overgrowths, but they do occur in lithic arenites and even in some quartz arenites. Microquartz (chert) cement can occur both as a syntaxial rim on quartz grains and as a mosaic of chert deposited within pore space. Although chert cement can conceivably precipitate directly from pore waters, more likely it precipitates initially as opal-A or opal-CT. Opal-CT has a lower solubility ($\sim$70 ppm) than opal-A, and thus in silica-enriched pore waters opal-CT may form in preference to quartz because it apparently precipitates more rapidly than quartz. Opal-CT subsequently transforms to microcrystalline quartz (chert). In highly enriched pore waters, opal-A may be the initial cement, which subsequently inverts and recrystallizes first to opal-CT and then to microquartz.

In volcanogenic sequences, the silica in opal and microquartz is derived locally by alteration of volcanic glass and other volcanic products, as mentioned. Microquartz (chert) cements in quartz arenites, which are rarely associated with volcaniclastic rocks, must have a different source. Pettijohn et al. (1987, p. 455) reason, however, that the source must nevertheless be nearby. Owing to the relative rapidity of precipitation of opaline silica phases, the silica would not remain in pore waters over long distances of transport. They suggest that the source may be soil waters in lateritic terrains, which are known to precipitate opal in the lower parts of soil zones. Precipitation of opal/microquartz cements may thus occur in sandstones at shallow depths below lateritic soils, i.e., below weathering surfaces that ultimately become unconformities.

Cementation by Other Minerals

Sulfate Cements: Gypsum, Anhydrite, Barite.   Gypsum or anhydrite occur as cements in some sandstones, especially those deposited in evaporitic environments. Evaporated seawater migrating downward under sabkha flats can precipitate gypsum cements in sands at shallow burial depths. At the higher temperatures encountered with burial to depths below about 700–1000 m, gypsum converts to anhydrite, with concomitant loss of water and decrease in volume. Anhydrite remains the stable phase during further burial unless the sediments are again uplifted to shallow depths where temperatures and salinities are lower. Late-stage anhydrite cement may precipitate directly during sediment burial if an adequate source of SO_4^{2-} ions are available in subsurface brines. Some subsurface brines are known to be rich in sulfate (Cheboratev, 1955). Ca^{2+} may be supplied for late-stage anhydrite precipitation by albitization of feldspars or dolomitization of limestones. Barite is a relatively uncommon mineral in siliciclastic rocks but occurs both as a cement and as concretions. The source of barium for barite cements may be associated volcanic rocks or deep-seated, barium-containing basement rocks.

Iron Oxides: Hematite and Goethite (Limonite).   These cements are common in some sandstones. The iron oxides act as a pigment that may coat detrital grains, form halos around ferromagnesian minerals, pervade and stain clay matrix, or be disseminated within calcite cement. Goethite occurs mainly in young sediments. With loss of water, goethite ages to hematite, which is the dominant iron oxide in most sediments. The source of the hematite that acts as a pigment in sandstones and other redbeds is controversial. Some iron oxides may be transported from subaerial weathering sites to depositional basins along with siliciclastic detritus. Walker (1967, 1974) maintains, however, that much if not most of the hematite in redbeds is authigenic. He cites convincing evidence, such as halos around iron-silicate grains and the irregular nature of the hematite coatings on feldspar and quartz grains, that suggests that the hematite is derived *in situ* by alteration of iron-bearing detrital grains such as hematite and biotite. Hematite cement appears to be particularly characteristic of sediments deposited in oxic alluvial or deltaic environments; however, diagenetic hematite can occur in sandstone deposited in any environment if ferromagnesian minerals and oxic pore waters are present. Dapples (1979) suggests that diagenetic reactions involving oxidation and reduction occur particularly during early burial (his redoxomorphic stage). Iron oxides may also form during telogenesis when sediments are brought into the presence of oxidizing meteoric waters.

Authigenic Feldspar Cements.   Feldspar cements occur primarily as overgrowths on detrital feldspar grains, either on K-feldspars or on albite grains. Authigenic feldspars are most common in feldspathic and volcaniclastic sandstones but occur also in some quartz arenites and lithic arenites. Overgrowths may form both during eogenesis and some stages of mesogenesis. The geochemical conditions that favor precipitation of feldspar over-growths include an adequate concentration of dissolved silica and Na^+ or K^+ in pore waters and moderately high temperatures (Kastner and Siever, 1979). Na^+ and K^+ may be supplied by alteration of kaolinite, smectite, muscovite, or zeolite minerals such as analcime and phillipsite, as well as by dissolution of volcanic rock fragments and potassium- and sodium-rich feldspars. See Sections 9.5.3, 9.5.5, and 9.5.6 for further discussion of these reactions. Many authigenic feldspars are small and may be difficult to identify by routine petrographic analysis. Table 4.3 lists analytical methods and criteria for distinguishing authigenic feldspars from detrital feldspars. See also Krainer and Spötl (1989). Authigenic K-feldspars have also been reported to form as pseudomorphs by precipitation into cavities created by dissolution of detrital K-feldspars and plagioclase (Morad et al., 1989).

Other Cements.   Other minerals that may act as cements in siliciclastic rocks include clay minerals, zeolites, tourmaline, zircon, and pyrite. Clay minerals (Section 9.5.5) and zeolites (Section 9.5.6) are important cements in some sandstones, particularly lithic arenites and volcaniclastic sandstones. Tourmaline, zircon, and pyrite are commonly of minor importance.

9.5.2 Dissolution and Porosity Enhancement

Introduction

Dissolution of certain mineral phases is mentioned in various parts of the preceding discussion of cementation. Dissolution of minerals, rock fragments, and skeletal materials during weathering and diagenesis accounts for much of the dissolved silica and carbonate required for cementation. Further, dissolution processes are responsible for producing most of the secondary porosity in sandstones, particularly at depths below 2–3 km, by removal of cements and framework grains. The development of secondary porosity in sandstones is a subject of paramount importance to petroleum geologists evaluating the oil potential of reservoir rocks, and extensive research has been done since the 1960s to evaluate the porosity effects of dissolution. As a result of dissolution processes, low-stability minerals such as calcium plagioclase and mafic minerals are destroyed, leading to mineral assemblages biased toward more-stable minerals. Interpretation of such biased assemblages is a major problem and challenge in provenance analysis, as discussed in Chapter 8. Therefore, an understanding of dissolution and other diagenetic pro-cesses is vital to provenance interpretation. Dissolution also plays a part in other diagenetic pro-cesses such as replacement and recrystallization. In this section, we look further at the processes of dissolution and their effect in particular on mineral composition and porosity.

The process of dissolution consists of removal in solution of all or part of previously existing minerals, leaving void spaces in the rocks. Solution of some minerals such as $CaCO_3$ and SiO_2 occurs by simple **congruent dissolution,** which involves complete, homogeneous dissolution of parts of a mineral such as quartz or calcite. This process leaves the remaining part unchanged in composition. That is, the material dissolved from the mineral has the same composition as the mineral. **Incongruent dissolution** is a selective solution process that causes the undissolved part of a mineral to be altered in composition. In other words, the more-soluble parts of a mineral dissolve in greater amounts. For example, dolomite is thought to dissolve congruently at surface temperatures (Equation 9.14) but dissolves incongruently, at least in part, at higher temperatures (Equation 9.15) (Krauskopf, 1979, p. 72):

$$CaMg(CO_3)_2 \rightarrow Ca^{2+} + Mg^{2+} + 2CO_3^{2-} \qquad (9.14)$$

$$CaMg(CO_3)_2 \rightarrow CaCO_3 + Mg^{2+} + CO_3^{2-} \qquad (9.15)$$

Note in Equation 9.15 that the original dolomite is eventually converted to calcite by selective dissolution and removal of Mg^{2+}. Other examples of incongruent dissolution include the alteration of K-feldspar to kaolinite (see Table 9.4) and the alteration of volcanic glass to smectite. Incongruent dissolution reactions are extremely important in the formation of authigenic clays. An excellent, short discussion of partial or incongruent dissolution is given in Pettijohn et al. (1987, p. 439–442).

Dissolution may take place in pure water, water with dissolved carbon dioxide, water of variable salinity containing dissolved cations and anions, and water containing various kinds of organic acids. Dissolution occurs when the solubility of a particular mineral is exceeded under a given set of Eh, pH, temperature, and salinity conditions. For example, the stability of most silicate minerals decreases with increasing temperature, whereas the stability of carbonate minerals increases with increasing temperature. The stability of some minerals, such as the carbonate minerals, is further controlled by pH. Calcite and aragonite, for example, are stable at pH values above about 7.8 but undergo solution under acidic conditions. Calcite and aragonite are also more soluble in waters of high salinity than in fresh water. Eh may exert an important control on the stability of some minerals such as hematite and magnetite. The presence of organic acids in pore waters may promote dissolution of minerals such as feldspars and mafic minerals by forming soluble chelates with the metal ions in the minerals (Table 9.4). See Garrels and Christ (1965) for details of solutions and mineral equilibria.

Intrastratal Solution

Dissolution can affect both minerals and rock fragments, including K-feldspars, plagioclase, muscovite, biotite, unstable heavy minerals, carbonate rock fragments, volcanic rock fragments, and chert. Even quartz may undergo solution along strained zones and microcrystalline boundaries. Only the ultrastable heavy minerals zircon, tourmaline, and rutile appear to resist dissolution (McBride, 1985). Dissolution of detrital grains or cements can be partial or complete. Partial dissolution of mineral grains can often be recognized by jagged or embayed boundaries. On the other hand, selective dissolution of more-soluble minerals in rock fragments can make identification of the rock fragments questionable. If grains are completely dissolved, no evidence remains of the grain identity unless the void left by the dissolved grain has a distinctive shape, such as the rhombic shape of dolomite, that is preserved by subsequent filling with clay minerals or other cements. Alteration of feldspars or glass shards to clay minerals by incongruent solution can also be recognized if the distinctive feldspar or shard shape is preserved.

The selective dissolution of less-stable framework grains or parts of grains during diagenesis is called **intrastratal solution**. The problem that complete dissolution of less-stable grains creates for provenance analysis has been recognized at least since the 1940s. Pettijohn (1941) reported a statistical tendency for older sedimentary rocks to have fewer heavy minerals, a less varied suite of heavy minerals, and a greater ratio of ultrastable (zircon, tourmaline, rutile) to less-stable heavy minerals than younger rocks. He attributed these characteristics to intrastratal solution. Although alternative explanations for this trend, involving progressive unroofing of a sedimentary-metamorphic-plutonic igneous source, have been offered (e.g., Van Andel, 1959), Pettijohn's views have generally been vindicated by subsequent work. For example, Morton (1984) suggests that intrastratal solution can alter a diverse suite of heavy minerals consisting of 20 or more species to one containing only zircon, rutile, and tourmaline. Similar observations are reported by Milliken and Mack (1990). Dissolution proceeds according to the relative stabilities of the minerals, with least-stable minerals undergoing dissolution first (see Table 8.4 for the relative stabilities of common heavy minerals). Morton suggests that the relative order of stability is controlled by composition of the interstitial waters but that the limits of persistence of particular heavy minerals is a function of pore-fluid temperatures, rate of water migration, and geologic time.

Potassium feldspars and plagioclase may also be destroyed by dissolution during diagenesis. For example, Milliken (1988) reports significant loss of K-feldspar, plagioclase, and heavy minerals in Plio-Pleistocene sandstones of the northern Gulf of Mexico owing to diagenesis. She found a sharp decrease in K-feldspar below depths of about 3300 m. Although some K-feldspar in samples examined by Milliken was lost by albitization, SEM evidence indicates dissolution of the K-feldspar along cleavage or other crystallographically controlled features. Also, K-feldspar in some deeper samples shows random dissolution. Milliken also reports loss of calcic plagioclase below depths of 3300 m. She concludes that substantial loss of provenance information has occurred in these Plio-Pleistocene sandstones in less than 5 m.y., with heavy mineral loss occurring at temperatures above 50°C and feldspar loss at temperatures above 80°C. See also Milliken (1989) and Milliken et al. (1989) for additional discussion of feldspar dissolution. Loss of sand- and silt-size K-feldspars has also been reported from Tertiary shales in the Gulf Coast (e.g., Hower et al., 1976). Dissolution of silicate framework grains can occur both at shallow burial depths and at low temperatures under the influence of meteoric waters and at high temperatures during deeper burial. See Helmold (1985) for further discussion of the effects of feldspar diagenesis on provenance interpretation.

Decementation

In addition to dissolution of framework grains in sandstones and shales, cements precipitated during an earlier stage of diagenesis may be removed during a later stage. This process is called **decementation**. Decementation operates primarily on carbonate cements and may result in complete or partial removal of previously formed cements. Obviously, dissolution of carbonate cements requires a change in the geochemical conditions of the diagenetic environment from that present at the time of precipitation.

Decrease in pH, decrease in temperature, or increase in salinity could be responsible for the change. Many workers have suggested that decrease in pH with greater burial occurs as a result of the formation of organic acids and decarboxylation reactions accompanying the maturation of organic matter (Section 9.4.3). The salinity of pore water is known to increase at an almost linear rate with increasing burial depth. Increase in salinity could cause carbonate decementation, provided that increase in carbonate stability owing to increased temperature did not overwhelm the effects of increased salinity. The reverse situation is true during uplift of sediments. Decreased temperature favors decementation, but decrease in pore-water salinity works against decementation. On the other hand, uplift may bring the rocks within the zone of meteoric recharge, where decementation can proceed rapidly in acidic meteoric waters. If cements are completely removed during decementation, no evidence of the former presence of the cement remains, and decementation cannot be documented. Remnants of the former cement may be preserved, however, as evidence of decementation. Petrographic studies suggest that multiple episodes of cementation and decementation may occur during burial and uplift of a sediment pile.

Porosity Evolution Owing to Dissolution

Aside from the potential loss of provenance information, dissolution also has a strong effect on the evolution of porosity during diagenesis. Decementation and destruction of framework grains in sandstones obviously create secondary porosity; however, not all dissolution reactions necessarily create porosity. For example, incongruent dissolution of K-feldspar to form kaolinite may result in *in situ* formation of kaolinite with little or no increase in porosity. Pressure solution of quartz grains at contact points, admittedly a special kind of solution, reduces rather than increases porosity. Pressure solution causes interpenetration of quartz grains and therefore increased compaction and overall thinning of beds. Compaction owing to pressure solution is often called chemical compaction. Furthermore, as discussed in the preceding section, much of the silica released by pressure solution may be reprecipitated virtually *in situ* as quartz overgrowths, further reducing porosity. Schmidt and McDonald (1979a) suggest that primary intergranular porosity in sandstones is reduced to essentially zero at burial depths of about 3 km (Fig. 9.18) by a combination of mechanical and chemical compaction. The amount of secondary porosity owing to intrastratal solution and decementation increases concomitantly with decrease in primary porosity and becomes especially important below about 3 km. Below this depth, secondary porosity gradually decreases to nearly zero at about 8 km depth. Criteria for recognizing secondary porosity are provided by Schmidt and McDonald (1989b) and Shanmugam (1984). Milliken (1989) reports as much as 12 percent secondary porosity in Oligocene Frio Formation sandstones of west Texas owing to grain dissolution. Milliken et al. (1989) suggest that 50–80 percent of the feldspars in Frio Formation sandstones has been lost to dissolution, although not all of this loss produced secondary porosity. That is, some of the pore space generated by dissolution was subsequently destroyed by compaction. Further details of the factors affecting secondary porosity development may be found in numerous references, mainly involving the petroleum industry. See, for example, Surdam, Dunn, et al. (1989) and the accompanying extensive bibliography.

9.5.3   Replacement

Nature of Process

Replacement involves the dissolution of one mineral and essentially simultaneous precipitation of another mineral in its place. This statement does not mean that a large mineral is dissolved rapidly and another instantly deposited in its place. Rather, dissolution and reprecipitation may take place over a long period of time, with the guest mineral gradually replacing the host. Replacement reactions appear to take place within

FIGURE 9.18

Changes in primary and secondary porosity as a function of burial depth. See Figure 9.4 for an explanation of the maturity stages. *(After Schmidt, V., and D. A. McDonald, 1979, Soc. Econ. Paleontologists and Mineralogists Spec. Pub. 26, Aspects of diagenesis. Fig. 12, p. 189, reprinted by permission of SEPM, Tulsa, Okla.)*

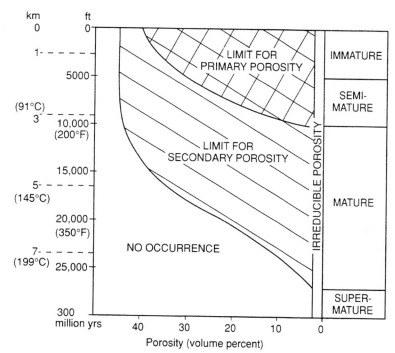

Lithology: medium-grained quartz arenite
Sedimentation rate: 30.5 m (100 ft)/1,000,000 yrs
Geothermal gradient: 2.7°C/100 m (1.5°F/100ft)

a very thin film or solution. Within this film, ions in a mineral undergoing solution diffuse outward into open pore water while the replacing ions diffuse from the pore water onto the freshly dissolved surface (precipitation surface).

It has commonly been assumed that replacement can occur only under geochemical conditions in which pore fluids are undersaturated with respect to the host mineral undergoing replacement and are supersaturated with respect to the guest mineral doing the replacing. Maliva and Siever (1988a) propose, however, that undersaturation with respect to the bulk host phase is not necessarily a prerequisite for replacement. They suggest that replacement can occur by a **force of crystallization-controlled mechanism** along thin films that have only diffusional access to bulk pore waters in equilibrium with the bulk host sediments. All that is required for replacement is a sufficient degree of local supersaturation with respect to the authigenic mineral phase, and suitable conditions in the rock for initial crystal nucleation. For additional details of this process, see the discussion under replacement in Section 12.4.3.

Replacement appears to take place without any volume change between the replaced and replacing minerals. Owing to the nature of the replacement process, delicate textures and structures present in the original mineral may be faithfully preserved in the replacement mineral. Well known examples of such preservation are petrified wood and carbonate fossils replaced by chert.

Replacement, as described, is not the same process as dissolution followed by precipitation. The latter process first produces a cavity, which is subsequently filled by precipitation of another mineral. Is there also a difference between replacement and alteration by incongruent solution? Incongruent solution may preserve the outline or shape of a grain but does not preserve its internal texture, e.g., alteration of K-feldspar to kaolinite. Clearly, many petrographers nonetheless consider the alteration of feldspars to clay minerals to be a type of replacement process, although kaolinitization of feldspar (and mica) may be accompanied by large increases in volume (McBride, 1984). Petrographic observations show that replacement of one mineral by another may begin along the outside edge of a grain, at the center of the grain, or in some position in between.

Common Replacement Minerals

Many kinds of minerals may be involved in replacement reactions, including the silica minerals, carbonates, feldspars, clay minerals, sulfates, halides, zeolites, pyrite, and iron oxides. Common replacement events include replacement of carbonate minerals by microquartz (chert), replacement of chert by carbonate minerals, replacement of feldspars and quartz by carbonate minerals (Fig. 9.19), replacement of clay matrix by carbonate minerals, replacement(?) of feldspars, rock fragments, and ferromagnesian minerals by clay minerals (Fig. 9.20), albitization of plagioclase and K-feldspar (see below), replacement of feldspars and volcanic rock fragments with zeolites in volcaniclastic sequences, and replacement of feldspars by evaporite minerals such as anhydrite in sandstone sequences associated with evaporites. Although less common, feldspars can also be replaced by quartz. For example, Morad and Aldahan (1987) report extensive replacement of feldspars by quartz in Proterozoic sandstones of Sweden and Norway. Also, some clay minerals, muscovite, and some zeolites can be replaced by Na- or K-feldspars. Probably the most characteristic replacement events in sandstones are replacement of silicate minerals by calcite or dolomite, replacement (loosely) of feldspars by clay minerals, and albitization of feldspars.

The major problems in studying paragenetic sequences of replacement minerals are (1) to recognize that replacement has occurred and (2) to determine which mineral has replaced which. That is to say, it is not always easy to tell which is the guest mineral and which is the host. Furthermore, if a grain is completely replaced, the identity of the original grain is lost unless the replacement process produces an identifiable pseudo-

FIGURE 9.19
K-feldspar (K) and plagioclase (P) partially replaced by calcite (C). Note the irregular embayments (arrows) in the K-feldspar owing to replacement by calcite. Bateman Formation (Eocene), southwest Oregon. Crossed nicols. Scale bar = 0.1 mm. *(Specimen courtesy D. Weatherby.)*

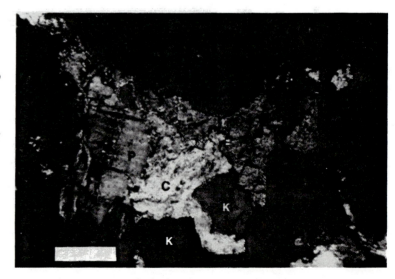

FIGURE 9.20
Feldspar (F) replaced along edge
(arrows) by dark-colored clay
minerals. Otter Point Formation
(Jurassic), southwest Oregon.
(Specimen courtesy R. L. Lent.)

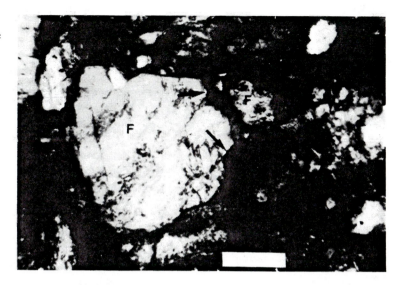

morph. Table 9.5 lists some useful criteria for identifying replacement textures and distinguishing between host and guest minerals. Readers are cautioned, however, that replacement events can be very complex and can involve several stages of replacement by different minerals as well as reversals in replacement between two minerals (Walker, 1962). Therefore, it may be very difficult to identify all stages of replacement, and extremely careful study may be required to sort out the replacement sequence.

Albitization

The albitization of feldspars is a special kind of replacement process that involves replacement of calcic plagioclase or K-feldspar with sodic plagioclase (albite). This process produces changes in feldspars that are not always detected during petrographic study. Although many workers have reported albitization of feldspars, the fact that no distinction is made between the different plagioclase feldspars during many routine petrographic studies means that albitization may go unrecognized. Unless a distinctive albitization texture is produced during diagenesis, or unless composition is analyzed by the electron probe microanalyzer, albitization may not be detected. Partial albitization of K-feldspar grains or the presence of "chessboard" texture (Walker, 1984) is an indication of albitization. Distinctive patterns of albitization (Fig. 9.21) may also be revealed by back-scattered electron microscopy (e.g., Morad et al., 1990). Exceptionally high abundance of albite in a sandstone also suggests albitization. More-routine use of the electron probe microanalyzer and backscattered electron microscopy in diagenesis studies is increasing the chance of identifying albitized feldspars.

Both Ca-bearing plagioclase and K-feldspar may undergo albitization. Most investigators suggest that albitization occurs at temperatures on the order of 100°–150°C (Milliken et al., 1981; Boles, 1982; Surdam, Crossey, et al., 1989a). On the other hand, albitization has also been reported to take place at temperatures as low as 70°–100°C (e.g., Morad et al., 1990; Aagaard et al., 1990). Albitization of feldspars may occur by direct replacement by albite, or through intermediate stages that involve initial replacement of feldspars by calcite, anhydrite, or other minerals, followed by replacement of these minerals with albite. Albitization may also occur by partial dissolution of detrital

TABLE 9.5

Criteria for recognizing replacement textures

Texture/composition	Criterion	Examples	Appearance
Illogical mineral composition (pseudomorphs)	Shape of grain does not fit mineralogy	Calcareous shell replaced by microquartz (chert) or anhydrite; volcanic glass shard replaced by smectite; feldspar replaced by kaolinite	Brachiopod / Glass shard / feldspar
Automorphic penetration	Cross-cutting relationship; one mineral clearly cuts across or transects the boundary of another	Quartz or feldspar grain cut by calcite or dolomite crystal	Q / Dolomite / Quartz
Preservation of unreplaced relic of host grain	Remnant of host mineral appears to "float" in the surrounding replacement mineral	Remnant of a feldspar grain within a larger mass of calcite; feldspar remnants have the same optical orientation	Calcite / Feldspar
Ghost or phantom structure	Recognizable "ghosts" of original grain (best seen in plane polarized light)	Ghosts of carbonate fossils in dolomite or calcite cement; most common in carbonate rocks	Q Q Q / Pelecypod
Caries texture	Bitelike embayments of guest mineral into host mineral	Calcite replacing quartz grain along edges (caution: don't confuse simple overlap of grains having irregular boundaries with replacement along grain boundaries)	Quartz / Q / Calcite
Abnormally loose packing	Abnormally large numbers of detrital grains that appear to be "floating"; i.e., large patches of a rock are not grain-supported	Quartz grains that appear to "float" in calcite; calcite has replaced the original mineral that occupied the space among the quartz grains	Q Q / Q Q / Calcite

FIGURE 9.21
Partial albitization of a large
plagioclase feldspar grain (center
of photograph) as revealed by
backscattered electron microscopy.
The dark patches are albitized
areas. Miocene deep-sea sandstone,
Japan Sea. Ocean Drilling program
Leg 127, Site 797; approximately
800 m below seafloor. Width of
image in photo is 0.2 mm.
*(Photograph courtesy David
Krinsley, Arizona State
University.)*

feldspars, followed by precipitation of albite in dissolution voids (Milliken, 1989). Table 9.6 shows replacement stages in the albitization of K-feldspar in the Fountain Formation of Colorado that were identified by Walker (1984). For description of additional feldspar textures created by albitization, see Gold (1987) and Saigal et al. (1988).

Effects of Replacement

Replacement processes, like dissolution processes, can affect the porosity of formations as well as influencing interpretations of provenance on the basis of detrital mineralogy. Porosity is affected mainly by alteration of feldspars, rock fragments, and micas to clay minerals, with a concomitant increase in volume, which tends to plug pores and reduce porosity. Loss of provenance information arises from complete replacement of grains so that the identity of the original grain is in doubt. Complete alteration of feldspars and other detrital grains to clay minerals, complete replacement of feldspars or other grains by carbonate minerals, and complete albitization of feldspars are probably the most important replacement processes that cause loss of provenance information and misinterpretation of provenance. Albitization of K-feldspars, for example, may result in the complete disappearance of K-feldspars from a formation or stratigraphic unit, with concomitant increase in the content of sodic plagioclase. If albitization goes undetected, interpretation of provenance on the basis of feldspars can give very misleading results.

9.5.4   Recrystallization and Inversion

The term **recrystallization,** as used in this book, refers to change in size or shape of crystals of a given mineral, without accompanying change in chemical composition or mineralogy. **Inversion** is a more complex kind of recrystallization in which one polymorph of a mineral changes to another polymorph without change in chemical composition. Folk (1965) suggested the term **neomorphism** to cover both recrystallization and inversion.

Recrystallization of a substance represents a change from a less stable to a more stable form of the substance. The driving force behind recrystallization is the tendency toward a minimum in the Gibbs free energy of the chemical system (Pettijohn et al., 1987,

TABLE 9.6
Stages of replacement involved in albitization of K-feldspars in the Pennsylvanian Fountain
Formation of Colorado

Precursor replacements:

1. Replacement of potassium feldspar by anhydrite:

$$KAlS_3O_8 + Ca^{++} + SO_4^{--} + 8H_2O = CaSO_4 + 3H_4SiO_4^0 + Al(OH)_4^- + K^+$$

2. Replacement of potassium feldspar by calcite:

$$KAlSi_3O_8 + Ca^{++} + HCO_3^- + 8H_2O = CaCO_3 + 3H_4SiO_4^0 + Al(OH)_4 + K^+ + H^+$$

3. Replacement of potassium feldspar by dolomite:

$$KAlSi_3O_8 + Ca^{++} + Mg^{++} + 2HCO_3^- + 8H_2O = CaMg(CO_3)_2 + 3H_4SiO_4^0 + Al(OH)_4^- + K^{++} 2H^+$$

Albitization:

4. Direct replacement of potassium feldspar by albite:

$$KAlSi_3O_8 + Na^+ = NaAlSi_3O_8 + K^+$$

5. Replacement of anhydrite by albite:

$$CaSO_4 + 3H_4SiO_4^0 + Na^+ + Al(OH)_4^- = NaAlSi_3O_8 + Ca^{++} + SO_4^{--} + 8H_2O$$

6. Replacement of calcite by albite:

$$CaCO_3 + 3H_4SiO_4^0 + Na^+ + Al(OH)_4^- + H^+ = NaAlSi_3O_8 + Ca^{++} + HCO_3^- + 8H_2O$$

7. Replacement of dolomite by albite:

$$CaMg(CO_3)_2 + 3H_4SiO_4^0 + Na^+ + Al(OH)_4^- + 2H^+ = NaAlSi_3O_8 + Ca^{++} + Mg^{++} + 2HCO_3^- + 8H_2O$$

Source: Walker, J. R., 1984, Diagenetic albitization of potassium feldspar in arkosic sandstones: Jour. Sed. Petrology, v. 54. Table 1, p. 15, reprinted by permission of Society of Economic Paleontologists and Mineralogists, Tulsa, Okla.

p. 438). Most commonly, this tendency drives recrystallization in the direction of increasing crystal size. That is, small crystals recrystallize to big crystals, a process sometimes referred to as **aggrading recrystallization** or **aggrading neomorphism**. The opposite process, recrystallization of large crystals to small crystals, has been reported, particularly in carbonate rocks, but **degrading recrystallization** is not common and is poorly understood. Recrystallization energetics also drive inversion of less-stable polymorphs to more-stable forms.

Simple recrystallization involving only a change in size or shape of grains is an important process in carbonate rocks (Chapter 12). Recrystallization is generally regarded to be relatively unimportant in siliciclastic rocks; however, Ji and Mainprice (1990) report extensive recrystallization in plagioclase crystals that have undergone metamorphism. A kind of inversion process that takes place in the presence of fluids occurs in both kinds of rocks. This process, which is actually a solution reprecipitation process, is most common in $CaCO_3$ and SiO_2 minerals. Alteration of metastable aragonite to calcite is particularly common in carbonate rocks (Chapter 12). Because such alteration takes place comparatively rapidly under aqueous conditions, aragonite rarely occurs in rocks older than the Tertiary. Nonetheless, aragonite has been reported in older rocks, e.g., Pennsylvanian rocks. It commonly occurs in older rocks under some kind of special conditions where the aragonite has been prevented from reacting with pore fluids. For example, aragonite in shells may be preserved by asphalt coatings (Brand, 1989). Polymorphic transformation of aragonite to calcite has minor significance in sandstones and shales, where it is limited mainly to alteration and recrystallization of aragonite fossils and cements.

Transformation of the SiO_2 polymorphs from opal-A to opal-CT to quartz (chert) is a common diagenetic process in muddy sediments that contain significant quantities of opaline skeletal remains, e.g., deep-sea diatom or radiolarian oozes. Sandstones commonly contain few opaline skeletal remains, but they may contain opal-A or opal-CT cements, particularly sandstones in volcaniclastic sequences. Opal cements subsequently transform to microcrystalline quartz (chert). The conversion of opal-A to microcrystalline quartz is controlled by both temperature and time. That is, conversion proceeds at a faster rate in areas having high geothermal gradients and high sedimentation rates than in areas with lower rates (Siever, 1983). The rate of conversion depends also upon the nature of the starting material and the solution chemistry, particularly the availability of Mg^{2+} and OH^- ions, which form magnesium hydroxide compounds that serve as a nucleus for crystallization of opal-CT (Kastner and Gieskes, 1983). Opal inversion is a solution-precipitation process. Opal-A first dissolves to yield solutions rich in silica. The silica then flocculates to yield opal-CT, which becomes increasingly ordered before dissolving to yield pore waters of low silica concentration. Slow growth of quartz takes place from these waters (Williams and Crerar, 1985). Identification of opal-A, opal-CT, and quartz phases commonly requires X-ray diffraction techniques.

9.5.5 Clay Mineral Authigenesis

Diagenetic reactions that affect clay minerals are discussed in several preceding sections; therefore, only a brief summary of the various processes that involve clay-mineral authigenesis is given here. Diagenetic changes that affect clay minerals include alteration of nonclay-mineral precursor minerals such as feldspars and volcanic glass to produce clay minerals, alteration of one kind of clay mineral to another, and precipitation of clay minerals into pore spaces where no obvious precursor mineral is present. Preceding discussion indicates that clay-mineral diagenesis is influenced by temperature and by organic and inorganic reactions that affect pore-water composition. Increasing temperature generally promotes clay-mineral diagenesis, although some clay minerals such as kaolinite are known to form at both low and high temperatures.

Table 9.7 summarizes some common clay-mineral diagenetic changes that occur in sandstones and shales. Reactions involving kaolinite and smectite are particularly important. For additional information on clay-mineral diagenesis in sandstones, see reviews by Wilson and Pittman (1977) and Pittman (1979).

The burial diagenesis of clay minerals is often depicted in burial-history curves such as that shown in Figure 9.17. Many studies of clay-mineral diagenesis are now being carried out as multidisciplinary studies that may include also investigation of other authigenic minerals such as quartz, feldspar, and carbonate cements. Techniques for such multidisciplinary studies include petrography, K/Ar dating, oxygen and carbon isotope measurements, and fluid-inclusion studies. These techniques provide information on the timing and temperature of formation of authigenic mineral phases and on the nature and origin of the diagenetic fluids. See Girard et al. (1989) for a good example of this approach.

Clay-mineral reactions are obviously most important in shales, which contain large amounts of clay minerals. Most sandstones probably contain little detrital clay, except possibly infiltered clay; however, alteration of feldspars, ferromagnesian minerals, and

TABLE 9.7
Some important clay-mineral diagenetic reactions in sandstones and shales

Precursor mineral*	Mineral formed	Chemical reaction	Diagenetic stage	Approx temp. of reaction	Remarks
K-feldspars*	Kaolinite	$2KAlSi_3O_8 + 2H^+ + 9H_2O \rightarrow Al_2Si_2O_5(OH)_4 + 4(H_4SiO_4) + 2K^+$	Eogenesis, telogenesis	$<\sim25°C$	Silicic acid and K^+ released
Ca-plagioclase	Kaolinite	$CaAl_2Si_2O_8 + 2H^+ + H_2O \rightarrow Al_2Si_2O_5(OH)_4 + Ca^{2+}$	Eogenesis, telogenesis	$<\sim25°C$	Ca^{2+} released
Kaolinite	K-feldspar	$Al_2Si_2O_5(OH)_4 + 4(H_4SiO_4) + 2K^+ \rightarrow 2KAlSi_3O_8 + 2H^+ + 9H_2O$	Mesogenesis	$<\sim150°C$ (?)	Silicic acid, K^+ added; H_2O, H^+ released
Kaolinite	Ca-plagioclase	$Al_2Si_2O_5(OH)_4 + Ca^{2+} \rightarrow CaAl_2Si_2O_8 + 2H^+ + H_2O$	Mesogenesis	$<\sim150°C$	Ca^{2+} added; H_2O, H^+ released
Kaolinite	Muscovite (or illite**)	$3Al_2Si_2O_5(OH)_4 + 2K \rightarrow 2KAl_3Si_3O_{10}(OH)_2 + 2H^+ + H_2O$	Mesogenesis	$\sim120°-150°C$	K^+ added; H^+, H_2O released
Smectite	Illite	Smectite $+ 4.5K^+ + 8Al^{3+} \rightarrow$ illite $+ Na^+ + 2Ca^{2+} + 2.5Fe^{3+} + 2Mg^{2+} + 3Si^{4+} + 10H_2O$	Mesogenesis	$\sim55°-200+°C$	Smectite dominant at $<\sim100°C$; mixed-layer S/I occurs at $\sim100°-200°C$; illite dominant at $>\sim200°C$; H_2O released
Smectite	Chlorite	Smectite $+ (Fe^{2+}, Fe^{3+}) \rightarrow (Mg,Al,Fe)_6[(Si,Al)_4O_{10}](OH)_8 + SiO_2 + Na^+ + Ca^{2+} + H_2O$ (not balanced)	Mesogenesis	$\sim55°-200+°C$	Chlorite dominant at $>200°C$

Reactant	Product	Reaction**	Diagenetic regime	Temperature	Comments
Volcanic glass	Na-smectite	$Na_2KCaAl_5Si_{11}O_{32} + MgSiO_3 + H_2O + 4H^+ + 4HCO_3^- \rightarrow Na(Al_5Mg)Si_{12}O_{30}(OH)_6 + Na^+ + Ca^{2+} + 4HCO_3^-$	Eogenesis, mesogenesis	$<25°{-}150+°C$	Mg^{2+}, SiO_2, H_2O added; Na^+, Ca^{2+} released
Illite (or muscovite)	Glauconite	Illite + $(Fe^{2+}, Fe^{3+}) \rightarrow$ glauconite + K^+ + Al_2O_3 (not balanced)	Eogenesis, mesogenesis	$<{\sim}50°C$	Fe^{2+}, Fe^{3+} added; K^+, Al_2O_3 released
Muscovite	K-feldspar	$KAl_3Si_3O_{10}(OH)_2 + 2K^+ + 6(H_4SiO_4) \rightarrow 3KAlSi_3O_8 + 12H_2O + 2H^+$	Mesogenesis	$<{\sim}150°C$ (?)	Silicic acid, K^+ added; H_2O, H^+ released
Na-Al smectite	Na-feldspar	$Na(Al_5Mg)Si_{12}O_{30}(OH)_6 + 4Na^+ + 3(H_4SiO_4) \rightarrow 5NaAlSi_3O_8 + 8H_2O + 2H^+ + Mg^{2+}$	Mesogenesis	$<{\sim}150°C$ (?)	Silicic acid, Na^+ added; H_2O, H^+, Mg^{2+} released
Na-Al smectite	K-feldspar	$Na(Al_5Mg)Si_{12}O_{30}(OH)_6 + 5K^+ + 3(H_4SiO_4) \rightarrow 5KAlSi_3O_8 + 8H_2O + 2H^+ + Na^+ + Mg^{2+}$	Mesogenesis	$<{\sim}150°C$	Silicic acid, K^+ added; H_2O, H^+, Na^+, Mg^{2+} released
Kaolinite + K-feldspar	Illite + quartz	$Al_2Si_2O_5(OH)_4 + KAlSi_3O_8 \rightarrow KAl_3Si_3O_{10}(OH)_2 + 2SiO_2 + H_2O$	Mesogenesis	$\sim120{-}150°C$	H_2O released
Pore waters	Kaolinite	$Al_2O_3 + 2SiO_2 + 2H_2O \rightarrow Al_2Si_2O_5(OH)_4$	Eogenesis, telogenesis	$<{\sim}25°C$	No obvious precursor; kaolinite precipitated in pore space

*Alteration of Na-feldspar proceeds by same reaction but with release of Na^+.

**Reaction shown is for muscovite; to form illite requires addition of more K^+ and SiO_2.

Source: Data from Bjørlykke (1983), Boles and Franks (1979), Kastner and Siever (1979), Pettijohn et al. (1987), and Lee et al. (1989).

volcanic glass can lead to formation of substantial quantities of authigenic clay minerals. Also, pore waters generated in shale sequences can migrate into associated sandstones and thereby affect the course of diagenesis in the sandstones, as noted. Thus, clay-mineral reactions are an important aspect of diagenesis in both shales and sandstones. These reactions proceed as a function of pore-water compositions and temperature. Time has also been mentioned as a factor in clay-mineral diagenesis, and several workers have pointed out, for example, that there is a statistical increase in illite and chlorite with increasing age (e.g., Garrels and McKenzie, 1971, p. 235; Blatt et al., 1980, p. 389). On the other hand, drastic diagenetic changes in clay minerals have been reported in rocks as young as late Tertiary (e.g., Ramseyer and Boles, 1986). The correlation between clay-mineral diagenesis and time may thus be more apparent than real. Older rocks may simply have had a statistically higher chance of being subjected to the kinds of temperatures required for diagenesis. The geothermal gradient and residence time of diagenetically active temperature may be more important than actual geologic age (Chamley, 1989).

9.5.6   Zeolite Authigenesis

Sedimentary zeolites occur in a variety of environments, including (1) closed hydrologic systems such as saline, alkaline lakes, (2) saline, alkaline soils and land surfaces, (3) open hydrologic systems where groundwaters can percolate freely, particularly in nonmarine settings, (4) deep-sea sediments, (5) hydrothermal alteration zones, and (6) burial diagenesis or metamorphic environments (Hay, 1981). The principal zeolites that occur in sedimentary rocks are shown in Table 9.8. As shown by Table 9.8, zeolites are hydrated aluminosilicates. Structurally, they are composed of a three-dimensional network of SiO_4^{4-} tetrahedra, like the feldspars, but the structure is much more open than that of the feldspars.

Zeolites that occur in alkaline lake deposits form at shallow depths and low temperatures. Only six zeolites are common in this environment: analcime, chabazite, clinoptilolite, erionite, mordenite, and phillipsite (Surdam, 1981). In deep-sea sediments, zeolites can also occur at shallow depths at relatively low temperatures in clays and volcaniclastic sediments. Phillipsite and clinoptilolite are most common, with analcime next in abundance. Erionite, natrolite, and mordenite occur only rarely. These zeolites form by reaction of volcanic glass with pore waters, possibly with addition of biogenic silica (Hay, 1981).

Volcaniclastic sediments buried to greater depths in open hydrologic systems, in either marine or nonmarine environments, typically undergo extensive alteration to zeolites. Zeolitization occurs particularly in thick sequences of volcaniclastic deposits, in which zeolites may occur in zones as a function of burial depth and temperature. Sediments at shallower depths are characterized by a diagenetic mineral assemblage that may include mordenite, clinoptilolite, heulandite, phillipsite, and smectite. Zeolites in moderately to deeply buried volcaniclastic sediments may include analcime, laumontite, and wairakite, together with lawsonite, prehnite, and pumpellyite. Zeolites may be absent from the most deeply buried volcanogenic deposits, which contain instead an albite-chlorite-quartz assemblage that formed by alteration of zeolites. Also, one zeolite may transform to another during burial diagenesis. For example, Ogihara and Iijima (1989) report that clinoptilolite has transformed to heulandite at burial depths of 1710–1770 m in some Cretaceous marine strata in Japan.

TABLE 9.8

Common zeolite minerals in sedimentary rocks

Zeolite	Typical formula	Crystal system	Habit in sedimentary rocks
Analcime	$Na_2(Al_2Si_4O_{12})\cdot2H_2O$	Cubic	Trapezohedra
Chabazite	$Ca(Al_2Si_4O_{12})\cdot6H_2O$	Hexagonal	Rhombs
Clinoptilolite	$(Na_2K_2)(Al_2Si_{10}O_{20})\cdot8H_2O$	Monoclinic	Laths and plates
Erionite	$(Na_2K_2,Ca,Mg)_{4.5}(Al_9Si_{27}O_{72})\cdot27H_2O$	Hexagonal	Hexagonal rods and bundles of rods
Ferrierite	$(Na_2Mg_2)(Al_6Si_{30}O_{72})\cdot18H_2O$	Orthorhombic	Laths and rods
Heulandite	$(Ca,Na_2)(Al_2Si_7O_{18})\cdot6H_2O$	Monoclinic	Laths
Laumontite	$Ca(Al_2Si_4O_{12})\cdot4H_2O$	Monoclinic	Laths
Mordenite	$(Na_2K_2Ca)(Al_2Si_{10}O_{24})\cdot7H_2O$	Orthorhombic	Needles and fibers
Natrolite	$Na_2(Al_2Si_3O_{10})\cdot2H_2O$	Orthorhombic	—
Phillipsite	$(1/2Ca,Na,K)_3(Al_3Si_5O_{16})\cdot6H_2O$	Orthorhombic	Rods and laths
Wairakite	$Ca(Al_2Si_4O_{12})\cdot2H_2O$	Monoclinic	—

Note: Zeolites are colorless or white when pure but may be colored by iron oxides or other impurities.
Source: Data from Deer et al. (1966) and Mumpton (1981).

The diagenetic formation of zeolites in volcaniclastic sequences takes place by alteration of plagioclase, volcanic glass, and volcanic lithic fragments. Clay minerals and biogenic silica may also be involved in some reactions. Plagioclase may alter partially or completely to zeolites or albite or may develop clay rims. Glass shards and volcanic lithic fragments may be replaced by zeolites or clay minerals. Zeolites may also precipitate as cements in pore spaces. Three principal kinds of reactions occur in the diagenetic process: (1) hydration, (2) formation of carbonates (carbonatization), and (3) dehydration (Surdam and Boles, 1979). Hydration and carbonate formation occur during the early stages of diagenesis, whereas dehydration occurs during later diagenesis. According to Surdam and Boles, zeolite diagenesis in volcanogenic sediments can take place over a broad range of temperatures from about 10°C to 200+°C, with the upper limit of pressure and temperature being about 3 kb pressure and 300°C temperature.

Hydration is commonly the earliest diagenetic reaction and may be recognized petrographically by features such as hydration rims on plagioclase or glass fragments. The reaction is either

$$\text{Glass} + H_2O \rightarrow \text{zeolite} \qquad (9.16)$$

or

$$\text{Plagioclase} + H_2O \rightarrow \text{zeolite} \qquad (9.17)$$

Hydration reactions may be accompanied by addition of SiO_2, which could be supplied by replacement of quartz. Hydration commonly causes significant reduction in porosity and permeability of the sediments owing to increase in volume of the zeolites. The hydration process may release Ca^{2+}, which can combine with HCO_3^- to form early carbonate cements if a source of bicarbonate is available. These carbonate cements further reduce porosity and permeability.

Dehydration reactions occur during later diagenesis and higher burial temperatures. These reactions may involve alteration of one zeolite to another or alteration of a zeolite to a different kind of mineral. For example, heulandite can dehydrate to laumontite (Surdam and Boles, 1979):

$$Ca_3K_2Al_8Si_{28}O_{72} \cdot 23H_2O \text{ (heulandite)} \rightarrow$$

$$3CaAl_2Si_4O_{12} \cdot 4H_2O + 10SiO_2 + 2KAlSi_3O_8 + 11H_2O \text{ (laumontite)} \quad (9.18)$$

or heulandite can dehydrate to albite:

$$8Na^+ + Ca_3K_2Al_8Si_{28}O_{72} \cdot 23H_2O \text{ (heulandite)} \rightarrow$$

$$8NaAlSi_3O_8 + 4SiO_2 + 23H_2O + 2K^+ + 3Ca^{2+} \text{(albite)} \quad (9.19)$$

Large quantities of water are released during these dehydration reactions, along with silica and cations such as calcium, sodium, and potassium.

Surdam and Boles (1979) suggest that temperature may be overemphasized in zeolite diagenesis and that fluid flow and pore-water composition may be as significant as depth of burial in controlling the distribution of diagenetic minerals in volcaniclastic deposits. Fluid effects are especially important during early diagenesis, when the fluid-to-grain ratio is high, and during later stages, when dehydration and fracturing are important.

Vavra (1989) provides an example of the influence of multiple variables in zeolite diagenesis. He reports that zeolite-facies mineral assemblages are an important authigenic component of feldspathic arenites and volcanic lithic arenites in Triassic sandstones of the central Transantarctic Mountains, Antarctica. The mineral assemblages include heulandite, mordenite, and analcime, as well as smectite, chlorite, and quartz. Vavra interprets that this mineral suite formed at shallow burial depths and moderately low temperatures by reaction between sediments and groundwater. With deeper burial, zeolite-facies minerals were subjected to higher temperatures associated with Jurassic diabase intrusions. Under these conditions, fluid–rock reactions resulted in conversion of smectite to mixed-layer illite/smectite and produced albite, laumontite, prehnite, epidote, and chlorite. Vavra suggests that the diagenetic reactions in this sandstone were controlled by a combination of parent-rock composition, fluid chemistry, permeability, and temperature.

The diagenetic changes that occur in volcanogenic sediments from the early to late stages of diagenesis are summarized in Figure 9.22.

9.5.7   Oxidation of Ferromagnesian Grains

Oxidation of iron-rich volcanic fragments and ferromagnesian minerals such as magnetite, biotite, and pyroxenes is a common diagenetic process. Oxidation initially produces goethite (limonite), which dehydrates with age to hematite. The iron oxides may alter and obscure the original detrital ferromagnesian grains so thoroughly that they cannot be recognized. Also, the oxides may migrate outward from the grain to form a surrounding halo, or they may invade and coat adjacent grains, making them difficult to identify. Walker (1967, 1974) suggests that much, if not most, of the hematite in ancient redbeds may be the product of diagenetic alteration of ferromagnesian minerals. Oxidation of ferromagnesian grains occurs particularly at shallow burial depths in the presence of oxic meteoric waters but can occur at any depth where oxidizing waters are present. Destruction of detrital grains by oxidation is thus most common during eogenesis and telogenesis.

FIGURE 9.22
Descriptive framework for zeolitization and other diagenetic changes in volcaniclastic sediments. *(From Surdam, R. C., and J. R. Boles, 1979, Aspects of diagenesis; Soc. Econ. Paleontologists and Mineralogists Spec. Pub. No. 26. Fig. 17, p. 241, reprinted by permission of SEPM, Tulsa, Okla.)*

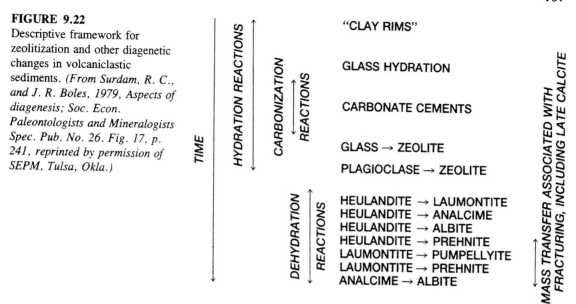

9.5.8 Ages of Diagenetic Events

In the preceding sections, the timing of diagenetic events is discussed in terms of relative stages of diagenesis: eogenesis, mesogenesis, telogenesis; early, middle, late diagenesis; shallow, moderate, deep burial; and so on. The relative timing of events in a diagenetic paragenetic sequence is usually established by petrographic study of compaction, cementation, and replacement textures as well as by a knowledge of the pressure-temperature stability fields of the various diagenetic minerals. The timing of porosity enhancement and loss is of special interest to petroleum geologists, who try to construct predictive models for evaluating the timing of diagenetic events (e.g., Surdam, Dunn, et al., 1989b).

In addition to relative dating of diagenetic events, geologists are increasingly interested in determining the absolute timing of such events. There is also increasing interest in relating the timing of diagenetic events to large-scale geologic phenomena such as periods of emergence, rapid sedimentation, and faulting (Lundegard, 1989). Determining the absolute age of a diagenetic event requires that the absolute age of characteristic diagenetic minerals be determined. Absolute ages are determined by isolating authigenic minerals whose ages can be established by using standard techniques of geochronology that involve dating by radiometric methods. A common technique is to determine the ages of authigenic illite and glauconite by using the potassium/argon method. The ages of overgrowths on potassium feldspars may also be determined by this technique. Alternatively, the $^{40}Ar/^{39}Ar$ method can be used to determine ages of these minerals, and the potassium/calcium method may be used for dating feldspar cements. Rubidium/strontium dating of illite and $^{87}Sr/^{86}Sr$ dating of cements such as marine carbonate cements are less common techniques. Paleomagnetic methods have also been applied to determining the ages of authigenic iron oxides in redbeds or in sedimentary units that have experienced solution of iron carbonates with subsequent reprecipitation of iron oxides. The age resolution of such methods is generally not very good, and the paleomagnetic scale for rocks older than the Jurassic has not yet been adequately developed. For a review of these diagenetic dating techniques and their limitations, see Lundegard (1989). As discussed in preceding sections of this chapter, ages determined by the above methods are often combined with information obtained by petrography, isotope analysis, and fluid-inclusion studies to produce burial-history curves. Figure 9.17 is an example of such curves.

9.6   Diagenesis of Organic Matter and Hydrocarbon Generation

The effects of organic-inorganic reactions on the diagenesis of shales and sandstones is discussed in several preceding sections of this chapter. Organic matter itself is changed during diagenesis. The diagenesis of organic matter and its conversion to hydrocarbons is a special topic of considerable economic and academic interest. The processes by which organic matter is altered during diagenesis, and maturation of organic matter to form petroleum, are discussed in Chapter 14. Readers are referred to Chapter 14 for these details.

Additional Readings

Berner, R. A., 1980, Early diagenesis: Princeton University Press, Princeton, NJ, 241 p.

Chilingarian, G. V., and K. H. Wolf, eds., 1975, Compaction of coarse-grained sediments, I: Elsevier, Amsterdam, 555 p.

——— 1988, Diagenesis, I, Developments in Sedimentology 41: Elsevier, Amsterdam, 591 p.

——— 1988, Diagenesis, II, Developments in Sedimentology 43: Elsevier, Amsterdam, 268 p.

Gautier, D. L., ed., 1986, Roles of organic matter in sediment diagenesis: Soc. Econ. Paleontologists and Mineralogists Spec. Pub. 38, 203 p.

Hutcheon, I. E., ed., 1989, Burial diagenesis: Mineral. Assoc. Canada, Short Course Handbook, v. 15, 409 p.

Larsen, G., and G. V. Chilingarian, eds., 1979, Diagenesis in sediments and sedimentary rocks, Developments in Sedimentology 25A: Elsevier, Amsterdam, 579 p.

——— 1983, Diagenesis in sediments and sedimentary rocks, 2, Developments in Sedimentology 25B: Elsevier, Amsterdam, 572 p.

Marshall, J. D., ed., 1987, Diagenesis of sedimentary sequences: Geol. Soc. Spec. Pub. 36, Blackwell, Oxford, 360 p.

McDonald, D. A., and R. C. Surdam, eds., 1984, Clastic diagenesis: Am. Assoc. Petroleum Geologists Mem. 37, 434 p.

Parker, A., and B. W. Sellwood, eds., 1983, Sediment diagenesis: D. Reidel, Dordrecht, Holland, 427 p.

Reicke, H. H., III, and G. V. Chilingarian, 1974, Compaction of argillaceous sediments: Elsevier, Amsterdam, 424 p.

Scholle, P. A., and P. R. Schluger, ed., 1979, Aspects of diagenesis: Soc. Econ. Paleontologists and Mineralogists Spec. Pub. 26, 443 p.

Taylor, J. C. M., 1978, Sandstone diagenesis, introduction. Jour. Geol. Soc. London, v. 135, p. 3–5, and following papers, p. 6–135.

PART 3
Carbonate Sedimentary Rocks

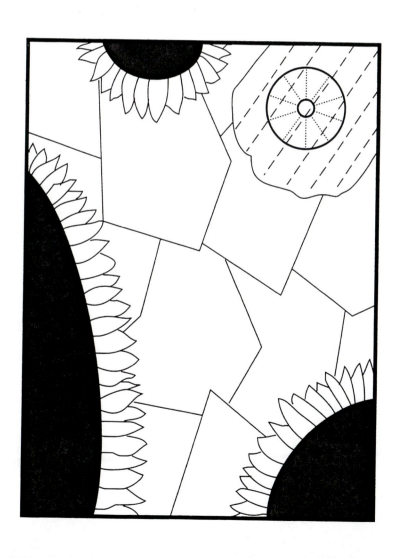

Chapter 10
Limestones

10.1 Introduction

As indicated in Chapter 1, carbonate rocks make up about one-fifth to one-quarter of all sedimentary rocks in the stratigraphic record. They occur in many Precambrian assemblages and in all geologic systems from the Cambrian to the Quaternary. Both limestone and dolomite are well represented in the stratigraphic record. Dolomite is the dominant carbonate rock in Precambrian and Paleozoic sequences, whereas limestone is dominant in carbonate units of Mesozoic and Cenozoic age (Ronov, 1983).

On the basis of their abundance alone, about the same as that of sandstones, carbonate rocks are obviously an important group of rocks. They are important for other reasons as well. They contain much of the fossil record of past life forms, and they are replete with structures and textures that provide invaluable insight into environmental conditions of the past. Aside from their intrinsic value as indicators of Earth history, they also have considerable economic significance. They are used for a variety of agricultural and industrial purposes, they make good building stone, they serve as reservoir rocks for more than one-third of the world's petroleum reserves, and they are hosts to certain kinds of ore deposits such as epigenetic lead and zinc deposits.

The microscopic study of carbonate rocks dates back to the beginning of petrographic analysis. The science of petrography was initiated by an English geologist named Henry Clifton Sorby, who began petrographic analysis about 1851 with the study of limestones. Other historically interesting early studies of carbonate rocks include investigation of carbonate sediments in the Bahamas by Black (1933) and Cayeux's (1935) classic work on the carbonate rocks of France. Modern study of carbonate sediments and depositional processes is generally regarded to have begun in the 1950s with the publications of Newell et al. (1951), Illing (1954), and Ginsburg (1956) dealing with modern carbonate sediments in the Bahamas and Florida Bay. Since the 1950s, the pace of research on carbonate rocks has accelerated. Dozens of books and many hundreds of research papers in many languages are devoted to carbonate rocks. A check of my bibliographic index turned up more than 75 full-length, English-language books that deal mainly or entirely with some aspect of carbonate rocks. A few of the more pertinent of these volumes are listed under Additional Readings at the end of this chapter.

With so much information available, a great deal is obviously known about the characteristics of carbonate rocks and the conditions under which they form. Nonetheless, field and laboratory research on carbonates continues to be extremely active. Microscopic petrography remains the primary tool for petrologic analysis of carbonate rocks; however, standard petrography is being increasingly supplemented by use of more sophisticated tools such as scanning electron microscopy (SEM), cathodoluminescence, X-ray diffraction (XRD) analysis, atomic absorption spectrophotometry (AA), X-ray fluorescence (XRF), and mass spectrometry (isotope analysis). For an excellent review of many of these techniques, see Tucker (1988). Computers and statistics are being increasingly used also for evaluation and interpretation of the vast amounts of data generated by these research tools.

In this chapter, we examine the mineralogy, chemical composition, and characteristic textures and structures of limestones. Also, we see how these features are used in classification. Further, we examine the petrographic characteristics of the principal kinds of limestones and explore these characteristics as a basis for interpreting the depositional environments of ancient limestones. The characteristics and origins of dolomite are discussed in Chapter 11, and the diagenesis of carbonate rocks is covered in Chapter 12.

10.2   Mineralogy

10.2.1   Principal Carbonate Groups

Carbonate rocks are so called because they are composed primarily of carbonate minerals. These minerals, in turn, derive their identity from the CO_3 anion, which is a fundamental part of their structure. The CO_3 carbonate ion combines with cations such as Ca, Mg, Fe, Mn, and Zn to form the common carbonate minerals (Table 10.1). A large number of other, little-known minerals also contain the CO_3 ion (see Frye, 1981). In many of these lesser known minerals, the carbonate anion is present along with other anions, such as Cl^-, SO_4^-, OH^-, F^-, and PO_4^{2-}, and many of the minerals are hydrated. Most of these uncommon minerals do not occur in sedimentary rocks, and they are not considered further in this book. A few of the carbonate phosphate minerals such as carbonate fluorapatite $[Ca_5(PO_4,CO_3)3F]$ and carbonate hydroxyapatite $[Ca_5(PO_4,CO_3)3OH]$ are discussed in Chapter 13.

The common carbonate minerals fall into three main groups: the **calcite group,** the **dolomite group,** and the **aragonite group** (Table 10.1). Minerals in the calcite and dolomite groups belong to the rhombohedral (trigonal) crystal system, and those in the aragonite group belong to the orthorhombic system. Dolomite-group minerals differ from calcite-group minerals in that they are double carbonates. That is, they contain Mg and/or Fe^{2+} in addition to Ca. The rhombohedral carbonates have 6-fold coordination. In calcite minerals, the atoms are arranged such that layers containing Ca, Mg, Mn, Fe, or Zn atoms alternate with layers of carbonate atoms, with the layers being equidistant along the z-axis. The triangular CO_3 groups have identical orientation within a given layer, but the orientation is reversed in alternate layers. Each oxygen ion in the calcite structure has two Ca (or other cation) ions as nearest neighbors. In dolomite minerals, the cation layers alternate; that is, layers of Ca ions alternate with layers of Mg (or Fe^{2+}) ions. Orthorhombic carbonates have a 9-fold coordination (3 cations for each oxygen). In aragonite-group carbonates, the CO_3 groups do not lie midway between Ca (or Pb, Sr, Ba) layers. Further, they are rotated 30 degrees to right or left so that each oxygen atom has three neighboring Ca (or other cation) atoms. For further details on the structure of the carbonate minerals, see Lippman (1973), Reeder (1983), Speer (1983), and Tucker and Wright (1990, p. 284).

Cations with smaller ionic radii (Mg, Zn, Fe, Mn) tend to favor 6-fold coordination, whereas those with larger ionic radii (Sr, Pb, Ba) favor 9-fold coordination. The radius of the calcium ion is intermediate in size between these smaller and larger ions; therefore, Ca can enter into a 6-fold coordination to form rhombohedral $CaCO_3$ (calcite) or into a 9-fold coordination to form orthorhombic $CaCO_3$ (aragonite). Aragonite is the less-stable polymorph and transforms in time to calcite. More details of aragonite-to-calcite transformation are given in Chapter 12.

Among the carbonate minerals listed in Table 10.1, only calcite, dolomite, and aragonite are volumetrically important minerals in limestones and dolomites. Furthermore, aragonite is important only in Cenozoic-age carbonate rocks and modern carbonate sediments. Siderite and ankerite are common as cements and concretions in some sedimentary rocks but rarely form important parts of carbonate units. The other carbonate minerals in Table 10.1 occur so rarely in carbonate rocks that they receive little further mention in this chapter.

10.2.2 Ion Substitution in Carbonate Minerals

Because the common cations in carbonate minerals have the same charge and similar ionic radii, substitution of cations is common. Solid-solution series exist between many of the end-member series (Table 10.1). Substitution of Mg (ionic radius 0.072 nm) for Ca (ionic radius 0.100 nm) is particularly common. On the other hand, the larger Ca ion does not readily substitute for Mg. In the calcite structure, disordered cation substitution of Mg for Ca can occur, up to several mole percent $MgCO_3$. Calcite containing more than about 4 mol % $MgCO_3$ (5 mol % according to some authors) is commonly called **magnesian calcite** or **high-magnesian calcite.** Some magnesian calcites may contain as much as 30 mol % $MgCO_3$ (Veizer, 1983). Calcite with less than about 4 mol % $MgCO_3$ is called **low-magnesian calcite** or simply **calcite.** Less commonly, Fe^{2+} can substitute for Ca or Mg in calcite to form **ferroan calcite,** and minor amounts of Mn can also substitute for Ca. High-magnesian calcite is metastable with respect to calcite and may lose its Mg in time and alter to calcite. Alternatively, if exposed to Mg-rich pore waters, high-magnesian calcite can gain additional Mg and be replaced by dolomite.

Mg does not commonly replace Ca in aragonite, although the aragonitic skeletons of some organisms incorporate Mg during growth. Small amounts of Sr or Pb may substitute for Ca in aragonite. In dolomite, ferrous iron (ionic radius 0.078 nm) may substitute for Mg in a limited solid solution series with up to 70 mol % $CaFe(CO_3)$, although the compound $CaFe(CO_3)$ does not occur naturally (Scoffin, 1987, p. 5). The more–iron-rich phase of dolomite is called ankerite and the less–iron-rich ferroan dolomite. The iron in dolomites may give the mineral a brownish color in outcrops and hand specimens. Small amounts of Mn can also replace Mg in dolomite.

10.2.3 Identification of Carbonate Minerals

Because calcite, aragonite, and dolomite are the dominant minerals in carbonate rocks, it is important to identify these minerals in petrologic study. Some distinguishing features

TABLE 10.1
Common carbonate minerals

Mineral	Crystal system	Formula	Substitutions	Indicatrix	Distinguishing characteristics	Remarks
Calcite group						
Calcite	Rhombohedral (trigonal)	$CaCO_3$	Mg for Ca common; also small amounts of Fe^{2+} and Mn for Ca	Uniaxial ($-$)	Lower birefringence than other rhombohedral carbonates; twin lamellae more common; lamellae parallel to edge or the long diagonal of the cleavage rhomb	Dominant mineral of limestones, especially in rocks older than the Tertiary
Magnesite	"	$MgCO_3$	Fe^{2+} for Mg; complete solid-solution series with siderite; minor Mn and Ca for Mg		Lacks twin lamellae; marked change in relief with rotation; Fe varieties yellow or brown	Uncommon in sedimentary rocks, but occurs in some evaporite deposits
Rhodochrosite	"	$MnCO_3$	Fe^{2+} for Mn; complete solid-solution series with siderite; also, Ca for Mn	"	Pink color (if present); association with other Mn-bearing minerals	Uncommon in sedimentary rocks; may occur in Mn-rich sediments associated with siderite and Fe-silicates
Siderite	"	$FeCO_3$	Complete solid solution series between siderite and magnesite and siderite and rhodochrosite	"	Yellow-brown or brown color; higher indices than other rhombohedral carbonates	Occurs as cements and concretions in shales and sandstones; common in ironstone deposits; also in carbonate rocks altered by Fe-bearing solutions

Mineral	Formula	Crystal system	Optical data	Chemical substitution	Distinguishing features	Occurrence
Smithsonite	$ZnCO_3$	"	"	Fe^{2+} and Mn for Zn; minor Ca, Mg, Cd, Cu, Co, Pb for Zn	Dirty, yellow-brown color	Uncommon in sedimentary rocks; occurs in association with Zn ores in limestones
Dolomite group Dolomite	$CaMg(CO_3)_2$	"	"	Fe^{2+} for Mg; forms solid-solution series with ankerite; minor Mn for Mg	Commonly forms euhedral rhombs; may be stained with Fe-oxides; higher indices than calcite; twin lamellae may be parallel to both long and short diagonals of rhomb	Dominant mineral of dolomites; commonly associated with calcite or evaporite minerals
Ankerite	$Ca(Mg,Fe,Mn)(CO_3)_2$	"	"	Limited solid-solution series with dolomite; also Mn for Mg or Fe^{2+}	Like dolomite; distinguished from magnesite by presence of twin lamellae	Much less common than dolomite; occurs in Fe-rich sediments as disseminated grains or concretions
Aragonite group Aragonite	$CaCO_3$	Orthorhombic	Biaxial (−) $2V = 18°$	Small amounts of Sr or Pb for Ca	Distinguished from calcite by lack of rhombohedral cleavage, biaxial character, and slightly higher indices	Common mineral in recent carbonate sediments; alters readily to calcite
Cerussite	$PbCO_3$	"	Biaxial (−) $2V = 9°$		White color; adamantine luster	Occurs in supergene lead ores
Strontianite	$SrCO_3$	"	Biaxial (−) $2V = 7-10°$	Ca, Ba for Sr	Higher 2V than aragonit-eOccurs in veins in some limestones	Occurs in veins in some limestones
Witherite	$BaCO_3$	"	Biaxial (−) $2V = 16°$	Minor Ca, Sr, Mg for Ba	Optically similar to aragonite	Occurs in veins associated with galena ore

Note: Common features of most carbonate minerals: high birefringence, change of relief with rotation, colorless in thin section.

Source: Data from Deer, Howie, and Zussman (1966), Nesse (1986), Reeder (1983).

of the carbonate minerals are listed in Table 10.1. Nonetheless, it is often difficult to distinguish among these minerals in hand specimens and thin sections. Identification can be greatly aided by staining and etching techniques. For example, aragonite is stained black with Fiegl's solution (Ag_2SO_4 + $MnSO_4$) whereas calcite remains unstained. Calcite is stained red in a solution of alizarin red S and dilute HCl, whereas dolomite remains unstained. Dolomite and high-magnesian calcite can be stained yellow in an alkaline solution of titan yellow. Also, dolomite can be distinguished from calcite by etching. For details of these methods, see Choquette and Trusell (1978), Dickson (1965, 1966), Friedman (1959), and Miller (1988). The carbonate minerals can be differentiated also by X-ray diffraction methods.

10.2.4   Mineralogy of Carbonate-secreting Organisms

The skeletal remains of calcium carbonate–secreting organisms are volumetrically important components of many limestones. These skeletal remains may consist of aragonite, calcite, or high-magnesian calcite containing as much as 30 mol % $MgCO_3$. For example, most pelecypods are composed of aragonite, and most echinoderms are composed of magnesian calcite. Benthonic foraminifers are also composed mainly of magnesian calcite, whereas most planktonic foraminifers are composed mainly of calcite. Table 10.2 gives the dominant mineralogy of the major groups of carbonate-secreting organisms. Note that some groups of organisms may build skeletons of both aragonite and calcite. Milliman (1974, p. 51–147) provides additional details on the mineralogy of skeletal components. The mineral composition of calcareous organisms may change with burial diagenesis. Aragonite in skeletal grains transforms to calcite with time and, as indicated, high-magnesian calcite may either lose Mg and alter to low-magnesian calcite or gain Mg to form dolomite.

10.2.5   Noncarbonate Components

Most carbonate rocks contain various amounts of noncarbonate minerals, but commonly less than about 5 percent. Noncarbonate minerals may include common silicate minerals such as quartz, chalcedony or microquartz, feldspars, micas, clay minerals, and heavy minerals. Clay minerals are particularly abundant constituents of some carbonates. Other minerals reported in carbonate rocks include fluorite, celestite, zeolites, iron oxides, barite, gypsum, anhydrite, and pyrite. Most noncarbonate minerals in limestones and dolomites are probably of detrital origin; however, some minerals such as chalcedony, pyrite, iron oxides, and anhydrite, may form during carbonate diagenesis. For study, noncarbonate minerals are commonly separated from carbonate constituents by acid treatment. Thus, they are usually referred to as **insoluble residues;** however, not all noncarbonate constituents are insoluble in acids. Carbonate rocks may also contain fine-size plant or animal organic matter. The mean organic content of carbonate rocks is commonly about 0.2 percent (Hunt, 1979, p. 265).

10.3   Chemical and Isotope Composition

10.3.1   Elemental Composition

Limestones contain one major cation, Ca^{2+}, and one minor cation, Mg^{2+}. Numerous other cations may be present in limestones in trace amounts (Turekian and Wedepohl, 1961). The most abundant of these trace cations are silicon, aluminum, iron, potassium,

TABLE 10.2

Mineralogy of major groups of carbonate-secreting organisms

Taxon	Aragonite	Calcite mol % MgCO₃ (0 10 20 30)	Both aragonite and calcite
Calcareous algae			
Red		× —— ×	
Green	×		
Coccoliths		×	
Foraminiferans			
Benthonic	0	× — ×	
Planktonic		× —— ×	
Sponges	0	× —— ×	
Coelenterates			
Stromatoporoids*	×	× ?	
Milleporoids	×		
Rugose*		×	
Tabulate*		×	
Scleractinian	×		
Alcyonarian	0	× —— ×	
Bryozoans	0	× —— ×	0
Brachiopods		× ×	
Molluscs			
Chitons	×		
Pelecypods	×	× — ×	×
Gastropods	×	× — ×	×
Pteropods	×		
Cephalopods (most)	×		
Belemnoids*		×	
Annelids (serpulids)	×	× —— ×	×
Arthropods			
Decapods		× — ×	
Ostracods		× — ×	
Barnacles		× — ×	
Trilobites*		×	
Echinoderms		× —— ×	

Note: ×, common; 0, rare; *not based on modern forms.
Source: Scholle, P. A., 1978, Carbonate rock constituents, textures, cements, and porosities. Am. Assoc. Petroleum Geologists Mem. 27, p. xi, reprinted by permission of AAPG, Tulsa, Okla.

manganese, strontium, sodium, phosphorus, titanium, and boron. The major anion in carbonate rocks is CO_3^{2-}, but significant amounts of SO_4^{2-}, OH^-, F^-, and Cl^- may also be present.

The major-element composition of carbonate rocks is a function of the kinds and amounts of carbonate minerals (Table 10.1), fossils (Table 10.2), and noncarbonate constituents present in the rocks. Major elements occur in all carbonate minerals and fossils. On the other hand, trace elements such as B, P, Mg, Ni, Cu, Fe, Zn, Mn, V, Na, U, Sr, Pb, K, and Ba are particularly concentrated in skeletal material. Cations with large ionic radii, such as strontium and to a lesser extent barium, lead, and uranium, commonly show higher concentrations in aragonite than in calcite. Elements with small ionic radii, such as magnesium, manganese, iron, nickel, and phosphorus, tend to be concentrated in calcite (Milliman, 1974, p. 142). Magnesium appears to interfere with precipitation of aragonite; therefore, the magnesium content of most aragonites is reported to be less than about 0.5 percent. As discussed in Section 10.2, the magnesium content of calcite may be low ($<\sim4$ mol % $MgCO_3$) or high (up to 30 or more mol % $MgCO_3$). For further details on the factors that affect incorporation of trace elements into carbonate materials see Veizer (1983). Trace elements such as Si, Al, Na, K, Ti, Mn, Fe, and Ba probably owe their presence in carbonate rocks mainly to the content of noncarbonate constituents, such as clay minerals, in these rocks.

10.3.2   Stable Isotope Composition

General Principles

The most important stable isotopes in carbonate rocks are isotopes of oxygen and carbon. The relative abundance of these isotopes is shown in Table 10.3. Note that oxygen has three stable isotopes; however, ^{18}O and ^{16}O are most abundant and are the principal oxygen isotopes used in isotope studies. Carbon has only two stable isotopes, ^{13}C and ^{12}C, but it has an additional radioactive isotope (^{14}C). The isotope composition of geologic materials is commonly expressed as ratios of the isotopes rather than actual abundances of the isotopes. The isotope ratio of a sample is compared to that in an arbitrary standard, and the isotope composition of the sample is expressed as per mil (parts per thousand) deviation from that of the standard. The standard for both oxygen and carbon isotopes is the University of Chicago PDB Standard, which is the isotope composition of a fossil belemnite from the Cretaceous Pee Dee Formation of South Carolina. Thus, the oxygen isotope composition of carbonates is given as the ratio of $^{18}O/^{16}O$ in a carbonate sample compared to that of the standard, and is expressed as

$$\delta^{18}O = \frac{[(^{18}O/^{16}O)\ \text{sample} - (^{18}O/^{16}O)\ \text{standard}]}{(^{18}O/^{16}O)\ \text{standard}} \times 1000 \qquad (10.1)$$

Some workers also express oxygen isotope composition relative to SMOW (Standard Mean Ocean Water). Coplen et al. (1983) show the relationship between V-SMOW (Vienna SMOW) and PDB to be

$$\delta^{18}O_{\text{V-SMOW}} = 1.03091\ \delta^{18}O_{\text{PDB}} + 30.91 \qquad (10.2)$$

and

$$\delta^{18}O_{\text{PDB}} = 0.97002\ \delta^{18}O_{\text{V-SMOW}} - 29.98 \qquad (10.3)$$

The carbon isotope composition of carbonates is expressed in the same manner as oxygen isotopes with reference to the PDB Standard.

Oxygen Isotopes

The oxygen isotope composition of carbonate minerals is a function primarily of the isotope composition and temperature of the water in which the minerals are precipitated. Additional oxygen isotope fractionation is dependent upon the mineralogical form of $CaCO_3$ minerals, their Mg content, and upon biological processes—called the vital effect.

TABLE 10.3

Relative abundance of carbon and oxygen isotopes and range of natural variations

Element	Atomic number	Isotope	Geochemical abundances (ppm)	Relative isotopic abundances (%)	Range of natural variations (‰)
C	6	12	230	98.99	
		13		1.108	90
		14		10^{-12}	
O	8	16	470,000	99.759	
		17		0.037	
		18		0.203	110

Source: Odin, G. S., M. Renard, and C. V. Grazzini, Geochemical events as a means of correlation, in Odin, G. S. ed., Numerical dating in stratigraphy, Table 2, p. 55, © 1982, John Wiley & Sons, Ltd. Reprinted by permission of John Wiley & Sons, Ltd., Chichester, England.

The oxygen isotope composition of meteoric waters can range widely from about $+10$ to $-40‰$. These variations are controlled mainly by evaporation/condensation processes related to geographic latitude and altitude. Because most carbonates are precipitated in warm regions, calcites precipitated in equilibrium with meteoric waters typically have $\delta^{18}O$ values of about $-4‰$ (Anderson and Arthur, 1983). The average $\delta^{18}O$ composition of seawater is approximately $0‰$; however, $\delta^{18}O$ values in surface waters may range from about $-1‰$ to $+1‰$ depending upon salinity. $\delta^{18}O$ values display an almost linear increase with increasing salinity (Fig. 10.1). In low-latitude, shallow waters, where most carbonate deposition occurs, variations owing to salinity are small and tend to correlate positively with the net evaporation/precipitation balance in a particular geographic area (Veizer, 1983).

When water evaporates at the surface of the ocean, the lighter ^{16}O isotopes are preferentially removed in the water vapor, leaving the heavier ^{18}O in the ocean. This isotopic fractionation process thus causes water vapor to be depleted in ^{18}O with respect to the seawater from which it evaporates. When ^{18}O-depleted moisture falls in polar regions, it is locked up as ice on land and is prevented from returning quickly to the ocean. Because of this retention of light-oxygen water in the ice caps, the ocean becomes progressively enriched in ^{18}O as ^{18}O-depleted ice caps build up during a glacial stage. Thus, marine carbonates that precipitate in the ocean during a glacial stage, particularly biogenic carbonate such as in foraminifers, will be enriched in ^{18}O relative to those that precipitate during times when the climate is warmer and ice caps on land are absent or smaller. Melting of ice caps, with consequent return of light-oxygen water to the ocean, will be reflected in a decrease in $\delta^{18}O$ values in marine biogenic carbonates. Savin and Yeh (1981) suggest that complete melting of global ice caps could produce a depletion in the average $\delta^{18}O$ of the whole ocean by -0.8 to $-1.3‰$ relative to that of the present ocean.

The equilibrium fractionation of oxygen isotopes is strongly affected by temperature. The relationship between ocean temperature and oxygen isotope composition has been shown by Shackleton (1967) to be

$$T(°C) = 16.9 - 4.38(\delta_c - \delta_w) + 0.10(\delta_c - \delta_w)^2 \qquad (10.4)$$

FIGURE 10.1

Relationship between salinity and $\delta^{18}O$ values of ocean water. *(From Berger, W. H., 1981, Paleooceanography: the deep record, in Emiliani, C., ed., The sea, v. 7, Fig. 6, p. 1448, © 1981, John Wiley & Sons, Inc. Reprinted by permission of John Wiley & Sons, Inc., New York.)*

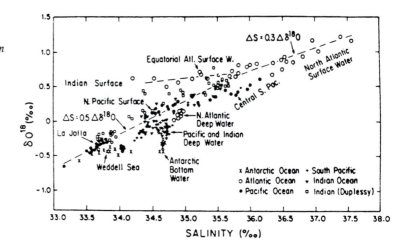

where δ_c = equilibrium oxygen isotope composition of calcite and δ_w = oxygen isotope composition of the water from which the calcite was precipitated. Because of this relationship, the oxygen isotope composition of carbonates can be used as a tool to determine ocean paleotemperatures, provided that the salinity of the water can also be determined. Figure 10.2 shows the relationship between $\delta^{18}O$ composition of calcite and dolomite and temperature. Note that $\delta^{18}O$ values decrease as temperature increases. Note also that the $\delta^{18}O$ values for dolomite are higher for a given temperature than those for coexisting calcite. Also, although not shown, aragonite tends to have higher $\delta^{18}O$ values than calcite, and high-magnesian calcites have higher values than aragonite (Anderson and Arthur, 1983).

FIGURE 10.2

The relationship between temperature and the $\delta^{18}O$ values of dolomite and calcite, assuming $\delta^{18}O$ water = 0‰ (SMOW). The wide band for calcite reflects variations in values reported by different investigators. *(After Land, L. S., 1983. The application of stable isotopes to studies of the origin of dolomite and to the problems of diagenesis of clastic sediments, in Arthur, M. A., et al., eds., Stable isotopes in geology, Soc. Econ. Paleontologists and Mineralogists Short Course Notes 10. Fig. 4-1, p. 4-4, reprinted by permission of SEPM, Tulsa, Okla.)*

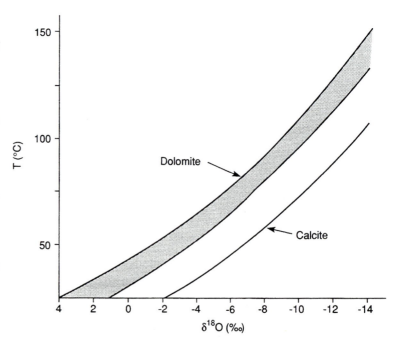

Organisms that secrete calcium carbonate can cause additional fractionation of oxygen isotopes, the so-called vital effect. These effects can bring about either positive or negative deviations from inorganic equilibrium (Veizer, 1983).

Carbon Isotopes

The carbon isotope composition of carbonate sediments is influenced primarily by the isotope composition of the water from which carbonate minerals precipitate and to a lesser extent by temperature and biogenic fractionation. The $\delta^{13}C$ values in carbonates reflect the $^{13}C/^{12}C$ ratio of CO_2 dissolved in water. This ratio, in turn, is a reflection of the source of carbon in the CO_2. Anderson and Arthur (1983) and Veizer (1983) suggest that the $\delta^{13}C$ of total dissolved carbon (TDC) in meteoric waters is a function of several factors: (1) the amount of CO_2 contributed by breakdown of organic carbon, which yields highly negative $\delta^{13}C$ values ($\sim -25\permil$), (2) the amount of carbon from dissolution of carbonates ($\delta^{13}C \sim +2\permil$), (3) the amount of carbon derived from atmospheric CO_2 ($\delta^{13}C \sim -6$ to $-7\permil$) and the extent of its equilibrium with water, which yields $\delta^{13}C$ values of $\sim +1$ to $+2\permil$, and (4) the extent of photosynthetic withdrawal of ^{12}C into organic carbon, which leads to heavier $\delta^{13}C$ in the residual total dissolved carbon reservoir. Depending upon the relative importance of each of these factors, the $\delta^{13}C$ of meteoric waters may range from highly negative values to positive values (Veizer, 1983), although negative values are most characteristic.

Values of $\delta^{13}C$ in the total dissolved carbon of ocean water are commonly higher than those in meteoric waters, with average values of approximately $+1\permil$. Surficial waters are generally isotopically heavier than this average, and deep waters are lighter (Kroopnick, 1980). The $\delta^{13}C$ of ocean water is affected by runoff of isotopically light water from the continents, but the change toward lighter carbon with depth probably reflects the residence time of deep-water masses in the ocean. Owing to oxidation of low-$\delta^{13}C$ marine organic matter that sinks from the surface, carbon-13 is depleted in deep-water masses that have long residence times near the ocean bottom. Oxidation of this low-$\delta^{13}C$ organic matter leads to production of low-$\delta^{13}C$ dissolved bicarbonate (HCO_3^-), which is then used by organisms to build shells. Respiration by bottom-dwelling organisms may also cause a decrease in $\delta^{13}C$ of deep-bottom waters. The $\delta^{13}C$ values of ocean waters may vary latitudinally as well as with depth. Surface waters in regions of upwelling in the zone of equatorial divergence tend to have lighter values and midlatitude areas have heavier values, reflecting variations in organic productivity, rates of upwelling and down-sinking of waters, CO_2 invasion and evasion, and temperature (Veizer, 1983).

The carbon isotope composition of carbonates is affected also by temperature in a manner similar to that by which oxygen isotopes are affected. The temperature fractionation effect is not nearly so strong, however, and carbon isotopes are not commonly used in paleotemperature studies. Biogenic fractionation can cause large deviations in $\delta^{13}C$ values from inorganic equilibrium values, commonly resulting in much lower than equilibrium values. Finally, kinetic factors may also influence carbon isotope composition. Turner (1982) demonstrated that the extent of fractionation between calcite and HCO_3^- decreases with increasing rate of precipitation.

10.3.3 Stable Isotope Composition of Carbonate Sediments and Fossils

Owing to the various factors that influence the oxygen and carbon isotope composition of skeletal and inorganic carbonates, wide variations in the isotopic composition of natural

FIGURE 10.3
Distribution of $\delta^{18}O$ and $\delta^{13}C$
values in various kinds of
Quaternary marine carbonates.
*(From Milliman, J. D., 1974,
Marine carbonates. Fig. 19, p. 83,
reprinted by permission of
Springer-Verlag, New York.)*

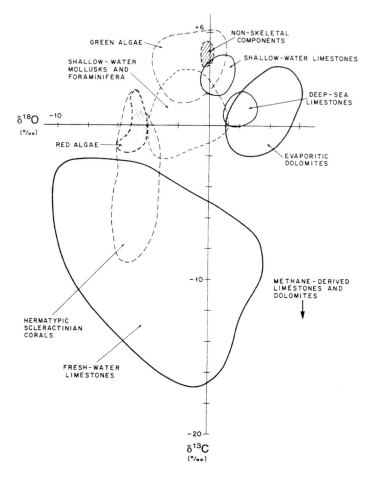

carbonates occur. For example, freshwater carbonates tend to have negative $\delta^{13}C$ and $\delta^{18}O$ values, whereas marine carbonates tend to have positive values. The distribution of $\delta^{13}C$ and $\delta^{18}O$ values in various kinds of Quaternary carbonates is summarized in Figure 10.3. Stable isotopes of carbon and oxygen, particularly oxygen isotopes, can be affected by isotopic exchange during diagenesis. Therefore, the isotopic composition of carbonates may change as a function of geologic age. In general, $\delta^{18}O$ values tend to decrease with increasing geologic age of carbonates (Veizer and Hoefs, 1976). $\delta^{13}C$ values also display a slight tendency to decrease with increasing age, particularly in rocks older than the Permian, but the trend is less clear. This subject is discussed further in Chapter 12.

10.3.4   Radiogenic Isotopes in Carbonate Rocks

In addition to the stable isotopes of oxygen and carbon, carbonates may contain several radiogenic isotopes, including carbon-14, thorium-230, protactinium-231, and strontium-87. Carbon-14, thorium-230, and protactinium-231 isotopes, in particular, have proven to be useful for direct determination of the ages of young carbonate sediments. For a brief discussion of the usefulness and limitations of these dating techniques, see Boggs (1987, p. 680–684).

10.4   Major Components of Limestones

10.4.1   General Statement

As discussed in Chapters 4 and 5, sandstones consist dominantly of various kinds of sand- and silt-size siliciclastic grains with various amounts of fine, siliciclastic mud matrix and secondary cements, including carbonate cements. The mineralogy of carbonate rocks is almost totally different from that of sandstones, but carbonate rocks resemble sandstones texturally in that they consist of various kinds of sand- and silt-size carbonate grains and various amounts of fine lime mud matrix and carbonate cements. Although carbonate rocks commonly contain only one or two dominant minerals, in contrast to sandstones, several distinct kinds of carbonate grains are recognized. Most of these grains are not single crystals but are composite grains made up of large numbers of small calcite or aragonite crystals. Folk (1962) proposed the term **allochem** to cover all of these organized carbonate aggregates that make up the bulk of many limestones. The kinds of carbonate grains in limestones and the relative abundance of grains, matrix, and cement have special significance in paleoenvironmental studies. Therefore, it is particularly important that carbonate workers understand and correctly identify these features.

10.4.2   Carbonate Grains

Lithoclasts

General Statement.   Carbonate lithoclasts are detrital fragments of carbonate rock produced by disintegration of preexisting carbonate rock or sediment, either within or outside a depositional basin. In other words, these particles are simply carbonate rock fragments (Figs. 10.4, 10.5). They are also sometimes called **limeclasts.** Lithoclasts may range in size from very fine sand to pebbles or even boulders. Lithoclasts tend to be well rounded, but may also be subrounded, subangular, or angular. Very small lithoclasts may be confused with large peloids.

Intraclasts.   Some lithoclasts originate within a depositional basin by fragmentation of penecontemporaneous, commonly weakly cemented carbonate sediment, which is eroded

FIGURE 10.4
Moderately rounded, micritic intraclasts cemented with sparry calcite cement (white). Ordovician limestone, Kentucky. Ordinary light. Scale bar = 0.5 mm.

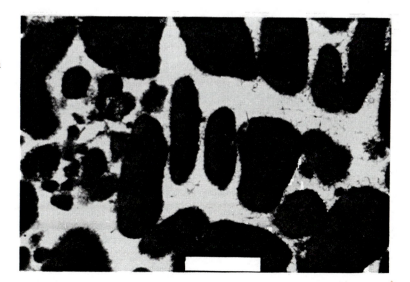

FIGURE 10.5
Probable extraclasts. Note fossil
fragments and other faint structures
preserved within the large angular
clast. Calville Limestone
(Permian), Nevada. Crossed nicols.
Scale bar = 0.5 mm.

from the seafloor and redeposited at or near the original area of deposition. Lithoclasts
having this origin are called intraclasts. Folk (1962) suggests that intraclasts may range in
origin from fragments reworked from the immediate seafloor perhaps a few days after
deposition to fragments torn by deeper erosion of sediment buried perhaps a few meters
below the surface, as long as the sediment is still unconsolidated. Figure 10.4 illustrates
typical intraclasts.

Although most intraclasts are probably produced by physical disruption of penecontemporaneous sediment by normal waves, storm waves, or currents, some intraclasts
may form by other mechanisms. These mechanisms could include organic activity on the
surface of sediment, burrowing or boring activity within sediment, and local sliding of
weakly consolidated sediment (Flügel, 1982, p. 164). Two mechanisms for producing
intraclasts that are regarded to be particularly common are (1) erosion of lithified beachrock within the intertidal and supratidal zones and (2) disruption of desiccated and cracked
supratidal, partially lithified calcareous muds to produce lime-mud clasts. Intraclasts
produced by erosion of surface sediment almost immediately after it is deposited are
plastic and easily deformed upon redeposition. Erosion of more deeply buried sediment
produces better-consolidated clasts that are more likely to preserve their shape when
redeposited.

Extraclasts.   Lithoclasts generated by erosion of much older, lithified carbonate rock
exposed on land outside the depositional basin in which the clasts accumulate are called
extraclasts. They are simply carbonate rock fragments (Fig. 10.5). Because intraclasts and
extraclasts differ fundamentally in origin, it is important in genetic studies to distinguish
between the two. Intraclasts yield information about the depositional basin, such as the
presence of bottom currents and the consolidation state of the sea bottom. Lithoclasts have
little environmental significance but yield information about the source area (provenance).
Table 10.4 lists some of the features of extraclasts that distinguish them from intraclasts.
In spite of this list of criteria, it is often extremely difficult to distinguish between
extraclasts and intraclasts. If the lithoclasts lack recognizable fossils (which is common)
or if they show no evidence of weathering or lack recrystallized veins (which is also
common), it may be impossible to unequivocally differentiate between lithoclasts. Nonetheless, it is important to try.

TABLE 10.4
Distinguishing characteristics of extraclasts

Characteristic	*Description*
Fossil content	May contain reworked, stratigraphically older fossils
Weathering features	Clasts have iron-stained, oxidized boundaries or rims
Inherited veins	May contain recrystallized veins inherited from parent rock
Nature of clast boundaries	Fossils or other grains within the clast possibly truncated at the extraclast boundary
Roundness	Clasts either conspicuously rounded or conspicuously angular
Compaction effects	Extraclasts lack compaction features, whereas intraclasts may show evidence of postdepositional compaction
Diagenetic textures	Extraclasts containing cementation or recrystallization textures that differ from associated carbonate grains
Associated noncarbonate grains	Tend to be associated with detrital chert or sandstone fragments

Source: Folk (1962), Zuffa (1980), Flügel (1982, p. 165).

Paleocurrent Significance of Lithoclasts.   Intraclasts generated by strong wave or current activity may become stacked in an imbricate arrangement. Imbricate, intraformational, carbonate conglomerates of this type are common in the geologic record, particularly in Lower Paleozoic carbonates. The imbricate structure of these conglomerates provides a useful method for determining paleocurrent directions. See Whisonant (1987) for an example of this kind of application.

Coated Grains: Ooids, Oncoids, and Cortoids

Definition.   Coated grain is a general term used for all carbonate grains composed of a **nucleus** surrounded by an enclosing layer or layers commonly called the **cortex.** Various kinds of coated grains are recognized, largely on the basis of the structure of the cortex, although some carbonate workers classify coated grains on the basis of presumed depositional environment. The terminology of coated grains is still evolving, and usage of terms is not consistent. In this book, I divide coated grains into three broad groups: ooids, oncoids, and cortoids.

Ooids.   Calcareous ooids are small, more or less spherical to oval carbonate particles, which are characterized by the presence of concentric laminae that coat a nucleus. The nucleus may be a skeletal fragment, peloid, smaller ooid, or even a siliciclastic grain such as a quartz grain. These particles have also been called **ooliths** and **oolites,** and a good deal of inconsistency has existed in usage of these names. Teichert (1970) makes a telling argument in favor of the term ooid rather than oolith; however, oolith is still in use. Although the term oolite has been used by some workers as a synonym for ooid, most now agree that oolite should be reserved as a name for a rock composed of ooids and should not be used for particles in the rock.

Most ooids are sand- to silt-size particles. They range in size from about 0.1 mm to more than 2 mm, with 0.5 mm to 1 mm being most common. Very large ooids are called **pisoids** (a rock composed of pisoids is a pisolite). Although the size boundary between ooids and pisoids is commonly regarded to be 2 mm (Choquette, 1978), such a rigid size distinction may not be justified or meaningful. Some authors (e.g., Flügel, 1982, p. 158) suggest that pisoids also differ from ooids in that they are more irregular in shape and are of nonmarine origin. On the other hand, Swett and Knoll (1989) report pisoids from Proterozoic carbonates of east Greenland and Spitzbergen that are very similar in shape and other characteristics to ooids. They suggest that these pisoids are simply oversized ooids and that they formed under marine conditions. Many occurrences are known in which ooids in a particular rock unit range in size from <2 mm to 10 mm or more without structural or mineralogical changes (Richter, 1983b). Ooids are white to cream in color and commonly have a pearly luster if formed in agitated water. Quiet-water ooids may have a dull luster. Ooids are distinguished especially by the presence of concentric, accretionary layers or laminae.

Ooids that consist of numerous laminae are regarded to be **mature** or **normal ooids** (Fig. 10.6). Those that have only a few laminae are called **superficial ooids** (Fig. 10.7), following the usage of Illing (1954). Carozzi (1960, p. 238) restricts the term superficial ooid to those ooids having only a single accretionary layer; however, most workers appear to prefer the definition of Illing. Richter (1983b) suggests that ooids are distinguished by the following general characteristics: (1) they are formed of a cortex and a nucleus of variable composition and size, (2) the cortex is smoothly laminated, (3) the laminae are either concentric or they are thinner on points of stronger curvature of the nucleus, and vice-versa, thus increasing the sphericity of the ooid during its growth, and (4) constructive biogenic structures are lacking.

With respect to structure of the cortex, carbonate workers distinguish three principal kinds of primary ooids: (1) ooids in which crystals are arranged tangentially within layers, (2) ooids with radially arranged crystals, and (3) micritic or microsparitic ooids with randomly oriented crystals (Fig. 10.8). In addition, secondary fabrics occur in some ooids. These diagenetic fabrics may include radial structures as well as oomoldic and *in*

FIGURE 10.6

Ooid (center of photograph) made up of numerous thin concentric layers surrounding an intraclast (nucleus). Tertiary limestone, Spain. Ordinary light. Scale bar = 0.1 mm. *(Specimen courtesy P. D. Snavely.)*

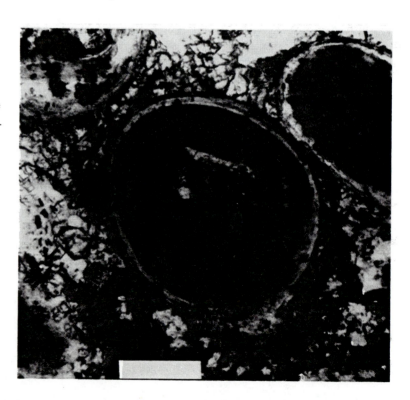

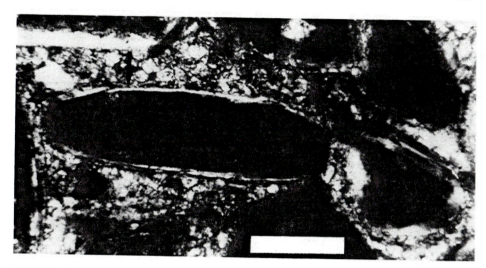

FIGURE 10.7
Superficial ooid (arrow) that displays a single thin layer or coat around a large micritic nucleus. Tertiary limestone, Spain. Crossed nicols. Scale bar = 0.5 mm. *(Specimen courtesy P. D. Snavely.)*

situ calcitized fabrics. Ooids vary also in mineral composition. Most modern-Holocene marine ooids are composed of aragonite, although some are composed of Mg-calcite. Most modern nonmarine ooids are composed of calcite. Most ancient ooids are also composed of calcite, presumably owing to diagenetic alteration of aragonite or Mg-calcite.

1. *Ooids with Tangential Structure.*   So-called Bahama-type ooids consist of elongated aragonite crystals oriented mainly tangentially, that is, parallel, to the ooid laminations (Fig. 10.6). These ooids commonly exhibit a distinct black cross under crossed polarizing prisms. They occur in marine environments, where they have been studied by numerous investigators. Tangential ooids have been reported also from some hypersaline lakes such as Great Salt Lake (Halley, 1977) and from some hot springs (Richter and Besenecker, 1983). The tangential orientation of aragonite needles in these ooids is commonly attributed to abrasion of the ooids in an agitated environment; however, the reason for this orientation is not fully understood. Tangential ooids composed of calcite occur in some caliche deposits. Clearly, the crystal orientation in these ooids is not related to water agitation. Organic matter may play a role in their orientation, although organisms themselves probably do not participate in the formation of ooids. Shearman et al. (1970) suggest that laminae of organic mucilage alternating with laminae of aragonite needles is an important element in causing orientation of the aragonite.

2. *Ooids with Radial Structure.*   Primary ooids composed of either aragonite or Mg-calcite in which elongated crystals are oriented in a radial arrangement (normal to concentric layers) occur in marine environments. Marine ooids of this type are much less common than tangential ooids. Radial orientation of crystals in marine ooids may occur within individual layers in an ooid with mainly tangential layers, or the ooid may be composed entirely of radially arranged aragonite or calcite. The latter type, which may show a black bar cross in the outer layers, occurs in only a few localities in the modern ocean. Radial-aragonitic ooids have been reported from the Persian Gulf, the Gulf of

STRUCTURAL TYPE	PRIMARY OOIDS			DIAGENETICALLY ALTERED OOIDS
	ARAGONITE	Mg-CALCITE	CALCITE	CALCITE
1  TANGENTIAL	marine lacustrine/hypersaline thermal	?	caliche	?
2  RADIAL	marine lacustrine/hypersaline	marine	cave pearls lacustrine fluvial	Mg-calcite → calcite
3  RANDOM	marine	?	lacustrine caliche	Mg-calcite → calcite or aragonite → calcite
4  OOMOLDIC				aragonite ↘ dissolution ↓ calcite cement
5  PSEUDOSPAR				aragonite → calcite

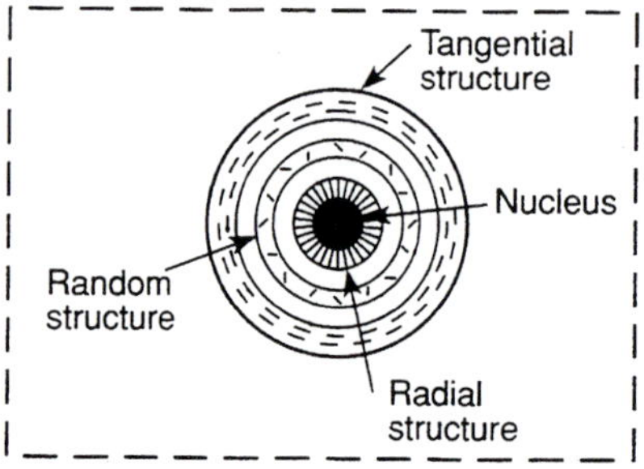

Inset (dashed rectangle):
Tangential structure — Nucleus — Random structure — Radial structure

FIGURE 10.8
Principal kinds of carbonate ooids. The inset (dashed rectangle) shows details of tangential, radial, and random structure. *(After Richter, D. K., 1983b, Calcareous ooids: a synopsis, in Peryt, T. M., ed., Coated grains. Fig. 1, p. 75, reprinted of permission of Springer-Verlag, Berlin.)*

Aqaba, and the Great Barrier Reef of Australia, and radial-Mg-calcite ooids from the Brazil-Guiana shelf north of the Amazon mouth and the Great Barrier Reef. Davies et al. (1978) synthesized radial ooids under quiet-water conditions and tangential ooids in agitated water, leading to the assumption that marine, radial ooids are formed in water with low agitation. Radial-aragonitic ooids can occur also in nonmarine environments and are known particularly from the Great Salt Lake (Halley, 1977). Nonmarine radial-Mg-calcitic ooids have not been reported, but nonmarine radial-calcitic ooids are known to occur in caves and mines and in some fluvial environments.

Radial-calcitic ooids can also form diagenetically by alteration of primary ooids (Fig. 10.9). These ooids are composed mainly of large, radially arranged calcite crystals that transect faintly preserved concentric laminae. These secondary radial-calcitic ooids may have formed by alteration of originally Mg-calcite or aragonite ooids, although some workers (e.g., Wilkinson, 1979) have suggested that Paleozoic ooids may have been

FIGURE 10.9
Radial ooids of marine origin.
Note that the ooids preserve some
of the concentric layers, suggesting
the radial fabric is secondary.
Devonian limestone, Canada.
Crossed nicols. Scale bar =
0.5 mm.

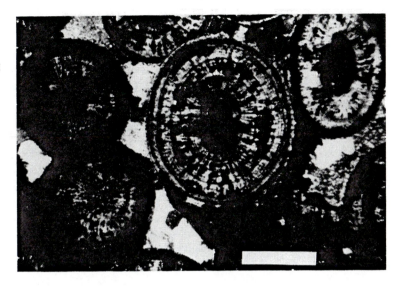

composed originally of calcite. If this is so, presumably the magnesium content of ocean water was lower in Paleozoic time.

3. *Ooids with Randomly Oriented Structure.*   Some ooids with dominantly tangential or radial structure contain layers within which the crystals consist of randomly arranged micrite or microspar (structural type 3, Fig. 10.8). Apparently, marine examples of ooids made up entirely of randomly oriented micrite are rare, although nonmarine occurrences are known in calcretes and lakes. The origin of micritic ooids is not well understood. Primary deposition of randomly oriented micrite crystals or pelagic particles such as coccoliths, aided by the presence of organic films or mucilage sheaths, has been postulated by some workers. Many others have suggested that the micrite forms by micritization of other kinds of structures in ooids owing to the action of boring algae or fungi. Some kind of degrading recrystallization of aragonitic or Mg-calcitic ooids to micrite may also be possible.

4. *Ooids with Oomoldic Structure.*   The fabric of this kind of ooid (structural type 4, Fig. 10.8) forms as a result of dissolution of the ooid cortex. This process causes the nucleus to fall to the bottom of the resulting void. The cavity or mold is subsequently filled with mosaic sparry calcite cement or coarse, fibrous calcite that grows inward from the periphery of the cavity.

5. *Ooids with Pseudospar Structure.*   These ooids (structural type 5, Fig. 10.8) consist of coarse pseudospar calcite (Folk, 1965) that has replaced the primary mineral, probably aragonite. Richter (1983b) refers to this type of ooid as *in situ* calcitized. Evidence of replacement may be indicated by relict concentric layers such as inclusion-rich zones that pass through individual calcite crystals. All gradations may exist from ooids with pseudostructure to those with micrite or microspar structure.

Ooids form by accretionary processes whereby $CaCO_3$ is precipitated onto the surface of a nucleus in water saturated to supersaturated with calcium bicarbonate. Water agitation appears to be important to growth of ooids, particularly tangential ooids, which commonly have highly spherical forms and polished surfaces. As discussed above, however, some ooids can form under quiet-water conditions. Quiet-water ooids are more likely to have a radial structure, less-spherical form, and less-polished surfaces. Ooids occur most commonly on shallow carbonate platforms but can be deposited in a wide variety of environments from fluvial to deep-marine, owing in some cases to retransport (Fig. 10.10). Ooids may contain organic matter such as the remains of blue-green algae (cyanobacteria), fungi, and bacteria. Although there has been much speculation about the possible role of organisms in the formation of ooids, there is no conclusive evidence that organisms do in fact influence their formation. Ooids appears to form primarily by inorganic processes.

Oncoids.   Coated grains more irregular in shape than ooids, and with more irregular laminae, are called oncoids (Fig. 10.11). Oncoids are also generally larger than ooids, commonly ranging from <2 mm to >10 mm. They form in both nonmarine and marine environments. Many authors restrict the term oncoid to grains of (green) algal, cyanobacterial, and bacterial origin (see Peryt, 1983a). Others (e.g., Richter, 1983a) include also as oncoids carbonate grains encrusted by red algae and bryozoans. Also, irregularly layered, coated grains without organic structures (such as spongiostromate oncoids) and nodules in vadose environments (so called vadoids in pedocretes) are considered by some workers to be oncoids. Most oncoids that form through the activities of encrusting organisms are a type of stromatolite, corresponding to Type SS (discrete spheroids) of Logan et al. (1964). Oncoids have a nucleus consisting of a shell fragment, lithoclast, ooid, and so on. The nucleus is surrounded by layers composed mainly of fine micrite but that may contain silt- or sand-size noncarbonate detrital grains. These layers form through the activities of organisms such as cyanobacteria that cause the formation of successive layers around the nucleus as the nucleus is shifted about in agitated water. The layers may be wavy or crinkly. Logan et al. recognize three subtypes of algal oncoids on the basis of layer structure and shape (Fig. 10.12):

1. **Type I—inverted stacked oncoids.** These oncoids form when growth of layers on one side of a nucleus is interrupted, and the oncoid is overturned. This step is followed by partial dissolution or erosion of the grain. A second oncoid then grows on top of the first.

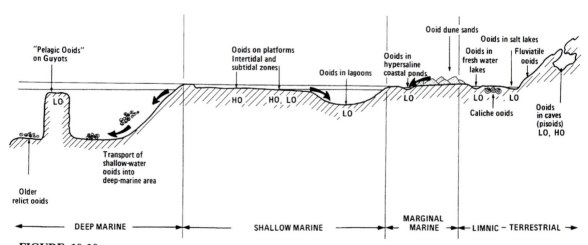

FIGURE 10.10

Depositional environments of ooids. HO refers to high-energy, agitated-water environments; LO refers to low-energy, quiet-water environments. *(From Flügel, E., 1982, Microfacies analysis of limestones. Fig. 21, p. 156, reprinted by permission of Springer-Verlag, Berlin.)*

FIGURE 10.11
Large oncoids exposed on a weathered limestone surface. Wasatch formation (Eocene), Utah. *(Photograph courtesy of H. P. Buchheim.)*

2. **Type C—concentrically stacked oncoids.** Layers are arranged concentrically around the nucleus; thus, these grains may resemble pisoids. The shape of the oncoid is governed by the shape of the nucleus.

3. **Type R—randomly stacked oncoids.** Dome-shaped layers grow at different rates as the oncoid is occasionally moved by waves or currents, producing the random stacking.

Leinfelder and Hartkopf-Fröder (1990) report a different kind of oncoid from Oligocene deposits of the Mainz Basin, Germany. These lacustrine oncoids have a concavo-convex shape, and Leinfelder and Hartkopf-Fröder refer to them as "swallow nests." They have a pustular surface and a characteristic cavity on their undersides (Fig. 10.12). Leinfelder and Hartkopf-Fröder suggest that these oncoids were formed by cyanobacteria and that they grew *in situ*. That is, they were never overturned during formation.

Oncoids of different origin may have somewhat different shapes from those illustrated in Figure 10.12. Oncoids may be further classified genetically on the basis of the kinds of organism that presumably formed them (cyanobacteria, green algae, red algae—which form "rhodoids," bryozoans, sessile foraminifers) or on the basis of their environment (caliche and cave vadoids that form in the vadose zone) (Flügel, 1982, p. 140–144; Peryt, 1983a; Richter, 1983a). For additional information on oncoids see the more than one dozen papers on this subject in Peryt (1983b).

Cortoids.   Some coated grains consist of fossils, ooids, or peloids (see pp. 433–34) coated with a thin envelope (Fig. 10.13) of generally dark colored micrite commonly

FIGURE 10.12

Principal types of algal oncoids.
Type I: inverted, stacked oncoids,
Type C: concentrically stacked
oncoids, Type R: randomly stacked
oncoids, swallow nest: concavo-
convex oncoids; the patterns
indicate successive growth stages.
*(Types I, C, and R after Logan,
R. W., et al., 1964, Classification
and environmental significance of
algal stromatolites: Jour. Geology,
v. 72. Fig. 4, p. 76, reprinted by
permission of University of
Chicago Press. Swallow-nest
oncoids after Leinfelder, R. R.,
and C. Hartkopf-Fröder, 1990, In
situ accretion mechanism of
concavo-convex lacustrine oncoids
('swallow nests') from the
Oligocene of the Mainz Basin.
Rhineland, FRG: Sedimentology,
v. 37. Fig. 2, p. 289, reprinted by
permission of Elsevier Science
Publishers, Amsterdam.)*

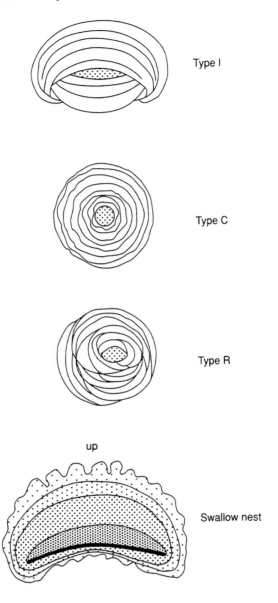

Type I

Type C

Type R

up

Swallow nest

consisting of crystals 2 to 5 μm in size. Flügel (1982, p. 160) calls these grains cortoids,
after the word *cortex* (Latin for bark). According to Flügel, the micrite envelopes form
primarily by micritization of the surface of the nucleus by boring organisms such as
cyanobacteria and red and green algae. Other processes that may form the micrite layers
include crystal growth on the surface of the particles, induced by organic films; selective
dissolution of the surface of shells; intergrowth of externally calcified filamentous algae;
and replacement of indigenous organic matter by micrite. Cortoids may resemble superficial ooids that have only a single layer around a nucleus; however, superficial ooids are
characterized by a thin concentric layer that forms by accretionary processes rather than
by micritization of the nucleus.

FIGURE 10.13
Micritic envelopes (arrows)
developed around echinoderm and
other fossil fragments. Renault
Formation (Mississippian),
Missouri. Crossed nicols. Scale
bar = 0.5 mm.

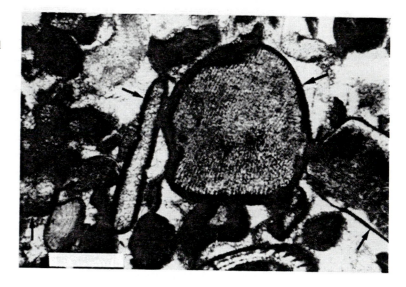

Peloids

Peloids are spherical, ovoid, or rod-shaped, mainly silt-sized carbonate grains that commonly lack definite internal structure (Fig. 10.14). They are generally dark gray to black owing to contained organic material and may or may not have a thin, dark outer rim. The most common size of peloids ranges from about 0.05 to 0.20 mm, although some are much larger. Peloids are composed mainly of fine micrite 2 to 5 μm in size, but larger crystals may be present. They are commonly well sorted and they may occur in clusters.

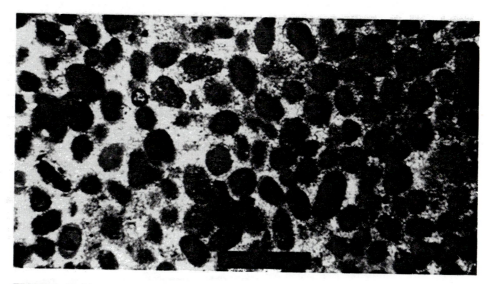

FIGURE 10.14
Peloids and a few small coated grains cemented by sparry calcite cement. Specimen source unknown. Crossed nicols. Scale bar = 0.5 mm.

Many peloids, especially those with well-rounded, symmetrical shapes, are thought to be of fecal origin. These peloids are commonly called **pellets.** Fecal pellets are produced by a variety of organisms that ingest fine, carbonate mud while feeding on organic-rich sediments. The pellets are shaped in the guts of these organisms and then extruded. Fecal pellets are produced by organisms such as crustaceans, holothurians, brachiopods, amphineurans, pelecypods, gastropods, ostracods, copepods, decepods, echinoderms, tunicates, worms, and even fish. Each kind of organism tends to produce pellets of a distinctive shape and size. Although most fecal pellets are structureless, some may have a sievelike internal structure formed by cylindrical holes parallel to the long dimension of the pellet. These holes may be filled with sparry cement, giving the pellets a speckled appearance in thin sections. Other pellets may have lamellaelike or ringlike internal structures. Although most fecal pellets in ancient carbonate rocks are small, the pellets of some modern organisms may exceed 5 mm. Most pellets appear to be produced by organisms living in quiet, marine water with muddy bottoms. Thus, pellets in ancient limestones occur most commonly in muddy limestones (micrites), and their presence suggests deposition in low-energy environments.

Not all marine peloids are fecal pellets. Some are believed to form by carbonate encrustation around filaments of cyanobacteria, endolithic algae (e.g., Schroeder, 1972), or fragments of other algae. Indistinct algal structures may still be present in these **algal peloids.** Still others are believed to form as a result of bacterial activity (Chafetz, 1986; Sun and Wright, 1989), by replacement (Shinn, 1969), by repeated nucleation (chemical precipitation) around centers of growth (Macintyre, 1985; Aissaoui, 1988), by crystal growth on nuclei of clastic origin (James et al., 1976), and by spontaneous nucleation of microcrystalline calcite (Marshall, 1983). Some peloids, referred to as **pseudopeloids,** may simply be small rounded intraclasts, formed by reworking of semiconsolidated carbonate mud or mud aggregates. Others may represent bioclasts or ooids that have been totally micritized by boring organisms or possibly by recrystallization. Compared to fecal pellets, these various nonfecal peloids tend to be characterized by poorer sorting, more-irregular shapes, and diverse sizes (0.02 mm to >2 mm).

Aggregate Grains and Lumps

Aggregate grains are irregularly shaped carbonate grains that consist of two or more carbonate fragments joined together by a micritic matrix. Aggregate grains were first described by Illing (1954) who called them lumps. Illing identified three major subtypes of lumps: (1) grapestones—aggregates resembling in shape a bunch of grapes, (2) botryoidal lumps—grapestones with superficial ooid coatings, and (3) encrusting lumps—lumps smoother than grapestones and with hollow interiors. Milliman (1974, p. 42) suggests that these three types of lumps form a continuum rather than representing discrete grain types, and he proposes use of the term **aggregate grain** to cover all of these types. Purdy (1963) previously suggested using the term **grapestone** to include all of Illings' lumps except botryoidal lumps, which he calls cryptocrystalline grains (lumps). Milliman (1974, p. 42) distinguished between aggregate grains and lumps on the basis of the micrite matrix content of the grains. That is, micrite matrix makes up less than half of aggregate grains but more than half of lumps.

Aggregate grains and lumps are mud aggregates that form *in situ* by agglutination of adjacent carbonate grains such as peloids, ooids, and skeletal fragments. Physical, chemical, and biologic processes all appear to have a role in the agglutination process. Winland et al. (1974) propose that initial binding is accomplished by algae and encrusting forams. Later, cementation and infilling of the aggregates occurs owing to continued growth of these organisms within the aggregates and to filling of internal voids by chemical or biochemical precipitation of cement. Tucker and Wright (1990, p. 12) suggest that lumps evolve from grapestones by continued cementation and micritization of grains (Fig. 10.15). (The term chasmoliths used in Figure 10.15 refers to microorganisms that live in holes not of their own creation. Endoliths are organisms that bore into grains and micritize them; see Chapter 12.)

Where the organism that encrusts aggregate grains is recognizable, such as coralline algae or encrusting foraminifers, Milliman (1974, p. 44) suggests that the aggregate grains be classified as skeletal. Winland et al. suggest that aggregate grains form under conditions where there is a supply of firm carbonate grains, uneven water turbulence, high water-circulation rates, and very low sedimentation rates. Some aggregate grains and lumps, particularly lumps, may closely resemble intraclasts. In fact, some workers regard these grains as a type of intraclast. Aggregate grains are particularly common in the modern ocean in the Bahama area. They have also been reported from shallow, brackish water of tidal zones and some sea-marginal hypersaline pools. Aggregate grains are not particularly common in ancient limestones, probably because of the limiting environmental conditions under which they form. Also, unless the aggregates undergo very early cementation, their shapes and textures may be obliterated by compaction during burial.

Skeletal Grains

Skeletal grains are among the most abundant and important kinds of grains that occur in limestones of Phanerozoic age. Skeletal grains may consist of whole fossil organisms, angular fragments of fossils, or fragments rounded to various degrees by abrasion. Skeletal grains may occur with other kinds of carbonate grains, or they may constitute the only kind of carbonate grains in a particular limestone. Some limestones are composed almost entirely of skeletal remains, which are cemented together with a small amount of micrite or sparry calcite cement. The kinds of skeletal grains that occur in limestones encompass essentially the entire spectrum of organisms that secrete hard parts; however, the relative abundance of various kinds of skeletal remains has varied through time. Figure 10.16 illustrates the relative importance of the principal kinds of calcareous marine organisms that formed carbonate sediments during Phanerozoic time. Note that some groups of organisms achieved their maximum development as carbonate sediment producers during the Paleozoic, others during the Mesozoic and Cenozoic. Figure 10.16 does not include all organisms that contribute to formation of carbonate sediment. Omitted here, for example, are the cyanobacteria that do not themselves calcify, but which play an extremely important role in trapping and binding fine carbonate sediment to form stromatolites.

Most skeletal grains are composed of aragonite, calcite, or magnesian calcite (Table 10.2). Vertebrate remains, fish scales, conodonts, and the remains of a few invertebrate organisms such as inarticulate brachiopods are composed of calcium phosphate. The original composition of skeletal grains may be altered during diagenesis. As mentioned, aragonite skeletons transform to calcite, and high-magnesian calcite grains may alter to calcite or become dolomitized. Carbonate skeletal grains may also undergo replacement by silica.

Each kind of organism that lived in the past was adapted to a particular set of ecological conditions. Because this was so, fossils yield vital information about environmental conditions such as water depth, salinity, turbidity, and energy levels. Therefore,

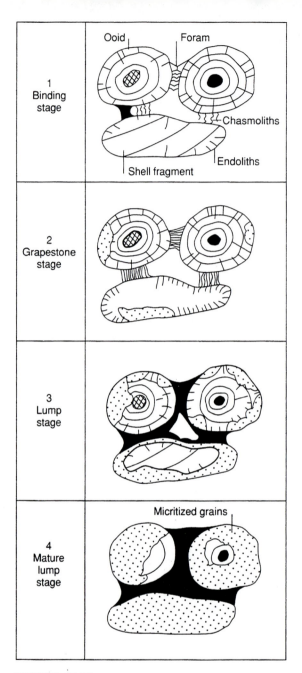

FIGURE 10.15

Stages in the formation of grapestones and lumps. Stage I: carbonate grains are bound together by foraminifers, microbial filaments, and mucilage. Chasmolithic microorganisms occur between the grains, whereas endolithic forms bore into the carbonate substrates. Stage 2: calcification of the microbial braces occurs, typically by high-magnesian calcite, to form a cemented grapestone. Progressive micritization of carbonate grains takes place. Stage 3: increased cementation at grain contacts, by microbially induced precipitation, fills depressions to create smoother relief. Stage 4: filling of any central cavity to form a dense, heavily micritized and matrix-rich aggregate. Some replacement of the high-magnesian calcite components by aragonite may also occur. *(After Tucker, M. E., and V. P. Wright, 1990, Carbonate sedimentology. Fig. 1.8, p. 12. Blackwell Scientific Publications, Oxford.)*

FIGURE 10.16

The approximate diversity, abundance, and relative importance of various calcareous marine organisms as sediment producers. P, Paleozoic; M, Mesozoic; C, Cenozoic. *(Modified from Wilkinson, B. H., 1979, Biomineralization, paleooceanography, and evolution of calcareous marine organisms: Geology, v. 7, Fig. 1, p. 526. Published by Geological Society of America, Boulder, Co.)*

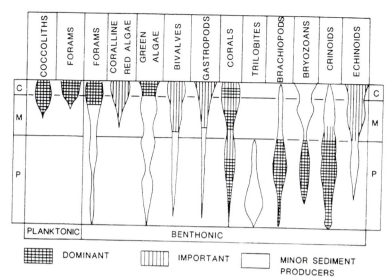

it is extremely important in petrologic studies of carbonate rocks to identify each skeletal grain. Identifying skeletal grains in thin sections can be an extremely challenging task, even for relatively experienced researchers. In many cases, only fragments of the original shell are preserved, and these fragments may be cut at various angles in the plane of the thin section. For example, a fragment of a brachiopod spine may have an outline that appears in thin section to be a circle, an oval, or an elongated rectangle, depending upon the orientation of the thin-section cut. Also, different organisms may appear quite similar in small fragments of thin-section size. Fragments of corals and bryozoans, for example, can be quite difficult to differentiate.

Beginning carbonate petrographers can generally benefit from some kind of identification guide to help cut through the maze of confusing forms that fossils present in thin section. The appendix at the end of the chapter, modified from Flügel (1982, p. 266–273), is a simple, hierarchal key that uses the skeletal outline and internal structure of fossil grains, as seen in thin section, to reduce the many possible choices to an identifiable few. This key is highly generalized, may not apply to all skeletal grains, and should not be considered a definitive guide to skeletal grain identification. Figure 10.17 provides additional details about skeletal microstructure that can be used in conjunction with the appendix. This appendix and Figure 10.17 should definitely be supplemented by reference to one or more of the photographic atlases that show plates of all the major fossil groups as they appear in thin section. See, for example, Flügel (1982), Horowitz and Potter (1971), Johnson (1971), Majewske (1969), and Scholle (1978). Space requirements in this book do not allow inclusion of numerous figures to illustrate all the major groups of fossil organisms. Figures 10.18 through 10.20 provide examples of a few kinds of skeletal grains that are common in limestones.

10.4.3   Microcrystalline Carbonate (Lime Mud)

Although many limestones are composed dominantly of carbonate grains (allochems), few if any limestones are made up entirely of sand/silt-sized grains. The remaining part of the rock is composed either of sparry calcite cement (described in the next section) or

FIGURE 10.17

Microstructure of skeletal grains as they appear in thin section. *(From Scoffin, T. P., 1987, An introduction to carbonate sediments and rocks. Fig. 4.2, p. 17, reprinted by permission of Blackie and Sons, Ltd., Glasgow.)*

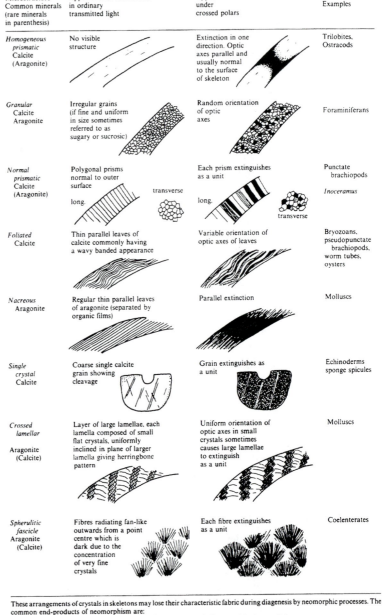

Microstructure Common minerals (rare minerals in parenthesis)	Appearance of thin section in ordinary transmitted light	Appearance of thin section under crossed polars	Examples
Homogeneous prismatic Calcite (Aragonite)	No visible structure	Extinction in one direction. Optic axes parallel and usually normal to the surface of skeleton	Trilobites, Ostracods
Granular Calcite Aragonite	Irregular grains (if fine and uniform in size sometimes referred to as sugary or sucrosic)	Random orientation of optic axes	Foraminiferans
Normal prismatic Calcite (Aragonite)	Polygonal prisms normal to outer surface long. transverse	Each prism extinguishes as a unit long. transverse	Punctate brachiopods *Inoceramus*
Foliated Calcite	Thin parallel leaves of calcite commonly having a wavy banded appearance	Variable orientation of optic axes of leaves	Bryozoans, pseudopunctate brachiopods, worm tubes, oysters
Nacreous Aragonite	Regular thin parallel leaves of aragonite (separated by organic films)	Parallel extinction	Molluscs
Single crystal Calcite	Coarse single calcite grain showing cleavage	Grain extinguishes as a unit	Echinoderms sponge spicules
Crossed lamellar Aragonite (Calcite)	Layer of large lamellae, each lamella composed of small flat crystals, uniformly inclined in plane of larger lamella giving herringbone pattern	Uniform orientation of optic axes in small crystals sometimes causes large lamellae to extinguish as a unit	Molluscs
Spherulitic fascicle Aragonite (Calcite)	Fibres radiating fan-like outwards from a point centre which is dark due to the concentration of very fine crystals	Each fibre extinguishes as a unit	Coelenterates

These arrangements of crystals in skeletons may lose their characteristic fabric during diagenesis by neomorphic processes. The common end-products of neomorphism are:

1. Microcrystalline calcite (aragonite) crystals $1-10\,\mu m$

 (ppl)   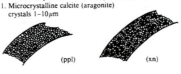 (xn)

2. Sparry calcite (aragonite) crystals $10-500\,\mu m$

 (ppl)   (xn)

FIGURE 10.18
Fusulinid foraminifers as they
appear in thin section. Both cross-
section and longitudinal-section
views are shown. Morgan
Formation (Pennsylvanian),
Colorado. Crossed nicols. Scale
bar = 0.5 mm.

FIGURE 10.19
Crinoid columnal (arrow) and other
echinoderm fragments as they
appear in thin section. All grains in
this photograph are echinoderm
fragments consisting of large,
single calcite crystals. Crossed
nicols. Scale bar = 0.5 mm.

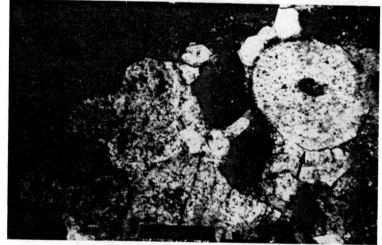

FIGURE 10.20
Gastropod shell and other mollusc
fragments surrounded by micrite.
Edwards Limestone (Cretaceous),
Texas. Crossed nicols. Scale bar =
0.5 mm.

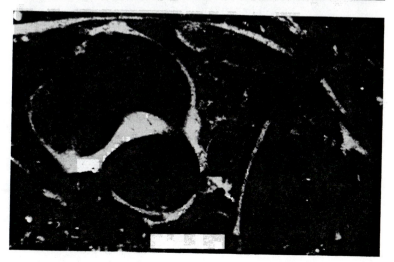

microcrystalline calcite or aragonite, commonly referred to as **lime mud.** Texturally, lime mud is analogous to the clay-size matrix in siliciclastic sedimentary rocks. In modern carbonate environments, lime muds are composed mainly of aragonite needles about 1–5 μm in length. In ancient limestones, they consists of similar-sized, but more equant, crystals of calcite. Lime muds may also contain a few percent clay-sized, noncarbonate impurities such as clay minerals, quartz, feldspar, and organic matter. Folk (1959) proposed the term **micrite** as a contraction for microcrystalline calcite, and this term has been universally adopted to signify very fine-grained carbonate sediment. It is used loosely to include all carbonate mud, including aragonite muds—that is, all microcrystalline carbonate. Micrite has a grayish to brownish, subtranslucent appearance under the microscope (Fig. 10.21). It is generally easily distinguished from carbonate grains by its finer size and from sparry calcite, which is coarser grained and more translucent. Although micrite typically occurs as a matrix among carbonate grains, some limestones are composed almost entirely of micrite. Such a limestone is texturally analogous to a siliciclastic shale or mudrock.

The presence of substantial micrite in a limestone is commonly interpreted to indicate deposition under fairly low energy conditions, where little winnowing of fine mud takes place. By contrast, deposition in agitated water is believed to cause removal of micrite, leaving mud-free carbonate grains that may later become cemented by sparry calcite. This long-standing dogma must be reexamined, however, in light of the discovery that micrite can precipitate internally in cavities and sediment pores below the sediment–water interface. The precipitation of cements with microcrystalline texture has been recognized since the mid-1960s (Reid et al., 1990). Initially, micrite cement was thought to precipitate internally in the same manner as spar cement (discussed below) by successive growth of crystals on the walls of cavities. It is now recognized that micrite can also nucleate in suspension within cavities and settle on cavity floors. Furthermore, repeated nucleation can result in formation of peloids. Thus, there are two kinds of micrite: **seafloor micrite,** which is deposited at the sediment–water interface, and **internal micrite,** which accumulates inside cavities and intergranular pores (Fig. 10.22). Carbonate deposits, including high-energy deposits such as oolites, may thus contain micrite that is not related to depositional conditions. Therefore, we must exercise considerable caution in interpreting environments of limestone deposition on the basis of micrite/grain ratios. Pore-filling micrite that exhibits peloidal or clotted textures, or that occurs in limestones

FIGURE 10.21
Photomicrograph of a limestone composed dominantly of micrite. The larger white areas are small fossil fragments; the smaller white specks are incipiently recrystallized micrite (microspar). Louisiana Limestone (Devonian), Illinois. Crossed nicols. Scale bar = 0.1 mm.

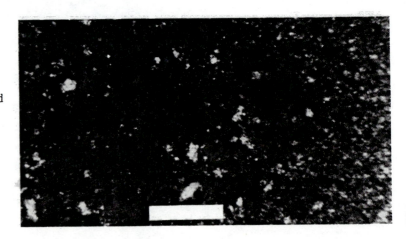

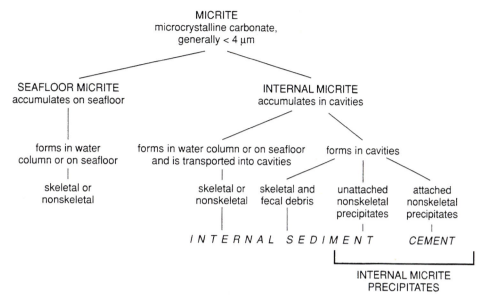

FIGURE 10.22

Classification of microcrystalline carbonate (micrite) on the basis of sites of deposition. Seafloor micrite accumulates at the sediment–water interface during deposition and reflects depositional energy and textural maturity of sediments. Internal micrite accumulates below the depositional interface during diagenesis and is unrelated to environment of deposition. *(From Reid, R. P., et al., 1990, Internal precipitation of microcrystalline carbonate: a fundamental problem for sedimentologists: Sed. Geology, v. 68. Fig. 1, p. 164, reprinted by permission of Elsevier Science Publishers, Amsterdam.)*

that exhibit evidence of submarine lithification (Chapter 12), may very likely be internal micrite.

The origin of seafloor micrite (microcrystalline aragonite and calcite mud) presents an interesting problem. Theoretical consideration of carbonate equilibria suggests that $CaCO_3$ should precipitate inorganically under conditions of supersaturation. The equilibrium relationship for $CaCO_3$ in water and dissolved carbon dioxide is

$$H_2O + CO_2 + CaCO_3 \text{ (solid)} \leftrightarrow Ca^{2+} \text{ (aq)} + 2HCO_3^- \text{ (aq)} \qquad (10.5)$$

Loss of CO_2 owing to heating, pressure decrease, photosynthesis, or other reasons disturbs this equilibrium, causing the reaction to shift to the left. This reaction is easily demonstrated in the laboratory. Nonetheless, it has now been adequately established that calcite does not precipitate freely in the presence of Mg in seawater. Aragonite nucleation is also inhibited, apparently by the presence of organo-phosphatic compounds (see Boggs, 1987, p. 78).

Questions about the inorganic precipitation of $CaCO_3$ have long been tied in with questions about the origin of whitings. **Whitings** are drifting clouds of seawater that are milky because of suspended carbonate (mainly aragonite and Mg-calcite). The origin of these whitings in modern carbonate environments such as the Bahama banks has been suggested by some authors to be proof of inorganic precipitation of $CaCO_3$. On the other hand, others have attributed the whitings to the action of bottom-feeding fish that stir up sediment or to resuspension of bottom mud by wave action. The most recent attempt to settle the question of whiting origin is reported by Shinn et al. (1989). On the basis of extensive observations made in the Bahama banks area, these authors conclude that whitings do indeed contain some (perhaps 25 percent) inorganic precipitates of aragonite and Mg-calcite. They were unable, however, to detect measurable differences in chemistry of the water within and outside the whitings or to identify and document the exact cause of inorganic precipitation. Thus, we are still left with a dilemma with regard to the overall importance of purely inorganic precipitation of calcium carbonate in the ocean.

A great deal of the lime mud in the geologic record may have been produced in some way through the activities of organisms. Photosynthesis by algae and other plants, and bacterial action and decomposition of organic matter, may promote micrite precipitation by changing the water chemistry. It is likely, however, that most organically related micrite is produced in a more direct manner by organisms. A principal source of lime mud may be the tiny aragonite needles contained within some green and red algae. When these organisms die, chemical and bacterial decomposition of their soft tissue releases the aragonite needles to the seafloor. For example, Neumann and Land (1975) studied lime mud deposition on Little Bahama Bank and concluded that most of the lime mud deposited in this area since Pleistocene time could have been supplied by disintegration of calcareous algae. This finding is somewhat at odds, however, with the more recent observations of Shinn et al. (1989), discussed above, who suggest that only about 10–20 percent of the carbonate mud in the Bahamas is algal carbonate.

Other mechanisms for production of fine carbonate sediments, many of which are organically related, include (1) disintegration of the hard parts of invertebrates, (2) breakdown of skeletal grains and other material owing to bioerosion by organisms such as parrot fish, (3) boring of carbonate grains or substrate by endolithic algae, fungi, and invertebrates, (4) physical abrasion of larger particles, and (5) production of micrite-size skeletal tests by microorganisms and nannoorganisms. Many of these processes, as they are thought to operate in the Bahamas, are illustrated diagrammatically in Figure 10.23. The widths of the arrows in Figure 10.23 suggest the relative lime-mud contribution by each of these processes, as interpreted by Neumann and Land (1975). Shinn et al. (1989) suggest that inorganic precipitation of lime mud is more important and disintegration of calcareous algae is less important than indicated by this diagram.

10.4.4   Sparry Calcite

The third major constituent of limestones, in addition to carbonate grains and micrite, is sparry calcite. Crystals of sparry calcite are large (0.02–0.1 mm) compared to micrite crystals and appear clear or white when viewed in plane light under a polarizing microscope. They are distinguished from micrite by their larger size and clarity and from carbonate grains by their crystalline shapes and lack of internal microstructures.

Much of the sparry calcite in limestones occurs as a cement that fills interstitial space among carbonate grains. Sparry calcite cement is particularly common in limestones such as oolites that were deposited in agitated water that prevented micrite from filling pore spaces. Therefore, as mentioned, the presence of significant amounts of sparry calcite cement in a limestone is commonly interpreted to indicate deposition of the limestone in agitated water. This criterion should be applied with caution, however, because much pore space in limestones can be secondary—produced by dissolution

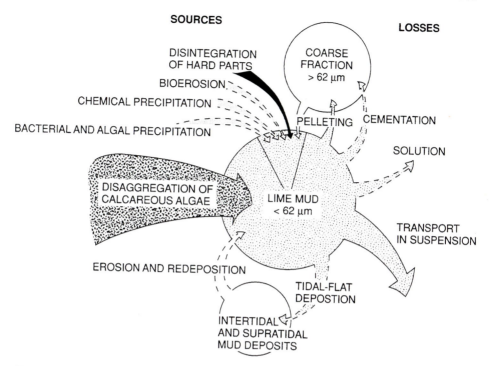

FIGURE 10.23

Multiple origin of lime mud in Bight of Abaco, Bahamas. *(After Neumann, A. C., and L. S. Land, 1975, Lime mud deposition and calcareous algae in the Bight of Abaco, Bahamas: a budget: Jour. Sed. Petrology, v. 45. Fig. 4, p. 766, reprinted by permission of the Society of Economic Paleontologists and Mineralogists, Tulsa, Okla.)*

during diagenesis. Sparry calcite cement that fills secondary pores has no relationship to depositional conditions.

Sparry calcite can form a variety of cementation fabrics, and several distinctive types of cement are recognized. The most common types are **granular** or **mosaic cement,** which is composed of nearly equant crystals; **fibrous cement,** either coarsely or finely fibrous; **bladed cement;** and **syntaxial cement** (overgrowths) (Fig. 10.24, 1). The most common syntaxial overgrowths are monocrystalline overgrowths around echinoderm fragments, which consist of plates composed of a single calcite crystal (Fig. 10.24, 1). Monocrystalline overgrowths are in optical continuity with these single-crystal echinoderm plates. These syntaxial cement rims are analogous to the overgrowths on quartz grains. Syntaxial overgrowths are also known to occur on brachiopod and mollusc fragments, foraminifer tests, and corals. Overgrowths of this type consist of bladed crusts, and the individual bladed crystals are in optical continuity with foundation crystals in the fossils (Scoffin, 1987, p. 127).

Although most sparry calcite is clear or limpid, calcite cements from many Paleozoic-age reefs and mudmounds have undulose extinction and abundant inclusions that make them cloudy or turbid. Three types of these cements, all of which are generally called radiaxial-fibrous, are recognized (Kendall, 1985). As shown in Figure 10.24, 2,

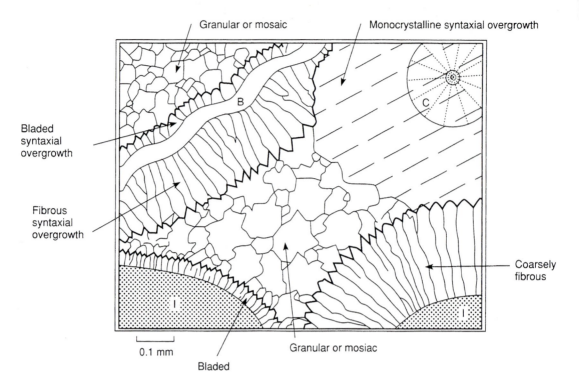

Granular or mosaic

Monocrystalline syntaxial overgrowth

Bladed syntaxial overgrowth

Fibrous syntaxial overgrowth

Coarsely fibrous

B

C

I

I

0.1 mm

Granular or mosiac

Bladed

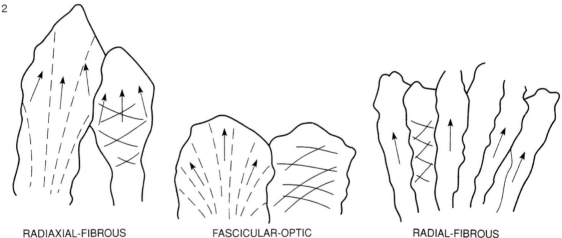

RADIAXIAL-FIBROUS

FASCICULAR-OPTIC

RADIAL-FIBROUS

FIGURE 10.24

Principal kinds of sparry calcite cement in limestones. In Figure 1, B = brachiopod, C = crinoid, I = intraclast. The change from bladed crystals to larger, granular crystals in the lower left corner of the figure illustrates "drusy" fabric. In Figure 2, radiaxial-fibrous and fasicular-optic calcite have undulose extinction and subcrystals have curved twin planes. Radial-fibrous calcite has unit extinction and straight twin planes. *(Figure 1 after Folk, R. L., 1965, Some aspects of recrystallization in ancient limestones, in Pray, L. C., and R. C. Murray, eds., Dolomitization and limestone diagenesis, Soc. Econ. Paleontologists and Mineralogists Spec. Pub. No. 13. Fig. 6, p. 27, reprinted by permission of SEPM, Tulsa, Okla. Figure 2 after Kendall, A. C., 1985, Radiaxial fibrous calcite: a reappraisal, in Schneidermann, N., and P. M. Harris, eds., Carbonate cement: Soc. Econ. Paleontologists and Mineralogists Spec. Pub. No. 36. Fig. 1, p. 60, reprinted by permission of SEPM, Tulsa, Okla.)*

these cement types are radiaxial fibrous, fascicular-optic, and radial-fibrous. **Radiaxial fibrous** calcite has a pattern within each crystal of subcrystals that diverge away from the substrate, an opposing pattern of distally convergent fast-vibration directions (optic axes), and corresponding curvature of cleavage, twin lamellae, and glide planes (Fig. 10.24, 2). **Fascicular-optic calcite** is distinguished by the presence of a divergent pattern of fast-vibration directions; this pattern coincides with the pattern of the subcrystals. To tell the difference between fascicular-optic and radiaxial fibrous calcite under the microscope, rotate the microscope stage and observe the direction of extinction swing or sweep in each crystal. In radiaxial-fibrous calcite, it is in the same direction that the stage turns. In fascicular-optic calcite, it is in the opposite direction. **Radial-fibrous** calcite does not exhibit marked undulose extinction but is distinguished by turbid crystals.

Although radiaxial-fibrous calcite has generally been attributed to replacement of acicular carbonate cement, Kendall (1985) reinterprets it as a primary precipitate. According to Kendall, it forms by a process of asymmetric growth within calcite crystals undergoing split growth. These composite crystals represent the attempts of the crystals to assume a spherulitic growth form in a diagenetic environment that favors the growth of length-slow calcite.

Euhedral to anhedral, bladed or coarsely fibrous crystals that are oriented perpendicular to carbonate grain surfaces may display an increase in grain size toward the center of the pore or cavity and a concomitant change to more-equant, granular crystals. This distinctive pore-filling fabric is commonly referred to as drusy cement. Drusy fabric is illustrated in Figure 10.24, 1 by the change from small bladed to larger granular crystals in the lower left corner of the figure. Figure 10.25 also illustrates drusy fabric. Because sparry calcite cements are a product of diagenesis, further discussion of these cements is reserved for Chapter 12.

Not all sparry calcite in limestones is cement that formed by precipitation into a pore or other cavity. Some sparry calcite originates through recrystallization of micrite or carbonate grains (Fig. 10.26). This kind of sparry calcite does not fill pore space; therefore, its presence has no value in interpreting depositional conditions. It is important in petrologic studies to differentiate sparry calcite formed by recrystallization processes from

FIGURE 10.25

Sparry calcite cementing rounded intraclasts. Note drusy texture; small oriented crystals line the margins of the cavity, and larger, unoriented crystals fill the center of the cavity. Ordovician limestone, Kentucky. Crossed nicols. Scale bar = 0.1 mm.

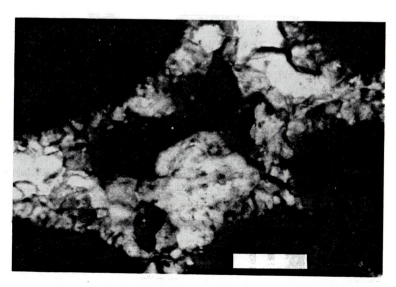

FIGURE 10.26
Coarse neospar (large, clear
crystals) formed by
recrystallization of micrite. Note
indistinct boundary between the
neospar and the ''dirty'' incipiently
recrystallized micrite. Robinson
Limestone (Pennsylvanian),
Colorado. Ordinary light. Scale
bar = 0.5 mm.

sparry calcite cement. Although recrystallized sparry calcite does not commonly form a drusy fabric, it can mimic most of the other types of cement mentioned above. Therefore, differentiating between sparry calcite cement and recrystallization spar is a major problem in carbonate petrology. Further details are given in Chapter 12.

10.5   Classification of Carbonate Rocks

10.5.1   Available Classifications

Classifications for carbonate rocks have not proliferated quite to the extent of sandstone classifications. Nonetheless, twenty or so carbonate classifications have appeared in print since the early 1960s. Somewhat more than half of these classifications are in English-language journals or books. If you are interested in that sort of thing, see Flügel (1982, p. 37) for a comparison of the strong and weak points of several of these classifications. Flügel (p. 381–382) also provides an extended bibliography of carbonate classifications, including classifications by Bulgarian, French, and German geologists.

Many of the English-language classifications are contained in Memoir 1 (''Classification of carbonate rocks'') of the American Association of Petroleum Geologists published in 1962 (W. E. Ham, editor). Two of the more widely used carbonate classifications in this memoir are the classifications of R. L. Folk and R. J. Dunham.

10.5.2   Folk Classification (1962)

Folk's classification requires definite knowledge of the kinds and abundances of carbonate grains (allochems) and the relative abundance of micrite and sparry calcite cement. The classification (Table 10.5) is hierarchal and requires that you proceed through the following definite steps: (1) Determine the relative abundance of carbonate grains (allochems) as a percentage of the total rock, that is, >10 percent grains or <10 percent grains. (2) Determine the relative abundance of micrite and sparry calcite cement; that is, sparry calcite abundance > micrite abundance, or micrite > sparry calcite. (3) Normalize

TABLE 10.5

Classification of carbonate rocks by Folk (1962)

Volumetric allochem composition / Normalized volume percent allochems	Limestones, partly dolomitized limestones, and primary dolomites					Replacement dolomites (V)	
	>10% Allochems — Allochemical rocks (I and II)		<10% Allochems — Microcrystalline rocks (III)		Undisturbed bioherm rocks (IV)		
Relative abundance of allochems, sparry calcite cement and micrite	Sparry calcite cement > microcrystalline ooze matrix — Sparry allochemical rocks (I)	Microcrystalline ooze matrix > sparry calcite cement — Microcrystalline allochemical rocks (II)	1–10% Allochems — Most abundant allochems	<1% Allochems		Evident allochem	No allochem ghosts / Allochem ghosts
>25% Intraclasts	Intrasparrudite Intrasparite	Intramicrudite Intramicrite	Intraclasts: intraclast-bearing micrite	Micrite: if disturbed, dismicrite; if primary dolomite, dolomicrite	Biolithite	Finely crystalline intraclastic dolomite	Allochem ghosts
<25% Intraclasts, >25% Ooids	Oosparrudite Oosparite	Oomicrudite Oomicrite	Oolites: oolite-bearing micrite			Coarsely crystalline oolitic dolomite	Medium crystalline dolomite
Volume ratio of fossils to pellets >3:1	Biosparrudite Biosparite	Biomicrudite Biomicrite	Fossils: fossiliferous micrite			Aphanocrystalline biogenic dolomite	Finely crystalline dolomite
Volume ratio of fossils to pellets 1:3–3:1	Biopelsparite	Biopelmicrite	Pellets: pelletiferous micrite			Very finely crystalline pellet dolomite	
Volume ratio of fossils to pellets <1:3	Pelsparite	Pelmicrite				etc.	
Volume ratio of fossils to oncoids <1:2 / ≤1:2	Oncosparite Oncosparrudite	Oncomicrite Oncomicrudite	Oncoids: oncoid-bearing micrite				

Source: After Folk, R. L., 1962, Spectral subdivision of limestone types, *in* Ham, W. E., ed., Classification of carbonate rocks: Am. Assoc. Petroleum Geologists Mem. 1, Table 1, p. 70, reprinted by permission of AAPG, Tulsa, Okla.

Note: The classification of oncoids in this figure is modified from Flügel (1980, p. 370).

the carbonate grain types to 100 percent grains; that is, determine the abundance of each major kind of carbonate grain (intraclasts, ooids, fossils, pellets, oncoids) as a percentage of total grains. (4) Enter the appropriate row in the classification: (a) If intraclasts exceed 25 percent of total grains, the rock is an intraclastic limestone. (b) If intraclasts make up less than 25 percent of total grains, but ooids exceed 25 percent, the rock is an oolitic limestone. (c) If the rock fits neither of these categories, it is a biogenic, pellet, or oncoid limestone, depending upon the ratio of fossils to pellets. (5) Select the appropriate name from classification on the basis of the above factors.

The rock name is derived by combining the appropriate abbreviation for the dominant allochem (**intra** for intraclast, for example) with the abbreviation for micrite (**mic**) or sparry calcite cement (**spar**), as appropriate. To this name is added a suffix denoting dominant size of the carbonate grains. This suffix is either **-ite,** shortened from arenite, or **rudite,** meaning gravel-size. (Note: Folk uses 1 mm as the boundary between arenites and rudites. This boundary differs from the more common 2 mm boundary used in the Wentworth size scale.)

Limestones that contain 10 percent or less carbonate grains are micritic limestones. For limestones containing 1–10 percent grains, Folk assigns names based on dominant grain type; e.g., oolite-bearing micrite, fossiliferous micrite. If the rock contains less than 1 percent grains, it is simply a **micrite,** or, if it is disturbed by burrowing organisms, a **dismicrite.** Biogenic limestones that were bound together at the time of deposition, e.g., coral reef rock, are called **biolithites.**

Folk's classification can be used also for dolomites. He suggests that fine-grained primary dolomites be classified with limestones. Such fine-grained dolomite is called **dolomicrite.** If the dolomite formed by replacement of a limestone, the assigned name depends upon the presence or absence of recognizable "ghosts" of carbonate grains, e.g., **oolitic dolomite,** or simply **medium-crystalline** dolomite (Table 10.6). See Chapter 11 for more details on dolomites.

As an adjunct to his classification, Folk also provides a textural maturity classification for limestones (Fig. 10.27) that is based on the percentage of carbonate grains in the total rock and the relative abundance of micrite and sparry calcite cement. Textural terms from Table 10.6 can be added to the main classification names to yield names such as **packed intrasparite** rather than simply intrasparite.

10.5.3   Dunham's Classification (1962)

Dunham's (1962) classification (Table 10.6A) focuses upon depositional textures rather than upon the identity of specific kinds of carbonate grains. To use this classification you must first determine if the original constituents of the limestone were or were not bound together at the time of deposition. For rocks composed of components not bound together during deposition (i.e., components deposited as discrete grains or crystals), the rocks are further divided into those that contain lime mud (micrite) and those that lack mud. Rocks that contain lime mud are either mud-supported or grain-supported. Determining the type of fabric support can be a problem, and the fabric must be visualized as it might appear in three dimensions and not simply as it appears in the two dimensions of a thin-section plane. Mud-supported limestones are **mudstones** (i.e., lime mudstones) if they contain less than 10 percent carbonate grains and **wackestones** if they contain more than 10 percent grains. Grain-supported limestones that contain some micrite mud matrix are

TABLE 10.6

Classification of limestones according to depositional textures

A

DEPOSITIONAL TEXTURE RECOGNIZABLE				DEPOSITIONAL TEXTURE NOT RECOGNIZABLE
Original components not bound together during deposition			Original components were bound together during deposition . . . as shown by intergrown skeletal matter, lamination contrary to gravity, or sediment-floored cavities are roofed over by organic or questionably organic matter and are too large to be interstices.	CRYSTALLINE CARBONATE
Contains mud (particles of clay and fine silt size)		Lacks mud and is grain-supported		
Mud-supported	Grain-supported			
Less than 10% grains / More than 10% grains				(Subdivide according to classifications designed to bear on physical texture or diagenesis.)
MUDSTONE / WACKESTONE	PACKSTONE	GRAINSTONE	BOUNDSTONE	

B

ALLOCHTHONOUS LIMESTONE — Original components not organically bound during deposition						AUTOCHTHONOUS LIMESTONE — Original components organically bound during deposition		
Less than 10% >2 mm components				Greater than 10% >2 mm components		By organisms that build a rigid framework	By organisms that encrust and bind	By organisms that act as baffles
Contains Lime mud (<0.03 mm)			No lime mud	Matrix-supported	>2 mm component-supported			
Mud-supported		Grain-supported	Grain-supported					
Less than 10% grains (>0.03 mm <2 mm)	Greater than 10% grains							
MUDSTONE	WACKESTONE	PACKSTONE	GRAINSTONE	FLOATSTONE	RUDSTONE	FRAMESTONE	BINDSTONE	BAFFLESTONE
						B O U N D S T O N E		

Source: A, Dunham (1962); B, as modified by Embry and Klovan (1972). After Dunham, R. J., 1962, Classification of carbonate rocks according to depositional textures, *in* Ham, W. E., ed., Classification of carbonate rocks: Am. Assoc. Petroleum Geologists Mem. 1. Table 1, p. 117, reprinted by permission of AAPG, Tulsa, Okla.; Embry, E. F., III and J. E. Klovan, 1972, Absolute water depth limits of late Devonian paleoecological zones: Geol. Rundschau, v. 61. Fig. 5, p. 676, reprinted by permission.

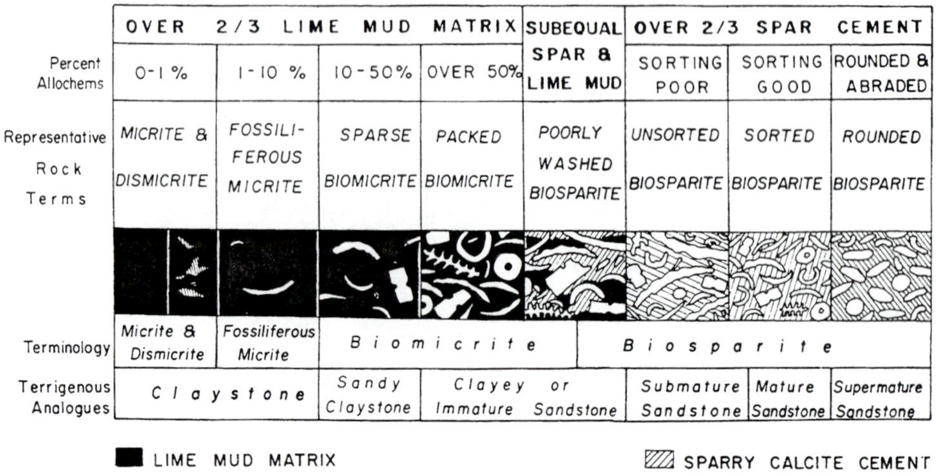

Percent Allochems	OVER 2/3 LIME MUD MATRIX				SUBEQUAL SPAR & LIME MUD	OVER 2/3 SPAR CEMENT		
	0-1 %	1-10 %	10-50%	OVER 50%		SORTING POOR	SORTING GOOD	ROUNDED & ABRADED
Representative Rock Terms	MICRITE & DISMICRITE	FOSSILI-FEROUS MICRITE	SPARSE BIOMICRITE	PACKED BIOMICRITE	POORLY WASHED BIOSPARITE	UNSORTED BIOSPARITE	SORTED BIOSPARITE	ROUNDED BIOSPARITE
Terminology	Micrite & Dismicrite	Fossiliferous Micrite	B i o m i c r i t e			B i o s p a r i t e		
Terrigenous Analogues	C l a y s t o n e		Sandy Claystone	Clayey or Immature Sandstone		Submature Sandstone	Mature Sandstone	Supermature Sandstone

■ LIME MUD MATRIX   ▨ SPARRY CALCITE CEMENT

FIGURE 10.27

Textural maturity classification of limestones. Limestones composed of other carbonate-grain types, such as intraclasts or ooids, are classified texturally in the same manner as shown in this example. *(After Folk, R. L., 1962, Spectral subdivision of limestone types, in Ham, W. E., ed., Classification of carbonate rocks: Am. Assoc. Petroleum Geologists Mem. 1. Fig. 4, p. 76, reprinted by permission of AAPG, Tulsa, Okla.)*

packstones. Grain-supported limestones that lack mud matrix are **grainstones.** Dunham uses the term boundstone for limestones composed of components bound together at the time of deposition. This name is roughly equivalent to Folk's biolithite and can include stromatolitic limestones as well as coral-reef and similar limestones.

Embry and Klovan (1972) modified Dunham's classification by subdividing limestones composed of originally unbound constituents on the basis of carbonate grain size. This modified classification scheme places more emphasis on limestone conglomerates (Table 10.6 B). Thus, they recognize two kinds of limestones on the basis of grain size: limestones that contain less than 10 percent >2 mm-size grains and those that contain greater than 10 percent >2 mm-size grains. Names for limestones containing less than 10 percent >2 mm-size grains are the same as those used by Dunham. Embry and Klovan added two new names for rocks composed of more than 10 percent >2 mm-size grains: **floatstone** for matrix-supported limestones and **rudstone** for component (grain)-supported limestones. They also divided Dunham's boundstone into three subcategories on the basis of the presumed kinds of organisms that built the limestones: **framestone**—built by organisms that construct rigid frameworks (i.e., *in situ,* massive organisms such as corals that build rigid, three-dimensional frameworks), **bindstone**—built by organisms that encrust and bind (these organisms do not build a three-dimensional framework), and **bafflestone**—formed by organisms that act as baffles (i.e., stalk-shaped organisms that acted as baffles at the time of deposition). This added subdivision of boundstone into three genetic categories constitutes mixing of genetic and descriptive parameters in the same classification. Such a subdivision may be difficult to make in practice, and it is certainly subjective. It may be preferable to stick with the name boundstone unless you are certain of the types of organisms that formed the boundstone.

10.5.4   Classification of Mixed Carbonate and Siliciclastic Sediments

Most carbonate rocks contain only a small percentage of noncarbonate, siliciclastic constituents. On the other hand, some carbonate rocks contain significant percentages of siliciclastic constituents, just as some siliciclastic sedimentary rocks contain high concentrations of carbonate constituents. Although such mixed sediments are not abundant in the geologic record, they are common enough (see Doyle and Roberts, 1988) to create problems when we try to classify them using classification schemes designed for limestones and sandstones. See Mount (1985) and Zuffa (1980) for examples of classification schemes for mixed carbonate and siliciclastic sediments.

10.6   Structures and Textures in Limestones

10.6.1   Structures Common to Siliciclastic Rocks

Although limestones are intrabasinal rocks, composed mainly of constituents precipitated in some manner from water, most carbonate grains in limestones undergo some transport before final deposition. Transport may occur by a variety of mechanisms, including tidal currents, longshore currents, wave action in shoaling zones, and turbidity currents. Accordingly, many ancient limestones display sedimentary structures that appear very similar to those that occur in siliciclastic sedimentary rocks (Chapter 3). These structures may include (1) physically produced structures such as cross-beds, lamination, ripple marks, and graded bedding, (2) deformation structures such as convolute lamination and synsedimentary folds, and (3) biogenically produced structures such as mottled bedding and other trace fossils. The textures of carbonate grains (size, shape, arrangement) may also be affected by transport processes, although shapes and sizes of some kinds of carbonate grains are determined mainly by their origin and not by transport and abrasion processes (e.g., whole fossils, ooids, pellets).

10.6.2   Special Structures

In addition to these common kinds of sedimentary structures and textures, limestones may contain some special structures and textures that are not common or that do not occur in siliciclastic rocks. The most important of these special characteristics are listed and briefly described in Table 10.7. Further details are provided below.

Cryptalgal Fabrics

Cryptalgal fabrics are fabrics in carbonate rocks believed to be biosedimentary structures that originated through the sediment-binding and/or mineral-precipitating activities of blue-green algae (cyanobacteria) and bacteria (Monty, 1976). Table 10.8 summarizes the principal kinds of cryptalgal fabrics. Layered cryptalgal structures are best known because they include **stromatolites**. Stromatolites are hemispherical structures (domal, club-shaped, columnar) characterized by generally nonplanar laminations ranging to a few millimeters in thickness (Table 10.9, Fig. 10.28). They generally have well-defined, three-dimensional boundaries and all except oncolites grow attached to the substrate. Logan et al. (1964) divide stromatolites into a few fundamental types on the basis of the

TABLE 10.7

Special kinds of structures and textures in limestones

Structure or texture	Description and origin
Cryptalgal fabric	Layered and unlayered biosedimentary structure that originated through the sediment-binding and/or mineral-precipitating activities of blue-green algae (cyanobacteria) and bacteria
Stromatactis	Evenly to irregularly distributed subhorizontal voids, generally in carbonate muds, several mm to cm in size, with a nearly flat bottom and a convex-upward upper surface with an irregular, digitate roof; filled with sparry calcite cement or micrite
Birdseye structure	Similar to stromatactis but smaller, commonly 1–3 mm, and shapes may be spherical, oval, elongated, or irregular
Geopetal fabric	A structure displayed inside the cavities of fossils or beneath convex-up shells, formed by partial infilling by micrite, creating a nearly flat floor, with the remaining, upper part of the cavity filled with sparry calcite cement; allows top and bottom of beds to be differentiated
Tepee structure	Structure that resembles an inverted, depressed V in cross section; occurs in marine, peritidal, lacustrine, and caliche environments; these arched-up antiforms are generated in carbonate hardgrounds owing to expansion caused by cementation or by fracturing and fracture fill by sediment or cement
Nodular structure	Nodules ranging in size to several cm that occur in a matrix of micrite and clay; particularly common in deep-water carbonates; origin probably early diagenetic, but poorly understood
Stylolites	Suturelike seams in carbonate rocks marked by irregular and interlocking penetration of the two sides; caused by pressure solution
Bioherm	Large-scale, moundlike or lenslike mass built by sedentary organisms; i.e., reefs and banks

TABLE 10.8

Principal kinds of cryptalgal structures and fabrics

I. LAMINATED FABRIC—characterizes stromatolites, oncolites, and cryptalgal laminites
  A. Laminated fabric—prominent laminations separated by distinct physical discontinuities
  B. Laminoid fabric—lacks distinct laminae, but contains more or less horizontal structures that outline an overall layering; three principal kinds of structures:
    1. Laminoid fenestral fabric—repetitive, elongate cavities or fenestrae
    2. Laminoid boundstone fabric—fabric arising from orientation of elongate particles parallel to the overall lineation that pervades the cryptalgal body; presumably formed by binding of particles by successive algal mat surfaces
    3. Lenticular laminoid fabric—fabric created by irregular juxtaposition and superposition of algal cushions or mats much smaller than the width of the stromatoid
II. NONLAYERED FABRIC
  A. Thrombolitic fabric—centimeter-sized patches or clots of micrite with rare clastic particles, separated by spaces filled with sparry calcite; may occur together with algal filaments; may be indistinctly laminated in part (Aitken, 1967)
  B. Massive fabric—shows no internal lineation or spatial organization of particles; may be difficult to identify as cryptalgal
  C. Radial fabric—shows radial growth of calcified algal filaments; may be interrupted in places by concentric layers, but overall fabric is nonlayered

Source: After Monty, 1976.

TABLE 10.9

Structure of hemispherical stromatolites showing examples of laterally linked hemispheroids, vertically stacked hemispheroids, and discrete spheroids

Types	Description	Vertical section of stromatolite structure
Laterally linked hemispheroids	Space-linked hemispheroids with close-linked hemispheroids as a microstructure in the constituent laminae	
Discrete, vertically stacked hemispheroids	Discrete, vertically stacked hemispheroids composed of close-linked hemispheroidal laminae on a microscale	
Discrete spheroids	Spheroidal structures consisting of inverted, stacked hemispheroids	
	Spheroidal structures consisting of concentrically stacked hemispheroids	
	Spheroidal structures consisting of randomly stacked hemispheroids	
Combination forms	Initial space-linked hemispheroids passing into discrete, vertically stacked hemispheroids with upward growth of structures	
	Initial discrete, vertically stacked hemispheroids passing into close-linked hemispheroids by upward growth	
	Alternation of discrete, vertically stacked hemispheroids and space-linked hemispheroids due to periodic sediment infilling of interstructure spaces	
	Initial space-linked hemispheroids passing into discrete, vertically stacked hemispheroids; both with laminae of close-linked hemispheroids	
	Initial discrete, vertically stacked hemispheroids passing into close-linked hemispheroids; both with laminae of close-linked hemispheroids	

Source: After Logan, B. W., R. Rezak, and R. N. Ginsburg, 1964, Classification and environmental significance of algal stromatolites: Jour. Geology, v. 72. Fig. 4, p. 76, and Fig. 5, p. 78, reprinted by permission of University of Chicago Press.

FIGURE 10.28
Algal stromatolites in the Snowslip Formation (Precambrian), west of Logan Pass, Glacier National Park, Montana. *(Photograph by G. J. Retallack.)*

shapes of the hemispheres and the presence or absence of linking structures between hemispheres (Table 10.9).

Logan et al. suggest that the change from laterally linked hemispheroids to discrete, vertically stacked hemispheroids to discrete hemispheroids shown in Table 10.9 indicates an increasing energy spectrum. That is, laterally linked hemispheroids are deposited under the lowest energy conditions and discrete spheroids under the highest energy conditions. Buekes and Lowe (1989) describe four morphological types of stromatolites from Archean carbonate rocks of South Africa, which they also relate to environmental conditions: stratiform, domal, columnar, and conical stromatolites. They suggest that stratiform stromatolites formed in upper intertidal, domal stromatolites formed in middle intertidal, columnar stromatolites formed in lower intertidal, and conical stromatolites formed in high-energy, subtidal channel environments.

Cryptalgal laminites lack the hemispherical form of stromatolites. They are laterally continuous stratiform deposits that are characterized by subcontinuous planar lamination ranging from less than a millimeter to several centimeters (Monty, 1976).

Monty (1976) discusses a variety of mechanisms that may be responsible for the development of laminated cryptalgal fabrics, including (1) diurnal differences in algal growth, (2) alternation between preferred vertical and preferred horizontal growth patterns of the algal filaments, (3) different growth rates of various algal species within the algal mat, (4) growth of algal layers inside a mat (rather than at the surface of the mat), (5) periodic calcification of algal mats by poorly understood processes, (6) precipitation of carbonate crystals with different crystal habits, (7) periodic enrichment in various mineral compounds (e.g., Fe and Mn), (8) periodic influx of detrital particles (sand, silt, mud, organic matter), (9) periodic inorganic cementation owing to changes in geochemical conditions, (10) different reaction of laminae during diagenesis, resulting in alternating dolomitic and calcitic laminae, (11) alignment of detrital particles and agglutination by algae, and (12) formation of fenestral fabrics owing to separation of algal mats because of gas bubble formation or shrinkage.

Best known of the **nonlayered cryptalgal fabrics** are **thrombolites.** Thrombolite fabrics are characterized by millimeter- and centimeter-size patches or clots of micrite separated by spaces filled either with sparry calcite cement or silt- and sand-sized sediment (Aitken, 1967; Kennard and James, 1986). Microscopic algal filaments may be present in the clots. Thrombolite fabrics are generally nonlayered but may be indistinctly laminated in part. Thrombolites are believed to form as a result of oxidation of dead colonies of algae, intergrowth and coalescence of colonies of algae, or internal dissolution of a calcified algal mass. According to Kennard and James (1986), "the individual mesoscopic clots (mesoclots) within thrombolites are interpreted as discrete colonies or growth forms of calcified, internally poorly differentiated, and coccoid-dominated microbial communities." Kennard and James suggest that thrombolites form a continuum with stromatolites. That is, thrombolites grade through **undifferentiated boundstones** to stromatolites. Thrombolites occur primarily in Cambrian and Lower Ordovician rocks.

Massive fabrics may develop as a result of more or less continuous entrapment and agglutination of detrital particles, rapid sedimentation of detrital particles compared to algal growth, invasion and binding of previously deposited sediment by algae, continuous precipitation of carbonate in growing mucilaginous algal masses or diatom mats, or obliteration of original laminated or thrombolitic fabrics during diagenesis. **Radial fabrics** apparently form by radial growth of calcified algal filaments. For further description of these structures and their modes of origin see Monty (1976).

Stromatactis

Stromatactis cavities are particularly common in Paleozoic carbonate mudstones and wackestones associated with reefs and banks. These spar-filled cavities have a smooth base and an irregularly digitate roof, tend to occur in swarms, and have a reticulate (network) distribution (Bathurst, 1982). Figure 10.29 illustrates the typical shapes of stromatactis cavities with their sediment-filled floors and spar-filled upper parts. Although well known from numerous limestone units, the origin of these structures is problematic and controversial. Wallace (1987) cites some of the different viewpoints on origin and provides a short bibliography of pertinent papers on this subject. Origin of the structures has variously been linked to organisms in some way, such as decay of soft-bodied organisms, burrowing and bioturbation activities of crustaceans, diagenetic alteration and cementation of sponge networks, recrystallization of algae and bryozoans, and microbial accretion. Numerous inorganic processes have also been postulated, including slumping, dewatering and collapse, dynamic metamorphism, recrystallization of carbonate mud, pressure solution, and submarine crust formation.

On the basis of studies of limestones in eastern Victoria, Australia, Wallace (1987) proposes that the structures he studied formed by a process of internal erosion and sedimentation (internal reworking) that took place immediately below the sediment–water interface. Carbonate mud was eroded from the roofs of existing cavities and redeposited on the floors. This process left a digitate, erosional roof in each cavity and created a smooth floor that migrated upward as the cavity was filled with internal sediment. The unfilled part of the cavity above the floor subsequently filled with sparry cement. This postulated erosional-depositional process requires that a precursor cavity system be present to initiate the process of internal erosion. How such a precursor system of cavities may have formed is not clear.

FIGURE 10.29
Thin-section photomicrographs and
equivalent line drawings of
stromatactis cavities in Devonian
limestones, eastern Victoria,
Australia. A. Lime mud with
abundant stromatactis cavities. B.
Skeletal wackestone with abundant
stromatactis and shelter cavities
(cavities under fossils). C.
Packstone-grainstone with
irregularly distributed lime mud. In
the photos, (i) refers to first-
generation, inclusion-filled cement,
and (e) refers to second-generation,
clear, equant cement. In the line
drawings, lime mud is patterned,
skeletal elements are black, and
cement-filled cavities are left
blank. The bar scale in all
photomicrographs = 1 cm. *(After
Wallace, M. W., 1987, The role of
internal erosion and sedimentation
in the formation of stromatactis
mudstones and associated
lithologies: Jour. Sed. Petrology,
v. 57. Fig. 3, p. 697, reprinted by
permission of Society of Economic
Paleontologists and Mineralogists,
Tulsa, Okla.)*

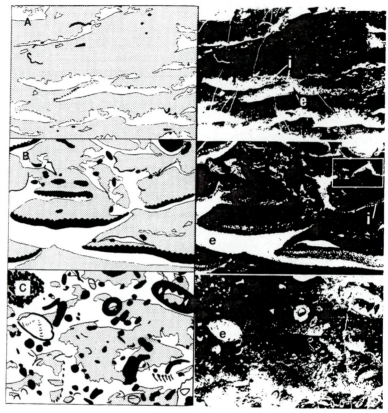

Birdseye Structure

This term was apparently introduced by Ham (1952) to describe small, spar-filled voids
in micritic limestones of Oklahoma. These structures (Fig. 10.30) superficially resemble
stromatactis but differ in some important respects. They are smaller (commonly about
1–3 mm), have shapes ranging from irregular to spherical or oval, lack internal sediment-
constructed floors, and generally occur irregularly distributed or arranged in patterns
parallel to bedding, rather than in reticulate patterns. Mechanisms for producing the
birdseye cavities suggested by various investigators include creation of voids in algal mats
by decay of the algae, escape of gas bubbles from decaying organic matter, bioturbation
by small organisms, formation of pores by desiccation of carbonate mud in tidal zones,
trapping of air bubbles in muds during temporary subaerial exposure, leaching of anhydrite nodules, and selective recrystallization during diagenesis.

Geopetal Fabric

Geopetal fabrics resemble stromatactis to the extent that they are cavities with nearly
smooth, sediment-constructed floors that are filled above the floors with sparry calcite
cement. They differ, however, in that the cavities occur either within whole fossils or

FIGURE 10.30
Birdseye texture in micritic
limestone. Devonian(?) limestone,
Alaska. Ordinary light. Scale
bar = 0.5 mm.

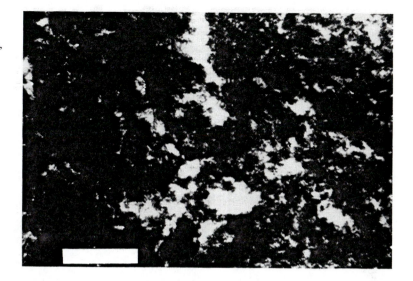

underneath convex-upward fossil shells such as mollusc and brachiopod shells (Fig. 10.31). These structures are also called shelter cavities and may occur in association with stromatactis. They form by precipitation or infiltration of micrite into or beneath a fossil shell, followed by precipitation of sparry calcite cement in the uppermost part of the cavity. Many geopetal structures are large enough to be easily seen in outcrops, where they provide useful indicators of tops and bottoms of beds.

Tepee Structures

Tepee structures are small- to large-scale polygons with upturned antiform margins that resemble a depressed inverted V (Fig. 10.32). These structures have been reported from submarine hardgrounds (a zone at the seafloor that has been lithified to form a hardened surface owing to early cementation and encrustation by sessile organisms and that is commonly bored by organisms and encrusted by solution) and from lacustrine and caliche

FIGURE 10.31
Geopetal fabric inside a brachiopod
shell. Micrite fills the lower part of
the shell, with sparry calcite
cement filling the uppermost part.
The top of the bed is toward the
top of the photograph. Devonian
limestone, Alaska. Crossed nicols.
Scale bar = 0.5 mm.

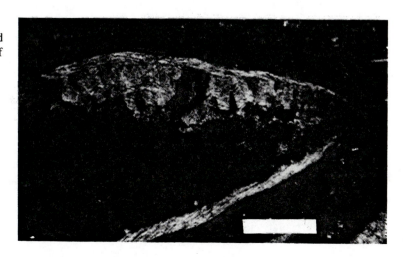

FIGURE 10.32

Tepee structure in limestone. Yates Formation (Permian), New Mexico. *(Photograph courtesy of E. F. McBride.)*

environments. Tepee structures occur as buckled margins of saucerlike megapolygons. They form when the surface area of carbonate crusts expands by some mechanism, causing the hardened sediment to crumple and form a pattern of megapolygons with upturned antiform margins. Submarine tepees occur mainly in shallow-water, subtidal, intertidal, and supratidal environments. They occur also in peritidal environments as well as in lacustrine and caliche environments. Although all tepees are apparently caused by expansion, the causes of expansion are different in different environmental settings. Table 10.10 lists the major setting of tepees and explains the postulated process of formation in each environmental setting. For a review of the origin and setting of tepee structures, see Kendall and Warren (1987).

Nodular Structures

Limestones composed of nodules set in a micritic or, less commonly, carbonate sand matrix occur in many ancient carbonate sequences. They appear to be particularly common in deeper-water deposits, but they have been reported also from shallower-water deposits (e.g., Noble and Howells, 1974). The nodules can range in size from less than a centimeter to tens of centimeters and in shape from rounded, subelliptical forms to irregular, elongated sausagelike forms. Nodular structures may be confined to relatively

TABLE 10.10

Origin and depositional setting of tepee structure

Environmental type	Process of formation
I. Submarine tepees	Expansion by pressure of crystallization in the sediment matrix and possibly in fractures in submarine crusts; indicates a hardground in shallow, carbonate-saturated waters
II. Peritidal tepees	Major expansion caused by fill by cement and sediment of fractures formed by minor expansion by force of crystallization or thermal contraction and expansion of cement fill of cracks; indicative of shallow water table and climate extremes about the margin of a shallow water body; a strandline indicator
III. Lacustrine tepees	
A. Groundwater tepees	Major expansion caused by fill by cement and sediment of fractures formed by thermal contraction and expansion; minor expansion caused by fluctuations of pore pressures in porous boxwork sediments below the tepee-affected crust; indicates periodic groundwater resurgence about the margins of shallow water body; a strandline indicator
B. Extrusion tepees	Major expansion caused by fill cement of fractures formed by thermal contraction and expansion; minor expansion by swelling and contraction of unconsolidated muds beneath tepee structures; indicative of ground resurgence about margins of nearshore salinas; a strandline indicator
IV. Caliche tepees	Expansion by a combination of the force of crystallization, wetting and drying, and rhizobrecciation; indicative of an exposure surface (paleosol) in a continental environment

Source: Kendall, C. G. St. C., and J. Warren, 1987, A review of the origin and setting of tepees and their associated fabrics: Sedimentology, v. 34. Table 2, p. 1010, reprinted by permission of Elsevier Science Publishers, Amsterdam.

thin units within a limestone body or they may make up a substantial fraction of the overall sequence, as shown in Figure 10.33. Many, but not all, nodular structures are associated with hardgrounds (see Section 12.2.2) or incipient hardgrounds (Kennedy and Garrison, 1975). Most theories advanced to explain the formation of nodular structure assume *in situ* early submarine cementation aided perhaps by compaction, pressure solution, or slumping. In a study of modern nodular carbonate sediments from the Bahamas, Mullins et al. (1980) call on a combination of physical (bottom currents), chemical, and biological (burrowing) processes and *in situ* submarine cementation to explain nodule formation.

Stylolites

Stylolites are suturelike seams in sedimentary rocks marked by irregular and commonly interlocking penetration of the two sides of the seam. They are most common in carbonate rocks but occur also in some sandstones and quartzites. Stylolite seams in limestones are characterized by concentrations of insoluble residues that may consist of clay minerals, other fine-size silicate minerals, iron oxides, or fine-size organic matter. Stylolites may occur on a microscale (grain-to-grain suturing) (Fig. 10.34) or on a much larger scale (Fig. 3.44; Fig. 12.12) where they cut across grains, micrite, and possibly cement. Stylolites are commonly oriented parallel to bedding planes, but they may occur at an angle to bedding planes or even perpendicular to bedding. Stylolites are believed to form

FIGURE 10.33
Rhythmically bedded nodular chalks, Middle Chalk (Cretaceous), Ballard Cliffs, Swanage, Dorset, England. *(From Kennedy, W. J., and R. E. Garrison, 1975, Morphology and genesis of nodular chalks and hardgrounds in the Upper Cretaceous of southern England: Sedimentology, v. 22. Fig. 13, p. 322, reprinted by permission of Elsevier Science Publishers, Amsterdam.)*

FIGURE 10.34
Microstylolite (arrow) in limestones of the Wadleigh Formation (Devonian), Alaska. Ordinary light. Scale bar = 0.5 mm.

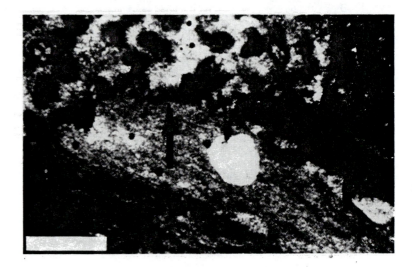

as a result of pressure solution, although the exact manner in which pressure solution works to produce the seams is poorly understood. Because stylolites are essentially diagenetic structures, they are discussed further in Chapter 12.

Bioherms

Bioherms are large, lenslike or moundlike bodies of carbonate carbonate rock that consist largely of the remains of sedentary organisms and that are enclosed in rocks of different lithology or character (Fig. 10.35). Some bioherms are composed of colonial organisms that constructed a rigid framework or core around which the bioherm accreted. Such a bioherm is called a **reef.** Some bioherms do not contain a framework core but are simply built of organisms that piled up in place as they died. Is this type of bioherm also a reef? Opinions differ. Many geologists call a bioherm that lacks a framework core a bank, mound, or mudmound, or they use the nongenetic term **carbonate buildup** rather than calling it a reef. See Dunham (1970) and Longman (1981) for some insight into this

FIGURE 10.35
Small bioherm (arrow) enclosed within well-bedded limestone. Point Peak Member of the Wilherns Formation (Cambrian), Texas. *(Photograph courtesy of W. M. Ahr.)*

problem. In any case, bioherms are common in Phanerozoic-age carbonate rocks, and they represent a type of structure not found in siliciclastic sedimentary rocks. Although most bioherms are composed in part of organic remains, the kinds of organisms that built bioherms have varied through time. The earliest bioherm-builders in Cambrian time were the archaeocyathids. These organisms have been succeeded through time as mound builders by organisms such as stromatoporoids, algae, bryozoans, rudists, and corals (James, 1983). Most modern framework-core reefs are built by corals.

10.7   Carbonate Microfacies and Marine Depositional Environments

10.7.1   Microfacies Analysis

The term **microfacies** refers to sedimentary facies that can be studied and characterized in small sections of a rock. The name is generally applied to characteristics that can be determined by study of thin sections with a petrographic microscope or by similar methods. Flügel (1982, p. 1) defines a microfacies as "the total of all the paleontological and sedimentological criteria which can be classified in thin-sections, peels, and polished slabs." According to Carozzi (1989, p. 24), a carbonate microfacies is "the total of the mineralogic, paleontologic, textural, diagenetic, geochemical, and petrophysical features of a carbonate rock." Carozzi (p. 24) also reviews the history of the microfacies concept. Microfacies analysis can be carried out by a variety of analytical techniques, including chemical, isotope, and X-ray diffraction analysis, and cathodoluminescence. Nonetheless, petrographic microscopic analysis has always been, and remains, the primary method for microfacies study.

The purpose of microfacies analysis is to provide a detailed inventory of carbonate rock characteristics (carbonate grain types, kinds and growth forms of fossils, size and shape of grains, nature of micrite, cement, particle fabrics) that can subsequently be related to depositional conditions. Thus, the ultimate aim of microfacies analysis is environmental interpretation. Microfacies analysis begins in the field with collection of samples. It then moves to the laboratory, where thin sections are prepared for study, which commonly includes both qualitative and detailed quantitative analysis. Such analysis commonly includes determination of (1) the relative abundance of the main constituents (grains, matrix, cements), (2) shapes, sizes, sorting, and other characteristics of the carbonate grains, including types of fossil organisms, (3) nature of the matrix, (4) kinds of cements, and (5) nature of the fabric (grain-support, mud-support, bioturbation, etc.). To insure that important characteristics of the rock are not overlooked during analysis, it is generally advisable to use a checklist of some type. You can make such a checklist yourself or refer to published checklists such as those of Flügel (1982, p. 394), Wilson (1975, p. 60), or Klovan (1964, Table 6).

Owing to the large number of criteria observed during microfacies analysis, statistical methods are commonly used to test the statistical validity of the data and to reduce the data to a more manageable level that can be used in environmental interpretation. Cluster and factor analysis are among the most commonly used statistical techniques. Flügel (1982, p. 388–394) provides an extended bibliography of such statistical applications. See also Carozzi (1989, p. 30–31). A final classification of each thin section into distinct microfacies types is reached at the end of statistical analysis.

10.7.2   Standard Microfacies Types

Flügel (1972, 1982) and Wilson (1975), in particular, have advocated use of carbonate microfacies data to establish a restricted number of major microfacies types that serve as models for all carbonate microfacies, regardless of the ages of the carbonate rocks. These microfacies are referred to as standard microfacies (SMF) types. They can be grouped into facies "belts," which are then used to build a generalized depositional model for carbonate rocks. Carozzi (1989, p. 33) disagrees with this approach and suggests that there is essentially an infinite number of depositional models. Thus, he regards the concept of standard microfacies types to be a "premature oversimplification" and suggests that such oversimplification even contradicts the purpose of microfacies techniques, "which is merely to unravel and explain the variety rather than to standardize the results into a rigid framework before its time."

Even though Carozzi is probably correct in stating that the standard microfacies concept is an oversimplification, it is nonetheless a useful simplification for beginning students and others not yet skilled in microfacies techniques. It provides a kind of standard to which we can compare our own analytical results; this can be exceedingly helpful in initial environmental interpretation. Even Flügel (1982, p. 403) cautions that standard microfacies types should not be overrated, and Wilson (1975, p. 64) acknowledges that organization of microfacies into a limited number of categories is an oversimplification. Nonetheless, if we recognize that these microfacies are not absolute standards, but simply norms for comparison, they serve a very useful purpose. Virtually everyone agrees that many depositional environments are complex and cannot be adequately represented by a simple model; however, if we exercise good judgment and common sense in use of models such as standard microfacies types, they can be invaluable aids in environmental interpretation. With these cautions in mind, let's take a look at Wilson's standard microfacies types.

Wilson (1975, p. 64) proposes 24 standard microfacies types, several of which are drawn from Flügel (1972). He selects names for these standard microfacies on the basis of major kinds of carbonate grains, paleontologic data, micrite abundance, and carbonate fabrics, using the textural classification of Dunham (1962) or of Embry and Klovan (1972), as appropriate (Table 10.11). Note also in Table 10.11 some terms that are not covered in these classifications, e.g., **calcisiltite** (a rock composed of carbonate grains of silt size) and **bioclasts** (fossil fragments). These standard microfacies types represent the most common kinds of carbonate microfacies; however, experience in working with carbonate rocks has shown that not every microfacies fits neatly into one of these categories. Students using these standard microfacies types as models should be prepared to accept the likelihood that some microfacies may contain elements of more than one of these standard types or may not fit any of the types exactly. Considering the wide variety and complexity of carbonate environments and the variations in fossil organisms and other geologic factors that have occurred with time, we could hardly expect anything else. These inconsistencies are likely to be frustrating for beginning students but become less of a problem as experience is gained in study of carbonate rocks.

10.7.3   Facies Belts and Depositional Environments

Wilson (1975, p. 26–27) suggests a general model for carbonate deposits that encompasses nine major **facies belts** corresponding to nine major carbonate environments. These facies belts are shown in Table 10.12. The major characteristics of the facies belts are as follows.

1. Facies belt 1—dark shales and carbonate mudstones; deposited in deeper-water basin environment, commonly below the oxygenation level
2. Facies belt 2—very fossiliferous limestones with shale interbeds; deposited on the open-sea shelf below storm-wave base but above the oxygenation level
3. Facies belt 3—fine-grained, graded to nongraded limestones, possibly containing exotic blocks derived from the foreslope; deposited on the toe of the foreslope
4. Facies belt 4—fine- to coarse-grained limestone with breccia and exotic blocks, deposited on the foreslope seaward of the platform edge; carbonate debris derived from facies belt 5
5. Facies belt 5—organic buildups (reefs and other bioherms) composed of various kinds of boundstones, particularly framestones; commonly make up the edge or rim of the carbonate platform but may not be present on all carbonate platforms
6. Facies belt 6—winnowed, sorted carbonate sands (calcarenites) composed particularly of skeletal grains derived from facies belts 4 and 5; ooids also common; deposited in very shallow water immediately landward of organic buildups or, if no buildups are present, at the very edge of the platform
7. Facies belt 7—mixed carbonate deposits that may include carbonate sands derived from belt 6, wackestones, mudstones; possible interbeds of shale or silt; patch reefs or other bioherms may be present; deposited in shallow water on the open-shelf platform where water circulation is normal

TABLE 10.11
Standard carbonate microfacies types

SMF No.	Name	Distinguishing characteristics
1	Spiculite	Dark, organic-rich, clayey mudstone or wackestone containing silt-sized spicules; spicules commonly oriented, generally siliceous monaxons, commonly replaced by calcite
2	Microbioclastic calcisiltite	Small bioclasts and peloids with a fine grainstone or packstone texture; mm-scale ripple cross-lamination common
3	Pelagic mudstone or wackestone	Micritic matrix with scattered fine sand- or silt-sized grain composed of pelagic microfossils (e.g., radiolarians or globigerinids) or megafauna (e.g., graptolites or thin-shelled bivalves)
4	Microbreccia or bioclastic-lithoclastic packstone	Worn grains of originally robust character; may consist of locally derived bioclasts and/or previously cemented lithoclasts; may also include quartz, chert, or other kinds of carbonate fragments; commonly graded
5	Bioclastic grainstone-packstone or floatstone	Composed mainly of bioclasts derived from organisms inhabiting reef top and flanks; geopetal fillings and infiltered fine sediment in shelter cavities common
6	Reef rudstone	Large bioclasts of reef-top or reef-flank organisms; no matrix material
7	Boundstone	Composed of *in situ* sessile organisms. May be called **framestone** if composed of massive upright and robust forms, **bindstone** if composed of encrusting lamellar mats enclosing and constructing cavities and encrusting micrite layers, and **bafflestone** if composed of delicate, complex, frond-like forms; commonly, micrite clotted or vaguely pelleted
8	Whole-fossil wackestone	Sessile organisms rooted in micrite, which contains only a few scattered bioclasts; well-preserved infauna and epifauna
9	Bioclastic wackestone or bioclastic micrite	Micritic sediment containing fragments of diverse organisms jumbled and homogenized through bioturbation; bioclasts may be micritized
10	Packstone-wackestone with coated and worn bioclasts in micrite	Sediment displays textural inversion; i.e., grains show evidence of formation in high-energy environment but contain mud matrix
11	Grainstone with coated bioclasts in sparry cement	Bioclasts cemented with sparry cement; bioclasts may be micritized

12	Coquina, bioclastic grainstone or rudstone	A shell hash in which certain types of organisms dominate (e.g., dasyclads, shells, or crinoids); lacks mud matrix
13	Oncoid biosparite grainstone	Composed mainly of oncoids in sparry cement
14	Lags	Coated and worn particles; may include ooids and peloids that are blackened and iron-stained; with phosphate; may also include allochthonous lithoclasts
15	Oolite, ooid grainstone	Well-sorted, well-formed, multiple-coated ooids ranging from 0.5 to 1.5 mm in diameter; fabric commonly overpacked; invariably cross-bedded
16	Pelsparite or peloidal grainstone	Probable fecal pellets; may be admixed with concentrated ostracod tests or foraminifers; may contain cm-thick graded laminae and fenestral fabric
17	Grapestone pelsparite or grainstone	Mixed facies of isolated peloids, agglutinated peloids, and aggregate grains (grapestones and lumps); may include some coated grains
18	Foraminiferal or dasycladacean grainstone	Consists of concentrations of tests, commonly mixed with peloids
19	Loferite	Laminated to bioturbated, pelleted lime mudstone-wackestone; may grade to pelsparite with fenestral fabric; ostracod-peloid assemblage common in mudstones; may also include micrite with scattered foraminifers, gastropods, and algae
20	Algal stromatolite mudstone	Stromatolites
21	Spongiostrome	Tufted algal fabric in fine lime mud sediment
22	Micrite with large oncoids	Wackestone or floatstone containing oncoids
23	Unlaminated, homogeneous, unfossiliferous pure micrite	Micrite; may contain crystals of evaporitic minerals
24	Rudstone or floatstone with coarse lithoclasts and/or bioclasts	Clasts commonly consist of unfossiliferous micrite or calcisiltite, and may have an edgewise or imbricate arrangement; may be cross-bedded; matrix sparse

Source: Wilson (1975, p. 64–69).

8. Facies belt 8—bioclastic wackestones, lithoclastic and bioclastic sands, pelleted carbonate mudstones, stromatolites, interbeds of shale or silt; deposited in shallow water on inner platform where water circulation may be restricted

9. Facies belt 9—nodular dolomites and anhydrites (on platforms where evaporative conditions exist); stromatolites; siliciclastic muds or silts; deposited in intertidal to supratidal zone

Note from Table 10.12 that each facies belt is characterized by a few standard microfacies types, but that a particular standard microfacies type can appear in more than one facies belt.

The facies belts shown in Table 10.12 are not drawn to scale. Belts 3, 4, 5, and 6, in particular, are much narrower than illustrated in Table 10.12. Also, belts 5 and 6 may not both be present on all carbonate shelves. Figure 10.36 depicts the facies belts and their associated carbonate environments at a more realistic scale.

As mentioned, Carozzi (1989, p. 33) disagrees with Wilson's concept of a few standard microfacies types and a limited number of depositional models. He maintains

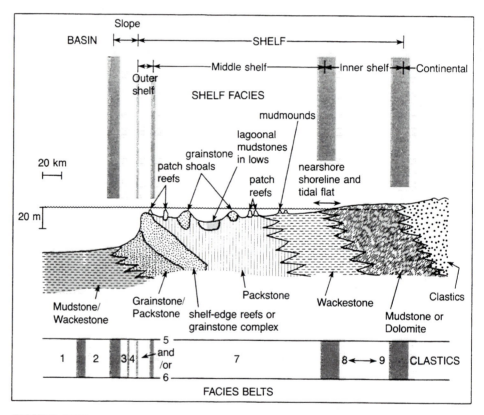

FIGURE 10.36
Schematic profile of a typical carbonate-shelf environment. Wilson's (1975) facies belts are shown to approximate scale in this figure. The facies belt numbers correspond to the facies belt numbers in Table 10.12. *(After Wilson, J. L., and C. Jordan, 1983, Middle shelf environment, in Scholle, P. A., et al., Carbonate depositional environments: Am. Assoc. Petroleum Geologists Mem. 33. Fig. 1a, p. 298, reprinted by permission of AAPG, Tulsa, Okla.)*

that there is an "infinite variety of depositional models, as infinite as the number of possible interactions between biological and physical factors in a carbonate environment." Carozzi describes numerous detailed microfacies that characterize a wide variety of carbonate depositional environments. Readers are referred to his extensive work on this subject (Carozzi, 1989, "Carbonate Rock Depositional Models: a Microfacies Approach") for details.

As suggested by Table 10.12, limestones are deposited in a variety of marine environments. From the shoreline outward, these environments may range from the supratidal through the shallow subtidal, open-shelf lagoon, platform margin (carbonate sands or carbonate buildups), and slope to the deep-marine basin environment. Within these environments, a variety of depositional processes operate to generate, transport, and deposit sediment. These processes may include those that are mainly chemical (e.g., inorganic precipitation of ooids and lime mud), mainly biochemical (e.g., secretion of $CaCO_3$ by organisms to build skeletal elements, decomposition of calcareous algae to release aragonite needles, organic production of pellets, micritization of skeletal grains by endolithic algae and fungi), and mainly physical (e.g., erosion of semiconsolidated lime mud to generate intraclasts, physical breakdown of larger particles, transport and deposition of terrigenous grains). Thus, depending upon the environment and the process or processes involved, the many kinds of limestones discussed in preceding sections of this chapter are generated. Environmental analysis is not the major focus of this book; therefore, I do not provide here greater coverage of marine carbonate depositional environments and depositional processes than that discussed above. The subject of carbonate environments is so extensive in scope that the entire book could be devoted to that subject alone. In fact, several books dealing with carbonate environments are available. See, for example, Bhattacharyya and Friedman (1983), Friedman (1969), and Scholle et al. (1983). For a shorter summary of carbonate environments and depositional processes, see Boggs (1987, Ch. 13) or Sellwood (1986).

10.7.4   Depositional Environments and Compositional Maturity of Limestones

The compositional maturity of siliciclastic sedimentary rocks is measured by the abundance of highly stable grains (quartz, chert, quartzite/quartz arenite fragments, ultrastable heavy minerals) relative to metastable and unstable grains. High compositional maturity reflects the intensity and duration of the chemical and physical processes involved in weathering, transportation, and deposition (and to a limited extent diagenesis) that combine to eliminate less-stable grains and selectively concentrate the more-stable constituents. This compositional maturity concept cannot easily be applied to limestones, which are made up primarily of $CaCO_3$ minerals (calcite, Mg-calcite, aragonite). These minerals are deposited in the same general, stable, marine environment, which does not change chemically very much before final burial.

Smosna (1987) makes the interesting suggestion, however, that compositional maturity of limestones can be expressed by the relative abundance of the various kinds of carbonate grains (allochems), micrite, and terrigenous grains that make up limestones. Specifically, he groups these constituents into six categories: (1) intraclasts (including aggregate grains and lumps), (2) ooids, (3) fossils, (4) peloids, (5) micrite matrix, and (6) terrigenous grains. Smosna proposes that the relative abundance of these compositional end members in a carbonate sediment is a function of relative rates of production by

TABLE 10.12

Generalized depositional environments, facies belts, and standard microfacies types

FACIES BELTS	1	2	3	4	5	6	7	8	9
ENVIRONMENT	BASIN	OPEN SEA SHELF (NERITIC)	DEEP SHELF MARGIN	FORESLOPE	PLATFORM EDGE ORGANIC BUILDUPS	WINNOWED PLATFORM EDGE	OPEN PLATFORM (SHELF LAGOON)	RESTRICTED PLATFORM	EVAPORITE PLATFORM (SABKHA)
SCHEMATIC FACIES PROFILE	Sea level — Normal wave base — Storm wave base — Oxygenation level ← WIDE FACIES BELTS →		← VERY NARROW BELTS →				← WIDE FACIES BELTS →		
SMF TYPES	1,2,3	2,8,9,10	2,3,4	4,5,6	7,11,12	11,12,13,14,15	8,9,10,16,17,18	16,17,18,19,21,22,23,24	20,23
MAJOR FACIES	Shale, limestone	Fossiliferous limestone with minor siliciclastics	Fine-grained graded to nongraded limestone with exotic blocks	Fine- to coarse-grained limestone with breccia and exotic blocks	Boundstone (framestone, bindstone, bafflestone)	Winnowed carbonate sands	Carbonate sand wackestone-mudstone; bioherms; fine siliciclastics	Bioclastic wackestone, lithoclastic and bioclastic sand, carbonate mudstone	Nodular dolomite and anhydrite; possible laminated evaporites

	1	2	3	4	5	6	7	8	9	
LITHOLOGY	Dark shale, siltstone, thin-bedded carbonate mudstone; even mm-scale lamination; rhythmic bedding; cross-bedding	Limestone (bioclastic and whole fossil wackestone, calcisiltite); interbedded marls bioturbated; wavy to nodular beds; shale or siltstone interbeds	Carbonate mudstone calcisiltite; cherty; massive; lenses of graded beds, lithoclasts, exotic blocks; rhythmic bedding, shale or silt interbeds	Carbonate breccia, calcisiltite; bioclastic wackestone-packstone; clasts of various sizes; slump structures, inclined bedding, exotic blocks, bioherms; shale-silt interbeds	Massive limestone, dolomite; boundstone and pockets of grainstone-packstone; massive organic structures or open framework with roofed cavities; no siliciclastics	Calcarenite (well-sorted grainstone) with ooids; skeletal grains, etc.; dolomite, admixed quartz sands; cross-bedding	Various kinds of carbonates (grainstones to mudstones); bioturbation tracks common; interbeds of siliciclastics	Dolomitic limestone; pelleted mudstone and grainstone, laminated in part; coarse intraclastic wackestone; birdseye texture, stromatolites, mm-laminations; some graded beds; siliciclastic interbeds	Dolomite and anhydrite, nodular; irregular lamination; caliche; may grade to red-beds; siliciclastics may be important	LITHOL-OGY
FOSSILS	Exclusively nektonic-pelagic faunas; locally abundant on bedding planes	Very diverse shelly fauna, including both infauna and epifauna	Redeposited shallower-water faunas; some indigenous benthonic and pelagic faunas	Colonies of whole-fossil organisms; bioclastic debris redeposited from belt 5	Major frame-building colonies with ramose forms in pockets; growth form determined by water energy	Worn, abraded organisms derived from belts 4 and 5; few indigenous organisms such as gastropods, foraminifers, dasyclads	Molluscs, sponges, foraminifers, algae abundant; open marine faunas (echinoderms, cephalopods, brachiopods, etc.) lacking; patch reefs	Very limited fauna, mainly gastropods, algae, certain foraminifers (e.g., miliolids) and ostracods	Stromatolites; almost no indigenous faunas	FOSSILS

Source: Redrawn from Wilson, J. L., 1975, Carbonate facies in geologic history. Fig. II-4, p. 26–27, reprinted by permission of Springer-Verlag, New York.

various environmental processes. Thus, the percentage of a given constituent in a carbonate deposit may be low, not because the environment was unfavorable to production of that constituent but because the constituent is diluted by others that formed more rapidly.

Smosna defines carbonate compositional maturity as "a type of sedimentary maturity in which the lime sediment approaches the compositional end-member (one of the six constituents listed above) to which it is driven by the environmental processes that operate upon it." As discussed in the preceding section, these processes can include a wide array of chemical, physical, and biological activities. Carbonate sediments are defined as compositionally immature, mature, or supermature as follows:

(1) **Compositionally immature sediment** consists of a mixture of several of the six major constituents, four or more of which exceed 10 percent by volume. Immature sediment is produced in an environment in which several formative processes are operating simultaneously, and more or less at the same rate.

(2) **Compositionally mature sediment** contains two or three dominant constituents, each in excess of 10 percent.

(3) **Compositionally supermature sediment** contains only one dominant constituent, which makes up more than 90 percent of the sediment. Supermature sediment is produced in an environment where a single formative process has operated to completion. Note that a lime mud, largely free of carbonate or terrigenous material, is considered to be compositionally supermature by this definition.

10.7.5   Chemistry of Marine Carbonate Deposition

Regardless of the specific marine depositional environment, the formation of carbonate sediment involves chemical processes in some way. Although some of these processes may be biogenically induced, and mechanical processes may erode and transport carbonate sediment, chemistry remains a first-order control on the formation of carbonate deposits. As discussed in Section 10.2, aragonite, calcite, and Mg-calcite are the dominant minerals in limestones. What chemical factors are particularly important to precipitation of these minerals? Let's begin an answer to this question by examining the **calcium carbonate solubility product,** which is related to CO_2 partial pressure, temperature, and salinity.

The equilibrium solubility product (K) of calcium carbonate is the ionic activity solubility product of calcium and carbonate in any solution that is in equilibrium with calcium carbonate, as shown by the relationship

$$K = \frac{{}^{\alpha}Ca^{2+} \times {}^{\alpha}CO_3{}^{2-}}{{}^{\alpha}CaCO_3} \qquad (10.6)$$

The terms ${}^{\alpha}Ca^{2+}$ and ${}^{\alpha}CO_3{}^{2-}$ refer to the activities of Ca^{2+} and $CO_3{}^{2-}$ ions in the aqueous solution, and ${}^{\alpha}CaCO_3$ refers to the activity of $CaCO_3$ in the carbonate phase. Activities are a function of concentration, and the relation between activity and concentration is

$$\alpha = \gamma X \qquad (10.7)$$

where X for solid phases (e.g., $CaCO_3$) is the mole fraction (number of moles of a component divided by the total number of moles of all components in solid solution) of the pure end member in the solid solution. For aqueous species, it is expressed as the concentration of the ion in the electrolyte solution (e.g., mol/kg or mol/L). γ is the rational activity coefficient (see Berner, 1971, p. 21, for an equation for calculating calcium activity). For pure end-member composition, $\gamma = 1$ and $X_{CaCO_3} = 1$. For very dilute solutions, activity and concentration are approximately the same.

The solubility product (K) is the value measured at equilibrium or calculated by use of the equation

$$\Delta G = -RT \ln K \tag{10.8}$$

where ΔG refers to change in Gibbs free energy, R is the gas constant (8.314 joules·mol^{-1}·K^{-1}), T is absolute temperature (Kelvin), and ln is natural logarithm. Gibbs free energy (G) = $H - TS$, where H is enthalpy = $U + PV$, U is internal energy, P is pressure, V is volume, T is temperature (Kelvin), and S is entropy.

The calcium carbonate solubility product is a constant for a given set of conditions but varies as a function of CO_2 partial pressure, temperature, and salinity. The solubility product increases with rising partial pressure of CO_2 and increasing salinity but decreases with increasing temperature. In practical terms, this relationship means that aragonite and calcite are most likely to precipitate under conditions of (1) low CO_2 partial pressure, (2) low salinity, and (3) high temperature. Further, the solubilities of aragonite and calcite under a given set of conditions are not the same. Aragonite (and Mg-calcite) is generally more soluble than calcite; however, calcite may not precipitate in preference to aragonite owing to the presence of specific inhibitors in seawater (e.g., Mg).

The amount of dissolved carbon dioxide in seawater exerts a major control on the solubility of calcium carbonate, as indicated by the following equilibrium relations:

$$CO_2 + H_2O \leftrightarrow H_2CO_3 \tag{10.9}$$

$$H_2CO_3 \leftrightarrow H^+ + HCO_3^- \tag{10.10}$$

$$HCO_3^- \leftrightarrow H^+ + CO_3^{2-} \tag{10.11}$$

The interaction between CO_2, H_2O, and **solid** $CaCO_3$ can be summarized as

$$CO_2 + H_2O + CaCO_3 \text{ (solid)} \leftrightarrow Ca^{2+} \text{ (aq)} + 2HCO_3^- \text{ (aq)} \tag{10.12}$$

The dissociation of H_2CO_3 and HCO_3^- in seawater can be expressed approximately by the following equations:

$$K'1 = \frac{[H^+] [HCO_3^-]}{[H_2CO_3]} \tag{10.13}$$

$$K'2 = \frac{[H^+] [CO_3^{2-}]}{[HCO_3^-]} \tag{10.14}$$

where $K'1$ and $K'2$ are apparent dissociation constants (which change with temperature, pressure, and salinity), and brackets denote concentration. The following equations show the relation of pH to the carbon dioxide/water system (Milliman, 1974, p. 9):

$$pH = pK'1 + \log \frac{[HCO_3^-]}{[H_2CO_3]} \tag{10.15}$$

$$pH = pK'2 + \log \frac{[CO_3^{2-}]}{[HCO_3^-]} \tag{10.16}$$

where p refers to negative logarithm. At pH values less than 7.5, the main CO_2 species are H_2CO_3 and HCO_3^- and Equation 10.15 approximates the system. At pH values greater than 7.5, the dominant species are HCO_3^- and CO_3^{2-} and Equation 10.16 applies.

Clearly, the amount of CO_2 dissolved in seawater affects the pH of the water. The decrease in pH with increasing CO_2 partial pressure is not, however, a linear function because the high alkalinity of seawater tends to buffer the seawater system. Carbonate alkalinity *(CA)* is defined as

$$CA = [HCO_3^-] + 2[CO_3^{2-}] \qquad (10.17)$$

For each mole of water and dissolved CO^2 that react, some yield H^+ and HCO_3^-. Owing to the high alkalinity of seawater, however, a high proportion of the dissolved CO_2 may form undissociated H_2CO_3. Also, some of the CO_2 is held in CO_3^{2-}. Thus, the addition of n moles of CO_2 may not release $2n$ moles of H^+, as predicted by Equations 10.9–10.11, but a much smaller quantity. Therefore, the pH of seawater may change less than theoretical considerations predict with respect to changes in partial pressure of CO_2. According to Bathurst (1975, p. 233), values of pH for seawater rarely fall outside the range of 7.8 to 8.3. On the other hand, fresh waters are poorly buffered and may show much larger changes of pH with variations in CO_2 partial pressure.

Nonetheless, CO_2 partial pressure exerts a very important control on carbonate solubility, and any process that removes CO_2 from seawater alters its carbonate equilibrium and may bring about carbonate precipitation (Equation 10.12). Thus, loss of CO_2 owing to increased temperature, decreased pressure (water agitation), or photosynthesis can theoretically cause precipitation of $CaCO_3$.

Although a great deal of the carbonate sediment in the geologic record is probably related to organic activity in some way (Section 10.4.3), an important fraction of this sediment may be of abiotic (inorganic) origin. Conventional wisdom has it that both the composition and crystal habit of abiotic carbonate minerals are influenced by the Mg/Ca ratio of the water. Precipitation from waters with low Mg/Ca ratios, such as meteoric waters, is thought to produce equant crystals of low-magnesium calcite. By contrast, precipitation from waters with high Mg/Ca ratios, such as seawater or hypersaline brines, is thought to form acicular or elongate crystals of mainly high-magnesian calcite and/or aragonite. Given and Wilkinson (1985) maintain, however, that many exceptions to this pattern occur in modern carbonate environments.

All this is by way of pointing out that the influence of Mg/Ca ratios on the mineralogy, composition, and morphology of abiotic carbonate minerals is an important and controversial problem. Some investigators (e.g., Mucci and Morse, 1983) suggest that the Mg/Ca ratio of the water is the primary factor that determines whether abiotic carbonate minerals are calcite, high-magnesium calcite, or aragonite. Several workers in addition to Mucci and Morse (e.g., Berner, 1975; Reddy and Wang, 1980) have suggested that Mg^{2+} strongly inhibits the precipitation of calcite and that aragonite or Mg-calcite preferentially precipitates in the presence of Mg^{2+} in seawater. Some growth inhibitors may also affect aragonite precipitation. For example, Berner et al. (1978) demonstrated experimentally that organic compounds found in natural humic and fulvic acids or in phosphatic compounds can form thin organophosphatic coatings on aragonite seed nuclei, inhibiting nuclei growth and preventing or delaying aragonite precipitation.

On the other hand, Given and Wilkinson (1985) believe that mineralogy, composition, and morphology are controlled largely by the kinetics of surface nucleation and the amount of reactants, principally carbonate ions, at growth sites. That is, Mg/Ca ratios only indirectly control the precipitated phases. They suggest that equant crystals form under conditions where CO_3^{2-} ion concentration and/or rates of fluid flow are low. Under these conditions, growth sites on precipitating crystals are starved for carbonate ions, c-axis growth is retarded, and an equant crystal results. Where carbonate ions are abundant and/or rates of fluid flow are high, c-axis growth is enhanced and an acicular or elongate crystal results. Furthermore, they maintain that the Mg content of calcite is also controlled by crystal-growth rates. At high CO_3^{2-} concentration and rapid crystal-growth rates, acicular high-Mg calcite forms. At lower CO_3^{2-} concentrations and lower crystal-growth rates, low-Mg calcite forms. Finally, they suggest that aragonite precipitation is favored when rates of reactant supply are high, whereas calcite forms when rates are low. Morse (1985) takes strong exception to Given and Wilkinson's stance that carbonate-ion concentration and crystal-growth rates are the dominant factors controlling the carbonate mineral phases precipitated. He regards this hypothesis as largely incorrect and unsupported by experimental and observational data. In retrospect, it seems likely that both kinetics and Mg/Ca ratios influence the precipitation of abiotic carbonates. It appears, however, as it so often does in science, that more research is needed before we can determine the relative importance of each of these factors.

10.8   Nonmarine Carbonates
10.8.1   Introduction

Because most carbonate rocks in the stratigraphic record are of marine origin, the preceding sections of this chapter have focused almost exclusively on marine limestones. Carbonate rocks also form in a variety of nonmarine settings, including lakes, streams, springs, caves, soils, and dune environments. The volume of these nonmarine or terrestrial carbonates is small, but they are an interesting addition to the overall carbonate record. Also, when they can be identified in ancient deposits, they make useful paleoenvironmental indicators. Therefore, a very brief description of the principal kinds of nonmarine carbonate rocks is given here.

10.8.2   Lacustrine Carbonates

Carbonate sediments occur in some freshwater lakes such as Lake Zürich and Lake Constance in Europe, as well as in saline lakes such as Great Salt Lake, Utah, and ephemeral playa lakes of Death Valley, California. The principal carbonate mineral formed in freshwater lakes is low-Mg calcite. The deposits of saline lakes may include low-Mg calcite, high-Mg calcite, and protodolomite (Chapter 11). Aragonite and magnesite ($MgCO_3$) may also occur in lake sediments but are uncommon minerals. Differences in the mineralogy and composition of these lacustrine carbonate minerals has commonly been regarded to reflect variations in Mg/Ca ratios of the lake water. That is, low-Mg calcite forms at at low Mg/Ca ratios (<2), high-Mg calcite forms at high Mg/Ca ratios (2–12), and aragonite and magnesite form at very high Mg/Ca ratios (>12). As discussed above, however, Given and Wilkinson (1985) maintain that carbonate ion availability and crystal-growth rate may also be important considerations. Deposits of saline lakes with very high alkalinity may include some rare carbonate minerals that commonly do not occur in marine environments; e.g., vaterite ($v\text{-}CaCO_3$), monohydrocalcite ($CaCO_3 \cdot H_2O$), trona ($NaHCO_3 \cdot Na_2CO_3 \cdot 2H_2O$), nahcolite ($NaHCO_3$), and natron ($Na_2CO_3 \cdot 10H_2O$).

Lacustrine carbonate sediments may include **abiotic precipitates, algal carbonates,** and **carbonate shell accumulations.** These carbonate materials may be mixed to various degrees with organic matter, biogenic silica (mainly diatom frustules), fine detrital siliciclastics, and evaporite minerals. Abiotic precipitation is probably important mainly in saline lakes in areas where evaporation rates are high. Under these conditions, both loss of water and loss of CO_2 can trigger precipitation of calcite and Mg-calcite. Important quantities of detrital silt and clay may occur with the carbonate minerals, as well as evaporitic minerals such as gypsum and halite. In ephemeral lakes such as playa lakes, variations in influx of terrigenous detritus may be reflected in alternating layers of mud and salts, with various amounts of carbonate minerals. In freshwater lakes, precipitation of low-Mg calcite probably occurs largely as a result of CO_2 loss owing to high surface-water temperatures and the photosynthetic activities of planktonic algae. Again, influx of detrital clay and silt may be significant, and this detritus may mix with carbonate sediment to form chalky or marly deposits. Alternatively, terrigenous detritus may be so abundant that it completely masks or swamps carbonate sediment. In some lakes, varves consisting of alternating thin layers of calcite-rich and clay-rich sediments may be present, reflecting variations in depositional conditions from summer to winter. In addition to their role in photosynthesis, algae may also form stromatolites and oncolites in both freshwater and saline lakes. Some algae such as *Chara* produce calcareous tubes and round, calcified reproductive bodies that become part of the carbonate sediment. Rooted, aquatic plants may induce precipitation of carbonate, which may cover the plants with a heavy carbonate crust. Nonmarine invertebrates such as ostracods, molluscs, and gastropods are common in many lakes and may generate shell deposits in the marginal areas of the lakes. These deposits commonly make up only a small volume of the overall carbonate sediments. Ooids are present in some saline lakes, e.g., Great Salt Lake, Utah, and one occurrence of low-Mg ooids in a freshwater lake (Higgins Lake, Michigan) has also been reported. For an excellent short review of carbonate deposition in lakes, see Dean and Fouch (1983).

10.8.3 Carbonates in Rivers, Streams, and Springs

Only a very small volume of carbonate sediments forms in the flowing water of rivers, streams, and springs. Commonly, such water is too undersaturated with calcium carbonate to precipitate carbonate minerals; however, saturation may occur in waters that drain regions underlain by carbonate rocks. Carbonate precipitated from streams and cold-water springs commonly consists of low-Mg calcite, whereas precipitates from hot springs may also include aragonite. The terminology of carbonate sediments formed in these freshwater environments is a bit confusing. The general name for all of these deposits is **travertine.** On the other hand, some authors apply the name travertine only to the more massive, dense, finely crystalline varieties of these deposits. These carbonates range in color from tan to white or cream, and some may be banded (Fig. 10.37). The more porous, spongy, or cellular varieties of these freshwater carbonates (Fig. 10.38), which commonly form as encrustations on plant remains, are called **tufa.** Pedley (1990) defines tufa as ''cool water deposits of highly porous or 'spongy' freshwater carbonate rich in microphytic and macrophytic growth, leaves and woody tissue.'' Some workers use the term **calcareous sinter** to refer to both travertine and tufa.

FIGURE 10.37
Crudely banded travertine.
Unknown location. *(Specimen
furnished by R. A. Linder.)*

FIGURE 10.38
Tufa formed in a hot spring,
Nevada. *(University of Oregon
collection.)*

Precipitation of travertine occurs predominantly in springs and spring-fed lakes and at waterfalls or cascades. Chafetz and Folk (1984) suggest that, morphologically, travertines can occur as (1) waterfall or cascade deposits, (2) lake-fill accumulations, (3) sloping mounds, fans, or cones, (4) terraced mounds, and (5) fissure ridges. Precipitation of travertines requires that groundwaters or streams be supersaturated with calcium carbonate with respect to calcite and supersaturated in CO_2 with respect to air. Precipitation can occur in cold-water springs owing to loss of CO_2 resulting from higher temperatures at the mouth of the spring and exposure of spring water to the atmosphere (decrease in pressure). Agitation of water in the springs and photosynthesis by plants are additional factors in CO_2 loss. In hot springs, evaporation of water around the mouths of the springs may also play a role in CO_2 loss. Away from the mouths of springs, photosynthesis by blue-green algae (cyanobacteria) and mosses is a more important factor in CO_2 loss. Tufas apparently form as a result of precipitation of calcite onto plants such as mosses and algae. Chafetz and Folk (1984) conclude that bacteria are also important agents in precipitation of travertines. Precipitation of travertine at waterfalls results from CO_2 loss as a result of decreased pressure arising from strong water agitation (Julia, 1983) and possibly also from photosynthesis by algae and mosses. These plants become calcified by the travertine, resulting in an irregular tangle of plant molds and inorganic crusts and cements. For further discussion of tufas and travertines, see Pedley (1990) and Heimann and Sass (1989).

Thus, depending upon their specific mode of origin, travertines form a wide variety of deposits. These deposits may range from vertical sheets built by encrustations of mosses in hanging springs and cascades to horizontal sheets formed in pools and dams. Travertines include varieties that are relatively dense and massive, relatively dense and banded, and porous, spongy, cellular, or laminated owing to encrustation on algae, mosses, and other plants. In addition to layered travertines, ooids have been reported from the agitated waters of some hot springs. Ancient travertines commonly occur in association with sedimentary strata or alluvium but occur also in some volcanic and granitic terrains. A common association of travertines with faults has also been reported.

10.8.4   Speleothem (Cave) Carbonates

Carbon dioxide–rich groundwaters that migrate through carbonate formations dissolve away parts of the formations and create solution pipes, sinks, and caves. The geomorphological features resulting from such solution activity are referred to as **karst** features (Fig. 10.39). When carbonate- and CO_2-saturated groundwaters enter air-filled caves, carbonate precipitation occurs on a massive scale owing to loss of CO_2, probably as a result of decreased pressure and evaporation. Therefore, caves in carbonate terrains are the sites of extensive carbonate deposition, which may take the form of **stalactites** (conical projections hanging from the roof, formed by dripping water), **stalagmites** (conical dripstone projecting upward from the floor), laminated **flowstones** (formed by flowing water), and **globoids** (cave pearls or cave pisolites formed by precipitation of carbonate as concentric layers around a nucleus). For example, Jones and MacDonald (1989) report cave pisolites up to 8 cm long that formed in rimstone pools in a cave on Grand Cayman. They suggest that these pisoids formed by a combination of inorganic precipitation and organic processes (calcification of filamentous microorganisms). Caves may also contain collapse breccia consisting of carbonate blocks that have fallen from cave roofs (Fig. 10.39).

10.8.5   Caliche (Calcrete) Carbonates

Soils in arid to semiarid regions, especially those developed on underlying carbonate rocks, may become so enriched in calcium carbonate that they form a caliche or calcrete deposit. Genetically, caliche is defined as a fine-grained, chalky to well-cemented, low-magnesian calcite deposit that formed as a soil in or on preexisting sediments, soils, or

FIGURE 10.39

Features of a mature karst profile. The vadose zone is the groundwater zone of aeration, and the phreatic zone is the zone of saturation where pore space is filled with water. Note that speleothems occur especially in the vadose zone, whereas collapse breccias and other cave sediments are common in the phreatic zone. *(After Esteban, M., and C. F. Klappa, 1983, Subaerial exposure environment, in Scholle, P. A., et al., eds., Carbonate depositional environments: Am. Assoc. Petroleum Geologists Mem. 33, Fig. 5, p. 4, as modified by Scoffin, T. P., 1987, Carbonate sediments and rocks. Fig. 12.1, p. 147, reprinted by permission of Blackie, Glasgow.)*

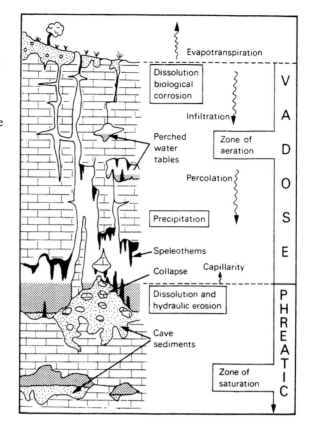

rocks. Esteban and Klappa (1983) suggest, however, as a more useful descriptive definition that "caliche is a vertically zoned, subhorizontal to horizontal carbonate deposit, developed normally with four rock types: (1) massive-chalky, (2) nodular-crumbly, (3) platy or sheet-like, and (4) compact crust or hardpan." An idealized caliche profile is shown in Figure 10.40. Esteban and Klappa stress that many variations of this profile exist and that the only consistent relation is that the massive-chalky rock grades downward into the original rock or sediment through a transition zone. These caliche profiles may extend downward below the soil surface for distances ranging from <1 m to a few tens of meters.

Soils are formed on carbonate bedrock by physical and chemical weathering processes that result in concentration of insoluble residues from the bedrock. $CaCO_3$ accumulates in the upper soil profile owing to evaporation of carbonate-saturated pore waters drawn to the surface by capillary action. Continued carbonate precipitation and downward mobilization result in formation of a **hardpan,** which is well indurated, lacks visible porosity, and is more resistant to weathering than underlying horizons. This hardpan may contain rhizoliths (root traces) and pisoids and macroscopically may appear massive or structureless, laminated, brecciated, or nodular. **Platy caliche,** which is characterized by horizontal to subhorizontal, platy, wavy, or thinly bedded habit, commonly occurs immediately below the hardpan. Platy caliche may also contain rhizoliths but is more friable and has greater porosity than the hardpan. It probably represents an immature stage in hardpan development. Searl (1989) reports an unusual occurrence of extensive sheetlike and fan developments of columnar calcite within and immediately below hardpan layers

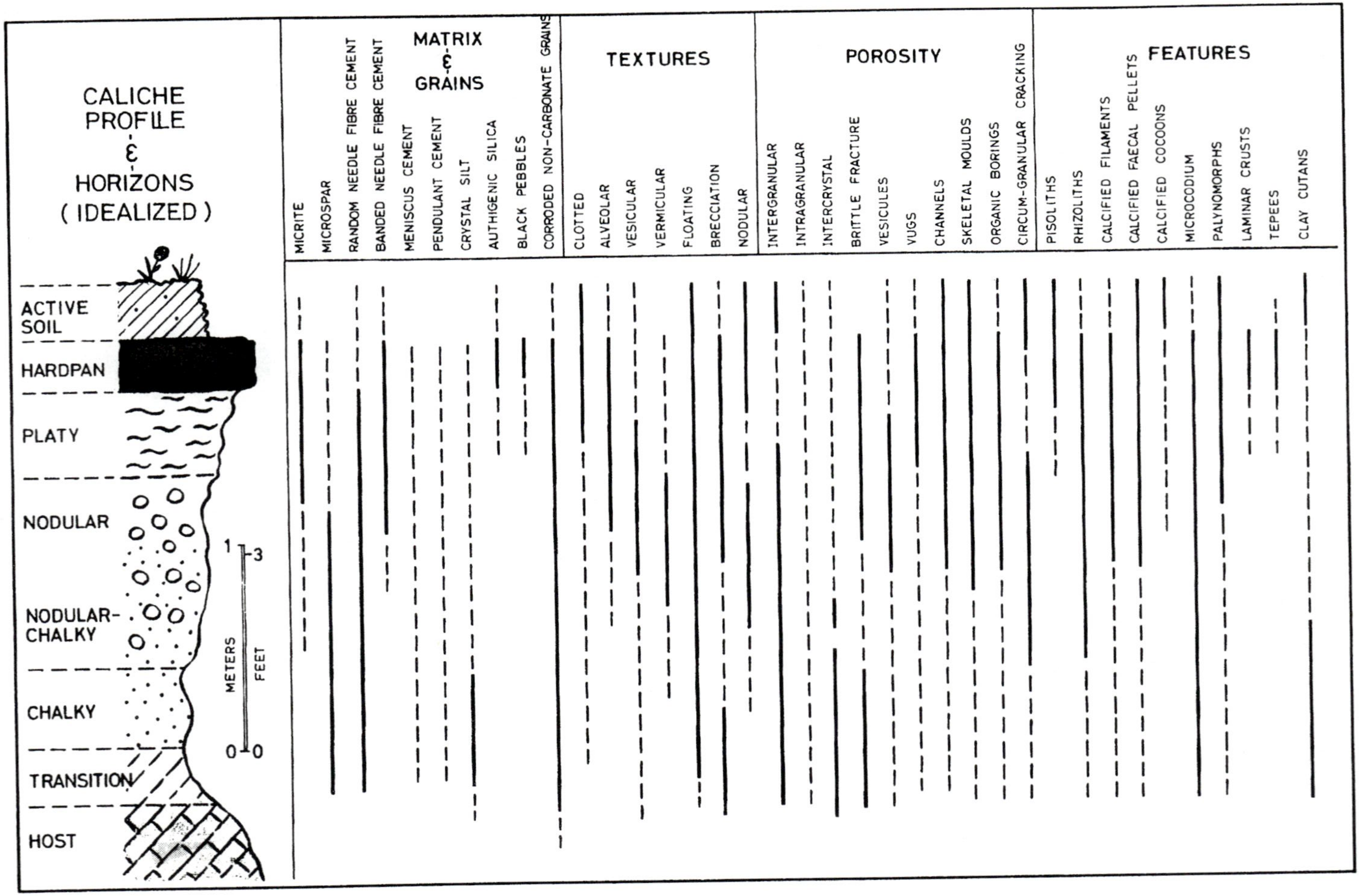

FIGURE 10.40
Idealized caliche profile and the distribution of major characteristics within the profile. (*After Esteban, M., and C. F. Klappa, 1983, Subaerial exposure environment, in Scholle, P. A., et al., eds., Carbonate depositional environments: Am. Assoc. Petroleum Geologists Mem. 33. Fig. 7, p. 5, reprinted by permission of AAPG, Tulsa, Okla.*)

in some south Wales calcretes. Individual crystals in this columnar calcite range to 30 cm long. **Nodular caliche** is made up of silt- to pebble-sized nodular structures called glaebules, which consist of concentrations of $CaCO_3$ intermixed in a less carbonate-rich matrix. **Chalky caliche** is generally unconsolidated (noncemented) and is characterized by the presence of cream to white, silt-sized calcite grains. Chalky caliche tends to be homogeneous and structureless. It grades upward into nodular caliche and downward through a transition zone to the host material. Vertical variations in carbonate grains and matrix, textures, porosity, and other features of caliches are summarized in Figure 10.40.

Although the carbonate mineral in most calcretes is calcite, soil profiles cemented with dolomite are also known (e.g., Khalaf, 1990). Such dolomite-cemented profiles are referred to as **dolocretes.**

10.8.6   Eolian Carbonates

Wind-blown carbonate sands, referred to as carbonate **eolianites,** occur adjacent to carbonate shorelines. Less commonly, they occur also along the margins of some saline lakes. They consist mainly of sand-size skeletal grains and ooids that were blown inland from carbonate beaches. Much of the land surface of Bermuda, for example, consists of carbonate dune sands, and they are also common on the Bahama islands. Carbonate dune sands exhibit the same general kinds of textures and structures as those occurring in siliciclastic dune deposits. Thus, they are characterized by generally well-rounded, well-sorted grains and may display large-scale, high-angle cross-stratification. The carbonate grains may be bonded with calcite cement. Ancient carbonate eolinites may be difficult to differentiate from carbonate beach sands. If present, terrestrial fossils (land snails, pollen, other plant remains), root traces (rhizoids or rhizoconcretions), and the remains of paleosols are diagnostic.

▬▬▬ *Additional Readings*

Bathurst, R. G. C., 1975, Carbonate sediments and their diagenesis, 2nd ed.: Elsevier, Amsterdam, 658 p.

Carozzi, A. V., 1989, Carbonate depositional models—A microfacies approach: Prentice Hall, Englewood Cliffs, N.J., 604 p.

Chilingar, G. V., H. J. Bissell, and R. W. Fairbridge, ed., 1967, Carbonate rocks. Developments in Sedimentology 9B: Elsevier, Amsterdam, 413 p.

Flügel, E., 1977, Fossil algae: Springer-Verlag, Berlin, 375 p.

Flügel, E., 1982, Microfacies analysis of limestones: Springer-Verlag, Berlin, 633 p.

Ham, W. E., ed., 1962, Classification of carbonate rocks: Am. Assoc. Petroleum Geologists Mem. 1, 279 p.

Horowitz, A. S., and P. E. Potter, 1971, Introductory petrography of fossils: Springer-Verlag, New York, 302 p.

Johnson, J. H., 1971, An introduction to the study of organic limestones: Colorado School Mines Quart., v 66, 185 p.

Majewske, O. P., 1969, Recognition of invertebrate fossil fragments in rocks and thin sections: E. J.

Brill, Leiden, 101 p.

Milliman, J. D., 1974, Marine carbonates: Springer-Verlag, New York, 375 p.

Monty, C., ed., 1981, Phanerozoic stromatolites: Springer-Verlag, Berlin, 249 p.

Peryt, T., ed., 1983. Coated grains: Springer-Verlag, Berlin, 655 p.

Reijers, T. J. A., and K. J. Hsü, ed., 1986, Manual of carbonate sedimentology: A lexicographical approach: Academic Press, London, 302 p.

Scholle, P. A., 1978, Carbonate rock constituents, textures, cements, and porosities: Am. Assoc. Petroleum Geologists Mem. 27, 241 p.

Scoffin, T. P., 1987, An introduction to carbonate sediments and rocks: Blackie, Glasgow, 274 p.

Toomey, D. F., and M. H. Nitecki, eds., 1985, Paleoalgology: Springer-Verlag, Berlin, 376 p.

Tucker, M. E., and V. P. Wright, 1990, Carbonate sedimentology: Blackwell Scientific Publications, Oxford, 482 p.

Walter, M. R., ed., 1976, Stromatolites: Elsevier, Amsterdam, 790 p.

APPENDIX

Key to identification of common skeletal grains in thin sections of carbonate rocks

I. SKELETAL GRAINS WITH CIRCULAR AND ELLIPTICAL SHAPES
  A. Circular or elliptical elements occur as scattered grains
    1. Grains contain a central opening
      a. Central opening plus outer openings
        Charophytes (algae)—large central opening surrounded by an outer ring of smaller openings
      b. One small central opening (grains <0.5 mm)
        Calpionellids (Protozoa; similar to tintinnids) (oblique sections)—central opening <0.1 mm; calcite rings without microstructure; may occur with U- and V-shaped forms
        Calcitized radiolarians—microcrystalline or sparry circular figures; original inner boundary commonly indistinct; outer boundary may be obscured by jagged edges; generally <0.5 mm
        Calcispheres (problematic organisms)—commonly circular sections with unclear periphery wall; wall may consist of various kinds of layers, some with fine radial pores; generally <0.5 mm
      c. One large central opening (grains >0.5 mm)
        Charophyte oogonia—thin calcitic wall elongated to regular pointed spines; some may have spiral structure and an outer ring of smaller openings; diameter about 3 mm
        Foraminifers—transparent, commonly yellowish; some may have spiky outline around a circular figure; maximum diameter 1 mm
        Echinoid spines—very complex geometrical pattern of radial and tangential elements around a central opening; may be ornamented; some are finely perforated single crystals
        Dasycladacean algae or **gymnocodiacean algae**—central opening with walls of various thickness that may be transversed with pore canals that commonly appear dark; the pore canals may be of uniform width or may taper to either the outside or inside boundaries
        Tentaculitids (organisms of uncertain affinity)—large central opening; dentate outline (outside ornamentation); diameter to about 1 mm
        Sponge spicules—sharply defined circular and elliptical figures with small central opening; $CaCO_3$ or SiO_2; commonly together with rod-shaped figures; some may display a pointed end; no internal structure
        Brachiopod spines (e.g., **productids**)—circular and ellipitical figures with a narrow central opening; wall with lamellar microstructure
        Trilobite fragments—calcitic phosphate; commonly dark-colored circular figures; diameter mm-sizes; some may be irregularly elongated to an elliptical form; commonly occur together with forms displaying ''shepherd's crooks''
        Serpulids (worm tubes)—large, calcitic figures of encrusted shells; microstructure divided into two parts: outer concentric lamellar structure, clear inner layer; figures commonly in clusters; diameter commonly several mm
        Scaphopods—large, calcitic figures; microstructure divided into three parts: differentiated prismatic middle layer and clear inner and outer layers; figures commonly overlap each other like the tubes of a telescope
        Solitary corals (Rugosa and **Scleractinia)**—Large central opening with radiating elements that generally do not extend to the center; variously differentiated walls that may or may not be well defined; may have a dark spot in the center (axial structure); microstructures (of septa) like bundles of fibers (herringbone structure) or with dark lines down the middle
        Nautiloids—large central opening; thick shell wall that is commonly recrystallized; longitudinal sections show cylindrical figures containing partitions with a single pore in the center
        Gastropods (oblique section of columella)—large circular figure indented on one side, with an element somewhat thicker toward the center; shell commonly recrystallized or cross-lamellar
        Calcareous sponges—circular figure with poorly defined wall exhibiting vermiform structures
    2. Grains contain no central opening, or a very small opening (grains >1 mm)
        Crinoid plates—small central lumen with various shapes (circular, pentagonal, etc.) surrounded by very fine meshwork; commonly yellowish single crystal

Belemnite rostrum—large circular figure (3–10 mm) with very small or absent central opening, surrounded by alternating layers of light- and dark-colored rings and/or large radiating prisms of calcite

Vertebrate remains—subrounded particles; very fine meshwork, dark brown to black in transmitted light; commonly differentiated into a thick peripheral part and a gauzelike central part (spongy material); phosphatic

Recrystallized sponge spicules (rhaxes) which cannot be more closely identified, **radiolarians,** planispiral **foraminifers,** and **calcispheres**

  B. Circular or elliptical elements occur interconnected

    1. A few circular figures or segments of circles arranged linearly or in spherical aggregates

    Planktonic foraminifers—chambers round or angular; commonly with spikelike ornamentation; diameter commonly 0.5—1.0 mm

    2. Circular segments with mirror-image, matching protuberances, lying adjacent to each other

    Ophiuroids (echinoderms)—protuberances are two calcite plates that become extinct at different positions

    3. Chambers arranged symmetrically in a line; a round chamber in the middle with concentric chambers becoming larger toward the ends

      a. Microscopic size:

      Foraminifers (planispiral and trochospiral)

      b. Macroscopic size:

      Juvenile ammonites and **gastropods**

    4. Unsymmetrically arranged chambers

      a. Microscopic size:

      Encrusting foraminifers

      b. Macroscopic size:

      Serpulid tubes

II. SKELETAL GRAINS WITH U AND V SHAPES

  A. Length commonly <0.20 mm

    Calpionellids/tintinnids (pelagic protozoans)—U- and V-shaped figures with thorn-shaped elongation or with flange or constriction near open end; walls calcitic or with dark-colored (organic) lamellae

  B. Length commonly >0.50 mm

    1. V-shaped figures

    Tentaculitids—narrow V figures; some may have serrated outer ornamentation or inner ornamentation; may be segmented; length ranges from mm to cm

    Pelagic crinoids—broad, V-shaped figures with symmetrical bulbous thickenings at the base or along the side; single crystals; commonly smaller than **tentaculitids**

    Belemnites—long, narrow V-shaped figures with narrow V-shaped inner cavity; walls of light-colored and dark-colored lamellae; diameter ranges to cm size

    2. Broad U- (and V-) shaped figures

    Dasycladacean algae—large central opening; distinct wall with pores that are uniformly thick or taper toward the inside or outside, particularly as viewed in oblique longitudinal sections

    Gymnocodiacean algae—serial association of barrel-shaped or ovoid segments; commonly thin walls with numerous very fine pores that radiate obliquely outward; inner space wide, generally filled with calcite cement or sediment; rarely primary longitudinal structures

III. CURVED SKELETAL GRAINS

  A. Commonly solid

    1. Slightly curved

    Filaments—very thin (~0.03 mm) and long; commonly arranged parallel to each other and in great abundance; lamellar and fine prismatic microstructure

    Pelecypod valves—thicker than filaments; may be corrugated; shell may contain lamellar, prismatic, or crossed-prismatic microstructures

Inarticulate brachiopods—lamellar microstructure; composed of calcium phosphate (brownish, low birefringence); shell thickness to 0.20 mm

Articulate brachiopods—thick, commonly very curved and bent elements with (and without) dark-colored vertical and oblique pores (~0.03 mm); conspicuous fine lamellar microstructures

Phylloid algae—generally slightly curved elements with peripheral inner structures (pores in the cortex zone); commonly recrystallized; coarse, sparitic central zone

Arm plates of crinoids or **dorsal and lateral shields of ophiuroids**—half-moon elements, single crystals

  2. Very curved, recurved, and coiled

    a. Spiral

Gastropod cross section—unpartitioned spiral; center commonly thickened; crossed-prismatic microstructure commonly destroyed by recrystallization

Foraminifers—partitioned spiral; mm-size; wall and surface elements variously differentiated

Juvenile ammonites—partitioned spiral; cm-size; partitions with an opening; walls commonly recrystallized

    b. Bent at one end

Trilobites (thorax segments)—figure bent at one end like a shepherd's crook; prismatic microstructure; generally distinct dark-colored edges

Ostracods—thin valves or shells, edges recurved; size 0.5–1.0 mm; fine prismatic microstructure

 B. Commonly hollow

Porostromate algae (e.g., *Girvanella*)—very small (0.10–0.05 mm) short, curved tubes, commonly occurring in clusters; generally embedded in micritic layers; may be linked in chains, which can be intertwined

Serpulid tubes—well-defined curved or coiled tubes, commonly cemented; microstructure divided into two parts (inside: parallel lamination; outside: cone-in-cone lamination); mm size, tube diameter about 1 mm

IV. SKELETAL GRAINS WITH NETLIKE STRUCTURE

 A. Regular, well-defined meshwork

  1. Meshwork rectangular, square, or polygonal

    a. Network structures

Coralline algae—very fine net structure of generally rectangular cells, which appear black in transmitted light; differentiated into areas with larger and smaller meshes; sparry calcite-filled circular figures embedded in the net (sporangia); cell diameter about 5–15 μm

Solenoporoid red algae—cm-sized colonies of long thin vertical threads; horizontal elements, if present, may be irregularly spaced (zonation); no other differentiation; mesh diameters 10–20 μm

Wood fragments—fine net structure of various-sized cells in the center and cyclic alternations of cell sizes on the periphery; cell walls may be quite thickened; may contain carbonized material, or wood may be replaced by SiO_2, FeS_2, or $CaCO_3$

Fragments of larger foraminifers—networks of fine, porous, very distinct horizontal elements that are either closely or widely spaced; commonly microgranular fine structure; serial association of chambers common

Stromatoporoids—larger colonies with net structures of horizontal, possibly undulating elements (laminae) and vertical pillars and spines of various lengths; extremely differentiated microstructures

    b. Open meshwork:

Sponges—open meshwork of thin, rod-shaped elements interlocked at distinct angles (SiO_2 or calcite)

    c. Bifurcated "tubes":

Porostomate filamentous algae—bifurcated "tubes" whose angles are all the same within a "colony"

    d. Colonies of tubes more or less parallel to each other

Bryozoans—colonies or fragments of colonies of uniform or various-sized tubes containing flat and curved transverse plates; colony shape commonly ramose, with tubes leading obliquely outward; tube diameter generally <0.5 mm; wall structure lamellar, rarely fibrous

    Colonial corals and **chaetetids**—reticulated colonies of uniformly large, angular meshes (may show radial elements or septa in tangential sections) or vertical tubes with transverse plates; walls may contain pores; fibrous microstructure that may consist of groups of radiating fibrous bundles; tube diameter commonly >0.5 mm, generally several mm

e. Shells composed of prisms lying parallel to each other

    Rudistid lamellibranch shells—shells with cellular prismatic structure; very regular rectangular and radial pattern; prism size ~0.05–0.10 mm

    Pelecypods (*Inoceramus*)—shells with ''honeycomb structure'' in tangential sections; diameter of prisms to 0.10 mm

  2. Cystose meshwork (curved plates)

    Stromatoporoids—colonies of stacked hemispherical cysts; some with spinelike columnal elements

    Bryozoans—cystose structure containing short vertical tubes with horizontal elements; tube diameter generally <0.50 mm

    Tabulate corals—similar to bryozoans but tube diameter generally >0.50 mm, commonly several mm

    Peripheral and longitudinal sections of corals and sphinctozoan sponges—vertically arranged cysts, outer periphery of the total structure very distinct

  3. Meshwork extremely regular

    Echinoderms—very fine network; each grain a single crystal

B. Irregular network of various shapes

    Calcareous sponges—wormlike elements in a micritic matrix that is commonly dark-colored; the entire structure may appear as a circular figure

    Reticulate foraminifers—flat fragments or conical figures with irregular network of dark-colored elements, some with distinct chambers

    Hydrozoans—sessile colonies; irregular network structure consisting chiefly of vertical and horizontal elements; commonly with star-shaped channel systems and tabulate tubes of various widths

    Bone remains—fragments of calcium phosphate, brownish or black-colored (low birefringence); meshes of various sizes

V. LAYERED STRUCTURES

A. Encrusted, planar distribution on the substrate

    Stromatolites—irregular, micritic fabric with spar-filled cavities parallel to bedding plane; basic geometric form comparable to a hemisphere; cm- to m-sized

B. Nonencrusted particles

  Fish scales—phosphatic; regular outline (commonly rhombic) particles with distinct very concentric lamellar structure; commonly dark in transmitted light

  Conodonts—phosphatic; toothlike, commonly irregular outline; dark- or light-gray color; distinct lamellar structure in various orientations; overall size 1 to 3 mm

VI. CHAINS OF SEGMENTS

A. Uniserial (arranged in a single row or series)

  1. Microscopic

    Foraminifers—small chambers becoming successively larger

  2. Macroscopic

    Orthoceratid nautiloids—straight chambered shells, cm-sized; transverse elements with pores

    Tangential sections of ammonites—ladderlike structure with slightly curved transverse elements

    Segmented calcisponges—arched, overlapping segments; some with perforated roofs; may contain vesicular structures in the segments

B. Biserial or triserial

    Foraminifers—small chambers becoming successively larger

Source: Flügel, E., 1982. Microfacies analysis of limestones, p. 266–273, reprinted by permission of Springer-Verlag, Berlin.

■■■ *Chapter 11*
Dolomites

■■■ *11.1 Introduction*

Carbonate rocks range in age from Holocene to Precambrian. The mass of Precambrian carbonate rocks is much smaller than that of Phanerozoic carbonates, which are particularly abundant in stratigraphic sequences of Paleozoic age. Carbonate rocks less than about 100 million years old are dominantly calcium carbonates with a low Mg/Ca ratio consistent with the ratio that would be expected if the rocks formed mainly by accumulation of carbonate skeletal debris. The Mg/Ca ratio rises sharply, but irregularly, with increasing age in carbonate rocks older than about 100 million years. Thus, it is a commonly accepted tenet that dolomites make up an increasing proportion of carbonate rocks with increasing age and that the average composition of Precambrian carbonate rocks approaches that of the mineral dolomite (Garrels and McKenzie, 1971, p. 237). This long-accepted view that dolomites increase in abundance relative to other carbonates with increasing age has recently been challenged by Given and Wilkinson (1987). These authors maintain that whereas dolomites do change in relative abundance through time, these changes do not correlate directly with age. Rather, they say that increased amounts of dolomite formation correlate to periods of sea-level highs and continental flooding. In turn, this viewpoint expressed by Given and Wilkinson has been challenged by Zenger (1989), who says that these authors have neither convincingly demonstrated that there is no increase in relative dolomite content with age nor provided more than the weakest suggestion that it correlates instead with highstands of sea level. So it goes!

Whatever the true story of relative abundance, dolomites are an extremely intriguing group of rocks, and they have considerable economic significance as reservoir rocks for petroleum. Certainly, few kinds of sedimentary rocks have generated so much interest and controversy among geologists and geochemists with regard to their origin. Most ancient dolomites are relatively thick, coarse-crystalline, porous to nonporous, massive rocks, many of which appear to have formed through pervasive dolomitization (replacement and recrystallization) of precursor limestones. On the other hand, dolomites that form in modern environments, as well as some ancient dolomites, tend to be thinner and finer-crystalline and otherwise lack distinctive evidence of massive dolomitization. The origin of these fine-grained dolomites, as well as the mechanisms responsible for large-scale, pervasive dolomitization of ancient limestones, has been hotly debated by geologists for well over half a century.

In this chapter, we examine first the mineralogy, textures, and structures of dolomites. We then take a look at some current ideas and controversies regarding the origin of dolomite.

11.2   Mineralogy of Dolomites

11.2.1   Stoichiometric vs. Nonstoichiometric Dolomite

As shown in Table 10.1, only two minerals belong to the dolomite group: **dolomite** [$CaMg(CO_3)_2$] and **ankerite** [$Ca(Mg,Fe,Mn)(CO_3)_2$]. Dolomite forms a limited solid solution series with ankerite. Perfectly ordered dolomite [$CaMg(CO_3)_2$], in which calcium and magnesium have equal molar proportions, is called **stoichiometric** dolomite. In this ideal dolomite, a plane of CO_3 ions alternates first with a plane of Ca ions, then with a plane of Mg ions (i.e., CO_3—Ca—CO_3—Mg—CO_3, and so on), with the c-axis of the crystal perpendicular to the stacked planes. Oxide analysis of an ideal dolomite yields 21.9 percent MgO, 30.4 percent CaO, and 47.7 percent CO_2 by weight (Blatt et al., 1980, p. 510). As described by Land (1985), ideal dolomite is the most stable form in which $CaCO_3$ and $MgCO_3$ can combine under sedimentary conditions. Thermodynamically, ideal dolomite has the lowest free energy possible for any combination of subequal amounts of $CaCO_3$ and $MgCO_3$ that can be combined under these conditions. Thus, ideal dolomite is the least soluble form in which subequal amounts of these materials can be combined. Any changes in the ideal structure or composition raise the free energy of the crystal and make it more soluble.

Ideal dolomite is rare in the geologic record. Many natural dolomites, particularly modern or Holocene dolomites, are poorly ordered **protodolomite** with an excess of calcium. Although less common, magnesium-rich dolomite is known also. The reported composition of naturally occurring dolomite ranges from about $Ca_{1.16}Mg_{0.84}(CO_3)_2$ to about $Ca_{0.96}Mg_{1.04}(CO_3)_2$ (Land, 1985). The structure of protodolomite may have fault-like steps or dislocations, and trace elements such as Na and Sr may be common in the crystal lattice owing to substitution for Ca or Mg. Isomorphous substitution of divalent ions, particularly ferrous iron, for calcium or magnesium (Table 10.1) is a further reason for departure of natural dolomites from ideal composition.

Land (1985) points out that the very young dolomite that occurs in Holocene environments is poorly ordered, but that it displays at least partial ordering. Crystals are only a few micrometers in size and consist of aggregates of submicrometer crystals that have generally similar orientation to that of neighboring crystals. The crystals have numerous structural defects of various kinds, and great variation in the degree of ordering and calcium enrichment can exist within the crystals. Thus, the crystal structures are highly strained and inhomogeneous, and they are further characterized by a high degree of trace-element substitution. This kind of structurally and compositionally inhomogeneous dolomite is unstable to metastable, and it dissolves or alters much more readily than more highly ordered dolomite. Metastable phases of dolomite have the potential to achieve a more stable state by undergoing structural or compositional changes with time. Thus, most older Phanerozoic dolomites and Precambrian dolomites are more highly ordered than Holocene protodolomite. According to Land, the most common kind of sedimentary dolomite, which is especially characteristic of post-Paleozoic sequences, is **calcium-rich dolomite** that exhibits a lamellar structure when examined by transmission electron microscopy and electron diffraction. This type of dolomite represents a more stable phase than Holocene dolomite but is still metastable and dissolves more rapidly than ideal dolomite. Continued stabilization of this dolomite can occur through time as a result of solution-reprecipitation, which gradually moves it toward a more ordered state. Finally, a third common kind of sedimentary dolomite is **"nearly stoichiometric,"** **well-ordered dolomite,** which is known mostly from Paleozoic and Precambrian sequences. Nonetheless, it is still less well ordered than ideal, stoichiometric dolomite, and it is characterized by numerous structural defects, especially various kinds of faultlike features. The density of these structural defects is much less common, however, than in the two preceding kinds of dolomite. Even this "nearly stoichiometric" dolomite is less stable than ideal dolomite owing to the strained and broken bonds associated with the structural defects.

11.2.2   Identifying Dolomite in Thin Section

Dolomite and calcite commonly occur together in many carbonate rocks. Dolomite can be readily identified by X-ray diffraction techniques, but because both dolomite and calcite are uniaxial negative minerals, they may be difficult to distinguish optically. Nesse (1986, p. 144) lists the following criteria that may be used to help make the distinction:

1. Dolomite is more likely to form euhedral crystals than is calcite.
2. Calcite is more likely to be twinned than is dolomite.
3. Twin lamellae in calcite may be parallel or oblique to the long diagonal of the crystal or parallel to the edges of cleavage rhombs but not parallel to the short diagonal. Twin lamellae in dolomite may be parallel to either the long or short diagonals of cleavage rhombs (Fig. 11.1).
4. Dolomite has higher refractive indices, higher specific gravity, and is less reactive in cold, dilute HCl.
5. Dolomite may be colorless, cloudy, or stained by iron oxides, whereas calcite is commonly colorless.

In addition to these criteria, several procedures are available for staining carbonate rocks that allow distinction between dolomite and calcite. For a review of these techniques, see Miller (1988).

FIGURE 11.1
Orientation of twin lamellae in calcite and dolomite. *(From Nesse, W. D., 1986, Introduction to optical mineralogy: Oxford University Press. Fig. 10.3, p. 139, reprinted by permission.)*

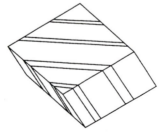

Calcite

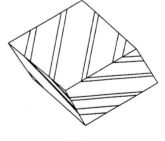

Dolomite

11.3   Dolomite Textures

Dolomites may be composed of crystals of nearly uniform size (**unimodal size distribution**) or crystals of various sizes (**polymodal size distribution**). Dolomite can occur either as rhomb-shaped euhedral to subhedral crystals or as nonrhombic, commonly anhedral crystals. Rhomb-shaped dolomite commonly displays straight compromise boundaries between crystals and is referred to as **planar dolomite** by Sibley and Gregg (1987), who originally called it **idiotopic dolomite** (Gregg and Sibley, 1984). (A compromise boundary is a surface of contact, not corresponding to a crystal face, between two mutually growing but differently oriented crystals.) Anhedral, nonrhombic dolomite is called **nonplanar dolomite** (originally called **xenotopic dolomite** by Gregg and Sibley, 1984). Boundaries between crystals in nonplanar dolomite are mostly curved, lobate, serrated, or indistinct. On the basis of textural characteristics, Gregg and Sibley (1984) and Sibley and Gregg (1987) divide planar dolomite (or idiotopic dolomite) into four subcategories and nonplanar dolomite (xenotopic dolomite) into three subcategories (Fig. 11.2).

Planar-euhedral dolomite is made up of loosely packed but crystal-supported, well-formed rhombs. The intercrystalline spaces among crystals may be filled with another mineral such as calcite, or the spaces may be empty (porous). The texture of porous dolomites of this type is sometimes referred to as **sucrosic** (sugary). **Planar-subhedral dolomite** has subhedral to anhedral crystals and very low porosity. Dolomites of this type have straight compromise boundaries and large numbers of preserved crystal-face junctions (Fig. 11.3). Crystal-face junctions form where straight compromise boundaries between two crystals meet at a distinct angle that is less than about 160 degrees (see arrows in Fig. 11.3). **Planar void-filling dolomite** consists of euhedral dolomite crystals with terminations that project into open spaces. Such dolomite may be a cement, but it may form also by replacement of the margins of a carbonate grain, with subsequent dissolution of the center of the grain. Alternatively, it could form by replacement of a precursor cement. Some dolomite rhombs may appear to ''float'' in a limestone (micrite) matrix. Such matrix-supported dolomites are called **planar-porphyrotopic.**

Nonplanar-anhedral dolomite is characterized by mainly anhedral crystals with curved, lobate, serrated, or indistinct intercrystalline boundaries. Inclusions may be abundant in the crystals, and they commonly display undulatory extinction. Nonplanar-anhedral dolomite may be confused with planar-subhedral dolomite; however, the presence of irregular crystal boundaries and the scarcity of crystal-face junctions help to distinguish the two. Gregg and Sibley (1984) also include replacement **saddle dolomite** crystals in this category of dolomite. Saddle dolomite, described by Radke and Mathis (1980), is a variety of dolomite with a warped crystal lattice that is characterized by curved crystal faces and cleavage and by sweeping extinction (Fig. 11.4). **Nonplanar void-filling dolomite** is irregular-shaped or saddle-shaped dolomite that fills open space. Saddle dolomite crystals of cement origin have long, curved edges leading to pointed terminations, resembling somewhat a Persian scimitar. **Nonplanar-porphyrotopic dolomite** is similar to planar-porphyrotopic dolomite except that the crystals are mainly anhedral. The crystals probably form by replacing micritic limestones or other limestones.

Dolomites of replacement origin may preserve original carbonate textures to various degrees. Therefore, description of dolomite texture may include characteristics of original carbonate grains, matrix (micrite), and void-filling crystals. As illustrated in Figure 11.2, carbonate grains may be unreplaced; may be dissolved, leaving molds; or may be replaced

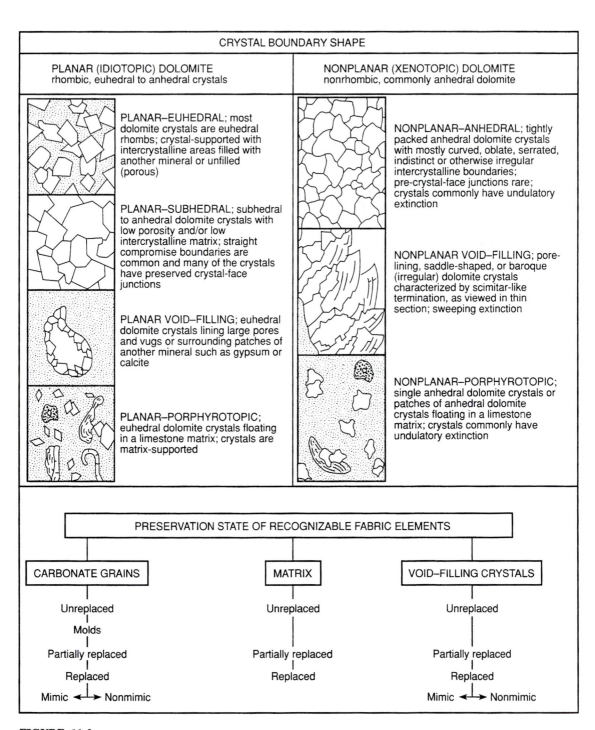

FIGURE 11.2

Classification of dolomite textures. See text for further explanation. *(After Gregg, J. M., and D. F. Sibley, 1984, Epigenetic dolomitization and the origin of xenotopic dolomite texture: Jour. Sed. Petrology, v. 54, Fig. 6, p. 913, and Sibley, D. F., and Gregg, J. M., 1987, Classification of dolomite rock textures: Jour. Sed. Petrology, v. 57, Fig. 1, p. 968; figures reprinted by permission of Society of Economic Paleontologists and Mineralogists, Tulsa, Okla.)*

FIGURE 11.3
Idiotopic-subhedral dolomite with crystal face junctions (arrows). Whitewood Formation (Ordovician), South Dakota. Crossed nicols. Scale bar = 0.05 mm.

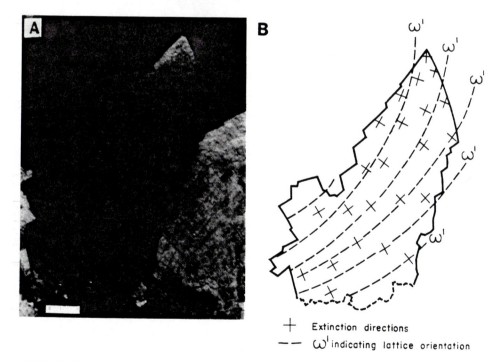

$+$ Extinction directions

$--$ ω' indicating lattice orientation

FIGURE 11.4
Saddle dolomite. A. Photomicrograph of spear-form dolomite with only part of crystal in complete extinction. The compromise boundary with adjacent crystal (right side) has a zigzag appearance. B. Sketch of same crystal. Directions of extinction (crosses) are plotted within the crystal with interpolated ω' traces (dashed lines), which indicate distorted lattice planes. The direction of optimum crystal growth coincides with the region of greatest lattice divergence. *(From Radke, B. M., and R. L. Mathis, 1980, The formation and occurrence of saddle dolomite: Jour. Sed. Petrology, v. 50. Fig. 3, p. 1153, reprinted by permission of Society of Economic Paleontologists and Mineralogists, Tulsa, Okla.)*

to various degrees. If carbonate grains are completely replaced, they may be replaced **mimically** or **nonmimically.** Mimic replacement refers to preservation of the form and internal structure of a carbonate grain and generally occurs only if replacement is by many small dolomite crystals. Nonmimic replacement may preserve the form but not the internal structure of a carbonate grain (Fig. 11.5). If the form of the grain is completely destroyed during replacement, of course no evidence of replacement remains. In many dolomites, only very faint outlines of original carbonate grains remain; these are referred to as **ghosts.** Figure 11.5 illustrates how ghost textures may form in dolomites. Matrix and original void-filling crystals may also be replaced partially or completely.

LIMESTONE

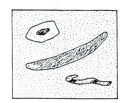

PARTIAL DOLOMITE

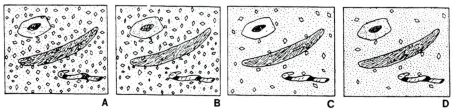

DOLOMITE

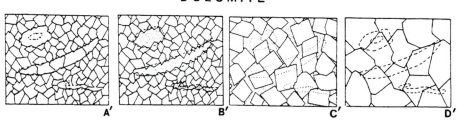

Saturation state with respect to dolomite decreases

$$\longrightarrow$$

FIGURE 11.5

Replacement of carbonate grains and the formation of carbonate grain ghosts in planar dolomite. Note that overall texture is related to the saturation state of the dolomitizing fluids with respect to dolomite, which decreases to the right in the diagram. D–D' represents a condition where saturation state is low but residence time in the dolomitizing solution is long, resulting in a low-porosity dolomite. Note also that A' has mimically replaced crinoid and nonmimically replaced brachiopod and trilobite fragments. B' has molds, and C' and D' have ghosts of the fossil grains. *(After Sibley, D. F., and J. M. Gregg, 1987, Classification of dolomite rock textures: Jour. Sed. Petrology, v. 57. Fig. 11, p. 974, reprinted by permission of Society of Economic Paleontologists and Mineralogists, Tulsa, Okla.)*

Gregg and Sibley (1984) propose that dolomite texture is related to temperature of formation. Below some critical temperature that lies between about 50°C and 100°C, called the **critical roughening temperature,** dolomite crystal growth produces dominantly euhedral crystals, resulting in planar textures. Growth of dolomite crystals above this temperature may produce anhedral crystals, resulting in nonplanar texture, although planar crystals can apparently also form above this temperature for reasons not clearly understood. By contrast, calcite has a critical roughening temperature below about 25°C. Therefore, calcite crystals are commonly anhedral. Gregg and Sibley (1986) suggest that the saturation state of dolomitizing fluids can also affect dolomite textures. Theoretically, nonplanar dolomite can form at low temperature (25°C) under conditions of high supersaturation. Shukla (1986) disagrees with Gregg and Sibley's (1984) conclusion that temperature is a primary factor that controls dolomite texture. He suggests that many other kinetic factors may also be important.

Subsequently, Sibley and Gregg (1987) added to their crystallization model by suggesting that unimodal crystal size distribution within a dolomite indicates a single dolomite nucleation event on a unimodal $CaCO_3$ substrate. Polymodal size distribution may result from multiple nucleation events on a unimodal or polymodal substrate or from differential nucleation on an originally polymodal substrate. Planar crystal boundaries develop when dolomite crystals undergo faceted growth (nucleation at active sites, and lateral migration of layers or growth spirals). Nonplanar crystal boundaries develop when crystals undergo nonfaceted growth (random addition of atoms to the crystal surface). Nonfaceted growth is characteristic of growth at elevated temperatures (>50°C) and/or high saturation. Sibley and Gregg suggest that both planar and nonplanar dolomite can form as a cement, by replacement of $CaCO_3$ minerals, or by neomorphism of a precursor dolomite.

Scoffin (1987, p. 132) proposes that dolomite fabrics may be related to rates of dolomite precipitation. When replacement by dolomite is fabric-preserving, precipitation must have been slow and the system relatively closed. Fabric-destructive dolomitization occurs with higher fluid-flow rates through the precursor calcium carbonate sediment. Large-scale dissolution of this precursor allows the dolomite to grow into the most energetically suitable forms, which is commonly planar-euhedral (sucrosic).

With respect to the origin of saddle dolomite, Radke and Mathis (1980) report that saddle dolomite is commonly associated with hydrocarbons, epigenetic base-metal mineralization, and sulfate-rich carbonates. They suggest that these associations imply late diagenetic formation by sulfate-reduction processes and that both temperature and the chemistry of the pore waters may be important to their formation. Machel (1987) concludes that saddle dolomites in Late Cretaceous Nisku deposits in Alberta, Canada, formed at temperatures in excess of at least 110°C from hypersaline brines as a byproduct of two processes: (1) chemical compaction of dolomitized reef rock and (2) increase in carbonate alkalinity as a result of thermochemical sulfate reduction. Pressure solution of reef rock supplied both magnesium and carbonate ions. Carbonate and/or bicarbonate was supplied also by organically driven oxidation of hydrocarbons during thermochemical sulfate reduction, resulting in increased carbonate alkalinity.

11.4   Spheroidal Dolomite

Von Der Borch and Jones (1976) published the first report of natural spheroidal or spherular dolomite, although spheroidal aggregates of protodolomite had been precipitated previously in laboratory experiments. Von Der Borch and Jones reported that sphe-roidal aggregates of relatively well-ordered micrometer-size dolomite occur in modern dolomite deposits from the Coorong area, South Australia. Subsequently, spheroidal dolomite has been found in other areas, including Sugarloaf Key, Florida (Carballo et al., 1987), Libya (Amiri-Garroussi, 1988), and Kuwait (Gunatilaka, 1989). The spheroidal dolomite described by Von Der Borch and Jones ranges from 0.2 to 1.0 μm in diameter and is composed of subunits measuring about 100 nm in diameter. Carballo et al. report subrounded microscopic crystallites 0.1 to 0.3 μm in size that are aggregated together to form rhombs about 6 μm across. Amiri-Garroussi describes spheroidal dolomite as ir-regularly spheroidal, ovoid or rounded bodies, about 0.5 to 2.0 μm in diameter, that are composed mostly of smaller, similarly rounded subunits with no definable shape. Spher-oids reported by Gunatilaka (1989) are larger, about 5 to 350 μm in diameter, and have concentric zones and nuclei of fluid inclusions with or without a radial fabric. Most of the above authors suggest that spheroidal dolomite is a primary precipitate (of protodolo-mite?), although explanations for its origin vary: flocculation in pore waters of Coorong lake sediments, tidal pumping of seawater through Florida Bay sediment, and association with hydrocarbons in Kuwait.

11.5   *Zoned Dolomite*

Many rhombic dolomite crystals have a cloudy, rhombic central zone surrounded by a clear rim (Fig. 11.6). These crystals are often referred to as zoned dolomite. The cloudy centers of the crystals result from the presence of inclusions, which may be bubbles or unreplaced inclusions of calcite or other minerals, whereas the clear rims are inclusion-free. These ''zoned crystals'' may form by replacement of a $CaCO_3$ precursor such as a micritic limestone, or they may grow into open pore space. Where they form within a precursor limestone, the cloudy centers represent replacement of the precursor $CaCO_3$. The clear rims must have formed in empty pore space around the margins of the cloudy rhombs. To account for this open space, Murray (1964) suggests that the empty space may be created by dissolution of $CaCO_3$ from just beyond the limits of the cloudy replacement rhombs. The $CaCO_3$ dissolved from the immediately surrounding area (referred to as ''local-source'' $CaCO_3$) is then reprecipitated syntaxially in the newly created space around the cloudy rhomb. Clear syntaxial rims may also form in optical continuity on a dolomite crystal that projects into voids. These syntaxial rims may either enlarge earlier-formed clear rims or produce clear dolomite rims on preexisting cloudy crystals.

In addition to these so-called zoned dolomites, some dolomites exhibit fine-scale internal zoning that results from differences in composition, particularly iron composi-tion. Ferrous iron is common in many dolomite crystals as a substitute for magnesium. If this ferrous iron is subsequently oxidized to ferric iron (hematite), it is visible with a standard petrographic microscope. Thus, some dolomite crystals may contain concentric alternating zones of red, iron-rich and clear, iron-poor dolomite that mark growth stages of the rhomb (Blatt, 1982, p. 313). Because iron does not so readily substitute for calcium, calcite does not shows this type of visible zoning. Dolomite crystals that may not display visible zoning under a petrographic microscope may show well-developed fine-scale zoning when viewed by cathodoluminescence (Fig. 11.7). Reeder and Prosky (1986) attribute this type of zoning to systematic compositional differences between nonequivalent growth sectors in dolomite rhombs and refer to it as **sector zoning.** They

FIGURE 11.6
Zoned dolomite crystal showing a
clouded center and an almost clear
rim. Paleozoic limestone, Great
Basin, USA. Crossed nicols. Scale
bar = 0.1 mm.

FIGURE 11.7
Compositional sector zoning in
dolomite crystals as revealed by
cathodoluminescence. *(From
Reeder, R. J., and J. L. Prosky,
1986, Compositional sector zoning
in dolomite: Jour. Sed. Petrology,
v. 56. Fig. 2, p. 239, reprinted by
permission of Society of Economic
Paleontologists and Mineralogists,
Tulsa, Okla.)*

report that growth sectors forming under {1120} crystal faces are enriched in Fe and Mg and slightly enriched in Mn relative to sectors forming under {1014} faces. Furthermore, transmission electron microscopy shows that growth microstructures differ within these different growth sectors. Reasons for these compositional and structural differences are poorly understood, but may be related to differences in growth rate, atomic configuration at the growing surface, and mechanisms of growth.

Dolomite crystals may also exhibit syntaxial overgrowths, although overgrowths are not as common on dolomite crystals as on calcite crystals (particularly on echinoderm fragments). For example, Jones (1989) describes abundant syntaxial overgrowths on dolomite crystals that line the walls of cavities in dolomites (dolostones) exposed along the coast of Grand Cayman, British West Indies. Jones attributes the formation of these dolomite overgrowths to evaporation of water films from the walls of cavities within the dolostones.

11.6   Mottled and Zebra Structure

Some dolomites display a distinctive **mottled structure,** particularly on weathered surfaces. The mottles range in shape from tubular to highly irregular, including anastomosing networks. In general, mottling appears to be the result of incomplete dolomitization; however, various kinds of processes or conditions have been suggested to account for the differential dolomitization. Mottles oriented approximately parallel to bedding surfaces of the precursor limestone probably represent selective dolomitization along zones of higher porosity and permeability. Tubular-shaped mottles may be relict borings or other organic traces. Very irregular mottles that cut indiscriminately across preserved stratification in limestones and that are not associated with obvious fracture systems are more difficult to explain. They are clearly diagenetic, but the mechanism responsible for isolation of the patches is not well understood.

Zebra structure is a term used to describe some dolomites that display conspicuous, somewhat irregular, light and dark bands, which are commonly oriented parallel to bedding. The light-colored bands range in color from light gray to white, are typically coarse-grained, and may be vuggy (porous). The darker bands are generally fine-grained. Zebra structure has commonly been reported from altered dolomites associated with sedimentary ore deposits and evaporites. Beales and Hardy (1980) suggest that zebra structure (Fig. 11.8) they describe in dolomites from Mississippi-Valley type lead/zinc ore deposits is a type of expansion structure. Presumably, displacive precipitation of gypsum resulted in expansion and brecciation of the host rock. Subsequent dissolution of the gypsum created open spaces that were later filled by white, sparry dolomite.

11.7   Origin of Dolomite

11.7.1   The Dolomite Problem

As mentioned in the introduction to this chapter, dolomites are common rocks in the geologic record and appear to form an increasing proportion of carbonate rocks with increasing age. Dolomites have been studied for two hundred years, but their origin is still enigmatic and controversial. The so-called "dolomite problem" arises in part from the fact that geochemists have not yet successfully precipitated well-ordered, stoichiometric dolomite in the laboratory at the normal temperatures ($\sim$25°C) and pressure ($\sim$1 atm) that

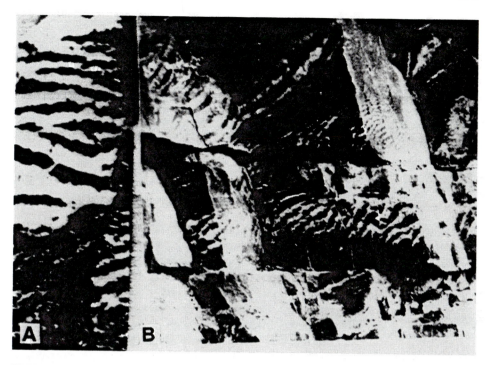

FIGURE 11.8
Zebra structure in dolomite, possibly caused by expansion during a precursor phase of evaporite precipitation. Width of specimen 35 mm. *(From Beales, F. W., and J. L. Hardy, 1980, Criteria for the recognition of diverse dolomite types with an emphasis on studies of host rocks for Mississippi Valley-type ore deposits, in Zenger, D. H., et al., eds., Concepts and models of dolomitization: Soc. Econ. Paleontologists and Mineralogists Spec. Pub. 28. Fig. 3, p. 206, reprinted by permission of SEPM, Tulsa, Okla.)*

occur at Earth's surface. Elevated temperatures in excess of 100°C are required to precipitate a well-ordered dolomite in the laboratory. At normal temperatures of about 25°C, only calcium-rich, poorly ordered protodolomite forms. Nonetheless, nearly stoichiometric dolomite is common in the geologic record. The reluctance of stoichiometric dolomite to precipitate under normal surface conditions has given rise to numerous theories to explain the occurrence of nearly stoichiometric dolomite in the geologic record. Some dolomites preserve relict or ghost limestone textures such as fossils or ooids. Other dolomites are known to grade fairly abruptly to limestone along a boundary that cuts across bedding planes. These dolomites clearly formed by replacement processes, with the mineral dolomite replacing calcite, Mg-calcite, or aragonite. Presumably, replacement occurred as Mg-bearing subsurface waters migrated through the carbonate sediments, possibly over periods of tens to hundreds of millions of years. Determining how extensive, thick sequences of ancient dolomites formed is also part of the dolomite problem.

In contrast to these obvious replacement dolomites, many dolomites do not contain visible relict textures, leaving their origin open to speculation. Did these dolomites form as a primary dolomite precipitate in spite of the failure of laboratory experiments to precipitate dolomite at surface temperatures? Alternatively, were they originally precipitated as $CaCO_3$, which quickly altered, partially or entirely, to dolomite before burial or at very shallow burial depths? Or did they form much later by replacing a precursor limestone such as a micrite that contained few recognizable textures or structures? A dolomite formed by replacing such a featureless limestone would presumably contain no relict limestone textures.

11.7.2   Origin of Penecontemporaneous Dolomite

Chemical Considerations

The chemical reactions of interest with respect to dolomite formation are

$$Ca^{2+} \text{ (aq)} + Mg^{2+} \text{ (aq)} + 2CO_3^{2-} \text{ (aq)} \leftrightarrow CaMg(CO_3)_2 \text{ (solid)} \qquad (11.1)$$

$$2CaCO_3 \text{ (solid)} + Mg^{2+} \text{ (aq)} \leftrightarrow CaMg(CO_3)_2 \text{ (solid)} + Ca^{2+} \text{ (aq)} \qquad (11.2)$$

Equation 11.1 represents primary precipitation of dolomite, and Equation 11.2 is an equation commonly used to represent dolomite replacement of calcite or aragonite. (Note: The reaction shown in Equation 11.2 is only one of a dozen or more reactions that have been proposed to represent the replacement of $CaCO_3$ by dolomite; see Machel and Mountjoy, 1986.) Some investigators have suggested that certain dolomites are indeed primary precipitates that formed by the reaction indicated in Equation 11.1, e.g., the spheroidal dolomites discussed in Section 11.4.

Is there any way to determine unequivocally if a particular dolomite formed as a primary precipitate? Unfortunately, the answer appears to be no; however, let's take a look at modern dolomites, some of which are suggested to be primary precipitates. Prior to about the middle 1940s, few occurrences of dolomite were known in modern environments. Thus, little support for the concept of primary precipitation could be gained from modern occurrences. Subsequently, dolomite has been reported from numerous modern localities, including some in Russia, South Australia, the Persian Gulf, the Bahamas, Bonaire Island off the Venezuela mainland, the Florida Keys, the Canary Islands, Deep Springs Lake, California, and Elk Lake, Minnesota. Ages of these dolomites range from about 0 to 10,000 years, although most are less than 3000 years old. Mole percent $MgCO_3$ ranges from about 30 to 50, but is commonly less than 45. Therefore, most modern dolomite is protodolomite. Experimental studies of oxygen isotopes in coexisting dolomite and calcite samples from some of these deposits have been made in an effort to confirm or disprove primary origin. Experimental data extrapolated from high temperature to low temperature suggest that at 25°C primary precipitates of dolomite should be enriched, with respect to coexisting calcite, by detectable amounts of heavy oxygen (higher $\delta^{18}O$ values). Unfortunately, isotope studies of modern dolomites have failed to consistently produce the results predicted by these experimental data. Consequently, many geologists and geochemists now apparently believe that only an insignificant amount of ancient dolomite is truly the product of primary precipitation at or above the sediment–water interface. The relative importance of primary dolomite in the geologic record is far from settled, but much of the recent work focuses on the problem of penecontemporaneous dolomitization of precursor calcium carbonate. That is, it attempts to determine the conditions that are required to bring about dolomitization of aragonite or calcite over short periods of time such as a few years to a few thousands of years.

The requirements for dolomite formation and possible mechanisms of dolomitization have been reviewed by numerous previous workers. See, in particular, Hardie (1987), Land (1985), and Machel and Mountjoy (1986). The formation of dolomite has to be considered in terms of both thermodynamic and kinetic considerations. Thermodynamic considerations include the Ca^{2+}/Mg^{2+} ratio, Ca^{2+}/CO_3^{2-} ratio, and temperature, which define the $CaCO_3$ and $CaMg(CO_3)_2$ thermodynamic stability fields in the system calcite-dolomite-water. According to Machel and Mountjoy (1986), dolomite formation is thermodynamically favored in solutions of (1) low Ca^{2+}/Mg^{2+} ratios (i.e., high Mg^{2+}/Ca^{2+} ratios), (2) low Ca^{2+}/CO_3^{2-} (or Ca^{2+}/HCO_3^-) ratios, and (3) high temperatures (Fig. 11.9A).

FIGURE 11.9

Thermodynamic (A) and kinetic (B) stability diagrams for the system calcite-dolomite-water. The ionic ratios (square brackets) in diagram A are activity ratios; seawater plots in the dolomite field, on the basal plane just outside the diagram (asterisk). The exact inclination of the stability field boundary (stippled) is not known. In diagram B, seawater plots just in the calcite field. At salinites higher than about 35‰, the field boundary (stippled) is probably bent toward higher $Ca^{2+}/$ Mg^{2+} ratios (in the direction of the arrows, wider stippled area). *(After Machel, H.-G., and E. W. Mountjoy, 1986, Chemistry and environments of dolomitization—a reappraisal: Earth Science Rev., v. 23. Figs. 2 and 3, p. 184, reprinted by permission.)*

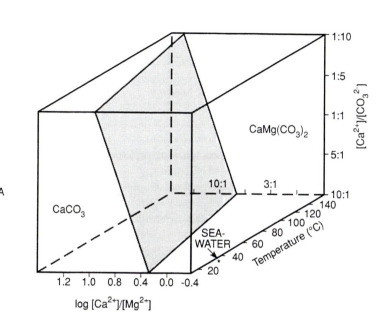

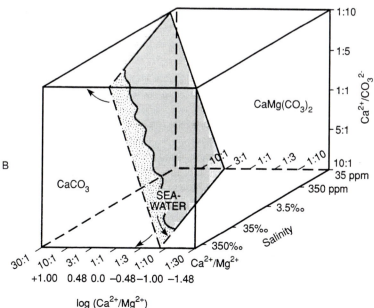

The failure of well-ordered, stoichiometric dolomite to precipitate from seawater at surface temperatures is generally attributed to kinetic factors. As summarized by Machel and Mountjoy (1986), kinetic factors that tend to inhibit formation of stoichiometric dolomite include but are not limited to the following: (1) Rapid crystallization from supersaturated, saline solutions impedes segregation of Ca^{2+} and Mg^{2+} ions into their respective layers because cations adhering to a ''wrong'' layer are more likely to become entombed. (2) The greater hydration energy of Mg^{2+} compared to Ca^{2+} reduces the chances of dehydration for Mg^{2+}, which favors Ca^{2+}-rich mineral phases. (3) The low activity of CO_3^{2-} in most natural solutions inhibits precipitation of Mg-rich carbonates because, owing to hydrated Mg^{2+} on crystal surfaces, only a few carbonate ions have enough kinetic energy to penetrate the hydration barrier. (4) Dilution of the solution decreases the rate of dolomite formation because it takes longer for the fewer ions present to travel to the reaction site; this can also be expressed as a lower degree of super-saturation. (5) The dissolution rate of the precursor carbonate (aragonite, calcite, Mg-calcite), which depends on grain size, an interplay of grain microstructure and saturation state of the solution, and the various ionic diffusion rates across crystal surface reaction zones with composition different from the bulk solution, determines the rate of dolomite formation. (6) Certain organic materials (e.g., aspartic acids, some soluble animal proteins), and dissolved SO_4^{2-} apparently inhibit dolomite precipitation in an unknown manner (surface adsorption?, complexing of Mg^{2+}). By contrast, certain other factors such as uric acid–fermenting bacteria, other organic compounds, and certain clay minerals may kinetically favor dolomite formation. It is the interaction of all these factors that determines the overall kinetics. Ca^{2+}/Mg^{2+} ratios, Ca^{2+}/CO_3^{2-} ratios, and salinity may have a particularly important effect on kinetics. These factors may be used to construct a kinetic stability diagram (Fig. 11.9B). This diagram suggests that dolomite is kinetically favored by low Ca^{2+}/Mg^{2+} ratios, low Ca^{2+}/CO_3^{2-} ratios, and low salinity. It is also favored by higher temperatures because the calcite-dolomite field boundary shifts toward higher Ca^{2+}/Mg^{2+} ratios with increasing temperature. In fact, at temperatures in excess of about 100°C most kinetic inhibitors become ineffective (Machel and Mountjoy, 1986).

Dolomite Models

Basic Models. Four basic models for penecontemporaneous dolomite formation in natural environments are currently under discussion by geologists: the hypersaline or sabkha model, the mixing-zone model, the low-sulfate model, and the shallow subtidal model. A brief summary of these models is given below.

The Hypersaline or Sabkha Model. Many, but certainly not all, of the known occurrences of modern or Holocene dolomites are in hypersaline environments such as the sabkhas of the Persian Gulf and the supratidal zones of arid climates. Under these strongly evaporative conditions, where rates of evaporation exceed rates of precipitation, seawater beneath the sediment surface becomes concentrated by evaporation. This concentration leads to precipitation of aragonite and gypsum, which preferentially removes Ca^{2+} from the water and increases the Mg/Ca ratio. The Mg/Ca ratio in normal seawater is about 5:1. When this ratio rises to a sufficiently high level owing to evaporation, possibly in excess of 10:1, dolomite is believed to form.

There is little disagreement among geologists that dolomite forms in modern hypersaline environments and that it likely formed contemporaneously in similar environments through geologic time. On the other hand, the question of exactly how dolomite forms in hypersaline environments and the overall volumetric importance of sabkha-type dolomite is still highly debatable. Many geologists (e.g., McKenzie, 1981; Patterson and Kinsman, 1982) assume that dolomite formation in the hypersaline environment proceeds by replacement of aragonite. Hardie (1987) takes exception to this viewpoint and maintains that the evidence cited for replacement origin by previous workers does not disprove the possibility that the dolomite forms as a primary precipitate of calcian dolomite or its disordered precursor. He suggests that direct precipitation of calcian dolomite from the sabkha brines be considered as an alternative to replacement. In support of this view, he points out that it is only from high-Mg, high-salinity brines that a dolomitic phase (disordered) can be precipitated in the laboratory at low temperatures. Further, he cites the presence in modern sabkha dolomites of euhedral aragonite crystals encrusting dolomite rhombs and dolomite rhombs encrusting aragonite needles with no visible evidence of absorption or replacement. Hardie does not maintain that replacement of aragonite sediment never occurs in sabkha environments, but suggests that direct precipitation and replacement may take place under different conditions and on different time scales. For example, direct precipitation may occur during periods of evaporative concentration in the vadose zone or in surface ponds, but replacement occurs on prolonged contact with slowly circulating groundwater brines.

Lasemi et al. (1989) lend support to Hardie's view. These authors report supratidal dolomite on Andros Island, Bahamas, that occurs as a cement in primary porosity. Their evidence suggests that the dolomite precipitated from solution in the pore waters. It did not form by replacement of the precursor carbonate sediment.

While conceding that dolomite can form in sabkha environments, Machel and Mountjoy (1986) conclude that sabkhas and similar hypersaline environments typically form only small quantities of distinctive, fine-crystalline protodolomite in thin beds, crusts, and patchy nodules. These dolomites are commonly associated with a variety of evaporite textures and minerals, notably calcium sulfates. Furthermore, they say that many modern sabkhas form little or no dolomite.

Whether or not dolomite forms by direct precipitation or replacement, a requirement commonly stipulated for the sabkha model is a high Mg/Ca ratio in the dolomite-forming fluid. As mentioned, Mg enrichment probably occurs through concentration of brines by evaporative processes that result in selective removal of Ca owing to precipitation of gypsum and aragonite. Concentration of brines may occur by evaporation of capillary water in the sediments of sabkhas. Under these conditions, upward flow of water from the saturated groundwater zone replaces the water lost by capillary evaporation. This process is called **evaporative pumping** (Hsü and Siegenthaler, 1969; McKenzie et al., 1980; Müller et al., 1990). Brines may also be concentrated in surface ponds or bays by surface evaporation of water. These concentrated waters have higher density than normal seawater, causing them to sink downward. Adams and Rhodes (1960) proposed that these dense brines would tend to sink down through earlier-deposited calcium-carbonate sediment and thus displace lighter seawater in the pores of the sediment. Flushing of large volumes of this Mg-rich brine through the sediment would putatively bring about dolomitization, a process they referred to as **seepage refluxion.** Although the seepage reflux model has been applied to explain various dolomite occurrences, Machel and Mountjoy (1986) state that no sabkhas have been found to form dolomite below a depth of 1–1.5 m. This finding does not prove that deeper reflux cannot occur, but it does suggest that reflux is not an important process for dolomitization in the sabkha environment.

The Mixing-Zone Model.   Some modern/Holocene dolomites and most ancient dolomites are not directly associated with evaporites. Therefore, the hypersaline model does not appear to be appropriate for these dolomites. Hanshaw and others (1971) proposed that dolomitization could occur by brackish groundwaters that were produced through mixing of seawater-derived brines and fresh water. Such low-salinity groundwaters could be saturated with respect to dolomite at Mg/Ca ratios as low as 1. Subsequently, this concept was further developed by Badiozamani (1973), Land (1973), and Folk and Land (1975). The mixing-zone model, or variations thereof, has been referred to also as the **Dorag model** (Badiozamani, 1973) and the **schizohaline model** (Folk and Land, 1975).

The fundamental concept underlying the Dorag model is explained in detail by Badiozamani on the basis of thermodynamic calculations. Mixing of meteoric waters with seawater causes undersaturation with respect to calcite, whereas dolomite saturation increases, resulting in replacement of $CaCO_3$ by dolomite. The schizohaline model of Folk and Land (1975) is illustrated in Figure 11.9B. These authors maintain that in solutions of low salinity and low ionic strength, dolomite can apparently form at Mg/Ca ratios as low as 1:1. When seawater or evaporated brine with high Mg/Ca ratios is diluted by mixing with fresh water (schizohaline environment), the mixture will retain the high Mg/Ca ratio (low Ca/Mg ratio) but not the high salinity of the saline water. Thus, these mixed waters putatively become special waters capable of forming ordered dolomite. According to Folk and Land, dolomite formed from dilute solutions is perfectly clear, with plane, mirrorlike faces (so called **limpid dolomite**), and is more resistant to solution than ordinary dolomite.

Judging from the published literature, the mixing-zone model has gained many supporters, and both **inland mixing models** and **coastal mixing models** have been proposed. On the other hand, Hardie (1987) levels some fairly devastating criticism at the model and the saturation index–percent seawater diagram that Badiozamani used as a basis for his model. Among other things, Hardie points out an inconsistency in the use of solubility values of ordered dolomite in Badiozamani's calculations. If the solubility values of less-soluble ordered dolomite are used in the calculation, as Badiozamani did, the Dorag zone of dolomite formation can extend through mixtures of meteoric water and seawater ranging from about 10 to 40 percent seawater. If, however, the solubility of more-soluble, less-ordered, Ca-rich dolomite (which is the kind of dolomite that actually forms under surface temperatures) is used, then the range of meteoric water–seawater mixtures that meet the Dorag requirement for dolomitization shrinks to a very narrow range of mixtures that fall between about 30 and 40 percent seawater (Fig. 11.10).

Machel and Mountjoy (1986) voice somewhat similar criticisms to Badiozamani's model and conclude that if dolomite precipitates in some natural mixing zones it is not because of a thermodynamic preference, but rather it is because the net effect of all kinetic inhibitors may be less potent. Further, they point out that dolomite does not form in most modern freshwater/seawater mixing zones, and where it does form, the volume of dolomite is small. With respect to the schizohaline model, Hardie stresses that there is no actual documentation that dolomite can form at Mg/Ca ratios of 1:1, nor is there hard evidence that demonstrates the special power of low-salinity conditions to produce cation-ordered dolomite.

On the other hand, Humphrey (1988) and Humphrey and Quinn (1989) suggest that massive dolomitization of platform-margin carbonates of Late Pleistocene age in southeastern Barbados, West Indies, is the result of mixing-zone dolomitization. Burns and

FIGURE 11.10

Theoretical saturation relations of dolomite and calcite in mixtures of seawater and meteoric water. Calculations were made using the starting water compositions of Badiozamani (1973). "Dorag zone" refers to that range of mixtures where the waters are supersaturated with respect to dolomite but undersaturated with calcite. Diagram A applies to ordered dolomite ($K = 10^{-17}$) and diagram B to disordered dolomite ($K = 10^{-16.5}$). *(From Hardie, L. A., 1987, Dolomitization: a critical view of some current views: Jour. Sed. Petrology, v. 57. Fig. 1, p. 169, reprinted by permission of Society of Economic Paleontologists and Mineralogists, Tulsa, Okla.)*

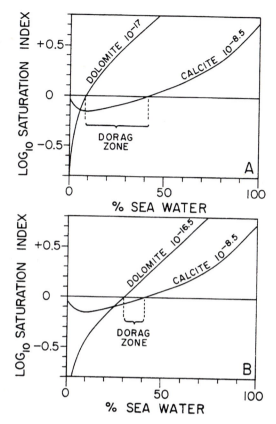

Rossinski (1989) disagree with some of Humphrey's conclusions. Gao et al. (1990) report the occurrence of minor quantities of limpid dolomite in Permian rocks in the Palo Duro Basin, Texas (USA), where it is associated with halite deposits. They propose that this dolomite formed as a result of mixing of diagenetic waters derived by dewatering of shales with brines evaporated from halite. On the basis of isotope composition and crystal zoning, Peryt and Magaritz (1990) suggest that dolomites in the Main Dolomite (Zechstein, Upper Permian) in northern Poland formed by a two-stage process that involved both hypersaline and fresh water. Dolomite growth was initiated in hypersaline water. It continued during progressive dilution by fresh water as continental waters invaded the basin following sea-level fall at the end of deposition of the Main Dolomite.

Perhaps the most widely studied modern occurrence of dolomites associated with fresh water is that of the dolomites of the Coorong area, South Australia, initially reported by Mawson (1929). These dolomites have subsequently been studied by numerous investigators (see review in Warren, 1990). Dolomitization in the Coorong lakes differs from that proposed for the mixing-zone model, however, because the Mg in the dolomitizing water is supplied by continental groundwater rather than by seawater. The purest and thickest Holocene dolomite occurs in ephemeral lakes located in the interdunal furrow immediately behind the dune ridge that forms the landward edge of the open Coorong lagoon (Warren, 1990). According to Warren, these are true primary dolomites. That is, the Coorong dolomite is precipitated where there was nothing before. It is not replacing an earlier-formed carbonate; however, the dolomite makes up no more than 10 percent of the carbonate minerals, forming surficial deposits across the coastal plain. Warren (1990) cautions that the hydrological model of the Coorong lakes should not be used to explain widespread dolomitization in ancient supratidal and shelf carbonates.

The Low-Sulfate Model.   Experimental work on the formation of dolomite at 200°C by Baker and Kastner (1981) demonstrated that the presence of dissolved SO_4^{2-} strongly inhibits the formation of dolomite. Apparently, dissolved SO_4^{2-} ions can inhibit the dolomitization of calcite at SO_4^{2-} values as low as 5 percent of their seawater value. Dolomitization of aragonite can occur at somewhat higher concentrations of SO_4^{2-}. Extrapolating these results to lower temperatures, Baker and Kastner suggest that the reason for the scarcity of dolomite in open-marine environments is the presence of dissolved SO_4^{2-} in seawater. They propose that dolomite can form rapidly in nature only where the SO_4^{2-} concentration is low. The most effective process for SO_4^{2-} removal in marine pore waters, according to Baker and Kastner, is microbial reduction of the ions in organic-rich sediments. Bacterial reduction of SO_4^{2-} promotes dolomitization by removal of the SO_4^{2-} inhibitor, by production of alkalinity, and by production of ammonia (NH^{4+}). The ammonia produced during reduction may exchange with Mg held in exchange sites in marine silicate-rich sediments, freeing the Mg for dolomitization. The concentration of SO_4^{2-} is decreased also by precipitation of calcium sulfate in evaporative environments during formation of gypsum. Mixing of fresh water with seawater in groundwater mixing environments likewise lowers the overall SO_4^{2-} concentration in the seawater. Baker and Kastner suggest that it is the reduction in SO_4^{2-} in these environments rather than the influence of the Mg/Ca ratio that allows the formation of dolomite.

Hardie (1987) raises a number of objections to the low-sulfate theory. Among other things, he points out several examples of Holocene dolomites that formed from brines with very high sulfate concentrations. He also suggests that if dolomite forms as a direct precipitate, the kinetic factors governing formation may be quite different from those governing a replacement reaction. Further, he mentions that it is possible to precipitate disordered protodolomite in the laboratory at 25°C from sulfate-rich solutions. Microbial reduction and methanogenesis reactions may promote dolomite formation, not because they remove sulfate but because they significantly increase supersaturation via massive HCO_3^- production.

Middleburg et al. (1990) report the formation of dolomite in anoxic sediments of Kau Bay, Indonesia, where they attribute it to both sulfate depletion and increase in alkalinity. Anaerobic decomposition of organic matter causes depletion of pore-water sulfate and concomitant generation of bicarbonate. These authors suggest that the dolomite formed by direct precipitation from solution and that precipitation was initiated by high-alkalinity (bicarbonate and carbonate) levels present in the zone of anoxic methane oxidation.

Morrow and Ricketts (1988) added additional experimental insight into the role of sulfates in dolomite formation. They carried out time series experiments relating to the dolomitization of calcite at 215°C–225°C in saline solutions of near-seawater salinity. Their experiments confirmed that concentrations of sulfate as low as 0.004M prevented the dolomitization of calcite. At concentrations less than 0.004M, dolomitization proceeded at a slower rate than in experiments where no sulfate was present. Their experiments demonstrated further, however, that the presence of sulfate in solution slowed the rate of calcite dissolution. Therefore, they suggest that sulfate in solution inhibits dolomitization in these experiments primarily by retarding the rate of calcite dissolution and by causing the precipitation of anhydrite. Anhydrite precipitation inhibits dolomitization by diminishing calcium-ion activity during calcite dissolution and possibly by coating calcite crystals with a thin $CaSO_4$ layer, which prevents the nucleation of dolomite on calcite crystal surfaces. Their experiments show that the rate of dolomitization of calcite is much more rapid in solutions with higher carbonate and bicarbonate activities, confirming suggestions to this effect by previous workers. Their experiments also showed that the presence of sulfate in solution did not prevent the direct precipitation of dolomite in solutions in which the solid reactants were carbonate minerals other than calcite ($BaCO_3$ and $2PbCO_3 \cdot PbOH$).

The Shallow Subtidal Model.   In all of the above models, some kind of "special" water (i.e., seawater modified in some way) is required for dolomitization. On the other hand, a few authors have proposed that penecontemporaneous dolomitization can take place in normal, unmodified seawater if a sufficient volume of seawater can be passed through the sediment (e.g., Carballo and Land, 1984; Carballo et al., 1987; Saller, 1984; Sass and Katz, 1982). Many geologists believe that dolomitization can occur only when active pumping of magnesium-bearing water takes place. Simply immersing carbonate sediments in static fluids, whether they be hypersaline, normal marine, or mixed marine-meteoric, rarely appears to nucleate dolomite, much less generate significant volumes of dolomite (Land, 1985). On the other hand, if large volumes of water are forced through the sediment so that each pore volume of water in the sediment is constantly being renewed with new water, then dolomitization can presumably occur. Movement of large volumes of water through the sediment provides a constant source of Mg and removes replaced Ca and other ions that might "poison" the dolomite crystal structure. Thus, any mechanism that provides a means of forcing large amounts of water through the sediment can presumably bring about dolomitization.

As an example, Carballo et al. (1987) report an area of Sugarloaf Key, Florida, where seawater is forced upward and downward through Holocene carbonate mud during rise and fall of seawater accompanying spring tides, a process they call **tidal pumping.** Owing to the large volume of seawater driven through the sediment by this mechanism, large quantities of Mg are imported into the sediment, and pore fluids are constantly being replaced by new fluids. Under these conditions, dolomite is forming in the sediment even though little or no evaporation of the seawater has occurred. Carballo et al. suggest that dolomite forms both by precipitation as a cement and by later replacement of preexisting crystallites. The results of this study thus clearly imply that normal seawater can act as a dolomite-forming fluid, without the requirement of Mg concentration through evaporation, if enough seawater is forced through sediment. If hypersaline waters are available, movement of these waters could also bring about dolomitization. Thus, a continuous spectrum of dolomitization may exist from normal-marine subtidal to hypersaline-subaerial (Machel and Mountjoy, 1986).

Deep-Sea Dolomite.   All of the above models for dolomite apply to dolomites formed in supratidal to shallow-subtidal environments. Dolomite in minor amounts is also geographically widespread in deep-marine sediments. It is most abundant in sites from the Atlantic, from small ocean basins, and near continental slopes and shelves, and it is less common in the Pacific and ocean-basin centers (Lumsden, 1988). It is present at depths ranging from about 1 m below the seafloor to more than 1000 m. According to Lumsden, dolomite makes up an average of about 1 percent of deep-marine sediment. The mean crystal size is about 6 μm and the dolomite is nonstoichiometric (average 56 percent

$CaCO_3$). The dolomite does not increase in crystal size or abundance downhole, or with increasing age of the sediment, nor does it change in crystal order or stoichiometry with increasing age. Lumsden estimates that as much as 10 percent (possibly more) of this deep-marine dolomite may be detrital and may be derived from supratidal dolomite sources. The remaining dolomite must have been precipitated postdepositionally in the sediment pores. The exact conditions that favored dolomite precipitation are poorly understood, but Mg was presumably derived from marine pore waters trapped in the sediment, with additional Mg advected and convected in from the adjacent sea.

11.7.3   Subsurface (Later-Stage) Dolomitization

The Problem of Massive Dolomitization

Let's return to the problem of explaining the relatively thick, massive, widespread dolomite that constitutes most of dolomite in the geologic record. Did this dolomite form penecontemporaneously, either by direct precipitation or dolomitization, by one or more of the mechanisms discussed in the above models? Alternatively, did it form much later, perhaps millions to hundreds of millions years later, by subsurface dolomitization of precursor limestones? There is reason to believe that many dolomites did indeed form by subsurface dolomitization; that is, dolomitization was not contemporaneous with sedimentation. If this is true, how did these dolomites form? Part of the ''dolomite problem'' involves coming up with a satisfactory explanation to account for the large amounts of Mg that had to be imported into preexisting calcium carbonate sediments to bring about dolomitization. Further, the mechanism has to account also for removal of the Ca and other ions released by the dolomite replacement process. Because seawater is the only common and abundant magnesium-rich fluid on Earth, seawater must have furnished the primary source of Mg for dolomitization. Under some circumstances, seawater may have been modified in some way (e.g., surface evaporation; mixing with meteoric waters; mixing with basement hydrothermal fluids; subsurface evolvement by salt-sieving, dissolution of buried salt, reaction with mineral phases, clay mineral dewatering to produce connate brines); however, such modification to increase Mg concentration may not be necessary. In fact, some subsurface waters derived from seawater may be depleted in Mg relative to seawater. If a sufficient volume of essentially normal seawater can be forced through carbonate sediment, dolomitization can probably occur.

Thus, seemingly, the problem of subsurface dolomitization boils down to finding suitable mechanisms for large-scale, mass transport of Mg-rich seawater through subsurface carbonate formations. This problem is complicated by the fact that subsurface formations exhibit significant variations in porosity and permeability. Effective, large-scale dolomitization presumably could occur only where an effective seawater circulation mechanism was present and where the carbonate precursor rocks had sufficient permeability to allow entry and passage of the dolomitizing waters. Perhaps these requisite conditions provide an explanation for the fact that not all carbonate rocks are dolomitized. It may be reasonable to ask, given hundreds of millions of years of burial history, why aren't all ancient limestones dolomitized? Machel and Mountjoy (1986) point out that the problem of proving subsurface dolomitization is twofold. First, we have to come up with a water-circulation mechanism that can account for the magnesium supplied to the limestones. Then we have to exclude near-surface dolomitization as the mechanism of dolomitization and, instead, establish that dolomitization took place in the subsurface. The latter step may be very difficult to do.

Land (1985) suggests that most dolomite must form relatively early (penecontemporaneously?) in the depositional and burial history of sediment when seawater or seawater-derived fluids can be actively pumped. On the other hand, several investigators have proposed a subsurface origin for some dolomites on the basis of petrographic, geochemical, or other evidence. For example, some dolomitization coincides with the formation of stylolites by pressure solution. Also, some dolomite may undergo more than one stage of dolomitization. For example, Gregg and Shelton (1990) propose two stages of dolomitization in dolomites of the Bonneterre and Davis formations of southeastern Missouri (USA). Planar dolomite formed by early diagenesis of cryptalgalaminites, and nonplanar dolomite formed by late-stage neomorphism (recrystallization) of planar dolomite and by dolomitization of peloid mudstones.

In any case, if subsurface dolomitization (as opposed to penecontemporaneous dolomitization) was important in the past, some mechanism or mechanisms had to operate that forced large quantities of water through precursor carbonate sediment. The principal mechanisms proposed for moving Mg-bearing waters through subsurface carbonates include (1) burial compaction, (2) topography-driven (or gravity-driven) flow, (3) thermal convection, and (4) hydrothermal flow. Hydrothermal flow is thought to be important only locally and in association with sedimentary ore deposits; however, one or more of the other mechanisms conceivably can account for larger-scale fluid movement.

Subsurface Fluid-Flow Models

Flow Resulting from Burial Compaction. Flow resulting from sediment compaction occurs at fairly shallow depths, probably less than 1000 m, above the zone where porosity and permeability are reduced by cementation and solution-compaction. Fluids expelled by compaction move largely upward, although some lateral movement is also possible. Although compaction flow may account for some dolomitization, many authors have pointed out that such flow cannot form massive dolomites over large regions. Owing to the limited volume of these fluids, there is simply not enough Mg in the fluids per unit volume of overlying strata to account for widespread dolomitization. Locally, where faults or other characteristics of the strata may funnel compaction flow into limited bodies of carbonates such as reefs, such flow may be responsible for dolomitization. Because burial compaction takes place at relatively shallow depths, dolomitization by compaction-driven fluids must occur relatively early during basin subsidence.

Topography-Driven (or Gravity-Driven) Flow. Topography-driven flow refers to flow of subsurface fluids downward and upward within a basin by gravity owing to the presence of a hydraulic head, the magnitude of which is determined by the elevation of the meteoric recharge area for the subsurface formations (Garven and Freeze, 1984). Such flow probably takes place to some extent in all sedimentary basins. The depth within the basin to which flow extends, the lateral distance of flow (possibly as much as several hundreds of kilometers), and the velocity of flow depend upon both the hydraulic head and the permeability distribution in the subsurface formations. The general conditions for such flow are shown schematically in Figure 11.11. By this mechanism of flow, enormous quantities of subsurface waters could be driven to considerable depths down along one side of a basin, to the extent allowed by permeability of the formations, and up again along the other side. Such flow could continue over a long period of time and might eventually result in massive dolomitization.

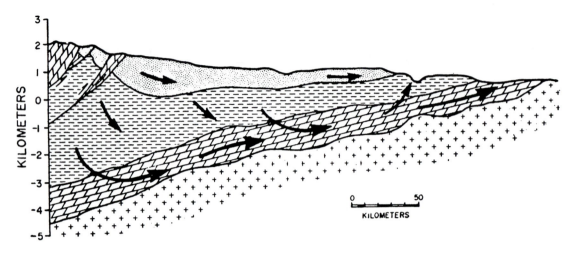

FIGURE 11.11
Conceptual model of topography-driven (or gravity-driven) fluid flow in sedimentary basins.
Flow is driven by the hydraulic head of fluid resulting from meteoric recharge. The geometry of
the flow system (arrows) is controlled by the water-table configuration and the permeability
distribution in subsurface formations. *(From Garvin, G., and R. A. Freeze, 1984, Theoretical
analysis of the role of groundwater flow in the genesis of stratabound ore deposits: Am. Jour.
Science, v. 284. Fig. 3, p. 1091, reprinted by permission.)*

Presumably, the fluids involved in such flow could continue to bring about dolomitization as long as they retained the chemical potential to form dolomite (adequate Mg
and CO_3?). Eventually, however, the chemical potential of the waters might become
exhausted as Mg is depleted owing to formation of dolomite. Also, the residual seawater
in the subsurface formations would eventually be expelled from the formations and
replaced by meteoric water with low Mg concentrations. Can additional Mg be added to
the subsurface waters from subsurface formations as the waters move downward and
upward through a basin? Some Mg might be released by pressure solution of dolomite or
Mg-calcite, but this Mg would not be new Mg. Solution of salt deposits containing
Mg-bearing salts (polyhalite, carnallite, kieserite) could also supply Mg, but the volume
of these salts is probably small. Magnesium may also be released during clay mineral
diagenesis; however, much of the released Mg may be taken up locally in forming new
authigenic minerals such as chlorite. Thus, it is possible that gravity-driven fluid flow is
a self-limiting process with respect to its dolomitization potential. Massive, late-stage
dolomitization by gravity-driven fluids may be limited simply because the fluids have lost
their chemical potential to form dolomite. The overall importance of this kind of fluid flow
to bring about massive dolomitization remains to be established; however, some recent
studies lend support to the potential importance of this process. For example, Gregg
(1985) has invoked topography-driven flow to explain dolomitization of a sheet of dolomite about 6 m thick in the Cambrian Bonneterre Formation of Missouri, which extends
over an area of about 17,000 km^2. As a final note, the Ca released by massive dolomitization must be removed in the expelled water that moves ahead of the dolomitization
front. What happens to this Ca? Does it precipitate as calcite cement in the precursor
limestones ahead of the front? If so, does this cement plug pore space and reduce per-

meability to the point that dolomitizing fluids cannot enter? Interesting questions to ponder!

It is also possible for fluids to be driven downward in a basin by gravity owing to density differences in fluids, rather than by a topography-controlled fluid head (i.e., the seepage reflux model of Adams and Rhodes, 1960). Ruppel and Cander (1988) suggest, for example, that some dolomitization in the Permian San Andres Formation of West Texas may have occurred by seepage refluxion of seawater brines evaporated to near calcium sulfate–saturation.

Fluid Flow Driven by Thermal Convection.   A thermal fluid convection model proposed by Kohout (Kohout, 1967; Kohout et al., 1977) is receiving increasing attention as a possible model for subsurface dolomitization (Fanning et al., 1981; Simms, 1984). This model, shown graphically in Figure 11.12, is particularly attractive because it is a kind of "half-cell," with seawater acting as a virtually infinite reservoir of Mg. The model is based on inferred circulation patterns beneath the Florida Platform. Cold seawater enters the platform at depths of as much as 3000 m under the Straits of Florida. This water is prevented from moving downward by a geothermal barrier, which forces the seawater to move laterally toward the platform interior. There, it displaces platform water upward and is itself forced upward, where it eventually discharges back to the sea through numerous geothermal springs. This model for thermal convection, commonly referred to as **Kohout convection,** has been proposed also as a mechanism for dolomitization under the Florida Platform and some Pacific atolls and in Jurassic-Cretaceous limestones of the southern Alps (Cervato, 1990). As mentioned, Kohout convection is an attractive model for dolomitization because of the virtually limitless supply of seawater that could potentially circulate through the shallow subsurface by this method. On the other hand, the model is largely based on inference, and definite proof is lacking that Kohout convection can actually produce massive dolomitization.

Combined Fluid-Flow Model.   It is possible under some geologic conditions that more than one mechanism may be operating to produce subsurface circulation of dolomitizing waters. For example, Whitaker and Smart (1990) discuss active circulation of saline groundwaters in carbonate platforms of the Great Bahama banks, which may be due to a combination of flow mechanisms. Factors responsible for fluid circulation may include elevation head, density, and temperature (Fig. 11.13).

The hydraulic head is created by differences in the sea-surface buildup across the carbonate platform owing to tides, wave conditions, and currents. Density-driven mechanisms involve both buoyant circulation and reflux. Buoyant circulation is caused by circulation within freshwater lenses along the mixing zone with saline waters. Resulting discharge of brackish water at the coast causes a compensating inflow of saline waters at depth (Fig. 11.13). Reflux involves downward and outward movement of seawater that has undergone an increase in density owing to evaporation, as described in the preceding discussion. Also, geothermal heat below the platform may be sufficient to induce thermal (Kohout) circulation.

Significance of Temperature and Time

In a discussion of subsurface dolomitization, Hardie (1987) emphasizes the importance of temperature. He points out that calcite can be converted to ordered dolomite in a matter of days at 100°C. Furthermore, as temperature increases, the Mg/Ca ratio of the waters

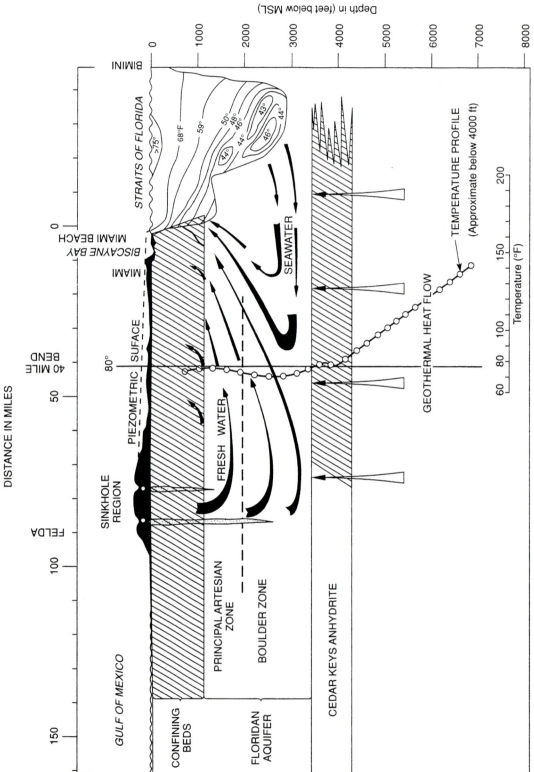

FIGURE 11.12

"Kohout convection" as illustrated by the flow system in the Florida subsurface. The flow pattern, shown by black arrows, is inferred and is believed to be the result of interaction of thermally heated seawater with a to geothermal conditions in the Floridan Plateau: *Florida Dept. Nat. Resources, Bur. Geology Spec. Pub. 21, p. 1–34, reprinted by permission.*)

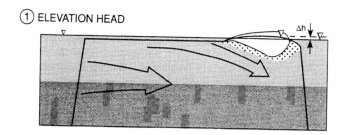

① ELEVATION HEAD

② DENSITY
a. Buoyant circulation

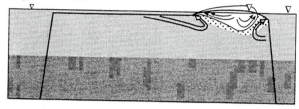

b. Reflux   Evaporation

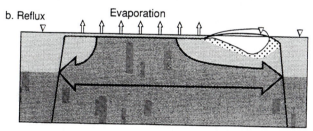

Fresh water

Transition zone

Lower-density saline water

Higher-density saline water

③ TEMPERATURE

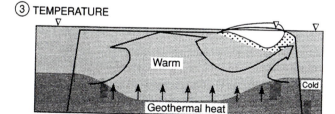

Warm

Cold

Geothermal heat

FIGURE 11.13

Schematic diagrams illustrating processes that may drive circulation of saline waters in carbonate platforms. (1) Circulation driven by elevation head of seawater above platform. (2a) Buoyant circulation created by circulation within freshwater lenses and mixing with saline water. (2b) Downward and outward reflux of ocean water concentrated by evaporation. (3) Circulation driven by geothermal heat from below. *(After Whitaker, F. F., and P. L. Smart, 1990, Active circulation of saline ground water in carbonate platforms: Evidence from the Great Bahama Bank: Geology, v. 18. Fig. 1, p. 200, reprinted by permission of Geological Society of America, Boulder, Co.)*

FIGURE 11.14

Plot of the expected stability field of ordered dolomite as a function of temperature and Ca/Mg ratio in 1M and 2M chloride brines. At the top of the plot is a histogram showing the Ca/Mg ratios of natural groundwaters in contact with a variety of sedimentary rocks. Note that at temperatures exceeding about 70°C dolomite can form in Ca-rich waters. Thus, most Ca-rich groundwaters can be dolomitizing fluids above this temperature. *(From Hardie, L. A., 1987, Dolomitization: a critical view of some current views: Jour. Sed. Petrology, v. 57. Fig. 7, p. 177, reprinted by permission of Society of Economic Paleontologists and Mineralogists, Tulsa, Okla.)*

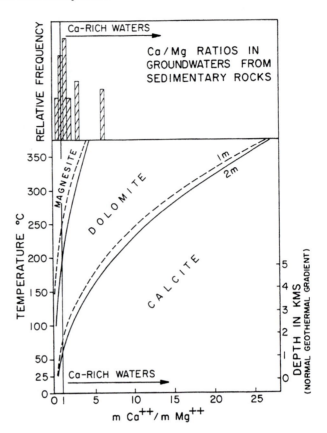

that is needed to convert calcite to dolomite decreases. At temperatures in excess of 100°C dolomitization can theoretically occur in Ca-rich waters (Fig. 11.14). Therefore, if adequate permeability exists at burial depths corresponding to temperatures exceeding 100°C, dolomite could form in waters with very low Mg/Ca ratios. Hardie further emphasizes that time may have an important effect on dolomitization at lower temperatures. He proposes, as others have suggested, that normal seawater can cause dolomitization if circulation through calcium carbonate sediment is sustained for a long enough time, perhaps 10^4 or 10^5 years.

Evidence for Late-Stage, Subsurface Dolomitization

As suggested, proving that dolomitization took place in the (deep) subsurface as opposed to near-surface is quite difficult. Evidence for deep-burial dolomitization commonly includes such criteria as coarse crystal size and the presence of microstylolites. In a recent study, for example, Lee and Friedman (1987) suggest deep-burial dolomitization for dolomites in the Ordovician Ellenburger Group in the Permian Basin of West Texas. As evidence for this interpretation, they cite (1) the presence of coarse, crystalline dolomite, (2) abundant xenotopic texture, (3) remnants of depositional textures in coarse-crystalline, xenotopic dolomite, (4) homogeneous cathodoluminescence, (5) the presence of saddle dolomite, (6) high temperature of dolomite formation as indicated by fluid inclusions, and (7) light oxygen isotope composition. Lee and Friedman suggest that the coarser the crystal size of dolomite the later the dolomitization and that coarse, crystalline dolomite with remnants of stylolite seams is indicative of moderate burial depths. They interpret the abundance of xenotopic (nonplanar) anhedral crystals to indicate burial temperatures above the critical roughening temperature of about 50°–80°C suggested by Gregg and Sibley (1984). They further suggest that the xenotopic texture indicates either recrystallization of earlier-formed dolomite or the replacement of limestone by dolomite at elevated temperatures. Homogeneous cathodoluminescence is suggested to indicate homogenization by neomorphism or replacement by the same kind of solution; they favor an origin for the dolomite by replacement of limestone by solutions of a similar chemistry. The presence of saddle dolomite is interpreted to indicate formation at temperatures ranging from 60° to 150°C (Radke and Mathis, 1980). Crystallization temperatures were calculated from fluid-inclusion data to range from about 50°C to >300°C. Lee and Friedman suggest that these ranges indicate progressive burial through time. Similarly, light oxygen isotope composition (low $\delta^{18}O$ values) suggests crystallization at elevated temperatures.

Kupecz et al. (1988) disagree with much of Lee and Friedman's interpretation. They state that none of the arguments for deep burial presented by Lee and Friedman are proof of a burial origin for the dolomite. They concede that some of the arguments may be consistent with elevated temperatures, but maintain that none place constraints on depth of burial. They argue that hot fluids could have been transported in (from below?) through some kind of natural conduit (not specified) to account for the elevated temperatures. This disagreement is clearly a case of different investigators drawing quite different conclusions from the same data. (See also the reply by Lee and Friedman, 1988.)

Machel and Anderson (1989) report pervasive subsurface dolomitization of the Devonian Nisku Formation in Alberta (Canada). On the basis of petrographic, geochemical, and isotopic evidence, they interpret that this dolomite formed from upward-flowing fluids at temperatures of 40°–50°C and at burial depths of 300–1000 m. They suggest that two types of subsurface flow could have caused pervasive dolomitization: (1) expulsion of burial-compaction waters, that is, chemically modified water, with or without addition of Mg-bearing water from deeper parts of the basin, and (2) thermal convection of formation fluids originally underlying and/or overlying the Nisku Formation.

11.7.4   Discussion and Summary

Field studies show that dolomite occurs in a wide range of modern carbonate environments. Some penecontemporaneous dolomites are reported from sabkha-type environments, others from presumed mixing zones, and still others from shallow, intertidal environments. Minor amounts of dolomite occur also in deep-marine sediments. Most of these dolomites are nonstoichiometric protodolomites, and their overall volume in modern environments is small. It has not been proven conclusively that these dolomites are either all primary precipitates or all dolomitized calcium carbonates. In fact, it seems likely that dolomites of both origins occur. Whether or not dolomite formation in sabkhas, shallow mixing zones, and shallow subtidal environments can account for all or most of the massive dolomite in the geologic record is an unresolved question. Some geologists believe that these massive dolomites are the product of later-stage, subsurface dolomitization during deeper burial.

Although some dolomites form in "special" waters (hypersaline, mixed meteoric-marine, low-sulfate), many geologists now believe that special waters are not necessary for dolomite formation if a constant supply of magnesium-bearing water (e.g., normal seawater) is forced through sediment. Higher temperatures tend to overcome the effects of the various factors that inhibit dolomite formation. Thus, temperature clearly has a strong effect on dolomitization. Machel and Mountjoy (1986) sum this all up by saying that if thermodynamic and kinetic considerations are combined, the following environments and conditions are conducive to dolomite formation: (1) environments of any salinity above thermodynamic and kinetic saturation with respect to dolomite (freshwater/seawater mixing zones or schizohaline environments, normal saline to hypersaline subtidal environments, hypersaline supratidal environments), (2) environments with high CO_3^{2-} and HCO_3^- waters (alkaline environments), and (3) environments with temperatures greater than about 50°C (subsurface and hydrothermal environments).

With respect to the stoichiometry of dolomite, both the stoichiometry and ordering of dolomite appear to increase in a general way with increasing age. Presumably this increase indicates greater stabilization of dolomite owing to solution-reprecipitation and recrystallization phenomena. On the other hand, Sass and Bien (1988) suggest that the stoichiometry of dolomite may be related also to salinity and environments. Their studies indicate that (1) marine (nonevaporitic) dolomites range in their calcium content from stoichiometric to calcian (57 mole percent), and their sodium content is 150–350 ppm; (2) dolomites in association with gypsum cover the same range of stoichiometry as the marine dolomites, but their sodium content reaches much higher values (as high as approximately 2700 ppm); furthermore, a negative correlation is present between excess calcium and sodium; and (3) dolomites in association with halite have an almost ideal stoichiometric composition and a sodium content overlapping the lower range of marine dolomites (150–270 ppm). They suggest that the ionic ratios in dolomites are related to the respective ratios in the mother solution, which evolved through a combination of surface evaporation and modifications in porewaters, mainly by dolomitization. Also, many investigators (e.g., Vahrenkamp and Swart, 1990) have suggested that the Sr^{2+}/Ca^{2+} ratio in dolomites is related to the Sr^{2+}/Ca^{2+} ratio in the dolomite-forming fluid.

This suggested relationship between composition of dolomitizing fluids and composition of the resulting dolomite is not entirely supported by experimental work. Sibley (1990) studied the formation of stoichiometric dolomite by experimental runs at 218°C in hydrothermal bombs. He reports that ordering of the dolomite occurred in stages. The first carbonate phase that formed was high-Mg calcite with approximately 35 percent $MgCO_3$. This phase was subsequently replaced by Ca-rich, poorly ordered dolomite, which, in turn, was replaced by ordered, stoichiometric dolomite. Sibley determined that the rate of dolomitization and the rate at which the unstable phases transformed to stoichiometric dolomite was related to the Mg^{2+}/Ca^{2+} ratio of the solution. That is, the rate of dolomitization was faster in solution with higher Mg^{2+}/Ca^{2+} ratios. On the other hand, he found no simple relationship between the Mg^{2+}/Ca^{2+} ratios of the solution and the composition of the growing dolomite rhombs. Sibley reports that (1) crystals continued to nucleate throughout the reaction, (2) overgrowths of the more-stable phases grew on the less-stable phases, and (3) the initial unstable crystals underwent intercrystalline-scale dissolution and reprecipitation replacement, which resulted in large rhombs with irregular, corroded interiors. Although the results of these high-temperature experiments cannot necessarily be extrapolated directly to natural, low-temperature systems, they suggest that both solution chemistry and reaction time need to be evaluated when interpreting the composition of ancient dolomites. The results appear to be somewhat at odds with Sass and Bien's (1988) findings that stoichiometry is related to salinity of the environment. The results also suggest that the transformation of nonstoichiometric to stoichiometric dolomite can be a stepwise process.

Additional Readings

Pray, L. C., and R. C. Murray, eds., 1965, Dolomitization and limestone diagenesis: Soc. Econ. Paleontologists and Mineralogists Spec. Pub. 13, 180 p.

Shukla, V., and P. A. Baker, eds., 1988, Sedimentology and geochemistry of dolostones: Soc. Econ. Paleontologists and Mineralogists Spec. Pub. 43, 266 p.

Zenger, D. H., J. B. Dunham, and R. L. Ethington, 1980, Concepts and models of dolomitization: Soc. Econ. Paleontologists and Mineralogists Spec. Pub. 28, 320 p.

Chapter 12
Diagenesis of Carbonate Rocks

Chapter 9 discusses the processes that bring about diagenesis of siliciclastic sediments. Significant differences exist between siliciclastic and carbonate sediments; therefore, the course of diagenesis differs in these two fundamentally different kinds of rocks. To illustrate, carbonate sediments are intrabasinal deposits, which are precipitated in some manner from the water in which they are deposited. Thus, initially, carbonate minerals are more or less in chemical equilibrium with the waters of their depositional environment. By contrast, siliciclastic sediments are brought into the depositional basin from outside. Further, carbonate sediments are composed of only a very few major minerals (aragonite, calcite, dolomite) in contrast to a much larger variety of minerals and rock fragments that may be present in siliciclastic sedimentary rocks. Carbonate minerals are more susceptible in general to diagenetic changes such as dissolution, recrystallization, and replacement than are most silicate minerals. Also, they are generally more easily broken down by physical processes, and they are much more susceptible to attack by organisms that may crush or shatter shells or that may bore into carbonate grains or shells.

Nonetheless, carbonate sediments proceed in a general way through the same diagenetic regimes as siliciclastic sediments. That is, carbonate sediments go through early (shallow-burial), middle (deep-burial), and possibly late (uplift and unroofing) stages of diagenesis. In terms of time and burial depth, these stages are similar to the eodiagenetic, mesodiagenetic, and telodiagenetic stages of siliciclastic diagenesis. If fact, the terms eogenetic, mesogenetic, and telogenetic were introduced by Choquette and Pray (1970) to designate zones of carbonate diagenesis. There are, however, significant differences between carbonate and siliciclastic sediments with respect to the nature of the diagenetic processes that occur in each of these stages. We cannot simply import the concepts developed for siliciclastic diagenesis to explain carbonate diagenesis. The ultimate result of burial diagenesis of siliciclastic sediments is to move initially incompatible assemblages of largely silicate minerals toward a state of greater equilibrium with their burial conditions of pressure, temperature, and pore-fluid compositions. These processes may thus produce important changes in mineral composition, but they rarely, if ever, result in wholesale, pervasive alteration or replacement of depositional mineral assemblages. By contrast, carbonate sediments may undergo pervasive changes involving aragonite calcitization, recrystallization, and replacement that may bring about complete or nearly complete change in depositional mineralogy. For example, an initial aragonite mud may alter entirely to calcite (micrite) during early diagenesis and burial. In turn, the calcite may conceivably be replaced completely or nearly completely by dolomite at a later time. Also, cementation tends to be a more important process in carbonate rocks than in siliciclastic rocks.

Pervasive alteration of carbonate sediments not only changes the mineralogy of the sediments, but it may also destroy or severely modify depositional textures (e.g., carbonate grains, micrite). Thus, important environmental information may be lost. Further, the diagenesis of carbonate sediments may produce marked changes in porosity. Porosity is enhanced by solution processes and reduced by compaction and cementation. Because of the marked susceptibility of carbonate sediments to diagenetic alteration, special care must be taken in the study of carbonate rocks to recognize and identify features of diagenetic origin. Unless diagenetic features are differentiated from depositional features, the validity of genetic interpretations may be severely compromised.

In this chapter, we take an in-depth look at carbonate diagenesis. We begin by discussing early diagenesis in the seafloor environment, where carbonate sediments may be affected by biologic as well as by chemical and physical processes. Many carbonate sediments are subsequently exposed to meteoric conditions, either before or after deep burial. Significant diagenetic change takes place in carbonates in this environment under the influence of dilute, chemically aggressive meteoric pore waters. We then follow the course of diagenesis during deep burial. During burial diagenesis, increased pressure, increased temperature, and changed pore-water compositions are the important factors that bring about diagenetic change.

12.2   Regimes of Carbonate Diagenesis

As discussed in Chapter 10, most carbonate sediments originate in marine environments. Therefore, this discussion of carbonate diagenesis focuses on the diagenesis of marine carbonates. Nonmarine carbonates also undergo diagenesis; however, diagenetic effects are generally less severe in nonmarine carbonates because they are composed of more-stable carbonate minerals (mainly low-Mg calcite) than are marine carbonates. Figure 12.1 illustrates schematically the principal environments of carbonate diagenesis. We recognize three major regimes of diagenesis (James and Choquette, 1983a):

1. **The seafloor and shallow-marine subsurface regime** includes the seafloor and the very near-surface environment, that is, the eogenetic zone of Choquette and Pray (1970). It is characterized mainly by marine waters of normal salinity, although hypersaline waters are present in evaporative environments. Mixed marine-meteoric waters may be present also at the strandline and in the shallow subsurface at the mixing interface between the marine realm and the meteoric realm (Fig. 12.1).

2. **The meteoric regime** is distinguished by the presence of fresh water. It includes the unsaturated, vadose, zone above the water table and the phreatic zone, or saturated

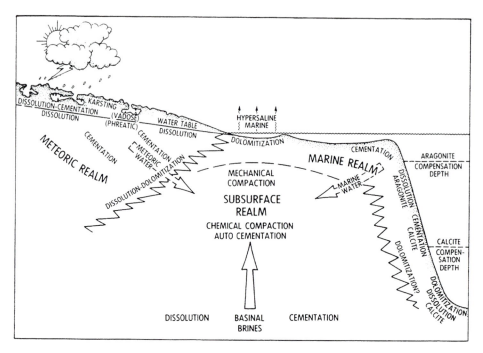

FIGURE 12.1

The principal environments in which postdepositional modification of carbonate sediments occur. The dominant diagenetic processes that occur in each of the major diagenetic realm are also indicated. See text for details. *(From Moore, C. H., 1989, Carbonate diagenesis and porosity. Fig. 3.1, p. 44, reprinted by permission of Elsevier Science Publishers, Amsterdam.)*

zone, below the water table. As mentioned, a zone of mixed marine-meteoric water exists between the meteoric and marine realms. Nonmarine carbonates originate in the meteoric regime. Marine carbonates may be brought into this regime, at least for a short time, in three ways: (a) by falling sea level, (b) by progressive sediment filling of a shallow carbonate basin (which produces a shallowing-upward sequence) until the sediment interface is at or above sea level, and (c) by late-stage uplift and unroofing of a deeply buried carbonate complex. Thus, the meteoric regime may include both the eogenetic and telogenetic zones of Choquette and Pray.

3. **The deep subsurface** is referred to as the **marine phreatic** or **deep phreatic** zone by some authors. In the deep subsurface, sediment pores are filled with waters that were either marine or meteoric waters in the beginning. The composition of deep pore waters is commonly different, however, from either marine or meteoric waters owing to burial modifications (Chapter 11). The deep subsurface is the mesogenetic zone of Choquette and Pray.

12.3 Diagenesis on the Seafloor

12.3.1 Biogenic Processes

As discussed in Chapter 10, organisms participate in a variety of ways in generating carbonate deposits. After carbonate sediments are deposited, however, organisms can degrade and break down skeletal grains and other carbonate materials. This organic degradation is actually a kind of sediment-forming process because it results in the production of finer-grained sediment. Nonetheless, it is included here as a type of very early diagenesis because it brings about modification of previously formed sediment.

The most important kind of biogenetic modification of sediment is caused by the boring activities of organisms. Milliman (1974, p. 253) points out that a variety of organisms are capable of boring into carbonate substrates. Microborings are produced by endolithic fungi, bacteria, and green, blue-green, and possibly red algae. Macroborings are formed by larger organisms, including sponges, molluscs, worms, echinoids, and crustaceans. Boring by algae, fungi, and bacteria is a particularly important process for modifying skeletal material and carbonate grains. Boring by photosynthesizing algae is confined to the photic zone and operates most effectively in the upper part of this zone at water depths less than about 70–100 m. On the other hand, the activities of fungi are believed to extend to depths of 500 m or more, and heterotrophic (nonphotosynthesizing) algae and bacteria may be present to abyssal depths (Friedman et al., 1971).

Boring of carbonate grains by endolithic fungi and algae is commonly most intense in shallow-water tropical areas. Very fine-grained aragonite and/or Mg-calcite is then precipitated into the holes left by these organisms. If boring activities are prolonged and intense, the entire surface of a grain may become infested by these aragonite- or Mg-calcite-filled borings, resulting in the formation of a thin coat of micrite around the grain. This coating is called a **micrite envelope** (Bathurst, 1966). The cortoids described in Chapter 10 probably formed in this way. Even more intensive boring may result in complete micritization of the grain, with the result that all internal textures are destroyed and a kind of peloid is created. In colder and deeper waters, boring by endoliths may also be fairly intense. Because these waters are not commonly saturated with $CaCO_3$, however, the vacated borings do not fill with precipitated $CaCO_3$. Under these conditions, rather than forming a micrite envelope, continued boring eventually results in breakdown of the skeletal materials into finer-grained particles (Alexandersson, 1979).

In addition to boring activities, organisms may contribute to carbonate destruction in the marine environment in some other ways: (1) predators such as fish and rays that eat shelled invertebrates break down the shells into smaller fragments, (2) predators attack and kill living substrate such as coral polyps, hastening the post-mortem invasion by boring organisms, (3) grazing organisms such as gastropods, echinoids, and fish attack and erode calcareous substrate, including reefs, and (4) browsing organisms such as holothurians and gastropods ingest carbonate sediment and pass it through their digestive mechanisms (Milliman, 1974, p. 258). This last process may cause some erosion and dissolution of carbonate particles, although its overall effect is believed to be quite minimal. The bioturbation activities of organisms can churn up carbonate sediment, resulting in textural mixing and destruction of sedimentary structures such as laminations. Furthermore, mixing of sediments near the sediment–water interface may cause some change in pore-water chemistry by mixing interstitial pore waters with overlying bottom waters. If the bottom water is undersaturated, such mixing could result in dissolution or etching of carbonate grains brought in contact with these waters as a result of bioturbation.

12.3.2   Carbonate Cementation and Dissolution

General Conditions Affecting Cementation and Dissolution

Carbonate sediments precipitated directly in the marine environment consist primarily of aragonite and high-magnesian calcite (Mg-calcite), although small amounts of low-magnesian calcite may also form. Dolomite also occurs in some modern marine environments. Whether or not this dolomite is a primary precipitate, or forms by rapid replacement of aragonite or Mg-calcite, is an unresolved problem (Chapter 11). Much of the carbonate in marine environments consists of the remains of $CaCO_3$-secreting organisms; however, the carbonate mineralogy of skeletal tests is different for different groups of organisms (Chapter 10). Most organisms have tests composed of aragonite (e.g., molluscs, green algae, corals) or Mg-calcite (e.g., echinoderms, benthonic foraminifers, trilobites). A few kinds of skeletal grains (e.g., brachiopod shells, coccoliths, planktonic foraminifers) are composed of low-Mg calcite.

In the depositional environment, marine carbonates are exposed to either normal marine or hypersaline waters. Most carbonate sediments are deposited in tropical waters where temperatures commonly exceed about 18°C; however, carbonate deposits composed mainly of shell remains occur also in temperate waters and even in some colder waters. Some carbonate sediments (e.g., pelagic oozes, carbonate turbidites) are deposited in deep water where bottom waters are quite cold and pressures are high. The diagenetic behavior of carbonate minerals in marine waters is influenced by the chemistry of the water (saturated or undersaturated) and by the relative solubilities of the carbonate minerals that make up skeletal grains and other carbonates. The relative solubilities of the carbonate minerals is a function of mineralogy and the Mg content of magnesium calcite. Very high-magnesian calcite (>12 mol % $MgCO_3$) is most soluble, followed by aragonite and Mg-calcite with about 12 mol % $MgCO_3$. In turn, aragonite and 12 mol % Mg-calcite are more soluble than Mg-calcite with less than 12 mol % $MgCO_3$, and low-Mg calcite (4 mol % $MgCO_3$ or less) is least soluble (James and Choquette, 1984). Dolomite is even less soluble than the $CaCO_3$ minerals. The rate of dissolution of dolomite is about 100 times slower than that of calcite and aragonite (Busenberg and Plummer, 1986).

The solubilities of all these minerals vary as a function of temperature and pressure (water depth). Solubilities decrease with increasing temperature and increase with increasing pressure of seawater. Therefore, seawaters can range in composition from those that are highly saturated with respect to calcium carbonate (surface waters in tropical regions) to those that are undersaturated (cold surface waters in high latitudes and deep bottom waters). Typical variations in temperature with depth in the modern tropical ocean are shown in Figure 12.2, which also shows relative solubility curves for aragonite and calcite in terms of percent carbonate in sediments. Note that the aragonite **lysocline,** the zone of abrupt increase in aragonite solubility, occurs at a much shallower depth than the calcite lysocline. The aragonite and calcite **compensation depths** refer to the depths below which these minerals dissolve faster than they are accumulating. Therefore, these minerals do not occur on the ocean floor below these depths. On the average, the aragonite compensation depth in the modern ocean is shallower by about 3 km than the calcite compensation depth (James and Choquette, 1983b).

James and Choquette (1983b) differentiate four carbonate "diagenetic" depth zones in the modern ocean. The conditions for carbonate diagenesis are approximately the same within each of these zones, but differ between zones. Zone I (Fig. 12.2) is the **zone of precipitation,** which occurs mainly in shallow tropical to subtropical settings. The average lower depth limit of this zone in the tropics is about 1000 m, and the lower limit rises to the surface at about 35 degrees north and south latitude. In this zone, precipitation of cements (primarily aragonite and Mg-calcite) predominate and little or no dissolution occurs. Zone II is a **zone of partial dissolution** of carbonates. It lies below the aragonite lysocline in tropical to subtropical regions, but occurs at the ocean surface at latitudes higher than about 35 degrees. Little inorganic precipitation of aragonite or Mg-calcite

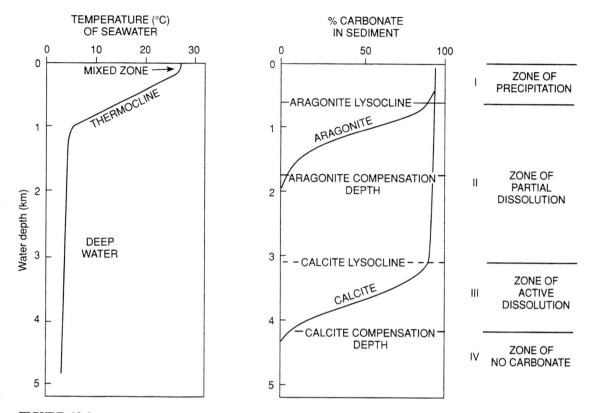

FIGURE 12.2
Generalized plots showing variations in seawater temperature with water depth and the relative positions of the aragonite and calcite lysoclines and compensation depths. Major zones of seafloor diagenesis are plotted to the right. *(From, James, N. P., and P. W. Choquette, 1983, Diagenesis 6. Limestones—The sea floor diagenetic environment: Geoscience Canada, v. 10. Fig. 1, p. 163, reprinted by permission of Geological Association of Canada.)*

takes place in this zone; but dissolution, particularly of aragonite and Mg-calcite, can occur. Zone III is the **zone of active dissolution,** which lies between the calcite lysocline and the calcite compensation depth. It may shallow to near the surface in polar seas. Zone IV is the **zone of no carbonate,** which occurs below the calcite compensation depth. Dissolution is predominant in this zone, and no carbonate is accumulating.

Cementation

Sites of Marine Cementation. The major sites of carbonate cementation on the seafloor are indicated in Figure 12.3. Early marine cementation occurs primarily in the warm, shallow waters of Zone I. Some cements may form in cold shelf waters and deeper waters of Zone II, but cementation appears to be very minor. Mg-calcite is apparently the most common marine cement overall, but aragonite is common also in waters of slightly elevated salinity (James and Ginsburg, 1979). Early cementation can occur within pores or cavities in carbonate grains lying loose on the seafloor, or buried to very shallow depths below the sediment interface, without cementing these particles together. The inner cav-

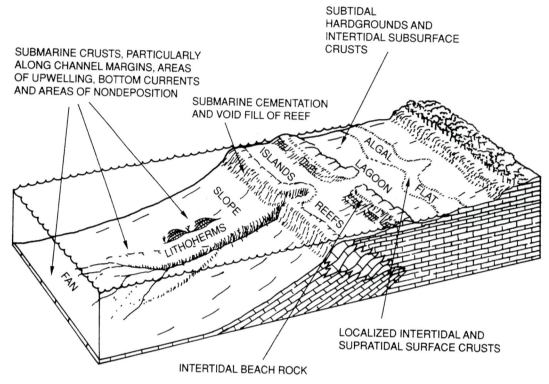

FIGURE 12.3

Major sites of carbonate cementation on the seafloor. *(Modified slightly from Harris, P. M., et al., 1985, Carbonate cementation—A brief review: Soc. Econ. Paleontologists and Mineralogists Spec. Pub. 36. Fig. 3, p. 83, reprinted by permission of SEPM, Tulsa, Okla.)*

ities of shells and other skeletal grains and microborings in the surfaces of skeletal grains appear to be particularly favorable sites for such early cementation. For example, early cementation inside pellets, grapestones (lumps), and possibly ooids results in early hardening of these grains.

Cements may also be precipitated between carbonate grains to create lithified substrate. The best conditions for cementation of grains appear to be in areas where good water circulation or turbulence furnishes a constant supply of $CaCO_3$-saturated fluid that can be pumped or forced through the surface and near-surface sediments by waves, currents, or tides. Because each pore volume of seawater contains only a minute amount of dissolved $CaCO_3$, and only a small percentage of this $CaCO_3$ may precipitate, the pore fluids must be constantly renewed for a pore to fill with cement. Thus, as many as 10,000 to 100,000 pore volumes of seawater may be required to fill a pore with carbonate cement (see discussion by Scholle and Halley, 1985). An additional requirement for cementation of grains is that the substrate must be well enough stabilized to allow cementation to occur before grains are remobilized by waves or currents.

Platform-margin reefs, especially those on windward margins, are areas of the carbonate platform where these requisite conditions, particularly strong water turbulence, are met particularly well. Thus, in a matter of tens to thousands of years, some reefs may become well cemented (James and Choquette, 1983b). Platform-margin carbonate sand shoals, where grains inside the shoals are at rest or are bound by algae, may also become well cemented to form **hardgrounds** (Bathurst, 1975, several sections). Hardgrounds in modern carbonate environments have been reported in numerous areas, including the margins of the Persian Gulf, the Bahama Platform, and Sharks Bay in Western Australia. Hardgrounds have also been reported in ancient carbonate sequences, where they may occur as part of ooid-bearing platform-margin sequences, within coarse grainstones in normal marine cratonic sequences, and as the terminal phases of shallowing-upward sequences (Moore, 1989, p. 88). Typical characteristics of hardgrounds are illustrated in Figure 12.4, and include the presence of an abraded, commonly irregularly bored surface. This surface may be marked also by the presence of encrusting fossils and intraclasts derived from the hardground and may be stained by manganese.

Some strandline sands may also become lithified by cementation to form beachrock (Milliman, 1974, p. 278; Bathurst, 1975, p. 367). Beachrock consists of carbonate-cemented layers of beach sand that dip seaward at about the same angle as the beach sediments. It may be composed mainly of cemented carbonate sands, but quartz or volcanic sands may also form beachrock. Presumably, the beach sand is stabilized owing to binding by algae, fungi, or roots long enough for cementation to occur (Strasser et al., 1989). High turbulence in the littoral zone furnishes a constant supply of $CaCO_3$-rich waters. Mixing of meteoric and seawater in strandline sediments may be an additional factor causing cement precipitation to form beachrock, although CO_2 degassing owing to water turbulence is probably a more important factor. Localized formation of carbonate crusts may occur also in algal mats and on other sediment in the intertidal and supratidal zones (Fig. 12.3).

Thus, reefs, platform-margin sand shoals, and strandline deposits are favored areas for early seafloor cementation. On the other hand, most sediments of shallow carbonate platforms are not cemented. This lack of cementation is especially evident in muddy

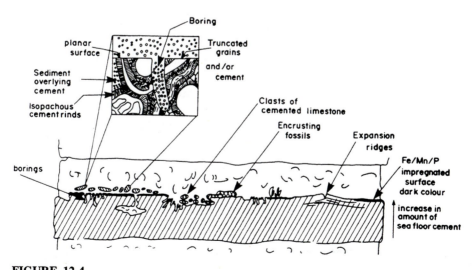

FIGURE 12.4
Distinguishing characteristics of seafloor hardgrounds. *(From James, N. P., and P. W. Choquette, 1983, Diagenesis 6. Limestones—The sea floor diagenetic environment: Geoscience Canada, v. 10. Fig. 12, p. 172, reprinted by permission of Geological Association of Canada.)*

sediments in shelf lagoons, where sluggish water movements, continuous bioturbation, and reducing conditions apparently discourage cementation.

Calcite– or magnesian-calcite–cemented hardgrounds may occur also in deeper water on the upper parts of platform-margin slopes and on the deeper seafloor in areas swept by bottom currents. These areas are within the zone of aragonite dissolution or partial dissolution, but above the zone of calcite dissolution. For example, Neumann et al. (1977) describe lithified carbonate mounds in the Straits of Florida that occur at water depths of about 600–700 m. These mounds, which they refer to as **lithoherms** (Fig. 12.3), rise as much as 50 m above the bottom. They are composed of ahermatypic corals and crinoids that are lithified by numerous superimposed, magnesian-calcite–cemented crusts. An additional example of a deep-marine lithified sediment was reported by Schlanger and James (1978). These authors describe lithified Bahamian peri-platform oozes in the Tongue of the Ocean at water depths of 700–1960 m. Lithification apparently took place by alteration of aragonite to low-magnesian calcite, dissolution of aragonite, and subsequent precipitation of calcite in the resulting voids. Submarine carbonate crusts may also occur on other parts of the deeper seafloor along some submarine-channel margins, in areas of upwelling, and in areas of nondeposition (Fig. 12.3). The tops of some seamounts are also sites of carbonate cementation.

Types of Marine Cements.   Magnesian calcite and aragonite are the dominant kinds of shallow-marine cements (Fig. 12.5), although dolomite may act as a cement in some environments (Chapter 11). Magnesian calcite can occur as **micrite-size crystals** that form very thin rinds around grains or fill pore spaces among grains. As a pore filler, micrite is not strictly a cement, as it may not bind grains together. Some micrite cements may have a peloidal structure. Micrite cements have been reported from beachrock, within Bahamian grapestones, in Bahamian hardgrounds (Moore, 1989, p. 77), and in reefs. Magnesian calcite can occur also as **fibrous** to **bladed crystals** that form much thicker rinds (Fig. 12.5). Aragonite cement commonly occurs as **fibrous rinds** around skeletal grains and other carbonate grains. These crusts are often called **isopachous rinds** because they have approximately equal thickness. Fibrous to bladed aragonite or Mg-calcite cement is common in beachrock (Fig. 12.6). Some fibrous cement in beachrock occurs only at grain contacts, where a meniscus of water would be held by capillary forces as interstitial water drained from the beach sand during low tide. This cement is called **meniscus cement** (Dunham, 1971). Nonisopachous fibrous cement in beachrock that appears to sag downward under grains is called **pendant** or **gravitational cement.** See Figure 12.9 for an illustration of meniscus and pendant cement, and see the many articles in Bricker (1971) for further discussion of beachrock cements. Aragonite may also occur as an intergranular **mesh** of **needlelike crystals** that grow with random orientation into pore space or as fibrous-radial crystals that have a **botryoidal** form (Fig. 12.7). Both of these forms can occur in forereef environments, for example. Even clear, irregular, very complex polyhedral aragonite crystals that resemble calcite spar have been reported from this environment (Land and Moore, 1980).

Isotope and Trace-Element Composition of Seafloor Cements.   The $\delta^{18}O$ composition of modern marine cements is generally high and ranges from about $-2‰$ to $+‰$, which is roughly in the same range as modern ooids. Values of $\delta^{13}C$ range from about $+2‰$ to $+4‰$, which is in the general range of other modern marine precipitates (Moore, 1989, p. 78). These cements tend to be enriched with respect to Mg, Sr, and Na and depleted with respect to Fe and Mn.

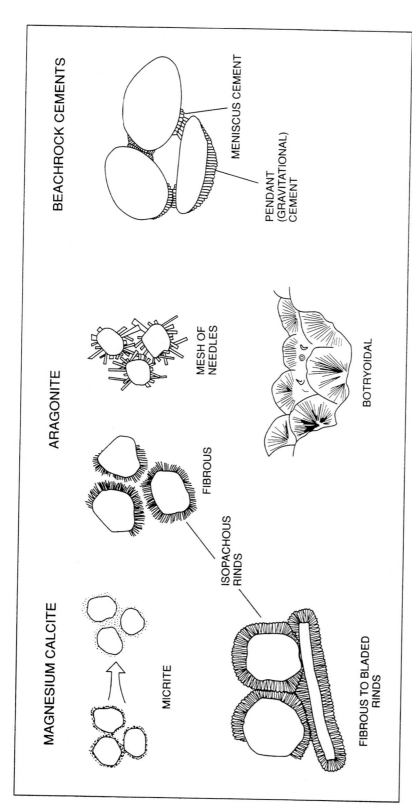

FIGURE 12.5

Major kinds of modern seafloor cements. (Modified from James, N. P., and P. W. Choquette, 1983, Diagenesis 6. Limestones—The sea floor diagenetic environment: Geoscience Canada, v. 10. Fig. 3, p. 165, reprinted by permission of Geological Association of Canada.)

FIGURE 12.6
Fibrous aragonite cement in
beachrock. Holocene, Berry Island,
Bahamas. Crossed nicols. *(From
Bathurst, R. G. C., 1975,
Carbonate sediments and their
diagenesis. Fig. 2, p. 367,
reprinted by permission of Elsevier
Science Publishers, Amsterdam.
Photograph by H. Buchanan.)*

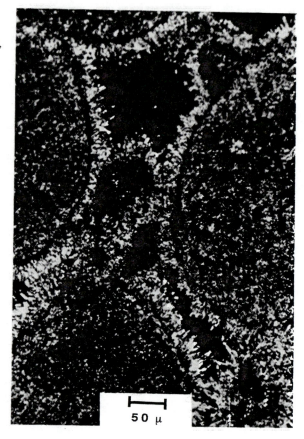

50 μ

Dissolution

As discussed above, seawater becomes undersaturated at depth in the ocean and at colder
temperatures in shallow water at high latitudes. Thus, dissolution may begin in diagenetic
Zone II (Fig. 12.2). It accelerates in Zone III and becomes complete in Zone IV (below
the calcite compensation depth). The relative rate of dissolution is determined by the
water geochemistry and the carbonate mineralogy. As mentioned, very high–Mg-calcite
is more soluble than aragonite, which is more soluble than Mg-calcite. In turn, Mg-calcite
is more soluble than calcite, and dolomite is least soluble. Thus, selective solution of
carbonate materials can take place, bringing about relative enrichment of the more-stable
carbonate phases, particularly calcite.

Obviously, most dissolution of carbonate on the seafloor takes place outside the
warm shallow waters of Zone I, which is the zone of precipitation and cementation. It
follows also that very little carbonate sediment is generated outside of Zone I except
skeletal grains. Therefore, most carbonate sediments in deeper water and at higher latitudes in shallow water are composed mainly of skeletal grains. Deeper waters may be
undersaturated with respect to aragonite and possibly high-magnesian calcite. Therefore,
depending upon their composition, selective dissolution of skeletal grains and other carbonate materials can occur. In the case of planktonic organisms, pteropods are composed
of aragonite, but foraminifers and coccoliths are composed of calcite. Thus, pteropods
will dissolve at a shallower depth than foraminifers and coccoliths, but these organisms

FIGURE 12.7
Coalesced aragonite botryoids, now replaced by calcite but preserving original fibrous aragonite texture. From a reef mount, Nansen Formation (Early Permian), Ellesmere Island, N.W.T. Scale bar = 2 mm. *(From James, N. P., and P. Choquette, 1983, Limestones—the sea floor diagenetic environment: Geoscience Canada, v. 10. Fig. 19, p. 174, reprinted by permission of Geological Association of Canada. Photograph by G. Davis.)*

will also disappear below the calcite compensation depth. Organisms in temperate waters tend to be composed mainly of calcite. Therefore, their remains are relatively stable on the seafloor. Nonetheless, shells can become visibly etched in these waters. If shells are heavily bored by endoliths, solution can aid in breaking down these shells to smaller pieces. Very little clay-size carbonate (micrite) occurs in temperate and high-latitude waters, probably because very little if any inorganic precipitation of fine carbonate takes place in these waters. Fine-size carbonate no doubt does occur by skeletal breakdown, but its scarcity in cold temperate waters suggests that it is destroyed by dissolution. Carbonate grains and lime mud transported from shallow platforms to deeper waters are also subject to dissolution. The relative importance of this dissolution obviously depends upon the mineralogy of the sediment and the saturation state of the water in which redeposition occurs. For example, an aragonite mud transported into deep, undersaturated water may dissolve, followed by precipitation of low-magnesian calcite. On the other hand, fragments of corals or bryozoans, composed mainly of calcite, would have a better chance of surviving dissolution unless they were transported below the calcite compensation depth. Owing to its slow rate of solubility, dolomite can presumably form and survive dissolution below the calcite compensation depth.

12.3.3   Marine Neomorphism

Neomorphism is a term used by Folk (1965) to cover the combined processes of inversion and recrystallization. **Inversion** is the change of one mineral to its polymorph (e.g., aragonite to calcite). The term, as applied to the transformation of aragonite to calcite, appears to have been used in the past with somewhat different meanings. Apparently, its correct meaning is the transformation of aragonite to calcite in the solid (dry) state. Transformation in the solid state proceeds by Ca-O bond destruction and reformation and requires temperatures on the order of 300°C to 600°C (Carlson, 1983). Inversion (in the solid state) probably does not occur in the diagenetic environment. Most diagenetic transformation of aragonite takes place in the presence of fluids. Under these conditions, transformation appears to proceed by solution of aragonite and essentially simultaneous precipitation of calcite. The transformation takes place across a thin film of water nanometers to micrometers in thickness (Wardlaw et al., 1978). As submicroscopic voids are created on one side of the film by dissolution of aragonite, calcite moves through the film to precipitate in these voids. Thus, the alteration of aragonite to calcite by this process is a special kind of replacement. Many geologists today refer to this fine-scale process simply as **calcitization.** For clarity in this book, I refer to the transformation of aragonite to calcite in the wet state as **polymorphic transformation.**

Trace elements, particularly Sr, tend to be partitioned into and become concentrated in calcite during the process of polymorphic transformation. This type of calcitization may preserve relict aragonite textures, which may be outlined by relics of organic matter or other insoluble material (Sandberg, 1983). For example, relict shell structure of aragonite fossils may still be visible, even though a mosaic of calcite crystals cross-cut the original fabric. Calcitization of aragonite can be accomplished also by larger-scale aragonite dissolution that produces visible pores, owing to solution of fossils, carbonate grains, or cements, followed by precipitation of calcite cement in the resulting molds. In this case, cementation may occur soon after formation of the molds or much later than the time of mold formation. This process, strictly speaking, is not polymorphic transformation but dissolution/cementation. It produces essentially the same mineralogical results as polymorphic transformation, but primary aragonite textures are not preserved.

Recrystallization is primarily a change in size or shape of crystals (e.g., increase or decrease in the size of calcite crystals) where no change in mineral composition takes place. Thus, the "replacement" of small calcite crystals by larger calcite crystals (no change in mineralogy) is recrystallization. The change of high-magnesian calcite to low-magnesian calcite is also recrystallization according to Folk (1965), but not according to Bathurst (1975, p. 475). Further, Bathurst (1975, p. 476) states that recrystallization in the dry state is unknown in carbonate diagenesis and that only "wet recrystallization" is important.

Because it is often very difficult to tell the difference (with a normal polarizing microscope) between solution/precipitation processes and recrystallization, Folk (1965) proposed what he called a comprehensive term of ignorance, **neomorphism,** to cover all transformations between one mineral and itself or a polymorph. Thus, neomorphism includes the polymorphic transformation of aragonite to calcite, alteration of Mg-calcite to calcite, and recrystallization. Note that neomorphism takes place without change in the major chemical composition, although trace-element and isotope composition may change. Typically, neomorphism results in an increase in crystal size of carbonate minerals (aggrading neomorphism).

Neomorphism appears to be a relatively unimportant process in the shallow, warm waters of diagenetic Zone I on the seafloor. Rare occurrence of Mg-calcite foraminifers and coralline algae that alter to aragonite have been reported. Also, aragonite cements may partially replace aragonitic mollusc shells, and parts of aragonitic cements and skeletons may change to Mg-calcite (see discussion by James and Choquette, 1983b);

however, neomorphism in this zone seems to be uncommon. On the other hand, carbonate grains and sediments transported into deeper water or pelagic shells settling into such water may undergo some aggrading neomorphism, which commonly involves alteration of Mg-calcite or aragonite to low-magnesian calcite. An example is the alteration of aragonite to calcite in Bahamian peri-platform oozes reported by Schlanger and James (1978). Examples of neomorphic alteration of skeletal debris have been reported also from the deep forereef off North Jamaica and on deep banks in the Gulf of Mexico (Scoffin, 1987, p. 105). Most of these examples appear to occur in areas of low sedimentation and prolonged exposure to cold seawater.

12.4 Diagenesis in the Meteoric Environment

12.4.1 Geochemical Constraints

As mentioned, carbonate sediments initially deposited in the marine environment can be brought into contact with meteoric waters in at least three ways: (1) complete sediment filling of a shallow carbonate basin (platform setting) to or above sea level, i.e., shoaling upward, (2) falling sea level, which exposes previously formed carbonates, and (3) late-stage uplift and unroofing of older carbonates, which brings them into the zone of meteoric water. Because of the common tendency for shallow-marine carbonates to shoal upward to sea level, or to be exposed by falling sea level, meteoric diagenesis dominates the diagenesis of most shallow-marine carbonate sequences (Land, 1986).

Although meteoric waters may exhibit a wide range of saturation states, they are commonly undersaturated with respect to most carbonate minerals. Thus, aragonite and Mg-calcite deposited under marine conditions can undergo significant alteration when brought under the influence of dilute, acidic meteoric waters. This alteration can include both dissolution and neomorphism to low-magnesian calcite. Although low-magnesian calcite is more stable (less soluble) than aragonite and Mg-calcite in meteoric waters, owing to the low Mg/Ca ratios and salinities of these waters, calcite may also undergo partial or complete dissolution in meteoric waters. Many meteoric waters are strongly acidic and undersaturated with respect to $CaCO_3$ owing to their high content of dissolved CO_2 derived by atmospheric interchange and solution from the soils through which they pass. Thus, even calcite may not escape alteration in such strongly "aggressive" meteoric waters. Dissolution may occur by **simple corrosion** (dissolution by rainwater), biogenic corrosion (dissolution owing to increased CO_2 from decaying organic matter in soils), **mixing corrosion** (resulting from mixing of different meteoric waters or mixing of meteoric water with connate water or seawater), and **hydrostatic corrosion** (owing to increased hydrostatic pressure with depth below the water table). See discussion by James and Choquette (1984).

Although low-magnesian calcite can dissolve in meteoric waters under some conditions, as stated, precipitation of low-magnesian calcite is a more common process than dissolution in the meteoric diagenetic environment. Precipitation occurs when waters become oversaturated with respect to calcite. According to James and Choquette (1984), precipitation may be either water-controlled or mineral-controlled. **Water-controlled precipitation** takes place under conditions of oversaturation owing to removal of CO_2 from the system. CO_2 may be removed by heating (including evaporation), pressure loss (e.g., waters emerging into a cave or the atmosphere), or photosynthesis. Calcite will precipitate from oversaturated waters if it can but may be inhibited from precipitation by kinetic factors or the presence of ions such as Mg^{2+}, SO_4^{2-}, or PO_4^{3-}. If inhibited, the saturation state will rise until a thermodynamic drive is reached that is sufficient to overcome the kinetic problem. **Mineral-controlled precipitation** involves dissolution of more-soluble aragonite and Mg-calcite, resulting in oversaturation with respect to calcite. Precipitation of calcite follows. In this process, calcite is never dissolved because oversaturation is achieved and maintained by dissolution of aragonite and Mg-calcite. On the other hand, kinetics may prevent calcite from precipitating as rapidly as the rate of dissolution of aragonite or Mg-calcite. For example, Schmalz (1967) estimated that the rate of aragonite dissolution in fresh meteoric water is 100 times greater than the rate of calcite precipitation. For a more recent and more detailed look at the dissolution and growth kinetics of calcite and aragonite, see Busenberg and Plummer (1986). Thus, meteoric waters may become several times oversaturated with respect to calcite owing to the slowness of calcite to precipitate. The new calcite can precipitate in open voids as cement between grains or within grains, or it can precipitate on a microscale inside particles, replacing the original aragonite or Mg-calcite.

This discussion suggests that mineral-controlled precipitation can eventually lead to wholesale alteration in carbonate mineralogy, with metastable aragonite and Mg-calcite being "replaced" by calcite. On the other hand, carbonate sediments that originally contained little aragonite may undergo only slight cementation during meteoric diagenesis. For example, James and Bone (1989) report that the Gambler Limestone of southern Australia, a cool-water limestone composed dominantly of bryozoans, has undergone very little cementation in 10 million years of diagenesis in the meteoric environment. They conclude from the characteristics of this limestone that meteoric cementation is effective only for carbonate sediments that originally contained abundant aragonite particles.

12.4.2   Diagenesis in the Vadose Zone

General Conditions

Longman (1980) suggests that the vadose zone (Fig. 12.1) can be divided into two parts, which represent the end members of a continuous spectrum: (1) the soil zone, or zone of solution, and (2) the zone of precipitation, or capillary fringe zone. Thickness of these zones may vary considerably, particularly as a function of climate. The boundaries between the zones are gradational and may fluctuate abruptly in response to rainfall. Also, the zone of precipitation may be absent under some conditions. Solution and neomorphism tend to proceed rapidly under humid conditions but much more slowly in dry regions. Much rainwater moves through the vadose zone fairly slowly by gravity percolation or vadose seepage. Water can also move through the zone by vadose flow, which involves relatively rapid flow by way of joints, large fissures, or sinkholes.

Dissolution

In the vadose zone of the meteoric environment (Fig. 12.1), both air and water may be present in the pores, but the pores are not completely filled (saturated) with water. Chemically aggressive meteoric water percolating (seeping) or flowing down through the vadose zone within a body of emerged marine carbonate sediment will initially cause extensive solution of aragonite and Mg-calcite and may also dissolve calcite. If the sediments are composed entirely of calcite, little further dissolution or precipitation may occur once the vadose waters reach saturation with respect to $CaCO_3$. Exceptions to this generality are possible if the calcite grains are very small, mixing of meteoric waters (vadose seepage and vadose flow) occurs, or degassing of CO_2 takes place owing to exposure to the atmosphere (in caves) or increased temperatures. If, however, the sediments are composed of a mixture of aragonite, Mg-calcite, and calcite, the vadose waters continue to bring about dissolution owing to mineral-controlled processes that stem from differences in solubility (James and Choquette, 1984). Thus, extensive dissolution of aragonite and Mg-calcite can occur, with concomitant precipitation of calcite.

Alteration of Mg-calcite to Calcite

Owing to its lower solubility, Mg-calcite alters to calcite through a solution reprecipitation process. Mg-calcite dissolves incongruently (Bathurst, 1980) and, presumably, the most $MgCO_3$-rich parts of skeletal grains dissolve first. As $MgCO_3$ is dissolved, calcite is simultaneously precipitated onto the existing calcite crystal lattice and preserves the original crystallographic orientation. Because this process is fabric-preserving, calcite crystals formed by alteration of Mg-calcite appear essentially identical to the precursor Mg-calcite under a polarizing microscope. Analysis of the iron content of the calcite is required to determine if it formed by alteration of Mg-calcite. Calcite precipitated from seawater does not contain iron, whereas Mg-calcite does. Thus, an iron-rich low-magnesian calcite must have formed by alteration of a Mg-calcite precursor (Richter and Füchtbauer, 1978).

Neomorphism of Aragonite to Calcite

Aragonite can invert to calcite in the solid (dry) state at high temperatures; however, as discussed, this process probably does not occur during sediment diagenesis. In the presence of fluids in the vadose zone, aragonite readily alters to calcite by solution-reprecipitation. Alteration can take place either on a microscale (replacement), which is fabric-preserving, or on a macroscale that involves solution and mold formation and possible filling of the molds by calcite cement. The difference between these two processes is illustrated diagrammatically in Figure 12.8. This figure shows that microscale dissolution and reprecipitation is fabric preserving and results in no secondary porosity. By contrast, macroscale dissolution may create molds (moldic porosity) that are not all immediately filled by calcite precipitation. In this case, the $CaCO_3$ in solution may be precipitated as calcite cement in other pore space farther downflow. It is commonly believed, however, that extensive moldic porosity does not occur in the vadose zone. Because aragonite is generally more abundant than Mg-calcite in shallow-water carbonate deposits, alteration of aragonite to calcite in the vadose zone is a volumetrically important process.

Calcite Cementation

In addition to precipitation of calcite in the process of microscale alteration of Mg-calcite and aragonite in the vadose zone, calcite is precipitated as a coarse cement into larger holes created by dissolution of these minerals. Calcite may also precipitate inside grains such as fossils and in open pore space among grains (James and Choquette, 1984). This calcite cement tends to form equant crystals. Because pore spaces are not completely filled with water, however, the cement commonly does not surround grains or completely fill pores. Typically, it takes the form of meniscus or pendant (stalactitic) cement (Fig. 12.9). Thus, it may resemble in form cements precipitated in the marine vadose zone (i.e., in beachrock); however, vadose cement is composed of calcite whereas beachrock cement is

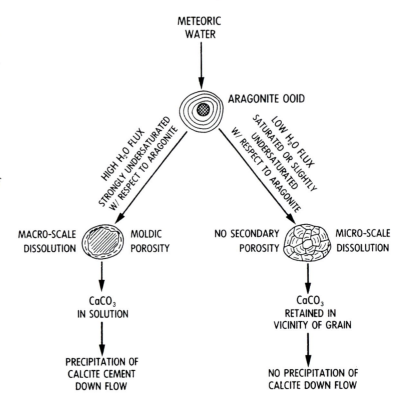

FIGURE 12.8
Schematic diagram illustrating the difference between microscale dissolution (fabric-preserving) and macroscale dissolution (fabric-destructive). Hydrologic and geochemical conditions that influence these pathways of diagenesis are indicated. *(From Moore, C. H., 1989, Carbonate diagenesis and porosity. Fig. 6.6, p. 170, reprinted by permission of Elsevier Science Publishers, Amsterdam.)*

composed of Mg-calcite or aragonite. If cementation is prolonged, pores may be completely filled with cement, and the characteristic meniscus or pendant shapes may be lost.

Micrite-envelope cement forms in voids created by complete dissolution of carbonate grains (probably aragonite grains) that had an original low-magnesian calcite micrite envelope. During dissolution, this envelope or rind resists solution and thus acts as a mold (Scoffin, 1987, p. 111; Tucker and Wright, 1990, p. 13). Subsequent filling of the mold by blocky, low-magnesian calcite cement gives the micrite cement pattern illustrated in Figure 12.9. Syntaxial (epitaxial) overgrowths on echinoderm fragments may also occur in the vadose environment. In finer-grained sediment, where capillary forces are greater and vadose water is held longer, calcite cementation may be more pervasive than in coarser sediment.

12.4.3   Diagenesis in the Phreatic Zone

Hydrologic Characteristics

The freshwater phreatic zone lies below the vadose zone, and all pore space in this zone is filled with meteoric water. The top of the zone is the water table, and the base is a zone of mixing with underlying water. This underlying water is normal seawater in areas near the coast, but it is subsurface formation water (connate water) in more-inland areas. Water enters the phreatic zone either by vadose seepage or by direct vadose flow through fractures or sinks from lakes, streams, or runoff. Thus, waters of quite different chemical character may be present (and may mix) in the phreatic zone. Water in the phreatic zone moves in a subhorizontal direction toward some outlet or discharge area.

METEORIC CEMENTS

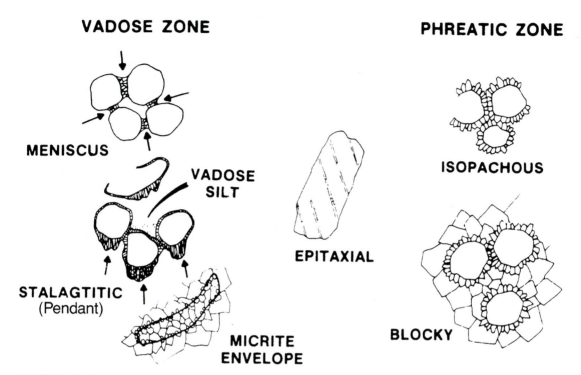

FIGURE 12.9
Principal kinds of cements precipitated in the vadose and phreatic meteoric diagenetic environment. Note that epitaxial cements (e.g., syntaxial rims on echinoderm fragments) can form in either the vadose or phreatic zone. *(After James, N. P., and P. W. Choquette, 1984, Diagenesis 9—The meteoric diagenetic environment: Geoscience Canada, v. 11. Fig. 24, p. 177, reprinted by permission of Geological Association of Canada.)*

Diagenetic Processes

Because the phreatic zone is buried below the water table and is not accessible for direct examination, observation of carbonate sediments in this zone can be had only from cores. Therefore, much less is known about diagenesis in the phreatic zone than in the vadose zone. Longman (1980) proposed an ideal model for freshwater phreatic diagenesis, in which he assumed that phreatic waters become increasingly saturated with $CaCO_3$ with increasing depth. According to this model, dissolution predominates in the shallower parts of the phreatic zone, and cementation (or no diagenetic activity) occurs in the deeper part. Because this model is based on inference, it remains to be tested.

Diagenetic processes are more intense in the phreatic zone, where grains are constantly surrounded by water and where water flow through sediment is more constant than in the vadose zone, where water flow is ephemeral. Furthermore, the water table itself is a zone of particularly intense activity (James and Choquette, 1984; Moore, 1989, p. 181). If aggressive CO_2-rich waters are still undersaturated when they reach the water table, particularly those arriving directly from the surface by vadose flow, dissolution will be concentrated along the water table. Many caves in carbonate deposits occur at the water table. On the other hand, if CO_2 degassing is active along the water table, cementation can occur. Furthermore, below the water table, mixing of waters having different carbon dioxide levels and $CaCO_3$ saturation states may bring about either dissolution or precipitation, depending upon the characteristics of the mixed waters. Thus, it is not easy to pinpoint specific zones of dissolution or precipitation in the phreatic zone.

In any case, metastable sediments appear to alter rapidly in the phreatic zone and to become better calcitized and lithified than the same kinds of sediments in the overlying vadose zone (James and Choquette, 1984). Furthermore, the rate of stabilization of aragonite and Mg-calcite appears to be faster in regional meteoric aquifer systems than in those hydrologic systems on small carbonate islands where there is a local floating meteoric water lens. Presumably, this difference exists because most of the water on small islands arrives by seepage from the vadose zone, whereas large volumes of water can enter regional aquifer systems by vadose flow. Also, there is more active movement of water through the system. With respect to island phreatic systems, Moore (1989, p. 185) suggests that dissolution of aragonite, creation of moldic porosity, and precipitation of calcite are all most active in the upper part of the phreatic zone. Aragonite dissolution diminishes downward toward the mixing zone, and the percentages of calcite cement and moldic porosity also decrease downward. Moore (1989, p. 195) proposes a much more complex model for diagenesis within a regional aquifer system where the aquifer is confined between nonpermeable layers (aquacludes). According to this model (Fig. 12.10), aragonite is initially present (immature stage) in the recharge area, and aragonite dissolution and concomitant calcite precipitation are very active processes. In the mature stage, aragonite is gone from the recharge area (but is still present in downflow areas), and aragonite dissolution and calcite precipitation are greatly reduced. Considerable porosity is present in the sediments in the mature stage.

Phreatic Cements

Because phreatic cements are precipitated in pores that are completely filled with water, in contrast to vadose pores, meniscus and pendant cements do not occur. Instead, isopachous cements and blocky cements are dominant (Fig. 12.9). As suggested by Figure 12.9, crystal size of phreatic cements tends to be somewhat larger than that of vadose cements. Syntaxial overgrowths of echinoderms may also be very common.

12.4.4   Diagenesis in the Mixing Zone

The base of the freshwater phreatic zone is the interface between freshwater and underlying more-saline waters (Fig. 12.1). In coastal regions and small islands, this underlying water is normal seawater. In more-inland settings, it is connate formation waters that may be greatly modified from original seawater or freshwater. Because there is mixing of freshwater and more-saline water across the interface, this zone is referred to as a mixing zone.

Geologists have generated considerable interest in the mixing zone as a likely site for dolomitization. As discussed in Chapter 11, however, the original thermodynamic underpinnings of the mixing-zone dolomite theory are now being questioned. Furthermore, little dolomite actually occurs in most modern mixing zones. With respect to the behavior of $CaCO_3$ in the mixing zone, dissolution appears to be more important than precipitation. The presence of caves that form in carbonate rocks at the mixing zone supports this conclusion. James and Choquette (1984) suggest that solution is likely

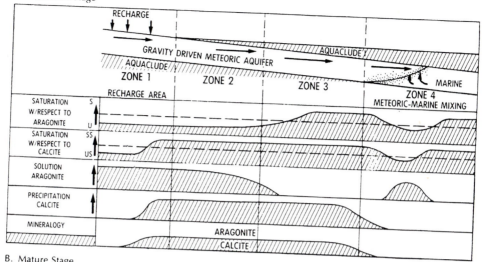

A. Immature Stage

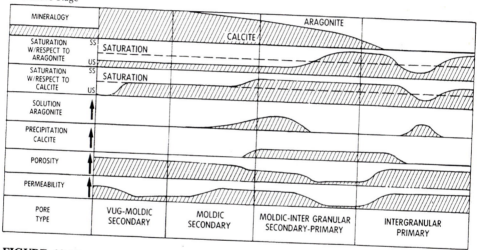

B. Mature Stage

FIGURE 12.10

Postulated diagenetic model for a gravity-driven, confined meteoric aquifer system (top of diagram). Water moves from zone 1 (recharge area) toward zone 4 (mixing zone). Note that in the early, immature stage of diagenesis aragonite is still present in the recharge area. In the more advanced, mature stage, aragonite has been destroyed by dissolution and calcite is present in the recharge area. *(From Moore, C. H., 1989, Carbonate diagenesis and porosity. Fig. 7.12, p. 195, reprinted by permission of Elsevier Science Publishers, Amsterdam.)*

favored in the upper, more-dilute parts of the mixing zone. Precipitation of calcite, if it occurs, is more likely to take place in the lower part of the zone. Furthermore, calcite cement may become more Mg-rich farther down in the mixing zone.

12.4.5 Diagenesis in the Telogenetic Meteoric Environment

The preceding discussion of meteoric diagenesis deals with marine carbonates exposed to meteoric waters owing to shoaling upward (basin filling) or falling sea level. These young

sediments consist mainly of Mg-calcite and aragonite; therefore, they are highly susceptible to dissolution and calcification in the meteoric environment owing to mineral-
controlled dissolution-precipitation reactions (James and Choquette, 1984). Old carbonate
sediments that have been deeply buried and subsequently exhumed by uplift and unroofing
may also be exposed to and altered by meteoric waters (telodiagenesis). These older
sediments may have been exposed initially to meteoric diagenesis, or they may have been
buried directly without exposure to meteoric waters. In either case, subsequent burial
diagenesis (next section) likely converted original Mg-calcite or aragonite sediments to
calcite or possibly dolomite. Therefore, when these mineralogically stabilized carbonates
arrive in the telogenetic zone, their late-stage diagenesis by meteoric waters differs from
that of unstabilized Mg-calcitic and aragonitic sediment.

During telodiagenesis, dissolution in both the freshwater vadose and phreatic zones
is the dominant diagenetic process. Dissolution probably takes place mainly in chemically
aggressive waters charged with soil-derived CO_2 but may also occur in mixed meteoric
and saline waters. In relatively humid climates, dissolution can lead to extensive karsti-
fication (Fig. 10.41). Solution during karst development generates openings of all sizes,
from tiny solution pores called **vugs** to large caves and caverns. For further details on the
formation of karst profiles, see Esteban and Klappa (1983) and the numerous papers in
Paleokarst, edited by James and Choquette (1988).

Because carbonates in the telogenetic zone are already stabilized, mineral-
controlled solution-precipitation does not occur. Aggressive meteoric waters will dissolve
calcite (or dolomite) until they become saturated with respect to calcite (or dolomite).
Once the waters are saturated, little further dissolution or precipitation will occur (Moore,
1989, p. 211). If saturation is achieved at shallow depth, cavity formation and porosity
enhancement will extend only to shallow depths. Also, little additional cementation will
occur within sediment pores in the downflow direction as saturated waters move through
the rock. Of course, precipitation may occur in caves, springs, etc., as described in
Chapter 10. Thus, other than extensive solution in the zone of unsaturated, aggressive
meteoric waters, the effects of diagenesis of old, stabilized carbonates in the meteoric
zone is much less pronounced than the diagenesis of immature, unstabilized carbonates
described in the preceding section. Carbonate telodiagenesis, as well as the diagenesis of
unstabilized carbonates, is also much less pronounced in arid and semiarid climates than
in humid climates.

12.4.6   Isotope Geochemistry and Trace-Element Composition of Carbonates in the Meteoric Zone

$\delta^{18}O$ Values

The $\delta^{18}O$ values of meteoric waters can extend over a wide range, from about 0 to -20
(SMOW). These values are strongly latitude-dependent, but they also reflect altitude and
temperature and even the isotope composition of coeval marine water. The isotope composition of diagenetically altered carbonates and concomitantly precipitated cements in
the meteoric zone is a function of the isotope composition of both the meteoric water and
the marine carbonates with which these waters react. The isotope values resulting from
rock–water reactions will depend upon the ratio O_L/O_W in the meteoric groundwater,
where O_L is the number of moles of oxygen derived from limestone or metastable
carbonate and O_W is the number of moles of oxygen derived from water (Allan and

Matthews, 1982). Because the solubility of calcium carbonate is low in nearsurface meteoric waters, O_L/O_W in recharge areas is much less than 1. Therefore, Allan and Matthews suggest that repeated solution-reprecipitation reactions will not noticeably change the O_L/O_W ratio of meteoric water bicarbonate or the $\delta^{18}O$ values of altered limestones and precipitated cements. In other words, meteoric water is such a ubiquitous source of oxygen for the formation of new calcium carbonate during the diagenetic processes that the $\delta^{18}O$ value of this new carbonate is determined primarily by the isotope composition of the meteoric water. Thus, the oxygen isotope composition of diagenetic carbonates remains virtually constant with increasing depth below subaerial surface exposures (e.g., under isolated islands) and with increasing distance down the hydrologic gradient in confined, meteoric aquifer systems.

$\delta^{13}C$ Values

The carbon isotope composition of the bicarbonate in meteoric waters, and thus the diagenetic carbonates formed in these waters, depends upon the ratio C_L/C_{SG}, where C_L is the number of moles of carbon derived from metastable carbonate sediments and C_{SG} is the number of moles of carbon derived from soil-gas CO_2 (Allan and Matthews, 1982). This ratio in the recharge area is initially equal to about 1. Because soil-derived CO_2 has highly depleted $\delta^{13}C$ values, diagenetic carbonates in the recharge area, which are strongly affected by soil-gas CO_2, also have low $\delta^{13}C$ values. As meteoric waters move deeper, beyond the influence of soil-derived CO_2, the C_L/C_{SG} ratio increases as rock–water reaction (dissolution of metastable carbonates) continues. Thus, the $\delta^{13}C$ values of diagenetic carbonates progressively increase with increasing depth. This increase may be particularly significant at the boundary of the vadose and phreatic zones.

Combined Variations in Oxygen and Carbon Isotopes

Because oxygen isotope composition in meteoric waters varies as a function of latitude, it is not meaningful to discuss absolute $\delta^{18}O$ values for carbonates altered diagenetically in the meteoric zone. At any given geographic locality, however, the oxygen isotope content of meteoric waters is relatively constant for that locality. The discussion above indicates that the oxygen isotope composition of diagenetic carbonates in any given meteoric zone is essentially invariant with depth. On the other hand, the $\delta^{13}C$ composition is highly variable with increasing depth. This constancy of $\delta^{18}O$ composition combined with the variable $\delta^{13}C$ composition defines a trend that Lohmann (1988) refers to as the **meteoric calcite line** (Fig. 12.11). Note from Figure 12.11 that, although $\delta^{18}O$ values typically are invariant with depth (distance from soil zone), variations can occur in at least two zones. One of these zones is very near the surface, where intense evaporation may cause preferential loss of ^{16}O from the meteoric water, thereby increasing $\delta^{18}O$ values in the water. Lohmann (1988) suggests that a second zone where increase in $\delta^{18}O$ values may occur is in unstable mineral assemblages (aragonite and Mg-calcite) undergoing rock–water interaction at sites distal from the surface recharge surface. Here, cements may precipitate that have oxygen isotope composition nearly equivalent to that of the dissolving mineral phases (R in Fig. 12.11). Carbon isotope composition may also show some deviation from the typical depth trend predicted by Figure 12.11. The carbon isotope content of cements precipitated at sites far removed from the recharge area tends to become more depleted with time. This progressive depletion results from diminished rock–water interaction with time as the carbonate system matures and the abundance of

FIGURE 12.11

Idealized plot of variations in $\delta^{18}O$ and $\delta^{13}C$ characteristics of meteoric vadose and phreatic carbonates. The almost invariant oxygen isotope composition and variable carbon isotope composition define a trend termed the meteoric calcite line. Although oxygen composition is dominantly invariant, deviations from this line can occur where there is increased rock–water interaction with polymineralic suites distal from meteoric recharge. At these distal sites, precipitating cements may have an isotope composition similar to that of dissolving mineral phases (R), as well as increasing Mg^{2+} or Sr^{2+} content. Deviations in oxygen isotope values can occur also at surface exposures, where evaporation drives the composition toward more positive values. Carbon isotopes tend to become more depleted with time; that is, the youngest cements have the lowest

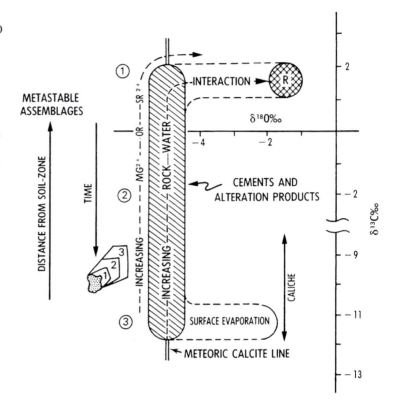

values. *(From Lohmann, R. C., 1988, Geochemical patterns of meteoric diagenetic systems and their application to studies of paleokarst, in James, N. P., and P. W. Choquette, eds., Paleokarst. Fig. 12.11, p. 72, reprinted by permission of Springer-Verlag, New York.)*

unstable minerals diminishes (Lohmann, 1988). In other words, less rock-water interaction (dissolution of unstable carbonates) means a greater influence of soil-gas carbon, which is strongly depleted in ^{13}C.

Note also in Figure 12.11 that Mg^{2+} and Sr^{2+} content of diagenetic carbonates tends to increase with increasing depth or distance below the soil zone. This trend reflects enrichment of these ions in meteoric waters owing to dissolution of aragonite and Mg-calcite. Calcite precipitated from these enriched waters will likewise be enriched in Mg^{2+} and Sr^{2+} over the original aragonite and Mg-calcite.

What happens to the carbon and oxygen isotope composition of diagenetic carbonates in the mixing zone of meteoric and more-saline waters? In the first place, the mixing zone, as mentioned, is dominantly a zone of dissolution rather than precipitation. For calcite that is precipitated in this zone, Allan and Matthews (1982) suggest that carbon and isotope composition will exhibit positive covariance. That is, $\delta^{13}C$ values will tend to increase as $\delta^{18}O$ values increase. On the other hand, Lohmann (1988) believes that diagenetic calcite is likely to precipitate only during partial mixing; that is, precipitation will occur in regions dominated by either meteoric or marine waters. Assuming such partial mixing, he suggests that the isotopic characteristics of mixed-water carbonates should define a trend parallel to the meteoric calcite line. In other words, the trend may be displaced from the meteoric calcite line, presumably toward more positive values, but will lie parallel to this line. Thus, it will not display positive covariance between carbon and oxygen isotope composition.

12.5   Diagenesis in the Deep-Burial Environment

12.5.1   Introduction

As discussed in preceding sections, many carbonate sediments undergo early diagenesis on the seafloor, owing largely to cementation by aragonite and Mg-calcite. These sediments may subsequently be subjected to additional diagenesis in the meteoric environment, where dissolution of aragonite and Mg-calcite and precipitation of calcite are dominant processes. Carbonate sediments previously modified by diagenesis on the seafloor or in the meteoric zone, as well as carbonates that may have escaped seafloor lithification or meteoric diagenesis, are ultimately buried. It is in this deeper-burial zone (Fig. 12.1) below the reach of surface-related processes—the mesogenetic diagenetic regime—that carbonates have their longest residence times. Therefore, there is ample opportunity in the deep-burial environment for additional diagenesis to occur. Choquette and James (1987) point out that lithification on the seafloor can take place very rapidly in as little as 10^1 to 10^4 years, and diagenesis in the meteoric environment can involve time intervals on the order of 10^3 to 10^5 years. On the other hand, diagenesis in the deep-burial environment may continue over tens to hundreds of millions of years (10^6–10^8 years), although the rate of diagenetic change drops off rapidly with increasing depth and time.

Deep burial subjects carbonate sediments to increased temperature, lithostatic pressure, and hydrostatic pressure and brings them into contact with pore waters that differ in composition from those on the seafloor or in the meteoric environment. Under these changed conditions, carbonate sediments undergo compaction and physical reorientation of grains, as well as chemical alteration, through the processes of cementation, dissolution, and neomorphism. The exact course of diagenesis in the deep-burial environment depends upon several factors, including the mineralogy of the starting material (aragonite, Mg-calcite, or low-magnesian calcite), grain size and texture of the sediment, porosity and permeability, composition of pore fluids, and the temperature and pressure of the deep-burial regime. These factors are explored further below.

12.5.2   Factors Influencing Burial Diagenesis

Initial Mineralogy and Grain Size

Carbonate sediments previously altered in the meteoric zone are mineralogically stabilized to various degrees; that is, aragonite and Mg-calcite have already been converted totally or mainly to low-magnesian calcite or dolomite. Such sediments have a relatively low diagenetic potential to further generate calcite during diagenesis (Scholle and Halley, 1985). By contrast, carbonate sediments composed mainly of aragonite and Mg-calcite have a much higher potential to generate calcite cements through dissolution and reprecipitation processes (Schlanger and Douglas, 1974). Thus, initially, chemical diagenesis of mineralogically stabilized carbonate sediments in the burial environments likely proceeds at a slower rate than that of unstabilized sediments. Because rates of diagenesis decrease with increasing time and burial depths, however, these rates may tend to equalize after a time.

Fine-grained sediments have higher surface-area to volume ratio than do coarse-grained sediments. Because of this higher reactive surface area, chemical diagenetic processes involving solution–reprecipitation and recrystallization (neomorphism) tend to proceed faster in fine-grained sediments than in coarser sediments. Fine-grained sediment may also undergo compaction more readily than do coarser sediments. On the other hand, fine-grained sediments have low permeability (although porosity may be high), which limits the passage of fluids through the sediments. Coarser-grained sediments may have lower porosity than do finer sediments, but commonly have much better permeability. Thus, grain size and permeability tend to have inverse effects on chemical diagenesis.

Pore-Fluid Composition

Marine carbonate sediments that escape diagenesis in the meteoric zone have initial pore fluids that are normal seawater. The pore fluids of carbonate sediments in the meteoric environment are composed of fresh water in the vadose and phreatic zones; the fresh water becomes mixed with more-saline waters in the mixing zone. With deeper burial, the chemical composition, pH, and Eh of pore waters may change significantly, as discussed in Chapters 9 and 11. These changes can be brought about by several processes, including rock–mineral reactions, dissolution of evaporites, shale dewatering (with waters derived from associated shales imported into carbonate sediments), decarboxylation of organic matter, and salt-sieving. Most waters in the deep-burial environment are more saline than seawater and may have different chemistries (Chapter 9). In many depositional basins, there is a general trend of increasing salinity with depth. In other basins, the waters of some basins are less saline than seawater, apparently owing to mixing of meteoric, marine, and perhaps basinal waters (Stoesser and Moore, 1983).

Pore-water composition and pH have an extremely important influence on diagenetic processes such as dissolution, pressure solution, cementation, dolomitization (Chapter 11), and neomorphism. As discussed in preceding parts of this book, pore fluids must move continuously through sediments for significant dissolution or precipitation to occur. In the shallower parts of the deep-burial environment, gravity-driven fluid flow predominates. In deeper parts of basins, large-scale basin circulation driven by thermal processes such as Kohout circulation may be important (Chapters 9 and 11). Thus, depending upon specific conditions of the basin, fluids may move either downward or upward through sediments. In the later stages of carbonate diagenesis, porosity tends to be destroyed by compaction and cementation, and the volume of water available for diagenesis is diminished. As the volume is decreased and flux through the rock is slowed, the diagenetic system changes from an open water-dominated system to a closed rock-dominated system (Prezbindowski, 1985). In some carbonate systems, liquid hydrocarbons may be present in pore waters at some stage. These hydrocarbons are believed to inhibit pressure solution as well as the formation of cements.

Temperature

The average geothermal gradient in sedimentary basins is about 25°C/km (Fig. 9.3). Wide variations in geothermal gradients are possible in different basins, ranging from as low as 10°C/km to as high as 35°C/km. As suggested in Chapter 9, increasing temperature causes an increase in the rate of most chemical reactions. On the other hand, increasing temperature causes a significant decrease in the solubility of calcite and other carbonate minerals (Bathurst, 1975, p. 273). The effect of temperature increase is to favor precipitation of carbonate cements and dolomitization during deeper burial (Bathurst, 1986). By contrast, the solubility of most silicate minerals, including chalcedony and quartz, increases with increasing temperature.

Increasing temperature combined with increasing pressure can bring about the dewatering (dehydration) of certain minerals, releasing waters that may affect carbonate diagenesis. Examples include the dewatering of smectite (Chapter 9), the conversion of gypsum to anhydrite, the transformation of opal-A to quartz, and the dehydration of hydrous iron oxides (limonite) to hematite. Finally, temperature is involved in the conversion of organic matter to organic acids, kerogen, and hydrocarbons (thermal maturation) during burial diagenesis (Section 9.7). Although many of these dewatering processes are more likely to occur in siliciclastic sedimentary rocks than in carbonates, waters released from siliciclastic rocks may move laterally or vertically into carbonate sediments.

Pressure

Both geostatic pressure and hydrostatic pressure increase linearly with increasing burial depth (Fig. 9.3). The geostatic pressure gradient in a limestone basin would be slightly less steep than the average gradient shown in Figure 9.3 because limestone is less dense than siliciclastic rocks. Sediments undergoing burial may be affected also by directed pressures related to tectonic stresses. The net pressure on sedimentary particles in the burial environment is the difference between lithostatic pressure and hydrostatic pressure. This pressure difference creates strain, which is relieved by dissolution and is the force that drives pressure solution or chemical compaction (Choquette and James, 1987).

Hydrostatic pressures at a particular depth are typically much lower than geostatic pressures (commonly less than about one-half). On the other hand, pore fluids trapped in sediments during compaction or in sediments subjected to tectonic stresses may develop abnormally high hydrostatic pressures. Petroleum geologists refer to such sediments as being **overpressured** or **geopressured.** Abnormally high fluid pressures can affect the diagenesis of carbonates in at least two ways. They may retard both mechanical and chemical compaction, thereby preventing porosity loss. Overpressured zones may also isolate pore fluids from surrounding diagenetic waters, retarding fluid and ion transfer and possibly preventing or significantly slowing cementation. These factors may explain the preservation of unusually high porosities reported in some deeply buried carbonate oil reservoirs (Feazel and Schatzinger, 1985).

12.5.3   Diagenetic Processes in the Deep-Burial Environment

Physical Compaction

Grain Packing and Reorientation.   Newly deposited, watery carbonate sediments commonly have very high initial porosities. Depending upon the grain size and nature of the carbonate particles, initial porosity may range from about 40 to more than 80 percent. Unless these sediments undergo very early cementation and lithification on the seafloor, significant porosity reduction and concomitant thinning of beds occur during burial diagenesis. Muddy carbonate sediments (lime muds and wackestones) commonly have the highest porosity; therefore, compaction effects are typically greatest for these sediments. A considerable amount of porosity reduction occurs very early during burial simply by grain settling, repacking, and reorientation that accompany early dewatering. This process continues until a grain-supported framework is established, which may occur within a few meters or so below the seafloor (Choquette and James, 1987). The actual amount of porosity lost in this process depends upon the nature of the sediment. Porosity loss may be as little as 10 percent (e.g., from 80 to 70 percent) in grain-rich sediments such as foraminiferal oozes to 40 percent (e.g., 80 to 40 percent) in mud-rich sediments.

Grain Deformation.   With deeper burial and progressively higher overburden pressure, carbonate grains are packed even more tightly. Eventually, overburden stresses result in grain deformation by brittle fracturing and breakage and by plastic or ductile squeezing. Significant additional reduction in porosity accompanies this process. To illustrate, Shinn and Robbin (1983) conducted *in-situ* compaction experiments on cores of modern carbonate sediments that have porosities ranging from 47 to 83 percent (mostly 60–75 percent). Porosities after compaction were reduced to as little as 29 percent (mostly 35–45 percent) of original porosities. Further, thickness of the cores was reduced to as little as 27 percent (mostly 30–65 percent) of original thickness. Shinn and Robbin report that most compaction effects occurred at pressures equivalent to burial depths less than about 1000 ft (305 m). Increasing loads to an equivalent of more than 10,000 ft (3,400 m) of burial depth did not significantly increase compaction or reduce sediment core length but did produce some chemical compaction (pressure solution). These experimental results suggest that significant compaction of carbonate sediments, with accompanying porosity loss and thinning of beds, can occur at burial depths less than 300 m. In fact, Shinn and Robbin (1983) conclude that at burial depths of as little as 100 m, compaction of marine lime sediments can reduce their depositional thickness by one-half, with accompanying porosity losses of 50–60 percent of original pore volumes.

Effects of Physical Compaction.   In addition to porosity loss and bed thinning, numerous other effects accompany physical compaction. As enumerated by Shinn and Robbin (1983) and Choquette and James (1987) these effects include (1) squeezing and deformation of organic matter into wispy, "stylolite-like" layers that drape over rigid grains, (2) mashing of sediment-filled circular burrows to produce ellipsoidal structures, (3) flattening of pellets or other grains, (4) rotation of shells or other grains into tighter packing, (5) crushing of shells, ooids, or other grains and fracturing of micrite envelopes (particularly in grain-supported carbonates), (6) obliteration of pellets and birdseye or fenestral voids in sediments that were not lithified by early cementation, (7) thinning of laminae between and draping over early concretions, (8) production of swirling structures, and (9) conversion of grain-poor lime mud or wackestone (mud-supported) to packstone (grain-supported). This latter effect occurs as a result of grains in an originally mud-supported fabric being forced together by compaction until they touch (Shinn and Robbin, 1983).

   The extent of porosity loss, bed thinning, and other mechanical compaction effects is related to the degree of early cementation in the sediments. If extensive intergranular cementation takes place in either the seafloor or shallow meteoric diagenetic environments, compaction effects are moderated. Soft grains such as pellets may preserve their shapes, and less overall thinning of beds occurs. Porosity loss owing to compaction is also less significant, although porosity is obviously reduced by cementation.

Chemical Compaction (Pressure Solution)

Nature of the Process.   Chemical compaction of siliciclastic sediments owing to pressure solution is discussed in Chapter 9. I point out in that chapter that grain-to-grain pressure solution at the contact points of individual grains can result in the formation of interpenetrating or sutured contacts. On a larger scale, pressure solution seams called

stylolites can develop. Pressure solution appears to be a much more prevalent process in carbonate rocks than in siliciclastic rocks, and a great variety of both microstylolites and larger-scale stylolites are common features of carbonates.

Pressure solution begins after mechanical compaction is essentially complete and a stable grain framework has been created. Load or tectonic stress transmitted to grain-contact points or surfaces causes dissolution at the contact. Solid calcium carbonate is changed to liquid, creating a solution film (Robin, 1978). Calcium and bicarbonate ions released into solution move away from the stressed contact point or surface by solution transfer or diffusion toward adjacent areas of lower stress (i.e., pores). These ions may be reprecipitated locally as calcite cement, or they may remain dissolved in the pore waters and move to more distant sites. Because pressure solution following mechanical compaction results in additional porosity loss and bed thinning, the process was referred to by Lloyd (1977) as **chemical compaction.**

Factors Affecting Chemical Compaction.   As suggested, pressure solution begins after mechanical compaction has established a stable grain framework, so load or tectonic stresses can be transmitted from grain to grain. Pressure solution is most easily visualized in grain-rich limestones such as oolites; however, it occurs also in mud-dominated limestones. Numerous factors are believed to influence the magnitude of chemical compaction, either to enhance or to retard compaction. These factors include burial depth and/or tectonic stress, carbonate mineralogy, the presence of insolubles such as clay minerals, composition of pore waters, the presence of liquid hydrocarbons, and elevated pore pressures (Choquette and James, 1987; Feazel and Schatzinger 1985; Moore, 1989, p. 252).

Geologists frequently state that pressure solution begins at burial depths as shallow as 200 m. Moore (1989, p. 251) points out, however, that the depth at which pressure solution begins depends upon many factors in addition to load stress, including tectonic stress arising from folding or faulting. The onset of chemical compaction has actually been reported by various authors to occur at depths ranging from less than 200 m to as much as 1500 m. One reason for such variation may be carbonate mineralogy. Carbonates composed of metastable aragonite or Mg-calcite should be more susceptible to chemical compaction than stable calcite sediments such as deep-sea pelagic carbonates or dolomite. Insolubles such as clay minerals and fine organic matter apparently also play an important role in pressure solution (Choquette and James, 1987; Sassen et al., 1987, Wanless, 1979), and stylolites appear unlikely to develop in carbonates that do not contain any insolubles. Magnesium-poor meteoric waters are regarded to be chemically more aggressive in promoting pressure solution than are marine pore waters; thus, reports of very shallow chemical compaction may be related to meteoric pore waters. By contrast, the early introduction of hydrocarbons into pore space may retard or shut down chemical compaction (Feazel and Schatzinger, 1985). Finally, elevated pore pressures in geopressured zones reduce stress at grain contacts and also retard chemical compaction.

Effects of Chemical Compaction.   The most obvious effect of pressure solution is the formation of stylolites and solution seams. Creation of these features is accompanied by reduction in bulk volume of the rocks and resultant loss of porosity. For example, Coogan and Manus (1975) suggest that grain-rich limestone with about 40 percent original porosity may experience as much as a 30 percent volume loss owing to pressure solution (volume loss calculated by measuring accumulated amplitudes of megascopic stylolites). Original porosity is thus reduced to about 25 percent. If approximately 4 percent cement originating from pressure solution is formed during this process, the final porosity is about 21 percent. These figures may not apply to all limestones, but they indicate that volume reduction and porosity loss can be significant. Furthermore, volume reduction by pressure solution may continue even after essentially all porosity is lost (Choquette and James, 1987).

Stylolites and related features have long been recognized as the product of pressure solution, and several early classifications for stylolites have been proposed (e.g., Park and Schot, 1968; Trurnit, 1968). A stylolite is a kind of **sutured seam** that is characterized by a jagged surface, is generally coated by insolubles such as clay minerals or organic matter, and is made up of interlocking pillars, sockets, and variously shaped teeth (Choquette and James, 1987). Sutured seams include both megastylolites (Fig. 12.12) and microstylolites between individual grains (Fig. 12.13). Stylolites occur as single or multiple features (swarms of stylolites), and their amplitude may range from small to large (Fig. 12.14). They are typically oriented parallel to depositional bedding; however, they can occur also at various angles to bedding and thus create reticulate or nodule-bounding patterns.

More recently, additional solution features have been recognized, including **non-sutured seams,** and new classifications have been suggested by several authors (Buxton and Sibley, 1981; Garrison and Kennedy, 1977; Logan and Semeniuk, 1976; Koepnick, 1984; Wanless, 1979). For example, Wanless (1979) suggests that relatively clean limestones with resistant units will form sutured seams (stylolites) and grain-contact sutures. On the other hand, pressure solution of limestones enriched in insolubles (>10 percent

FIGURE 12.12
Well-developed, sutured stylolites in limestone. Cretaceous limestone, Calcare Massicio, Tuscany, Italy. *(Photograph courtesy of E. F. McBride.)*

FIGURE 12.13
Microstylolites (arrows) in
limestone. Microstylolites
commonly occur in "swarms," as
seen here. Paleozoic limestone,
Alaska. Ordinary light. Scale
bar = 0.5 mm.

clay, quartz, organic matter) will concentrate insolubles as a film, producing nonsutured
seams (Fig. 12.15). These nonsutured seams have been given various names, including
clay seams, wispy laminae, wavy laminae, microstylolites, and pseudostylolites. Nonsutured seams may occur as single clay film surfaces or as anastomosing swarms of wispy
seams oriented parallel to bedding. They may occur also in nonparallel, reticulate orientation, producing a pattern of closely spaced nodules referred to as **fitted fabric** (Wanless,
1979; Buxton and Sibley, 1981). Logan and Semeniuk (1976) refer to this reticulate fabric
as **stylobreccia.** Fitted fabric appears to occur particularly in poorly cemented grainstones
but may occur also in packstones and wackestones (Buxton and Sibley, 1981). Pressure
solution apparently starts at grain contacts; microstylolites then propagate irregularly
through the rock to create a fitted-fabric network (Choquette and James, 1987). Figure
12.14 is a very simple classification that illustrates the principal kinds of stylolites,
microstylolites, and nonsutured seams that occur in carbonate rocks. For additional information on stylolites, see the references mentioned above and the special volume on
stylolites published by the Abu Dhabi National Reservoir Research Foundation (1984).

Cementation

Nature of the Process.   Much of the research on carbonate diagenesis carried out since
the 1950s has concentrated on study of Holocene and Pleistocene carbonates. Significant
evidence of cementation in these rocks, believed to have taken place in the seafloor and

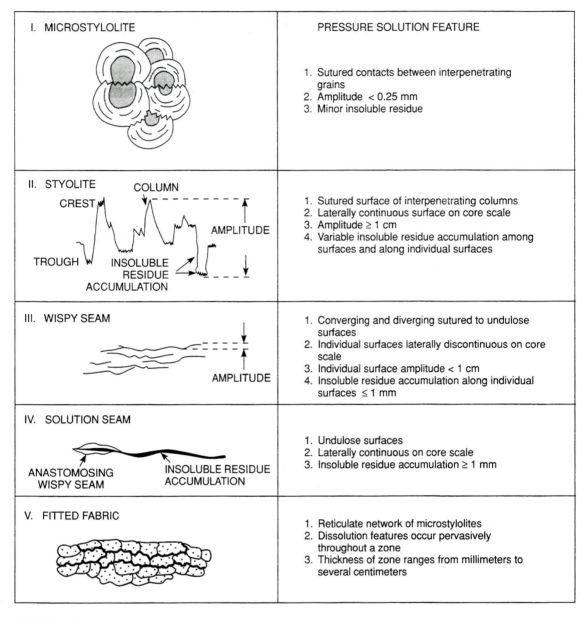

I. MICROSTYLOLITE	PRESSURE SOLUTION FEATURE 1. Sutured contacts between interpenetrating grains 2. Amplitude < 0.25 mm 3. Minor insoluble residue
II. STYOLITE CREST COLUMN AMPLITUDE TROUGH INSOLUBLE RESIDUE ACCUMULATION	1. Sutured surface of interpenetrating columns 2. Laterally continuous surface on core scale 3. Amplitude ≥ 1 cm 4. Variable insoluble residue accumulation among surfaces and along individual surfaces
III. WISPY SEAM AMPLITUDE	1. Converging and diverging sutured to undulose surfaces 2. Individual surfaces laterally discontinuous on core scale 3. Individual surface amplitude < 1 cm 4. Insoluble residue accumulation along individual surfaces ≤ 1 mm
IV. SOLUTION SEAM ANASTOMOSING WISPY SEAM INSOLUBLE RESIDUE ACCUMULATION	1. Undulose surfaces 2. Laterally continuous on core scale 3. Insoluble residue accumulation ≥ 1 mm
V. FITTED FABRIC	1. Reticulate network of microstylolites 2. Dissolution features occur pervasively throughout a zone 3. Thickness of zone ranges from millimeters to several centimeters

FIGURE 12.14

Principal kinds of pressure solution features in carbonate rocks. (*Modified from Koepnick, R. B., 1984, Distribution and vertical permeability of stylolites within a lower Cretaceous carbonate reservoir, Abu Dhabi, United Arab Emirates, in Stylolites and associated phenomena— Relevance to hydrocarbon reservoirs: Abu Dhabi National Reservoir Research Foundation Spec. Pub., Fig. 1, p. 262; with additions by C. H. Moore, 1989, p. 248; reprinted by permission.*)

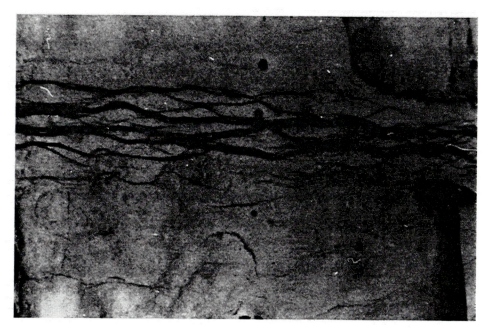

FIGURE 12.15
Nonsutured stylolite seams in limestone, Middle Chalk (Cretaceous), Compton Bay, Isle of Wight, Hampshire, southern England. *(From Garrison, R. G., and W. J. Kennedy, 1977, Origin of solution seams and flaser structures in upper Cretaceous chalks of southern England: Sed. Geology, v. 19. Fig. 8, p. 117, reprinted by permission of Elsevier Science Publishers, Amsterdam.)*

meteoric burial environments, has been reported in the literature. Consequently, geologists frequently assert that most cementation must occur at shallow depths. On the other hand, Scholle and Halley (1985), as well as many other geologists (see discussion by Choquette and James, 1987), argue that much of the cementation in carbonate rocks takes place in moderately deep to very deep subsurface settings (mesodiagenetic environments). Clearly, all cements in deep-water carbonates such as chalks that were never exposed to meteoric waters must be of deep-burial origin. A growing body of evidence obtained by petrographic, cathodoluminescence, stable-isotope, and fluid-inclusion study suggests that much of the cement in ancient peritidal, shallow marine, and platform-margin limestone may also have formed during deep-burial diagenesis.

The factors that control dissolution and precipitation of carbonates at depth are complex and poorly understood. Unless carbonate sediments escape mineralogical stabilization in the seafloor or meteoric environment, the subsurface fluids in contact with carbonate minerals may be more or less in equilibrium with $CaCO_3$. Thus, there may be no strong geochemical "drive" to bring about either dissolution or cementation. Nonetheless, much available evidence indicates that deep-burial dissolution and cementation must occur. What, then, drives these processes, particularly cementation? Temperature, pressure (particularly as it affects P_{CO_2}), salinity, and ion composition of pore waters must be major factors that influence dissolution and cementation. Other factors that may affect these processes include grain size, fabric, mineralogy, porosity, and permeability. In turn, these variables affect rates of fluid flow (flux). It is difficult to know exactly which combination of factors actually leads to cementation. Factors that favor cementation (or solution–reprecipitation) are suggested to include metastable mineralogy, highly over-saturated pore waters, high porosity and permeability (which enable high rates of fluid flow), increase in temperature, decrease in carbon dioxide partial pressure (P_{CO_2}), and pressure solution (Feazel and Schatzinger, 1985; Choquette and James, 1987).

Increasing temperature in the range of 25°C to 200°C may decrease the solubility of calcite in water by up to two orders of magnitude (Sharp and Kennedy, 1965) and thus promote cementation. On the other hand, accompanying increase in pressure probably counteracts some of this decrease. Nonetheless, temperature increase must be important. In a discussion of chalk cementation in DSDP cores, Wetzel (1989) reports that cement content of chalks doubles in sediments that have experienced temperatures four times higher. Fluid-inclusion studies suggest that deep-burial cement may be precipitated from $CaCO_3$-saturated pore waters ranging from brackish to highly saline (10–100‰). As discussed elsewhere in this book, tens of thousands to hundreds of thousands of pore volumes of water must pass through a pore to precipitate a significant amount of cement. Thus, oversaturation with respect to calcite and relatively high fluid flux favor cementation. It is probable, however, that fluid flux under many subsurface conditions is quite small. Degassing of CO_2, owing to escape from deep-burial zones to the shallow sub-surface along faults or fractures, is a potentially important but unproven mechanism for enabling cementation (Choquette and James, 1987).

If we accept the premise that a significant volume of calcite (or dolomite) cement may be precipitated in the deep-burial environment, then we must identify an adequate source of $CaCO_3$ to provide the cement. The mechanism for supplying deep-burial $CaCO_3$ most frequently mentioned (e.g., Scholle and Halley, 1985) is pressure solution. Moore (1989, p. 261) points out, however, that an enormous amount of carbonate section must be removed by pressure solution to furnish a small amount of cement. He suggests that $CaCO_3$ derived by local pressure solution must be supplemented from other sources. Possibly, it may be imported from other, adjacent subsurface carbonate units likewise undergoing pressure solution. Alternatively, late dissolution of limestones or dolomites by pore fluids made aggressive by organic acids and CO_2 generated during organic matter diagenesis (see Chapter 9) may furnish additional $CaCO_3$.

Characteristics of Burial Cements.   As mentioned, burial cements occur in both deep-water chalks and coarser-grained shallow-water carbonates. Cements in chalks are com-posed of very fine crystals (1–10 μm) and are best studied with the scanning electron microscope (Scholle, 1977). Burial carbonate cements in coarser-grained carbonates can include clear, coarse calcite as well as dolomite. The general characteristics of these cements are summarized by Choquette and James (1987). They are commonly enriched in iron and manganese, depleted in strontium, and depleted in ^{18}O compared to shallow-burial cements. Cathodoluminescence is dull, the cements may or may not show com-positional zonation, and fluid inclusions (containing both a gas and liquid phase) and hydrocarbon inclusions are common (Sellwood et al., 1989). The principal kinds of cement are as follows.

1. **Bladed-prismatic calcite** cement consists of elongate, scalenohedral crystals a few tens of micrometers across and up to a few hundreds of micrometers long. They typically grow directly on grain surfaces or on top of an earlier generation of marine cements (Figs. 12.16, 12.17).

FIGURE 12.16
Schematic representation of
principal kinds of deep-burial
cements. Cements subjected to
cathodoluminescence may display
luminescent banding, with
luminescence of the bands ranging
from nonluminescent through
bright to dull. These bands reflect
differences in Fe and Mn content
of the cements that may indicate
different generations of cement.
See subsequent discussion of
cement stratigraphy for details.
*(From Choquette, P. W., and
N. P. James, 1987, Diagenesis 12.
Diagenesis in limestones—3. The
deep burial environment:
Geoscience Canada, v. 14. Fig.
21, p. 16, reprinted by permission
of Geological Association of
Canada.)*

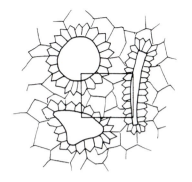

BURIAL
CALCITE
CEMENTS

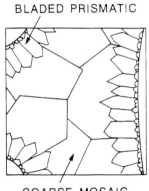

BLADED PRISMATIC

COARSE MOSAIC
CALCSPAR

PLANE LIGHT

NON-LUMINESCENT

BRIGHT

DULL

CATHODOLUMINESCENCE

FIGURE 12.17
Bladed, prismatic calcite lining
margins of a cavity and grading
inward to mosaic calcite toward
center of cavity. Cavity
incompletely filled with cement.
Robinson Limestone
(Pennsylvanian), Colorado.
Crossed nicols. Scale bar =
0.5 mm.

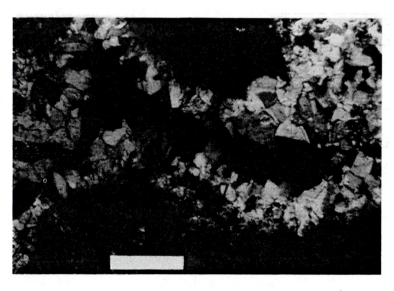

2. **Coarse mosaic calcispar,** commonly much coarser than bladed-prismatic calcite, typically consists of plane-sided, equant crystals (Fig. 12.16). Where present with bladed-prismatic calcite, it is typically younger than the bladed calcite.

3. **Poikilotopic calcite** cement consists of mm-size crystals that are large enough to enclose several carbonate grains (Fig. 12.18).

4. **Coarse dolomite** cements consist of clear-to-turbid, coarse crystals that may include planar (xenotopic) dolomite, saddle dolomite, and baroque dolomite. See Chapter 11 for additional description of these dolomites and the conditions under which they form.

5. **Coarse anhydrite** cements occur in some buried limestones. These cements generally occur in limestones that are associated with evaporite deposits.

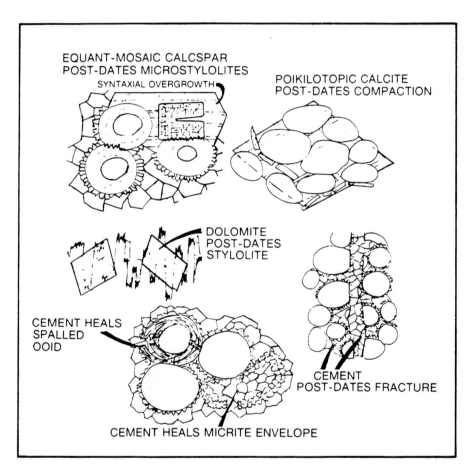

FIGURE 12.18

Examples of criteria that suggest late-burial cementation. A. Syntaxial overgrowth encloses microstylolites formed during a preceding stage of chemical compaction. B. Calcite poikiotopically encompasses grains packed and broken during physical compaction. C. Cement crystals cut across stylolites formed during chemical compaction. D. Ooids or micrite envelopes broken during compaction are healed by cement. E. Cement fills late-state tectonic fractures. *(After Choquette, P. W., and N. P. James, 1987, Diagenesis 12. Diagenesis in limestones —3 The deep burial environment: Geoscience Canada, v. 14. Fig. 24, p. 18, reprinted by permission of Geological Association of Canada.)*

It may be possible to differentiate burial cements from earlier-formed cements on the basis of careful petrographic examination. Evidence for late-burial origin are discussed by Moore (1985), Choquette and James (1987), and Sellwood et al. (1989). Cements are considered to be "late-stage" if cement crystals (1) cross-cut other features known to be formed by burial diagenesis, such as stylolites or other pressure solution seams (Fig. 12.18), (2) heal fractured grains or spalled ooid cortices, (3) enclose compacted grains (the poikilotopic calcite in Fig. 12.18) or fill compaction pores, indicating that cementation postdates physical or chemical compaction, (4) fill late-stage tectonic fractures or fill dissolution pore spaces that were created by solution of grains and early burial cements, and (5) enclose solid hydrocarbons such as asphalt or pyrobitumens that themselves were formed during deep-burial diagenesis.

Differentiating Cement Fabrics from Neomorphic Fabrics.   In addition to the problem of differentiating between early and late cement, it is also necessary to differentiate between cement and neomorphic calcite. Differentiating sparry calcite cement from neomorphic spar is one of the most troublesome problems in carbonate petrology. The fabric of neomorphic spar, which is formed by polymorphic transformation and recrystallization (discussed below), can strongly resemble that of some cements. Because misidentification of neomorphic spar for cement can lead to errors in interpreting depositional and diagenetic history, it is important to distinguish between the two. Unfortunately, avoiding misidentification is easier said than done. Bathurst (1975, p. 417) lists some 17 different characteristics of cements that help to distinguish them from neomorphic spar. These criteria, with additions from other sources, are summarized in Table 12.1. Table 12.1

TABLE 12.1
Distinguishing characteristics of sparry calcite cement

- Two or more distinct generations of spar are present (unlikely in neomorphic spar); the spar is commonly clear and free of relict structures or impurities.
- The cement encrusts free surfaces such as the surfaces of carbonate grains or molds.
- Crystals of the cement mosaic have a preferred orientation of longest diameters normal to the initial substrate on which they grew; size of the crystals increases away from the initial substrate, and crystals tend to become larger, more blocky, and less well oriented (drusy fabric).
- Cement fabrics are characterized by a high percentage of enfacial junctions (an enfacial junction is a triple junction between three crystals where one of the angles is 180 degrees), and boundaries between crystals are mainly plane interfaces.
- Spar fabrics that occur in association with particles composed of micrite (e.g., pellets or micrite envelopes) that are not themselves altered to neomorphic spar are likely to be cements.
- Contacts between spar and carbonate particles are sharp.
- Spar lines a cavity that is incompletely filled, or spar occupies the upper part of a cavity, the whole lower part of which is filled with more or less flat-topped, mechanically deposited micrite (geopetal structure).
- The presence of distinctly banded zones within crystals, as revealed by staining or cathodoluminescence; such zones would be uncommon in neomorphic spar.

Source: Bathurst, R. G. C., 1975, and other sources. Compare with Table 12.2.

should be compared with Table 12.2, which summarizes the distinguishing characteristics of neomorphic calcite.

Geochemistry and Isotope Composition of Deep-Burial Cements

Burial cements, including both calcite and late-burial saddle dolomite, are commonly enriched in Fe and Mn but have low Sr concentrations. Fe and Mn enrichment apparently reflects the progressive enrichment of Fe and Mn in concentrated subsurface brines owing to rock–water interactions with siliciclastic rocks and evaporites, and low Sr concentrations may reflect the slow rate of precipitation of deep-burial calcite and dolomite (Moore, 1989, p. 262).

Because oxygen isotope fractionation is affected by temperature, oxygen isotopes should theoretically show progressively decreasing $\delta^{18}O$ values with increasing burial (Choquette and James, 1987; Moore, 1985). Choquette and James (1987) suggest, for example, that $\delta^{18}O$ isotope compositions should hypothetically show a more or less steady decrease from values of about 0‰ in the meteoric zone to values as low as -10 to $-12‰$ with deep burial. This hypothetical trend reflects increasing temperatures of the precipitating waters. A number of empirical studies lend support to this general hypothetical trend. As subsurface waters become increasingly hotter with burial, however, the $\delta^{18}O$ composition of the waters themselves becomes increasingly higher. Heydari and Moore (1988) have shown that the progressive decrease in $\delta^{18}O$ values in cements may ultimately be buffered, and even reversed, by the increasingly high $\delta^{18}O$ values of deep pore fluids. Thus, the isotopic composition of cements precipitated at temperatures in excess of about 150°C, may be higher than for those precipitated at lower burial temperatures (e.g., 40°C–100°C).

Carbon isotope composition changes also with deep burial; however, most studies indicate that $\delta^{13}C$ values decline only slightly with burial depth. Moore (1989, p. 265) suggests that the reason most subsurface cements show little variation in carbon isotope composition is that the cement is rock-buffered owing to chemical compaction. On the other hand, increasing volumes of light carbon (^{12}C), generated as a byproduct of hydrocarbon maturation, may bring about slight depletion in $\delta^{13}C$ values.

Cement Stratigraphy

As discussed, carbonate cements can be precipitated in the seafloor, meteoric, or deep-burial diagenetic environments. During the progressive burial and diagenesis of a given carbonate deposit, one generation of cement may be deposited in the seafloor environment, another in the meteoric environment, and still another in the deep-burial environment. With careful study, it may be possible to detect each of these generations of cement. For example, it may be possible within single pores or adjacent pores in a limestone to identify a marine seafloor cement (oldest), a vadose cement (younger), and a meteoric phreatic cement (youngest). Alternatively, the cement sequence might consist of a beach-rock cement (oldest), a vadose cement (younger), and a deep-burial cement (youngest). This technique of studying successive generations of cement, referred to as **cement stratigraphy,** was introduced by Evamy (1969) and was first applied in cathodoluminescence studies by Meyer (1974). The technique has subsequently been employed by numerous investigators (e.g., Grover and Read, 1983; Meyer and Lohmann, 1985; Goldstein, 1988; Kaufman et al., 1988) as a tool for unraveling the burial history of limestones.

In some samples, different generations of cements may be tentatively identified simply by petrographic analysis; for example, a cloudy seafloor cement (oldest) may be present with a clear meteoric cement (youngest). Most cement stratigraphy requires more-detailed study that may involve staining, cathodoluminescence, ultraviolet microscopy, trace-element analysis, and fluid-inclusion studies. To illustrate, staining with potassium ferrocyanide can reveal differences in ferrous iron content of calcite or dolomite. These differences may reflect changes in oxidizing conditions of the precipitating fluid, such as a change from the meteoric to the deep-burial environment.

The most heavily used tool in cement stratigraphy is cathodoluminescence. Cathodoluminescence in carbonates is caused mainly by the presence of trace elements, particularly Mn^{2+}, and to a lesser extent by other factors such as distorted crystal structures and compositional inhomogeneities. On the other hand, Fe^{2+} (especially) tends to quench the luminescent reaction. It is commonly reported in cement stratigraphy studies that cements display a trend from oldest to youngest cements of (1) nonluminescence to (2) bright luminescence to (3) dull luminescence (Fig. 12.16). This trend is interpreted to signify progressive change during burial from well-oxidized conditions that (1) inhibit the uptake of manganese and iron (no luminescence) to (2) reducing conditions that favor uptake of Mn^{2+} but keep Fe^{2+} locked up in interactions with organic matter and sulfates (bright luminescence) to (3) deeper-burial reducing conditions where Fe^{2+} is available to quench luminescence (dull luminescence). Thus, these trends are commonly suggested to be due simply to concentrations of iron and manganese as a function of Eh and pH. Machel (1985) maintains, however, that many trace elements (e.g., Pb) in addition to Mn^{2+} and Fe^{2+} can affect cathodoluminescence. That is, cathodoluminescence in carbonates is due to the association of many trace elements and/or crystal-lattice defects and not just to the relative abundance of Mn^{2+} and Fe^{2+}. Therefore, he cautions that environmental and stratigraphic interpretation of diagenetic carbonates on the basis of their cathodoluminescence be undertaken with extreme care. On the other hand, Hemming et al. (1989) report that Fe/Mn ratio and the absolute concentrations of Fe and Mn control cathodoluminescence intensity in their studies. They suggest that rare earth elements and Pb do not play a significant role in cathodoluminescence intensity.

See Machel (1985) for a good general discussion of cathodoluminescence in carbonates, and see Goldstein (1988) for a more extended bibliography of research papers dealing with cement stratigraphy. The techniques for studying zoned cements and the applications and implications of such studies are covered in a special issue of *Sedimentary Geology* (Volume 65) edited by B. W. Sellwood (1989).

Neomorphism

As discussed in Section 12.2.3, neomorphism refers to diagenetic changes that take place owing to polymorphic transformation (solution–reprecipitation) and/or recrystallization (Folk, 1965). Neomorphism does not include formation of a solution cavity (mold) and subsequent refilling by cement. Neomorphism typically produces an increase in crystal size, referred to as **aggrading neomorphism.** Folk proposes that aggrading neomorphism can occur either by gradual enlargement of all crystals (coalescive neomorphism) or by growth of a few large crystals in a static groundmass (porphyroid neomorphism). The end result is the same (Fig. 12.19).

Folk suggests that a decrease in crystal size is also possible and refers to such a decrease as **degrading neomorphism** (grain diminution of other authors). A putative

FIGURE 12.19

Schematic illustration of
porphyroid and coalescive
aggrading neomorphism. *(From
Folk, R. L., 1965, Some aspects of
recrystallization in ancient
limestones, in Pray, L. C., and
R. C. Murray, eds., Soc. Econ.
Paleontologists and Mineralogists
Spec. Pub. 13. Fig. 3, p. 22,
reprinted by permission of SEPM,
Tulsa, Okla.)*

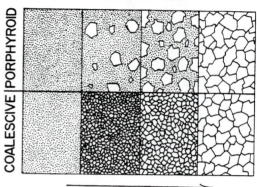

example of degrading neomorphism is the partial or complete alteration of an echinoderm
fragment (single calcite crystal) to a mosaic of small crystals. On the other hand, Bathurst
(1975, p. 478) raises serious doubts about the various reported examples of degrading
neomorphism and urges caution about the whole concept of grain diminution.

Three kinds of aggrading neomorphism are important in carbonates (Folk, 1965):
(1) change of aragonite crystals in a lime mud to $1-3$ μm size calcite (micrite), (2) more
advanced neomorphism of mud-size carbonate to $5-20$ μm size **microspar,** and (3)
further neomorphism of mud, micrite, fossils, or other carbonate grains to coarse sparry
or fibrous calcite, called **pseudospar,** with crystal size ranging to several tenths of
millimeters. Presumably, wet recrystallization is the dominant process causing formation
of microspar and pseudospar; however, wet polymorphic transformation of aragonite to
calcite or solution–reprecipitation involving Mg-calcite are probably also involved for
those carbonates that have not been completely stabilized mineralogically. Such metastable materials might include marine carbonate sediments that were never exposed to
meteoric diagenesis, fossils containing aragonite or Mg-calcite, or previously formed
aragonitic or Mg-calcitic cements. Burial neomorphism is probably affected by the same
factors discussed under cementation, that is, increasing temperature and pressure, metastable mineralogy, and pore-fluid flux. Folk (1965) suggests that the driving force behind
recrystallization may involve several factors, including (a) energy of a phase transformation, (b) solution of tiny grains or projections, and precipitation about a larger nucleus, (c)
pressure solution, (d) energy of a strained crystal lattice, and (f) surface tension caused by
crystal boundary curvature. In a discussion of the instability of micrometer-sized fabrics
and the drive toward formation of coarser fabrics, Bathurst (1975, p. 349) suggests that
some mud-size crystals are so small as to be supersoluble. Small crystals have a greater
density of edges per unit surface area and thus a greater concentration of unneutralized
bonds, and many crystals have residual elastic strain and are thus more soluble. Also, the
presence of metastable aragonite and Mg-calcite is an important factor.

As mentioned in Section 12.2.3, neomorphism appears to be a relatively minor
process in the seafloor diagenetic environment. Considerable neomorphism does occur in
the meteoric burial environment where aragonite and Mg-calcite are altered to calcite.
Change in crystal shape and increase in crystal size accompany this stabilization process.
Neomorphism is probably an important process also in the deep-burial diagenetic environment, given the long periods of time involved in burial diagenesis, although relatively
few hard data on the magnitude of deep-burial neomorphism appear to be available. As

an example of deep-burial neomorphism, Kendall (1975) reports replacement of aragonite in mollusc shells by coarse calcite. The relationship of the coarse calcite crystals to fracturing and pressure solution features in the molluscs shows that calcitization must have occurred after compaction of the shells. Bathurst (1983) also reports neomorphism of molluscan shells and early-formed cement after carbonate sediments had undergone compactive fracture and grain–grain pressure solution.

A major problem in evaluating the overall significance of deep-burial neomorphism is the difficulty in differentiating neomorphic calcite from calcite cement, as mentioned. With respect to neomorphism, Bathurst (1975, p. 483) states that "the process yields a sparry calcite that is always difficult to distinguish, sometimes impossible to distinguish, from the sparry calcite of cement." Neomorphism can produce at least three kinds of crystal fabrics:

1. **Granular fabrics** appear as a mosaic of crystals, with crystal size ranging upward from about 4 to 50–100 μm (Fig. 12.20). Crystal size may vary irregularly and patchily from place to place, although equigranular fabrics can also occur.

2. **Radial-fibrous fabrics** occur in which crystals are arranged in subspherical, stellate masses (Fig. 12.21). Commonly, the center of the masses consists of equigranular microspar (5–10 μm in size) surrounded by coarser elongate crystals with their axes radially distributed; elongate crystals may taper toward the center of the stellate structure (Bathurst, 1975, p. 484).

3. **Syntaxial rims** may occur on skeletal grains owing to neomorphism of surrounding micrite. Syntaxial rims are most common on echinoderm fragments but may occur also on coarse crystals in the walls of brachiopod or mollusc shells. These syntaxial rims may closely resemble syntaxial rim cements.

Neomorphic spar may replace detrital micrite, carbonate grains, or an earlier generation of aragonite, Mg-calcite, or calcite cement. The problem of differentiating between neomorphic spar and sparry calcite cement is a very difficult one in carbonate petrology because many cement fabrics mimic neomorphic fabrics. As mentioned by Bathurst above, it may be impossible in some cases to make the distinction. Nonetheless,

FIGURE 12.20
Fusulinid foraminifer cut in half owing to recrystallization to coarse, neomorphic spar (bottom half of photograph). Pennsylvanian limestone, Paradox Basin, Utah. Crossed nicols. Scale bar = 0.5 mm.

FIGURE 12.21
Stellate neomorphic fabric formed by alteration of a biomicrite. Note equigranular microspar in the core of the stellate mass. Carboniferous limestone, Denbigshire, England. Ordinary light. *(From Bathurst, R. G. C., 1975, Carbonate sediments and their diagenesis. Fig. 332, p. 479, reprinted by permission of Elsevier Science Publishers, Amsterdam.)*

100 μ

Bathurst (1975, p. 484–493; 1983) provides a number of criteria that he believes characterize neomorphic spar. The principal criteria are summarized in Table 12.2. Compare this table with Table 12.1, which lists the distinguishing features of cements. No single criterion in Table 12.2 can be regarded as unequivocal proof of the neomorphic origin of spar, however, and it is wise to apply as many of the criteria as possible.

Interpretation of syntaxial rims as neomorphic replacements of micrite is particularly troublesome. Petrographic criteria may support a neomorphic origin for syntaxial rims in a particular deposit, whereas cathodoluminescence studies may indicate a different origin. For example, Walkden and Berry (1984) restudied the origin of syntaxial rims in Lower Carboniferous limestones in Great Britain that were previously considered to be of neomorphic origin. On the basis of cathodoluminescence analysis, they conclude that the rims formed by passive cementation in solution voids that formed around echinoderm fragments during meteoric diagenesis. In another cathodoluminescence study of syntaxial rims in limestones, Maliva (1989) concludes that the rims formed by displacive crystal growth that simply pushed aside the enclosing micrite.

Replacement

Replacement involves the dissolution of one mineral and the nearly simultaneous precipitation of another mineral **of different composition** in its place. Thus, the process of replacement is similar to the process of wet polymorphic transformation, but it involves minerals of different composition. Many kinds of mineral may replace carbonate minerals during diagenesis, including chert, pyrite, hematite, apatite, anhydrite, and dolomite. As

TABLE 12.2

Distinguishing characteristics of neomorphic calcite

- Radial-fibrous fabrics are present.
- Neomorphic spar embays (nibbles) detrital micrite; embayments tend to be sawtoothed; relicts of micrite may appear as wisps or threads in spar.
- Spar transects skeletal grains, ooids, or other carbonate grains; grains may be only partially replaced by spar, preserving some of the internal structure of the grain (e.g., the fibrous structure of aragonitic fossils), or they may be replaced entirely, thus preserving only the outline or form of the grain.
- Crystal size of spar may vary irregularly and patchily from place to place; however, equigranular fabrics also occur.
- The presence of undigested silt, clay, or organic matter trapped in spar, or the presence of dusty relics of earlier cement fringes; sparry calcite cement is commonly free of such cloudy impurities.
- Abnormally loose packing (floating relics), as shown by patches of micrite, skeletal grains, or other carbonate grains entirely surrounded by spar; this criterion must be used with caution because grains that do not appear to be in contact in the two-dimensional plane of a thin section may actually be in contact in three dimensions, in which case the spar may be cement.
- Patches of spar in the midst of homogeneous micrite may indicate neomorphic spar; however, such patches may also form by filling of small burrows, gas bubble holes, etc. by sparry calcite cement.
- Intercrystalline boundaries tend to be curved or wavy, in contrast to the plane boundaries typical of sparry calcite cements.
- Neomorphic spar displays few enfacial junctions between crystals in contrast to cement spar, which commonly contains high percentages of enfacial junctions.
- Neomorphic syntaxial rims may contain cloudy impurities, display sawtoothed margins, or transect adjacent carbonate grains; syntaxial rims surrounded by micrite may appear to be neomorphic; however, cathodoluminescence study may reveal that the rims formed by displacive crystallization or by filling of solution cavities around grains.
- Cathodoluminescence may show relics of earlier cement fabrics or other replaced fabrics.
- Crystals may be separated by concentrations of impurities that were expelled by growing crystals and displaced to line compromise boundaries.

Source: Bathurst R. G. C. (1975, 1983) and other sources. Compare with Table 12.1.

discussed in Chapter 11, dolomite can replace aragonite or calcite on a massive scale, resulting in pervasive dolomitization of limestone sequences. Other than dolomite, chert is probably the most common replacement mineral in carbonate rocks. Bodies of replacement chert may range in size from micrometer-size patches within carbonate grains to centimeter-size nodules and lenses that engulf both carbonate grains and micrite. Replacement may be either fabric-destructive or fabric-preserving. For example, replacement of fossils by chert very commonly preserves the fine textural details of the fossils. On the other hand, much replacement of aragonite or calcite by dolomite is fabric-destructive (Chapter 11).

Replacement can occur in all diagenetic environments. Dolomite, for example, may replace calcium carbonate in intertidal-supratidal environments, mixing-zone environments, and deep-burial environments. Replacement of carbonates by chert is not common on the seafloor but can occur in both the meteoric and deep-burial environments. Maliva and Siever (1989) report chertification of Paleozoic carbonates at burial depths ranging from 30 to 1000 m. In carbonate-evaporite sequences, replacement of carbonate minerals by anhydrite at depth is known to be a common process.

Conventional wisdom holds that replacement can occur in any environment that favors the dissolution of calcium carbonate and the precipitation of the guest mineral and where an adequate source of magnesium, silica, iron, sulfate, etc. is available to form the precipitating mineral. Most previous investigators have assumed that replacement occurs in a diagenetic environment in which the bulk pore waters of the sediment or rock undergoing replacement are supersaturated with respect to the guest (replacing) mineral and undersaturated with respect to the bulk host (replaced) mineral, whose free surfaces are in contact with pore waters. Maliva and Siever (1988a) make the extremely interesting suggestion, however, that undersaturation with respect to the bulk host phase is not necessarily a prerequisite for replacement. They propose that diagenetic replacement can occur by a **force of crystallization-controlled mechanism** along thin films that have only diffusional access to bulk pore waters in equilibrium with the bulk host sediments. According to this model, "nonhydrostatic stresses resulting from authigenic crystal growth are principally responsible for local host-phase dissolution at authigenic-host phase contacts." All that is required for replacement is a sufficient degree of local supersaturation with respect to the authigenic mineral phase and suitable conditions in the rock for initial crystal nucleation. The degree of saturation of the bulk pore water with respect to the host mineral being replaced is of secondary importance. Maliva and Siever maintain that nonhydrostatic stresses resulting from authigenic crystal growth are sufficient for host-mineral dissolution. They developed this model originally to explain silicification of fossils in limestones (Maliva and Siever, 1988b) but suggest that it may also be applied to replacement of detrital quartz by carbonate minerals and replacement of carbonate minerals by dolomite.

Replacement of carbonate minerals by silica may be highly selective, with silica quite commonly replacing fossils and other carbonate grains such as ooids in preference to micrite (Newell et al., 1953; Maliva and Siever, 1988b). Maliva and Siever (1989) report that chert nodules can occur in grainstone, packstone, wackestone, and mudstone lithologies, although, owing to depositional environmental controls, the nodules tend to occur most commonly in mudstones and wackestones.

Significant silicification of carbonates requires an adequate source of silica. Dissolution of siliceous sponge spicules is commonly postulated to be the principal mechanism for supplying silica. Otherwise, silica must come from sources outside carbonate formations. Generally, it is extremely difficult to identify the source of silica in a given carbonate unit. The silica-carbonate replacement reaction is known to be a reversible one (Walker, 1962; Hesse, 1987). At some stages during diagenesis, silica can replace carbonate and subsequently be itself replaced by carbonate. In turn, the replacing carbonate may be partially or completely replaced by silica at a later stage. Assuming that Maliva and Siever's (1988a) "force of crystallization" theory is correct, these reversals may be governed mainly by varying silica concentrations in pore waters.

Carbonate replacement can generally be recognized by careful petrographic examination. Criteria for recognizing replacement textures are discussed in Chapter 9 and illustrated in Table 9.4. Illogical mineral composition (pseudomorphs), cross-cutting relationships, and caries texture (bitelike embayments into the host mineral) are particularly useful criteria. Figure 12.22 shows an example of illogical mineral composition (a fossil shell replaced by anhydrite), and Figure 12.23 illustrates caries texture.

FIGURE 12.22
Section of a brachiopod shell almost completely replaced by coarse anhydrite. An example of illogical mineral composition as a clue to recognition of replacement fabrics. Pennsylvanian limestone, Paradox Basin, Utah. Crossed nicols. Scale bar = 0.5 mm.

FIGURE 12.23
Chert (CH) replacing an echinoderm fragment (E) along a scalloped replacement front (arrow)—an example of caries texture. Paleozoic limestone, Alaska. Crossed nicols. Scale bar = 0.5 mm.

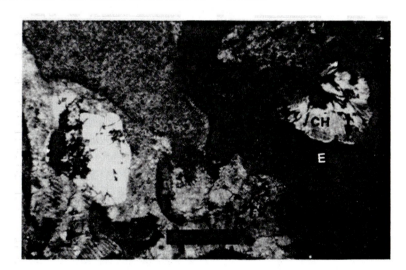

Dissolution

Deeply buried carbonate rocks have presumably been mineralogically stabilized owing to dolomitization, neomorphism, and cementation. Therefore, few highly soluble phases should be present. Nonetheless, dissolution in the deep-burial environment has been suggested by several investigators (e.g., Elliot, 1982; Moore and Druckman; 1981, Moore, 1985). Two factors may account for deep-burial dissolution: (1) Subtle differences in the relative solubilities of stabilized carbonates may exist owing to factors such as crystal size, trace element and organic matter content, and microporosity. In general, biogenic calcite is more soluble than calcitized aragonite or Mg-calcite, which are more soluble than calcite cement (Choquette and James, 1987). (2) Decarboxylation of organic matter during maturation may release sufficient CO_2 and organic acid into pore fluids to cause dissolution (see Chapter 9, Section 9.4.3).

Thus, dissolution may be fabric-selective, with preferential removal of less-stable biogenic calcite and possibly calcitized carbonate grains. On the other hand, dissolution may be non-fabric-selective if CO_2 and organic acid levels in pore waters are high enough. In non-fabric-selective dissolution, dissolution begins in interparticle or intergranular pores, which are enlarged by dissolution into large vugs. These enlarged vugs are rounded, and they cut across all textural elements, including carbonate grains, late-burial cements, and stylolites. This cross-cutting relationship of solution pores is the best evidence of deep-burial dissolution. Because decarboxylation processes take place at burial temperatures just below those required to form liquid hydrocarbons, decarboxylation-driven dissolution may create important amounts of secondary porosity in carbonate reservoirs for storage of petroleum.

12.6   Evolution of Porosity in Carbonate Rocks

12.6.1   Introduction

Because carbonate rocks contain more than one-third of the world's petroleum reserves, the porosity of carbonate rocks is an important subject to petroleum geologists. The porosity and permeability of sedimentary rocks influence the migration of petroleum, and the porosity of reservoir rocks determines the amount of petroleum that can be stored in a given petroleum trap. Thus, the origin, evolution, and preservation of porosity in carbonate rocks has considerable economic significance in addition to being a subject of scientific interest. Analysis and interpretation of carbonate porosity require some knowledge of porosity classification as well as an understanding of the origin and evolution of porosity. These topics are discussed very briefly in this section. Students and other interested workers who want more detail on carbonate porosity should consult the excellent book by Moore (1989) dealing with carbonate diagenesis and porosity, as well as a special issue of *Sedimentary Geology* (Volume 63), edited by Hanford et al. (1989), devoted to microporosity in carbonate strata.

12.6.2   Classification and Origin of Carbonate Porosity

Most geologists worldwide use the porosity classification of Choquette and Pray (1970) to describe carbonate porosity. This classification (Fig. 12.24) divides porosity into three basic types: fabric-selective, not fabric-selective, and fabric-selective or not. **Fabric-selective** porosity includes interparticle porosity (pore space among grains), intraparticle porosity (pore space within grains), fenestral porosity (irregular, elongated openings commonly oriented parallel to bedding), intercrystalline porosity (pore space among crystals), moldic porosity (solution molds), shelter porosity (pore space beneath umbellalike fossils or other grains), and growth-framework porosity (porosity within reef frameworks or similar structures).

Most of these fundamental kinds of fabric-selective porosity are generated as a result of depositional processes and are thus kinds of **primary porosity.** Porosity that results from diagenetic processes is secondary porosity. Moldic porosity is entirely of

FIGURE 12.24
Classification of porosity in carbonate rocks. *(From Choquette, P. W., and L. C. Pray, 1970, Geologic nomenclature and classification of porosity in sedimentary carbonates: Am. Assoc. Petroleum Geologists Bull., v. 54. Fig. 2, p. 224, reprinted by permission of AAPG, Tulsa, Okla.)*

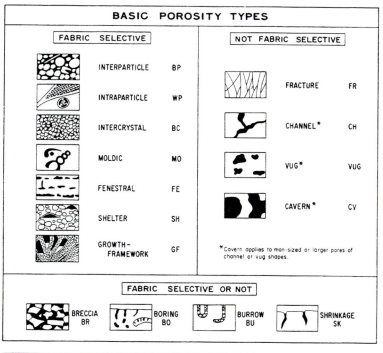

secondary origin, and most intercrystalline porosity is probably also secondary. Interparticle pores are initially primary but can be enlarged by secondary solution processes. Some intraparticle porosity may also be generated by dissolution, and dissolution processes are probably involved to some extent in the formation of fenestral porosity (Chapter 10). Porosity that is **not fabric-selective** is entirely of secondary origin and forms as a result of fracturing and/or solution. **Fabric-selective or not porosity** includes both primary and secondary types. Breccia porosity probably forms mainly by secondary processes that involve solution collapse or tectonism. Boring and burrow porosity form by primary (biogenic) processes, but porosity can subsequently be modified by cementation and decementation (solution). Shrinkage porosity is formed primarily by secondary processes.

Choquette and Pray (1970) include modifying terms in their classification (Fig. 12.24) that allow expression of pore size, abundance, and origin. Although the genetic modifiers are useful, some are so subjective that they may be difficult to apply (e.g., epigenetic, mesogenetic, telogenetic).

12.6.3   Changes in Carbonate Porosity with Burial

As discussed in preceding sections of this chapter, carbonate sediments undergo various diagenetic modifications during burial that either reduce or enhance porosity. Mechanical compaction, chemical compaction, and cementation reduce porosity, whereas dissolution increases porosity. Uncompacted carbonate sediments may have initial porosities ranging from about 40 to 80 percent. By contrast, the final porosity of carbonate rocks may range from nearly zero to a few percent. Values of 5–15 percent are common in reservoir facies (Choquette and Pray, 1970). Clearly, the overall effect of burial diagenesis is to reduce the porosity of carbonate sediments to low levels. Deeply buried carbonate rocks that undergo late-stage uplift into the telogenetic zone may, of course, develop enhanced porosity owing to dissolution of carbonate grains or previously formed cements by chemically aggressive meteoric waters.

Several workers have constructed empirical porosity-depth curves on the basis of data derived from deep wells. See, for example, the porosity-depth curves for chalks in North America and Europe by Scholle (1977) and the porosity-depth plots for South Florida limestones by Schmoker and Halley (1982). These curves typically show a nonlinear, exponential trend of decreasing porosity with increasing depth. Typically, they show that initially high porosity values are reduced to values of about 10–20 percent at burial depth of about 3 km and to less than 5 percent at depths of about 6 km. The shapes of these curves are normally considered to be due mainly to mechanical and chemical compaction; however, many of the factors discussed in preceding sections of this chapter, especially cementation, must also affect the shapes of the curves.

James and Choquette (1987) generated a hypothetical porosity-depth curve that schematically illustrates how these various factors can affect porosity during burial (Fig. 12.25). Note that considerable deviation from a "normal" exponential curve (Curve 1) can occur as a result of meteoric zone cementation, overpressuring, and deep-burial dissolution. Curve 2 is a curve for chalk or lime mud subjected to early meteoric or possibly marine cementation, 3 is a curve that might be expected if solution is created at considerable depth, and 4 is a curve that might result from overpressuring.

Ricken (1987) reports that percent compaction of carbonate sediments (obtained by measurements of deformed burrows and fossil casts) plotted against percent carbonate in carbonate sediments shows a nonlinear trend of increasing compaction with decreasing carbonate content. Ricken uses this observed relationship to develop a method for calculating the original porosity of carbonate sediments on the basis of what he calls the **carbonate compaction law.** This law utilizes the principle that in a given calcareous sediment or rock sample the volume of the noncarbonate fraction (clays and silt-size

FIGURE 12.25

Hypothetical curves illustrating (1) a so-called "normal" porosity-depth relationship for fine-grained sediments with marine pore waters, (2) cementation in the meteoric zone (horizontal segments) alternating with burial in marine pore waters, (3) reversal of normal porosity-depth trend owing to dissolution in the deep subsurface, followed by resumption of normal burial, and (4) arrested porosity reduction owing to abnormally high pore pressure. *(From Choquette, P. W., and N. P. James, 1987, Diagenesis 12. Diagenesis in limestones—3. The deep burial environment: Geoscience Canada, v. 14. Fig. 33, p. 23, reprinted by permission of Geological Association of Canada.)*

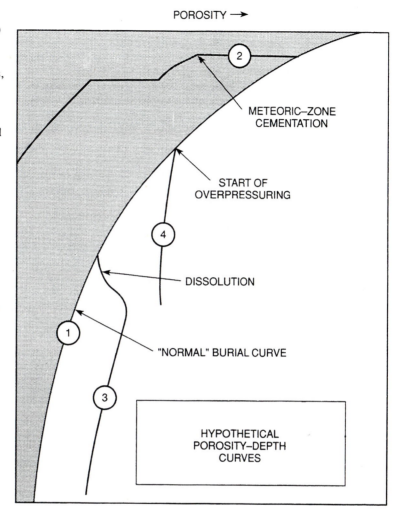

POROSITY →

METEORIC–ZONE CEMENTATION

START OF OVERPRESSURING

DISSOLUTION

"NORMAL" BURIAL CURVE

HYPOTHETICAL POROSITY–DEPTH CURVES

silicates) normalized to the primary decompacted sediment volume must essentially be constant during carbonate diagenesis. In other words, the volume of carbonate sediment decreases during compactive diagenesis, whereas the volume of noncarbonate sediment remains essentially constant. The present carbonate content in nonporous or nearly nonporous rocks depends only on the degree of compaction and the amount of noncarbonate contained in the primary sediment volume. Compaction can be calculated numerically by using the present carbonate content of the rock and the standardized noncarbonate content, which can be calculated from empirical measurements of the noncarbonate content. Ricken says that the compaction law can be used to evaluate the most important diagenetic variables that affect sediment rock transformations (e.g., mechanical compaction, chemical compaction, cementation). He goes on to derive decompaction formulas, which allow him to calculate from compacted carbonate rocks both the original volume of carbonate and the amount of original porosity. Anyone interested in this theoretical approach to evaluating porosity in carbonate sediments should see Ricken (1987) for details.

Additional Readings

Abu Dhabi National Reservoir Research Foundation, 1984, Stylolites and associated phenomena—Relevance to hydrocarbon reservoirs: Abu Dhabi National Reservoir Research Foundation Spec. Pub., 304 p.

Bathurst, R. G. C., 1975, Carbonate sediments and their diagenesis, 2nd ed.: Elsevier, Amsterdam, 658 p.

Bricker, O. P., ed., 1971, Carbonate cements: The Johns Hopkins Press, Baltimore and London, 376 p.

Choquette, P. W., and N. P. James, 1987, Diagenesis 12. Diagenesis in limestones—3. The deep burial environment: Geoscience Canada, v. 14, p. 3–35.

Hanford, C. R., R. G. Loucks, and S. O. Moshier, eds., 1989, Nature and origin of micro-rhombic calcite and associated microporosity in carbonate strata: Sed. Geology, v. 63, p. 187–325.

James, N. P., and P. W. Choquette, 1983, Diagenesis 6. Limestones—The sea floor diagenetic environment: Geoscience Canada, v. 10, p. 162–179.

——— 1984, Diagenesis 9. Limestones—The meteoric diagenetic environment: Geoscience Canada, v. 11, p. 161–194.

James, N. P., and P. W. Choquette, eds., 1988, Paleokarst: Springer-Verlag, New York, 416 p.

Moore, C. H., 1989, Carbonate diagenesis and porosity: Elsevier, Amsterdam, 338 p.

Schneidermann, N., and P. M. Harris eds., 1985, Carbonate cements: Soc. Econ. Paleontologists and Mineralogists Spec. Pub. 36, 379 p.

Scoffin, T. P., 1987, Carbonate sediments and rocks, Chapter 4: Diagenesis: Blackie, Glasgow, p. 90–145.

Tucker, M. E., and V. P. Wright, 1990, Carbonate sedimentology: Blackwell, Oxford, 482 p.

PART 4

Other Chemical/Biochemical and Carbonaceous Sedimentary Rocks

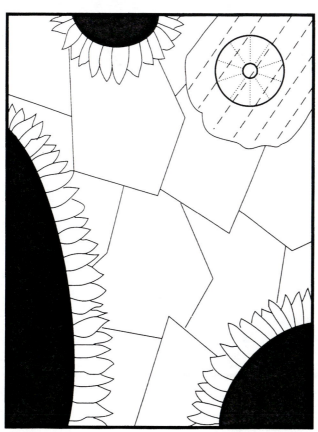

Chapter 13

Evaporites, Cherts, Iron-Rich Sedimentary Rocks, and Phosphorites

13.1 Introduction

This chapter deals with the chemical/biochemical rocks other than carbonates. Thus, it discusses evaporites, the siliceous sedimentary rocks (cherts), iron-rich sedimentary rocks (iron-formations and ironstones), and phosphorites. Volumetrically, these rocks are far less abundant than siliciclastic sedimentary rocks and carbonate rocks, although accurate estimates of their volume as a percentage of the total stratigraphic record are unavailable. If we add the percentages of shale, sandstone, and carbonate rocks calculated or measured by various authors such as Ronov et al. (1980), the unaccounted volume of total sedimentary rocks that can be attributed to these chemical/biochemical rocks appears to be no more than about 1–2 percent. The small volume of these rocks is not, however, a measure of their importance or the interest that we have in them. All of the sedimentary rocks discussed in this chapter have considerable economic significance. Evaporite deposits such as gypsum, halite (rock salt), and trona are mined for a variety of industrial and agricultural purposes, iron-rich sedimentary rocks are the source of most of our iron ores, phosphorites are extremely important sources of fertilizers and other chemicals, and the siliceous sedimentary rocks have some economic value, e.g., in the semiconductor industry.

In addition to their economic importance, all of the rocks discussed in this chapter are intrinsically interesting owing to their compositions and origins. The origin of many of these rocks is still enigmatic even after many years of study. For example, we are still unsure of the depositional mechanisms that account for the formation of the iron-rich sedimentary rocks, nor do we have a fully satisfactory explanation for the source or sources of the vast amount of iron locked up in these deposits. Also, there are many unanswered questions about the origin of phosphorites and the processes by which phosphorus becomes so highly concentrated in phosphate deposits. We understand somewhat better the origin of evaporites and cherts; nonetheless, we certainly do not have all the answers.

In this chapter we take a look at each of these interesting and economically significant kinds of sedimentary rocks. Space limitations do not permit analysis and discussion of each of these groups of rocks in the detail accorded the siliciclastic sedimentary rocks and carbonates. Accordingly, I do not place undue stress on the petrographic characteristics of these rocks. Nonetheless, I have tried to provide enough details about their mineralogy, chemical compositions, and other properties to give readers a reasonable feel for their principal characteristics. I also discuss some of the more interesting and enigmatic aspects of their origin, and alternate points of view about origin are presented where appropriate. An extensive list of additional readings is provided at the end of the chapter for those readers who wish to explore the properties of these rocks in greater detail.

13.2   Evaporites

13.2.1   Introduction

We use the term evaporites to include all those sedimentary rocks formed by evaporation of saline waters. Sedimentary rocks containing evaporite minerals are common in the geologic record. Evaporite deposits occur in rocks as old as early Precambrian, where they are preserved mainly as pseudomorphs rather than as the actual salt. Extensive accumulations of evaporite minerals are very common in Phanerozoic stratigraphic sequences. Evaporite deposition was especially widespread and important during the Late Cambrian, Permian, Jurassic, and Miocene. Lesser accumulations took place during the Silurian, Devonian, Triassic, and Eocene (Ronov et al., 1980). Although evaporites make up less than about 2 percent of the sediments deposited on the world's platforms during Phanerozoic time, evaporites were deposited quite rapidly when they were forming. As much as 100 m of evaporites could be deposited in about 1000 years when conditions were right (Schreiber and Hsü, 1980); this rate is two or three orders of magnitude higher than the rate at which most other shelf sediments are deposited. One of the thickest known deposits of evaporites is the Late Miocene Mediterranean Messinian evaporite sequence, which may exceed 2 km in thickness. The Messinian evaporites are believed to have been deposited in less than 200,000 years.

Although the bulk of ancient evaporite deposits were probably deposited under marine or marginal-marine conditions, evaporites were also deposited at various times in the past in nonmarine settings. Deposition in continental settings was particularly important during early phases of the Triassic-Jurassic rifting of Pangaea, when the basal deposits in the Atlantic and Gulf Coast basins, and in many other rift basins, were formed (Schreiber, 1988a). Evaporites are also forming today in many parts of the world where rates of evaporation exceed water input. Many of these areas lie in nonmarine settings within continental masses. Although evaporite deposition is most common in warm areas of the world, evaporites are forming at present in arid portions of the Arctic and Antarctic. On the other hand, evaporite deposition is much slower in colder regions than in warmer regions. Also, depositional rates for evaporites in nonmarine settings are slower than for those in marine-fed basins because the feed waters for nonmarine basins are initially very dilute (Shreiber, 1988a). Marine evaporites tend to be thicker and more laterally extensive than nonmarine evaporites and are generally of greater geologic interest.

Evaporite salts have been mined and used by humans for the last 6000 years (Warren, 1989, p. 2). Halite, gypsum, trona, and other salts are currently used for a variety of industrial and agricultural purposes. In addition to their commercial value, evaporites are associated with carbonate rocks in many major oil fields of the world. Squeezing and remobilization of salt deposits creates petroleum traps in association with salt domes, subsurface solution of evaporite cements in carbonate and siliciclastic rocks can create important amounts of secondary porosity in these rocks, and evaporite caprocks form seals over many petroleum traps that prevent the petroleum from escaping. This relationship between evaporites and hydrocarbons has been explored in at least two recent books (Schreiber, 1988b; Warren, 1989).

13.2.2 Composition

Major Minerals

Evaporite deposits are composed dominantly of varying proportions of halite (NaCl), anhydrite ($CaSO_4$) and gypsum ($CaSO_4 \cdot 2H_2O$), although numerous other minerals may be present in minor amounts. Gypsum is by far the most abundant calcium sulfate mineral in modern evaporite deposits (Dean, 1982); however, anhydrite is more abundant than gypsum in deposits buried to depths exceeding about 2000 ft (610 m) owing to dewatering of gypsum and conversion to anhydrite. Halite crystals occur in a variety of forms that include skeletal hopper crystals (cubes with depressed faces), pyramidal hopper crystals, and toothlike forms (Fig. 13.1; Shearman, 1982). Gypsum and anhydrite crystals also take a variety of forms that include single crystals, radial clusters of crystals, and numerous types of complex twinned crystals (Schreiber, 1982). An example of anhydrite crystals as they appear in thin section is illustrated in Figure 13.2. Evaporite deposits range from those composed dominantly of anhydrite and gypsum to those that are mainly halite; however, approximately 80 minerals in total have been reported from evaporite deposits (Stewart, 1963). About 40 of the more important minerals are listed, together with their chemical compositions, in Table 13.1. Of these 40 minerals, only about a dozen are common and abundant enough to be considered major rock formers. As mentioned, evaporites can form in both marine and nonmarine environments. Owing to differences in the dissolved mineral composition of the feed waters, different suites of minerals tend to form in the two environments.

Marine Evaporite Minerals

Because marine evaporites are precipitated from seawater, the mineral composition of marine evaporites in various deposits tends to be relatively constant. Seawater has an average salinity of about 35 parts per thousand (‰), and 12 elements are present in seawater in amounts greater than 1 ppm (0.001‰) (Braitsch, 1971). The major elements Cl^- (18.98‰), Na^+ (10.56‰), SO_4^{2-} (2.65‰), Mg^{2+} (1.27‰), Ca^{2+} (0.40‰), K^+ (0.38‰), and HCO_3^- (0.14‰) make up the bulk of the dissolved solids in seawater.

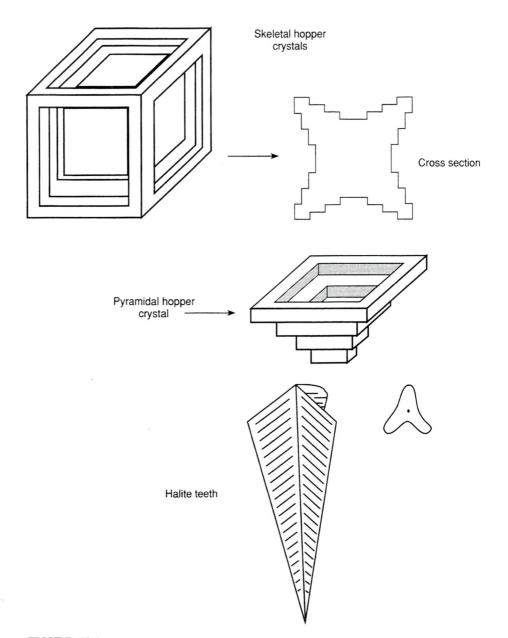

Skeletal hopper
crystals

Cross section

Pyramidal hopper
crystal

Halite teeth

FIGURE 13.1
Some halite crystal habits. A. Skeletal hopper crystal (cube with depressed faces). B. Pyramidal hopper crystal. C. Toothlike forms. *(After Shearman, D. J., 1982, Evaporites of coastal sabkhas, in Dean, W. E., and B. C. Schreiber, eds., Marine evaporites: Soc. Econ. Paleontologists and Mineralogists Short Course 4. Figs., p. 32 and 35, reprinted by permission of SEPM, Tulsa, Okla.)*

FIGURE 13.2

Large anhydrite crystals (A) in a fine-crystalline dolomite (note dolomite rhombs), Beaver Hill Lake Formation (Devonian), Alberta, Canada. Crossed nicols. Scale bar = 0.5 mm.

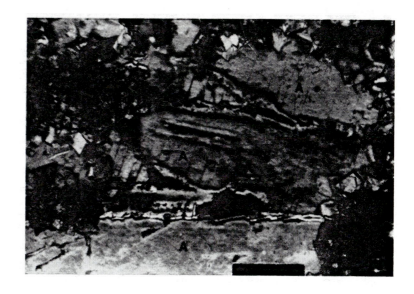

TABLE 13.1

Chemical composition of evaporite minerals deposits

Mineral	Composition	Mineral	Composition
anhydrite	$CaSO_4$	leonhardtite	$MgSO_4 \cdot 4H_2O$
aphthitalite		leonite	$MgSO_4 \cdot K_2SO_4 \cdot 4H_2O$
(glaserite)	$K_2SO_4 \cdot (Na,K)SO_4$	loewite	$2MgSO_4 \cdot 2Na_2SO_4 \cdot 5H_2O$
antarcticite	$CaCl_2 \cdot 6H_2O$	magnesian	
aragonite	$CaCO_3$	calcite	$(Mg_xCa_{1-x})CO_3$
bassanite	$CaSO_4 \cdot \frac{1}{2}H_2O$	mirabilite	$Na_2SO_4 \cdot 10H_2O$
bischofite	$MgCl_2 \cdot 6H_2O$	nahcolite	$NaHCO_3$
bloedite		natron	$Na_2CO_3 \cdot 10H_2O$
(astrakanite)	$Na_2SO_4 \cdot MgSO_4 \cdot 4H_2O$	pentahydrite	$MgSO_4 \cdot 5H_2O$
burkeite	$Na_2CO_3 \cdot 2Na_2SO_4$	pirssonite	$CaCO_3 \cdot Na_2CO_3 \cdot 2H_2O$
calcite	$CaCO_3$	polyhalite	$2CaSO_4 \cdot MgSO_4 \cdot K_2SO_4 \cdot 2H_2O$
carnallite	$MgCl_2 \cdot KCl \cdot 6H_2O$	rinneite	$FeCl_2 \cdot NaCl \cdot 3KCl$
dolomite	$CaCO_3 \cdot MgCO_3$	sanderite	$MgSO_4 \cdot 2H_2O$
epsomite	$MgSO_4 \cdot 7H_2O$	schoenite	
gaylussite	$CaCO_3 \cdot Na_2CO_3 \cdot 5H_2O$	(picro-	
glauberite	$CaSO_4 \cdot Na_2SO_4$	merite)	$MgSO_4 \cdot K_2SO_4 \cdot 6H_2O$
gypsum	$CaSO_4 \cdot 2H_2O$	shortite	$2CaCO_3 \cdot Na_2CO_3$
halite	$NaCl$	sylvite	KCl
hanksite	$9Na_2SO_4 \cdot 2Na_2CO_3 \cdot KCl$	syngenite	$CaSO_4 \cdot K_2SO_4 \cdot H_2O$
hexahydrite	$MgSO_4 \cdot 6H_2O$	tachyhydrite	$CaCl_2 \cdot 2MgCl_2 \cdot 12H_2O$
kainite	$MgSO_4 \cdot KCl \cdot 1\frac{1}{4}H_2O$	thenardite	Na_2SO_4
kieserite	$MgSO_4 \cdot H_2O$	thermonatrite	$Na_2CO_3 \cdot H_2O$
langbeinite	$2MgSO_4 \cdot K_2SO_4$	trona	$NaHCO_3 \cdot Na_2CO_3 \cdot 2H_2O$
		van'thoffite	$MgSO_4 \cdot 3Na_2SO_4$

Source: Hardie, L. A., 1984, Evaporites: Marine or non-marine? Am. Jour. Science, v. 284. Appendix, p. 233, reprinted by permission.

Trace elements present in amounts ranging from about 65 ppm to 1 ppm include Br^-, Sr^{2+}, B, F^-, and H_4SiO_4. Excluding the carbonate minerals, most of which are not evaporites, the most common marine evaporite minerals are the calcium sulfate minerals gypsum and anhydrite. Halite is next in abundance, followed by the potash salts (sylvite, carnellite, langbeinite, polyhalite, kainite) and the magnesium sulfate kieserite. The common marine evaporite minerals can be grouped on the basis of chemical composition into chlorides, sulfates, and carbonates (Table 13.2). Note that the anions (Cl^-, SO_4^{2-}, CO_3^{2-}) are combined with Na^+, K^+, Mg^{2+}, or Ca^{2+} to form the various minerals. Evaporite deposits may also contain various amounts of impurities such as clay minerals, quartz, feldspar, or sulfur.

Nonmarine Evaporites

Nonmarine evaporites form from waters that were originally river water or groundwater. The chemistries of these original waters can be highly variable, depending upon the lithology of the rocks with which they interact. For example, rivers that flow across limestones are commonly enriched in Ca^{2+} and HCO_3^-, whereas those that flow across igneous and metamorphic rocks tend to be enriched in silica, Ca^{2+}, and Na^+. Because of these differences in original water chemistry, more-diverse suites of minerals typify nonmarine evaporite deposits than typify marine deposits (Harvie et al., 1982). Thus, many nonmarine deposits contain evaporite minerals that are not common in marine evaporites. These minerals may include trona [$Na_3H(CO_3)_2 \cdot 2H_2O$), mirabilite [$Na_2SO_4 \cdot 10H_2O$], glauberite [$Na_2Ca(SO_4)$], borax [$Na_2B_4O_5(OH)_4 \cdot 8H_2O$], epsomite [$MgSO_4 \cdot 7H_2O$], thenardite [$NaSO_4$], gaylussite [$Na_2CO_3 \cdot CaCO_3 \cdot 5H_2O$], and bloedite [$Na_2SO_4 \cdot MgSO_4 \cdot 4H_2O$]. On the other hand, nonmarine deposits may also contain anhydrite, gypsum, and halite and may even be composed dominantly of these minerals.

TABLE 13.2
Classification of marine evaporites on the basis of mineral composition

Mineral class	Mineral name	Chemical composition	Rock name
Chlorides	Halite	NaCl	Halite; rock salt
	Sylvite Carnallite	KCl $KMgCL_3 \cdot 6H_2O$	Potash salts
Sulfates	Langbeinite Polyhalite Kainite	$K_2Mg_2(SO_4)_3$ $K_2Ca_2Mg(SO_4)_4 \cdot 2H_2O$ $KMg(SO_4)Cl \cdot 3H_2O$	
	Anhydrite Gypsum Kieserite	$CaSO_4$ $CaSO_4 \cdot 2H_2O$ $MgSO_4 \cdot H_2O$	Anhydrite; anhydrock Gypsum; gyprock —
Carbonates	Calcite Magnesite Dolomite	$CaCO_3$ $MgCO_3$ $CaMg(CO_3)_2$	Limestone — Dolomite; dolostone

Source: Data from Stewart, F. H., 1963, Marine evaporites, *in* Fleischer, M., ed., Data of geochemistry: U.S. Geol. Survey Prof. Paper 440-Y; Borchert, H., and R. O. Muir, 1964, Salt deposits: The origin, metamorphism, and deformation of evaporites: Van Nostrand, London.

Therefore, it may not always be easy to distinguish between marine and nonmarine evaporites on the basis of mineralogy.

13.2.3   Classification of Evaporites

Evaporites can be classified informally into marine and nonmarine types on the basis of origin. They can be further classified as chlorides, sulfates, and carbonates on the basis of mineralogy, as shown in Table 13.2. Other than these simple, informal classifications, no general classification scheme for evaporite rocks as a whole appears to exist. Only a few rock names have been given to evaporite deposits, and these names are applied on the basis of the dominant mineral in the deposits. Rocks composed mainly of halite are called halite or **rock salt.** Rocks made up dominantly of gypsum or anhydrite are simply called gypsum or anhydrite, although some geologists use the names **rock gypsum** or **rock anhydrite.** Less commonly they are called gyprock and anhydrock. Few evaporite beds are composed dominantly of minerals other than the calcium sulfates and halite. No formal names have been proposed for rocks enriched in other evaporite minerals, although the term potash-salts is used informally for potassium-rich evaporites.

As indicated, most evaporite deposits consist of either anhydrite/gypsum or halite. I know of no formal scheme for subdividing halite (rock salt) into subtypes; however, Maiklem et al. (1969) proposed a useful scheme for the structural classification of anhydrites. Many anhydrites are characterized by distorted fabrics that result from volume changes owing to dehydration of gypsum and rehydration of anhydrite during diagenesis. Maiklem et al. (1969) proposed a structural classification for anhydrite on the basis of fabric and bedding. This classification divides anhydrites into about two dozen structural types, which can be lumped into three fundamental structural groups: nodular anhydrites, laminated anhydrites, and massive anhydrites.

Nodular anhydrite consists of irregularly shaped lumps of anhydrite that are partly or completely separated from each other by a salt or carbonate matrix (Fig. 13.3). **Mosaic anhydrite** is a type of nodular anhydrite in which the anhydrite masses or lumps are approximately equidimensional and are separated by very thin stringers of dark carbonate mud or clay. The formation of nodular anhydrite begins by displacive growth of gypsum in carbonate or clayey sediments. Subsequently, gypsum crystals alter to anhydrite pseudomorphs, which continue to enlarge by addition of Ca^{2+} and SO_4^{2-} from an external source. This displacive growth ultimately results in segregation of the anhydrite into nodular masses. **Chickenwire structure** is a term used for a particular type of mosaic or nodular anhydrite that consists of slightly elongated, irregular polygonal masses of anhydrite separated by thin dark stringers of other minerals such as carbonate or clay minerals (Fig. 13.4). This structure apparently forms when growing nodules ultimately coalesce and interfere. Most of the enclosing sediment is pushed aside and what remains forms thin stringers between the nodules (Shearman, 1982).

Laminated anhydrites, sometimes called laminites, consist of thin, nearly white anhydrite or gypsum laminate that alternate with dark gray or black laminae rich in dolomite or organic matter (Fig. 13.5). The laminae are commonly only a few millimeters thick and rarely attain a thickness of 1 cm. Many thin laminae are remarkably uniform, with sharp planar contacts that can be traced laterally for long distances. Some are reported to be traceable for distances of 290 km (Anderson and Kirkland, 1966). Laminites may comprise vertical sequences hundreds of meters thick in which hundreds

FIGURE 13.3
Nodular anhydrite in a core sample of the Buckner Anhydrite (Jurassic), Texas. Dark-colored carbonate separates and surrounds the lighter colored anhydrite nodules.

of thousands of laminae are present. Laterally persistent laminae are believed to form by precipitation of evaporites in quiet water below wave base. They could presumably form either in a shallow-water area protected in some manner from strong bottom currents and wave agitation or in a deeper-water environment. Alternating light and dark pairs of bands in laminated evaporites may represent annual varves resulting from seasonal changes in water chemistry and temperature, or they may represent cyclic changes or disturbances of longer duration. Laminae of anhydrite can occur with thicker layers of halite, producing laminated halite.

Some laminated anhydrite may form by coalescence of growing anhydrite nodules, which expand laterally until they merge into a continuous layer. Layers formed by this mechanism are thicker, less distinct, and less continuous than laminae formed by precipitation. A special type of contorted layering that has resulted from coalescing nodules has been observed in some modern sabkha deposits where continued growth of nodules creates a demand for space. The lateral pressures that result from this demand cause the layers to become contorted, forming ropy bedding or **enterolithic structures** (Fig. 13.6).

Massive anhydrite is anhydrite that lacks perceptible internal structures. True massive anhydrite is less common than nodular and laminated anhydrite, and its origin is poorly understood. Presumably, it represents sustained, uniform conditions of deposition. Haney and Briggs (1964) suggest that massive anhydrite forms by evaporation at brine

FIGURE 13.4
Chickenwire structure in anhydrite.
Evaporite series of the Lower Lias
(Jurassic), Aquitaine Basin,
southwest France. *(From
Bouroullec, J., 1981, Sequential
study of the top of the evaporitic
series of the Lower Lias in a well
in the Aquitaine Basin (Auch 1),
southwestern France, in Chambre
Syndical de la Recherche et de la
Production due Pétrole et du Gaz
Naturel, eds., Evaporite deposits:
Illustration and interpretation of
some environmental sequences. Pl.
36, p. 157, reprinted by
permission of Editions Technip,
Paris, and Gulf Publishing Co.,
Houston, Tex. Photography
courtesy of J. Bouroullec.)*

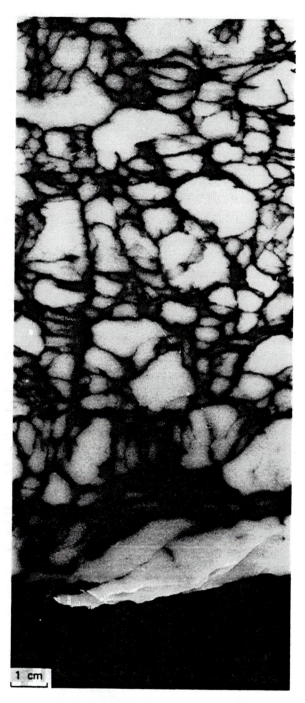

1 cm

FIGURE 13.5
Laminated anhydrite from the
Prairie Evaporite (Devonian),
Canada.

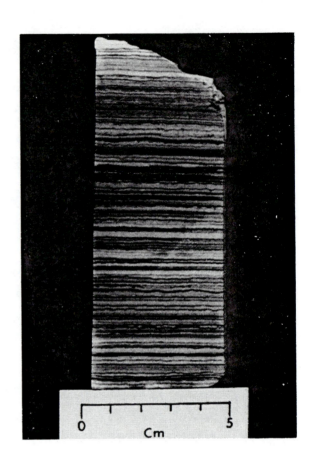

FIGURE 13.6
Enterolithic structure in nodular
gypsum of the Grenada Basin,
southern Spain. *(Photograph
courtesy of J. M. Rouchy.)*

salinities of approximately 200–275‰, just below the salinities at which halite begins to precipitate.

13.2.4   Deposition of Evaporites

Evaporation Sequence

Ocean water has an average salinity of about 35‰. When ocean water is evaporated in the laboratory, evaporite minerals are precipitated in a definite sequence that was first demonstrated by Usiglio in 1848 (reported in Clarke, 1924). Minor quantities of carbonate minerals begin to form when the original volume of seawater is reduced by evaporation to about one-half and brine concentration is about twice that of seawater. Gypsum appears when the original volume has been reduced to about 20 percent. At this point, the brine has a concentration of about four to five times that of normal seawater (130–160‰). Halite begins to form when the water volume reaches approximately 10 percent of the original volume, or 11 to 12 times the brine concentration of seawater (340–360‰). Magnesium and potassium salts are deposited when less than about 5 percent of the original volume of water remains; at that point, brine concentration may be more than 60 times that of seawater.

The same general sequence of evaporite minerals occurs in natural evaporite deposits, although many discrepancies exist between the theoretical sequences predicted on the basis of laboratory experiments and the sequences actually observed in the rock record. In general, the proportion of $CaSO_4$ (gypsum and anhydrite) is greater and the proportion of late-stage Na-Mg-K sulfates and chlorides is less in natural deposits than predicted from theoretical considerations (Borchert and Muir, 1964). This early-sulfate excess and late-sulfate deficiency is commonly attributed to incomplete cycles of evaporation, with brines containing most of the dissolved Na, Mg, and K continually being refluxed out of the basin. Diagenesis may also contribute to differences between theoretical and observed successions of evaporites. For more-extended discussion of this subject, see Dean (1982) and Hardie (1984).

Many marine evaporite deposits are quite thick, some exceeding 2 km, yet it has long been recognized that evaporation of a column of seawater 1000 m thick will produce only about 14 to 15 m of evaporites. Evaporation of all the water of the Mediterranean Sea, for example, would yield a mean thickness of evaporites of only about 60 m. Obviously, special geologic conditions operating over a long period of time are required to deposit thick sequences of natural evaporites. The basic requirements for deposition of marine evaporites are a relatively arid climate, where rates of evaporation exceed rates of precipitation, and partial isolation of the depositional basin from the open ocean. Isolation is achieved by means of some type of barrier that restricts free circulation of ocean water into and out of the basin. Under these restricted conditions, the brines formed by evaporation are prevented from returning to the open ocean, allowing them to become concentrated to the point where evaporite minerals are precipitated. Nonmarine evaporites form, of course, under different conditions. Nonetheless, they also require strongly evaporative conditions and a plentiful supply of water for thick deposits to form.

Physical Processes in Deposition of Evaporites

Although we tend to regard evaporite deposits as simply the products of chemical precipitation owing to evaporation, many evaporite deposits are not just passive chemical precipitates. The evaporite minerals have, in fact, been transported and reworked in the same way as the constituents of siliciclastic deposits and carbonate deposits. Transport can occur by normal fluid-flow processes or by mass-transport processes such as slumps and turbidity currents. Therefore, evaporite deposits may display clastic textures, including both normal and reverse size grading, and various types of sedimentary structures such as cross-bedding and ripple marks.

Depositional Models for Evaporite

Dominance of Modern Subaerial and Shallow-Water Environments.   Modern evaporites are accumulating in a variety of nonmarine and marginal-marine settings, as indicated in Figure 13.7. Note from this figure that modern evaporites form in sabkhas, salinas, and interdune environments. All of these settings are in subaerial or shallow subaqueous environments. This subaerial to very shallow-water origin of modern evaporites stands in sharp contrast to the apparent environment of many ancient evaporites. Many thick sequences of ancient marine evaporites appear to have formed in laterally extensive, shallow- to deep-water basins, for which there are no modern equivalents. Thus, the principle of uniformitarianism cannot be strictly applied in the case of evaporites because the present range of evaporite environments is clearly not an adequate guide to evaporite environments of the past.

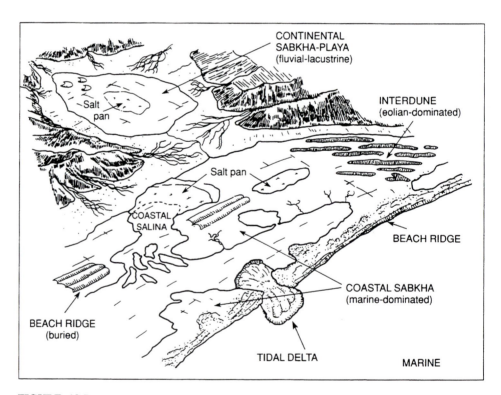

FIGURE 13.7
Principal settings in which modern evaporite deposits are accumulating. *(From Kendall, A. C., 1984, Evaporites, in Walker, R. G., ed., Facies models: Geoscience Canada Reprint Ser. 1. Fig. 1, p. 260, as modified slightly by Warren, 1989, reprinted by permission of Geological Association of Canada.)*

Subaerial Evaporites.   Many modern evaporites accumulate in **sabkhas,** or salt flats. Sabkhas are for the most part subaerial mud flats in which evaporites form, although they may be covered at times by ephemeral shallow water. Modern sabkhas have been described in many parts of the world but are particularly well known in the Persian Gulf. Warren (1989, p. 38) defines sabkhas as "marine and continental mudflats where displacive and replacive evaporite minerals are forming in the capillary zone above a saline water table." According to Hanford (1981), sabkhas can occur in both continental settings (nonmarine sediments and continental groundwaters) and marginal-marine settings (marine sediments and mainly marine-derived groundwaters). In continental environments, they occur in inland areas in fluvial-lacustrine (playa)–dominated settings and in eolian-dominated, interdune settings (Fig. 13.7). In the marginal-marine environment, they occur as coastal sabkhas in the intertidal and supratidal zones. Sabkhas commonly do not consist entirely or even dominantly of evaporites. Sabkha sediments are composed of a combination of evaporite minerals and carbonate and/or terrigenous clastic sediments. Evaporite minerals may form both at the surface of sabkhas and displacively within sabkha sediments, creating distorted fabrics as mentioned. Figure 13.8 shows some typical characteristics of sabkha sediments based on sabkhas in the Arabian Gulf.

Shallow Subaqueous Evaporites.   Shallow subaqueous evaporites accumulate in the marginal-marine environment in coastal lakes called **salinas** (Warren and Kendall, 1985). Modern salinas are particularly common in southern and western Australia, but they occur also around the margins of the Mediterranean, Black, and Red seas. Salinas typically occur in carbonate environments, commonly in depressions within coastal carbonate dunes or in association with carbonate reefs. Gypsum appears to be the most common evaporite mineral in most salinas; however, some (e.g., Lake Macleod, Australia) are filled with halite. Figure 13.9 shows typical evaporite facies in the salinas of southern Australia. The stratigraphic record suggests that many ancient shallow subaqueous evaporites were deposited on broad platforms that were far more extensive laterally than any evaporite environments known today. Thus, modern salinas are probably not adequate models for these ancient platform evaporites. Subaqueous evaporites occur also in lacustrine settings in continental basins. Lake Magadi in the African Rift is probably the best-documented modern example. Trona ($NaHCO_3 \cdot Na_2CO_3 \cdot 2H_2O$) and sodium silicate minerals such as magadiite [$NaSi_7O_{13}(OH)_3 \cdot H_2O$] are characteristic deposits of this lake.

Deep-Water Evaporites.   With the possible exception of the Dead Sea in the Middle East, no modern examples of deep-water evaporite basins exist. Therefore, models for deep-water evaporites are based on theoretical considerations and study of presumed ancient deep-water evaporites. Deep-water evaporites are putatively characterized by thin bedding and lamination and lateral continuity of beds and laminae. Strata are composed predominantly of laminar evaporitic carbonate, sulfate, and halite. They occur in sections tens to hundreds of meters thick that can be correlated over tens to hundreds of kilometers (Schreiber et al., 1986). Although most deep-water evaporites probably formed mainly by *in situ* precipitation, some appear to be turbidites. These evaporite turbidites are composed dominantly of gypsum or carbonate and gypsum, although rare halite turbidites have also been reported. Evaporite turbidites occupy the deepest portions of ancient evaporite basins. They apparently formed as a result of rapid precipitation in the shallow areas of evaporite basins, resulting in unstable marginal accumulations that were subsequently redeposited downslope by turbidity currents.

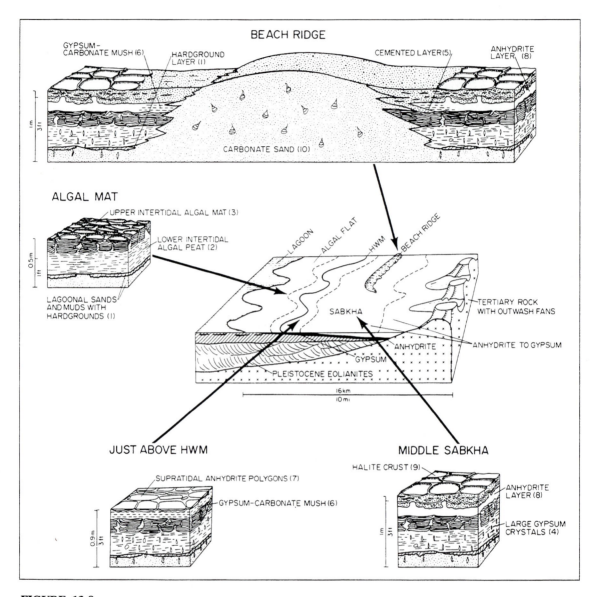

FIGURE 13.8

Schematic representation of vertical and lateral facies relationships in sabkhas of the Arabian Gulf. HWM = high-water mark. *(From Warren, J. K., and G. St. C. Kendall, 1985, Comparison of sequences formed in marine sabkha (subaerial) and salina (subaqueous) settings—modern and ancient: Am. Assoc. Petroleum Geologists Bull., v. 69. Fig. 2, p. 1015, reprinted by permission of AAPG, Tulsa, Okla.)*

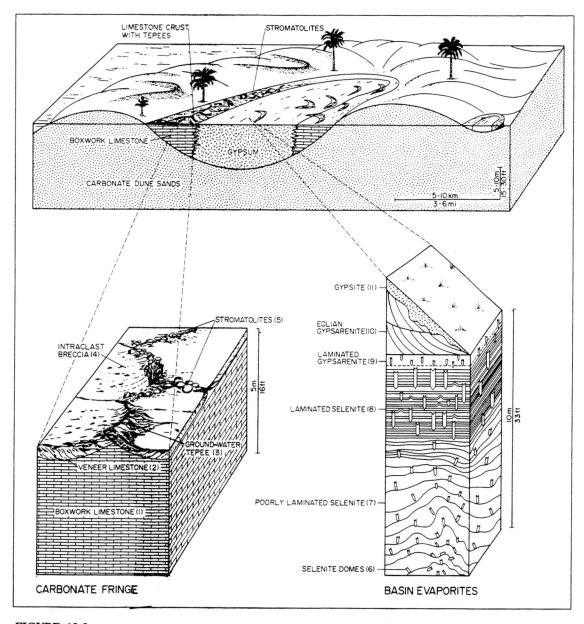

FIGURE 13.9
Environmental setting and typical evaporite facies in southern Australia salinas. *(From Warren, J. K., and G. St. C. Kendall, 1985, Comparison of sequences formed in marine sabkha (subaerial) and salina (subaqueous) settings—modern and ancient: Am. Assoc. Petroleum Geologists Bull., v. 69. Fig. 4, p. 1017, reprinted by permission of AAPG, Tulsa, Okla.)*

Shallow-Water vs. Deep-Water Evaporites.   As mentioned, most ancient evaporite deposits appear to have accumulated in platform or basinwide settings that extended for tens to hundreds of kilometers. Basinwide evaporites are thick, and they contain textural evidence indicating deposition in a variety of environments ranging from deep-water to platform to continental (Warren, 1989, p. 33). Some textural and structural differences in continental, coastal sabkha, and subaqueous evaporites are illustrated in Figure 13.10. These textural criteria should be used with caution, however, because similar kinds of features can form in different environments. For example, nodular anhydrites have been observed in many modern coastal sabkhas; therefore, many geologists assume that nodular anhydrites found in ancient evaporite deposits also signify deposition under subaerial, sabkha conditions. Dean et al. (1975) point out, however, that nodular anhydrites and laminated anhydrites, which both form in standing water, occur together in some environments. This association is taken to mean that nodular anhydrites can also form in deeper-water environments. Apparently, all that is needed for formation of nodular anhydrite is growth of crystals in mud in contact with highly saline brines, which can occur in deep or shallow standing water as well as in a sabkha environment. Also, note from Figure 13.10 that laminated evaporites can form in both shallow and deep water.

Kendall (1984) suggests that ancient basins containing thick sequences of bedded evaporites are of three subtypes (Fig. 13.11). The **deep-water, deep-basin model** for ancient basin evaporites assumes existence of a deep basin separated from the open ocean by some type of sill. The sill acts as a barrier to prevent free interchange of water in the basin with the open ocean but allows enough water into the basin to replenish that lost by evaporation. Seaward escape of some brine allows a particular concentration of brine to be maintained for a long time, leading to thick deposits of certain evaporite minerals such as gypsum. The **shallow-water, shallow-basin model** assumes concentration of brines in a shallow, silled basin but allows for accumulation of great thicknesses of evaporites owing to continued subsidence of the floor of the basin. The **shallow-water, deep-basin model** requires that the brine level in the basin be reduced below the level of the sill, a process called **evaporative drawdown,** with recharge of water from the open ocean taking place only by seepage through the sill or by periodic overflow of the sill. Total desiccation of the floors of such basins could presumably occur periodically, allowing the evaporative process to go to completion, thereby depositing a complete evaporite sequence, including magnesium and potassium salts.

Application of these basin models to interpretation of ancient evaporite deposits is a challenging task. Different interpretations about ancient deposits are likely to be made by different geologists—and at different times. A recent Penrose Conference report (Sonnenfeld and Kendall, 1989) suggests that concepts of evaporite models have swung away from deep-water to shallow-water deposition, and the pendulum has again swung away from the tidal sabkha regime to very moderate depths.

13.2.5   *Diagenesis of Evaporites*

Evaporites are particularly prone to burial alteration and undergo a variety of diagenetic modifications that may include dehydration of gypsum and rehydration of anhydrite, dissolution, cementation, recrystallization, replacement, calcitization of sulfates owing to bacterial processes, and deformation (Hardie, 1984; Schreiber, 1988c; Warren, 1989, p. 9). The conversion of gypsum to anhydrite and back again is particularly important. With burial, gypsum is transformed to anhydrite at temperatures above 60°C, with a volume

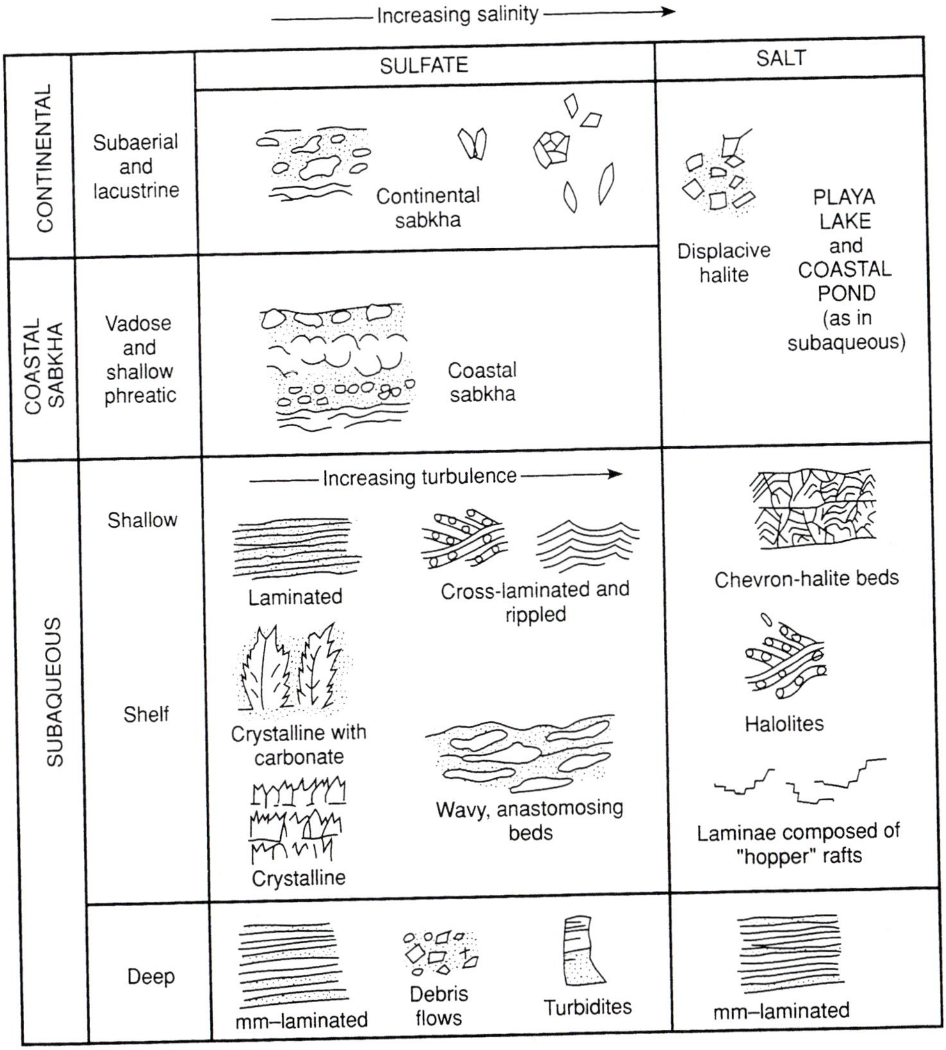

FIGURE 13.10
Summary of evaporite textures and structures characteristic of major evaporite environments. *(After Kendall, A. C., Evaporites, in Walker, R. G., ed., Facies models: Geoscience Canada Reprint Ser. 1. Fig. 2, p. 261, with subsequent modifications by Schreiber et al., 1976, and Warren, 1989, reprinted by permission of Geological Association of Canada.)*

loss in water of about 38 percent. Exhumation of buried anhydrite results in rehydration with accompanying increase in volume. These volume changes and hydration and rehydration reactions account for a considerable amount of the deformation in evaporites (development of nodules, enterolithic structures, etc.), as discussed. Owing to their low yield strengths, evaporites respond to burial and tectonic pressures by plastic deformation. Deformation may take place as a result of pressure solution, by the formation of spaced cleavage, or by folding and diapirism (Schreiber, 1988c). The scale of deformation can range from millimeter-scale ptygmatic folds to kilometer-scale salt diapirs.

FIGURE 13.11
Schematic diagram illustrating three models for deposition of marine evaporites in basins where water circulation is restricted by the presence of a topographic sill. *(From Kendall, A. C., 1979, Subaqueous evaporites, in Walker, R. G., ed., Facies models: Geoscience Canada Reprint Ser. 1. Fig. 17, p. 170, reprinted by permission of Geological Association of Canada.)*

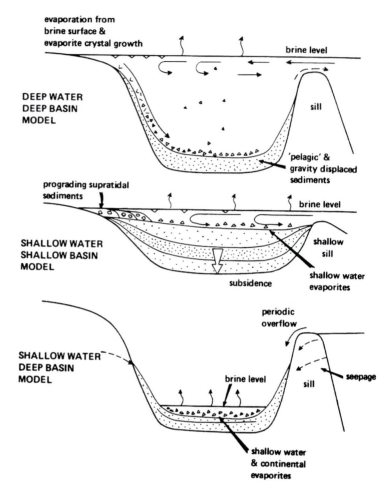

Although considerable diagenesis of evaporite deposits may take place at moderate burial depths, some diagenetic changes can occur very early during diagenesis and at very shallow depths. For example, Casas and Lowenstein (1989) report that dissolution and cementation reactions in some modern California and Mexico saline-pan halites are essentially complete within the first 45 m of burial. Also, Hussain and Warren (1989) describe the early diagenetic formation of nodular and enterolithic gypsum in the vadose zone (unsaturated groundwater zone above the water table) in playa deposits of West Texas. Diagenetic processes include dissolution and replacement of gypsum by halite or dolomite at depths of only 1–2 m, early-diagenetic enlargement and coalescence of gypsum crystals to create nodules within laminated host sediment, diffuse growth of nodules in the lower vadose zone to create chickenwire textures, and growth of gypsum in the upper vadose zone to form enterolithic bands.

13.2.6   Ancient Evaporite Deposits

As mentioned, evaporite deposits are common in the rock record and are particularly abundant in stratigraphic sequences of Late Cambrian, Permian, Jurassic, and Miocene ages. As also mentioned, most ancient evaporites appear to be of marine origin. They accumulated on a vertical and lateral scale far exceeding anything known in modern evaporite deposits. Many of these extensive platform and basinwide ancient evaporite sequences have been intensively studied in sediment cores and in mine exposures. Particularly well-studied examples of ancient basinal evaporites include the Miocene Messinian of the Mediterranean region, exceeding 2 km in thickness; the Permian Zechstein of the North Sea area, also exceeding 2 km in thickness; Upper Silurian evaporites in the Michigan Basin (USA) exceeding 600 m in thickness; and evaporite deposits of the Permian Delaware Basin of West Texas and New Mexico that exceed 1 km in thickness. Evaporite deposits of the Delaware Basin are especially characterized by their striking lateral continuity.

Although not as common and abundant as marine evaporites, ancient nonmarine evaporites are also recognized. Putative ancient sabkha evaporites have been reported from many formations, including the Permian San Andreas Formation of West Texas, the Permian Lower Clear Fork Formation in the Palo Duro Basin of the Texas Panhandle, the Ordovician Red River Formation in the Williston Basin along the U.S.–Canadian border, the Permian Upper Minnelusa Formation in Wyoming, and formations in rift basins such as those of the East African and Baikal rifts. The literature on ancient evaporite deposits is voluminous. Readers who wish more information about ancient evaporites should consult the list of additional readings at the end of this chapter. These books contain extensive lists of references to published articles on evaporites. *Evaporite Sedimentology* by J. K. Warren (1989) is a particularly useful place to begin.

13.3   Siliceous Sedimentary Rocks (Cherts)

13.3.1   Introduction

Siliceous sedimentary rocks are fine-grained, dense, commonly very hard rocks composed dominantly of the SiO_2 minerals quartz, chalcedony, and opal. Most siliceous rocks also contain minor amounts of impurities such as clay minerals, hematite, calcite, dolomite, and organic matter. **Chert** is the general group name used for siliceous sedimentary rocks. Cherts are common but not abundant rocks in the geologic record. Overall, they probably make up less than 1 percent of all sedimentary rocks, but they are represented in stratigraphic sequences ranging in age from Precambrian to Quaternary. Also, they are forming today as siliceous oozes in some parts of the modern ocean. Most ancient chert occurs as bedded deposits in association with generally deep-water sediments such as pelagic shales and turbidites. Bedded cherts are common in ophiolite and subduction complexes, and many bedded chert sequences are quite thick. Chert occurs also as nodules and stringers in shallow-water limestones of all ages, where it forms diagenetically as a replacement for carbonate minerals. Rocks such as shales that are highly cemented with silica are also a kind of siliceous sedimentary rock; however, these rocks are not considered in this discussion.

In this section, we examine the mineralogy and chemical composition of cherts, briefly discuss the principal kinds of cherts, consider some of the problems involved in their origin, and investigate chert diagenesis. The origin, diagenesis, and common occurrence of bedded cherts in ancient, so-called geosynclinal and subduction complexes is of great interest to geologists concerned with aspects of Earth history such as paleogeography, paleoceanographic circulation patterns, and plate tectonics. Siliceous rocks also have some economic significance. Silicon is used in the semiconductor and computer industries, and it is used also for making glass and related products such as fire bricks. Furthermore, siliceous deposits occur in association with important economic deposits of other minerals, including Precambrian iron ores, uranium deposits, manganese deposits, and phosphorite deposits. Many petroleum deposits also occur in association with siliceous rocks, which may be source rocks and possibly even reservoir rocks for petroleum. Developing a better understanding of the origin of chert may help to understand the origin of these other, economically significant, deposits.

13.3.2   Mineralogy and Texture

Quartz is the primary mineral of siliceous deposits; however, other SiO_2 minerals in these deposits can include chalcedonic quartz, amorphous silica (opal-A), and disordered cristobalite and tridymite (opal-CT). Opal-CT is low-temperature cristobalite disordered by interlayered tridymite lattices (terminology of Jones and Segnit, 1971). As indicated, we commonly use the group name chert to cover all rocks formed of these minerals. Opal-A is primarily of biogenic origin and forms the tests of siliceous plankton and the spines of some sponges. Skeletal opal A is metastable and converts in time to opal-CT and finally quartz. Nonetheless, unaltered opal-A organic remains are present in some cherts, particularly those of Cenozoic age, suggesting a biogenic origin for these cherts. All gradations may be present in chert deposits—from pure opal to pure quartz chert, depending upon the age of the deposits and the conditions of burial.

Texturally, the quartz that forms cherts can be divided into three main types: (1) **microquartz,** consisting of nearly equidimensional grains of quartz less than about 20 μm in size (Fig. 13.12), (2) **megaquartz,** composed of equant to elongated grains greater than 20 μm (Fig. 13.12), and (3) **chalcedonic quartz,** forming sheaflike bundles of radiating, extremely thin crystals about 0.1 mm long (Fig. 13.13; Folk, 1974). In most chalcedonic quartz, the elongated fibers have negative elongation, with the c-axis being perpendicular to the length of the fibers, and the crystals are said to be length-fast. In some chalcedony, however, the c-axes are parallel to the fibers, and the chalcedony is thus length-slow. Folk and Pittman (1971) found a strong correlation between length-slow chalcedony and the

FIGURE 13.12
Fine-textured, nearly equigranular microquartz (chert) (CH) cut by a vein of much coarser megaquartz (Q). Source of specimen unknown. Crossed nicols. Scale bar = 0.1 mm.

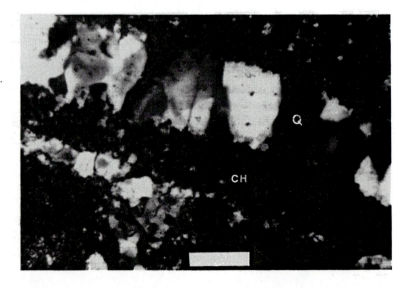

FIGURE 13.13
Fibrous, chalcedonic quartz that formed in a cavity within microquartz (upper right part of photograph). Source of specimen unknown. Crossed nicols. Scale bar = 0.1 mm.

presence of evaporites or former evaporites. The chalcedony can occur as pseudomorphs of gypsum and anhydrite or in nodules that closely resemble the chickenwire structure of nodular evaporites. Folk and Pittman point out, however, that not all chalcedony in evaporites is length-slow and also that some length-slow chalcedony has been found in basic igneous rocks. Therefore, length-slow chalcedony is not an infallible indicator of vanished evaporites. More recently, Hattori (1989) reports length-slow chalcedony from the Mino Terrane in Japan. In this terrane, it occurs as vein minerals or replaces carbonates that Hattori interprets to have formed in an evaporative environment. Hattori cautions, however, that length-slow chalcedony can be confused in thin section with length-slow colorless chlorite or flamboyant quartz. He recommends that it be identified by determining the chemical composition, crystal structure, and fibrous nature by electron microprobe, X-ray diffraction, and scanning electron microscope, respectively.

As mentioned, many cherts contain recognizable remains of siliceous organisms, including radiolarians (Fig. 13.14), diatoms, silicoflagellates, and sponge spicules. Some

FIGURE 13.14
Poorly preserved radiolarians (small round bodies) in an organic-rich chert. Otter Point formation (Jurassic), southwest Oregon. Ordinary light. Scale bar = 0.1 mm. *(Specimen courtesy of Shelia Monroe.)*

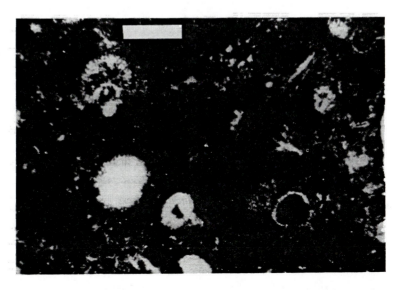

cherts contain no preserved remains of siliceous organisms but may contain constituents such as detrital clays and other siliciclastic minerals, pyroclastic particles, and organic matter. They may also contain authigenic minerals such as silica cement, clay minerals, hematite, pyrite, and magnetite.

13.3.3   Chemical Composition

Cherts are composed dominantly of SiO_2 but can include minor amounts of Al, Fe, Mn, Ca, Na, K, Mg, Ni, Cu, Ti, Sr, and Ba. The amount of SiO_2 varies markedly in different types of cherts, ranging from more than 99 percent in very pure cherts such as the Arkansas Novaculite to less than 65 percent in some nodular cherts (Cressman, 1962). Aluminum is commonly the second most abundant element in cherts, followed by Fe, Mg or K, Ca, and Na. Cherts may also contain trace amounts of rare earth elements such as cerium (Ce) and europium (Eu).

     Jones and Murchey (1986) suggest that the chemical elements in cherts are derived from four possible sources: biogenic, detrital, hydrogenous (precipitated or absorbed from seawater), and hydrothermal. Siliceous organisms furnish the major source of Si, and Ca may be derived in part from calcareous organisms. Detrital impurities furnish additional Si, as well as Al, Ti, Ca, Mg, K, and Na. In areas of high volcanic activity such as backarc basins and seamounts, significant amounts of K and Mg may be furnished in detrital components (Hein et al., 1983). The hydrogenous elements may include Fe, Mn, Ni, and Cu. Elements that may be contributed from hydrothermal fluids in areas of high heat flow such as oceanic spreading centers include Fe, Mn, and Ba.

13.3.4   Principal Kinds of Cherts

Varieties of Chert

Although chert is the general group name for siliceous sedimentary rocks composed dominantly of SiO_2 minerals, several names are applied to various varieties of chert. **Flint** is used both as a synonym for chert and as a varietal name for chert, particularly chert that occurs as nodules in Cretaceous chalks. **Jasper** is a variety of chert colored red by impurities of disseminated hematite. Jasper that is interbedded with hematite in Precambrian iron-formations is called **jaspilite.** **Novaculite** is a very dense, fine-grained, even-textured chert that occurs mainly in mid-Paleozoic rocks of the Arkansas, Oklahoma, Texas region of south-central United States. **Porcellanite** is a term used for fine-grained siliceous rocks with a texture and fracture resembling that of unglazed porcelain. The term is often used by chert workers for cherts having this character that are composed mainly of opal-CT. **Siliceous sinter** is porous, low-density, light-colored siliceous rock deposited by waters of hot springs and geysers. Although most siliceous rocks consist dominantly of chert, some have a high content of detrital clays or micrite. These impure cherts grade compositionally into siliceous shales or siliceous limestones.

Bedded Chert vs. Nodular Chert

General Statement.   Cherts can be divided on the basis of gross morphology into bedded cherts and nodular cherts. Bedded cherts are further distinguished by their content of siliceous organisms of various kinds. Mineralogy is not used as a basis for classifying cherts because these rocks are all composed mainly of SiO_2 minerals. The principal distinguishing characteristics of bedded and nodular cherts are described below.

Bedded Chert.   Bedded chert consists of layers of nearly pure chert, ranging to several centimeters in thickness, that are commonly interbedded with millimeter-thick partings or laminae of siliceous shale (Fig. 13.15). Bedding may be even and uniform or may show pinching and swelling. These rhythmically bedded deposits are also referred to as **ribbon cherts.** Many chert beds lack internal sedimentary structures; however, graded bedding, cross-bedding, ripple marks, sole markings, convolute layers, and soft-sediment folds have been reported in some cherts. The presence of these structures indicates that mechanical transport was involved in the deposition of these rocks, quite possibly transport by turbidity currents. Bedded cherts are commonly associated with ophiolitic rocks such as submarine volcanic flows or pillowed greenstones, tuffs, pelagic limestones, shales or argillites, and siliciclastic or carbonate turbidites. As indicated, many bedded cherts are composed dominantly of the remains of siliceous organisms, which are commonly altered to various degrees by solution and recrystallization. Bedded cherts can be subdivided on the basis of type and abundance of siliceous organic constituents into four principal types:

1. **Diatomaceous deposits** include both diatomites and diatomaceous cherts. **Diatomites** are light-colored, soft, friable siliceous rocks composed chiefly of the opaline frustules of diatoms, unicellular aquatic plants related to the algae. Thus, they are fossil diatomaceous oozes. Diatomites of both marine and lacustrine origin are recognized. Marine diatomites are commonly associated with sandstones, volcanic tuffs, mudstones or clay shales, impure limestones (marls), and, less commonly, gypsum. Lacustrine diatomites are almost invariably associated with volcanic rocks. **Diatomaceous chert** consists of beds and lenses of diatomite that have well-developed silica cement or groundmass that has converted the diatomite into dense, hard chert. Beds of diatomaceous chert comprising strata several hundred meters in thickness have been reported from some sedimentary

FIGURE 13.15
Thin, well-bedded cherts in the Mino Belt Group (Triassic), near Inuyama, Honshu, Japan.

sequences such as the Miocene Monterey Formation of California (Garrison et al., 1981), which may reach a thickness in some areas of as much as 2000 m. Marine diatomaceous deposits occur in rocks as old as the Cretaceous, and nonmarine deposits are reported from rocks as old as the Eocene (Barron, 1987).

Note that these so-called diatomaceous cherts or diatom cherts are composed of diatoms cemented by silica and that the diatoms still consist mainly of opal-A. When diatomaceous deposits are converted to quartz chert during diagenesis (discussed below), the diatoms commonly do not survive. Therefore, quartz cherts containing recognizable diatoms are rare. Hein et al. (1990) reported the first known example of such chert from Eocene deposits on Adak Island, Alaska. They propose that diatoms were preserved in this quartz chert because early and rapid alteration of ubiquitous volcanic glass in the section released silica and saturated the pore waters with respect to opal-A.

2. **Radiolarian deposits** consist dominantly of the remains of radiolarians, which are marine planktonic protozoans with a latticelike skeletal framework. Radiolarian deposits can be divided into radiolarite and radiolarian chert. **Radiolarite** is the comparatively hard, fine-grained, chertlike equivalent of radiolarian ooze, that is, indurated radiolarian ooze. **Radiolarian chert** is well-bedded, microcrystalline radiolarite that has a well-developed siliceous cement or groundmass. Because radiolarians are more robust and contain less surface area than diatoms, radiolarians tend to survive silica diagenesis. Therefore, they are common components of many bedded quartz cherts (Hein et al., 1990). Radiolarian cherts are commonly associated with tuffs, mafic volcanic rocks such as pillow basalts, pelagic limestones, and turbidite sandstones that are believed to indicate a deep-water origin. On the other hand, some radiolarian cherts are associated with micritic limestones and other rocks that suggest deposition at shallower depths of perhaps 200–1000 m (Iijima et al., 1979).

3. **Siliceous spicule deposits** include spicularite (spiculite), a siliceous rock composed principally of the siliceous spicules of invertebrate organisms, particularly sponges. Spicularite is loosely cemented in contrast to spicular chert, which is hard and dense. Spicular cherts are mainly marine in origin and occur associated with glauconitic sandstones, black shales, dolomite, argillaceous limestones, and phosphorites. They are not generally associated with volcanic rocks and are probably deposited mainly in relatively shallow water a few hundred meters deep.

4. **Bedded cherts containing few or no siliceous skeletal remains** have been described by many authors. Some of these reported occurrences of barren cherts may simply be the result of inadequate microscopic examination of the cherts, which might be found upon closer examination to contain siliceous organisms. Others have been examined closely and clearly contain few siliceous organisms. Cherts in this latter group include most cherts associated with the Precambrian iron-formations, as well as many Phanerozoic-age cherts such as the Mississippian-Devonian–age Arkansas Novaculite of Arkansas and Oklahoma and the Caballos Novaculite of Texas. Except for the absence of skeletal remains, these cherts resemble radiolarian cherts both megascopically and in their lithologic associations (Cressman, 1962). Some of these cherts are probably radiolarian cherts that have undergone such severe diagenesis that no recognizable radiolarians remain. Others, particularly Precambrian cherts, may have some other origin.

Nodular Cherts

Nodular cherts are subspheroidal masses, lenses, or irregular layers or bodies that range in size from a few centimeters to several tens of centimeters (Fig. 13.16). They commonly

FIGURE 13.16
Nodular chert (arrows) in limestones of the Rico Formation (Permo-Pennsylvanian) near Hite, Utah. The length of the largest (saddle-shaped) nodule is about 12 cm. *(Photograph courtesy of Bruce Bartleson.)*

lack internal structures, but some nodular cherts contain silicified fossils or relict structures such as bedding. Colors of these cherts range from green to tan and black. They typically occur in shelf-type carbonate rocks, where they tend to be concentrated along certain horizons parallel to bedding. More rarely, they occur in sandstones and mudrocks, lacustrine sediments, and evaporites. They have also been encountered in cores of deep sea sediment recovered during Deep Sea Drilling Program (DSDP) and Ocean Drilling Program (ODP) coring in the open ocean. Nodular cherts originate mainly by diagenetic replacement. In carbonate rocks, nodular chert can replace micrite as well as fossils and other carbonate grains; such chert occurs in both limestones and dolomites. Diagenetic origin is clearly demonstrated in many nodules by the presence of partly or wholly silicified remains of calcareous fossils or ooliths. Nodular cherts can also replace anhydrite, pelagic clays, and, rarely, sandstones.

13.3.5   Deposition of Chert

Silica Solubility
The solubility of silica in seawater varies with different silicate minerals (Fig. 13.17). The solubility of solid SiO_2 at 25°C ranges from ~6–10 ppm for quartz to ~60–130 ppm for amorphous or noncrystalline varieties of silica such as opal (Krauskopf, 1959; Morey et al., 1962, 1964; Iller, 1979). The solubility of disordered cristobalite-tridymite (opal-CT) is somewhat lower than that of amorphous silica, but it is considerably higher than that of quartz and is inferred to lie somewhere between that of amorphous silica and α-cristobalite (Kastner et al., 1977). Silica solubility is affected by both temperature and pH. Solubility increases with increasing temperature in essentially a linear fashion, and solubility at 100°C is approximately three to four times that at 25°C (Fig. 13.17). Change in solubility of silica with pH is illustrated in Figure 13.18. Solubility changes only slightly with increase in pH up to about 9, but rises sharply at pH values above 9.

Silica Concentrations in Seawater
Silica is transported in river water to the modern ocean as silicic acid (H_4SiO_4) in concentrations averaging about 13 ppm SiO_2. In addition, silica is added to the oceans

FIGURE 13.17
Solubility of silica in water as a
function of temperature. *(After
Fournier, R. O., 1970, Silica in
thermal waters: Laboratory and
field investigations: Proc. Internat.
Symposium on Hydrochemistry and
Biochemistry, Tokyo, p. 122–139,
reprinted by permission.)*

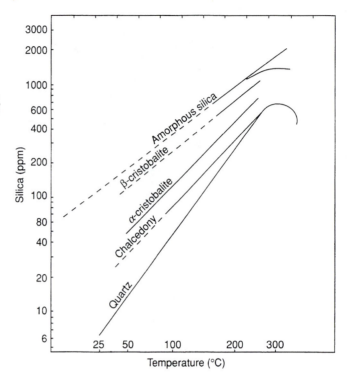

FIGURE 13.18
Solubility of silica as a function of
pH at 25°C. The solid line shows
the variation in solubility of
amorphous silica as determined
experimentally. The upper dashed
curve shows the calculated
solubility of amorphous silica,
based on an assumed constant
solubility of 120 ppm SiO_2 at pH
below 8. The lower dashed line is
the calculated solubility of quartz
based on the approximately known
solubility of 6 ppm SiO_2 in neutral
and acid solutions. *(From
Krauskopf, K. B., Introduction to
geochemistry. © 1979. Fig. 6.3,
p. 133, reprinted by permission of
McGraw-Hill, New York.)*

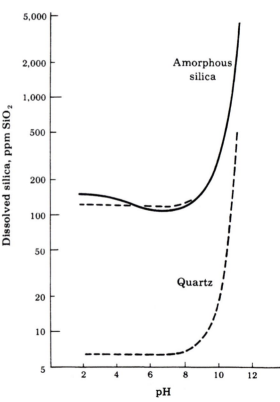

through reaction of seawater with hot volcanic rocks along midocean ridges and by low-temperature alteration of oceanic basalts and detrital silicate particles on the seafloor. Some silica may also escape from silica-enriched pore waters of pelagic sediments on the seafloor. These silica sources are summarized in Figure 13.19. Despite contributions of silica from these various sources, the silica concentration in seawater is quite low. Surface waters are particularly depleted in silica (commonly <0.01 ppm SiO_2). Silica concentration increases downward to a maximum of about 11 ppm below a depth of about 2 km. The average dissolved silica content of the ocean is only 1 ppm (Heath, 1974). The variation in silica concentration with depth reflects uptake of silica near the surface by phytoplankton and regeneration of silica at depth owing to dissolution of the silicon tests of phytoplankton. The difference between the average silica content of rivers and that of modern seawater thus reflects biogenic removal of silica to construct skeletal tests of diatoms, radiolarians, and other siliceous organisms. Most of the silica in these tests redissolves upon the death of the organisms; however, a small amount (perhaps 1–10 percent) reaches the bottom sediments (Calvert, 1974).

Precipitation of Chert from Seawater

Seawater with an average dissolved silica content of only 1 ppm is grossly undersaturated with respect to silica. Once silica is in solution under a given set of temperature and pH conditions, it does not appear to crystallize readily to form quartz even from solutions that have silica concentrations greatly exceeding the solubility of quartz (6–10 ppm). Therefore, it is unlikely that chalcedony or microquartz (chert) can be precipitated directly by inorganic processes from highly undersaturated ocean water. Microcrystalline quartz might be precipitated in some local basins where waters are saturated with silica owing perhaps to dissolution of volcanic ash. Some silica may be removed from seawater in the

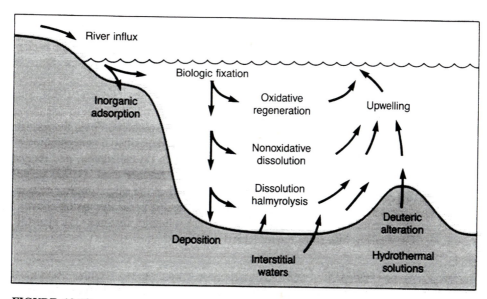

FIGURE 13.19

Sources of dissolved silica in ocean water. *(After Heath, G. R., 1974, Dissolved silica and deep-sea sediments, in Hay, W. W., ed., Studies in paleooceanography: Soc. Econ. Paleontologists and Mineralogists Spec. Pub. 20. Fig. 7, p. 81, reprinted by permission of SEPM, Tulsa, Okla.)*

open ocean by adsorption onto clay minerals or other silicate particles (Heath, 1974); however, such processes cannot account for the many bedded sequences of nearly pure chert that occur in the geologic record.

Removal of silica from ocean water by silica-secreting organisms to build opaline skeletal structures appears to be the only mechanism capable of large-scale silica extraction from undersaturated seawater. This biologic process has operated since at least early Paleozoic time to regulate the balance of silica in the ocean. **Radiolarians** (Cambrian/ Ordovician–Holocene), **diatoms** (Jurassic(?)–Holocene), and **silicoflagellates** (Cretaceous–Holocene) are microplankton that build skeletons of opaline silica. These siliceous microplankton have apparently been abundant enough in the ocean during Phanerozoic time to extract most of the silica delivered to the oceans by rock weathering, etc. Diatoms are probably responsible for the bulk of silica extraction from ocean waters in the modern ocean (Calvert, 1983); however, radiolarians were the most important silica-secreting organisms in the Phanerozoic oceans of Jurassic and older age. Heath (1974) calculates that the residence time for dissolved silica in the ocean ranges from 200 to 300 years for biologic utilization to 11,000– 16,000 years for incorporation into the geologic record—a very short time from a geologic point of view.

While silica-secreting organisms are alive, their siliceous skeletons are protected by an organic coating that prevents them from dissolving in highly undersaturated and corrosive seawater. After death, this coating is destroyed by biochemical decomposition and the opaline skeletons began to undergo dissolution. In areas of the ocean where siliceous organisms flourish (zones of upwelling), the rate of production of siliceous skeletons may be so high that they cannot all be dissolved as rapidly as they are produced. Under such conditions, a sufficient number of the siliceous skeletons may survive total dissolution to accumulate on the seafloor as siliceous oozes (sediments containing >30 percent siliceous skeletal material). After burial by additional siliceous ooze or clayey sediment, these opaline skeletal materials continue to undergo solution; however, the dissolving silica is trapped in the pore spaces of the sediment and cannot all escape back to the open ocean. The pore waters thus become increasingly enriched in silica. Silica concentrations exceeding 120 ppm are reported in pore fluids extracted from some deep-sea cores recovered at some Deep Sea Drilling Program (DSDP) sites (Gieskes, 1983). Such pore waters are saturated to supersaturated with respect to silica, and cherts are thought to slowly precipitate from these concentrated interstitial solutions. Thus, the formation of cherts is in part a sedimentation process involving the depositional concentration of biogenic opaline tests and in part a diagenetic process with crystallization, and recrystallization, of the chert taking place after sediment burial. Kastner et al. (1977) suggest that quartz can crystallize directly from solutions only when the silica concentration in solution is below the inferred equilibrium solubility for opal-CT (see discussion above). When silica concentrations in pore waters exceed the solubility for opal-CT, opal-CT is likely to crystallize instead of quartz (chert).

Nonbiogenic Cherts

The origin of bedded, ribbon cherts that contain no siliceous organic remains is poorly understood. Direct, inorganic precipitation of amorphous silica has been reported in some ephemeral Australian lakes (Peterson and von der Borch, 1965). Pleistocene cherts from alkaline Lake Magadi, Kenya, are also believed to form inorganically by early replacement of sodium silicate precursors—magadiite, $NaSi_7O_{13}(OH)_3\cdot3H_2O$, and/or kenyaite, $NaSi_{11}O_{20}\cdot5(OH)\cdot3H_2O$ (Schubel and Simonson, 1990). No similar occurrences have been reported in the open-marine environment that could help explain the presence of widespread nonfossiliferous chert deposits such as the Arkansas Novaculite. The scarcity of radiolarians and sponge spicules in the Arkansas Novaculite and similar Phanerozoic-age cherts does not, however, preclude the possibility that these cherts were formed by organisms. They could have been derived from siliceous oozes that were subsequently almost completely dissolved and recrystallized, leaving few or no recognizable siliceous organic remains (Weaver and Wise, 1974). The origin of Precambrian bedded cherts is a particular problem. Siliceous organisms are not definitely known to have existed in Precambrian time. Scattered remnants of possible siliceous planktonic organisms have been reported from Precambrian cherts by some investigators (e.g., LaBerge, 1973); however, geologists disagree about the biogenic origin of these structures. More recently, LaBerge et al. (1987) reported the common presence of spheroids in many Precambrian cherts associated with iron-formations. One type is about 30 μm in diameter and has a double-wall structure (Fig. 13.20). LaBerge et al. believe that these structures may represent the remains of an organism that had a siliceous frustule. If these structures indeed represent the remains of siliceous organisms, the origin of Precambrian cherts takes on a new dimension. Until the existence of Precambrian siliceous organisms is definitely proven, however, we have to assume that Precambrian cherts were formed mainly by inorganic precipitation. The source of the silica that was required to saturate Precambrian ocean waters to the point of quartz precipitation is not known, but is presumably related to volcanism.

FIGURE 13.20
Spheroidal structures of a possible silica-secreting organism in 3.5 b.y. old iron-formations from the Pilbara region, Western Australia. The structure is pigmented by submicrometer hematite dust. *(From LaBerge, G. L., et al., 1987, A model for the biological precipitation of Precambriam iron-formations, in Appel, P. W. V., and G. L. LaBerge, eds., Precambrian iron-formations. Fig. 11C, p. 86, reprinted by permission of Theophrastus Publications, S. A., Athens, Greece.)*

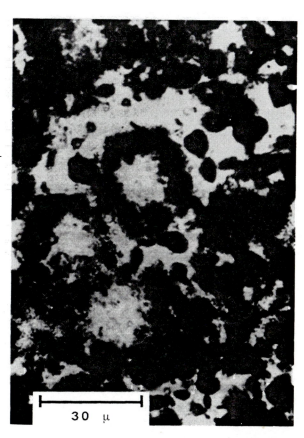

30 μ

13.3.6   Diagenesis of Chert

The formation of quartz cherts from the remains of siliceous organisms composed of opal-A is a solution–reprecipitation process. As indicated, biogenic opal-A typically does not convert directly to quartz chert. Instead, it commonly goes through an intermediate phase in which metastable opal-CT precipitates from pore waters that are saturated with silica owing to dissolution of opal-A. Skeletal opal-A first dissolves to yield solutions high in silica. The silica then flocculates to yield opal-CT, which becomes increasingly ordered before dissolving to yield pore waters of low silica concentration. Slow growth of quartz takes place from these waters (Williams and Crerar, 1985). Opal-CT may occur in open spaces in sediments as **lepispheres,** which are 5–20 μm spherulitic aggregates of blade-shaped crystals (Wise and Kelts, 1972). It can occur also as nonspherulitic blades, as rim cements and overgrowths, and as a massive cement (Maliva and Siever, 1988c). As mentioned, chert composed of opal-CT is referred to by many silica workers as porcellanite.

The rates of diagenetic evolution of silica from biogenic opal-A to opal-CT and finally to quartz chert are controlled by several physico-chemical factors. Temperature is commonly considered to be a particularly important control, with increasing temperature promoting an increased rate of transformation. Siever (1983) stresses the importance of temperature and sedimentation rates on silica diagenesis. He suggests that the conversion process can require a few millions of years to tens of millions of years depending upon conditions. Faster rates of transformation occur where rates of sedimentation and burial are high because sediment is quickly buried to depths where high temperatures bring about rapid transformation. The effects of time and burial temperatures on the diagenesis of silica are illustrated in Figure 13.21. The diagrams in this figure show the time–temperature regions for conversion of opal-A to opal-CT and quartz for two geographic areas with different sedimentation and burial rates and geothermal gradients. Note that a much longer time is required for transformation of opal-A to opal-CT under conditions of slow sedimentation and a low geothermal gradient (Fig. 13.21B) than under conditions of rapid sedimentation and a high geothermal gradient (Fig. 13.21A).

Kastner and Gieskes (1983) have demonstrated that the rate of transformation of opal-A to quartz also depends upon the nature of the opal starting material and the presence of magnesium hydroxide compounds, which serve as a nucleus for the crystallization of opal-CT. Williams et al. (1985) conclude that increasing surface-to-volume ratio of siliceous particles—a ratio that increases with decreasing particle size—results in greater solubility of opal and an increase in the rate of transformation. These authors also suggest that under some conditions opal-A can transform directly to quartz chert without going through the intermediate opal-CT stage.

13.3.7   Replacement Chert

In addition to its occurrence as bedded chert, chert can also occur in the form of small nodules, lenses, or thin, discontinuous beds. Nodular cherts are especially common in carbonate rocks but occur also in evaporites and siliciclastic rocks. Relict textures in these nodular cherts suggest that most are formed by diagenetic replacement. For example, Hein and Karl (1983) suggest that open-ocean cherts recovered from DSDP cores formed by volume-for-volume replacement of carbonates (chalk) and clay, as shown by preservation of burrows and other sedimentary structures. Chert nodules and lenses nucleate among local concentrations of magnesium compounds (Kastner et al., 1977), apparently because magnesium hydroxide nuclei aid in the formation of opal-CT lepispheres. The

FIGURE 13.21

A. Time–temperature diagram for the diagenetic alteration of siliceous sediments under conditions of variable, but moderately rapid, sedimentation in a rift zone with a high geothermal gradient. The diagram shows a time–temperature curve that is based on estimates of sedimentation rates and the geothermal gradient. Time–temperature regions for the conversion of opal-A to opal-CT and quartz (chert) are also shown. B. Time–temperature diagram for the diagenetic alteration of siliceous sediments under conditions of slow sedimentation on a spreading seafloor where the geothermal gradient is low. *(From Siever, R., 1983, Evolution of chert at active and passive continental margins, in Iijima, A., et al., eds., Siliceous deposits in the Pacific regions. Fig 6, p. 17, and Fig. 8, p. 20, reprinted by permission of Elsevier Science Publishers, Amsterdam.)*

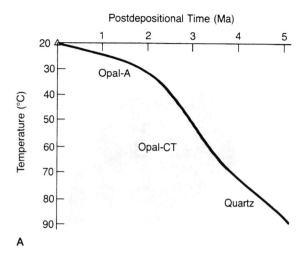

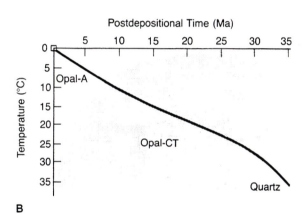

nodules grow until all the local biogenic silica in the host sediments is dissolved. Silica dissolved from opaline skeletal grains is precipitated initially as opal-CT, which is converted to quartz chert with increasing time and temperature. Maliva and Siever (1988) refer to this mechanism for forming deep-sea nodular cherts as a maturation process whereby chert evolves from originally dispersed biogenic opal-A through an intermediate stage to quartz porcellanite or chert. These cherts are typically characterized by quartz-replaced lepispheres, which provide good textural evidence for an opal-CT precursor. Hein and Karl (1983) stress that these open-ocean or deep-ocean replacement cherts are not analogous to bedded cherts or ribbon cherts (described above), which form in orogenic belts near continental margins.

Replacement cherts also occur in shallow, platform carbonate rocks. Knauth (1979) suggests that the mixing zone where meteoric groundwaters mix with seawater in a coastal area may be a particularly favorable geochemical environment for chert replacement of carbonates. Silica is supplied in this environment by the dissolution of sponge spicules or other forms of biogenic opal-A within the sediment pile and is then transported into the zone of mixing, where replacement of $CaCO_3$ occurs. Presumably, opal-A is first transformed to opal-CT, which then replaces the carbonate and is later diagenetically altered to quartz chert. Alternatively, the carbonate could be replaced directly by quartz. Maliva and Siever (1988c) report the presence of quartz-replaced lepispheres in several Mesozoic and Paleozoic nodular cherts in shelf limestone formations. They suggest that quartz-replaced lepispheres in these cherts indicate formation by a maturation process similar to that for deep-sea cherts. Nodular cherts formed by direct replacement of limestone by quartz are distinguished by absence of quartz-replaced lepispheres and a coarser (>10 μm) crystal size. The necessary geochemical conditions for replacement of carbonate by chert are discussed in Chapter 12. Nodular cherts have also been reported to form by the silica replacement of anhydrite, with the silica again being derived by dissolution of biogenic opal (Chowns and Elkins, 1974).

13.3.8   Ancient Cherts

Bedded cherts or ribbon cherts occur in sedimentary sequences ranging in age from Precambrian to Neogene. Hein and Parrish (1987) provide a particularly useful compilation of bedded chert deposits from all parts of the world. In this compilation, they inventory a total of 306 chert deposits and indicate that this list is not completely comprehensive. Although bedded chert deposits occur in geologic systems of all ages, they are particularly abundant in Jurassic to Neogene rocks, moderately abundant in Devonian and Carboniferous rocks, and lowest in Silurian and Cambrian deposits (Fig. 13.22). Ancient chert deposits are also unevenly distributed with respect to paleolatitude. The greatest majority of them are concentrated between 0 and 30 degrees latitude (Fig. 13.23). This latitude range encompasses the equatorial-oceanic divergence and west-coast upwelling zones. Presumably, the greatest productivity of siliceous plankton took place in these zones of upwelling. Note from Figure 13.23, however, that the distribution by latitude is

FIGURE 13.22

Distribution of chert through time. *(After Hein, J. R., and J. T. Parrish, 1987, Distribution of siliceous deposits in space and time, in Hein, J. R., Siliceous sedimentary rock-hosted ores and petroleum. Fig. 2–1, p. 15. Copyright 1987 by Van Nostrand Reinhold. All rights reserved.)*

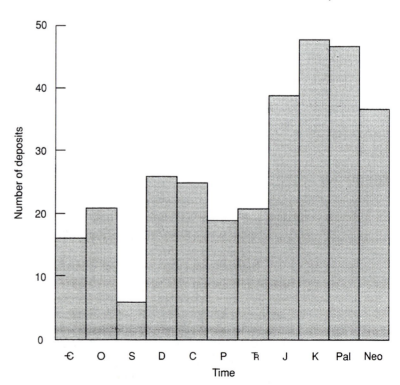

FIGURE 13.23

Paleolatitudinal distribution of chert through time. *(After Hein, J. R., and J. T. Parrish, 1987, Distribution of siliceous deposits in space and time, in Hein, J. R., ed., Siliceous sedimentary rock-hosted ores and petroleum. Fig. 2–3, p. 17. Copyright 1987 by Van Nostrand Reinhold. All rights reserved.)*

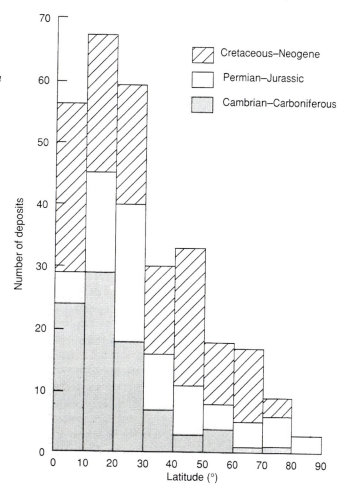

not uniform through time. Apparently, different paleoceanographic circulation patterns existed before the assembly of Pangaea, during Pangaea's existence, and after its breakup (Hein and Parrish, 1987). Because chert deposits are apparently related to paleoceanographic circulation patterns and zones of upwelling, there is considerable interest in chert deposits as clues to ancient ocean-circulation patterns.

Chert deposits are particularly common around the margins of the Pacific and in the Tethys region (the Alpine-Himalayan orogenic belt extending from the Caribbean region to the Indonesian region). Hein and Parrish's (1987) compilation of chert deposits indicates that the greatest number of chert deposits occur in the United States (about 22 percent of total deposits) and the USSR (14 percent of total deposits). Chert deposits are moderately abundant in Japan, Canada, Malaysia, Mexico, Indonesia, and Australia. Cherts occur also in many other countries including China, Great Britain, Peru, Chile, Antarctica, France, Italy, Greece, Spain, New Zealand, Central and South America, Turkey, Yugoslavia, Bulgaria, India, the Philippines, and Germany. *Siliceous Deposits in the Pacific Region*, edited by Iijima et al. (1983), and *Siliceous Deposits of the Tethys and Pacific Regions*, edited by Hein and Obradović (1989), are useful publications that describe many of the chert deposits of these regions.

The ancient chert deposits inventoried by Hein and Parrish (1987) include many examples of diatomites, radiolarites, radiolarian cherts, and some spiculites. Also included are cherts that contain less-abundant remains of siliceous organisms. Bedded chert deposits occur in association with a wide variety of sedimentary rocks, including limestone, dolomite, shale, sandstone and siltstones (including graywackes), conglomerate, phosphates, iron-formations, argillite, slate, schist, marble, quartzite, greenstone, tuff, basalt, ophiolite, pillow lavas, and ultramafic rocks.

Although most ancient bedded cherts, with the possible exception of Precambrian cherts, are believed to be of biogenic origin, the origin of bedded chert or ribbon chert is still something of an enigma. As mentioned, it seems reasonably clear now that the deep-sea cherts encountered during the Deep Sea Drilling Project and the more recent Ocean Drilling Project are not true analogs of ancient bedded cherts. These deep-sea or open-sea cherts form by replacement of carbonates or clays, with silica furnished by dissolution of local siliceous organisms. They lack the rhythmic bedding of ribbon cherts, make up only a small percentage of the carbonate or clay sequences in which they occur, and occur as lenses, nodules, or stringers on a millimeter or centimeter scale. By contrast, ribbon cherts make up at least 50 percent of sequences composed also of basalt, shale, sandstone, or other rock and consist of beds ranging to tens of centimeters thick and that extend laterally for tens of meters or more (Hein and Karl, 1983). Ribbon chert sequences must have been deposited initially as siliceous oozes or as turbidites composed mostly of siliceous organisms with various amounts of admixed clay. The siliceous oozes or turbidites were subsequently converted to chert by diagenetic recrystallization, as described. A key difference between ancient bedded or ribbon cherts and DSDP and ODP cherts appears to be that ribbon cherts do not form by diagenetic replacement of carbonate or clay. Instead, the whole mass of siliceous ooze deposits is transformed to chert by the opal-A to opal-CT to quartz conversion process. Also, DSDP and ODP cherts occur in the open ocean, whereas ribbon cherts appear to have formed mainly in continental-margin settings in orogenic-belt sequences. These settings may include backarc basins such as the Bering, Philippine, and Japan seas; small ocean basins with transform-fault–dominated spreading centers (spreading basins rather than spreading ridges) such as the Gulf of California; and silled basins that are locally isolated from terrigenous material such as the wrench basins of the California continental borderland (Hein and Parrish, 1987).

13.4   Iron-Rich Sedimentary Rocks

13.4.1   Introduction

Small amounts of iron occur in nearly all sedimentary rocks—about 4.8 percent in the average shale, 2.4 percent in the average sandstones, and 0.4 percent in the average limestones (Blatt, 1982). Sedimentary rocks that contain more than about 15 percent iron, corresponding to 21.3 percent Fe_2O_3 or 19.4 percent FeO, are considered to be iron-rich. Although iron-bearing minerals occur in sedimentary rocks of all ages, true iron-rich rocks are very unevenly distributed by age in the geologic record. Most iron-rich sedimentary rocks were deposited during three major time periods: the Precambrian, the early Paleozoic, and the Jurassic and Cretaceous.

Iron-rich sedimentary rocks make up only a very small fraction (<1 percent) of the sedimentary rocks in the geologic record; however, their enormous economic significance greatly overshadows their relatively minor abundance. Most of the world's currently mined iron ore comes from sedimentary deposits, primarily Precambrian banded, cherty iron-formations. These banded iron deposits, often referred to as **taconite,** have iron contents ranging from about 24 to 40 percent (James and Trendall, 1982). They are located primarily in the shield areas of the major continental masses, including North and South America, Africa, India, Russia, and Australia.

The iron-rich sedimentary rocks are an enigmatic group of rocks from the standpoint of their temporal and spatial distribution and origin. We do not fully understand why these rocks were deposited mainly during the three time periods mentioned, and we have many other questions concerning their origin. For example, what special depositional conditions existed during these time periods that were not present during other times? What was the source of the vast quantities of iron now locked up in sedimentary rocks of these ages? What was the mechanism of iron transport, and under what conditions was the iron deposited? We will examine these questions in this section, along with discussion of the physical and chemical characteristics of the iron-rich sedimentary rocks.

13.4.2 Principal Kinds of Iron-Rich Sedimentary Rocks

The classification of iron-rich sedimentary rocks is a contentious subject (Trendall, 1983), and a variety of names, many of local origin, are applied to these rocks. Dimroth (1979) suggests that all iron-rich rocks can be grouped into three broad categories of deposits on the basis of composition and physical characteristics: detrital chemical iron-rich sediments, iron-rich shales, and miscellaneous iron-rich deposits (Table 13.3). The iron-rich

TABLE 13.3

Principal classes of iron-rich sedimentary rocks

I. Detrital chemical iron-rich sediments
  A. Cherty iron formation
      Texture: analogous to limestone texture
      Composition: iron-rich chert containing hematite, magnetite, siderite, ankerite, or (predominantly alumina-poor) silicates as predominating iron minerals; relatively poor in Al and P
  B. Minette-type ironstone (aluminous iron formation)
      Texture: analogous to limestone texture
      Composition: aluminous iron silicates (chamosite, chlorite, stilpnomelane), iron oxides, and carbonates; relatively rich in Al and P
II. Iron-rich shales
  C. Pyritic shales
      Bituminous shales containing nodules or laminae of pyrite; grade into massive pyrite bodies by coalescence of pyrite laminae and nodules
  D. Siderite-rich shales
      Bituminous shales with siderite concretions; grade into massive siderite bodies by coalescence of concretions
III. Miscellaneous iron-rich deposits
  E. Iron-rich laterites
  F. Bog iron ores
  G. Manganese nodules and oceanic iron crusts
  H. Iron-rich muds precipitated from hydrothermal brines, Lahn-Dill–type iron oxide ores, and stratiform, volcanogenic sulfide deposits
  I. Placers of magnetite, hematite, or ilmenite sand

Source: Modified slightly from Dimroth, E., 1979, Models of physical sedimentation of iron formations, *in* Walker, R. G., ed., Facies models: Geoscience Canada Reprint Ser. 1. Table I, p. 175, reprinted by permission of Geological Association of Canada.

shales and miscellaneous types of iron-rich rocks rarely form deposits of economic significance and are volumetrically unimportant compared to the detrital chemical sediments. The nomenclature problem applies primarily to these last rocks. James (1966) proposed that these rocks be divided into two principal types: iron formation and ironstone. He used the term **iron formation** to include iron-rich rocks of largely Precambrian age and **ironstone** for rocks of mainly Phanerozoic age. Differences in age, physical characteristics, and mineralogy of iron formation and ironstone, as identified by James, are indicated in Table 13.4. Note in particular that iron-formation rocks are thinly bedded and well banded, whereas ironstones are massive to poorly banded.

Trendall (1983) points out, however, that the term ironstone has been used in contexts other than that suggested by James. He recommends that the term be avoided and that iron-formation be used instead as the general lithological and stratigraphic term for iron-rich sedimentary rocks. He further recommends that we drop the arbitrary lower limit of 15 percent iron for iron-rich sedimentary rocks and define them instead as "any sedimentary rock whose principal chemical characteristics are an anomalously high content of iron." Trendall suggests the abbreviation **IF** for iron-formation (presumably nonbanded) and **BIF** for banded iron-formation. Dimroth (1979) uses the term **aluminous iron-formation** (Minette-type ironstone) for the nonbanded iron-rich rocks (Table 13.3).

TABLE 13.4
Differences between ironstone and iron-formation

	Ironstone	Iron formation
Age	Pliocene to Middle Precambrian; principal beds from Lower Paleozoic and Jurassic	Cambrian to Early Precambrian; principal formations approximately 2000 million years old
Thickness	Major units a few meters to a few tens of meters	Major units 50–600 m
Original areal extent	Individual depositional basins rarely more than 150 km in maximum dimension	Difficult to determine; some deposits with continuity over many hundreds of kilometers
Physical character	Massive to poorly banded; silicate and oxide facies oolitic	Thinly bedded; layers of dominantly hematite, magnetite, siderite, or silicate alternating with chert, which makes up approximately half the rock; oolites rare
Mineralogy	Dominant oxide goethite; hematite very common; magnetite relatively rare; chamosite primary silicate; calcite and dolomite common constituents	No goethite; magnetite and hematite about equally abundant; primary silicate greenalite; chert a major constituent; dolomite present in some units, but calcite rare or absent
Chemistry	Except for high iron content, no distinctive aspects	Remarkably low content of Na, K, and Al; low P
Associated rocks	Both typically interbedded with shale, sandstone, or graywacke; yet the iron formation has few or no clastics compared to the ironstone.	
Relative abundance of facies	No gross differences apparent; probable order of abundance for ironstone: oxide, silicate, siderite, sulfide; for iron formation the order is similar, but siderite facies may be more abundant than silicate facies	

Source: James, H. L., 1966, Chemistry of the iron-rich sedimentary rocks, *in* Fleischer, M., ed., Data of geochemistry: U.S. Geol. Survey Prof. Paper 440-W.

Although Trendall's reservations about use of the term ironstone are well founded, the name appears to be entrenched in the geological literature. Furthermore, it is a convenient term for distinguishing between mainly nonbanded, noncherty, Phanerozoic iron-rich rocks and mainly well banded, cherty, Precambrian iron-rich sedimentary rocks. Therefore, I have retained James's (1966) terminology in this book. In any case, the ironstones represent much smaller concentrations of iron than in iron-formations, although locally they have considerable economic value (James and Trendall, 1982). Most of the literature on iron-rich sedimentary rocks deals with the Precambrian iron-formations.

13.4.3   Iron-Formations

Distribution

Iron-formations range in age from early Precambrian to Devonian (James and Trendall, 1982), although they are primarily of Precambrian age. Three periods of peak deposition are recognized: mid-Archean (3400–2900 m.y.), early Proterozoic (2000–2500 m.y.), and late Proterozoic (750–500 m.y.). Some of the more important deposits are listed in Table 13.5. Note that these deposits occur on the major Precambrian cratons of the Earth and that one deposit of banded iron characterized as a **very large** deposit occurs in each of the continents listed in Table 13.5. Approximately 90 percent of the preserved iron-formations in the world are contained in these five large deposits (James and Trendall, 1982).

Characteristics

Sedimentary Structures. Banded iron-formations consist of distinctively banded sequences (Fig. 13.24) composed of layers enriched in iron alternating with layers rich in chert. Banding occurs at a variety of scales. In some iron-formations, millimeter-scale concentrations of iron-bearing minerals can define **microbands** within centimeter-scale **mesobands** of chert (terminology of Trendall and Blockley, 1970). In turn, chert mesobands are separated by mesobands of iron-rich materials called **chert matrix.** These microbands and mesobands can occur within larger-scale bands (macrobands) ranging from 20 m to 500 m or more (Fig. 13.25). The bands tend to be laterally continuous for long distances. In the Hamersley Basin of Western Australia, for example, the banding is laterally continuous over the entire estimated 150,000 km^2 original depositional area of the basin (James and Trendall, 1982).

Gole and Klein (1981) suggest that microbanding, in the strict sense described above (Fig. 13.26E, F), is not common; lamination is more commonly expressed by alternating 0.1–2.0 mm thick laminae of different iron-bearing minerals (Fig. 13.26C, D). Laminated iron formations may also include other small-scale structures such as chert pods, macules (spots or knots), and small spherical nodules. Some iron-formations have oolitic and granular textures (Fig. 13.26A, B), which James and Trendall (1982) refer to as **granule iron-formation.** These iron-formations, which occur mainly in Proterozoic sequences (Gole and Klein, 1981), appear to be less well laminated than other iron-formations, although they may be laminated in part. The textures of these iron-formations resemble those of limestones. Dimroth (1979) recognizes textural types in iron-formations equivalent to micritic, pelleted, intraclastic, oolitic, pisolitic, and stromatolitic limestone textures (Table 13.6). Other sedimentary structures reported from banded iron-formations include cross-bedding, graded bedding, load casts, ripple marks, erosion channels,

TABLE 13.5
Estimated initial size and age of selected cherty, banded iron-formations

	Area (may include more than one stratigraphic unit)	Class	Estimated age* (m.y.)
Africa	Damara Belt, Namibia	Moderate	650 (590–720)
	Shushong Group, Botswana	Small	1875 (1750–2000)
	Ijil Group, Mauritania	Moderate	2100 (1700–2500)
	Transvaal-Griquatown, S. Africa	Very large	2263 (2095–2643)
	Witwatersrand, S. Africa	Small	2720 (2643–2800)
	Liberian Shield, Liberia-Sierra Leone	Large	3050 (2750–3350)
	Pongola beds, Swaziland-S. Africa	Moderate	3100 (2850–3350)
	Swaziland Supergroup, Swaziland-S. Africa	Small	3200 (3000–3400)
Australia	Nabberu Basin	Large	2150 (1700–2600)
	Middleback Range	Moderate	2200 (1780–2600)
	Hamersley Range	Very large	2500 (2350–2650)
Eurasia	Altai region, Kazakhstan-W. Siberia	Moderate	375 (350–400)
	Maly Khinghan-Uda, Far East USSR	Large	550 (500–800?)
	Central Finland	Moderate	2085 ± 45
	Krivoy Rog-KMA, USSR	Very large	2250 (1900–2600)
	Bihar-Orissa, India	Large	3025 (2900–3150)
	Belozyorsky-Konski, Ukraine, USSR	Moderate	3250 (3100–3400)
North America	Raritan Group, western Canada	Moderate	700 (550–850)
	Yavapai Series, southwest USA	Small	1795 (1775–1820)
	Lake Superior, USA	Large	1975 (1850–2100)
	Labrador Trough and extensions, Canada	Very large	2175 (1850–2500)
	Michipicoten, Canada-Vermilion, USA	Moderate	2725 (2700–2750)
	Beartooth Mountains, Montana, USA	Small	2920 (2700–3140)
	Isua, Greenland	Small	>3760 ± 70
South America	Morro du Urucum-Mutun, Brazil-Bolivia	Moderate	600? (450–900)
	Minas Gerais, Brazil	Very large	2350 (2000–2700)
	Imataca Complex, Venezuela	Large	3400 (3100–3700)

*In the absence of other data, the assigned age is the arithmetic mean of the age limits given in brackets.

Source: James, H. L., and A. F. Trendall, 1982, Banded iron formation: Distribution in time and paleoenvironmental significance, *in* Holland, H. D., and M. Schidlowski, eds., Mineral deposits and the evolution of the biosphere. Table 1, p. 202, reprinted by permission of Springer-Verlag, Berlin.

shrinkage cracks, and slump structures. These structures show that many of the constituents of iron-formations must have undergone mechanical transport and deposition.

Mineralogy.   On the basis of relative abundance of major iron-bearing minerals, James (1966) defined four principal kinds of iron-rich sedimentary rocks, which he referred to as facies: (1) oxides, (2) silicates, (3) sulfides, and (4) carbonates (Table 13.7). **Oxides** are characterized by the presence of hematite, goethite, and magnetite. Hematite is present in both iron-formations and ironstones. Goethite occurs in ironstones but is absent in Precambrian iron deposits. Magnetite is most abundant in Precambrian iron deposits, but occurs also in Phanerozoic deposits. **Silicates** are distinguished particularly by the iron silicate minerals chamosite and greenalite but may also contain glauconite, stilpnomelane, and iron talc. Chamosite is the principal iron silicate mineral in ironstones, whereas greenalite is predominant in Precambrian iron-formations. **Sulfides** may contain both

FIGURE 13.24
Banded iron-formation from the
Negaunee Iron Formation
(Precambrian), Michigan. The
light-colored bands are chert; dark
layers are the iron-rich units.
Length 8 cm. *(Specimen furnished
by M. H. Reed.)*

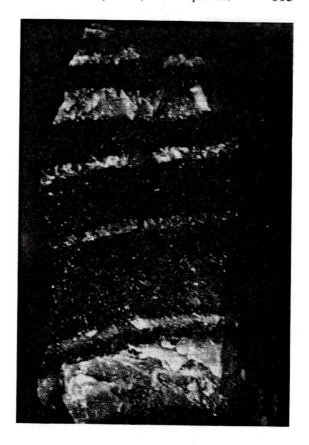

pyrite and marcasite, although pyrite is commonly the predominant sulfide mineral in iron-rich sedimentary rocks. Sulfide minerals are rarely the dominant iron minerals in iron-rich sedimentary rocks, but locally they can predominate in some thin beds. **Carbonates** are common minerals in iron-rich deposits and may include siderite, ankerite, dolomite, and calcite. Siderite is an important constituent of both Precambrian and Phanerozoic iron-rich rocks, where it commonly consists of flattened nodules or more or less continuous beds. Dolomite is common also in both iron-formations and ironstones. Calcite is common in ironstones but rare in iron-formations, and ankerite is most common in iron-formations.

Chemical Composition. Iron-formations consist mainly of SiO_2 and Fe, although the chemical composition of iron-rich sedimentary rocks varies over a wide range, depending upon the type of deposit. Iron expressed as Fe_2O_3, FeO, or FeS is the dominant chemical constituent in some iron-rich sediments; however, the iron content of iron-rich rocks is frequently exceeded by that of silica. Although it is probably not possible to establish a representative average composition for all iron-formations, chemical data compiled by Gole and Klein (1981) for eight major iron-formations of Archean and Proterozoic age show that average chemical composition of these formations is about 45–50 percent SiO_2, 13–27 percent Fe_2O_3, 17–25 percent FeO (29–32 percent total Fe), 3–6 percent MgO, 2–7 percent CaO, 1–2 percent Al_2O_3, and less than 1 percent each of TiO_2, MnO, Na_2O, K_2O, P_2O_5, S, and C.

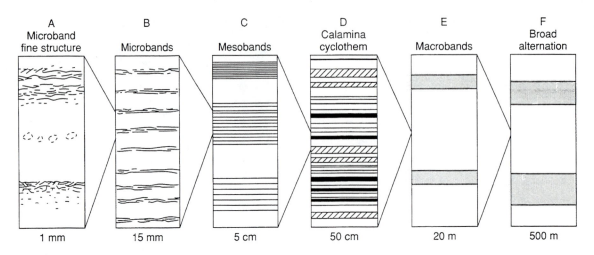

FIGURE 13.25
Various scales of banding in banded iron-formations of the Hamersley Basin, Western Australia.
A shows the fine structure within the microbands defined by iron-rich minerals shown in B. In
C, microbanded chert mesobands of different microband interval are represented, separated by
chert matrix (blank pattern). D shows a regular cyclicity of different chert types, which occur
within the larger cyclicites shown in E and F. *(From James, H. L., and A. F. Trendall, 1982,
Banded iron formation: Distribution in time and paleoenvironmental significance, in Holland,
H. D., and M. Schidlowski, eds., Mineral deposits and the evolution of the biosphere. Fig. 2,
p. 209, reprinted by permission of Springer-Verlag, Berlin.)*

Eichler (1976) points out that chemical composition can vary considerably in the
different mineral facies of iron-formations. Iron content tends to be highest in the oxide
and silicate facies and lower in the carbonate and sulfide facies (Table 13.8). Although the
chemical composition of different iron-formations is rather variable, Gole and Klein
maintain that similarities between Proterozoic and Archean banded iron-formations are
more significant than their differences. For a more detailed look at the chemical composition of mesobands and macrobands in selected iron-formations, see Davy (1983). Fryer
(1983) adds additional details on the rare earth elements in iron-formations, and Perry
(1983) discusses the oxygen isotope geochemistry of these formations.

Associated Facies.   Iron-formations occur in association with a wide variety of other
rock types, and there appears to be no consistent lithologic association in rock units either
below or above iron-formations (Gole and Klein, 1981). Iron-formations commonly occur
within sequences of siliciclastic sedimentary rocks, and cherty iron-formations can grade
into slightly cherty iron-rich sandstones, siltstones, and shales. They may also be present
in association with limestones and dolomites, quartzite, argillite, schist, volcanic rocks,
and ultramafic rocks. Volcanic rocks appear to be particularly common in Archean banded
iron-formation sequences, although they occur also in Proterozoic sequences. On the
other hand, carbonate rocks are comparatively rare in Archean sequences.

Principal Kinds of Iron-Formations.   Various attempts have been made to classify
iron-formations into a few specific, representative types. For example, Gross (1965)
divided iron-formations into Algoma-type and Superior-type. According to him, **Algoma-
type** iron-formations are intimately associated with various volcanic rocks. They are

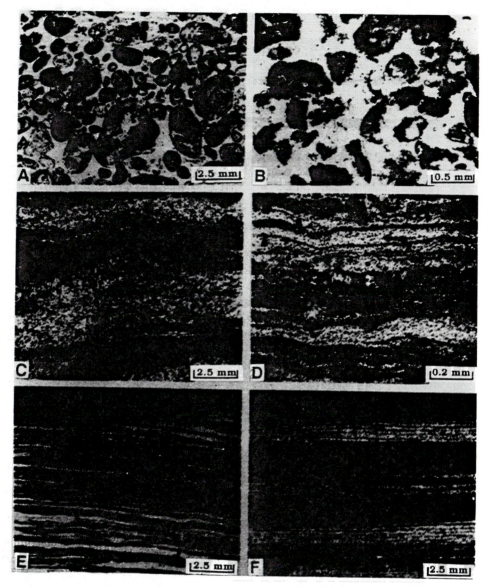

FIGURE 13.26

Granule textures and small-scale banding in Precambrian banded iron-formations. A. Ooids and granules composed of magnetite, chert, minor carbonate, and stilpnomelane in a chert-rich matrix (Sokoman Iron Formation, Labrador Trough). B. Granules of minnesotaite, greenalite, stilpnomelane, siderite, and minor magnetite in a chert-rich matrix (Biwabik Iron Formation, Minnesota). C. Laminae in a minnesotaite–stilpnomelane–quartz–minor magnetite assemblage (Negaunee Iron Formation, northern Michigan). D. Microbands defined by quartz, magnetite, and minor siderite, ankerite, and chlorite (Negaunee Iron Formation). E. Microbanding in a quartz-magnetite mesoband (Joffre Member of the Hamersley Group, Western Australiak). F. Alternating massive and microbanded mesobands in a magnetite–quartz–minor grunerite-bearing iron-formation, metamorphosed to amphibolite grade (Forrestania area, Archean Yilgarn Block, Western Australia). *(From Gole, M. J., and C. Klein, 1981, Banded iron-formation through much of Precambrian time: Jour. Geology, v. 89. Fig. 2, p. 174, reprinted by permission of University of Chicago Press.)*

TABLE 13.6
Textural types of cherty iron-formation equivalent to limestone textural types

Type	Description
Micrite type	Deposited as a mud whose particles are too fine grained to survive diagenesis; only lamination and stratification visible as depositional structures; small-scale cross-beds here and there prove deposition as particulate, noncohesive matter
Pelleted	Fine pellet texture of silt or very fine sand
Intraclastic	Containing gravel-size fragments (intraclasts) whose internal textures prove derivation from penecontemporaneous sediment; fragments embedded in a micrite-type matrix or bound by a cement introduced during diagenesis
Peloidal	Containing sand-size fragments (peloids) without internal textures; peloids embedded in a micrite-type matrix or bound by a clear chert cement introduced during diagenesis
Oolitic	Containing concentrically laminated ooids, either set in a micrite-type matrix or, more commonly, bound by a clear chert cement introduced during diagenesis
Pisolitic	Containing pisolites set either in a micrite-type matrix or cemented by clear chert
Stromatolitic	Wavy, columnar, or digitating stromatolites

Source: After Dimroth, E., 1979, Models of physical sedimentation of iron formations, *in* Walker, R. G., ed., Facies models: Geoscience Canada Reprint Ser. 1. Table II, p. 176, reprinted by permission of Geological Association of Canada.

TABLE 13.7
Principal iron-bearing minerals in iron-rich sedimentary rocks

Mineral class	Mineral	Chemical formula
Oxides	Goethite*	$FeOOH$
	Hematite	Fe_2O_3
	Magnetite	Fe_3O_4
Silicates	Chamosite	$3(Fe,Mg)O \cdot (Al,Fe)_2O_3 \cdot 2SiO_2 \cdot nH_2O$
	Greenalite	$FeSiO_3 \cdot nH_2O$
	Glauconite	$KMg(Fe,Al)(SiO_3)_6 \cdot 3H_2O$
	Stilpnomelane	$2(Fe,Mg)O \cdot (Fe,Al)_2O_3 \cdot 5SiO_2 \cdot 3H_2O$
	Minnesotaite (iron talc)	$(OH)_2(Fe,Mg)_3Si_4O_{10}$
Sulfides	Pyrite	FeS_2
	Marcasite	FeS_2
Carbonates	Siderite	$FeCO_3$
	Ankerite	$Ca(Mg,Fe)(CO_3)_2$
	Dolomite	$CaMg(CO_3)_2$
	Calcite	$CaCO_3$

*Not found in Precambrian iron-formations

TABLE 13.8
Chemical composition of sedimentary facies of iron-formations

	Oxide facies	*Silicate facies*	*Carbonate facies*	*Sulfide facies*
Fe	37.80	26.5	21.23	20.0
FeO	2.10	28.9	22.22	2.35
Fe_2O_3	51.69	5.6	5.74	—
FeS_2	—	—	—	38.70
SiO_2	42.89	50.7	48.72	36.67
Al_2O_3	0.42	0.4	0.15	6.90
Mn	0.3	0.4	0.50	0.001
P	0.03	—	0.07	0.09
CaO	0.1	0.1	4.60	0.13
MgO	+	4.2	0.84	0.65
K_2O	+	—	+	1.81
Na_2O	+	—	0.01	0.26
TiO_2	+	+	+	0.39
CO_2	—	5.1	14.10	—
S	—	+	2.76	—
SO_3	—	—	—	2.60
C	—	+	++	7.60
H_2O^+	0.43	5.2	2.67	1.25

+ = trace; + + = larger trace; — = not present
Source: Eichler, J., 1976, Origin of Precambrian banded iron-formations, in Wolf, K. H., ed., Handbook of strata-bound and stratiform ore deposits, v. 7. Table VI, p. 187, reprinted by permission of Elsevier Science Publishers, Amsterdam.

thinly banded or laminated, lack oolitic or granular textures, and commonly extend laterally for only a few kilometers. By contrast, Gross suggested that **Superior-type** iron-formations are not associated with volcanic rocks but commonly occur with quartzite, black carbonaceous shale, conglomerate, dolomite, massive chert, chert breccia, and argillite. They contain granular and oolitic textures and commonly extend laterally for hundreds of kilometers. Granular/oolite-facies rocks are ferruginous chert arenites. The chert clasts, which contain finely disseminated inclusions of hematite or iron silicates and ooids may have been transported some distance from shallower sources. Alternatively, clasts might be ripped up locally by current scour (Simonson and Goode, 1989). Subsequent workers have made some modifications in the putative characteristics of these two kinds of deposits. For example, Eichler (1976) suggests that both Algoma-type and Superior-type iron-formations have granular and oolitic textures. Other distinguishing characteristics of these two types of formations are shown in Table 13.9. Eichler includes as an example of Superior-type deposits the huge Hamersley deposits of Australia (Simonson and Goode, 1989, recently reported clastic facies, i.e., chert arenites, in the Hamersley Group). On the other hand, Dimroth (1976) includes the Hamersley deposits as an example of Algoma-type deposits.

Button et al. (1982) list **Raritan-type** deposits as a third kind of iron-formation. These deposits occur in several basins of late Precambrian age and are associated in part with glacial deposits. Deposits of significant dimensions occur in the Raritan Group of northwestern Canada and in the Morro du Urucum region of western Brazil and eastern Bolivia. Button et al. suggest that these deposits are somewhat richer in iron than are normal iron-formations. Again, Dimroth (1976) includes deposits of the Raritan Group in

TABLE 13.9

Distinguishing features of Algoma- and Superior-type banded iron-formations

	Algoma-type (Archean-type)	*Superior-type (Animikie-type)*
Age	Pre-2600 m.y. (also Proterozoic and Phanerozoic)	Pre-1800 m.y.
Sedimentary environment	Eugeosynclinal tectonic basins of several 100 km diameter; iron-formation in marginal parts in connection with greenstone belts	Miogeosynclinal; iron-formation along margins of stable continental shelves; shallow water; restricted intracratonic basins
Extent	Commonly lenticular bodies of a few km	Extensive formations persistent over some 100 km to more than 1000 km
Thickness	0.1 m to some 10 m	Several m to more than 100 m (1000 m)
Location in sedimentary sequence	Irregular, lenticular bodies within Archean "basement" rocks	In bottom and middle parts of sedimentary sequences as sheet deposits, transgressive on older "basement" rocks
Associated rocks	Graywackes and shales; carbonaceous slates; mafic volcanics; felsic pyroclastics; rhyolitic flows; pillowed andesites	Coarse clastics; quartzites, conglomerates, dolomites, black shales (graphitic)
Volcanics	Close association to volcanism in time and space	No direct association with contemporaneous volcanism; volcanics normally absent
Sedimentary facies	Oxide facies predominant; carbonate and sulfide facies thin and discontinuous; silicate facies; all facies frequently closely associated	Oxide facies most abundant; silicate and carbonate facies frequently intergradational
	Sulfide and carbonate facies near the center of volcanism; oxide facies on the margins	Sulfide facies insignificant or absent
	Heterogeneous lithological assemblages with fine-grained clastic beds	More homogeneous (especially oxide facies); little or no detritus
	Granular and oolitic textures	Granular and oolitic textures
Examples	Canada: Archean basins (e.g., Michipicoten)	Labrador Trough
	USA: Vermilion District, Minnesota	Lake Superior region
	S. Africa: greenstone belts, Kaapvaal and Rhodesia cratons	Transvaal and Witwatersrand supergroups
	Brazil: Rio das Velhas Series	Minas Series (itabirites); Carajás/Pará
	India: southern Mysore	Bihar, Orissa, Goa, Mysore
	Australia: Yilgarn and Pilbara blocks	Hamersley Group
	USSR: Taratash/Urals	Krivoy Rog, Kursk; Ukranian Shield

Source: Eichler, E., 1976, Origin of the Precambrian banded iron-formations, *in* Wolf, K. H., ed., Handbook of strata-bound and stratiform ore deposits. Table 1, p. 173, reprinted by permission of Elsevier Science Publishers, Amsterdam.

Canada as an example of Algoma-type deposits. Clearly, there is lack of agreement among the experts with respect to classification of major iron-formations. Furthermore, Trendall (1983) does not agree with the practice of classifying most iron-formations into two major types: Algoma and Superior. He maintains that "this demonstrably subjective distinction is not only inadequate to accommodate, without distortion, many major iron-formations of the world, but imposes an artificial two-fold division not present in the complex spectrum of rocks it seeks to cover."

13.4.4 Ironstones

Ironstones are dominantly Phanerozoic-age sedimentary deposits. They occur mainly in early Paleozoic and Jurassic-Cretaceous rocks, but they range in age from Pliocene to middle Precambrian. The geographic and age distribution of Phanerozoic-age oolitic ironstones is shown in Figure 13.27. Note that they occur on all the major continents but that few deposits of Cambrian age or Carboniferous to Triassic age occur on any continent. The time intervals represented by rocks of these ages must have been unfavorable for deposition of iron-rich sediments.

The volume of preserved ironstones is much less than that of the iron formations, although they are locally important, as mentioned. Ironstones form thin, massive, or poorly banded sequences (Fig. 13.28), a few meters to a few tens of meters thick, in sharp contrast to the much thicker, well-banded iron-formations. They generally have an oolitic texture (Fig. 13.29), and they may contain fossils that have been partly or completely replaced by iron minerals. Van Houten (1982) suggests two major ironstone facies: (1) sandy and shelly shallow-marine sediments with abundant pellets of glauconite and (2) ferric oxide-chamosite (mostly 7 Å iron-rich clay mineral) oolite; ferric oxides can include both hematite and goethite. These two facies can be associated and, locally, oolitic deposits may replace glauconite facies vertically or interfinger with them laterally. Most of the ferric oxide-chamosite oolites accumulated during early Ordovician to late Devonian and early Jurassic to middle Cenozoic time (Van Houten, 1982). Nonferrous minerals in ironstones may include detrital quartz, calcite, dolomite, authigenic phosphorite, and authigenic chert. As mentioned, ironstones are commonly poorly banded; however, they may contain a variety of other sedimentary structures. These structures can include cross-bedding, ripple marks, scour-and-fill structures, rip-up clasts, and burrows. Structures such as cross-beds and ripple marks indicate that mechanical transport of grains was involved in the origin of these deposits.

Ironstones are commonly interbedded with carbonates, mudrocks, and fine-grained sandstones of shelf to shallow-marine origin. Many occur in deltaic sequences. On the other hand, some ironstones occur in association with deeper-water mudstones or marly limestones.

13.4.5 Iron-Rich Shales

Pyritic black shales occur in association with both Precambrian iron-formations and Phanerozoic ironstones. They commonly form thin beds in which sulfide content may range as high as 75 percent. Pyrite occurs disseminated in these black carbonaceous shales and in some limestones. It occurs also as nodules, laminae, and as a replacement of fossil fragments and other iron minerals. Pyrite-rich layers have also been reported in some limestones. **Siderite-rich shales** (clay ironstones) occur primarily in association with

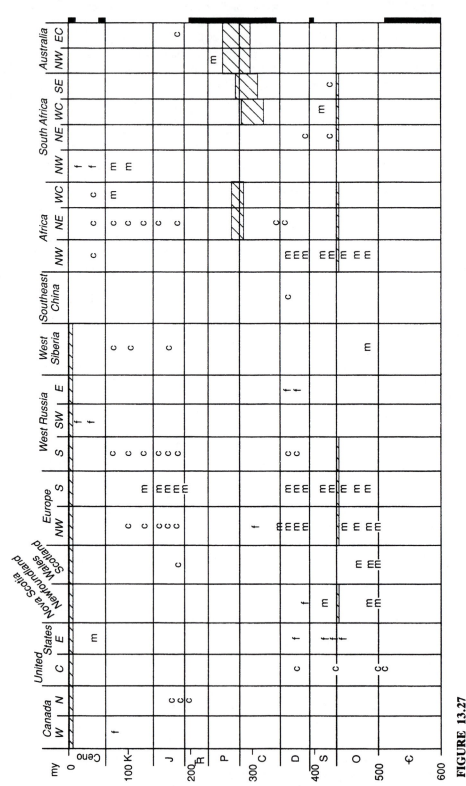

FIGURE 13.27

Geographic and age distribution of Phanerozoic oolitic ironstones; c = intercratonic basin, f = foredeep, and m = cratonic margin. Unfavorable episodes and intervals—thick line along right side of chart. Diagonal pattern—continental glaciation. *(After Van Houten, F. B., 1982, Phanerozoic oolitic ironstones: Ann. Rev. Earth and Planetary Sciences, v. 10. Fig. 2, p. 444, reprinted by permission.)*

FIGURE 13.28
Hematitic oolitic Clinton ironstone
(lower part of photograph) of the
Keefer Member of the Mifflintown
Formation (Silurian), Pennsylvania.
Limestone beds overlie the
ironstones. *(Photograph courtesy
of T. Schuster.)*

FIGURE 13.29
Ironstone ooids with quartz nuclei,
cemented with sparry calcite
cement. Clinton Formation
(Silurian), New York. Ordinary
light. Scale bar = 0.5 mm.

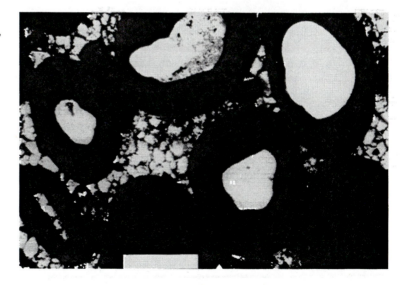

other iron-rich deposits. They are also present in the coal measures of both Great Britain and the United States. Siderite occurs disseminated in the mudrocks or as flattened nodules and more or less continuous beds.

13.4.6   Miscellaneous Iron-Rich Deposits

Bog-iron ores are minor accumulations of iron-rich sediments that occur particularly in small freshwater lakes of high altitude. They range from hard, oolitic, pisolitic and concretionary forms to soft, earthy types. **Iron-rich laterites** are residual deposits that form as a product of intense chemical weathering. They are basically highly weathered soils in which iron is enriched. **Manganese crusts** and **nodules** are widely distributed on the modern seafloor in parts of the Pacific, Atlantic, and Indian oceans in areas where sedimentation rates are low. They have also been reported from ancient sedimentary deposits in association with such oceanic sediments as red shales, cherts, and pelagic limestones. Both iron-rich (15–20 percent Fe) and iron-poor (less than about 6 percent Fe) varieties of manganese nodules are known. These nodules contain variable amounts of Cu, Co, Ni, Cr, and V in addition to manganese and iron.

Metalliferous sediments occur in oceanic settings, particularly near active mid-ocean spreading ridges. They are believed to form by precipitation from metal-rich hydrothermal fluids that have become enriched through contact and interaction with hot basaltic rocks. These sediments are enriched in Fe, Mn, Cu, Pb, Zn, Co, Ni, Cr, and V. Metal-enriched sediments have also been reported from some ancient sedimentary deposits in association with submarine pillow basalts and ophiolite sequences of ocean crustal rocks.

Heavy mineral placers are sedimentary deposits that form by mechanical concentration of mineral particles of high specific gravity, commonly in beach or alluvial environments. Magnetite, ilmenite, and hematite sands are common constituents of placers, particularly beach and marine placers. Placers are local accumulations that occur mainly in Pleistocene- to Holocene-age sediments and that commonly do not exceed 1–2 m in thickness. Marine placer deposits containing about 5 percent iron ore have been mined off the southern tip of Kyushu, Japan, for many years (Mero, 1965), and offshore placers containing up to 10 percent magnetite and ilmenite have been reported off the southeastern coast of Taiwan (Boggs, 1975). Beach placers containing ilmenite have been exploited commercially in Australia since about 1965 (Hails, 1976). "Fossil" placer deposits are comparatively rare, although thin, heavy-mineral laminae are common in some ancient beach deposits. Hails (1976) reports that outcrops of ilmenite- and magnetite-bearing placers of Cretaceous age are exposed discontinuously through New Mexico, Colorado, Wyoming, and Montana subparallel to the Rocky Mountains.

13.4.7   Origin of Iron-Rich Sedimentary Rocks

Introduction

The origin of iron-rich sedimentary rocks is an enigmatic topic that has been debated by geologists for well over a century. The number of papers written about the origin of iron-rich sediments is probably exceeded only by the number of papers written about the origin of dolomite, which is equally enigmatic. Because there is no modern analog for the iron-rich sediments, depositional models for these sediments have to be formulated mainly on the basis of the ancient record. Geologists with access to the same data quite commonly draw different inferences from these data; therefore, numerous models for the origin of iron-rich sedimentary rocks have been advanced—none of which has gained complete acceptance. Most of the controversy focuses on the origin of iron-formations, which form the bulk of the iron-rich rocks.

A number of important questions are involved in the debate about the origin of iron-formations and ironstones. Why, for example, are cherty, banded iron-formations confined primarily to the Precambrian? Why are iron-formations and ironstones not forming today? What is the source of the vast amount of iron locked up in the known iron-rich deposits? Was volcanism involved in some way in the formation of some or all of these iron deposits? How was iron transported from its source to the depositional site, and what was the depositional environment in which iron-rich sediments accumulated? Finally, are some or even all iron-rich deposits of secondary (replacement) origin?

Depositional Environments of Modern Iron-Bearing Minerals

There are no modern counterparts to the ancient environments that presumably favored widespread deposition of iron-rich sediments to produce iron-formations and ironstones, but iron-bearing minerals are being deposited on a small scale in a variety of modern environments. Iron sulfides, particularly pyrite (FeS_2), are forming in black muds that accumulate under reducing conditions in stagnant ocean basins, tidal flats, and organic-rich lakes. Iron sulfides are also accumulating around the vents of hot springs located on the crests of midocean ridges. Chamosite, a complex Fe-Mg-Al silicate, has been reported in modern sediments at water depths as great as 150 m in the Orinoco and Niger deltas and on the ocean floor off Guinea, Gabon, Sarawak, and in the Malacca Straits. Glauconite is a K-Mg-Fe-Al silicate mineral that has been reported from Monterey Bay, California, and from various other parts of the ocean at water depths ranging down to about 2000 m. Iron oxides such as goethite ($FeOOH$) are accumulating in some modern lakes and bogs, as oolites on the floor of the North Sea, and in manganese nodules in both seawater and fresh water. Manganese nodules, which contain manganese, copper, cobalt, nickel, and other metals in addition to iron, are particularly widespread in the Pacific Ocean at water depths of about 4–5 km.

The Problem of Iron Solubility

The solution of iron in source areas and its subsequent transport and deposition to form iron-rich deposits are governed by the Eh and pH of the environment. Iron in the oxidized or ferric (Fe^{3+}) state is much less soluble than iron in the reduced or ferrous (Fe^{2+}) state (Fig. 13.30). Ferric iron is soluble only at pH values less than about 4, which rarely occur under natural conditions. Eh-pH diagrams such as Figure 13.30 can be used in a general way to predict the stability of iron-bearing minerals and serve to illustrate that Eh is commonly more important than pH in determining the solubility of iron and the kind of iron-bearing mineral that will be deposited under given conditions. For example, hematite (Fe_2O_3) is precipitated under oxidizing conditions at the pH levels commonly found in the ocean and surface waters, siderite ($FeCO_3$) forms under moderately reducing conditions, and pyrite (FeS_2) forms under moderate to strong reducing conditions.

On the other hand, the iron geochemistry of natural systems is far more complex than the simplified conditions assumed in constructing Eh-pH diagrams, and such diagrams are of only limited use in environmental interpretations. Many problems are associated with the formation of sedimentary iron deposits, and the mechanisms by which transport and deposition of iron occurred in the past to form iron-formations and ironstones are still poorly understood and controversial. One of the principal problems stems

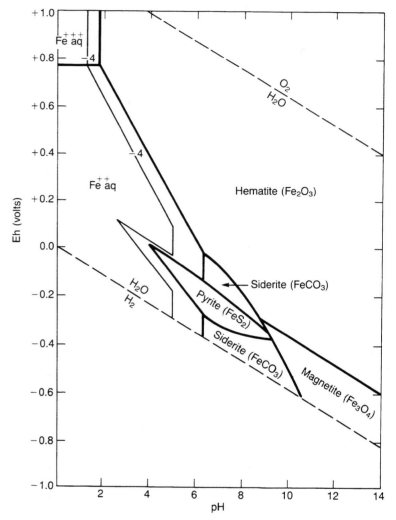

FIGURE 13.30

Eh-pH diagram showing the stability fields of the common iron minerals, sulfides, and carbonates in water at 25°C and 1 atmosphere total pressure. Total dissolved sulfur = 10^{-6}; total dissolved carbonate = 10^{0}. *(From Garrels, R. M., and C. L. Christ, Solutions, minerals, and equilibria. © 1965. Fig. 7.21, p. 224, reprinted by permission of Harper & Row, New York.)*

from the fact that iron tends to be insoluble in the oxidized or Fe^{3+} state. Assume for a moment that subaerial weathering of iron-rich minerals is the major source of iron for iron-formations. Given that oxidized iron is largely insoluble, how can large quantities of iron be taken into solution and transported from subaerial weathering sites under the oxidizing conditions that commonly prevail in streams and rivers?

This problem of solution and transport of iron under oxidizing conditions prompted some workers (Lepp and Goldich, 1964; Cloud, 1973; Lepp, 1987) to postulate that low oxygen levels existed during the Precambrian which allowed great quantities of iron to be transported in the soluble, reduced state (Fe^{2+}) to marine basins. Presumably, local oxidizing conditions existed within some parts of broad, shallow-marine basins, owing to photosynthetically generated oxygen, where the iron was oxidized to the insoluble ferric state (Fe^{3+}) and precipitated. The concept of low oxygen levels during the late Precambrian has been questioned by other investigators (e.g., Dimroth and Kimberley, 1976). In any case, this argument cannot be used to explain the solution and transport of iron during the Phanerozoic, when an oxidizing atmosphere clearly existed. Some workers have suggested that iron may have been transported as colloids by physical processes rather than in true solution or that it was sorbed to clay particles or organic materials and transported along with these substances. It is generally assumed by most investigators, however, that mechanisms such as colloidal transport cannot account for transport of the large quantities of iron that occur in iron-formations or ironstones (Ewers, 1983). Reducing conditions seem to be required for transport of large amounts of iron in solution, hence the dilemma. On the other hand, Schieber (1987) suggests that large quantities of iron might be moved in groundwater, where the iron could be dissolved in the reduced state. When groundwater is discharged to the drainage the iron oxidizes; however, instead of precipitating immediately the iron may form a stable sol and be transported on the last leg of its journey as a colloid.

Models for Deposition of Iron-Rich Sediments

Three principal aspects of the iron problems have to be addressed to account for the formation of iron-rich rocks: (1) the source of the iron, (2) transport of iron to the depositional basin in a soluble state, presumably under reducing conditions, and (3) episodic precipitation of the iron within the basin. Presumably, precipitation occurs under oxidizing conditions, in environments free of conspicuous clastic components, to produce rhythmic banding and lamination. A fourth problem with respect to origin of the banded iron-formations (BIFs) is the source of the chert and its mechanism of deposition. Many different hypotheses have been offered to get around the difficulties inherent in these problems. Garrels (1987) briefly summarizes some current opinions. Majority opinion indicates that the iron-rich rocks are of marine origin, with a strong minority opting for a freshwater origin. Many workers assume that BIFs originated under anoxic or very-low oxygen conditions, but that they are somehow related to the rise of atmospheric oxygen to present-day levels. Most believe that iron-rich rocks formed under climatic conditions similar to those of today, but some think the climate was hotter and some think it was cooler. Many early models for iron-formations assumed deposition in restricted environments in restricted or isolated basins rather than in the open ocean; however, several recent workers (e.g., Simonson, 1985) prefer an open-ocean environment. Some researchers believe that silica-secreting organisms were important in forming the cherts of BIFs; others do not. Still other researchers believe that bacteria may have been important in precipitating iron compounds.

Many workers originally assumed that the iron was derived by subaerial weathering of iron-silicate minerals; however, the problem of transport of ferric iron to depositional basins remains an obstacle if iron is derived from land. Also, the immense volume of rock that would have to be weathered to furnish the staggering amount of iron in large deposits such as the Hamersley deposits of Australia is mind-boggling. To get around these difficulties, some workers have suggested that the source of the iron was within the depositional basin close to the depositional site. For example, Drever (1974) proposed an upwelling model in which iron and silica are precipitated from marine bottom waters as a result of upwelling of bottom waters into an oxidizing environment. That is, localization of iron deposition is related to areas of upwelling of deep anoxic water that are charged with ferrous iron as a result of reduction of ferric oxide in terrigenous siliciclastic bottom sediments.

A somewhat similar model is proposed by Button et al. (1982). This model (Fig. 13.31) involves a Precambrian ocean with an upper oxic layer overlying a much larger volume of anoxic water. Ferrous iron accumulated from terrigenous and submarine

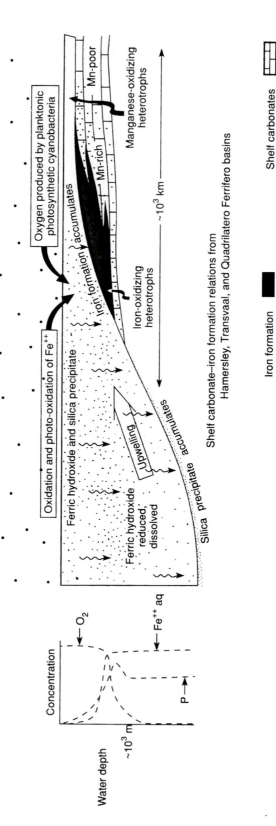

FIGURE 13.31
Conceptual model for the deposition of large iron-formations. *(After Button, A., et al., 1982, Sedimentary iron deposits, evaporites, and phosphorites, in Holland, H. D., and M. Schidlowski, eds., Mineral deposits and the evolution of the biosphere. Fig. 1, p. 261, reprinted by permission of Springer-Verlag, New York.)*

sources over a long period of time and is stored in basinal bottom waters. Lepp (1987) still maintains that the iron in BIFs must have been transported from land in the Fe^{2+} state under low-oxygen conditions in the Precambrian. Presumably, this transport took place during early Precambrian and the iron was then thoroughly mixed in the world ocean before later precipitation occurred. In any case, according to Button et al. (1982), even relatively low concentrations of ferrous iron in the anoxic water would be adequate, owing to the immense volume of water, to form BIFs. Upwelling of the Fe^{2+}-bearing anoxic waters would bring ferrous iron onto shelves or continent-margin basins where oxidation to Fe^{3+} and subsequent precipitation would occur, perhaps aided by iron-oxidizing bacteria. Oxygen in the shallow waters was presumably furnished by photodissociation of water vapor and photosynthesis by planktonic organisms. Seasonal variations in upwelling and organic activity might thus account for banding in the BIFs. Owing to absence of silica-secreting organisms in the Precambrian, Button et al. (1982) suggest that silica must have concentrated in the Precambrian ocean, building to concentrations close to saturation with amorphous silica. They visualize a more or less constant rain of silica-rich precipitates to the seafloor.

An intriguing recent hypothesis proposes that iron-formations are primarily exhalative or hydrothermal in origin (Gross, 1980; Simonson, 1985). According to this hypothesis, the source of the iron is located basinward from the depositional site, not landward. Hydrothermal activity from hot springs situated along midocean spreading ridges provides a source for iron and silica. The effluents from these hot springs are postulated to be supersaturated with iron and silica, which would commonly be dispersed from the deeper basins upward and outward to shallower water, where precipitation could take place. A major objection to this model is the great distance over which the hydrothermal solutes would have to be dispersed. Also, Lepp (1987) objects to Simonson's hypothesis of hydrothermal activity as a source for the iron. He maintains that the source had to be a well-mixed (ocean) reservoir and that volcanic exhalations do not fit this requirement.

The difficulties in explaining large-scale solution transport and precipitation of iron have stimulated some investigators to propose that both iron-formations and ironstones may be secondary deposits that were created by iron replacement of original carbonate minerals or other minerals during burial and diagenesis. It has long been recognized that some ironstone deposits are at least partly secondary in origin because of the presence in these deposits of relict calcium carbonate constituents such as oolites and fossils, which have clearly been replaced by iron. Dimroth (1979) suggests that iron-formations as well as ironstones were most likely precipitated initially as $CaCO_3$. The calcium carbonate deposits were subsequently replaced during diagenesis by silica and iron to form the cherty-iron deposits. The sources of the iron needed to bring about such massive replacement and the mechanisms of selective iron and silica replacement that would be required to produce the alternating bands of silica-rich and iron-rich sediments are still highly speculative. This postulated replacement mechanisms would probably require that ferrous iron be introduced uniformly into the carbonate deposits over a wide area under reducing conditions, with subsequent intermittent change to oxidizing conditions to allow precipitation of the iron to occur. Ewers (1983) considers this process unlikely, at least for many iron-formations. Other models for origin of iron-formations have also been proposed. For example, Garrels (1987) proposes that the microbands in Australian BIFs formed by evaporation of water in restricted basins.

No model for the origin of iron-formations has been fully accepted by all research-ers, and many puzzling aspects of the formation of sedimentary iron deposits remain. Among other things, any general model for the origin of iron-rich sedimentary rocks should apply to the Phanerozoic-age ironstones as well as to Precambrian BIFs, although there must have been some important differences in the conditions under which these two kinds of iron-rich rocks formed. We still do not have a completely adequate explanation for the banding in iron-formations and the absence of banding in ironstones. Further, why was chert deposition confined primarily to the Precambrian? Also, we are unsure of the role that organisms may have played in the deposition of iron-formations and ironstones. Was the local production of oxygen by photosynthesizing organisms such as algae im-portant? Did low forms of life such as bacteria and algae catalyze or initiate precipitation in some manner? If so, how did they cause precipitation and how important was such biologic activity? LaBerge et al. (1987) believe that the iron in iron-formations is pre-cipitated as ferric hydrate and that the precipitation was effected mainly by bacterial stripping and algal photosynthesis. Furthermore, these authors believe that the chert associated with iron-formations may be the result of organic precipitation of the silica in the form of siliceous frustules; in other words, that organisms capable of secreting sili-ceous tests may have existed in the Precambrian. Clearly, there is still considerable difference of opinion about the origin of iron-formations, and it seems highly probable that the debate over the origin of iron-rich sedimentary rocks will continue for many years.

13.5   Sedimentary Phosphorites

13.5.1   Introduction

Phosphorites are sedimentary deposits containing more than about 15–20 percent P_2O_5. Thus, they are significantly enriched in phosphorus over most other types of sedimentary rocks. The phosphorus content of average shales is about 0.11–0.17 percent P_2O_5, the average sandstone contains about 0.08–0.16 percent P_2O_5, and the average limestone about 0.03–0.7 percent P_2O_5 (McKelvey, 1973). Shales, sandstones, or limestones that contain less than 20 percent P_2O_5 but which are enriched in phosphorus over that found in average sediments are referred to as phosphatic, e.g., phosphatic shale. Phosphorus-rich sedimentary rocks are called by a variety of names—**phosphate rock, rock phos-phate, phosphates, phosphatites,** and **phosphorites.** Not all of these terms have exactly the same meaning, however, and geologists do not agree about which is most appropriate as a name for phosphorus-rich sedimentary rocks. In this book, the term phosphorite is used with the meaning given at the beginning of this paragraph.

The total volume of sedimentary phosphates in the geologic record is quite small; however, sedimentary phosphorites are of special economic interest. They contribute about 82 percent of the world's production of phosphate rock and make up about 96 percent of the world's total resources of phosphate rock (Notholt et al., 1989a). The remaining phosphate resources are in igneous rocks. The most important production at this time comes from rocks of Miocene–Pliocene, Late Cretaceous–Eocene, and Permian ages in the USA and from Cambrian rocks in the USSR and China. Sedimentary phos-phates occur in rocks of all ages from Precambrian to Holocene. Modern phosphorites also occur as nodules on some parts of the ocean floor. The origin of phosphorites is perhaps less enigmatic than that of iron-rich sedimentary rocks; nonetheless, many aspects of their origin are still poorly understood. For example, modern ocean water contains only about 70 parts per billion (ppb) phosphorus, but some ancient phosphorite deposits contain 30–40 percent P_2O_5. These values represent a concentration factor of up to two million times as great as the phosphorus content of the ocean and 200–300 times greater than the P_2O_5 content of average sedimentary rock. We are left to ponder the special depositional conditions that must have existed to produce such significant enrichment. Also, we do not have a clear understanding of why phosphorites accumulated in so much greater volumes at certain times in the geologic past than they are accumulating in the modern ocean.

13.5.2   Occurrence and Distribution

Most phosphate deposits occur in marine sedimentary sequences. They have been reported in rocks of all ages and on all continents, although they are unevenly distributed in space and time. Figure 13.32 shows the distribution of major phosphate deposits. Phosphorite deposition appears to have been particularly prevalent during the Late Precambrian and Cambrian in central and southeast Asia, the Permian in North America, the Jurassic and Early Cretaceous in eastern Europe, the Late Cretaceous to Eocene in the

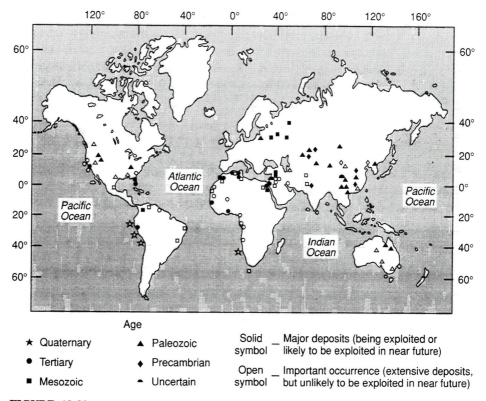

Age

★ Quaternary	▲ Paleozoic	Solid symbol — Major deposits (being exploited or likely to be exploited in near future)
● Tertiary	◆ Precambrian	Open symbol — Important occurrence (extensive deposits, but unlikely to be exploited in near future)
■ Mesozoic	▲ Uncertain	

FIGURE 13.32

Worldwide distribution of major sedimentary phosphorite deposits. *(After Cook, P. J., 1976, Sedimentary phosphate deposits, in Wolf, K. H., ed., Handbook of strata-bound and stratiform ore deposits. Fig. 1, p. 505, reprinted by permission of Elsevier Science Publishers, Amsterdam.)*

Tethyan Province of the Middle East and North Africa, and the Miocene of southeastern North America (Cook, 1976). In the United States, for example, deposits containing more than 20 percent P_2O_5 occur in Florida, Georgia, and North Carolina (Miocene-Pliocene), in the Western Phosphate Field (Permian) in Idaho, and in Utah (Carboniferous). Cook and McElhinney (1979) show that most ancient phosphorite deposits accumulated at latitudes between about 0 and 40 degrees. It is not fully understood why phosphorite deposition was favored at low latitudes, although it is widely believed that upwelling in the ocean played a major role in the origin of these deposits.

Deposits of phosphorite nodules and phosphatic sediments occur also on the present ocean floor, mainly at depths less than about 400 m and mostly in the vicinity of coastlines. They are concentrated particularly off the coasts of southern California and Baja California, eastern United States, Peru and Chili, and southwest Africa. They occur also on the floors of the Pacific, Indian, and Atlantic oceans and on some seamounts (Baturin, 1982; Fig. 13.33). Most of these phosphate nodule accumulations are deposits older than the Holocene that formed on the seafloor more than 700,000 years ago (Kolodny and Kaplan, 1970). Some modern phosphorite nodules are now forming on the ocean floor in a few places such as the Peru-Chile continental margin (Burnett and Froelich, 1988) and the Namibian shelf off southwest Africa. Holocene-age phosphorite nodules have also been reported from estuarine environments (Cullen et al., 1990).

In ancient phosphorite deposits, phosphate-rich layers typically occur interbedded with carbonate rocks, shales, or chert. A characteristic feature of many major phosphorite accumulations is the triple association of phosphate, chert, and sediments containing abundant organic carbon. The phosphatic beds of some deposits are lenticular over short distances, whereas others are very uniform in lithology and thickness over hundreds of square kilometers (Cook, 1976). Phosphatic rocks commonly grade regionally into non-

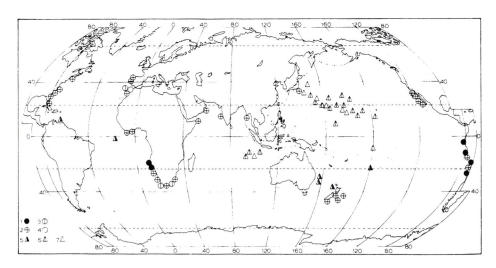

FIGURE 13.33

Distribution of phosphorite nodules and phosphatic sediments on the ocean floor: 1–4 show the locations of phosphorites on continental shelves; 5–7, phosphorites on seamounts. Ages of the phosphorites are 1, Holocene; 2 and 5, Neogene; 3 and 6, Paleogene; 4 and 7, Cretaceous. *(From Baturin, G. N., 1982, Phosphorites on the sea floor: Developments in Sedimentology 33. Fig. 2.1, p. 56, reprinted by permission of Elsevier Science Publishers, Amsterdam.)*

phosphatic sedimentary rocks of the same age. The stratigraphic associations of some major phosphate deposits are illustrated in Figure 13.34. One of the best-studied examples of an ancient phosphorite deposit is the Permian-age Phosphoria Formation of western Wyoming and Idaho (Fig. 13.34). This formation is several hundred meters thick. Phosphatic members of the formation reach 30–35 m in thickness and cover an area of thousands of square kilometers. Much thicker phosphate units occur in some phosphatic sequences, such as in the Cambrian Karatou deposits of the USSR (Fig. 13.34). For an excellent review of the distribution and characteristics of the major phosphate deposits of the world, readers should consult *Phosphate Deposits of the World*, edited by Notholt et al. (1989b). Additional details of the distribution of sedimentary phosphorites are given in Bentor (1980) and in Vol. 136 of the *Journal of Geological Society* (London) (1980).

13.5.3 Composition of Phosphorites

Sedimentary phosphates are composed of phosphate minerals, all of which are varieties of apatite. The principal varieties are fluorapatite $[Ca_5(PO_4)_3F]$, chlorapatite $[Ca_5(PO_4)_3Cl]$, and hydroxyapatite $[Ca_5(PO_4)_3OH]$. Most sedimentary phosphates are carbonate hydroxyl fluorapatites in which up to 10 percent carbonate ions can be substituted for phosphate ions to yield the general formula $Ca_{10}(PO_4,CO_3)_6F_{2-3}$. Many other substitutions of both cations and anions in the fluorapatite structure are possible. See Nathan (1984) for details. These carbonate hydroxyl fluorapatites are commonly called **fran-**

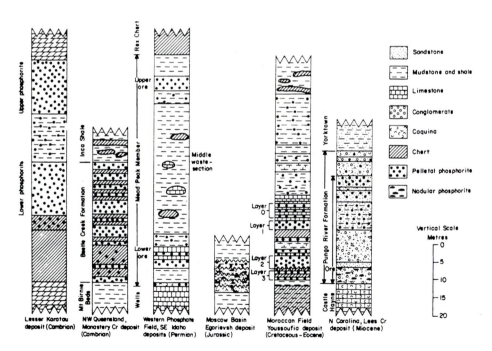

FIGURE 13.34

Representative strategic sections of some major phosphorite deposits showing various kinds of sedimentary facies associated with phosphorites. *(After Cook, P. J., 1976, Sedimentary phosphate deposits, in Wolf, K. H., ed., Handbook of strata-bound and stratiform ore deposits. Fig. 2, p. 506, reprinted by permission of Elsevier Science Publishers, Amsterdam.)*

colite. Although more than 300 phosphate minerals are known, francolite is virtually the only phosphate mineral that occurs in unweathered marine phosphorites. The wastebasket term **collophane** is often used for sedimentary apatites for which the exact chemical composition has not been determined.

Phosphorites commonly also contain some detrital quartz and authigenic chert. Opal-CT may also be present. The common association of cherts, carbonates, and organic-rich shales with phosphorites has already been mentioned. Both calcite and dolomite may occur in phosphorites, and dolomite may be particularly abundant. Glauconite, illite, montmorillonite, and zeolites, especially clinoptilolite, may also be present in some deposits. Organic matter is a characteristic constituent of many phosphorites (Nathan, 1984).

The chemical composition of phosphorites is dominated by P, Si, and Ca. Moderate amounts of Al, Fe, Mg, Na, K, S, F, and Cl may also be present. Representative analyses of phosphorites from various parts of the world are given in Table 13.10. Many trace elements appear to be enriched in phosphorites over their average composition in seawater, the crust, and the average shale. In particular, Ag, Cd, Mo, Se, Sr, U, Y, Zn, and the rare earth elements (R.E.E.) show this enrichment. According to Nathan (1984), enrichment in these trace elements stems from the marine origin of phosphorites, their paragenetic association with organic matter, and the crystallographic properties of francolite.

13.5.4 Petrography of Phosphates

Phosphorites have textures that resemble those in limestones. Thus, they may contain ooids, peloids, fossils (bioclasts), and clasts. These grains are, of course, composed of apatite rather than calcite. Some phosphorites lack distinctive granular textures and are composed instead of fine, micritelike textureless collophane. Cook (1976) refers to this kind of phosphorite as **pholutite.** Sedimentary apatites are quite commonly cryptocrystalline; that is, the crystals are too small to be visible under the petrographic microscope. Thus, in polarized light they appear as brownish isotropic material; however, when crystal size increases above a few micrometers, their birefringence appears (Slansky, 1986, p. 41). Phosphatic grains such as pellets may contain inclusions of organic matter, clay minerals, silt-size detrital grains, and pyrite. The abundant organic matter in phosphorites contributes to their color, which is commonly dark brown to black.

Peloidal or pelletal phosphorites are particularly common, e.g., the phosphates of the Permian Phosphoria Formation, USA (Sheldon, 1989), and many other phosphorite deposits. Some peloids are essentially structureless, whereas others contain a nucleus of some type. Still others are compound, grapestonelike particles. Some peloids may be of fecal origin (i.e., they are fecal pellets); however, such an origin apparently remains uncertain. Peloids may contain various kinds of fossils such as foraminiferal or radiolarian tests, as well as quartz, clay minerals, and organic matter, as mentioned. Peloidal phosphates may also include ooids that contain a nucleus and display well-developed concentric layering (Fig. 13.35). Ooids are less common in phosphorites than in peloids but may make up most of the phosphatized grains in some deposits, e.g., phosphatic ooids in the Lower Jurassic of central England (Horton et al., 1980). Soudry and Southgate (1989) report pelletal phosphates from the Georgina Basin, Australia, which they interpret as the remains of solitary and colonial unicells and microbial sheaths that once formed part of a microbial mat community in a vadose environment. Phosphatized fossils or fragments of

TABLE 13.10

Chemical composition of selected phosphorite samples of various ages

Origin	Age	P_2O_5	SiO_2	Al_2O_3	Fe_2O_3	CaO	MgO	Na_2O	K_2O	CO_2	SO_3	F^-	Cl^-	H_2O^-	H_2O	C org.	$\dfrac{CaO}{P_2O_5}$	$\dfrac{F}{P_2O_5}$	$\dfrac{CO_2}{P_2O_5}$	$\dfrac{SO_3}{P_2O_5}$
Peru-Chile (offshore)	Holocene	22.61	22.13	5.15	2.85	33.93	1.07	0.85	1.30	3.07		2.22			1.88		1.50	0.098		
Florida ('pebble')	Pliocene	32.07	9.31	1.29	1.57	46.98	0.19	0.21	0.13	3.00	0.59	3.68	0.013	0.42	1.81	0.053	1.46	0.115	0.09	0.018
Venezuela (Riecito)	Miocene	34.28	7.05	1.00	0.69	48.05	0.23	0.70	0.08	3.00	0.70	2.23	0.08				1.40	0.065	0.08	0.02
Christmas Island	Miocene	38.50	0.79		0.41	52.10	0.10	0.20	0.03	1.20		1.14		1.43	1.97		1.35	0.029	0.03	
		37.40	3.10	3.10	1.17	48.60	0.15	0.22	0.06	2.00		2.17		1.73	3.19	0.09	1.29	0.058	0.05	
Senegal (Taiba)	Mid. Eocene	33.30	7.30	3.20	3.60	45.10	0.60	0.30	0.21	1.40		3.75			0.60		1.35	0.112	0.04	
Benin	Mid. Eocene	28.15	13.15	5.40	0.17	40.94	0.52	0.27	0.05	1.79		2.65	0.48	1.77	3.45	0.07	1.45	0.094	0.06	
Togo (concentrate)	Mid. Eocene	36.85	2.99	1.00	1.30	51.69	0.03					3.75	0.12		1.44		1.40	0.101		
Morocco (Khouribga)	Ear. Eocene	34.26	0.03	0.37	0.26	52.78	0.48	0.84	0.09	3.59	1.59	3.05	0.03		2.34		1.54	0.089	0.10	0.04
(Bou Craa)		34.70	8.07	0.71	0.37	50.45	0.19	0.07	0.10	1.98	0.66	3.56			0.74		1.45	0.102	0.05	0.019
Tunisia (Metlaoui)	Ear. Eocene	26.09	8.90	1.53	0.60	42.85	0.50	1.45	0.38	4.62	3.90	2.98	0.09	3.16	3.00	0.93	1.64	0.114	0.17	0.149
	Ear. Eocene	24.70	7.80	0.52	0.71	44.16	3.04	1.23	0.19	10.95		2.92		0.42	1.46	0.34	1.78	0.118	0.44	
Egypt (Nile Valley)	La. Cretaceous	26.00	9.25	0.53	1.77	45.14	0.77	0.78	0.12	8.92		2.57	0.24	0.66	1.73	1.32	1.73	0.098	0.34	
(Quseir)	La. Cretaceous	25.20	12.50	0.84	1.86	40.66	1.75	0.68	0.10	5.58		2.57	1.06	2.11	2.19	1.20	1.61	0.101	0.22	
Israel (Oron)	La. Cretaceous	25.20	2.00	0.50	0.30	52.50	0.20	0.80	0.03	13.00	1.80	2.90	0.40				2.08	0.115	0.51	0.07
Syria	La. Cretaceous	24.50	6.90	0.08	0.25	44.16	0.27	0.52	0.03	2.92	11.20	2.65	0.26	4.65	1.90	0.26	1.80	0.108	0.12	0.45
Colombia	La. Cretaceous	28.04	17.40	3.10	1.72	40.08	0.10	0.10		2.77		1.98					1.42			
Thailand (Maeta)	Permian	38.84	0.01	0.51	0.97	51.10		0.18	0.03	2.40	0.13	0.78	0.06	0.49	2.26	0.06	1.31	0.020	0.06	0.003
India (Mussoorie)	Permian	22.50	7.05	0.66	2.56	40.55	6.00		0.24	15.10	1.35	2.15	0.06	0.24	0.75	0.76	1.80	0.095	0.67	0.06
U.S.A. (Rocky Mountains)	Permian	30.50	11.90	1.70	1.10	44.00	0.30	0.60	0.50	2.20	1.80	3.10		0.60	1.60	2.10	1.44	0.101	0.07	0.059
Australia (Duchess)	Cambrian	37.20	2.60	0.85	0.94	51.70	0.15	0.28	0.09	1.63	0.56	3.20		0.53	1.42	0.60	1.38	0.086	0.04	0.015
(Lady Annie)		35.00	10.30	1.59	0.14	48.20	0.15	0.04	0.09	1.20	0.07	3.16		0.53	0.70	0.60	1.37	0.090	0.03	0.002
China	La. Cambrian	23.41	19.74	3.12	1.90	41.98	0.21	0.54		5.52		2.72	0.67	0.33	0.24	0.62	1.79	0.116	0.23	
Benin (Mekrou)	Cambrian	28.25	24.65	1.28	2.14	38.55	0.08	0.18	0.10	1.12		2.60	0.05	0.21	1.29	0.07	1.36	0.092	0.03	
India (Udaipur)	Precambrian	26.12	20.93	3.02		40.28				4.03		2.01			0.46		1.54	0.076	0.15	
		35.46	10.56	1.18		47.75	0.28			0.94	0.21	1.60			1.66		1.34	0.045	0.02	0.05

Source: Slansky, M., 1986, Geology of sedimentary phosphorites. Table 9, p. 70, reproduced by permission of North Oxford Academic Publishers, Essex, Great Britain.

FIGURE 13.35
Brown, phosphatic ooids in a phosphorite. Phosphoria Formation (Permian), northern Rocky Mountains. Ordinary light. Scale bar = 0.1 mm.

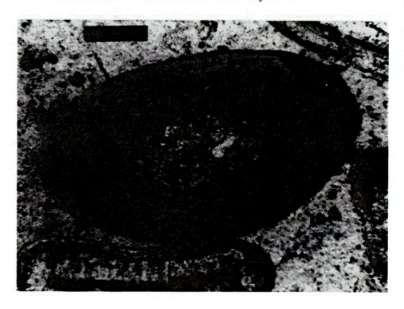

originally phosphatic shells may (in some cases) form a significant part of some phosphorite deposits. Fossils may include both invertebrate and vertebrate remains. Lithoclasts consisting of reworked fragments of preexisting phosphorites also occur in some deposits, e.g., Miocene phosphorites of Florida (Riggs, 1980).

Most grains in phosphorites are sand size; however, particles greater than 2 mm may occur. These large grains are commonly referred to as nodules and can range in size from 2 mm to several tens of centimeters. Like nodules in limestones, they can be spherical, lenticular, or flattened parallel to bedding.

No scheme for classification of phosphorites on the basis of texture appears to be in wide use at this time, although Mabie and Hess (1964) developed a very comprehensive classification for peloidal phosphorites in which they describe more than thirty different textures in phosphate ores of the Western Phosphate Field, USA. Their classification system is based on the internal structure of the phosphate particles, the apatite crystal size, and the degree of compaction of the rock as determined by the types of grain-to-grain contact. Slansky (1986) advocates use of a classification system, based to some extent on Folk's limestone classification, that combines grain size and composition terms. This classification yields names such as **biophospharenite,** a phosphorite composed of sand-size phosphatized fossils, and **intraphospharudite,** a phosphorite composed of gravel-size phosphatized clasts.

13.5.5   *Principal Kinds of Phosphate Deposits*

Phosphate-rich sedimentary rocks may occur in layers ranging from thin laminae a few millimeters thick to beds a few meters thick. No widely accepted scheme for classifying phosphorite deposits appears to be in general use. Cook (1976) mentions use of a threefold classification based essentially on analogy with the characteristics of phosphorite deposits in three well-studied areas of the United States: (1) **geosynclinal** or **west-coast type** (e.g., the Western Phosphate Fields of the United States), (2) **platform** or **east-coast type** (e.g., the North Carolina deposits), and (3) **weathered** or **residual types** (e.g., the brown rock deposits of Tennessee). The principal characteristics of the west-coast and east-coast types are shown in Table 13.11. Residual phosphorites form as a result of concentration of phosphate by chemical or mechanical weathering and are thus not primary sedimentary phosphorites. Cook (1976) points out that phosphorite deposits may have characteristics of more than one of these types of deposits and that there are also undesirable genetic implications involved in using terms such as geosynclinal and platform that detract from the usefulness of this classification scheme.

A second approach to classification of phosphorite deposits is to divide them into groups based on bedding characteristics and the principal types of phosphate materials that make up the deposits. Five principal groups can be identified:

1. **Bedded phosphorites** form distinct beds of variable thickness, commonly interbedded and interfingering with carbonaceous mudrocks, cherts, and carbonate rocks. The phosphorite occurs as pellets, oolites, pisolites, phosphatized brachiopods and other skeletal fragments, and cements. A classic example of a bedded phosphate deposit is the Permian Phosphoria Formation of the northwestern United States. This formation has a total thickness of 420 m and extends over an area of about 350,000 km^2 (McKelvey et al., 1959; Sheldon, 1989). Bedded phosphates are believed to form on shelf areas associated with zones of upwelling in the ocean.

2. **Bioclastic phosphorites** are a special type of bedded phosphate deposit composed largely of vertebrate skeletal fragments such as fish bones, shark teeth, fish scales, coprolites, etc. Deposits composed mainly of invertebrate fossil remains such as phosphatized brachiopod shells are also known. These phosphate-bearing organic materials commonly become further enriched in P_2O_5 during diagenesis and may be cemented by phosphate minerals.

3. **Nodular phosphorites** are brownish to black spherical to irregular-shaped nodules ranging in size from a few centimeters to a meter or more. Internal structure of phosphate nodules ranges from homogeneous (structureless) to layered or concentrically banded. Phosphatic grains, pellets, shark teeth, and other fossils may occur within the nodules. Phosphate nodules are forming today in zones of upwelling in the ocean, and many ancient nodular phosphorites may have a similar origin; however, some ancient

TABLE 13.11

Differences in west-coast and east-coast types of phosphorite deposits

Feature	Geosynclinal or west-coast type	Platform or east-coast type
Phosphorite type	Generally pelletal; minor oolitic	Pelletal, nodular, and nonpelletal
Matrix to pellets	Argillaceous or siliceous	Quartzose (sandy) or calcareous
Grade of the phosphorite	High grade	Low grade
Nature of the deposit	Thick extensive continuous beds	Discontinuous beds
Fauna	Pelagic	Shallow water
Inferred water depth	Hundreds of meters	Tens of meters
Sediment association	Black shale-chert	Carbonates, sands
Influence of synsedimentary structures	Generally lacking	Deposits commonly in synclines or the flanks of anticlines
Tectonics	Strongly folded	Gently folded

Source: Cook, P. J., 1976, Sedimentary phosphate deposits, *in* Wolf, K. H., ed., Handbook of strata-bound and stratiform ore deposits, v. 7. Table I, p. 511, reprinted by permission of Elsevier Science Publishers, Amsterdam.

phosphorite nodules may be of diagenetic origin. Some are relict features of disconformity and unconformity surfaces. For a thorough description of phosphatic nodules in one modern environment, the Peru continental margin, see Burnett and Froelich (1988).

4. **Pebble-bed phosphorites** are composed of phosphatic nodules, phosphatized limestone fragments, or phosphatic fossils that have been mechanically concentrated by reworking of earlier-formed phosphate deposits.

5. **Guano deposits are** composed of the excrement of birds and possibly bats that has been leached to form an insoluble residue of calcium phosphate. Guano occurs today on small oceanic islands in the Eastern Pacific and the West Indies. Guano deposits are not important in the geologic record.

13.5.6 Deposition of Phosphorites

Chemical and Biochemical Processes

As mentioned, the principal phosphate minerals in sedimentary rocks are various varieties of apatites, of which carbonate-apatite $[Ca_{10}CO_3(PO_4)_6]$ is most important. The conditions that favor precipitation of calcium carbonate also favor formation of carbonate-apatite, although carbonate-apatite can precipitate at values of pH possibly as low as 7.0, whereas calcium carbonates generally do not precipitate below a pH of about 7.5 (Bentor, 1980). Carbonate-apatite precipitation also appears to be favored by slightly reducing conditions. The factors affecting phosphate solubility are not thoroughly understood, however, and the solubility of carbonate-apatite has not been definitely established. It is not known if the ocean is saturated with carbonate-apatite, although it is believed that it is very near saturation (Kolodny, 1981). The average concentration of phosphorus in the ocean is 70 ppb (parts per billion) (Gulbrandsen and Roberson, 1973) compared to 20 ppb in average river water. The concentration of phosphorus in ocean water ranges from only a few ppb in surface waters, which are strongly depleted by biologic uptake, to values of 50–100 ppb at depths greater than about 200–400 m.

Phosphorus is removed from seawater in several ways. Some phosphorus is precipitated along with calcium carbonate minerals during deposition of limestones; however, the average limestone contains only about 0.04 percent P_2O_5. Phosphorus is also removed from ocean water by concentration in the tissues of organisms, but this phosphorus is returned to the ocean when organisms die unless they are quickly buried by sediment before decay of the organic tissue is complete. Some phosphorus may be removed from seawater by incorporation into metalliferous sediments as a result of adsorption onto metallic minerals such as iron hydroxides. Finally, phosphorus is removed from ocean water in some manner to form marine apatite deposits. We are particularly interested in this latter process because the major problem associated with phosphorite deposition is identifying a mechanism that can explain how trace amounts of phosphorus in seawater can be concentrated to form phosphorite deposits.

A secondary or replacement origin has been suggested for some phosphorite deposits to account for this enrichment. On the other hand, the preservation in many other phosphorite deposits of clastic textures and primary sedimentary structures such as cross-bedding and laminations indicates that these sediments are primary deposits. Several geologists who made early studies of phosphorite deposits suggested an association between phosphorite deposition and areas of upwelling in the oceans. Studies of the distribution of ancient phosphorites show that most occur in lower latitudes in the trade wind belts along one side of a basin where deeper water could have upwelled adjacent to a continent. Most phosphate nodule deposits on the modern ocean floor also occur in areas of upwelling.

Early ideas on upwelling and phosphorite deposition assumed that inorganic precipitation of apatite occurred as cold, deep, phosphate-rich waters upwelled onto a shallow shelf. Under these postulated conditions, carbon dioxide would be lost from the upwelling waters owing to pressure decrease, warming, or photosynthesis, causing pH to increase and carbonate-apatite precipitation to occur. It has now been established, however, that Mg^{2+} ions in seawater have an inhibiting effect on the growth of carbonate-apatite crystals in much the same way that they inhibit the precipitation of calcite (Martens and Harris, 1970). Also, most or all of the phosphorus brought to surface waters by upwelling currents is quickly used up by organisms that utilize phosphate as one of the essential nutrients needed for organic growth. Rapid biologic utilization prevents phosphate levels in the ocean from rising to the point of saturation. These two factors taken together thus appear to rule out the direct inorganic precipitation of apatite from open-ocean water.

Nonetheless, biologic utilization of phosphate to build soft body tissue appears to provide the answer to the problem of phosphate concentration in sediments. Modern phosphate nodules are forming in areas of oceanic upwelling where a steady supply of phosphate brought from the large deep-ocean reservoir allows continuous growth of organisms in large numbers. After death, organisms and organic debris not consumed by scavengers pile up on the ocean floor under reducing conditions, where decay is inhibited. These organic materials include the remains of phytoplankton and zooplankton, coprolites (feces), and the bones and scales of fish. Under the toxic, reducing conditions of the seafloor, some of the soft body tissue is thus preserved long enough to be buried and incorporated into accumulating sediment. Perhaps about 1-2 percent of the total phosphorus involved in primary productivity in upwelling zones is ultimately incorporated into the sediments in this way (Baturin, 1982).

Slow decay of body tissue after burial releases phosphorus to the interstitial waters of the sediment. Studies of the chemistry of interstitial waters in sediments where modern phosphate nodules are forming, and in other areas of the seafloor where organic-rich sediments are accumulating under reducing conditions, have turned up phosphorus concentrations in the interstitial waters of these sediments ranging from 1400 ppb to as much as 7500 ppb (Bentor, 1980; Froelich et al., 1988). At such high concentrations of phosphorus in interstitial waters, the waters are supersaturated with respect to calcium phosphate. The phosphate thus begins to precipitate on the surfaces of siliceous organisms, carbonate grains, particles of organic matter, fish scales and bones, siliciclastic mineral grains, or older phosphate particles (Baturin, 1982). Phosphorite nodules thus form **within** the sediments by diagenetic reactions between organic-rich sediments and their phosphate-enriched interstitial waters. Mg^{2+}/Ca^{2+} ratios are possibly lowered below the threshold values, inhibiting apatite formation owing to magnesium iron replacements in clay minerals in the anoxic marine sediments (Drever, 1971).

Froelich et al. (1988) report that carbonate fluorapatite is precipitating from pore waters in the upper few centimeters of organic-carbon-rich muds on the Peru continental margin. They suggest that the mechanism of phosphate release may be linked to dissolution of fish debris or the presence of microbial mats in surficial sediments. Mg concentrations in the pore waters are nearly the same as that of seawater. Thus, for reasons that are poorly understood, Mg depletion is apparently not a prerequisite for carbonate fluorapatite precipitation in these sediments. In the Peru continental-margin sediments, phosphatic minerals precipitate only at very shallow depths. Below a depth of a few centimeters, excessive carbonate-ion concentration and diminished reactive iron and sulfate compositions favor dolomite formation, while precluding further precipitation of phosphatic minerals (Glenn and Arthur, 1988). Glenn and Arthur suggest that precipitation of phosphatic minerals in the Peru sediments may be influenced in some poorly understood way by filamentous bacteria and chemical absorption of clays into organic compounds.

Physical Processes

The presence of clastic textures and primary depositional sedimentary structures in some phosphorite deposits seems inconsistent with this proposed diagenetic concentration mechanism. Therefore, Kolodny (1980) suggested a two-stage process for the origin of ancient phosphorite deposits. In the first stage, apatite forms diagenetically in stagnant, reducing basins by phosphorus mobilization in interstitial waters in the manner postulated for formation of modern phosphorite nodules. The final stage involves reworking and enrichment of these diagenetically formed nodules by mechanical concentration processes. This stage is characterized by oxidizing conditions, and concentration presumably takes place in a high-energy environment during lower stands of sea level. This final stage, during which the original diagenetically formed phosphorite sediments are mechanically reworked under shallow-water conditions, accounts for the clastic textures and primary sedimentary structures found in many ancient phosphorites.

Support for this concept of mechanical concentration is provided by Glenn and Arthur (1990) in a study of major phosphorite deposits of Egypt. They report that phosphorite grains that were precipitated initially in reducing shales have been mechanically reworked and concentrated. Reworking occurred as a result of sea-level fall and delta progradation. Further, phosphorites in some of the deposits were concentrated into giant phosphorite sand waves owing to storm waves and tidal processes.

In summary, upwelling of phosphate-rich waters from deeper parts of the ocean and biologic utilization of the phosphate in soft body tissue appear to be important factors in the origin of phosphorite deposits. Phosphorus is deposited on the seafloor in organic detritus and buried with accumulating sediment. Phosphate becomes concentrated in the pore waters of sediment during slow decay of the phosphate-bearing, soft-bodied organisms and other organic detritus. Carbonate-apatite precipitates diagenetically from these phosphate-enriched pore waters by some process not yet fully understood to form gel-like, gradually hardening, phosphate concretions. Subsequently, these diagenetic deposits are reworked mechanically owing to lowered sea levels, allowing final concentration and deposition of phosphatic sediments by waves and currents. These processes are summarized diagrammatically in Figure 13.36.

This postulated multistage process for formation of phosphorite deposits is a plausible hypothesis, but it has some flaws. It does not, for example, explain why phosphorites accumulated on a much vaster scale at some times in the geologic past than at present. Bentor (1980) suggests that some additional mechanism, not realized or understood at the present time, may have been important in the past. Research on the origin of phosphorites is still very much in a state of flux. Interested readers may wish to consult Slansky (1986, p. 83–161) for other ideas on the origin of phosphorites.

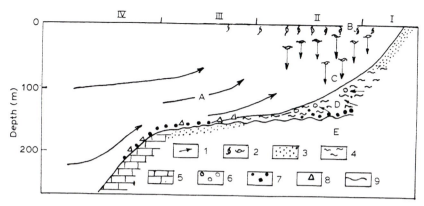

FIGURE 13.36

Schematic diagram illustrating formation of phosphorites in areas of upwelling on ocean shelves. The letters and numbers refer to A, supply of phosphorus to the shelf by upwelling waters; B, consumption of phosphorus by organisms; C, deposition of phosphorus on the bottom in biogenic detritus and burial by accumulating sediment; D, formation of phosphate concretions in the biogenic sediment by diagenetic processes; E, mechanical reworking of sediments and concentration of phosphate concretions when sea level is low. Zone I is the zone of shallow-water clastic deposits; Zone II is the zone where high contents of phosphate-rich biogenic detritus accumulate in the sediment; Zone III is the zone of reworking of phosphate-rich sediments during lowered sea level; and Zone IV is a deeper-water zone where carbonate sediments with local phosphate concentrations occur. The arrows refer to 1, paths of movement of phosphorus in the ocean and interstitial waters; 2, plankton; 3, clastic sediments; 4, biogenic siliceous, siliceous-clastic, and siliceous-carbonate sediments; 5, carbonate sediments; 6, unconsolidated phosphate concretions; 7, dense phosphate concretions; 8, glauconite; 9, erosional surface. *(From Baturin, G. N., 1982, Phosphorites on the sea floor: Developments in Sedimentology 33. Fig. 5.4, p. 227, reprinted by permission of Elsevier Science Publishers, Amsterdam.)*

Additional Readings

Evaporites

Borchert, H., and R. O. Muir, 1964, Salt deposits: The origin, metamorphism and deformation of evaporites: Van Nostrand, New York, 338 p.

Braitsch, O., 1971, Salt deposits: Their origin and composition: Springer-Verlag, New York, 279 p.

Chambre Sydnical de la Recherche et de la Production due Petrole et du Gaz Naturel, 1980, Evaporite deposits (Illustrations and interpretation of some environmental sequences): Editions Technip, Paris, 284 p.

Dean, W. E., and B. C. Schreiber, 1978, Marine evaporites: Soc. Econ. Paleontologists and Mineralogists Short Course Notes 4, Tulsa, 193 p.

Kirkland, D. W., and R. Evans, eds., 1973, Marine evaporites: Origin, diagenesis and geochemistry: Dowden, Hutchinson and Ross, Stroudsburg, Pa., 444 p.

Peryt, T. M., ed., Evaporite basins. Lecture notes in earth science: Springer-Verlag, Berlin, 188 p.

Renault, R. W., ed., 1989, Sedimentology and diagenesis of evaporites: Special issue of Sed. Geology, v. 64, p. 207–298.

Schreiber, B. C., ed., 1988, Evaporites and hydrocarbons: Columbia University Press, New York, 475 p.

Sonnenfeld, P., 1984, Brines and evaporites: Academic Press, London, 624 p.

Warren, J. K., 1989, Evaporite sedimentology: Prentice Hall, Englewood Cliffs, N. J., 285 p.

Cherts

Aston, S. R., ed., 1983, Silicon geochemistry and biochemistry: Academic Press, London, 248 p.

Calvert, S. E., 1974, Deposition and diagenesis of silica in marine sediments, *in* Hsü, K. J., and H. C. Jenkyns, eds., Pelagic sediments: On land and under the sea: Internat. Assoc. Sedimentologists Spec. Pub. 1, p. 273–300.

Cressman, E. R., 1962, Nondetrital siliceous sediments: U.S. Geol. Survey Prof. Paper 440-T, 22 p.

Garrison, R. E., R. G. Douglas, K. E. Pisciotta, C. M. Isaacs, and J. C. Ingle, eds., l981, The Monterey Formation and related siliceous rocks of California: Pacific Sect. Soc. Econ. Paleontologists and Mineralogists, Tulsa, 327 p.

Heath, G. R., 1974, Dissolved silica and deep-sea sediments, *in* Hay, W. W., ed., Studies in paleooceanography: Soc. Econ. Paleontologists and Mineralogists Spec. Paper 20, p. 77–94.

Hein, J. R., ed., Siliceous sedimentary rock-hosted ores and petroleum: Van Nostrand Reinhold, New York, 304 p.

Hein, J. R., and J. Obradović, 1989, eds., Siliceous deposits of the Tethys and Pacific regions: Springer-Verlag, New York, 244 p.

Iijima, A., J. R. Hein, and R. Siever, eds., l983, Siliceous deposits in the Pacific region: Elsevier, Amsterdam, 472 p.

Iller, R. K., 1979, Chemistry of silica: John Wiley & Sons, New York, 866 p.

Ireland, H. A., 1959, Silica in sediments: Soc. Econ. Paleontologists and Mineralogists Spec. Pub. 7, Tulsa, 185 p.

Linder, van der, G. J., ed., l977, Diagenesis of deep-sea biogenic sediments: Benchmark Papers in Geology, v. 40, Dowden, Hutchinson and Ross, Stroudsburg, Pa., 385 p.

McBride, E. F., ed., l979, Silica in sediments: Nodular and bedded cherts: Soc. Econ. Paleontologists and Mineralogists Reprint Ser. 8, 184 p.

Iron-Rich Sedimentary Rocks

Appel, P. W. U., and G. L. LaBerge, 1987, Precambrian iron-formations: Theophrastus Pub., S.A., Athens, Greece, 674 p.

Dimroth, E., 1976, Aspects of the sedimentary petrology of cherty iron-formation, *in* Wolf, K. H., ed., Handbook of strata-bound and stratiform ore deposits, v. 7: Elsevier, New York, p. 203–254.

Eichler, J., 1976, Origin of the Precambrian iron-formation, *in* Wolf, K. H., ed., Handbook of strata-bound

and stratiform ore deposits, v. 7: Elsevier, New York, p. 157–202.

James, H. L., and P. K. Sims, eds., 1973, Precambrian iron formations of the world: Econ. Geology, v. 68, p. 913–1179.

James, H. L., 1966, Chemistry of the iron-rich sedimentary rocks: Data of geochemistry, 6th ed., U.S. Geol. Survey Prof. Paper 440-W, 6l p.

Lepp, H., ed. 1975, Geochemistry of iron: Benchmark Papers in Geology, v. 18: Dowden, Hutchinson and Ross, Stroudsburg, Pa., 464 p.

Melnik, Y. P., l982, Precambrian banded iron-formations: Developments in Precambrian geology 5, Elsevier, Amsterdam, 310 p. Translated from the Russian by Dorothy B. Vitaliano.

Trendall, A. F., and R. C. Morris, eds., 1983, Iron-formation facts and problems: Developments in Precambrian geology 6, Elsevier, Amsterdam, 558 p.

Van Houton, F. B., and D. P. Bhattacharyya, 1982, Phanerozoic oolitic ironstone-facies models and distribution in space and time: Rev. Earth and Planetary Sci., v. 10.

Phosphorites

Baturin, G. N., and C. W. Finkl, Jr., l982, Phosphorites on the sea floor, Origin, composition and distribution: Developments in sedimentology 33, Elsevier, Amsterdam, 343 p. Translated from the Russian by Dorothy B. Vitaliano.

Bentor, Y. K., ed., 1980, Marine phosphorites: Geochemistry, occurrence, genesis: Soc. Econ. Paleontologists and Mineralogists Spec. Pub. 29, 249 p.

Burnett, W. C., and P. N. Froelich, eds., 1988, The origin of marine phosphorites: The results of the R.V. *Robert D. Conrad* Cruise 23–06 to the Peru shelf: Special issue of Marine Geology, v. 80, p. 181–346.

Cook, P. J., 1976, Sedimentary phosphate deposits, *in* Wolf, K. H., ed., Handbook of strata-bound and stratiform ore deposits, v. 7: Elsevier, New York, p. 505–536.

Journal Geological Society (London), 1980, v. 136, pt. 6: An issue devoted to phosphatic and glauconitic sediments, p. 657–805.

Nriagu, J. O., and P. B. Moore, ed., 1984, Phosphate minerals: Springer-Verlag, New York, 434 p.

Slansky, M., 1986, Geology of sedimentary phosphates: North Oxford Academic Pub., Essex, Great Britain, 210 p.

Chapter 14
Carbonaceous Sedimentary Rocks

14.1 Introduction

Most sedimentary rocks, including rocks of Precambrian age, contain at least a small amount of organic matter consisting of the preserved residue of plant or animal tissue. The average content of organic matter in sedimentary rocks is about 2.1 weight percent in mudrocks, 0.29 percent in limestones, and 0.05 percent in sandstones (Degens, 1965). The average in all sedimentary rocks is about 1.5 percent. Organic matter contains about 50–60 percent carbon; therefore, the average sedimentary rock contains about 1 percent organic carbon. A few special types of sedimentary rocks have significantly more organic material than these average rocks. Black shales and other organic-rich and bituminous mudrocks typically contain 3 to 10 or more percent by weight of organic matter. Some oil shale contains even higher percentages, ranging to 25 percent or more, and coals may be composed of more than 70 percent organic matter. Certain solid hydrocarbon accumulations, such as asphalt and bitumen deposits formed from petroleum by oxidation and loss of volatiles, constitute another example of a sedimentary deposit greatly enriched in organic carbon. In fact, even liquid petroleums can be thought of a special kind of organic-rich deposit, although they are clearly not rocks. Sedimentary rocks containing significant enrichment in organic matter over average sediments are called carbonaceous sedimentary rocks. The amount of organic matter required to designate a sediment as carbonaceous is not firmly established. Shales or carbonate rocks with at least 10 percent organic matter are probably called carbonaceous sedimentary rocks by many geologists, and rocks with 3–10 percent organic matter are regarded as organic-rich. The carbon in carbonaceous sedimentary rocks is believed to be almost entirely of organic origin, although a few scientists maintain that petroleum may originate by inorganic processes.

The overall abundance of carbonaceous sedimentary rocks is not accurately known, although the volume of economically significant carbonaceous rocks is probably less than

1 percent of the total sediment volume. These rocks are, however, extremely valuable. Most kinds of highly organic-rich carbonaceous sediments are either currently exploited for fossil fuels, or they have potential value as an energy source. Currently, coal and petroleum (including natural gas) are the principal kinds of fossil fuels. Oil shales or kerogen shales contain significant volumes of organic matter that can be converted to petroleum through heating. Oil shales may thus be exploited in the future as coal and petroleum resources diminish. Less–organic-rich rocks such as black shales that contain about 3–10 percent organic matter currently have little economic potential.

In this chapter, we examine briefly the distribution, characteristics, and origins of the major kinds of carbonaceous sedimentary rocks. Because of the extraordinary economic significance of these deposits, the literature on carbonaceous rocks is voluminous, particularly the literature concerning petroleum and coal. I make no attempt in this short chapter to provide comprehensive treatment of the occurrence and characteristics of coal, oil shale, and petroleum. Several references listed under additional readings provide interested readers with a starting point for further literature research.

14.2   Characteristics of Organic Matter in Carbonaceous Sedimentary Rocks

14.2.1   Principal Kinds of Organic Matter

Organic content of carbonaceous sedimentary rocks is the characteristic that particularly distinguishes them from other sedimentary rocks. Therefore, in the study of carbonaceous rocks, it is essential to develop some understanding of the physical and chemical (including isotope) attributes of this organic matter. Three basic kinds of organic matter accumulate in subaerial and subaqueous environments: humus, peat, and sapropel. Soil humus is plant organic matter that accumulates in soils to form a number of decay products such as humic and fulvic acids. Most soil humus is eventually oxidized and destroyed; thus, little is preserved in sedimentary rocks. Peat is also humic organic matter, but peat accumulates in freshwater or brackish-water swamps and bogs where stagnant, anaerobic conditions prevent total oxidation and bacterial decay. Therefore, some humus that accumulates under these conditions can be preserved in sediments, e.g., in coals and shales. Sapropel is fine organic matter that accumulates subaqueously in lakes, lagoons, and marine basins where oxygen levels are moderately low. It consists largely of the remains of phytoplankton and zooplankton and of spores, pollen, and macerated fragments of higher plants.

It is often difficult to differentiate accurately between the types of organic matter found in ancient sediments, but both humic and sapropelic types are recognized. Humic organic matter is the chief constituent of most coals, although a few coals are composed of sapropel. The organic material that occurs in shales is sapropelic; however, it is so finely disseminated and altered that is difficult to identify. The organic matter in shales is not the same as the original organic material deposited in the parent mud. The original organic material has been changed by a complex diagenetic process involving chemical and biochemical degradation, yielding an insoluble organic substance called **kerogen.** Among other things, kerogen is believed to be the precursor material of petroleum, which is derived from kerogen by natural thermal or thermocatalytic cracking (Section 14.3.4). Kerogen is insoluble in both aqueous alkaline solvents and common organic solvents. Some organic material in sediments is soluble in and extractable by organic solvents. This material is called **bitumen.** In ancient shales, kerogen makes up 80 to 99 percent of the organic matter; the rest is bitumen (Tissot and Welte, 1984, p. 132). For additional discussion of the depositional conditions that favor the production and preservation of organic matter in sediments, see Chapter 7, Section 7.4.3.

14.2.2   Chemical Structure of Kerogen

Kerogen consists of masses of almost completely macerated organic debris, which is chiefly plant remains such as algae, spores, spore coats, pollen, resins, and waxes. On the basis of the kinds of organic remains from which it was derived, kerogen has been classified into five principal types: (1) algal—composed dominantly of the remains of algae, (2) amorphous—composed largely of sapropelic organic matter from plankton and other low forms of life, (3) herbaceous—composed of pollen, spores, cuticles, etc., (4) woody, and (5) coaly (inertinite) (Hunt, 1979, p. 276).

The chemical structure of kerogen is complex. According to Tissot and Welte (1984, p. 148), the structure consists of nuclei crosslinked by chainlike bridges. The nuclei themselves consist of stacks, made up of two to four essentially parallel aromatic sheets, each containing less than 10 condensed aromatic rings (see Fig. 14.1 for an example of an aromatic ring). Some of the rings may be heterocycles containing nitrogen, sulfur, and possibly oxygen. A heterocyclic compound is one in which one or more of the carbon atoms in the ring are replaced by an atom of another element such as nitrogen, sulfur, or oxygen. The nuclei may also contain napthenic rings, alkyl chains substituted on the aromatic rings, and various functional groups (Fig. 14.1). The bridges that are crosslinked to the nuclei may consist of a variety of organic structures, including linear or branched aliphatic chains, which are attached to the nuclei as substituents, and oxygen or sulfur functional groups such as ketones, esters, ether, sulfides, and disulfides. Alkyl groups, which are formed from aliphatic chains by removal of a hydrogen atom, are commonly represented by the symbol R (Fig. 14.1). An aliphatic chain R may be combined with a functional group such as an ester to form part of a bridge structure. Also, various functional groups such as hydroxyl, carboxyl, or methoxyl may be substituted on nuclei or chains.

On the basis of chemical composition, Tissot and Welte (1984, p. 151) classify kerogen into three major types. **Type I kerogen** has high initial H/C and low O/C ratios. It is rich in aliphatic structures and thus in hydrogen. Algal kerogens belong in this group. **Type II kerogen** has relatively high H/C and low O/C ratios, but polyaromatic nuclei and heteroatomic ketone and carboxylic acid groups are more prevalent than they are in type I kerogen. Ester bonds are important. Sulfur is present also in substantial amounts, located in heterocycles and as a sulfide bond. Type II kerogen is apparently derived from phytoplankton, zooplankton, bacteria, etc. deposited in marine environments under reducing conditions. **Type III kerogen** has relatively low initial H/C and high initial O/C ratios. Polyaromatic nuclei and heteroatomic ketone and carboxylic acid groups are more important than in type II. No ester groups occur, but noncarbonyl oxygen may be included in ether bonds. Type III kerogen is believed to be derived mainly from terrestrial plants and contains much identifiable vegetal debris. Tissot and Welte (1984, p. 155) identify also a **residual type** of organic matter characterized by abnormally low H/C ratios associated with high O/C ratios. This type of organic matter has abundant aromatic nuclei and oxygen-containing groups, but aliphatic chains are absent. It is characterized by coaly fragments of oxidized organic matter or charcoal. It is the coaly type of kerogen men-

Aromatic ring

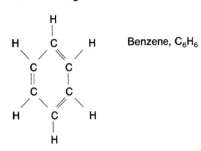

Benzene, C_6H_6

Napthenic ring

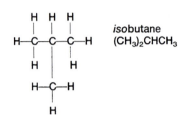

Cyclopropane
$CH_2CH_2CH_2$

Branched aliphatic chain

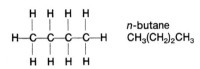

*iso*butane
$(CH_3)_2CHCH_3$

Linear aliphatic chain

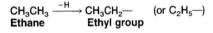

n-butane
$CH_3(CH_2)_2CH_3$

Alkyl group

$CH_3CH_3 \xrightarrow{-H} CH_3CH_2$—   (or C_2H_5—)
Ethane   Ethyl group

Alkyl groups represented by symbol R

CH_3—	Methyl	All of
CH_3CH_2—	Ethyl	these
$CH_3CH_2CH_2$—	Propyl	can be
CH_3CHCH_3	Isopropyl	designated by R

Functional groups

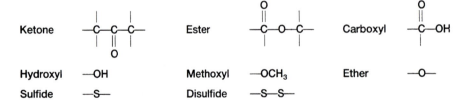

Ketone

Ester

Carboxyl

Hydroxyl   —OH   Methoxyl   —OCH_3 Ether   —O—

Sulfide   —S—   Disulfide   —S—S—

Aliphatic chain R with a functional group

$$ \overset{O}{\overset{\|}{-C}}-O-\overset{|}{\underset{|}{C}}-R \quad (e.g., ester + R) $$

FIGURE 14.1
Examples of organic structural groups that may occur in kerogen. Note that alkyl groups are formed from aliphatic chains by removal of a hydrogen atom. For convenience, alkyl groups are often represented by the symbol R. A functional group such as an ester can be combined with an aliphatic chain R.

tioned by Hunt (1979). For additional description of the chemical and structural makeup of kerogen and the methods used in isolating and studying kerogen, see Durand (1980).

14.2.3   Elemental Chemistry of Organic Matter

The elemental chemistry of organic matter is relatively simple. It consists dominantly of carbon, with lesser amounts of hydrogen, oxygen, nitrogen, and sulfur. Data shown in

TABLE 14.1
Average elemental composition of kerogen in shales from North America and Europe and the composition of coals from Germany and Yugoslavia

Kerogen type	*Weight percent*				
	C	*H*	*O*	*N*	*S*
Type I	78.8	8.8	7.7	2.0	2.7
Type II	77.8	6.8	10.5	2.2	2.7
Type III	82.5	4.6	10.5	2.1	0.2
C O A L Lignite	68.6	5.1	21.2	2.6	2.5
Bituminous	88.3	5.0	3.9	2.0	0.8
Anthracite	91.3	3.2	3.2	1.8	0.5

Source: Tissot and Welte, 1984, p. 141.

Table 14.1 suggest that the average weight percent carbon in typical kerogen ranges between about 75 and 85. Hydrogen abundance ranges from about 4 to 9 percent, oxygen from 7 to 11 percent, nitrogen about 2 percent, and sulfur from about 0.2 to 3 percent. Ranges among individual kerogens are greater than those shown by these average values.

Elemental composition is related to kerogen type and diagenetic history. Type I kerogen tends to be high in hydrogen and relatively low in oxygen, whereas Type III kerogen is high in oxygen and relatively low in hydrogen. Type II kerogen has high oxygen and intermediate hydrogen abundance. Thermal maturation or coalificaton of organic material results in a relative loss of oxygen and hydrogen and a relative enrichment in carbon.

14.2.4   Stable Isotope Chemistry

The elemental chemistry of organic matter provides some insight into the origin of kerogen, but it is not of significant value in evaluating the depositional environment or diagenetic history of the organic matter in carbonaceous sedimentary rocks. On the other hand, geologists and geochemists are showing growing interest in use of the isotope geochemistry of organic matter as a possible tool for characterizing organic matter and for analyzing its depositional and diagenetic history.

The general principles of isotope geochemistry are discussed briefly in Section 10.3. Potentially useful stable isotopes in the study of organic matter include those of carbon, oxygen, hydrogen, nitrogen, and sulfur. Carbon isotopes are expressed as $\delta^{13}C$ (the deviation in parts per mil [‰] of the ratio of $^{13}C/^{12}C$ in a sample compared to that of a standard). Similarly, oxygen isotopes are expressed as $\delta^{18}O$ (ratio of $^{18}O/^{16}O$ in a sample compared to a standard), hydrogen isotopes are expressed as δD (ratio of $^{2}D/^{1}H$ in a sample compared to a standard, where D is deuterium), nitrogen is expressed as $\delta^{15}N$ (ratio of $^{15}N/^{14}N$ in a sample compared to a standard), and sulfur isotopes are expressed as $\delta^{34}S$ (ratio of $^{34}S/^{32}S$ in a sample compared to a standard).

Comparatively few data are available on the isotope composition of nonmarine plants and aquatic marine plants or the isotope composition of organic matter in modern nonmarine and marine sediments. Initial studies suggest, however, that isotope fractionation in nonmarine and marine plants is different and that these differences are reflected in different isotope compositions of nonmarine and marine organic matter. Unfortunately, initial isotope ratios may be changed during diagenesis, and available data are still too few to provide a reasoned assessment of the potential usefulness of stable isotopes in organic matter as a paleoenvironmental tool. Nonetheless, interest in stable isotope studies of organic matter will surely increase in the future. For additional information on this subject, readers may wish to consult Anderson and Arthur (1983), Dean et al. (1986), Deniro (1983), Deniro and Epstein (1981), Faure (1986), Hoefs (1987), Kaplan (1983), and Rigby and Batts (1986).

14.3   Major Kinds of Carbonaceous Sedimentary Rock

14.3.1   General Statement

The dominant organic constituents of carbonaceous sediments are humic and sapropelic organic matter. The nonorganic constituents are either siliciclastic grains or carbonate materials. Carbonaceous sediments can be classified on the basis of relative abundance of these constituents and the kinds of organic matter that compose the constituents (humic vs. sapropelic) into three basic types of organic-rich rocks: coal, oil shale, and asphaltic substances (Fig. 14.2). Asphaltic substances are formed from sapropelic organic material that has undergone a complex process of change and maturation during burial. The origin of petroleum and asphaltic substances is discussed in Sections 14.3.4 and 14.3.5.

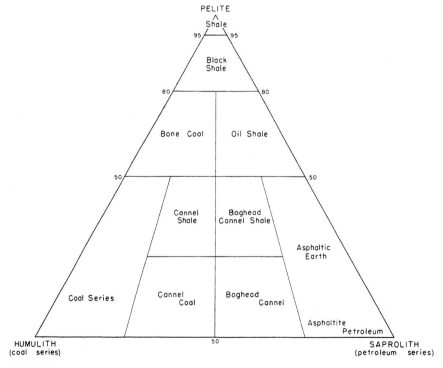

FIGURE 14.2

Classification and nomenclature of carbonaceous sediments on the basis of relative abundance of humic organic constituents (humulith), sapropelic organic constituents (saprolith), and fine-grained terrigenous constituents (pelite). *(From Pettijohn, F. J., 1975, Sedimentary rocks, 3rd ed. Fig. 11.37, p. 445, reprinted by permission of Harper & Row, New York.)*

14.3.2   Coals

Nature and Distribution

Coals are the most abundant type of carbonaceous sediment. They are composed dominantly of combustible organic matter but contain various amounts of impurities (ash) that are largely siliciclastic materials. The amount of ash that coals can contain and still retain the name of coal is not precisely fixed. Some very impure coals (bone coals) may contain 70–80 percent ash, but most coals have less than 50 percent ash by weight. Most coals are humic coals, although a few are sapropelic coals that are made up mostly of spores, algae, and fine plant debris. Cannel coals and boghead coals are sapropelic coals. Coals are defined in various ways, but a commonly accepted definition is that of Schopf (1956):

> Coal is a readily combustible rock containing more than 50 percent by weight and more than 70 percent by volume of carbonaceous material, formed from compaction or induration of variously altered plant remains similar those of peaty deposits. Differences in the kinds of plant materials (type), in degree of metamorphism (rank), and range of impurities (grade), are characteristic of the varieties of coal.

Although the overall volume of coal is quite small, coals are widely distributed geographically. They occur in sedimentary rocks ranging in age from Precambrian to Tertiary, and peat analogs of coal occur in Quaternary sediments. Coal is reported in Precambrian rocks as old as 1700 million years in the Michigamme Slate of Michigan (Tyler et al., 1957). This coal is probably algal coal. Coals did not become common until the development of woody land plants in the Devonian (Pettijohn, 1975, p. 451). The first extensive coal deposits occur in rocks of Carboniferous age. Throughout the world as a whole, coals occur in scattered localities in all of the post-Devonian geologic systems, although they tend to be more abundant in some systems, particularly the Carboniferous. In the United States, coals are most common in rocks of Pennsylvanian, Cretaceous, and Tertiary ages. Data compiled by Fettweis (1979) show that the major coal resources of the world, in descending order, occur in the USSR, the United States, China, Europe, and Canada. The remaining coal resources are scattered among the other countries of the world.

Classification

Fourteen or more approaches to classification of coals are in use (Elliott and Yohe, 1981). Perhaps the most common and important method is classification by **rank** (Table 14.2), which is based on volatile matter and calorific value. Rank is a function of the degree of coalification or carbonification (increase in organic carbon) attained by a given coal owing to burial and metamorphism. **Peat** is often included in a tabulation of coals but is actually not a true coal. Peat consists of unconsolidated, semicarbonized plant remains with high moisture content. **Lignite,** or brown coal, is the lowest-rank coal. Lignites are brown to brownish-black coals that have high moisture content and commonly retain many of the structures of the original woody plant fragments. They are dominantly Cretaceous or Tertiary in age. **Bituminous** coals are hard, black coals that contain lower amounts of volatiles and less moisture than lignite and have a higher carbon content. They commonly display thin layers consisting of alternating bright and dull bands (Fig. 14.3). **Subbituminous** coal has properties intermediate between those of lignite and bituminous coal. **Anthracite** is a hard, black, dense coal that contains more than 90 percent carbon. It is a bright, shiny rock that breaks with conchoidal fracture, such as the fractures in broken

TABLE 14.2
ASTM classification of coals by rank

Class	Group	Fixed carbon limits (percent) (dry, mineral-matter–free basis)		Volatile matter limits (percent) (dry, mineral-matter–free basis)		Calorific value limits (Btu per pound) (moist,[B] mineral-matter–free basis)		Agglomerating character
		Equal or greater than	Less than	Greater than	Equal or less than	Equal or greater than	Less than	
I. Anthracitic	1. Meta-anthracite	98	...	...	2	...	...	nonagglomerating
	2. Anthracite	92	98	2	8	...	...	
	3. Semianthracite[C]	86	92	8	14	...	...	
II. Bituminous	1. Low-volatile bituminous coal	78	86	14	22	...	...	commonly agglomerating[E]
	2. Medium-volatile bituminous coal	69	78	22	31	...	...	
	3. High-volatile A bituminous coal	...	69	31	...	14 000[D]	...	
	4. High-volatile B bituminous coal	...	...	...	...	13 000[D]	14 000	
	5. High-volatile C bituminous coal	...	...	...	...	11 500	13 000	
						10 500	11 500	agglomerating
III. Subbituminous	1. Subbituminous A coal	...	...	...	...	10 500	11 500	nonagglomerating
	2. Subbituminous B coal	...	...	...	...	9 500	10 500	
	3. Subbituminous C coal	...	...	...	...	8 300	9 500	
IV. Lignitic	1. Lignite A	...	...	...	...	6 300	8 300	
	2. Lignite B	...	...	...	...	...	6 300	

[A]This classification does not include a few coals, principally nonbanded varieties, which have unusual physical and chemical properties and which come within the limits of fixed carbon or calorific value of the high-volatile bituminous and subbituminous ranks. All of these coals either contain less than 48% dry, mineral-matter–free fixed carbon or have more than 15,500 moist, mineral-matter-free British thermal units per pound.

[B]Moist refers to coal containing its natural inherent moisture but not including visible water on the surface of the coal.

[C]If agglomerating, classify in low-volatile group of the bituminous class.

[D]Coals having 69% or more fixed carbon on the dry, mineral-matter–free basis shall be classified according to fixed carbon, regardless of calorific value.

[E]It is recognized that there may be nonagglomerating varieties in these groups of the bituminous class, and that there are notable exceptions in high-volatile C bituminous group.

Source: American Society for Testing Materials (ASTM), 1981, Annual book of ASTM standards. Part 26, American Society for Testing Materials, Philadelphia, Pa., reprinted by permission.

FIGURE 14.3
Banded coal (Pennsylvanian) from the Illinois Basin. Specimen about 20 cm long. *(Photograph courtesy of D. Bensely.)*

glass. Bituminous coals and anthracite are largely of Mississippian and Pennsylvanian (Carboniferous) ages. **Cannel coal** and **boghead coal** are nonbanded, dull, black coals that also break with conchoidal fracture; however, they have bituminous rank and much higher volatile content than does anthracite. Cannel coal is composed of conspicuous percentages of spores. Boghead coals are composed dominantly of nonspore algal remains. **Bone coal** is very impure coal with high ash content.

Petrography of Coals

Coals are not homogeneous materials, but instead are made up of various components somewhat analogous to the minerals of inorganic rocks. Coals can thus be classified also on the basis of megascopic textural appearance and their recognizable petrographic or microscopic constituents. Transmitted-light microscopes can be used to study coals up to the rank of low-volatile bituminous. At higher-rank levels, the organic material becomes opaque, and effective study requires use of incident (reflected) light. Incident light microscopy is most commonly used in coal petrography because it can be used to analyze coals over the complete rank from peat to anthracite and graphite (Bustin et al., 1985).

Stopes (1919) recognized four types of coal, now called lithotypes, on the basis of megascopic appearance. These lithotypes comprise millimeter-thick bands or layers of humic coal.

1. **Vitrain** is brilliant, glossy, vitreous, black coal that occurs in thin horizontal bands that are commonly 3–5 mm thick. It breaks with a conchoidal fracture and is clean to the touch (Fig. 14.4).

2. **Clarain** has a smooth fracture that displays a pronounced gloss or shine and is distinguished from vitrain by the presence of dull intercalations or striations. Small-scale sublaminations that occur within the layers or bands give the surface a silky luster (Fig. 14.4). Clarain is the most widely distributed and common macroscopic constituent of humic coals.

3. **Durain** occurs in bands up to a few centimeters in thickness that have a close, firm texture that appears somewhat granular; broken surfaces are not smooth but have a fine lumpy or matte texture (Fig. 14.4). Durain is characterized by its lack of luster, gray to brownish-black color, and earthy appearance.

4. **Fusain** is soft, friable, and black. It resembles common charcoal and has been described as "mineral charcoal." It occurs chiefly as irregular wedges and is friable and porous if not mineralized.

FIGURE 14.4
Bituminous coal showing examples of three different lithotypes: V, vitrain; C, clarain; and D, durain. The small divisions on the scale = 1 cm. *(From Bustin, R. M., et al., 1985, Coal petrology, its principles, methods, and applications: Geol. Assoc. Canada Short Course Notes, v. 3. Pl. 6A, p. 51, reprinted by permission.)*

Under the microscope, coal can be seen to consist of several kinds of organic units that are single fragments of plant debris, or in some cases the organic units are fragments consisting of more than one type of plant tissue. Stopes (1935) suggested the name **maceral** for these organic units as a parallel word for the term mineral used for the constituents of inorganic rocks. The starting materials for macerals are woody tissues, bark, fungi, spores, and so on; however, these materials are not always recognizable in coals. Macerals are divided into three major groups: vitrinite, inertinite, and liptinite. The characteristics of these macerals and their most common subtypes are summarized in Table 14.3, and they are further described below:

1. **Vitrinites** are macerals that originated as wood or bark (Fig. 14.5). The term **huminite** is also used for these macerals in the description of lignites and bituminous coals. Vitrinites are a major humic constituent of bright coals. Two principal types of vitrinites are known. **Collinite** is a structureless or nearly structureless maceral that commonly occurs as a matrix or impregnating material for fragments of other macerals. **Telinite** is derived from cell-wall material of bark and wood and preserves some of the cellular texture.

2. **Inertinites** are macerals composed of woody tissues, fungal remains, or fine organic debris of uncertain origin and are characterized by relatively high carbon content (Fig. 14.6). They can be divided into five subtypes:
a. **Fusinite**—displays cell structures composed of carbonized or oxidized cell walls and hollow lumens (the space bounded by the wall of an organ) that are commonly mineral-filled; characteristic of fusain
b. **Semifusinite**—a transition state between fusinite and vitrinite
c. **Sclerotinite**—composed of the remains of fungal sclerotia (a hardened mass of tubular filaments or threads) or altered resins; characterized by oval shape and varying size

TABLE 14.3
Coal macerals: classification, origin, properties

Maceral group	Maceral	Origin		Appearance		Chemistry	Carbonization	Liquefaction
		Source	Process	Color in thin section	Reflected light			
Vitrinite	Telinite (cell walls) Collinite (cell fillings, structureless vitrinite)	Woody tissues, bark, leaves, etc.	Mummification	Red-brown	Intermediate gray	Intermediate hydrogen content and volatiles	Principal reactive constituent in coking coal	Susceptible to liquefaction
Liptinite	Resinite Sporinite Cutinite Alginite	Resins Spore exines Cuticles Algae	Resistant	Yellow	Dark gray	Higher hydrogen content and volatiles; more aliphatic	Reactive during carbonization	Susceptible to liquefaction
Inertinite	Micrinite (massive and granular)	Plant materials	Degradation products	Opaque	White, yellowish, light gray	Lower hydrogen content and volatiles; more aromatic	Inert during carbonization	Resistant to liquefaction
	Fusinite	Woody tissues, etc.	Fire; or biochemical oxidation					
	Semifusinite	Woody tissues; etc.	Intermediate between vitrinite and fusinite					
	Sclerotinite	Fungal sclerotia, hyphae	Resistant remains					

Source: Davis, A., W. Spackman, and P. Givens, 1976, The influence of the properties of coals and their conversion to clean fuels: Energy Sources, *v.* 3, No. 1, p. 55−81.

FIGURE 14.5

Example of telinite, a huminite/ vitrinite maceral, in early Cretaceous bituminous coal from British Columbia, Reflected light. Scale bar = 50 μm. *(From Bustin, R. M., et al., 1985, Coal petrology, its principles, methods, and applications: Geol. Assoc. Canada Short Course Notes, v. 3. Pl. 3, Fig. 5, p. 41, reprinted by permission.)*

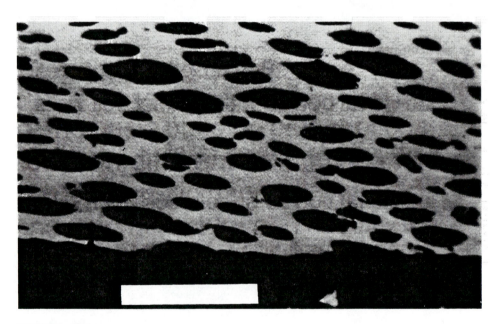

FIGURE 14.6

Example of fusinite, an inertinite maceral, in early Cretaceous bituminous coal from British Columbia. Reflected light. Scale bar = 50 μm. *(From Bustin, R. M., et al., 1985, Coal petrology, its principles, methods, and applications: Geol. Assoc. Canada Short Course Notes, v. 3. Pl. 5, Fig. 1, p. 47, reprinted by permission.)*

d. **Micrinite and macronite**—structureless, granular macerals derived from fine-grained organic detritus; opaque and generally less than 10 μm in size (micrinite) but ranges up to 100 μm (macronite)

e. **Inertodetrinite**—finely divided, structureless clastic form of inertinite in which fragments of various kinds of inertinite maceral occur as dispersed particles

3. **Liptinites** (exinites) originate from spores, cuticles, resins, and algae (Fig. 14.7). They can be recognized by their shapes and structures, although original constituents may be compacted and squashed. The major types of liptinites are the following:

a. **Sporinite**—composed of the remains of yellow, translucent bodies (spore exines) that are commonly flattened parallel to the bedding

b. **Cutinite**—formed from macerated fragments of cuticles (layers covering the outer wall of a plant's epidermal cells)

c. **Resinite**—the remains of plant resins and waxes; occurs as isolated rounded to oval or spindle-shaped, reddish, translucent bodies; also occurs as diffuse impregnations or as fillings in cell cavities

d. **Alginite**—macerals composed of the remains of algal bodies; have a serrated, oval shape; the characteristic maceral of boghead coal

The macerals in coals are identified on the basis of reflectivity, anisotropy, morphology, relief, and size. Under the reflecting microscope, liptinites appear dark gray with low reflectivity. Vitrinites are medium-gray, medium-reflecting, and inertinites are light-gray to white, high-reflecting. Anisotropy is defined as maximum reflectance minus minimum reflectance. Macerals of the liptinite and inertinite groups are, in general, isotropic. With increasing rank, coal become distinctly anisotropic (Bustin et al., 1985). Shape or form, size, and internal structures are used to distinguish macerals if reflectance is too similar to permit identification. Macerals may also have differences in relief when viewed in incident light owing to differences in hardness. Inertinite macerals commonly show positive relief, and liptinite macerals may also show positive relief compared to the vitrinite matrix. Etching the surface of coals with a strong oxidizing agent is also a

FIGURE 14.7
Example of resinite (R), a liptinite maceral, in a Tertiary lignite from Greece. Reflected light. Scale bar = 50 μm. *(From Bustin, R. M., et al., 1985, Coal petrology, its principles, methods, and applications: Geol. Assoc. Canada Short Course Notes, v. 3. Pl. 4, Fig. 2a, p. 45, reprinted by permission.)*

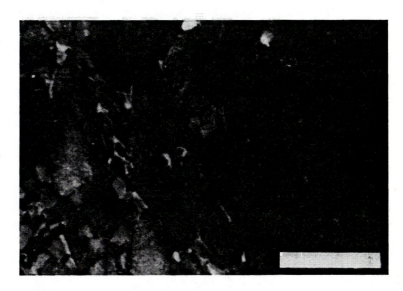

common technique in coal petrography. Etching enhances differences among macerals and makes them easier to identify. Fluorescence microscopy involves irradiation of coals with blue light or ultraviolet light, which causes the coals to luminesce or fluoresce. The presence or absence of fluorescence and the fluorescence colors can be used to identify the different macerals.

Coal petrology is a highly specialized kind of petrography and cannot be pursued in detail here. Several references are available to readers who may wish more information about this complex subject. See, for example, Bustin et al. (1985), Crelling and Dutcher (1980), International Committee for Coal Petrology (1971, 1975), Stach et al. (1982), Ting (1982), and Ward (1984). A more extensive bibliography of books published prior to 1980 that deal with coal and coal petrology is provided by Crelling and Dutcher (1980, p. 93).

Origin of Coal

As suggested, coals consist primarily of Type 3 (humic) organic material. Thus, they consist mainly of the degraded, but not highly degraded, remains of higher plants. Rare coals are composed of sapropels, that is, mainly spores and pollen. Because land plants did not become well established until the Devonian, older coals, such as Precambrian coals, must be composed largely of algal remains, as mentioned. Land plants did not become truly abundant until the Carboniferous; therefore, the first major coal deposits occur in rocks of Carboniferous age.

The origin of coal requires two major conditions: a favorable climate for plant growth and an environment in which preservation of organic matter is favored. Climate influences the rate of plant growth, and warm, wet, tropical to subtropical climates are most favorable to rapid growth. Rates of accumulation of peat of as much as 4 mm per year are reported under some tropical climates (e.g., J. A. R. Anderson, 1964). On the other hand, the decomposition rate of plant debris is quite fast under these climatic conditions. Plants grow slower in cool, dry climates, but organic matter decomposes at a slower rate. Coal can form only where the rate of plant accumulation is greater than the rate of decomposition. Although we tend to think of coals as being the deposits of warm, humid climates, coals can also form in cold climates. Ancient coals were deposited at all latitudes from the equator to polar regions, with the majority forming in midlatitudes (McCabe, 1984). Accumulating plant debris is most likely to be preserved in depositional environments where oxidation of the organic matter is inhibited owing to rapid burial in a subsiding swampy area where the water table is close to the peat surface. For thick coal seams to develop, these conditions must exist for a geologically long period of time. That is, the coal swamps must not be destroyed by major environmental changes such as a marine transgression or a rapid influx of terrigenous clastics. The most favorable depositional environments appear to be transitional marine (paralic) settings such as back-barrier environments, deltas, and coastal and interdelta plains.

The conversion of humic organic matter to coal results from both biochemical and geochemical processes. Organic matter is transformed progressively from peat through lignite, subbituminous, and bituminous coals to semianthracite and anthracite coal. Biochemical processes are involved in the coalification process up to the rank of bituminous coals (Bustin et al., 1985). At very shallow burial depths, aerobic bacteria and fungi initiate decomposition of organic materials. According to Bustin et al., the order of susceptibility of organic matter to decomposition is (1) protoplasm, (2) chlorophyll, (3) oils, (4) carbohydrates, (5) epidermis, seed coats, (6) pigments, (7) cuticles, (8) spore and pollen exines, (9) waxes, and (10) resins. Thus, organic residues such as cuticles, spores and pollen, waxes, and resins are most likely to be present in coals. With deeper burial and loss of oxygen, anaerobic microbes take over the process of biochemical decomposition, which continues to depths of tens of meters. During the biochemical stage of coalification, the more easily hydrolizable organic compounds are converted to carbon dioxide, ammonia, methane gas, and water and are subsequently lost. Some of the remaining compounds are oxidized, apparently with the aid of microorganisms, to form humic acids. The remaining, nonoxidized material is incorporated into the peat, where it undergoes further biochemical change under reducing conditions. This process results ultimately in the formation of humin. Readers concerned with the details of organic matter conversion during burial may wish to consult Tissot and Welte (1984) or Huc (1980).

Some compaction occurs during biochemical conversion of organic matter, resulting in loss of moisture. With deeper burial, compaction and moisture loss continue, together with loss of oxygen, whereas carbon content increases. Overall volatile content (water, carbon dioxide, hydrogen) of coals decreases with increasing coalification; however, coals retain about 5–6 percent hydrogen into the semianthracite stage. A number of additional chemical changes occur that affect the structure of organic molecules. Compaction and volatile loss accompanying deep burial result in thinning of coal beds by a factor of as much as 30 to 1 (Ryer and Langer, 1980); that is, 30 m of original peat might produce only 1 m of coal. Temperature has a particularly important effect on the coalification process. The rank of coal tends to increase with depth owing to increase in temperature with depth. The formation of anthracite coals requires temperatures in excess of ~200°C (Daniels et al., 1990). The time over which heating occurs appears to be important also. The rank of coal tends to increase with increasing time of heating.

As mentioned, coals form under climatic conditions that favor plant growth and in depositional environments where burial conditions favor preservation of organic material. Furthermore, the coal-forming environment had to be protected in some way from significant influx of detrital sediments. These conditions appear to be met best in swamps, where organic matter can accumulate below water that is relatively stagnant and deficient in oxygen. To explain the low detrital content of coals, McCabe (1984) suggests three possibilities: (1) detrital minerals were leached away owing to the highly acid waters in some swamps, (2) the peat accumulated in raised or floating swamps, which are protected from clastic deposition, or (3) peats accumulated at times when the regional clastic supply was cut off for some reason during swamp development.

Coals occur dominantly in clastic depositional sequences, although thin limestones may be associated with some coals. In some coal-bearing sequences, the coals display a cyclic pattern of occurrence. Pennsylvanian coal sequences of the United States midcontinent region, for example, typically begin with an underclay or seatrock followed by coal. The coal is overlain by marine limestone and/or shale as well as by siltstone or sandstone, which, in turn, are overlain by another underclay and coal, and so on. Some coals are overlain by lacustrine shales and freshwater limestones rather than by marine deposits, and still others are overlain by fluvial clastic sediments or even volcanic deposits.

Observation of cyclic coal sequences led in the early 1930s to the concept of coal **cyclothems,** which were tied to marine transgressions and regressions. Considerable controversy arose over the tectonic vs. eustatic origin of these transgressions and regressions, a debate that continued particularly through the 1960s. Since the 1960s, interest in the cyclothem concept appears to have abated somewhat. Much of the more recent research on coal stratigraphy has focused on depositional environments and local controls on sedimentation patterns. Some geologists have proposed, for example, that we may be able to explain simple transgression and regression processes in cyclothems by sedimentation processes associated with depositional events such as deltaic deposition. Thus, so the suggestion goes, it may not be necessary to invoke the classic concepts of major transgressions and regressions to explain coal cycles. Recently, Klein and Willard (1989) reexamined the origin of Pennsylvanian coal-bearing cyclothems of North America. They suggest that the transgressive and regressive cycles that affected North American cyclothems were controlled by several factors. These factors include (1) flexural deformation during plate accretion into a supercontinent, (2) concomitant glaciation and eustatic sea-level change, and (3) associated episodic thrust loading and foreland-basin subsidence of small magnitude on progressively more rigid crust. Many ancient coals appear to have formed in coastal areas in deltaic, back-barrier, and fluvial settings. Others formed in intermontane fluvial-dominated or lacustrine-dominated environments (Rahmani and Flores, 1984).

Swampy environments large enough to form major coal deposits have existed since the Carboniferous (Fig. 14.8). Only the Triassic Period appears to have been a time when coal-forming processes were at a minimum. Note from Figure 14.8 that most major coal deposits occur in the Northern Hemisphere; however, owing to plate movements, some coals may have formed originally in different latitudes than those in which they now exist. The locations of the most important coal occurrences on Earth are shown graphically in Figure 14.9. As indicated in this figure, the major coal deposits of the world occur in the USSR, the United States, China, Europe, and Canada. Discussion of individual coal deposits in these countries is beyond the scope of this book. Readers may wish to consult Fettweiss (1979) for a discussion of the coal resources of individual countries.

14.3.3   Oil Shale

Definition and Importance

The term "oil shale" is applied to fine-grained sedimentary rocks from which substantial quantities of oil can be derived by heating. The term is actually a misnomer in the sense that relatively little free oil occurs in these rocks, although small blebs, pockets, or veins of asphaltic bitumens may be present. This soluble bitumen fraction constitutes as much as 20 percent of the organic content of typical oil shales. The remaining 80 percent or more of organic substances is present in the form of kerogen, which yields oil when heated to a temperature of about 350°C. The composition of kerogen is discussed below. The organic content of oil shales commonly does not exceed about 25 percent; the remaining part of the rock is composed of inorganic constituents, which are mainly silicate grains or carbonate minerals. Thus, an "oil shale" can actually be an organic-rich limestone as well as a shale.

Oil shales are of special interest to geologists and economists because of their potential to generate oil when refined into fuel at sufficiently high temperatures. Several countries have attempted to develop small oil shale industries, including Australia, Brazil, New Zealand, Switzerland, Sweden, Estonia (USSR), Spain, China, Scotland, France, and South Africa. Peak development was reached immediately after World War II. Most oil shale plants have now closed, with the exception of some plants in the USSR and China (Tissot and Welte, 1984, p. 254). Relatively little commercial exploitation of oil shales has occurred, primarily because oil produced from oil shales cannot compete

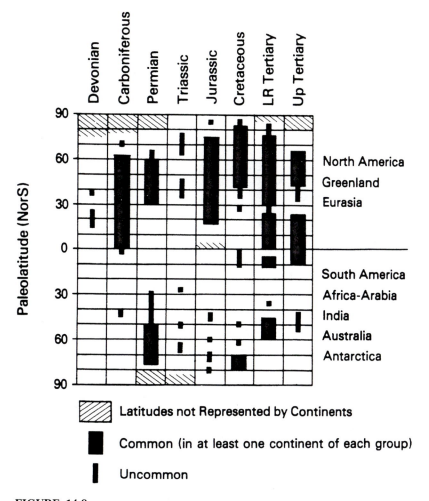

FIGURE 14.8

Distribution of coal by paleolatitude. Since the Permian, the upper group of continents has been mainly in the Northern Hemisphere and the lower group in the Southern Hemisphere. *(From McCabe, P. J., 1984, Depositional environment of coal and coal-bearing strata, in Rahmani, R. A., and R. M. Flores, eds., Sedimentology of coal and coal-bearing sequences: Internat. Assoc. Sedimentologists Spec. Pub. 7, Fig. 7, p. 23, as modified from Habicht, J. K. A., 1979, Paleoclimate, paleomagnetism, and continental drift: Am. Assoc. Petroleum Geologists Studies in Geology 9, Fig. 11, p. 15. Reprinted by permission of Blackwell Scientific Publications Limited, Oxford, and AAPG, Tulsa, Okla.)*

economically with petroleum. Also, many technological problems are associated with mining and extraction processes, as well as with disposal of the gigantic amounts of waste rock that would be created by surface retorting processes. Some observers believe that oil shales will be mined and utilized as a major source of fuel in the future as petroleum and natural gas supplies inevitably dwindle. Others are not so sure that large-scale exploitation of oil shales will ever become economically feasible. Only time will tell. At least 50 countries of the world have reserves of oil shale that have the potential to be exploited in the future (Russell, 1990, p. 4).

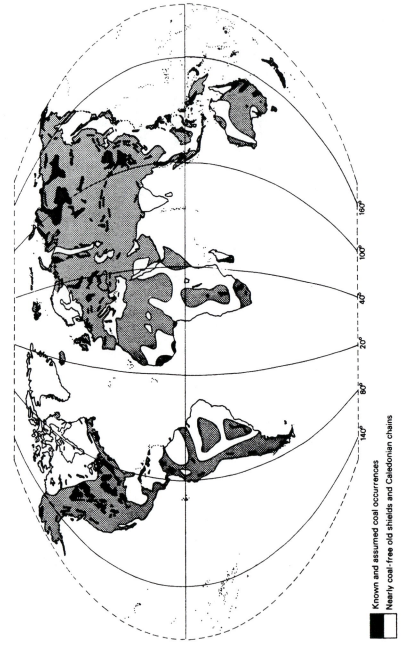

FIGURE 14.9

Geographical location of the most important coal occurrences on Earth. *(After Fettweiss, G. B., 1979, World coal resources, Appendix H., p. 415, reprinted by permission of Elsevier Science Publishers, Amsterdam.)*

■ Known and assumed coal occurrences
□ Nearly coal-free old shields and Caledonian chains

Oil shales are also of interest because of the information they convey about ancient depositional environments. To preserve such high levels of organic matter in shales and carbonate rocks requires unusual environmental conditions. Many oil shale deposits are believed to have been deposited in large lakes. Others formed in small lakes, bogs, and lagoons that may have been associated with coal-forming swamps. Still others probably formed in shallow seas on continental platforms and on continental shelves where water circulation was restricted. See also Chapter 7, Section 7.4.3. Although geologists commonly assume that preservation of large quantities of organic matter requires both high rates of organic productivity and anoxic conditions, Pederson and Calvert (1990) dispute the requirement for anoxia in the case of organic-rich marine sediments. They suggest that high rate of organic productivity is the fundamental control on accumulation of organic-rich facies in the ocean and marginal seas—not the presence or absence of anoxia. This contention is sure to stimulate considerable discussion among geologists.

Age, Distribution, and Abundance

Oil shales range in age mainly from Cambrian to Tertiary. To my knowledge, the only oil shales of Precambrian age are the Nonesuch shales of Michigan and Wisconsin (USA), which contain very minor amounts of organic matter (Russell, 1990, p. 98). In addition, some Quaternary-age deposits contain organic debris that can yield oil (Duncan, 1976). Oil shales are widespread geographically. They occur in at least 50 countries (Russell, 1990, p. 4), although they are concentrated especially in the United States, Europe, and the USSR. Cambrian and Ordovician deposits occur particularly in northern Europe, northern Asia, and east-central United States and Canada. Silurian-Devonian deposits are best developed in eastern and central United States, North Africa, and the central European USSR. Late Paleozoic oil shales occur on all the continents. Large deposits occur in southern Brazil, and smaller deposits are known in Scotland, France, Spain, South Africa, Australia, the USSR, Canada, Uruguay and southern Argentina, the northwestern United States, and Alaska. Mesozoic oil shales also occur on all continents, except Australia, in such widely scattered areas as central Africa (Zaire), northern and eastern Asia, the Middle East, Europe, Alaska, central Canada, and the western United States. Tertiary oil shales occur particularly in the western United States (Green River shales) as well as in New Zealand, Europe, the Andes Mountains of South America, the southern USSR and eastern China, and some other countries.

Estimates of the recoverable reserves of oil in oil shales range to trillions of barrels. For example, Yen and Chilingar (1976) estimate that approximately 1.8 trillion barrels of crude shale oil is contained within the Green River Formation alone in Colorado, Utah, and Wyoming. The greatest reserves of oil shale occur in the United States, followed by Brazil, the USSR, and Zaire. Much smaller but potentially economic reserves occur in a number of other countries, including Jordan, Morocco, Italy, China, Thailand, and northern and western Europe. Russell (1990, p. 4) estimates that the potential world supply of oil from oil shale is 2000 trillion barrels (10^{12} barrels) of in-place oil. Russell provides a brief description of the geology, reserves, and oil potential of all major world oil shale deposits.

Composition and Classification

The principal constituents of oil shales are shown in Figure 14.10, although the percentages of these constituents may vary considerably in different deposits. The detrital minerals are the common detrital minerals in all shales, that is, clay minerals, quartz, feld-

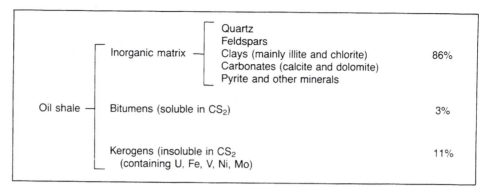

FIGURE 14.10

Principal constituents in oil shales. *(After Yen, T. F., and G. V. Chilingar, 1976. Introduction to oil shales, in Yen, T. F., and G. V. Chilingarian, eds., Oil shales: Developments in petroleum science 5. Fig. 1.2. p. 3, reprinted by permission of Elsevier Science Publishers, Amsterdam.)*

spars, and perhaps volcanic debris. Some oil shales may also contain siliceous or calcareous organic remains, which are not generally abundant. Authigenic minerals may include pyrite and other metal sulfides, carbonates (calcite, dolomite, siderite, magnesite), chert, phosphates, and saline minerals such as trona, gypsum, and halite (Shanks et al., 1976).

The organic content of oil shales is the property that particularly distinguishes them from other shales. Organic constituents commonly do not exceed about 25 percent of the rock, but much higher values are present in some deposits. It is this content of organic matter that gives oil shales their economic significance. As mentioned, some of the organic material consists of soluble bitumens; however, commonly more than 80 percent consists of kerogen. The elemental chemistry, isotope chemistry, and chemical structure of kerogen are described in preceding Sections 14.2.2–14.2.4. The origin of kerogen is discussed below under petroleum.

Not all oil shales are actually shales. Some are organic-rich siltstones, mudrocks, limestones, and impure coals. Three basic types are recognized: (1) **Carbonate-rich oil shales** are those in which the principal nonkerogen constituents are calcite, dolomite, ankerite, and variable amounts of siliciclastic silt. They are generally hard, tough, and resistant to weathering (Duncan, 1976). (2) **Silica-rich oil shales** are shales in which the main constituents apart from kerogen are fine-grained quartz, feldspar, and clay minerals. They may also contain chert, opal, and phosphatic nodules. Siliceous oil shales are generally dark brown or black and are less resistant to weathering than the carbonate-rich shales. (3) **Cannel shale** is an oil shale that consists predominantly of organic matter that completely encloses other mineral grains. The organic matter is composed largely of algal remains. Cannel shales are sometimes classified as impure cannel coals and are referred to as **torbanites.** Many oil shales are characterized by distinct lamination caused by alternations of millimeter-thick organic laminae with either siliciclastic or carbonate laminae. The amount of oil that can be extracted from oil shales ranges from about 4 percent to more than 50 percent of the weight of the rock; that is, between 10 and 150 gallons of oil per ton of rock (Duncan, 1976). See Russell (1990) for specific yields from various oil shale deposits of the world.

Origin

Oil shales form in environments where organic matter is abundant and anaerobic or reducing conditions prevent oxidation and total bacterial decomposition. They are deposited in both lacustrine (lake) and marine environments where the above conditions are met. As mentioned, the principal environments are (1) large lakes, (2) shallow seas or continental platforms and continental shelves in areas where water circulation was restricted and reducing or where weakly oxidizing conditions existed, and (3) small lakes, bogs, and lagoons associated with coal-producing swamps. The deposits of large lakes may range in composition from marls to argillaceous limestones. Probably the largest deposit of oil shale in the world, the Green River shale in west-central United States, is a lake deposit. Sediments associated with lake deposits may include volcaniclastic sediments and evaporites. Oil shale deposits in shallow seas consist mainly of siliciclastic sediments, although carbonates may also be present. In the United States, the Devonian black shales that extend over several states in the eastern midcontinent region are a good example of oil shales deposited in shallow seas. Oil shale deposits in small lakes or bogs may be associated with or overlie coal beds.

The main source of kerogen in major oil shale deposits appears to be algal remains. The microstratification in many oil shales suggests recurring variations in the supply of organic remains and fine siliciclastic detritus. Oil shales formed in lakes or swamps may be associated with impure cannel- or boghead-type coal, tuffs and other volcanic rocks, or even evaporites. Many oil shales deposited in large lakes are carbonate-rich types and tend to have high oil yields. Oil shales deposited in marine environments are characteristically the silica-rich type and have lower oil yields, although some Tertiary- and Mesozoic-age siliceous oil shales have rich oil yields. Oil shales extend over wide geographic areas and are commonly associated with limestones, cherts, sandstones, and phosphatic deposits. Additional details of the origin, distribution, and composition of oil shales are given in Yen and Chilingarian (1976).

14.3.4 Petroleum

Nature and Composition

Petroleum is not a carbonaceous sedimentary rock, but it is a carbon-rich organic substance that occurs as liquid and gas accumulations predominantly in sandstones and carbonate rocks. Therefore, a very brief discussion of petroleum is included here with the carbonaceous sedimentary rocks. Petroleum and natural gas constitute the most important sources of energy in the world today. For that reason, the factors that govern the availability of petroleum are of much more than academic interest. Major oil companies, world governments, politicians, and the average citizen are all vitally concerned with the petroleum supply. To say that the course of world political and economic events in the near future will be governed in a major way by the availability of petroleum is not too strong a statement. Until an adequate alternative source of energy is developed, the fate of the world's nations may very well depend upon who has and who does not have access to abundant, cheap supplies of petroleum.

Petroleum is composed dominantly of carbon (about 84 percent) and hydrogen (about 13 percent). It also contains an average of about 1.5 percent sulfur, 0.5 percent nitrogen, and 0.5 percent oxygen (Hunt, 1979). Despite its simple elemental chemical composition, the molecular structure of petroleum can be exceedingly complex. The molecules in petroleum range from the simple methane gas molecule (CH_4) with a mo-

FIGURE 14.11

Schematic structure of paraffin hydrocarbons having the general formula C_nH_{2n+2}, where n refers to the number of carbon or hydrogen atoms. A. Butane. B. Pentane.

A Butane, C_4H_{10}
$CH_3(CH_2)_2CH_3$

$$H-\overset{\displaystyle H}{\underset{\displaystyle H}{C}}-\overset{\displaystyle H}{\underset{\displaystyle H}{C}}-\overset{\displaystyle H}{\underset{\displaystyle H}{C}}-\overset{\displaystyle H}{\underset{\displaystyle H}{C}}-H$$

B Pentane, C_5H_{12}
$CH_3(CH_2)_3CH_3$

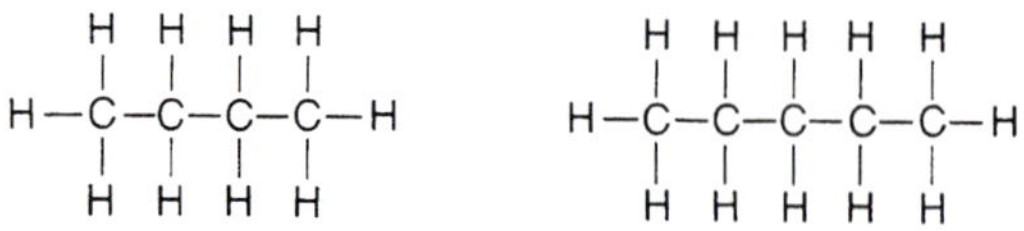

$$H-\overset{\displaystyle H}{\underset{\displaystyle H}{C}}-\overset{\displaystyle H}{\underset{\displaystyle H}{C}}-\overset{\displaystyle H}{\underset{\displaystyle H}{C}}-\overset{\displaystyle H}{\underset{\displaystyle H}{C}}-\overset{\displaystyle H}{\underset{\displaystyle H}{C}}-H$$

lecular weight of 16 to molecules with molecular weights in the thousands. Several hundred different hydrocarbons have been recorded in natural crude oils; however, all hydrocarbons can be grouped into a few basic classes or series having common molecular structural form. These structural forms are complex and are not explained in detail here, but the main hydrocarbon series are the following:

1. **Paraffins (alkanes)**—open-chain molecules with single covalent bonds between carbon atoms (Fig. 14.11).
2. **Napthenes (cycloparaffins)**—closed-ring molecules with single covalent bonds between carbon atoms (Fig. 14.12)
3. **Aromatics (arenes)**—one or more benzene-ring structures with double covalent bonds between some carbon atoms (Fig. 14.13)

Most natural gases as well as many liquid petroleums belong to the paraffin series of hydrocarbons. Most napthene hydrocarbons are liquid petroleums, although two napthenes occur as gases at normal temperatures. The aromatics, which are named for their strong aromatic odor, are liquid petroleums. They commonly make up only a small percentage of the petroleum in natural crude oils.

Occurrence and Distribution of Petroleum

Commercial accumulations of petroleum can be present in any kind of rock that contains sufficient porosity to permit storage of economically significant quantities of petroleum. Most oil deposits occur in sedimentary rocks, particularly sandstones and carbonate rocks. Roughly 55 percent of the world supply of petroleum and 75 percent of its natural gas occur in sandstones. About 45 percent of world petroleum reserves and 25 percent of natural gas reserves occur in carbonate rocks. Very small amounts of petroleum may occur in other kinds of sedimentary rocks and even in some fractured igneous or metamorphic rocks. Petroleum deposits are located in all the world's continents, but they are

FIGURE 14.12

Schematic structure of naphthene (cycloparaffin) hydrocarbons having the general formula C_nH_{2n}. A. Cyclopentane. B. Cyclohexane.

A Cyclopentane, C_5H_{10}
$CH_2CH_2CH_2CH_2CH_2$

B Cyclohexane, C_6H_{12}
$CH_2CH_2CH_2CH_2CH_2CH_2$

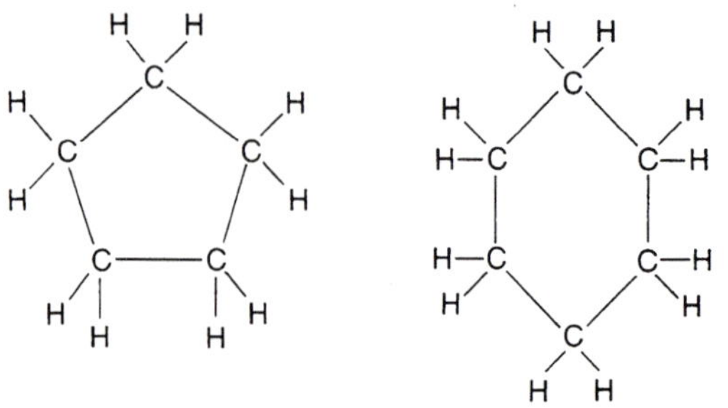

FIGURE 14.13

Schematic structure of aromatic hydrocarbons having the general formula C_nH_{2n-6}. A. Benzene. B. Toluene.

A Benzene, C_6H_6

B Toulene, $C_6H_5CH_3$

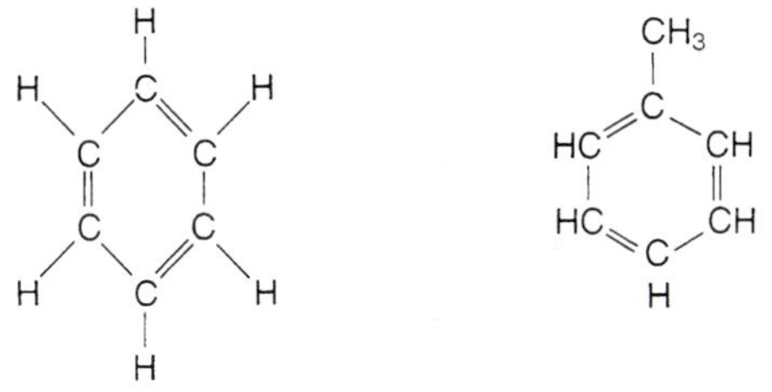

concentrated in four main areas: (1) the Middle East (Persian Gulf Basin), (2) the northern and western sides of the Gulf of Mexico and the southern side of the Caribbean, (3) the flanks of the Ouachita Mountains in the United States, and (4) the flanks of the Ural Mountains in the Soviet Union. These four areas account for nearly three-fourths of the world's oil and two-thirds of its natural gas. The major oil-producing countries of the world in terms of proven reserves of petroleum are the Middle East countries of Saudi Arabia, Iraq, Iran, Abu Dhabi, and Kuwait. These Middle East Countries are followed by the USSR, Venezuela, Mexico, the United States, and Libya. Petroleum occurs in rocks of all ages, although very little occurs in Precambrian or Quaternary rocks. Roughly 60 percent of the world's reserves of petroleum are in rocks of Mesozoic (especially Cretaceous) age. About 30 percent occurs in rocks of Tertiary age and 8–10 percent in rocks of Paleozoic age.

Petroleum deposits accumulate within some kind of trap. A trap is a structural or stratigraphic feature of a rock into which oil or gas can easily migrate but from which it is difficult to escape. Three common types of petroleum traps are anticlines, faults, and stratigraphic pinchouts (Fig. 14.14). Petroleum occurs within the pores of sedimentary strata along with water. Owing to its lower specific gravity, it tends to rise upward and accumulate above the water. Thus, petroleum will rise to the crest of a structure such as an anticline. If the sedimentary layer in which petroleum accumulates is overlain by an impermeable layer such as shale or evaporites, the petroleum cannot escape upward beyond the crest of the trap. Thus, it will accumulate until it fills the closed area of the trap (Fig. 14.14). The search for petroleum deposits centers particularly around locating traps, which may or may not be filled with petroleum. Because traps are located deep beneath the surface, seismic prospecting methods must commonly be used to locate the traps.

Origin of Petroleum

Most geologists believe that petroleum is of organic origin and that it formed from organic matter by a complex maturation process during sediment diagenesis. On the other hand, a small but respected number of scientists firmly believe that petroleum originates by inorganic processes, probably within the upper mantle. Decades of oil well drilling have not completely resolved this difference of opinion. Owing to the overwhelming weight of evidence that favors an organic origin, however, I discuss the origin of petroleum in this book only from the organic origin point of view. It is highly probable that petroleum forms from plant and animal organic matter by a complex maturation process during sediment burial. The transformation of organic matter to petroleum involves initial microbial alteration and subsequent thermal alteration and cracking. Most organic matter produced by plants and animals is destroyed by oxidation and organisms. Only a small

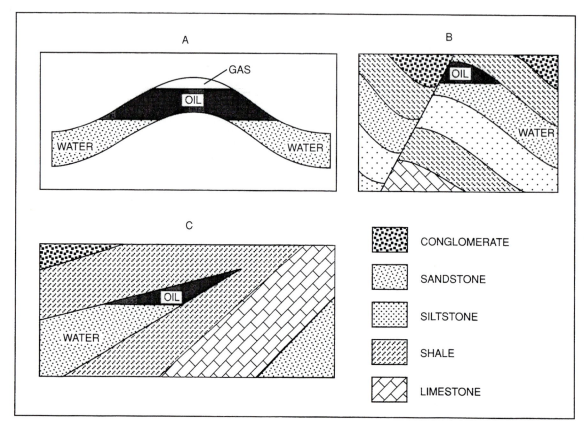

FIGURE 14.14
Schematic representation of three kinds of petroleum traps: A, anticlinal trap, B, fault trap, with
an impermeable fault gouge or mineral seal along the fault, and C, stratigraphic (pinchout) trap.

percentage of the organic matter created through time, perhaps 1 percent, needs to have
been preserved, however, to account for the known petroleum deposits. Organic matter
that is deposited along with fine-grained sediments in reducing environments has commonly been regarded to have the best chance of preservation, although Pederson and
Calvert (1990) suggest that anoxic conditions are not required if rates of organic productivity are high. Thus, the source materials for petroleum are believed to be contained
primarily in organic-rich shales and fine-grained carbonate rocks that were deposited
under conditions of high organic productivity and possibly in some kind of restricted or
low-oxygen environment.

During burial, organic matter undergoes complex changes as it proceeds through
three main stages of transformation, commonly referred to by petroleum geologists as
diagenesis, catagenesis, and metagenesis. During **diagenesis,** which takes place at burial
depths of tens to hundreds of meters and at temperatures below about 50°C (Hunt, 1979,
p. 100), organic **biopolymers** (carbohydrates, proteins, lipids, lignin, and other highly
organized organic substances) are degraded by microbial activity. This biochemical process results in formation of individual organic compounds, called **monomers,** such as

sugars, amino acids, fatty acids, and phenols. Microbial degradation also results in production of a small amount of methane gas. These microbial reactions occur particularly in clays and muds deposited in either reducing or oxic environments, where rapid burial prevents the organic matter from being destroyed completely by oxidation. With deeper burial, these monomers undergo additional change owing to polymerization and condensation. Polymerization is the combining of simple molecules to form more-complex molecules, and condensation is the combining of two or more molecules to form a larger molecule, with loss of another molecule such as H_2O. Through condensation and polymerization, monomers are gradually converted to **geopolymers,** which are humic and fulvic acids and humin. (Humic and fulvic acids are complex, high-molecular-weight organic compounds. Humic acid is soluble in alkaline solutions but is precipitated by acidification. Fulvic acid is soluble in both acid and base, and humin is insoluble in both acid and base.) A small amount of aromatic and napthenic hydrocarbons may also be generated during this stage owing to additional chemical processes such as decarboxylation and hydrogen disproportionation (loss of hydrogen from some molecules and enrichment in others). With further condensation and insolublization (conversion of soluble humic and fulvic acids to insoluble humin), the geopolymers are converted into kerogen. This diagenetic process is illustrated schematically in Figure 14.15. As shown, some hydrocarbonlike organic substances such as lipids may go through the diagenesis stage with only minor degradation.

Once formed, kerogen is highly stable and can persist at moderately low temperatures for hundreds of millions of years. As discussed above, kerogen is the principal kind of organic matter preserved in oil shales. On the other hand, with increasing temperature accompanying deeper burial, kerogen can undergo thermal degradation to form liquid hydrocarbons. This process, referred to as **catagenesis,** takes place at burial depths exceeding about 1000 m and at temperatures of about 50°–150°C and pressures of 300–1500 bars. Catagenesis involves the cracking (breaking of carbon-to-carbon bonds) of kerogen and other complex, high-molecular-weight compounds to form hydrocarbons. At the low end of the temperature range (up to about 125°C), thermocatalytic cracking (in the presence of a catalyst such as smectite clay) probably predominates. At higher temperatures, thermal cracking, which does not require a catalyst, predominates. Thus, during catagenesis, low- to medium-molecular-weight hydrocarbons are formed at lower temperatures, with lighter hydrocarbons and some methane forming at the high end of the temperature range above about 120°C. Above 120°C, some previously formed liquid hydrocarbons may also be cracked to methane.

In many oil-producing basins, the maturation of organic matter may not proceed beyond the catagenesis stage. If deeper burial occurs, with temperatures exceeding about 150°–200°C, the maturation process enters the **metagenesis stage.** In the temperature range of metagenesis, liquid hydrocarbons are no longer formed. Some methane may continue to be generated to burial depths corresponding to temperatures of about 250°C. At higher temperatures, any remaining organic matter is converted to anthracitic and graphitic materials. Also, liquid petroleum may be destroyed to form methane plus solid hydrocarbons (discussed below) and graphite. The above discussion is drawn mainly from Tissot and Welte (1984, p. 69–92) and Hunt (1979, Ch. 4). Readers may wish to consult these references for further details.

After petroleum has formed from organic source materials at substantial burial depths, it migrates by processes not yet fully understood from the fine-grained source

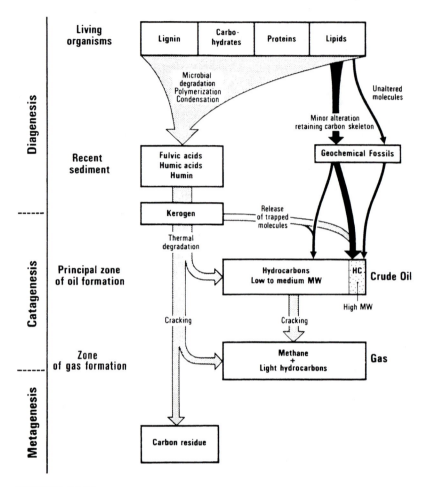

FIGURE 14.15
Schematic illustration of the sequential conversion of organic matter in living substances to
fulvic and humic acids, kerogen, and hydrocarbons. Note that some petroleumlike hydrocarbons
in lipids undergo only minor alteration during diagenesis and are thus referred to as geochemical
fossils. HC = hydrocarbons; MW = molecular weight. *(From Tissot, B. P., and D. H. Welte,
1984, Petroleum formation and occurrence, 2nd ed. Fig. II.3.1, p. 94, reprinted by permission
of Springer-Verlag, Berlin.)*

rocks into associated coarser-grained, porous, and permeable sandstone or carbonate
rocks. The petroleum then migrates, possibly along with water, up the regional dip of the
basin within these ''carrier beds'' until it reaches a trap. There, it slowly accumulates until
it may fill the trap.

The above discussion outlines the fundamental concept of petroleum origin that
appears to be most widely accepted by petroleum geologists today. Nonetheless, many
problems and questions remain regarding the origin of petroleum and its migration from
shale source rocks into permeable and porous sandstone or carbonate reservoir rocks. The
subject of petroleum geology constitutes an entire field of research in itself, and many
ideas regarding the origin of petroleum are controversial.

14.3.5   Solid Hydrocarbons

Introduction

These substances are hydrocarbons such as natural asphalts and mineral waxes that occur in a semisolid or solid state. Most solid hydrocarbons probably formed from liquid petroleums that were subjected to loss of volatiles, oxidation, and biologic degradation after seepage to the surface. Others may never have existed as light oils. Solid hydrocarbons occur as seepages, as surface accumulations, as impregnations occupying the pore spaces of sandstones or other sedimentary rock, and in veins and dikes. They are black or dark brown and have a characteristic odor of pitch or paraffin. Some solid hydrocarbons are combustible and thus have commercial value as fuels. Others are infusible. All solid hydrocarbons are of interest to petroleum geologists, however, because their presence at the surface is an indication of petroleum at depth in a region. Also, study of their occurrence may help to solve the problems related to the origin and alteration of petroleum.

Composition and Occurrence

Solid hydrocarbons have roughly the same elemental chemical composition as liquid petroleum, but the percentage of carbon and hydrogen tend to be somewhat lower and the content of sulfur, nitrogen, and oxygen somewhat higher. They are divided into four main varieties or series based on fusibility (melting temperature) and solubility in carbon disulfide (CS_2), an organic solvent (Fig. 14.16):

FIGURE 14.16
Terminology of principal kinds of naturally occurring solid hydrocarbons. *(From Petroleum geochemistry and geology, by J. M. Hunt. W. H. Freeman and Company, © 1979, Fig. 8.28, p. 400.)*

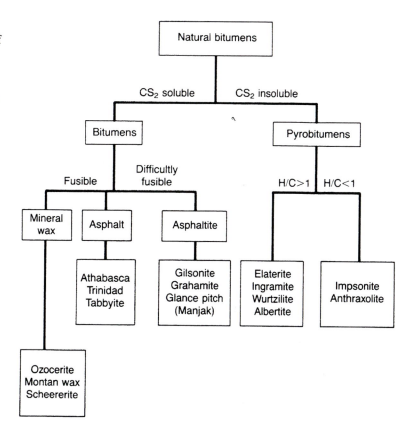

1. **Asphalts** are soft, semisolid bitumens that occur as seeps, as surface pools, as viscous impregnations in sediments (tar sands). They are dark colored, plastic to fair hard, easily fusible, and soluble in carbon disulfide. Varietal names for asphalts fro different areas are shown in Figure 14.16. Asphalts are commonly associated with acti oil seeps. The most important economic accumulations of asphalts are in tar sands. T sand deposits occur in many parts of the world and a few are large enough (e.g., t Athabasca tar sands of Canada) to have considerable economic significance. Major d posits occur in Canada, Venezuela, the Malagasy Republic, the United States, and A bania. Less important deposits occur in Trinidad, Romania, and the USSR.

2. **Asphaltites** occur primarily in dikes and veins that cut sediment beds. They a harder and denser than asphalts and melt at higher temperatures. They are largely solub in carbon disulfide. Names applied to varieties of asphaltites that differ slightly in densit fusibility, and solubility are **gilsonite, glance pitch, and grahamite.** In the United State deposits of asphaltites are known from Utah and Oklahoma. Other deposits occur in Per Argentina, Cuba, Trinidad, Mexico, the Dead Sea area, and the USSR.

3. **Pyrobitumens** occur in dikes and veins like asphaltites but are infusible a largely insoluble in carbon disulfide. They also tend to have higher sulfur content th other solid hydrocarbons. Several varieties of pyrobitumens are recognized. Softer for include **elaterite,** a soft elastic substance rather like India rubber, and **wurtzlite.** Mo indurated forms are **albertite,** a black, solid bitumen with a brilliant jetlike luster a conchoidal fracture; **ingramite;** and the metamorphosed pyrobitumens **imsonite** a **anthraxolite.** Owing to their infusibility, pyrobitumens do not have commercial value fuels.

4. **Native mineral waxes** are solid, waxy, light-colored substances that cons largely of paraffinic hydrocarbons of high molecular weight. They represent the residuu of high-wax oils exposed at the surface. The most important native mineral wax **ozocerite,** which consists of veinlike deposits of greenish or brown wax. **Montan wax** an extract obtained from some kinds of brown coals or lignites. Mineral waxes are fusit and soluble in carbon disulfide.

See Hunt (1979, p. 398–404) for additional discussion of the characteristics of t solid hydrocarbons. Details of the geochemistry, origin, distribution, and exploitation bitumen and other solid hydrocarbons are discussed in a series of papers in Chilingari and Yen (1978).

═══════ *Additional Readings*

Bustin, R. M., A. R. Cameron, D. A. Grieve, and W. D. Kalkreuth, 1985, Coal petrology, its principles, methods, and applications, 2nd ed.: Geol. Assoc. of Canada Short Course Notes, v. 3, 230 p.

Crelling, J. C., and R. R. Dutcher, 1980, Principles and applications of coal petrology: Soc. Econ. Paleontologists and Mineralogists Short Course Notes 8, 127 p.

Chilingarian, G. V., and T. F. Yen, 1978, Bitumens, asphalts and tar sands: Elsevier, New York, 331 p.

Fettweiss, G. B., 1979, World coal resourc Elsevier, Amsterdam, 415 p.

Hunt, J. M., 1979, Petroleum geochemistry a geology: W. H. Freeman, San Francisco, 617 p.

International Committee for Coal Petrology, 196 International Hand-Book of Coal Petrology, 2nd ed.: Ce tre National de la Recherche Scientifique, Paris. Suppl ments published in 1971, 1975.

Meyers, R. A., ed., 1982, Coal structure: A demic Press, New York, 340 p.

Petrakis, L., and D. W. Grandy, 1980, Coal analysis, characterization and petrography: Jour. Chem. Education, v. 57, p. 689–694.

Rahmani, R. A., and R. M. Flores, eds., 1984, Sedimentology of coal and coal-bearing sequences: Internat. Assoc. of Sedimentologists Spec. Pub. 7, 412 p.

Ross, C. A., and J. R. P. Ross, 1984, Geology of coal: Hutchinson and Ross, Stroudsburg, Pa., 344 p.

Russell, P. L., 1990, Oil shales of the world: Their origin, occurrence and exploitation: Pergamon Press, Oxford, 736 p.

Stach, E., M.-Th. Mackowsky, M. Teichmüller, G. H. Taylor, D. Chandra, and R. Teichmüller, 1982, Handbook of coal petrology, 3rd ed.: Gebrüder Borntraeger, Berlin-Stuttgart, 535 p.

Ward, C. R., ed., 1984, Coal geology and coal technology: Blackwell, Melbourne, 345 p.

Yen, T. F., and G. V. Chilingarian, eds., 1976, Oil shale: Elsevier, New York, 292 p.

References

Aagaard, P., P. K. Egeberg, G. C. Saigal, S. Morad, and K. Bjørlykke, 1990, Diagenetic albitization of detrital K-feldspars in Jurassic, lower Cretaceous and Tertiary clastic reservoir rocks from offshore Norway, II. Formation water chemistry and kinetic considerations: Jour. Sed. Petrology, v. 60, p. 575–581.

Abbott, P. L., and G. L. Peterson, 1978, Effect of abrasion durability on conglomerate clast populations: examples from Cretaceous and Eocene conglomerates of the San Diego area, California: Jour. Sed. Petrology, v. 48, p. 31–42.

Abu Dhabi National Reservoir Research Foundation, 1984, Stylolites and associated phenomena—Relevance to hydrocarbon reservoirs: Abu Dhabi National Reservoir Research Foundation Spec. Pub., 304 p.

Adams, A. E., W. S. Mackenzie, and C. Guilford, 1984, Atlas of sedimentary rocks under the microscope: John Wiley & Sons, New York, 104 p.

Adams, J. E., and M. L. Rhodes, 1960, Dolomitization by seepage refluxion: Am. Assoc. Petroleum Geologists Bull., v. 44, p. 1912–1921.

Ahmad, R., 1981, Stratigraphy, structure, and petrology of the Lookingglass and Roseburg formations, Agness-Illahe area, southwestern Oregon: M.S. Thesis, University of Oregon, Eugene, 150 p.

Ahmed, F., and D. C. Almond, 1983, Field mapping for geology students: George Allen & Unwin, London, 72 p.

Aissaoui, D. M., 1988, Magnesium calcite cements and their diagenetic dissolution and dolomitization, Mururoa

Atoll: Sedimentology, v. 35, p. 821–841.

Aitken, J. D., 1967, Classification and environmental significance of cryptalgal limestones and dolomites, with illustrations from the Cambrian and Ordovician of southwestern Alberta: Jour. Sed. Petrology, v. 37, p. 1163–1179.

Alexandersson, T., 1979, Marine maceration of skeletal carbonates in the Skagerrak North Sea: Sedimentology, v. 26, p. 845–852.

Ali, A. D., and P. Turner, 1982, Authigenic K-feldspar in the Bromsgrove sandstone Formation (Triassic) of Central England: Jour. Sed. Petrology, v. 52, p. 187–197.

Allan, J. R., and R. K. Matthews, 1982, Isotope signatures associated with early meteoric diagenesis: Sedimentology, v. 29, p. 797–817.

Allen, G. P., D. Laurier, and J. Thouvenin, 1979, Etude sédimentologique du delta de la Mahakam: Comp. Franc. Pétroles, Notes Mém. 15, 156 p.

Allen, J. R. L., 1962, Petrology, origin and deposition of the highest lower Old Red Sandstone of Shropshire, England: Jour. Sed. Petrology, v. 32, p. 657–697.

————1968, Current ripples: Their relation to patterns of water and sediment motion: North Holland Publishing, Amsterdam, 433 p.

————1982, Sedimentary structures—Their character and physical basis: Elsevier, Amsterdam, v. I, 593 p.; v. II, 663 p.

————1984, Laminations developed from upper-state plane beds: A model based on the larger coherent structures of the turbulent boundary layer: Sed. Geology, v. 39, p. 227–242.

Allen, P., 1945, Sedimentary variations: Some new facts and theories: Jour. Sed. Petrology, v. 15, p. 75–83.

————1947, Correlation between allogenic grade size and allogenic frequency in sediments: Jour. Sed. Petrology, v. 17, p. 3–7.

Allen, P. A., and P. Homewood, eds., 1986, Foreland basins: Internat. Assoc. Sedimentologists Spec. Pub. 8, Blackwell Scientific Publications, Oxford, 453 p.

Alling, H. L., 1945, Use of microlithologies as illustrated by some New York sedimentary rocks: Geol. Soc. America Bull., v. 56, p. 737–756.

Amiri-Garroussi, K., 1988, Eocene spheroidal dolomite from the western Sirte Basin, Libya: Sedimentology, v. 35, p. 577–585.

Anderson, F. F., and M. A. Arthur, 1983, Stable isotopes of oxygen and carbon and their application to sedimentologic and paleoenvironmental problems: Soc. Econ. Paleontologists and Mineralogists Short Course 10, p. 1-1–1.151.

Anderson, J. A. R., 1964, The structure and development of peat swamps of Sarawak and Brunei: Jour. Tropical Geography, v. 18, p. 7–16.

Anderson, J. B., 1983, Ancient glacial-marine deposits: their spatial and temporal distribution, *in* Molnia, B. F., ed., Glacial-marine sedimentation: Plenum Press, New York, p. 3–92.

Anderson, R. Y., and D. W. Kirkland, 1966, Intrabasin varve correlation: Geol. Soc. America Bull., v. 77, p. 241–256.

Anderson, T. F., and Arthur,

M. A., 1983, Stable isotopes of oxygen and carbon and their application to sedimentologic and paleoenvironmental problems, *in* Arthur, M. A., et al., eds., Stable isotopes in sedimentary geology: Soc. Econ. Paleontologists and Mineralogists Short Course Notes 10, p. 1-1-1-151.

Arche, A., 1983, Coarse-grained meander lobe deposits in the Jarama River, Madrid, Spain, *in* Collinson, J. D., and J. Lewin, eds., Modern and ancient fluvial systems: Internat. Assoc. Sedimentologists Spec. Pub. 6, p. 313–321.

Archer, J. B., 1984, Clastic intrusions in deep-sea fan deposits of the Rosroe Formation, Lower Ordovician, Western Ireland: Jour. Sed. Petrology, v. 54, p. 1197–1205.

Argast, S., and T. W. Donnelly, 1986, Compositions and sources of metasediments in the Upper Dhawar Supergroup, South India: Jour. Geology, v. 94, p. 215–231.

———1987, The chemical discrimination of clastic sedimentary components: Jour. Sed. Petrology, v. 57, p. 813–823.

Armin, R. A., 1987, Sedimentology and tectonic significance of Wolfcampian (Lower Permian) conglomerates in the Pedregosa Basin: southeastern Arizona, southwestern New Mexico, and northern Mexico: Geol. Soc. America Bull., v. 99, p. 42–65.

Arnott, R. W. C., and B. M. Hand, 1989, Bedforms, primary structures and grain fabrics in the presence of suspended sediment rain: Jour. Sed. Petrology, v. 59, p. 1062–1069.

Arthur, M. A., and S. E. Schlanger, 1979, Cretaceous "oceanic anoxic events" as causal factors in development of reef-reservoired giant oil fields: Am. Assoc. Petroleum Geologists Bull., v. 63, p. 870–885.

Ashley, G. M., chairperson, 1990, Classification of large-scale subaqueous bedforms: a new look at an old problem: Jour. Sed. Petrology, v. 60, p. 160–172.

Ashley, G. M., J. Shaw, and N. D. Smith, 1985, Glacial sedimentary environments: Soc. Econ. Paleontologists and Mineralogists Short Course 16, 246 p.

Asquith, G., 1982, Basic well log analysis for geologists: Am. Assoc. Petroleum Geologists, Tulsa, 216 p.

Aubouin, J., 1965, Geosynclines: Developments in geotectonics: Elsevier, Amsterdam, 335 p.

Austin, G. S., 1974, Multiple overgrowths on detrital quartz sand grains in the Shakopee Formation (Lower Ordovician) of Minnesota: Jour. Sed. Petrology, v. 44, p. 358–362.

Ayalon, A., and F. J. Longstaffe, 1988, Oxygen isotope studies of diagenesis and pore-water evolution in the western Canada sedimentary basins: Evidence from the Upper Cretaceous basal Belly River Sandstone, Alberta: Jour. Sed. Petrology, v. 58, p. 489–505.

Aylmore, L. A. G., and J. P. Quirk, 1960, Domain or turbostratic structures of clays: Nature, v. 187, p. 1046–1048.

Baas Becking, L. G. M., I. R. Kaplan, and D. Moore, 1960, Limits of the natural environment in terms of pH and oxidation-reduction potentials: Jour. Geology, v. 68, pp. 243–284.

Badiozamani, K., 1973, The Dorag dolomitization model—application to the Middle Ordovician of Wisconsin: Jour. Sed. Petrology, v. 43, p. 965–984.

Bagnold, R. A., and O. Barndorff-Nielsen, 1980, The pattern of natural size distribution: Sedimentology, v. 27, p. 199–207.

Bain, D. C., and B. F. L. Smith, 1987, Chemical analysis, *in* Wilson, M. J., ed., A handbook of determinative methods in clay mineralogy: Blackie, Glasgow and London (Chapman and Hall, New York), p. 248–274.

Baker, P. A., and M. Kastner, 1981, Constraints on the formation of sedimentary dolomite: Science, v. 213, p. 214–216.

Baker, V. R., 1977, Streamchannel response to floods, with examples from central Texas: Geol. Soc. America Bull., v. 88, p. 1057–1071.

———1984, Flood sedimentation in bedrock fluvial systems, *in* Koster, E. H., and R. J. Steel, eds., Sedimentology of gravels and conglomerates: Canadian Soc. Petroleum Geologists Mem. 10, p. 87–98.

Baldwin, B., and C. O. Butler, 1985, Compaction curves: Am. Assoc. of Petroleum Geologists Bull., v. 69, p. 622–626.

Bally, A. W., 1984, Structural styles and the evolution of sedimentary basins: Am. Assoc. Petroleum Geologists Short Course, 238 p.

Bally, A. W., and S. Snelson, 1980, Realms of subsidence, *in* Miall, A. D., ed., Facts and principles of world petroleum occurrence: Canadian Soc. Petroleum Geologists Mem. 6, p. 9–44.

Baltuck, M., 1982, Provenance and distribution of Tethyan pelagic and hemipelagic siliceous sediments, Pindos Mountains, Greece: Sed. Geology, v. 31, p. 63–88.

Barats, G. M., T. H. Nilsen, and R. T. Golia, 1984, Conglomerate clast composition of the Upper Cretaceous Hornbrook Formation, Oregon and California, *in* Nilsen, T. H., ed., Geology of the Upper Cretaceous Hornbrook Formation, Oregon and California: Pacific Section, Soc. Econ. Paleontologists and Mineralogists, v. 42, p. 111–122.

Barker, C., 1979, Organic geochemistry in petroleum exploration: Am. Assoc. Petroleum Geologists Course Note Ser. 10, 159 p.

Barndorff-Nielsen, O., 1977, Exponentially decreasing distributions for the logarithm of particle size: Proc. Royal Society London A, v. 353, p. 401–419.

Barndorff-Nielsen, O., K. Dalsgaard, C. Halgreen, H. Kuhlman, J. T. Moller, and G. Schou, 1982, Variation in particle size distribution over a small dune: Sedimentology, v. 29, p. 53–65.

Barnes, J. W., 1981, Basic geologic mapping: Open University Press and Halsted Press (John Wiley & Sons), New York, 113 p.

Barrett, P. J., 1980, The shape of rock particles, a critical review: Sedimentology, v. 27, p. 291–303.

Barron, J. A., 1987, Diatomite: Environmental and geologic factors affecting its distribution, *in* Hein, J. R., ed., Siliceous sedimentary rock-hosted ores and petroleum: Van Nostrand Reinhold, New York, p. 164–178.

Barth, T. R. W., 1969, Feldspars: Wiley Interscience, New York, 267 p.

Bartleson, B. L., 1968, Stratigraphy and petrology of the Gothic Formation, Elk Mountains, Colorado: Ph.D. Thesis, University of Colorado, 126 p.

———1972, Permo-Pennsylvanian stratigraphy and history of the Crested Butte-Aspen region: Colorado School Mines Quart., v. 67, p. 187–248.

Basan, P. B., ed., 1978, Trace fossil concepts: Soc. Econ. Paleontologists and Mineralogists Short Course 5, 181 p.

Baskin, Y., 1956, A study of authigenic feldspars: Jour. Geology, v. 64, p. 132–155.

Basu, A., 1976, Petrology of Holocene fluvial sand derived from plutonic source rocks: implications to paleoclimate interpretation: Jour. Sed. Petrology, v. 46, p. 694–709.

——1985a, Influence of climate and relief on compositions of sand released at source areas, *in* Zuffa, G. G., ed., Provenance of arenites: D. Reidel, Dordrecht, p. 1–18.

——1985b, Reading provenance from detrital quartz, *in* Zuffa, G. G., ed., Provenance of arenites: D. Reidel, Dordrecht, p. 231–247.

Basu, A., and E. Molinaroli, 1989, Provenance characteristics of detrital opaque Fe-Ti oxide minerals: Jour. Sed. Petrology, v. 59, p. 922–934.

Basu, A., S. W. Young, L. J. Suttner, W. C. James, and G. H. Mack, 1975, Reevaluation of the use of undulatory extinction and polycrystallinity in detrital quartz for provenance interpretation: Jour. Sed. Petrology, v. 45, p. 873–882.

Bates, R. L., and J. A. Jackson, eds., 1980, Glossary of geology, 2nd ed.: American Geological Institute, Falls Church, Va., 749 p.

Bathurst, R. G. C., 1966, Boring algae, micrite envelopes and lithification of molluscan biosparites: Geol. Jour., v. 5, p. 15–32.

——1975, Carbonate sediments and their diagenesis, 2nd ed.: Elsevier, Amsterdam, 658 p.

——1980, Lithification of carbonate sediments: Science Progress, v. 66, p. 451–471.

——1982, Genesis of stromatactis cavities between submarine crusts in Paleozoic carbonate mud buildups: Jour. Geol. Soc. London, v. 139, p. 165–181.

——1983, Neomorphic spar versus cement in some Jurassic grainstones: Significance for evaluation of porosity evolution and compaction: Jour. Geol. Soc. of London, v. 140, p. 229–237.

——1986, Carbonate diagenesis and reservoir development: conservation, destruction and creation of pores, *in* Warme, J. E., and K. W. Shanley, eds., 1986, Carbonate depositional environments, modern and ancient. Part 5: Diagenesis I: Colorado School of Mines Quart., v. 81, p. 1–25.

Baturin, G. N., 1982, Phosphorites on the sea floor: Elsevier, Amsterdam, 343 p.

Beales, F. W., and J. L. Hardy, 1980, Criteria for the recognition of diverse dolomite types with an emphasis on studies on host rocks for Mississippi Valley-type ore deposits, *in* Zenger,

D. H., J. B. Dunham, and R. L. Ethington, eds., Concepts and models of dolomitization: Soc. Econ. Paleontologists and Mineralogists Spec. Pub. 28, p. 197–213.

Belderson, R. H., and A. H. Stride, 1966, Tidal current fashioning of a basal bed: Marine Geology, v. 4, p. 237–257.

Bellemin, G. J., and R. Merriam, 1958, Petrology and origin of the Poway Conglomerate, San Diego County, California: Geol. Soc. America Bull., v. 69, p. 199–220.

Bennett, R. H., W. R. Bryant, and G. H. Keller, 1981, Clay fabric of selected submarine sediments: Fundamental properties and models: Jour. Sed. Petrology, v. 51, p. 217–232.

Bennett, R. H., N. R. O'Brien, and M. H. Hulbert, 1991, Determinants of clay and shale microfabric signatures: Processes and mechanisms, *in* Bennett, R. H., W. R. Bryant, and M. H. Hulbert, eds., Microstructure of fine-grained sediments: Springer-Verlag, New York, p. 5–32.

Bentor, Y. K., ed., 1980, Marine phosphorites—Geochemistry, occurrence, genesis: Soc. Econ. Paleontologists and Mineralogists Spec. Pub. 29, 249 p.

Berner, R. A., 1971, Principles of chemical sedimentology: McGraw-Hill, New York, 240 p.

——1975, The role of magnesium in crystal growth of aragonite from sea water: Geochim. et Cosmochim. Acta, v. 39, p. 489–505.

——1980, Early diagenesis: Princeton University Press, Princeton, N.J., 241 p.

Berner, R. A., J. T. Westrich, R. Graber, J. Smith, and C. S. Martens, 1978, Inhibition of aragonite precipitation from supersaturated seawater: A laboratory and field study: Am. Jour. Science, v. 278, p. 816–837.

Bhatia, M. R., 1983, Plate tectonics and geochemical composition of sandstones: Jour. Geology, v. 91, p. 611–627.

——1985a, Composition and classification of Paleozoic flysch mudrocks of eastern Australia: Implications in provenance and tectonic setting interpretation: Sed. Geology, v. 41, p. 249–268.

——1985b, Rare earth element geochemistry of Australian Paleozoic graywackes and mudrocks: provenance and tectonic control: Sed. Geology, v. 45, p. 97–113.

Bhatia, M. R., and S. R. Taylor, 1981, Trace-element geochemistry and sedimentary provinces: a study from the Tasman geosyncline, Australia: Chem. Geology, v. 33, p. 115–125.

Bhattacharyya, A., and G. M. Friedman, eds., 1983, Modern carbonate environments: Benchmark Papers in Geology, v. 74, Hutchinson and Ross, Stroudsburg, Pa., 376 p.

Bjørlykke, K., 1983, Diagenetic reactions in sandstones, *in* Parker, A., and B. W. Sellwood, eds., Sediment diagenesis: D. Reidel, Dordrecht, Holland, p. 169–213.

Black, M., 1933, The precipitation of calcium carbonate on the Great Bahama Bank: Geol. Mag., v. 70, p. 455–466.

Blatt, H., 1967a, Original characteristics of clastic quartz grains: Jour. Sed. Petrology, v. 37, p. 401–424.

——1967b, Provenance determination and recycling of sediment: Jour. Sed. Petrology, v. 37, p. 1031–1044.

——1979, Diagenetic processes in sandstones, *in* Scholle, P. A., and P. R. Schluger, eds., Aspects of diagenesis: Soc. Econ. Paleontologists and Mineralogists Spec. Pub. 26, p. 141–157.

——1982, Sedimentary petrology: W. H. Freeman, San Francisco, 564 p.

——1985, Provenance studies of mudrocks: Jour. Sed. Petrology, v. 55, p. 69–75.

——1987, Oxygen-isotopes and the origin of quartz: Jour. Sed. Petrology, v. 57, p. 373–377.

Blatt, H., and J. R. Caprara, 1985, Feldspar dispersal patterns in shales of the Vanoss formation (Pennsylvanian), south-central Oklahoma: Jour. Sed. Petrology, v. 55, p. 548–552.

Blatt, H., and J. M. Christie, 1963, Undulatory extinction in quartz of igneous and metamorphic rocks and its significance in provenance studies of sedimentary rocks: Jour. Sed. Petrology, v. 33, p. 559–579.

Blatt, H., R. L. Jones, and R. G. Charles, 1982, Separation of quartz and feldspars from mudrocks: Jour. Sed. Petrology, v. 52, p. 660–662.

Blatt, H., G. Middleton, and R. Murray, 1980, Origin of sedimentary rocks, 2nd ed.: Prentice-Hall, Inc., Englewood Cliffs, New Jersey, 782 p.

Blatt, H., and D. J. Schultz, 1976, Size distribution of quartz in mudrocks: Sedimentology, v. 23, p. 857–866.

Blatt, H., and M. W. Totten, 1981, Detrital quartz as an indicator of distance from shore in marine mudrocks: Jour. Sed. Petrology, v. 51, p. 1259–1266.

Bluck, B. J., 1967, Sedimentation of beach gravels: examples from South Wales: Jour. Sed. Petrology, v. 37, p. 128–156.

Boggs, S., Jr., 1966, Petrology of Minturn Formation, east-central Eagle County, Colorado: Am. Assoc. Petroleum Geologists Bull., v. 50, p. 1399–1422.

————1967a, Measurement of roundness and sphericity parameters using an electronic particle size analyzer: Jour. Sed. Petrology, v. 37, p. 908–913.

————1967b, A numerical method for sandstone classification: Jour. Sed. Petrology, v. 37, p. 548–555.

————1968, Experimental study of rock fragments: Jour. Sed. Petrology, v. 38, p. 1326–1339.

————1969a, Distribution of heavy minerals in the Sixes River, Curry County, Oregon: The Ore Bin, v. 31, p. 133–152.

————1969b, Relationship of size and composition in pebble counts: Jour. Sed. Petrology, v. 39, p. 1243–1246.

————1972, Petrography and geochemistry of rhombic calcite pseudomorphs from mid-Tertiary mudstones of the Pacific Northwest, U.S.A.: Sedimentology, v. 19, p. 219–235.

————1975, Seabed resources of the Taiwan continental shelf: Acta Oceanographica Taiwanica, v. 5, p. 1–18.

————1984, Quaternary sedimentation in the Japan arc-trench system: Geol. Soc. America Bull., v. 95, p. 669–685.

————1987, Principles of sedimentology and stratigraphy: Merrill, Columbus, Oh., 784 p.

Boggs, S., Jr., and C. A. Jones, 1976, Seasonal reversal of flood-tide dominant sediment transport in a small Oregon estuary: Geol. Soc. America Bull., v. 87, p. 419–426.

Boggs, S., Jr., D. G. Livermore, and M. G. Seitz, 1985, Humic macromolecules in natural waters: Jour. Macromolecular Science, v. C25(4), p. 599–657.

Boggs, S., Jr., W.-C. Wang, and J.-C. Chen, 1974, Texture and composi-

tional patterns of Taiwan shelf sediment: Acta Oceanographica Taiwanica, v. 4, p. 13–56.

Bokman, J., 1952, Clastic quartz particles as indices of provenance: Jour. Sed. Petrology, v. 22, p. 17–24.

Boles, J. R., 1974, Structure, stratigraphy, and petrology of mainly Triassic rocks, Hokonui Hills, Southland, New Zealand: New Zealand Jour. Geology and Geophysics, v. 17, p. 337–374.

————1982, Active albitization of plagioclase, Gulf Coast Tertiary: Am. Jour. Sci., v. 282, p. 165–180.

Boles, J. R., and S. G. Franks, 1979, Clay diagenesis in Wilcox Sandstone of southwest Texas: Implications of smectite diagenesis on sandstone cementation: Jour. Sed. Petrology, v. 49, p. 55–70.

Bond, G. C., and J. C. Devay, 1980, Pre-upper Devonian quartzose sandstones in the Shoo Fly formation, northern California—Petrology, provenance and implications for regional tectonics: Jour. Geology, v. 88, p. 285–308.

Boothroyd, J. C., and G. M. Ashley, 1975, Processes, bar morphology and sedimentary structures on braided outwash fans, northeastern Gulf of Alaska, in Jopling, V. V., and B. C. McDonald, eds., Glaciofluvial and glaciolacustrine sedimentation: Soc. Econ. Paleontologists and Mineralogists Spec. Pub. 23, p. 63–83.

Borchert, H., and R. O. Muir, 1964, Salt deposits: The origin, metamorphism, and deformation of evaporites: Van Nostrand, London, 338 p.

Bornhold, B. D., and P. Giresse, 1985, Glauconitic sediments on the continental shelf off Vancouver Island, British Columbia, Canada: Jour. Sed. Petrology, v. 55, p. 653–664.

Boswell, P. G. H., 1933, Mineralogy of sedimentary rocks: Murby, London, 393 p.

Bouma, A. H., 1962, Sedimentology of some flysch deposits: Elsevier, Amsterdam, 168 p.

————1969, Methods for the study of sedimentary structures: John Wiley & Sons, New York, 458 p.

Bourgeois, J., and E. L. Leithold, 1984, Wave-worked conglomerates—depositional processes and criteria for recognition, in Koster, E. H., and R. J. Steel, eds., Sedimentology of gravels and conglomerates: Canadian Soc. Petroleum Geologists Mem. 10, p. 331–344.

Braithwaite, C. J. R., 1989, Displacive calcite and grain breakage in sandstones: Jour. Sed. Petrology, v. 59, p. 258–266.

Braitsch, O., 1971, Salt deposits: Their origin and composition: Springer-Verlag, Berlin, 279 p.

Bramlette, M. N., 1946, The Monterey Formation of California and the origin of its siliceous rocks: U.S. Geol. Survey Prof. Paper 212, 55 p.

Brand, J., 1989, Aragonite-calcite transformation based on Pennsylvanian molluscs: Geol. Soc. America Bull., v. 101, p. 377–390.

Bricker, O. P., ed., 1971, Carbonate cements: The Johns Hopkins Press, Baltimore and London, 376 p.

Brindley, G. W., and G. Brown, eds., 1980, Crystal structures of clay minerals and their X-ray identification: Mineralog. Soc. (London) Mon. 5, 495 p.

Brownfield, M. B., 1972, Geology of Floras Creek drainage, Langlois Quadrangle, Oregon: M.S. Thesis, University of Oregon, Eugene, 104 p.

Bryant, R., and D. J. A. Williams, 1982, Mineralogical analysis of suspended cohesive sediment using an analytical electron microscope: Jour. Sed. Petrology, v. 52, p. 299–306.

Buczynski, C., and H. S. Chafetz, 1987, Siliciclastic grain breakage and displacement due to carbonate crystal growth: an example from the Lueders Formation (Permian) of north-central Texas, U.S.A.: Sedimentology, v. 34, p. 837–843.

Buekes, N. J., and D. R. Lowe, 1989, Environmental control on diverse stromatolite morphologies in the 3000 Myr Pongola Supergroup, South Africa: Sedimentology, v. 36, p. 383–397.

Bull, P. A., 1986, Procedures in environmental reconstruction by SEM analysis, in Sieveking, G. De C., and M. B. Hart, eds., The scientific study of flint and chert: Cambridge Univ. Press, Cambridge, p. 221–226.

Bull, W. B., 1972, Recognition of alluvial fan deposits in the stratigraphic record, in Hamblin, W. K., and J. D. Rigby, eds., Recognition of ancient sedimentary environments: Soc. Economic Paleontologists and Mineralogists Spec. Pub. 16, p. 63–83.

Burger, H., and W. Skala, 1976, Comparison of sieve and thin-section technique by a Monte-Carlo-model: Computer Geoscience, v. 2, p. 123–139.

Burke, K., 1975, Atlantic

evaporites formed by evaporation of water spilled from Pacific, Tethyan, and southern oceans: Geology, v. 3, p. 613–616.

————1977, Aulocogens and continental breakup: Ann. Rev. Earth and Planetary Sciences, v. 5, p. 371–396.

————1978, Evolution of continental rift systems in the light of plate tectonics, *in* Ramberg, I. B., and E. R. Neumann, eds., Tectonics and geophysics of continental rifts: D. Reidel, p. 1–9.

Burley, S. D., J. D. Kantorowicz, and B. Waugh, 1985, Clastic diagenesis, *in* Brenchley, P. J., and B. P. J. Williams, eds., Sedimentology, recent developments and applied aspects: Blackwell, Oxford, p. 189–226.

Burnett, W. C., and P. N. Froelich, eds., 1988, The origin of marine phosphorites: The results of the R.V. *Robert D. Conrad* Cruise 23–06 to the Peru shelf: Special issue of Marine Geology, v. 80, p. 181–346.

Burns, S. J., and V. Rossinsky, Jr., 1989, Late Pleistocene mixing zone dolomitization, southeastern Barbados, West Indies: Sedimentology, v. 36, p. 1135–1142.

Burst, J. F., 1965, Subaqueously formed shrinkage cracks in clay: Jour. Sed. Petrology, v. 35, p. 348–353.

Busenberg, E., and L. N. Plummer, 1986, A comparative study of the dissolution and crystal growth kinetics of calcite and aragonite, *in* Mumpton, F. A., ed., Studies in diagenesis: U.S. Geol. Survey Bull. 1578, p. 139–168.

Bustin, R. M., A. R. Cameron, D. A. Grieve, and W. D. Kalkreuth, 1985, Coal petrology, its principles, methods, and applications: Geol. Assoc. Canada Short Course Notes, v. 3, 230 p.

Button, A., T. D. Brock, P. J. Cook, H. P. Euster, A. M. Goodwin, H. L. James, L. Margulis, K. H. Nealson, J. O. Nriagu, A. F. Trendall, and M. R. Walter, 1982, Sedimentary iron deposits, evaporites, and phosphorites, *in* Holland, H. D., and M. Schidlowski, eds., Mineral deposits and evolution of the biosphere: Springer-Verlag, New York, p. 259–273.

Buxton, T. M., and D.F. Sibley, 1981, Pressure solution features in shallow buried limestones: Jour. Sed. Petrology, v. 51, p. 19–26.

Buyce, M. R., and G. M. Friedman, 1975, Significance of authigenic K-feldspar in Cambro-Ordovician carbonate rocks of the proto-Atlantic shelf in North America: Jour. Sed. Petrology, v. 45, p. 808–821.

Byerly, G. R., J. V. Mrakovich, and R. J. Malcutt, 1975, Use of Fourier shape analysis in zircon petrogenetic studies: Geol. Soc. America Bull., v. 86, p. 956–958.

Byers, C. W., 1974, Shale fissility: relation to bioturbation: Sedimentology, v. 21, p. 479–484.

Cadigan, R. A., 1967, Petrology of the Morrison Formation in the Colorado Plateau Region: U.S. Geol. Survey Prof. Paper 556, 113 p.

————1971, Petrology of the Triassic Moenkopi Formation and related strata in the Colorado Plateau Region: U.S. Geol. Survey Prof. Paper 692, 70 p.

Calvert, S. E., 1974, Deposition and diagenesis of silica in marine waters, *in* Hsü, K. J., and H. C. Jenkyns, Pelagic sediments: on land and under the sea: Internat. Assoc. Sedimentologists, Spec. Pub. 1, Blackwell, Oxford, p. 273–299.

————1983, Sedimentary geochemistry of silicon, *in* Aston, R. R., ed., Silicon geochemistry and biogeochemistry: Academic Press, London, p. 143–186.

Cameron, E. M., and R. M. Garrels, 1980, Geochemical composition of some Precambrian shales from the Canadian Shield: Chem. Geology, v. 28, p. 181–197.

Cameron, K. L., and H. Blatt, 1971, Durabilities of sand size schist and "volcanic" rock fragments during fluvial transport, Elk Creek, Black Hills, South Dakota: Jour. Sed. Petrology, v. 41, p. 565–576.

Campbell, C. V., 1966, Truncated wave-ripple laminae: Jour. Sed. Petrology, v. 36, p. 820–828.

————1967, Lamina, laminaset, bed and bedset: Sedimentology, v. 8, p. 7–26.

Campbell, J. A., 1972, Petrology of the quartzose sandstones of the Parting Formation in west central Colorado: Jour. Sed. Petrology, v. 42, p. 263–269.

Cant, D. J., 1984, Development of shoreline-shelf sandbodies in a Cretaceous epeiric sea deposit: Jour. Sed. Petrology, v. 54, p. 541–556.

Carballo, J. D., and L. S. Land, 1984, Holocene dolomitization of supratidal sediments by active tidal pumping, Sugarloaf Key, Florida (abs.): Am. Assoc. Petroleum Geologists Bull., v. 68, p. 459.

Carballo, J. D., L. S. Land, and D. E. Miser, 1987, Holocene dolomitization of supratidal sediments by active tidal pumping, Sugarloaf Key, Florida: Jour. Sed. Petrology, v. 57, p. 153–165.

Carlson, W. D., 1983, The polymorphs of $CaCO_3$ and the aragonite-calcite transformation, *in* Reeder, R. J., ed., Carbonates: Mineralogy and chemistry: Mineralog. Soc. America Rev. Mineralogy, v. 11, p. 191–225.

Carothers, W. W., and Y. K. Kharaka, 1978, Aliphatic acid anions in oil-field waters—implications for origin of natural gas: Am. Assoc. Petroleum Geologists Bull., v. 62, p. 2441–2453.

Carozzi, A. V., 1960, Microscopic sedimentary petrography: John Wiley & Sons, Inc., New York, 485 p.

————1989, Carbonate rock depositional models—A microfacies approach: Prentice Hall, Englewood Cliffs, N.J., 604 p.

Carrigy, M. A., and G. B. Mellon, 1964, Authigenic clay mineral cements in Cretaceous and Tertiary sandstones of Alberta: Jour. Sed. Petrology, v. 34, p. 461–472.

Carstens, H., 1985, Early diagenetic cone-in-cone structures in pyrite concretions: Jour. Sed. Petrology, v. 55, p. 105–108.

Carter, R. M., 1975, A discussion and classification of subaqueous mass-transport with particular application to grain-flow, slurry-flow, and fluxoturbidites: Earth Science Rev., v. 11, p. 145–177.

Carter, R. W. G., and J. D. Orford, 1984, Coarse clastic barrier beaches: A discussion of the distinctive dynamic and morphosedimentary characteristics: Marine Geology, v. 60, p. 377–389.

Carver, R. E., 1971a, Heavy-mineral separation, *in* Carver, R. W., ed., Procedures in sedimentary petrology, John Wiley & Sons, New York, p. 427–452.

Carver, R. E., ed., 1971b, Procedures in sedimentary petrology: John Wiley & Sons, New York, 653 p.

Cas, R. A. F., and J. V. Wright, 1987, Volcanic successions: modern and ancient: George Allen & Unwin, London, 528 p.

Casas, E., and T. K. Lowenstein, 1989, Diagenesis of saline pan halites: comparison of petrographic features of modern, Quaternary, and Permian halites: Jour. Sed. Petrology, v. 59, p. 724–739.

Cavazza, W., 1989, Detrital modes and provenance of the Stilo-Capo d'Orlando Formation (Miocene), southern Italy: Sedimentology, v. 36, p. 1700–1090.

Cawood, P. A., 1990, Provenance mixing in an intraoceanic subduction zone: Tonga Trench–Louisville Ridge collision zone, southwest Pacific: Sed. Geology, v. 67, p. 35–53.

Cayeux, L., 1935, Les roches sédimentares de Frances: Roches carbonatees: Masson et Cie, Paris, 447 p.

Cervato, C., 1990, Hydrothermal dolomitization of Jurassic-Cretaceous limestones in the southern Alps (Italy): Relation to tectonics and volcanism: Geology, v. 18, p. 458–461.

Chafetz, H. S., 1986, Marine peloids: a product of bacterially induced precipitation of calcite: Jour. Sed. Petrology, v. 56, p. 812–817.

Chafetz, H. S., and R. L. Folk, 1984, Travertines: Depositional morphology and the bacterially constructed constituents: Jour. Sed. Petrology, v. 54, p. 289–316.

Chambers, R. L., and S. B. Upchurch, 1979, Multivariate analysis of sedimentary environments using grain size frequency distributions: Jour. Math. Geology, v. 11, p. 27–43.

Chamley, H., 1989, Clay sedimentology: Springer-Verlag, Berlin, 623 p.

Chappell, B. W., 1968, Volcanogenic graywackes from the Upper Devonian Baldwin Formation, Tamworth-Burraba District, New South Wales: Jour. Geol. Soc. Australia, v. 15, p. 87–102.

Charles, R. G., and H. Blatt, 1978, Quartz, chert, and feldspars in modern fluvial muds and sands: Jour. Sed. Petrology, v. 48, p. 427–432.

Chebotarev, I. I., 1955, Metamorphism of natural waters in the crust weathering: Geochim. et Cosmochim. Acta, v. 8, p. 198–212.

Cheel, R. J., 1990, Horizontal lamination and the sequence of bed phases and stratification under upperflow-regime conditions: Sedimentology, v. 37, p. 517–529.

Cheel, R. J., and G. V. Middleton, 1986, Horizontal laminae formed under upper flow regime plane bed conditions: Jour. Geology, v. 94, p. 489–504.

Cheeney, R. F., 1983, Statistical methods in geology: George Allen & Unwin, London, 169 p.

Chilingarian, G. V., 1981, Compactional diagenesis, *in* Parker, A. B., and B. W. Sellwood, eds., Sediment diagenesis: D. Reidel, Dordrecht, Holland, p. 57–168.

Chilingarian, G. V., and K. H. Wolf, 1975, Compaction of coarse-grained sediments, I: Elsevier, Amsterdam, 555 p.

————— 1976, Compaction of coarse-grained sediments, II: Elsevier, Amsterdam, 808 p.

Chilingarian, G. V., and T. F. Yen, 1978, Bitumens, asphalts and tar sands: Elsevier, Amsterdam, 331 p.

Choquette, P. W., 1978, Oolite, *in* Fairbridge, R. W., and J. Bourgeois, Encyclopedia of sedimentology, Hutchinson and Ross, Stroudsburg, Pa., p. 510–515.

Choquette, P. W., and N. P. James, 1987, Diagenesis in limestones—3. The deep burial environment: Geoscience Canada, v. 14, p. 3–35.

Choquette, P. W., and L. C. Pray, 1970, Geologic nomenclature and classification of porosity in sedimentary carbonates: Am. Assoc. Petroleum Geologists Bull., v. 54, p. 207–250.

Choquette, P. W., and F. C. Trusell, 1978, A procedure for making the titan-yellow stain for Mg-calcite permanent: Jour. Sed. Petrology, v. 48, p. 639–641.

Chowns, T. M., and J. E. Elkins, 1974, The origin of quartz geodes and cauliflower cherts through the silicification of anhydrite nodules: Jour. Sed. Petrology, v. 44, p. 885–903.

Christiansen, C., 1984, A comparison of sediment parameters from log probability plots and log-log plots of the same sediments: Geoskrifter, v. 20, 20 p.

Christiansen, C., P. Blaesild, and D. Dalsgaard, 1984, Re-interpreting ''segmented'' grain-size curves: Geol. Magazine, v. 121, p. 47–51.

Clark, M.W., 1981, Quantitative shape analysis: A review: Jour. Math. Geology, v. 13, p. 303–320.

————— 1987, Image analysis of clastic particles, *in* Marshall, J. R., ed., Clastic particles: Van Nostrand Reinhold, New York, p. 256–266.

Clarke, F. W., 1924, The data of geochemistry, 5th ed.: U.S. Geol. Survey Bull., v. 770, 841 p.

Cleary, W. J., and J. R. Conolly, 1971, Distribution and genesis of quartz in a piedmont-coastal plain environment: Geol. Soc. America Bull., v. 82, p. 2755–2766.

————— 1972, Embayed quartz grains in soils and their significance: Jour. Sed. Petrology, v. 42, p. 899–904.

Clifton, H. E., 1969, Beach lamination: nature and origin: Marine Geology, v. 7, p. 553–559.

————— 1973, Pebble segregation and bed lenticularity in wave worked versus alluvial gravel: Sedimentology, v. 20, p. 173–187.

————— 1976, Wave-formed sedimentary structures: A conceptual model, *in* Davis, R. A., and R. L. Ethington, eds., Beach and nearshore sedimentation: Soc. Econ. Paleontologists and Mineralogists Spec. Pub. 24, p. 126–148.

————— 1981a, Progradational sequences in Miocene shoreline deposits, southeastern Caliente Range, California: Jour. Sed. Petrology, v. 51, p. 165–184.

————— 1981b, Submarine canyon deposits, Point Lobos, California, *in* Frizzel, V., ed., Upper Cretaceous and Paleocene turbidites, central California coast: Pacific Section, Soc. Econ. Paleontologists and Mineralogists Field Trip Guide Book, Field Trip No. 6, unpaginated.

————— 1984, Sedimentation units in stratified resedimented conglomerate, Paleocene submarine canyon fill, Point Lobos, California, *in* Koster, E. H., and R. J. Steel, eds., Sedimentology of gravels and conglomerates: Canadian Soc. Petroleum Geologists Mem. 10, p. 429–441.

Clifton, H. E., and J. K. Thompson, 1978, *Macaronichnus segregatus*: a feeding structure of shallow marine polychaetes: Jour. Sed. Petrology, v. 48, p. 1293–1302.

Cloud, P. E., 1973, Paleoecological significance of banded iron formations: Econ. Geology, v. 68, p. 1135–1143.

Collinson, J. D., 1970, Bedforms of the Tana River, Norway: Geog. Annaler, v. 52A, p. 31–55.

Collinson, J. D., and D. B. Thompson, 1982, Sedimentary structures: George Allen & Unwin, London, 194 p.

Combellick, R. A., and R. H. Osborne, 1977, Sources and petrology of beach sands from southern Monterey Bay, California: Jour. Sed. Petrology, v. 47, p. 891–907.

Compton, R. R., 1962, Manual of field geology: John Wiley & Sons, New York, 378 p.

Condie, K. C., J. E. Macke, and T. O. Reimer, 1970, Petrology and geochemistry of Early Precambrian graywackes from the Fig Tree Group: Geol. Soc. America Bull., v. 81, p. 2759–2776.

Condie, K. C., and S. Snansieng, 1971, Petrology and chemistry of the Duzel (Ordovician) and Gazelle (Silurian) formations, northern California: Jour. Sed. Petrology, v. 41, p. 741–751.

Coney, P. J., 1970, The geotectonic cycle and the new global tectonics: Geol. Soc. America Bull., v. 81, p. 739–748.

Connally, G. G., 1964, Garnet ratios and provenance in the glacial drift of western New York: Science, v. 144, p. 1452–1453.

Conolly, J. R., 1965, The occurrence of polycrystallinity and undulatory extinction in quartz in sandstones: Jour. Sed. Petrology, v. 35, p. 116–135.

Conybeare, C. E. B., and K. A. W. Crook, 1968, Manual of sedimentary structures: Bureau of Mineral Resources, Geology and Geophysics, Canberra, A.C.T., Commonwealth of Australia, 327 p.

Coogan, A. H., and R. W. Manus, 1975, Compaction and diagenesis of carbonate sands, in Chilingarian, G. V., and K. H. Wolf, eds., Compaction of coarse-grained sediments I. Developments in sedimentology 18A: Elsevier, New York, p. 79–166.

Cook, P. J., 1976, Sedimentary phosphate deposits, in Wolf, K. H., ed., Handbook of strata-bound and stratiform ore deposits, v. 7: Elsevier, Amsterdam, p. 506–536.

Cook, P. J., and M. W. McElhinney, 1979, A reevaluation of the spatial and temporal distribution of sedimentary phosphate deposits in light of plate tectonics: Econ. Geology, v. 74, p. 315–330.

Coplen, T. B., C. Kendall, and J. Hopple, 1983, Comparison of stable isotope reference samples: Nature, v. 302, p. 236–238.

Crelling, J. C., and R. R. Dutcher, 1980, Principles and applications of coal petrology: Soc. Econ. Paleontologists and Mineralogists Short Course Notes 8, 127 p.

Cressman, E. R., 1962, Nondetrital siliceous sediments: U.S. Geol. Survey Prof. Paper 440-T, 22 p.

Crimes, T. P., and J. C. Harper, eds., 1970, Trace fossils: Seel House Press, Liverpool, 547 p.

————1977, Trace fossils 2: Seel House Press, Liverpool, 351 p.

Critelli, S., R. De Rosa, and J. P. Platt, 1990, Sandstone detrital modes in the Makran accretionary wedge, southwest Pakistan: Implications for tectonic setting and long-distance turbidite transportation: Sed. Geology, v. 68, p. 241–260.

Crook, K. A. W., 1960, Petrology of Parry Group, Upper Devonian–Lower Carboniferous, Tamworth-Nundel District, New South Wales: Jour. Sed. Petrology, v. 30, p. 538–552.

————1970, Graywackes, in Encyclopaedia Britannica 10: Encyclopaedia Britannica, Chicago.

————1974, Lithogenesis and geotectonics: the significance of compositional variations in flysch arenites (graywackes), in Dott, R. H., Jr., and R. H. Shaver, eds., Modern and ancient geosynclinal sedimentation: Soc. Econ. Paleontologists and Mineralogists Special Pub. 19, p. 304–310.

Crowell, J. C., 1957, Origin of pebbly mudstones: Geol. Soc. America Bull., v. 68, p. 993–1009.

Cullen, D. J., G. A. Challis, and G. W. Drummond, 1990, Late Holocene estuarine phosphogenesis in Raglan Harbour, New Zealand: Sedimentology, v. 37, p. 847–857.

Cummins, W. A., 1962, The greywacke problem: Liverpool and Manchester Geol. Jour., v. 3, p. 51–72.

Curran, H. A., ed., 1985, Biogenic structures: Their use in interpreting depositional environments: Soc. Econ. Paleontologists and Mineralogists Spec. Pub. 35, 347 p.

Curtis, C. D., 1977, Sedimentary geochemistry: Environments and processes dominated by involvement in an aqueous phase: Philos. Trans. Royal Society, v. A286, p. 353–371.

————1980, Diagenetic alteration in black shales: Jour. Geol. Soc. London, v. 137, p. 189–194.

————1987, Inorganic geochemistry and petroleum exploration, in Advances in petroleum geochemistry, Academic Press, London, v. 2, p. 91–140.

Curtis, C. D., and M. L. Coleman, 1986, Controls on the precipitation of early diagenetic calcite, dolomite and siderite concretions in complex depositional sequences, in Gautier, D. L., ed., Roles of organic matter in sediment diagenesis: Soc. Econ. Paleontologists and Mineralogists Special Pub. 38, p. 23–33.

Curtis, C. D., S. R. Lipshie, G. Oertel, and M. J. Pearson, 1980, Clay orientation in some Upper Carboniferous mudrocks, its relationship to quartz content and some inferences about fissility, porosity and compaction history: Sedimentology, v. 17, p. 333–339.

Damanti, J. F., and T. A. Jordan, 1989, Cementation and compaction history of synorogenic foreland basin sedimentary rocks from Huaco, Argentina: Am. Assoc. Petroleum Geologists Bull., v. 73, p. 858–873.

Dana, J. D., 1873, On some results of the earth's contraction by cooling, including a discussion of the origin of mountains, and the nature of the earth's interior: Am. Jour. Sci., Ser. 3, v. 5, p. 423–443; v. 6, p. 6–14, 104–115, 161–171.

Daniels, E. J., S. P. Altaner, S. Marshak, and J. R. Eggleson, 1990, Hydrothermal alteration in anthracite in eastern Pennsylvania: Implications for the mechanisms of anthracite formation: Geology, v. 18, p. 247–250.

Dapples, E. C., 1979, Diagenesis of sandstones, in Larsen, G., and G. V. Chilingar, eds., Diagenesis in sediments and sedimentary rocks: Elsevier, Amsterdam, p. 31–97.

Dapples, E. C., and J. F. Rominger, 1945, Orientation analysis of fine-grained clastic sediments: Jour. Geology, v. 53, p. 246–261.

Darby, D. A., 1975, Kaolinite and other clay minerals in Arctic Ocean sediments: Jour. Sed. Petrology, v. 45, p. 272–279.

————1984, Trace elements in ilmenites: a way to discriminate provenance or age in coastal sands: Geol. Soc. of America Bull., v. 95, p. 1208–1218.

Darby, D. A., and Y. W. Tsang, 1987, Variation in ilmenite element composition within and among drainage basins: implication for provenance: Jour. Sed. Petrology, v. 57, p. 831–838.

Davies, I. C., and R. G. Walker, 1974, Transport and deposition of resedimented conglomerates: The Cap Enrage Formation, Cambro-Ordovician, Gaspé, Quebec: Jour. Sed. Petrology, v. 44, p. 1200–1216.

Davies, P. J., B. Bubela, and J. Ferguson, 1978, The formation of ooids: Sedimentology, v. 25, p. 703–730.

Davy, R., 1983, A contribution on the chemical composition of Precambrian iron-formations, in Trendall, A. F., and R. C. Morris, eds., Iron-formation: facts and problems: Elsevier, Amsterdam, p. 325–343.

Dean, W. E., 1982, Theoretical versus observed successions from evaporation of seawater, *in* Dean, W. E., and B. C. Schreiber, eds., Marine evaporites: Soc. Econ. Paleontologists and Mineralogists Short Course 4, p. 74–85.

Dean, W. E., M. A. Arthur, and G. E. Claypool, 1986, Depletion of ^{13}C in Cretaceous marine organic matter: source, diagenetic, or environmental signal?: Marine Geology, v. 70, p. 119–157.

Dean, W. E., G. R. Davies, and R. Y. Anderson, 1975, Sedimentological significance of nodular and laminated anhydrite: Geology, v. 3, p. 367–372.

Dean, W. E., and T. D. Fouch, 1983, Lacustrine environment, *in* Scholle, P. A., D. G. Bebout, and C. H. Moore, 1983, Carbonate depositional environments: Am. Assoc. Petroleum Geologists Mem. 33, p. 97–130.

DeCelles, P. G., 1987, Variable preservation of Middle Tertiary, coarse-grained, nearshore to outer-shelf storm deposits in southern California: Jour. Sed. Petrology, v. 57, p. 250–264.

———— 1988, Lithologic provenance modeling applied to the Late Cretaceous synorogenic Echo Canyon Conglomerate, Utah: a case of multiple source areas: Geology, v. 16, p. 1039–1043.

Deer, W. A., R. A. Howie, and J. Zussman, 1963a, Rock-forming minerals, v. 4, Framework silicates: John Wiley & Sons, New York, 435 p.

———— 1963b, Rock-forming minerals, v. 2, Chain silicates: John Wiley & Sons, New York, 379 p.

———— 1966, An introduction to the rock-forming minerals: Longman Group Limited, Essex, England, 528 p.

Degens, E. T., 1965, Geochemistry of sediments: Prentice-Hall, Englewood Cliffs, N.J., 342 p.

Deniro, M. J., 1983, Distribution of the stable isotopes of carbon, nitrogen, oxygen, and hydrogen among plants, *in* Meinschein, W. G., ed., Organic geochemistry of contemporaneous and ancient sediments: Great Lakes Sect., Soc. Econ. Paleontologists and Mineralogists, p. 4-1–4-27.

Deniro, M. J., and S. Epstein, 1981, Isotopic composition of cellulose from aquatic organisms: Geochim. et Cosmochim. Acta, v. 42, p. 495–506.

Dennen, W. H., 1966, Stoichiomeric substitution in natural quartz: Geochim. et Cosmochim. Acta, v. 30, p. 1235–1241.

———— 1967, Trace elements in quartz as indicators of provenance: Geol. Soc. America Bull., v. 78, p. 125–130.

Dias, J. M. A., and M. J. Neal, 1990, Modal size classification of sands: An example from the northern Portugal continental shelf: Jour. Sed. Petrology, v. 60, p. 426–437.

Dickinson, W. R., 1970, Interpreting detrital modes of graywacke and arkose: Jour. Sed. Petrology, v. 40, p. 695–707.

———— 1971, Detrital modes of New Zealand graywackes: Sed. Geology, v. 5, p. 37–56.

———— 1974a, Plate tectonics and sedimentation, *in* W. R. Dickinson, ed., Tectonics and sedimentation, Soc. Econ. Paleontologists and Mineralogists Spec. Pub. 22, p. 1–27.

Dickinson, W. R., ed., 1974b, Tectonics and sedimentation: SEPM Spec. Pub. 22, Soc. Econ. Paleontologists and Mineralogists, Tulsa, 204 p.

———— 1985, Interpreting provenance relations from detrital modes of sandstones, *in* Zuffa, G. G., ed., Provenance of arenites: D. Reidel, Dordrecht, p. 333–361.

———— 1988, Provenance and sediment dispersal in relation to paleotectonics and paleogeography of sedimentary basins, *in* Kleinspehn, K. L., and C. Paola, eds., New perspectives in basin analysis: Springer-Verlag, New York, p. 3–25.

Dickinson, W. R., L. S. Beard, G. R. Brakenridge, J. L. Erjavec, R. C. Ferguson, K. F. Inman, R. A. Knepp, F. A. Lindberg, and P. T. Ryberg, 1983, Provenance of North American Phanerozoic sandstones in relation to tectonic setting: Geol. Soc. America Bull., v. 94, p. 222–235.

Dickinson, W. R., K. P. Helmold, and J. A. Stein, 1979, Mesozoic lithic sandstones of central Oregon: Jour. Sed. Petrology, v. 49, p. 501–516.

Dickinson, W. R., and R. V. Ingersoll, 1990, Physiographic controls on the composition of sediments derived from volcanic and sedimentary terrains on Barro Colorado Island, Panama—Discussion: Jour. Sed. Petrology, v. 60, p. 797–798.

Dickinson, W. R., and E. I. Rich, 1972, Petrologic intervals and petrofacies in the Great Valley Sequence, Sacramento Valley, California: Geol. Soc. America Bull., v. 83, p. 3007–3024.

Dickinson, W. R., and C. A.

Suczek, 1979, Plate tectonics and sandstone compositions: Am. Assoc. Petroleum Geologists Bull., v. 63, p. 2164–2182.

Dickinson, W. R., and R. Valloni, 1980, Plate settings and provenance of sands in modern ocean basins: Geology, v. 8, p. 82–86.

Dickson, J. A. D., 1965, A modified staining technique for carbonates in thin section: Nature, v. 205, p. 587.

———— 1966, Carbonate identification and genesis as revealed by staining: Jour. Sed. Petrology, v. 36, p. 491–505.

Diks, A., and S. C. Graham, 1985, Quantitative mineralogic characterization by back-scattered electron image analysis: Jour. Sed. Petrology, v. 55, p. 347–355.

Dimroth, E., 1976, Aspects of the sedimentary petrology of cherty iron-formation, *in* Wolf, K. H., ed., Handbook of strata-bound and stratiform ore deposits, v. 7: Elsevier, Amsterdam, p. 203–254.

———— 1979, Models of physical sedimentation of iron formations, *in* Walker, R. G., ed., Facies models: Geoscience Canada Reprint Ser. 1, p. 159–174.

Dimroth, E., and M. M. Kimberley, 1976, Precambrian atmospheric oxygen: Evidence in the sedimentary distribution of carbon, sulfur, uranium and iron: Canadian Jour. Earth Science, v. 13, p. 1161–1185.

Dixon, J., and V. P. Wright, 1983, Burial diagenesis and crystal diminution. The origin of crystal diminution in some limestones from South Wales: Sedimentology, v. 30, p. 537–546.

Dobkins, J. E., and R. L. Folk, 1970, Shape development on Tahati-Nui: Jour. Sed. Petrology, v. 40, p. 1167–1203.

Doeglas, D. J., 1962, The structure of sedimentary deposits of braided rivers: Sedimentology, v. 1, p. 167–190.

Dott, R. H., Jr., 1964, Wacke, graywacke and matrix—What approach to immature sandstone classification?: Jour. Sed. Petrology, v. 34, p. 625–632.

———— 1974, The geosynclinal concept, *in* Dott, R. H., and R. H. Shaver, eds., Modern and ancient geosynclinal sedimentation: Soc. Econ. Paleontologists and Mineralogists Spec. Pub. 19, p. 1–13.

———— 1978, Tectonics and sedimentation a century later: Earth Sci. Rev., v. 14, p. 1–34.

———— 1979, The geosyncline—first major geological concept "made in America": University Press of New England, published for the University of New Hampshire, p. 239–264.

Dott, R. H., and J. Bourgeois, 1982, Hummocky stratification: Significance of its variable bedding sequences: Geol. Soc. America Bull. v. 93, p. 663–680.

———— 1983, Hummocky stratification: Significance of its variable bedding sequences: Discussion and reply: Geol. Soc. America Bull., v. 94, p. 1249–1251.

Dott, R. H., and R. H. Shaver, ed., 1974, Modern and ancient geosynclinal sedimentation: Soc. Econ. Paleontologists and Mineralogists Spec. Pub. 19, 380 p.

Dowdeswell, J. A., 1982, Scanning electron micrographs of quartz sand grains from cold environments: Jour. Sed. Petrology, v. 2, p. 1315–1324.

Dowdeswell, J. A., M. J. Hambrey, and R. Wu, 1985, A comparison of clast fabric and shape in Late Precambrian and Modern glaciogenic sediments: Jour. Sed. Petrology, v. 55, p. 691–704.

Dowdeswell, J. A., L. E. Osterman, and J. T. Andrews, 1985, Quartz sand grain shape and other criteria used to distinguish glacial and nonglacial events in a marine core from Frobisher Bay, Baffin Island, N.W.T., Canada: Sedimentology, v. 32, p. 119–132.

Dowdeswell, J. A., and M. Sharp, 1986, Characterization of pebble fabrics in modern terrestrial glaciogenic sediments: Sedimentology, v. 33, p. 699–710.

Doyle, L. J. and H. H. Roberts, 1988, Carbonate-clastic transitions: Elsevier, Amsterdam, 304 p.

Doyle, L. J., K. L. Carder, and R. G. Stewart, 1983, The hydraulic equivalence of micas: Jour. Sed. Petrology, v. 53, p. 643–648.

Drever, J. I., 1971, Magnesium iron replacements in clay minerals in anoxic marine sediments: Science, v. 172, p. 1334–1336.

———— 1974, Geochemical model for the origin of Precambrian banded iron formations: Geol. Soc. America Bull., v. 85, p. 1099–1106.

———— 1982, The geochemistry of natural waters: Prentice-Hall, Englewood Cliffs, N.J., 388 p.

Drewery, S., R. A. Cliff, and M. R. Leeder, 1987, Provenance of Carboniferous sandstones from U-Pb dating of detrital zircons: Nature, v. 325, p. 50–53.

Dryden, L., and C. Dryden, 1946, Comparative rates of weathering of some common heavy minerals: Jour. Sed. Petrology, v. 16, p. 91–96.

Duke, W. L., 1985, Hummocky cross-stratification, tropical hurricanes, and intense winter storms: Sedimentology, v. 32, p. 167–194.

———— 1987, Hummocky cross-stratification, tropical hurricanes, and intense winter storms: Reply: Sedimentology, v. 34, p. 344–359.

Duncan, D. C., 1976, Geologic setting of oil-shale deposits and world prospects, in Yen, T. F., and G. V. Chilingarian, eds., Oil shale: Elsevier, Amsterdam, p. 13–26.

Dunham, R. J., 1962, Classification of carbonate rocks according to depositional texture, in Ham, W. E., ed., Classification of carbonate rocks. Am. Assoc. Petroleum Geologists Mem. 1, p. 108–121.

———— 1970, Stratigraphic reef versus ecologic reefs: Am. Assoc. Petroleum Geologists Bull., v. 54, p. 1931–1932.

———— 1971, Meniscus cement, in Bricker, O. P., ed., Carbonate cements: Johns Hopkins Univ. Studies in Geology, No. 19, p. 297–300.

Dunoyer de Segonzac, G., 1970, The transformation of clay minerals during diagenesis and low-grade metamorphism: a review: Sedimentology, v. 15, p. 281–346.

Dupré, W. R., 1984, Reconstruction of paleo-wave conditions during the Late Pleistocene from marine terrace deposits, Monterey Bay, California: Marine Geology, v. 60, p. 435–454.

Dupré, W. R., H. E. Clifton, and R. E. Hunter, 1980, Modern sedimentary facies of the open Pacific coast and Pleistocene analogs from Monterey Bay, California, in Fields, M. E., et al., eds., Quaternary depositional environment of the Pacific Coast: Pacific Coast Paleogeography Symposium 4, Pacific Sect., Soc. Econ. Paleontologists and Mineralogists, p. 105–120.

Durand, B., ed., 1980, Kerogen: Editions Technip, Paris, 519 p.

Dutton, S. P., and T. N. Diggs, 1990, History of quartz cementation in the Lower Cretaceous Travis Peak Formation, east Texas: Jour. Sed. Petrol-

ogy, v. 60, p. 191–202.

Dzevanshir, R. D., L. A. Buryakovskiy, and G. V. Chilingarian, 1986, Simple quantitative evaluation of porosity of argillaceous sediments at various depths of burial: Sed. Geology, v. 46, p. 169–175.

Dzulnyski, S., and E. K. Walton, 1965, Sedimentary features of flysch and greywackes: Developments in sedimentology, v. 7, Elsevier, Amsterdam, 274 p.

Easterbrook, D. J., 1981, Characteristic features of glacial sediments, in Scholle, P. A., and D. Spearing, eds., Sandstone depositional environments: Am. Assoc. Petroleum Geologists Mem. 31, p. 1–10.

Eberhart, J. P., 1982, High resolution electron-microscopy studies applied to clay minerals, in Fripiat, J. J., ed., Advanced techniques for clay mineral analysis: Elsevier, Amsterdam, p. 31–50.

Edwards, M., 1986, Glacial environments, in Reading, H. G., ed., Sedimentary environments and facies, 2nd ed: Blackwell Scientific Publications, Oxford, p. 445–470.

Ehrlich, R., 1983, Size analysis wears no clothes, or have moments come and gone?: Jour. Sed. Petrology, v. 53, p. 1.

Ehrlich, R., P. J. Brown, J. M. Yarus, and R. S. Przygocki, 1980, The origin of shape frequency distributions and the relationship between size and shape: Jour. Sed. Petrology, v. 50, p. 475–484.

Ehrlich, R., and M. Chin, 1980, Fourier grain-shape analysis: A new tool for sourcing and tracking abyssal silts: Marine Geology, v. 38, p. 219–231.

Ehrlich, R., S. J. Crabtree, S. K. Kennedy, and R. L. Cannon, 1984, Petrographic image analysis, I. Analysis of reservoir pore complexes: Jour. Sed. Petrology, v. 54, p. 1365–1378.

Ehrlich, R., S. K. Kennedy, and C. D. Brotherhood, 1987, Respective roles of Fourier and SEM techniques in analyzing sedimentary quartz, in Marshall, J. R., ed., Clastic particles, Van Nostrand Reinhold, New York, p. 292–301.

Ehrlich, R., and B. Weinberg, 1970, An exact method for characterization of grain shape: Jour. Sed. Petrology, v. 40, p. 205–212.

Eichler, J., 1976, Origin of the Precambrian banded iron-formations, in Wolf, K. H., ed., Handbook of strata-bound and stratiform ore deposits, v. 7: Elsevier, Amsterdam, p. 157–201.

Ekdale, A. A., R. G. Bromley, and S. G. Pemberton, 1984, Ichnology, trace fossils in sedimentology and stratigraphy: Soc. Econ. Paleontologists and Mineralogists Short Course 15, 317 p.

Elliott, M. A., and G. R. Yohe, 1981, The coal industry and coal research and development in perspective, *in* Elliott, M. A., ed., Chemistry of coal utilization, second supplementary volume: Wiley Interscience, New York, p. 1–54.

Elliott, T. L., 1982, Carbonate facies, depositional cycles and the development of secondary porosity during burial diagenesis, *in* Christopher, J. E., and J. Kaldi, eds., 4th International Williston Basin Symposium: Saskatchewan Geol. Soc. Spec. Pub. 6, p. 131–151.

Embry, A. F., and J. E. Klovan, 1972, Absolute water depth limits of late Devonian paleoecological zones: Geol. Rundschau, v. 61, p. 672–686.

Emery, J. R., and J. C. Griffith, 1954, Reconnaissance investigation into relationships between behavior and petrographic properties of some Mississippian sediments: Bull. Mineral Ind. Expt. Sta. Pennsylvania State University, v. 62, p. 67–80.

Eslinger, E. V., L. M. Mayer, T. L. Durst, J. Hower, and S. M. Savin, 1973, An X-ray technique for distinguishing detrital and secondary quartz in fine-grained fraction of sedimentary rocks: Jour. Sed. Petrology, v. 43, p. 540–543.

Eslinger, E. V., and D. Pevear, 1988, Clay minerals for petroleum geologists and engineers: Soc. Econ. Paleontologists and Mineralogists Short Course Notes 22.

Eslinger, E. V., and S. M. Savin, 1973, Oxygen isotope geothermometry of the burial metamorphic rocks of the Precambrian Belt Supergroup, Glacier National Park, Montana: Geol. Soc. America Bull., v. 84, p. 2549–2560.

Eslinger, E. V., S. M. Savin, and H.-W. Yeh, 1979, Oxygen isotope geothermometry of diagenetically altered shales, *in* Scholle, P. A., and P. R. Schluger, eds., Aspects of diagenesis: Soc. Econ. Paleontologists and Mineralogists Spec. Pub. 26, p. 113–124.

Eslinger, E. V., and H.-W. Yeh, 1986, Oxygen and hydrogen isotope geochemistry of Cretaceous bentonites and shales from the Disturbed Belt,

Montana: Geochim. et Cosmochim. Acta, v. 50, p. 59–68.

Esteban, M., and C. F. Klappa, 1983, Subaerial exposure environment, *in* Scholle, P. A., D. G. Bebout, and C. H. Moore, eds., Carbonate depositional environments: Am. Assoc. Petroleum Geologists Mem. 33, p. 1–95.

Ethridge, F. G., N. Tyler, and L. K. Burns, 1984, Sedimentology of a Precambrian quartz-pebble conglomerate, southwest Colorado, *in* Koster, E. H., and R. J. Steel, eds., Sedimentology of gravels and conglomerates: Canadian Soc. Petroleum Geologists Mem. 10, p. 165–174.

Ethridge, F. G., and W. A. Wescott, 1984, Tectonic setting, recognition and hydrocarbon potential of fandelta deposits, *in* Koster, E. H., and R. J. Steel, eds., Sedimentology of gravels and conglomerates: Canadian Soc. Petroleum Geologists Mem. 10, p. 217–236.

Evamy, B. D., 1969, The precipitational environment and correlation of some calcite cements deduced from artificial staining: Jour. Sed. Petrology, v. 39, p. 787–821.

Evans, L. J., and W. A. Adams, 1975, Chlorite and illite in some lower Palaeozoic mudstones of mid-Wales: Clay Minerals, v. 10, p. 387–397.

Ewers, W. W., 1983, Chemical factors in the deposition and diagenesis of banded iron-formation, *in* Trendall, A. F., and R. C. Morris, eds., Developments in Precambrian geology 6: Elsevier, Amsterdam, p. 491–512.

Eyles, C. H., 1987, Glacially influenced submarine-channel sedimentation in the Yakataga Formation, Middleton Island, Alaska: Jour. Sed. Petrology, v. 57, p. 1004–1017.

Fairbridge, R. W., 1983, Syndiagenesis-anadiagenesis-epidiagenesis: phases in lithogenesis, *in* Larsen, G., and G. V. Chilingar, eds., Diagenesis in sediments and sedimentary rocks, 2: Elsevier, Amsterdam, p. 17–113.

Fanning, K. A., R. H. Bryne, J. A. Breland, P. R. Betzer, W. S. Moore, and R. J. Elsinger, 1981, Geothermal springs of the West Florida continental shelf: evidence for dolomitization and radionuclide enrichment: Earth and Planetary Science Letters, v. 52, p. 345–354.

Farooqui, S. M., 1969, Stratigraphy and petrology of the Port Orford Conglomerate, Cape Blanco, Oregon: M.S. Thesis, University of Oregon, 57 p.

Faure, G., 1986, Principles of isotope geology, 2nd ed: John Wiley & Sons, New York, 589 p.

Faust, S. D., and O. M. Aly, 1981, Chemistry of natural waters: Ann Arbor Science Publishers, Ann Arbor, Mich., 400 p.

Feazel, C. T., and R. A. Schatzinger, 1985, Prevention of carbonate cementation in petroleum reservoirs, *in* Schneidermann, N., and P. M. Harris, eds., Carbonate cements: Soc. Econ. Paleontologists and Sedimentologists Spec. Pub. 36, p. 97–106.

Ferree, R. A., D. W. Jordan, R. S. Kertes, K. M. Savage, and P. M. Potter, 1988, Comparative petrographic maturity of river and beach sand, and origin of quartz arenites: Jour. Geol. Education, v. 36, p. 79–87.

Fettweis, G. B., 1979, World coal resources: Elsevier, Amsterdam, 415 p.

Fieller, N. R. J., D. D. Gilbertson, and W. Olbricht, 1984, A new method for environmental analsis of particle size distribution data from shoreline sediments: Nature, v. 311, p. 648–651.

Fisher, R. V., 1961, Proposed classification of volcaniclastic sediments and rocks: Geol. Soc. America Bull., v. 72, p. 1409–1414.

———— 1966, Rocks composed of volcanic fragments and their classification: Earth Science Reviews, v. 1, p. 287–298.

Fisher, R. V., and H.-U. Schmincke, 1984, Pyroclastic rocks: Springer-Verlag, Berlin, 472 p.

Flegmann, A. W., J. W. Goodwin, and R. H. Ottewill, 1969, Rheological studies on kaolinite suspensions: British Ceramics Soc. Proc., v. 13, p. 31–45.

Flint, R. F., 1971, Glacial and Quaternary geology: John Wiley & Sons, New York, 892 p.

Flint, R. F., J. E. Sanders, and J. Rodgers, 1960, Diamictite, a substitute term for symmictite: Geol. Soc. America Bull., v. 71, p. 1809–1810.

Flügel, E., 1972, Mikrofazielle Untersuchungen in der alpinen Trias. Methoden und Probleme. Mitt. Gesell. Geol. Bergbaustud, Österreich, v. 21, p. 7–64.

———— 1982, Microfacies analysis of limestones: Springer-Verlag, Berlin, 633 p.

Folk, R. L., 1951, Stages of textural maturity in sedimentary rocks: Jour. Sed. Petrology, v. 21, p. 127–130.

———— 1959, Practical petrographic classification of limestones: Am. Assoc. Petroleum Geologists Bull., v. 43, p. 1–38.

———1962, Spectral subdivision of limestone types, *in* Ham, W. E., ed., Classification of carbonate rocks: Am. Assoc. Petroleum Geologists Mem. 1, p. 62–84.

———1965, Some aspects of recrystallization in ancient limestones, *in* Pray, L. C., and R. C. Murray, eds., Dolomitization and limestone diagenesis: Soc. Econ. Paleontologists and Mineralogists Spec. Pub. 13, p. 14–48.

———1968, Bimodal supermature sandstones. Product of the desert floor: XXIII Internat. Geol. Cong. Proc. 8, p. 9–32.

———1974, Petrology of sedimentary rocks: Hemphill, Austin, Tex., 182 p.

———1977, Stratigraphic analysis of the Navajo Sandstone: A discussion: Jour. Sed. Petrology, v. 47, p. 483–484.

Folk, R. L., P. B. Andrews, and D. W. Lewis, 1970, Detrital sedimentary rock classification and nomenclature for use in New Zealand: New Zealand Jour. Geology and Geophysics, v. 13, p. 937–968.

Folk, R. L., and L. S. Land, 1975, Mg/Ca ratio and salinity: Two controls over crystallization of dolomite: Am. Assoc. Petroleum Geologists Bull., v. 59, p. 60–68.

Folk, R. L., and J. S. Pittman, 1971, Length-slow chalcedony: A new testament for vanished evaporites: Jour. Sed. Petrology, v. 41, p. 1045–1058.

Folk, R. L., and W. C. Ward, 1957, Brazos River bar: A study in the significance of grain size parameters: Jour. Sed. Petrology, v. 27, p. 3–26.

Forbes, D. L., 1983, Morphology and sedimentology of a sinuous gravel-bed channel system: lower Babbage River, Yukon coastal plain, Canada, *in* Collinson, J. D., and J. Lewin, eds., Modern and ancient fluvial systems: Internat. Assoc. Sedimentologists Spec. Pub. 6, p. 195–206.

Force, E. R., 1980, The provenance of rutile: Jour. Sed. Petrology, v. 50, p. 485–488.

Forrest, J., and N. R. Clark, 1989, Characterizing grain size distributions: Evaluation of a new approach using multivariate extension of entropy analysis: Sedimentology, v. 36, p. 711–722.

Foster, R. J., 1960, Tertiary geology of a portion of the central Cascade Mountains, Washington: Geol. Soc. America Bull., v. 71, p. 99–126.

Fournier, R. O., 1983, A method of calculating quartz solubility in aqueous sodium chloride solutions: Geochim. et Cosmochim. Acta, v. 47, p. 579–586.

Fournier, R. O., and J. J. Rowe, 1962, The solubility of cristobalite along the three-phase curve, gas plus liquid plus cristobalite: Am. Mineralogist, v. 47, p. 897–902.

Franks, P. C., 1969, Nature, origin, and significance of cone-in-cone structures in the Kiowa Formation (Early Cretaceous), North-Central Kansas: Jour. Sed. Petrology, v. 39, p. 1438–1454.

Franks, P. C., and A. Swineford, 1959, Character and genesis of massive opal in Kimball Member, Ogallala Formation, Scott County, Kansas: Jour. Sed. Petrology, v. 29, p. 186–196.

Franzinelli, E., and P. E. Potter, 1983, Petrology, chemistry, and texture of modern river sands, Amazon River system: Jour. Geology, v. 91, p. 23–39.

Freed, R. L., and D. R. Peacor, 1989, Geopressured shale and sealing effect of smectite to illite transition: Am. Assoc. Petroleum Geologists Bull., v. 73, p. 1223–1232.

Frey, R. W., ed., 1975, The study of trace fossils: Springer-Verlag, New York, 562 p.

Frey, R. W., 1978, Behavioral and ecological implications of trace fossils, *in* Basan, P. B., ed., Trace fossil concepts: Soc. Econ. Paleontologists and Mineralogists Short Course 5, p. 43–66.

Frey, R. W., and S. G. Pemberton, 1984, Trace fossil facies models, *in* Walker, R. G., ed., Facies models: Geoscience Canada Reprint Ser. 1, p. 189–207.

Frey, R. W., S. G. Pemberton, and J. A. Fagerstrom, 1984, Morphological, ethological and environmental significance of the ichnogenera *Scoyenia* and *Ancorichnus*: Jour. Paleontology, v. 58, p. 511–528.

Friedman, G. M., 1959, Identification of carbonate minerals by staining methods: Jour. Sed. Petrology, v. 29, p. 87–97.

———1962, Comparison of moment measures for sieving and thin-section data in sedimentary petrological studies: Jour. Sed. Petrology, v. 32, p. 15–25.

———1965, Terminology of crystallization textures and fabrics in sedimentary rocks: Jour. Sed. Petrology, v. 35, p. 643–655.

———1967, Dynamic processes and statistical parameters compared for size frequency distribution of beach and river sands: Jour. Sed. Petrology, v. 37, p. 327–354.

Friedman, G. M. ed., 1969, Depositional environments in carbonate rocks: Soc. Econ. Paleontologists and Mineralogists Spec. Pub. 14, 209 p.

Friedman, G. M., 1979, Address of the retiring president of the International Association of Sedimentologists: Differences in size distributions of populations of particles among sands of various origins: Sedimentology, v. 26, p. 3–32.

Friedman, G. M., C. D. Gebelein, and J. E. Sanders, 1971, Micrite envelopes of carbonate grains are not exclusively of photosynthetic algal origin: Sedimentology, v. 16, p. 89–96.

Friedman, G. M., and J. E. Sanders, 1978, Principles of sedimentology: John Wiley & Sons, New York, 792 p.

Friedman, G. M., and V. Shukla, 1980, Significance of authigenic quartz euhedra after sulfates: Example from the Lockport Formation (Middle Silurian) of New York: Jour. Sed. Petrology, v. 50, p. 1299–1304.

Fripiat, J. J., ed., 1982a, Advanced techniques for clay mineral analysis: Elsevier, Amsterdam, 235 p.

Fripiat, J. J., 1982b, Application of far infrared spectroscopy to the study of clay minerals and zeolites, *in* Fripiat, J. J., ed., Advanced techniques for clay mineral analysis: Elsevier, Amsterdam, p. 191–210.

Froelich, P. N., M. A. Arthur, W. C. Burnett, M. Deakin, V. Hensley, R. Jahnke, L. Kaul, K.-H. Kim, K. Roe, A. Soutar, and C. Vathakanon, 1988, Early diagenesis of organic matter in Peru continental margin sediments: phosphorite precipitation: Marine Geology, v. 80, p. 309–343.

Froude, D. O., T. R. Ireland, P. D. Kinny, I. S. Williams, and W. Compston, 1983, Ion microprobe identification of 4,100–4,200 Myr-old terrestrial zircons: Nature, v. 304, p. 616–618.

Frye, K., 1981, Encyclopedia of mineralogy: Hutchinson Ross, Stroudsburg, Pa., 794 p.

Fryer, B. J., 1983, Rare earth elements in iron-formation, *in* Trendall, A. F., and R. C. Morris, eds., Iron-formation: facts and problems: Elsevier, Amsterdam, p. 345–358.

Füchtbauer, H., 1967, Influence of different types of diagenesis on sandstone porosity: Proceedings Seventh World Petroleum Congress, Mexico, v. 2, p. 353–369.

Full, W. E., R. Ehrlich, and S. K. Kennedy, 1984, Optimal configuration and information content of sets of frequency distributions: Jour. Sed. Petrology, v. 54, p. 117–126.

Galehouse, J. S., 1971, Sedimentation analysis, *in* Carver, R. E., ed., Procedures in sedimentary petrology: John Wiley & Sons, New York, p. 69–94.

Gao, G. S., S. D. Hovorka, and H. H. Posey, 1990, Limpid dolomite in Permian, San Andres halite rocks, Palo Duro Basin, Texas Panhandle: Characteristics, possible origin, and implications for brine evolution: Jour. Sed. Petrology, v. 60, p. 118–124.

Garrels, R. M., 1987, A model for the deposition of the microbanded Precambrian iron formations: Am. Jour. Science, v. 287, p. 81–106.

Garrels, R. M., and C. L. Christ, 1965, Solutions, minerals, and equilibria: Harper & Row, New York, 450 p.

Garrels, R. M., and F. T. Mackenzie, 1971, Evolution of sedimentary rocks: W. W. Norton, New York, 397 p.

Garrison, R. E., R. B. Douglass, K. E. Pisciotto, C. M. Isaacs, and J. C. Ingle, eds., 1981, The Monterey Formation and related siliceous rocks of California: Soc. Econ. Paleontologists and Mineralogists, Pacific Section, Los Angeles, Calif., 327 p.

Garrison, R. E., and W. J. Kennedy, 1977, Origin of solution-seams and flaser structure in Upper Cretaceous chalks of southern England: Sed. Geology, v. 19, p. 107–137.

Garven, G., and R. A. Freeze, 1984, Theoretical analysis of the role of groundwater flow in the genesis of stratabound ore deposits: Am. Jour. Science, v. 284, p. 1085–1174.

Gautier, D. L., Y. K. Kharaka, and R. C. Surdam, 1985, Relationship of organic matter and mineral diagenesis: Soc. Econ. Paleontologists and Mineralogists Short Course 17, 279 p.

Gergen, L. D., and R. V. Ingersoll, 1986, Petrology and provenance of deep sea drilling project sand and sandstone from the North Pacific Ocean and the Bering Sea: Sed. Geology, v. 51, p. 29–56.

Gibbs, R. J., 1983, Coagulation rates of clay minerals and natural sediments: Jour. Sed. Petrology, v. 53, p. 1193–1203.

Gieskes, J. M., 1983, The chemistry of interstitial waters of deep sea sediments: Interpretations of deep sea drilling data. *in* Riley, J. P., and R. Chester, eds., Chemical oceanography 8: Academic Press, New York, p. 221–269.

Gieskes, J. M., H. Elderfield, and B. Nevsky, 1983, Interstitial water studies, Leg 65, Deep Sea Drilling Project, *in* Lewis, B. R. R., P. Robison, et al., Initial Reports, DSDP 65, p. 441–449, U.S. Government Printing Office, Washington, D.C.

Gilligan, A., 1919, The petrography of the Millstone Grit of Yorkshire: Geol. Soc. London Quarterly Jour., v. 75, p. 251–292.

Gilman, R. A., and W. J. Metzger, 1967, Cone-in-cone concretions from western New York: Jour. Sed. Petrology, v. 37, p. 87–95.

Ginsburg, R. N., 1956, Environmental relationships of grain size and constituent particles in some south Florida carbonate environments: Am. Assoc. Petroleum Geologists Bull., v. 40, p. 2384–2387.

Girard, J.-P., S. M. Savin, and J. L. Aronson, 1989, Diagenesis of the Lower Cretaceous arkoses of the Angola Margin: Petrologic, K/Ar dating and $^{18}O/^{16}O$ evidence: Jour. Sed. Petrology, v. 59, p. 519–538.

Girty, G. H., and A. Armitage, 1989, Composition of Holocene river sand: An example of mixed-provenance sand derived from multiple tectonic elements of the Cordilleran continental margin: Jour. Sed. Petrology, v. 59, p. 597–664.

Girty, G. H., B. J. Mossman, and S. D. Pincus, 1988, Petrology of Holocene sand, peninsular ranges, California, and Baja Norte, Mexico: Implications for provenance-discrimination models: Jour. Sed. Petrology, v. 58, p. 881–887.

Given, R. K., and B. H. Wilkinson, 1985, Kinetic control of morphology, composition, and mineralogy of abiotic sedimentary carbonates: Jour. Sed. Petrology, v. 55, p. 109–119.

———— 1987, Dolomite abundance and stratigraphic age: Constraints on rates and mechanisms of Phanerozoic dolostone formation: Jour. Sed. Petrology, v. 57, p. 1068–1078.

Glaister, R. P., and H. W. Nelson, 1974, Grain-size distributions, an aid to facies identifications: Canadian Petroleum Geologists Bull., v. 22, p. 203–240.

Glenn, G. R., and M. A. Arthur, 1988, Petrology and major element geochemistry of Peru margin phosphorites and associated diagenetic minerals: authigenesis in modern organic-rich sediments: Marine Geology, v. 80, p. 231–267.

———— 1990, Origin of a Cretaceous phosphorite-greensand giant, Egypt: Sedimentology, v. 37, p. 123–154.

Gloppen, R. G., and R. J. Steel, 1981, The deposits, internal structure and geometry in six alluvial fan–fan delta bodies (Devonian, Norway)—a study in the significance of bedding sequences in conglomerates, *in* Ethridge, F. G., and R. M. Flores, eds., Recent and ancient nonmarine depositional environments: Models for exploration: Soc. of Econ. Paleontologists and Mineralogists Spec. Pub. 31, p. 49–69.

Gold, P. B., 1987, Textures and geochemistry of authigenic albite from Miocene sandstones, Louisiana Gulf Coast: Jour. Sed. Petrology, v. 57, p. 353–362.

Goldich, S. S., 1938, A study in rock weathering: Jour. Geology, v. 46, p. 17–58.

Goldstein, R. H., 1988, Cement stratigraphy of Pennsylvanian Holder Formation, Sacramento Mountains, New Mexico: Am. Assoc. Petroleum Geologists Bull., v. 72, p. 425–438.

Gole, M. J., and C. Klein, 1981, Banded iron-formations through much of Precambrian time: Jour. Geology, v. 89, p. 169–183.

Goodfellow, R. W., 1987, Petrology and provenance of sandstones from the Otter Point Formation, southwestern Oregon: M.S. Thesis, University of Oregon, 158 p.

Gorai, M., 1951, Petrological studies on plagioclase twins: Am. Mineralogist, v. 36, p. 884–901.

Graham, S. A., R. V. Ingersoll, and W. R. Dickinson, 1976, Common provenance for lithic grains in Carboniferous sandstones from Ouachita Mountains and Black Warrier Basin: Jour. Sed. Petrology, v. 40, p. 620–632.

Graham, S. A., R. B. Tolson, P. G. DeCelles, R. V. Ingersoll, E. Bargar, M. Caldwell, W. Cavazza, D. P. Edwards, M. F. Follo, J. F. Handschy, L. Lemke, I. Moxon, R. Rice, G. A. Smith, and J. White, 1986, Provenance modelling as a technique for analysing source terrane evolution and controls on foreland sedimentation, *in* Allen, P. A., and P. Homewood, eds., Foreland basins: Blackwell, Oxford, p. 425–436.

Grantham, J. H., and M. A. Velbel, 1988, The influence of climate and topography on rock-fragment abundance in modern fluvial sands of the southern Blue Ridge Mountains, North Carolina: Jour. Sed. Petrology, v. 58, p. 219–227.

Graton, L. C., and H. J. Fraser, 1935, Systematic packing of spheres with particular relation to porosity and permeability: Jour. Geology, v. 43, p. 785–909.

Greenwood, B., 1969, Sediment parameters and environmental discrimination: An application of multivariate statistics: Canadian Jour. Earth Sci., v. 6, p. 1347–1358.

Greenwood, B., and D. J. Sherman, 1986, Hummocky cross-stratification in the surf zone: flow parameters and bedding genesis: Sedimentology, v. 33, p. 33–45.

Gregg, J. M., 1985, Regional epigenetic dolomitization in the Bonneterre Dolomite (Cambrian), southeastern Missouri: Geology, v. 13, p. 503–506.

Gregg, J. M., and K. L. Shelton, 1990, Dolomitization and dolomite neomorphism in the backreef facies of the Bonneterre and Davis formations (Cambrian), southeastern Missouri: Jour. Sed. Petrology, v. 60, p. 549–562.

Gregg, J. M., and D. F. Sibley, 1984, Epigenetic dolomitization and the origin of xenotopic dolomite texture: Jour. Sed. Petrology, v. 54, p. 908–931.

———1986, Epigenetic dolomitization and the origin of xenotopic dolomite texture—a reply: Jour. Sed. Petrology, v. 56, p. 735–736.

Griffin, G. M., 1971, Interpretation of X-ray diffraction data, *in* Carver, R. E., ed., Procedures in sedimentary petrology: John Wiley & Sons, New York, p. 541–569.

Griffith, J. C., 1961, Measurement of the properties of sediments: Jour. Geology, v. 69, p. 487–498.

———1967, Scientific methods in analysis of sediments: McGraw-Hill, New York, 508 p.

———1971, Problems of sampling in geoscience: Inst. Mining and Metallurgy Trans., v. 80, p. B346–B356.

Griggs, G. B., and J. R. Hein, 1980, Sources, dispersal, and clay mineral composition of fine-grained sediment off central and northern California: Jour. Geology, v. 88, p. 541–566.

Grigsby, J. D., 1990, Detrital magnetite as a provenance indicator: Jour. Sed. Petrology, v. 60, p. 940–951.

Gromet, L. P., R. F. Dymek, L. A. Haskin, and R. L. Korotev, 1984, The "North American shale composite": Its compilation and major and trace element characteristics: Geochim. et Cosmochim. Acta, v. 48, p. 2469–2482.

Gross, G. A., 1965, Geology of iron deposits in Canada. I: General geology and evaluation of iron deposits: Geol. Survey Canada Econ. Geology Rept. 22, 181 p.

———1980, A classification of iron formations based on depositional environments: Canadian Mineralogist, v. 18, p. 215–222.

Grover, G., Jr., and J. F. Read, 1983, Paleoaquifer and deep burial related cements defined by regional cathodoluminescent patterns, Middle Ordovician carbonates, Virginia: Am. Assoc. Petroleum Geologists Bull., v. 67, p. 1275–1303.

Gulbrandsen, R. A., and C. E. Roberson, 1973, Inorganic phosphorite in seawater: Environmental phosphorus handbook: John Wiley & Sons, Ch. 5, p. 117–140.

Gunatilaka, A., 1989, Spheroidal dolomites—origin by hydrocarbon seepage? Sedimentology, v. 36, p. 701–710.

Gupta, A., 1983, High-magnitude floods and stream channel response, *in* Collinson, J. D., and J. Lewin, eds., Modern and ancient fluvial systems: Internat. Assoc. Sedimentologists Spec. Pub. 6, p. 219–227.

Gupta, G. D., B. Parkash, and R. J. Garde, 1987, An experimental study of fabric development in plane-bed phases: Sed. Geology, v. 53, p. 101–122.

Gustavson, T. C., 1978, Bedforms and stratification types of modern gravel meander lobes, Nueces River, Texas: Sedimentology, v. 25, p. 401–426.

Hails, J. R., 1976, Placer deposits, *in* Wolf, K. H., ed., Handbook of strata-bound and stratiform ore deposits, v. 3, Elsevier, New York, p. 213–244.

Hails, J. R., and J. H. Hoyt, 1972, The nature and occurrence of heavy minerals in Pleistocene sediments of the lower Georgia coastal plain: Jour. Sed. Petrology, v. 42, p. 646–666.

Haines, J., and J. Mazzullo, 1988, The original shapes of quartz silt grains: a test of the validity of the use of quartz grain shape analysis to determine the source of terrigenous silt in marine sedimentary deposits: Marine Geology, v. 78, p. 227–240.

Hall, J., 1859, Description and figures of the organic remains of the lower Helderberg Group and the Oriskany Sandstone: New York Geological Survey, Natural history of New York, pt. 6, Paleontology: Jour. v. 3, 532 p.

Hallam, A., 1981, Facies interpretation and the stratigraphic record: W. H. Freeman, San Francisco, 291 p.

Halley, R. B., 1977, Ooid fabric and fracture in the Great Salt Lake and the geologic record: Jour. Sed. Petrology, v. 47, p. 1099–1120.

Ham, W. E., 1952, Algal origin of the "birdseye" limestones in the McLish Formation: Oklahoma Acad. Science Proc. 33, p. 200–203.

———1962, Classification of carbonate rocks: Am. Assoc. Petroleum Geologists Mem. 1, 279 p.

Hamblin, W. K., 1971, X-ray photography, *in* Carver, R. E., ed., Procedures in sedimentary petrology: John Wiley & Sons, New York, p. 251–284.

Haney, W. D., and L. I. Briggs, 1964, Cyclicity of textures in evaporite rocks of the Lucas Formation, *in* Merriam, D. F., ed., Symposium on cyclic sedimentation: Kansas Geol. Survey, p. 191–197.

Hanford, C. R., 1981, A process-sedimentary framework for characterizing recent and ancient sabkhas: Sed. Geology, v. 30, p. 255–265.

Hanford, C. R., R. G. Loucks, and S. O. Moshier, eds., 1989, Nature and origin of micro-rhombic calcite and associated microporosity in carbonate strata: Sed. Geology, v. 63, p. 187–325.

Hanshaw, B. B., W. E. Back, and R. G. Deike, 1971, A geochemical hypothesis for dolomitization of groundwater: Econ. Geology, v. 66, p. 710–724.

Häntzschel, W., 1975, Trace fossils and problematica, 2nd ed.: Treatise on invertebrate paleontology, pt. W, Misc., Supp. 1, C. Teichert, ed.: Geol. Soc. America and Univ. Kansas, Boulder, Colo., and Lawrence, Kan., v. XXI+, 269 p.

Hardie, L. A., 1984, Evaporites: marine or non-marine: Am. Jour. Science, v. 284, p. 193–240.

———1987, Dolomitization: A critical view of some current views: Jour. Sed. Petrology, v. 57, p. 166–183.

Harland, W. B., K. Herod, and

D. H. Krinsley, 1966, The definition and identification of tills and tillites: Earth Science Rev., v. 2, p. 225–256.

Harms, J. C., and R. K. Fahnestock, 1965, Stratification, bed forms and flow phenomena (with examples from the Rio Grande), in Middleton, G. V., ed., Primary sedimentary structures and their hydrodynamic interpretation: Soc. Econ. Paleontologists Spec. Pub. 12, p. 84–155.

Harms, J. C., J. B. Southard, D. R. Spearing, and R. G. Walker, 1975, Depositional environments as interpreted from primary sedimentary structures and stratification sequences: Soc. Econ. Paleontologists and Mineralogists Short Course 2, 161 p.

Harms, J. C., J. B. Southard, and R. G. Walker, 1982, Structures and sequences in clastic rocks: Soc. Econ. Paleontologists and Mineralogists Short Course 9.

Harshbarger, J. W., C. A. Repenning, and J. H. Irwin, 1957, Stratigraphy of the uppermost Triassic and Jurassic rocks of the Navajo country: U.S. Geol. Survey Prof. Paper 291, 74 p.

Hart, B. S., and A. G. Plint, 1989, Gravelly shoreface deposits: a comparison of modern and ancient facies sequences: Sedimentology, v. 36, p. 551–557.

Harvey, A. M., 1984, Debris flows and fluvial deposits in Spanish Quaternary alluvial fans: Implications for fan morphology, in Koster, E. H., and R. J. Steel, eds., Sedimentology of gravels and conglomerates: Canadian Soc. Petroleum Geologists Mem. 10, p. 123–132.

Harvie, C. E., H. P. Euster, and J. M. Weare, 1982, Mineral equilibria in the six-component seawater system, Na-K-Mg-Ca-SO_4-Cl-H_2O at 250° C, II: Composition of the saturated solutions: Geochim. et Cosmochim. Acta, v. 46, p. 1603–1618.

Hathaway, J. C., 1979, Clay minerals, in Ribbe, P. H., ed., Marine minerals: Mineralog. Soc. America Short Course Notes, v. 6, p. 123–150.

Hattori, I., 1989, Length-slow chalcedony in sedimentary rocks of the Mesozoic allochthonous terrane in central Japan and its use for tectonic synthesis, in Hein, J. R., and J. Obradović, eds., Siliceous deposits of the Tethys and Pacific regions: Springer-Verlag, New York, p. 201–215.

Haug, E., 1900, Les géosynclinaux et les aires continentales: contribution a l'étude des regressions et des trans-gressions marines: Soc. Géol. France Bull., v. 28, p. 617–711.

Hawkins, J. W., and J. T. Whetten, 1969, Greywacke matrix minerals: Hydrothermal reactions with Columbia River sediments: Science, v. 166, p. 868–870.

Hay, R. L., 1981, Geology of zeolites in sedimentary rocks, in Mumpton, F. A., ed., Mineralogy and geology of natural zeolites: Mineralog. Soc. America, Rev. Mineralogy, v. 4, p. 53–64.

Heath, G. R., 1974, Dissolved silica and deep-sea sediments, in Hay, W. W., ed., Studies in paleooceanography: Soc. Econ. Paleontologists and Mineralogists Spec. Pub. 20, p. 77–93.

Heiken, G., 1972, Morphology and petrography of volcanic ashes: Geol. Soc. America Bull., v. 83, p. 1961–1988.

———1987, Textural analysis of tephra from a rhyodacitic eruption sequence, Thira (Santorini), Greece, in Marshall, J. R., ed., Clastic particles: Van Nostrand Reinhold, New York, p. 67–78.

Heimann, A., and E. Sass, 1989, Travertines in the northern Hula Valley, Israel: Sedimentology, v. 36, p. 95–108.

Heimlich, R. A., L. B. Shotwell, T. Cookro, and M. J. Gawell, 1975, Variability of zircons from the Sharon Conglomerate of northeastern Ohio: Jour. Sed. Petrology, v. 45, p. 629–635.

Hein, F. J., 1984, Deep-sea and fluvial braided-channel conglomerates: a comparison of two case studies, in Koster, E. H., and R. J. Steel, eds., Sedimentology of gravels and conglomerates: Canadian Soc. Petroleum Geologists Mem. 10, p. 33–49.

Hein, J. R., A. O. Allwardt, and G. B. Griggs, 1974, The occurrence of glauconite in Monterey Bay, California: diversity, origins, and sedimentary environmental significance: Jour. Sed. Petrology, v. 44, p. 562–571.

Hein, J. R., and S. M. Karl, 1983, Comparisons between open-ocean and continental margin chert sequences, in Iijima, A., J. R. Hein, and R. Siever, eds., Siliceous deposits in the Pacific region. Elsevier, Amsterdam, p. 25–43.

Hein, J. R., E. P. Kuijers, P. Denyer, and R. E. Sliney, 1983, Petrology and geochemistry of Cretaceous and Paleogene cherts from western Costa Rica, in Iijima, A., J. R. Hein, and R. Siever, eds., Siliceous deposits in the Pacific region: Elsevier, Amsterdam, p. 143–174.

Hein, J. R., and J. Obradović eds., 1989, Siliceous deposits of the Tethys and Pacific regions: Springer-Verlag, New York, 244 p.

Hein, J. R., and J. T. Parrish, 1987, Distribution of siliceous deposits in space and time, in Hein, J. R., ed., Siliceous sedimentary rock-hosted ores and petroleum: Van Nostrand Reinhold, New York, p. 10–57.

Hein, J. R., H.-W. Yeh, and J. A. Barron, 1990, Eocene diatom chert from Adak Island, Alaska: Jour. Sed. Petrology, v. 60, p. 250–257.

Hein, F. J., and R. G. Walker, 1977, Bar evolution and development of stratification in the gravelly, braided Kicking Horse River, British Columbia: Canadian Jour. Earth Sciences, v. 14, p. 562–570.

Heller, P. L., Z. E. Peterman, J. R. O'Neil, and M. Shafiqullah, 1985, Isotopic provenance of sandstones from the Eocene Tyee Formation, Oregon Coast Range: Geol. Soc. America Bull., v. 96, p. 770–780.

Helmold, K. P., 1985, Provenance of feldspathic sandstones—the effect of diagenesis on provenance interpretations: a review, in Zuffa, G. G., ed., Provenance of arenites: D. Reidel, Dordrecht, p. 139–163.

Hemming, N. G., W. J. Meyer, and J. C. Grams, 1989, Cathodoluminescence in diagenetic calcites: The roles of Fe and Mn as deduced from electron probe and spectrophotometric measurements: Jour. Sed. Petrology, v. 59, p. 404–411.

Herbig, H.-G., and K. Stattegger, 1989, Late Paleozoic heavy mineral and clast modes from the Betic Cordillera (southern Spain): transition from a passive to an active continental margin: Sed. Geology, v. 63, p. 93–108.

Hesse, R., 1987, Selective and reversible carbonate-silica replacements in Lower Cretaceous carbonate-bearing turbidites of the eastern Alps: Sedimentology, v. 34, p. 1055–1077.

Hesse, R., and S. K. Chough, 1980, The Northwest Atlantic Mid-ocean Channel of the Labrador Sea: II. Deposition of parallel laminated levee-muds from the viscous sublayer of low density turbidity currents: Sedimentology, v. 27, p. 697–711.

Heusser, L. E., 1988, Pollen distribution in marine sediments on the continental margin off northern California: Marine Geology, v. 80, p. 131–147.

Heydari, E., and C. H. Moore, 1988, Oxygen isotope evolution of the Smackover pore waters, southeast Mississippi salt basin: Geol. Soc. America Abs. with Program, 20:A261.

Higgs, R., 1978, Provenance of Mesozoic and Cenozoic sediments from the Labrador and western Greenland continental margin: Canadian Jour. Earth Science, v. 15, p. 1850–1860.

Hiscott, R. N., 1978, Provenance of Ordovician deep-water sandstones, Tourelle Formation, Quebec, and implications for initiation of the Taconic Orogeny: Canadian Jour. Earth Science, v. 15, p. 1579–1597.

Hoefs, J., 1987, Stable isotope geochemistry, 3rd ed.: Springer-Verlag, Berlin, 241 p.

Hoffman, P., J. F. Dewey, and K. Burke, 1974, Aulacogens and their genetic relation to geosynclines, with a Proterozoic example from Great Slave Lake, Canada, in Dott, R. H., Jr., and R. H. Shaver, eds., Modern and ancient geosynclinal sedimentation: Soc. Econ. Paleontologists and Mineralogists Spec. Pub. 19, p. 38–55.

Hogg, W. E., 1982, Sheetfloods, sheetwash, sheetflow, or . . .?: Earth Science Rev., v. 18, p. 59–76.

Hoholick, J. D., T. A. Metarko, and P. E. Potter, 1982, Weighted contact packing—improved formula for grain packing of quartz arenites: The Mountain Geologist, v. 19, p. 79–82.

Holmes, A., 1965, Principles of physical geology: Ronald Press, New York, 1288 p.

Hoque, M. V., 1968, Sedimentologic and paleocurrent study of Mauch Chunk sandstones (Mississippian), south-central and western Pennsylvania: Am. Assoc. Petroleum Geologists Bull., v. 52, p. 246–263.

Horowitz, A. S., and P. E. Potter, 1971, Introductory petrography of fossils: Springer-Verlag, New York, 302 p.

Horton, A., H. C. Ivimey-Cook, R. K. Harrison, and B. R. Young, 1980, Phosphatic ooids in the Upper Lias (Lower Jurassic) of central England: Jour. Geol. Soc. (London), v. 137, p. 731–740.

Houghton, H. F., 1980, Refined technique for staining plagioclase and alkali feldspars in thin section: Jour. Sed. Petrology, v. 50, p. 629–631.

Houseknecht, D. W., 1984, Influence of grain size and temperature on intergranular pressure solution, quartz cementation, and porosity in a quartzose sandstone: Jour. Sed. Petrology, v. 54, p. 348–361.

———1987, Assessing the relative importance of compaction processes and cementation to reduction of porosity in sandstones: Am. Assoc. Petroleum Geologists Bull., v. 71, p. 633–642.

Howarth, M. J., 1982, Tidal currents of the continental shelf, in Stride, A. H., ed., Offshore tidal sands: Processes and deposits: Chapman and Hall, London, p. 10–26.

Howell, D. G., and M. H. Link, 1979, Eocene conglomerate sedimentology and basin analysis, San Diego and the southern California borderland: Jour. Sed. Petrology, v. 49, p. 517–540.

Howell, D. G., and W. R. Normark, 1982, Sedimentology of submarine fans, in Scholle, P. A., and D. Spearing, eds., Sandstone depositional environments: Am. Assoc. Petroleum Geologists Mem. 31, p. 365–404.

Hower, J., E. V. Eslinger, M. E. Hower, and E. A. Perry, 1976, Mechanism of burial metamorphism of argillaceous sediments: 1. Mineralogical and chemical evidence: Geol. Soc. America Bull., v. 87, p. 725–737.

Hsü, K. J., 1989, Physical principals of sedimentology: Springer-Verlag, Berlin, 233 p.

Hsü, K. J., and C. Siegenthaler, 1969, Preliminary experiments and hydrodynamic movement induced by evaporation and their bearing on the dolomite problem: Sedimentology, v. 12, p. 11–25.

Hubert, J. F., 1960, Petrology of the Fountain and Lyons Formations, Front Range, Colorado: Colorado School of Mines Quart. Jour., v. 55, no. 1, 242 p.

———1962, A zircon-tourmaline-rutile maturity index and the interdependence of the composition of heavy mineral assemblages with the gross composition and texture of sandstones: Jour. Sed. Petrology, v. 32, p. 440–450.

Hubert, J. F., and W. J. Neal, 1967, Mineral composition and dispersal patterns of deep-sea sand in the western North Atlantic petrologic province: Geol. Soc. America Bull., v. 78, p. 749–772.

Huc, A. Y., 1980, Origin and formation of organic matter in recent sediments and its relation to kerogen, in Durand, B., ed., Kerogen: Editions Technip, Paris, France, p. 445–474.

Huckenholz, H. G., 1963, Mineral composition and texture in gray-wackes from the Harz Mountains (Germany) and in arkoses from the Auvergne (France): Jour. Sed. Petrology, v. 33, p. 914–918.

Humphrey, J. D., 1988, Late Pleistocene mixing zone dolomitization, southeastern Barbados, West Indies: Sedimentology, v. 35, p. 327–348.

Humphrey, J. D., and T. M. Quinn, 1989, Coastal mixing zone dolomite, forward modeling, and massive dolomitization of platform-margin carbonates: Jour. Sed. Petrology, v. 59, p. 438–454.

Hunt, J. M., 1979, Petroleum geochemistry and geology: W. H. Freeman, San Francisco, 617 p.

Hunter, R. E., 1967, The petrography of some Illinois Pleistocene and Recent sands: Sed. Geology, v. 1, p. 57–75.

———1977, Basic types of stratification in small eolian dunes: Sedimentology, v. 24, p. 361–387.

———1980, Depositional environments of some Pleistocene coastal terrace deposits, southwestern Oregon—case history of a progradational beach and dune sequence: Sed. Geology, v. 27, p. 241–262.

Hunter, R. E., H. E. Clifton, and R. L. Phillips, 1979, Depositional processes, sedimentary structures, and predicted vertical sequences in barred nearshore systems, southern Oregon Coast: Jour. Sed. Petrology, v. 49, p. 711–726.

Hussain, M., and J. K. Warren, 1989, Nodular and enterolithic gypsum: the "sabkha-tization" of Salt Flat Playa, west Texas: Sed. Geology, v. 64, p. 13–24.

Hutchinson, C. S., 1974, Laboratory handbook of petrographic techniques: John Wiley & Sons, New York, 527 p.

Iijima, A., J. R. Hein, and R. Siever, 1983, Siliceous deposits in the Pacific region, Developments in sedimentology 36: Elsevier, Amsterdam, 472 p.

Iijima, A., H. Inagaki, and Y. Kakuwa, 1979, Nature and origin of the Paleogene cherts in the Setogawa Terrain, Shizuoka, central Japan: Jour. Fac. Sci., Univ. Tokyo, v. 20, p. 1–30.

Iller, R. K., 1979, Chemistry of silica: Wiley-Interscience, New York, 866 p.

Illing, L. V., 1954, Bahaman calcareous sands: Am. Assoc. Petroleum Geologists Bull., v. 38, p. 1–95.

Ingersoll, R. V., 1983, Petrofacies and provenance of Late Mesozoic forearc basin, northern and central California: Am. Assoc. Petroleum Geologists Bull., v. 67, p. 1125–114.

——— 1988, Tectonics of sedimentary basins: Geol. Soc. America Bull., v. 100, p. 1704–1719.

——— 1990, Actualistic sandstone petrofacies: discriminating modern and ancient source rocks: Geology, v. 18, p. 733–736.

Ingersoll, R. V., T. D. Bullard, R. L. Ford, J. P. Grimm, J. D. Pickle, and S. W. Sares, 1984, The effect of grain size on detrital modes: A test of the Gazzi-Dickinson point counting method: Jour. Sed. Petrology, v. 54, p. 103–116.

Ingram, R. L., 1953, Fissility of mudrocks: Geol. Soc. America Bull., v. 64, p. 869–878.

——— 1954, Terminology for thickness of stratification and parting units in sedimentary rocks: Geol. Soc. America Bull., v. 65, p. 937–938.

——— 1971, Sieve analysis, *in* Carver, R. E., ed., Procedures in sedimentary petrology: John Wiley & Sons, New York, p. 49–67.

International Committee for Coal Petrology, 1971, International handbook of coal petrography, 1st supp. to 2nd ed.: Centre National de la Recherche Scientifique, Paris, France.

——— 1975, Analysis subcommission, fluorescence microscopy and fluorescence photometry, *in* International handbook of coal petrography, 2nd supp. to 2nd ed.: Centre National de la Recherche Scientifique, Paris, France.

Jackson, R. G., II, 1978, Preliminary evaluation of lithofacies models for meandering alluvial streams, *in* Miall, A. D., ed., Fluvial sedimentology, Canadian Soc. Petroleum Geologists Mem. 5, p. 543–576.

James, H. E., Jr., and F. B. Van Houten, 1979, Miocene goethitic and chamositic oolites, northeastern Colombia: Sedimentology, v. 26, p. 125–133.

James, H. L., 1966, Chemistry of iron-rich sedimentary rocks: Data of geochemistry, 6th ed.: U.S. Geol. Survey Prof. Paper 440-W, 61 p.

James, H. L., and A. F. Trendall, 1982, Banded iron formation: Distribution in time and paleoenvironmental significance, *in* Holland, H. D., and M. Schidlowski, eds., Mineral deposits and the evolution of the biosphere: Springer-Verlag, Berlin, p. 199–218.

James, N. P., 1983, Reef environment, *in* Scholle, P. A., D. G. Bebout, and C. H. Moore, eds., Carbonate depositional environments: Am. Assoc. Petroleum Geologists Mem. 33, p. 345–440.

James, N. P., and Y. Bone, 1989, Petrogenesis of Cenozoic, temperate water calcarenites, south Australia: A model for meteoric shallow burial diagenesis of shallow water calcite sediments: Jour. Sed. Petrology, v. 59, p. 191–203.

James, N. P., and P. W. Choquette, 1983a, Diagenesis 5. Limestones: Introduction: Geoscience Canada, v. 10, p. 159–161.

——— 1983b, Diagenesis 6. Limestones—The sea floor diagenetic environment: Geoscience Canada, v. 10, p. 162–179.

——— 1984, Diagenesis 9. Limestones—The meteoric diagenetic environment: Geoscience Canada, v. 11, p. 161–194.

——— 1987, Diagenesis 12. Diagenesis in limestones—3. The deep burial environment: Geoscience Canada, v. 14.

James, N. P., and P. W. Choquette, eds., 1988, Paleokarst: Springer-Verlag, New York, 416 p.

James, N. P., and R. N. Ginsburg, 1979, The seaward margin of Belize barrier and atoll reefs: Internat. Assoc. Sedimentologists Spec. Pub. 3, 197 p.

James, N. P., R. N. Ginsburg, D. S. Marszalek, and P. W. Choquette, 1976, Facies and fabric specificity of early subsea cements in shallow Belize (British Honduras) reefs: Jour. Sed. Petrology, v. 46, p. 523–544.

James, W. C., G. H. Mack, and L. J. Suttner, 1981, Relative alteration of microcline and sodic plagioclase in semiarid and humid climates: Jour. Sed. Petrology, v. 51, p. 151–164.

James, W. C., and R. Q. Oaks, Jr., 1977, Petrology of the Kinnikinic Quartzite (Middle Ordovician), east-central Idaho: Jour. Sed. Petrology, v. 47, p. 1491–1511.

Ji, S., and D. Mainprice, 1990, Recrystallization and fabric development in plagioclase: Jour. Geology, v. 98, p. 65–79.

Johansson, C. E., 1965, Orientation of pebbles in running water: Geol. Foren. Stockholm Förh., v. 87, p. 3–61.

Johnson, J. H., 1971, An introduction to the study of organic limestones: Colorado School of Mines Quart., v. 66, 185 p.

Johnsson, M. J., and R. C. Reynolds, 1986, Clay mineralogy of shale-limestone rhythmites in Scaglia Rossa (Turonian-Eocene), Italian Apennines: Jour. Sed. Petrology, v. 56, p. 501–509.

Johnsson, M. J., and R. F. Stallard, 1990, Physiographic controls on the composition of sediments derived from volcanic and sedimentary terrains on Barro Colorado Island, Panama—Reply: Jour. Sed. Petrology, v. 60, p. 799–801.

Johnsson, M. J., R. F. Stallard, and R. H. Meade, 1988, First-cycle quartz arenites in the Orinoco River basin, Venezuela and Colombia: Jour. Geology, v. 96, p. 263–277.

Jones, B., 1989, Syntaxial overgrowths on dolomite crystals in the Bluff Formation, Grand Cayman, British West Indies: Jour. Sed. Petrology, v. 59, p. 839–847.

Jones, B., and R. W. MacDonald, 1989, Micro-organisms and crystal fabrics in cave pisoliths from Grand Cayman, British West Indies: Jour. Sed. Petrology, v. 59, p. 387–396.

Jones, D. L., and B. Murchey, 1986, Geologic significance of Paleozoic and Mesozoic radiolarian chert: Ann. Rev. Earth and Planetary Science Letters, v. 14, p. 455–492.

Jones, J. B., and E. R. Segnit, 1971, The nature of opal I. Nomenclature and constituent phases: Jour. Geol. Soc. Australia, v. 18, p. 57–68.

Jones, K. P. N., I. N. McCave, and P. D. Patel, 1988, A computer-interfaced sedigraph for modal size analysis of fine-grained sediment: Sedimentology, v. 35, p. 163–172.

Jones, N. W., and F. D. Bloss, 1980, Laboratory manual for optical mineralogy: Alpha Editions (Burgess International Group, Inc.), Edina, Minn., variously paginated.

Jones, P. C., 1972, Quartzarenite and litharenite facies in the fluvial foreland deposits of the Trenchard Group (Westphalian), Forest of Dean, England: Sed. Geology, v. 8, p. 177–198.

Jones, R. L., and H. Blatt, 1984, Mineral dispersal patterns in the Pierre Shale: Jour. Sed. Petrology, v. 54, p. 17–28.

Jopling, A. V., and B. C. McDonald, eds., 1975, Glaciofluvial and glaciolacustrine sedimentation: Soc. Econ. Paleontologists and Mineralogists Spec. Pub. 23, 320 p.

Jopling, A. V., and R. G. Walker, 1968, Morphology and origin of ripple-drift cross-lamination and examples from the Pleistocene of Massachusetts: Jour. Sed. Petrology, v. 38, 971–984.

Jordan, C. F., G. E. Freyer, and E. H. Hemmen, 1971, Size analysis of silt and clay by hydrophotometer: Jour. Sed. Petrology, v. 41, p. 489–496.

Julia, R., 1983, Travertines, *in* Scholle, P. A., D. G. Bebout, and C. H. Moore, eds., Carbonate depositional environments: Am. Assoc. Petroleum Geologists Mem. 33, p. 64–72.

Kahn, J. S., 1956, The analysis and distribution of the properties of packing in sand-size sediments 1. On the measurement of packing in sandstones: Jour. Geology, v. 64, p. 385–395.

Kaplan, I. R., 1983, Stable isotopes of sulfur, nitrogen and deuterium in recent marine environments, *in* Arthur, M. A., organizer, Stable isotopes in geology: Soc. Econ. Paleontologists and Mineralogists Short Course 10, p. 2-1–2-108.

Kaplan, I. R., K. O. Emery, and S. C. Rittenberg, 1963, The distribution and isotopic abundance of sulphur in recent marine sediments off southern California: Geochim. et Cosmochim. Acta, v. 27, p. 297–331.

Karrow, P. F., 1976, The texture, mineralogy, and petrography of North American tills, *in* Legget, R. F., ed., Glacial till: Royal Soc. Canada Spec. Pub. 12, p. 83–98.

Kastner, M., 1971, Authigenic feldspars in carbonate rocks: Am. Mineralogist, v. 56, p. 1403–1442.

Kastner, M., and J. M. Gieskes, 1983, Opal-A to opal-CT transformation: A kinetic study, *in* Iijima, A., J. R. Hein, and R. Siever, eds., Siliceous deposits in the Pacific region: Developments in Sedimentology 36, Elsevier, Amsterdam, p. 211–228.

Kastner, M., J. B. Keene, and J. M. Gieskes, 1977, Diagenesis of siliceous oozes—I. Chemical controls on the rate of opal-A to opal-CT transformation—an experimental study: Geochim. et Cosmochim. Acta, v. 41, p. 1041–1059.

Kastner, M., and R. Siever, 1979, Low temperature feldspars in sedimentary rocks: Am. Jour. Science, v. 279, p. 435–479.

Kaufman, J., H. S. Cander, L. D. Daniels, and W. J. Meyer, 1988, Calcite cement stratigraphy and cementation history of the Burlington-Keokuk Formation (Mississippian), Illinois and Missouri: Jour. Sed. Petrology, v. 58, p. 312–326.

Keene, J. B., 1983, Chalcedonic quartz and occurrence of quartzine (length-slow chalcedony) in pelagic sediments: Sedimentology, v. 30, p. 449–454.

Kelling, G., and P. F. Williams, 1967, Flume studies on the orientation of pebbles and shells: Jour. Geology, v. 75, p. 243–267

Kendall, A. C., 1975, Postcompactional calcitization of molluscan aragonite in a Jurassic limestone from Saskatchewan, Canada: Jour. Sed. Petrology, v. 45, p. 399–404.

———1984, Evaporites, *in* Walker, R. G., ed., Facies models, 2nd ed., Geoscience Canada Reprint Ser. 1, p. 259–296.

———1985, Radiaxial fibrous calcite: a reappraisal, *in* Schneidermann, N., and P. M. Harris, eds., Carbonate cements: Soc. Econ. Paleontologists and Mineralogists Spec. Pub. 36, p. 59–77.

Kendall, C. G. St. C., and J. Warren, 1987, A review of the origin and setting of tepees and their associated fabrics: Sedimentology, v. 34, p. 1007–1027.

Kennard, J. M., and N. P. James, 1986, Thrombolites and stromatolites: two distinct types of microbial structures: Palaios, v. 1, p. 492–503.

Kennedy, S. K., and R. Ehrlich, 1985, Origin of shape changes of sand and silt in a high-gradient stream system: Jour. Sed. Petrology, v. 55, p. 57–64.

Kennedy, W. J., and R. E. Garrison, 1975, Morphology and genesis of nodular chalks and hardgrounds in the Upper Cretaceous of southern England: Sedimentology, v. 22, p. 311–386.

Kennedy, W. Q., 1951, Sedimentary differentiation as a factor in the Moine-Torridonian correlation: Geol. Mag., v. 88, p. 257–266.

Kessler, L. G., II, and K. Moorhouse, 1984, Depositional processes and fluid mechanics of Upper Jurassic conglomerate accumulations, British North Sea, *in* Koster, E. H., and R. J. Steel, eds., Sedimentology of gravels and conglomerates: Canadian Soc. Petroleum Geologists Mem. 10, p. 383–397.

Ketner, K. B., 1966, Comparison of Ordovician eugeosynclinal and miogeosynclinal quartzites of the Cordilleran geosyncline: U.S. Geol. Survey Prof. Paper 550-C, p. C54–C60.

Khalaf, F. I., 1990, Occurrence of phreatic dolocrete within Tertiary clastic deposits of Kuwait, Arabian Gulf: Sed. Geology, v. 68, p. 223–239.

Kirk, R. M., 1980, Mixed sand

and gravel beaches: morphology, processes, and sediments: Progress in Physical Geography, v. 4, p. 189–210.

Klein, C., and C. S. Hurlbut, Jr., 1985, Manual of mineralogy, 20th ed.: John Wiley & Sons, New York, 596 p.

Klein, G. deV., 1963, Analysis and review of sandstone classifications in the North American geological literature: Geol. Soc. America Bull., v. 74, p. 555–576.

———1977, Clastic tidal facies: Continuing Education Publication Co., Champaign, Ill., 149 p.

———1987, Current aspects of basin analysis: Sed. Geology, v. 50, p. 95–118.

Klein, G. deV., and K. M. Marsaglia, 1987, Hummocky cross-stratification, tropical hurricanes, and intense winter storms: Discussion: Sedimentology, v. 34, p. 333–337.

Klein, G. deV., and D. A. Willard, 1989, Origin of Pennsylvanian coal-bearing cyclothems of North America: Geology, v. 17, p. 152–155.

Kleinspehn, K. L., and C. Paola, eds., 1988, New perspectives in basin analysis: Springer-Verlag, New York, 453 p.

Kleinspehn, K. L., R. J. Steel, E. Johannessen, and A. Netland, 1984, Conglomeratic fan-delta sequences, Late Carboniferous-Early Permian, Western Spitsbergen, *in* Koster, E. H., and R. J. Steel, eds., Sedimentology of gravels and conglomerates: Canadian Soc. Petroleum Geologists Mem. 10, p. 279–294.

Klovan, J. E., 1964, Facies analysis of the Redwater reef complex, Alberta, Canada: Bull. Canadian Petroleum Geology, v. 12, p. 1–100.

———1966, The use of factor analysis in determining environments from grain size distributions: Jour. Sed. Petrology, v. 36, p. 57–69.

Knauth, L. P., 1979, A model for the origin of chert in limestone: Geology, v. 7, 274–277.

Koch, G. S., Jr., and R. F. Link, 1970, Statistical analysis of geological data: John Wiley & Sons, New York, 368 p.

Koepnick, R. B., 1984, Distribution and vertical permeability of stylolites within a Lower Cretaceous carbonate reservoir, Abu Dhabi, U.A.E., *in* Stylolites and associated phenomena—relevance to hydrocarbon reservoirs: Abu Dhabi Reservoir Research Foundation Spec. Pub., Abu Dhabi, U.A.E., p. 261–278.

Kohout, F. A., 1967, Ground

water flow and the geothermal regime of the Floridan Plateau: Trans. Gulf Coast Assoc. Geol. Soc., v. 17, p. 339–354.

Kohout, F. A., H. R. Henry, and J. E. Banks, 1977, Hydrogeology related to geothermal conditions of the Floridan Plateau, *in* Smith, K. L., and G. M. Griffin, eds., The geothermal nature of the Floridan Plateau: Florida Dept. Nat. Resources Bur. Geology Spec. Pub. 21, p. 1–34.

Kolodny, Y., 1980, The origin of phosphorite deposits in light of occurrences of Recent sea-floor phosphorites, *in* Bentor, Y. K., ed., Marine phosphorites: Soc. Econ. Paleontologists and Mineralogists Spec. Pub. 29, p. 249.

——— 1981, Phosphorites, *in* Emiliani, C., ed., The ocean lithosphere: The sea, v. 7: John Wiley & Sons, New York, p. 981–1023.

Kolodny, Y., and I. R. Kaplan, 1970, Uranium isotopes in sea floor phosphorites: Geochim. et Cosmochim. Acta, v. 34, p. 3–24.

Komar, P. D., 1976, Beach processes and sedimentation: Prentice-Hall, Inc., Englewood Cliffs, N. J., 429 p.

Komar, P. D., and C. E. Reimers, 1978, Grain shape effects on settling rates: Jour. Geology, v. 86, p. 193–209.

Koster, E. H., and R. J. Steel, eds., 1984, Sedimentology of gravels and conglomerates: Canadian Soc. Petroleum Geologists Mem. 10, 441 p.

Krainer, K., and C. Spötl, 1989, Detrital and authigenic feldspars in Permian and early Triassic sandstones, eastern Alps (Austria): Sed. Geology: v. 62, p. 59–77.

Kranck, K., 1975, Sediment deposited from flocculated suspensions: Sedimentology, v. 22, p. 111–123.

——— 1980, Experiments on the significance of flocculation in the settling of fine-grained sediment in still water: Canadian Jour. Earth Science, v. 17, p. 1517–1526.

——— 1984, Grain-size characteristics of turbidites, *in* Stow, D. A. V., and D. J. W. Piper, eds., Fine-grained sediments: Deep-water processes and facies: Geol. Soc. Spec. Pub. 15, Blackwell, Oxford, p. 83–92.

Kraus, M. J., 1984, Sedimentology and tectonic setting of early Tertiary quartzite conglomerates, northwest Wyoming, *in* Koster, E. H., and R. J. Steel, eds., Sedimentology of gravels and conglomerates: Canadian Soc. Petroleum Geologists Mem. 10, p. 203–216.

Krauskopf, K. B., 1959, The geochemistry of silica in sedimentary environments, *in* Ireland, H. A., ed., Silica in sediments: Soc. Econ. Paleontologists and Mineralogists Spec. Pub. 7, p. 4–19.

——— 1979, Introduction to geochemistry, 2nd ed.: McGraw-Hill, New York, 617 p.

Kreisa, R., and Nøttvedt, A., 1986, Hummocky cross-stratification is trough cross-stratification (abst.): Soc. Econ. Paleontologists and Mineralogists Ann. Midyear Mtg. Prog. 3, p. 63.

Krinsley, D. H., and C. R. Manley, 1989, Backscattered electron microscopy as an advanced technique in petrography: Jour. Geol. Education, v. 37, p. 202–209.

Krinsley, D. H., and P. Trusty, 1986, Sand grain surface textures, *in* Sieveking, G. De C., and M. B. Hart, eds., The scientific study of flint and chert: Cambridge University Press, Cambridge, p. 201–207.

Kroopnick, P., 1980, The distribution of ^{13}C in the Atlantic Ocean: Earth and Planetary Science Letters, v. 49, p. 469–484.

Krumbein, W. C., 1941, Measurement and geological significance of shape and roundness of sedimentary particles: Jour. Sed. Petrology, v. 11, p. 64–72.

——— 1942, Flood gravels of Arroyo Seco, Los Angeles County, California: Geol. Soc. America Bull., v. 53, p. 1355–1402.

Krumbein, W. C., and F. A. Graybill, 1965, An introduction to statistical models in geology: McGraw-Hill, New York, 475 p.

Krumbein, W. C., and F. J. Pettijohn, 1938, Manual of sedimentary petrology: Appleton-Century, New York, 549 p.

Krumbein, W. C., and L. L. Sloss, 1963, Stratigraphy and sedimentation, 2nd ed.: W. H. Freeman, San Francisco, 660 p.

Krynine, P. D., 1940, Petrology and genesis of the Third Bradford sand: Pennsylvania Mineral Industries Expt. Sta. Bull. 27, 134 p.

——— 1943, Diastrophism and the evolution of sedimentary rocks: Pennsylvania Mineral Industries Tech. Paper 84-A, 21 p.

——— 1946, The tourmaline group in sediments: Jour. Geology, v. 54, p. 65–87.

Kuehl, S. A., C. A. Nittrouer, and D. J. DeMaster, 1988, Microfabric study of fine-grained sediments: Observations from the Amazon subaqueous delta: Jour. Sed. Petrology, v. 58, p. 12–23.

Kuenen, Ph. H., 1958, Experiments in geology: Glasgow Geol. Soc. Trans., v. 23, p. 1–28.

——— 1959, Experimental abrasion, part 3: Fluviatile action on sand: Am. Jour. Sci., v. 257, p. 172–190.

——— 1960, Experimental abrasion, part 4: Eolian action: Jour. Geology, v. 68, p. 427–449.

Kugler, R. L., 1979, Stratigraphy and petrology of the Bushnell Rock Member of the Lookingglass Formation, southwestern Oregon Coast Range: M.S. Thesis, University of Oregon, Eugene, 118 p.

Kupecz, J. A., C. Kerans, and L. S. Land, 1988, Deep-burial dolomitization in the Ordovician Ellenburger Group carbonates, west Texas and southeastern New Mexico—Discussion: Jour. Sed. Petrology, v. 58, p. 908–910.

Kvale, E. P., A. W. Archer, and H. R. Johnson, 1989, Daily, monthly, and yearly tidal cycles within laminated siltstones of the Mansfield Formation (Pennsylvanian) of Indiana: Geology, v. 17, p. 365–368.

LaBerge, G. L., 1973, Possible biological origin of Precambrian iron-formations: Econ. Geology, v. 68, p. 1098–1109.

LaBerge, G. L., E. I. Robbins, and T.-M. Han, 1987, A model for the biological precipitation of Precambrian iron-formations—A: Geological evidence, *in* Appel, P. W. U., and G. L. LaBerge, eds., Precambrian iron-formations: Theophrastus Pub. S. A., Athens, Greece, p. 69–96.

Lajoie, J., 1984, Volcaniclastic rocks, *in* Walker, R. G., ed., Facies models, 2nd ed.: Geoscience Canada Reprint Ser. 1, p. 39–52.

Land, L. S., 1973, Holocene meteoric dolomitization of Pleistocene limestones, North Jamaica: Sedimentology, v. 20, p. 411–424.

——— 1984, Frio sandstone diagenesis, Texas Gulf Coast: a regional isotopic study, *in* McDonald, D. A., and R. C. Surdam, eds., Clastic diagenesis: Am. Assoc. Petroleum Geologists Mem. 37, p. 47–62.

——— 1985, The origin of massive dolomite: Jour. Geol. Education, v. 33, p. 112–125.

——— 1986, Limestone diagenesis—some geochemical considerations, *in* Mumpton, F. A., ed., Studies in diagenesis: U.S. Geol. Survey Bull. 1578, p. 129–137.

Land, L. S., and S. P. Dutton, 1978, Cementation of a Pennsylvanian deltaic sandstone: Isotopic data: Jour. Sed. Petrology, v. 48, p. 1167–1176.

Land, L. S., and C. H. Moore, 1980, Lithification, micritization, and syndepositional diagenesis of biolithites on the Jamaican island slope: Jour. Sed. Petrology, v. 43, p. 614–617.

Lane, E. W., and E. J. Carlson, 1954, Some observations of the effect of particle shape on movement of coarse sediments: Am. Geophys. Union Trans., v. 35, p. 453–462.

Larsen, V., and R. J. Steel, 1978, The sedimentary history of a debris flow dominated alluvial fan: A study of textural inversion: Sedimentology, v. 25, p. 37–59.

Larue, D. K., and M. M. Sampayo, 1990, Lithic-volcanic sandstones derived from oceanic crust in the Franciscal Complex of California: 'Sedimental memories' of source rock geochemistry: Sedimentology, v. 37, p. 879–889.

Lasemi, Z., M. R. Boardman, and P. A. Sandberg, 1989, Cement origin of supratidal dolomite, Andros Island, Bahamas: Jour. Sed. Petrology, v. 59, p. 249–257.

Lasius, G., 1789, Beobachtungen im Harzgebirge: Helwing, Hannover, p. 132–152.

Lawson, D. E., 1979, A comparison of pebble orientations in ice and deposits of Matanuska Glacier, Alaska: Jour. Geology, v. 87, p. 629–645.

———1982, Mobilization, movement, and deposition of active subaerial sediment flows, Matanuska Glacier, Alaska: Jour. Geology, v. 90, p. 279–300.

Leckie, D., 1988, Wave-formed, coarse-grained ripples and their relationship to hummocky cross-stratification: Jour. Sed. Petrology, v. 58, p. 607–622.

Leckie, D. A., and R. G. Walker, 1982, Storm- and tide-dominated shorelines in Cretaceous Moosebar-Lower Gates interval-outcrop equivalents of Deep Basin gas traps in western Canada: Am. Assoc. Petroleum Geol. Bull., v. 66, p. 138–157.

Leder, F., and W. C. Park, 1986, Porosity reduction in sandstones by quartz overgrowths: Am. Assoc. Petroleum Geologists Bull., v. 70, p. 1713–1728.

Lee, I. Y., and G. M. Friedman, 1987, Deep-burial dolomitization in the Ordovician Ellenburger Group carbonates, western Texas and southeastern New Mexico: Jour. Sed. Petrology, v. 57, p. 544–557.

———1988, Deep-burial dolomitization in the Ordovician Ellenburger Group carbonates, west Texas and southeastern New Mexico—Reply: Jour. Sed. Petrology, v. 58, p. 910–913.

Lee, M., J. L. Aronson, and S. M. Savin, 1989, Timing and conditions of Permian Rotliegende sandstone diagenesis, southern North Sea: K/Ar and oxygen isotopic data: Am. Assoc. Petroleum Geologists Bull., v. 73, p. 195–215.

Lee, M., and S. M. Savin, 1985, Isolation of diagenetic overgrowths on quartz sand grains for oxygen isotopic analysis: Geochim. et Cosmochim. Acta, v. 49, p. 497–501.

Leinfelder, R. R., and C. Hartkopf-Fröder, 1990, In situ accretion mechanism of concavo-convex lacustrine oncoids ('swallow nests') from the Oligocene of the Mainz Basin, Rhineland, FRG: Sedimentology, v. 37, p. 287–301.

Leith, C. K., and W. J. Mead, 1915, Metamorphic geology: Henry Holt, New York, 337 p.

Leithold, E. L., and J. Bourgeois, 1984, Characteristics of coarse-grained sequences deposited in nearshore, wave-dominated environments—examples from the Miocene of southwest Oregon: Sedimentology, v. 31, p. 749–775.

Lent, R. L., 1969, Geology of the southern half of the Langlois Quadrangle, Oregon: Ph.D. Dissertation, University of Oregon, 189 p.

Lepp, H., 1987, Chemistry and origin of Precambrian iron formations, in Appel, P. W. U., and G. L. LaBerge, eds., Precambrian iron-formations: Theophrastus Pub., S. A., Athens, Greece, p. 3–30.

Lepp, H., and S. S. Goldich, 1964, Origin of Precambrian iron formation: Econ. Geology, v. 58, p. 1025–1061.

Lerbekmo, J. F., 1961, Genetic relationship among Tertiary blue sandstones in central California: Jour. Sed. Petrology, v. 31, p. 594–602.

Leroy, S. D., 1981, Grain-size and moment measures: A new look at Karl Pearson's ideas on distributions: Jour. Sed. Petrology, v. 51, p. 625–630.

Lewan, M. D., 1978, Laboratory classification of very fine grained sedimentary rocks: Geology, v. 6, p. 745–748.

Lewis, D. W., 1984, Practical sedimentology: Hutchinson and Ross, Stroudsburg, Pa., 229 p.

Lewis, D. W., M. G. Laird, and R. D. Powell, 1980, Debris flow deposits of Early Miocene age, Deadman Stream, Marlborough, New Zealand: Sed. Geology, v. 27, p. 83–118.

Lindholm, R. C., 1987, A practical approach to sedimentology: George Allen & Unwin, London, 276 p.

Lindholm, R. C., J. M. Hazlett, and S. W. Fagin, 1979, Petrology of Triassic-Jurassic conglomerates in the Culpepper Basin, Virginia: Jour. Sed. Petrology, v. 49, p. 1245–1262.

Lippman, F., 1973, Sedimentary carbonate minerals: Springer-Verlag, New York, 228 p.

Lloyd, R. M., 1977, Porosity reduction by chemical compaction-stable isotope model: Am. Assoc. Petroleum Geologists Bull., v. 61, p. 809.

Lobo, C. F., and R. H. Osborne, 1976, Petrology of Late Precambrian-Cambrian quartzose sandstones in the Eastern Mojave Desert, southeastern California: Jour. Sed. Petrology, v. 46, p. 829–846.

Logan, B. R., and V. Semeniuk, 1976, Dynamic metamorphism: processes and products in Devonian carbonate rocks, Canning Basin, Western Australia: Geol. Soc. Australia Spec. Pub. 16, 138 p.

Logan, B. W., R. Rezak, and R. N. Ginsburg, 1964, Classification and environmental significance of algal stromatolites: Jour. Geology, v. 72, p. 68–83.

Logvinenko, N. V., 1982, Origin of glauconite in the recent bottom sediments of the ocean: Sed. Geology, v. 31, p. 43–48.

Lohmann, K. C., 1988, Geochemical patterns of meteoric diagenetic systems and their application to studies of paleokarst, in James, N. P., and P. W. Choquette, eds., Paleokarst, Springer-Verlag, New York, p. 58–80.

Longman, M. W., 1980, Carbonate diagenetic textures from nearsurface diagenetic carbonates: Am. Assoc. Petroleum Geologists Bull., v. 64, p. 461–487.

———1981, A process approach to recognizing facies of reef complexes, in Toomey, D. F., ed., European fossil reef models: Soc. Econ. Paleontologists and Mineralogists Spec. Pub. 30, p. 9–40.

Longstaffe, F. J., 1989, Stable isotopes as tracers in clastic diagenesis, *in* Hutcheon, I. E., ed., Burial diagenesis: Mineralog. Assoc. Canada Short Course Handb., v. 15, p. 201–277.

Lonnie, T. P., 1982, Mineralogic and chemical comparison of marine, nonmarine, and transitional clay beds on south shore of Long Island, New York: Jour. Sed. Petrology, v. 52, p. 529–536.

Lowe, D. R., 1975, Water escape structures in coarse-grained sediment: Sedimentology, v. 22, p. 157–204.

———1976, Grain flow and grain flow deposits: Jour. Sed. Petrology, v. 46, p. 188–199.

———1979, Sediment gravity flows; their classification and some problems of application to natural flows and deposits, *in* Doyle, L. J., and O. H. Pilkey, eds., Geology of continental slopes: Soc. Economic Paleontologists and Mineralogists Spec. Pub. 27, p. 75–82.

———1982, Sediment gravity flows: II. Depositional models with special reference to the deposits of high-density turbidity currents: Jour. Sed. Petrology, v. 52, p. 279–297.

Lowe, D. R., and R. D. LoPiccolo, 1974, The characteristics and origin of dish and pillar structures: Jour. Sed. Petrology, v. 44, p. 484–501.

Luepke, G., 1980, Opaque minerals as aids in distinguishing between source and sorting effects on beach-sand mineralogy in southwestern Oregon: Jour. Sed. Petrology, v. 50, p. 489–496.

———1984, Stability of heavy minerals in sediments: Benchmark Papers in Geology 81, Van Nostrand Reinhold, New York, 305 p.

Lumsden, D. N., 1988, Characteristics of deep-marine dolomite: Jour. Sed. Petrology, v. 58, p. 1023–1031.

Lundegard, P. D., 1989, Temporal reconstruction of sandstone diagenetic histories, *in* Hutcheon, I. E., ed., Burial diagenesis: Mineralog. Assoc. Canada Short Course Handb., v. 15, p. 161–200.

Lundegard, P. D., and N. D. Samuels, 1980, Field classification of fine-grained sedimentary rocks: Jour. Sed. Petrology, v. 50, p. 781–786.

Mabie, C. P., and H. D. Hess, 1964, Petrographic study and classification of Western Phosphate ores: U.S. Bur. Mines Rept. Inv. 6468, 95 p.

MacGowan, D., and R. C. Surdam, 1988, Disfunctional carboxylic acid anions in oil-field waters: Organic Geochemistry, v. 12, p. 245–259.

Machel, H.-G., 1985, Cathodoluminescence in calcite and dolomite and its chemical interpretation: Geoscience Canada, v. 12, p. 139–147.

———1987, Saddle dolomite as a by-product of chemical compaction and thermochemical sulfate reduction: Geology, v. 15, p. 936–940.

Machel, H.-G., and J. H. Anderson, 1989, Pervasive subsurface dolomitization of the Nisku Formation in central Alberta: Jour. Sed. Petrology, v. 59, p. 891–911.

Machel, H.-G., and E. W. Mountjoy, 1986, Chemistry and environments of dolomitization—a reappraisal: Earth Science Rev., v. 23, p. 175–222.

Macintyre, I. G., 1985, Submarine cements—the peloidal question, *in* Schneidermann, N., and P. M. Harris, eds., Carbonate cements: Soc. Econ. Paleontologists and Mineralogists Spec. Pub. 36, p. 109–116.

Mack, G. H., 1981, Composition of modern stream sand in a humid climate derived from a low-grade metamorphic and sedimentary foreland fold-thrust belt of North Georgia: Jour. Sed. Petrology, v. 51, p. 1247–1258.

———1984, Exceptions to the relationship between plate tectonics and sandstone compositions: Jour. Sed. Petrology, v. 54, p. 212–220.

Mack, G. H., and T. Jerzykiewicz, 1989, Detrital modes of sand and sandstone derived from andesitic rocks as a paleoclimate indicator: Sed. Geology, v. 65, p. 35–44.

Mack, G. H., and L. J. Suttner, 1977, Paleoclimate interpretation from a petrographic comparison of Holocene sands and the Fountain Formation (Pennsylvanian) in the Colorado Front Range: Jour. Sed. Petrology, v. 47, p. 89–100.

Mackenzie, R. C., 1982, Thermoanalytical methods in clay studies, *in* Fripiat, J. J., ed., Advanced techniques for clay mineral analysis: Elsevier, Amsterdam, p. 5–29.

MacKenzie, W. S., and C. Guilford, 1980, Atlas of rock-forming minerals in thin section: John Wiley & Sons, New York, 98 p.

Mackie, W., 1896, The sands and sandstones of eastern Moray: Edinburg Geol. Soc. Trans., v. 7, p. 148–172.

Maejima, W., 1982, Texture and stratification of gravelly beach sediments, Enju Beach, Kii Peninsula, Japan: Osaka City University, Jour. Geosciences, v. 25, p. 35–51.

Maiklem, W. R., D. G. Bebolt, and R. P. Glaister, 1969, Classification of anhydrite—A practical approach: Canadian Petroleum Geology Bull., v. 17, p. 194–233.

Majewske, O. P., 1969, Recognition of invertebrate fossil fragments in rocks and thin sections: E. J. Brill, Leiden, 101 p.

Major, J. J., and B. Voight, 1986, Sedimentology and clast orientations of the 18 May 1980 southwest-flank lahars, Mount St. Helens, Washington: Jour. Sed. Petrology, v. 56, 691–705.

Maliva, R. G., 1989, Displacive syntaxial overgrowths in open marine limestones: Jour. Sed. Petrology, v. 59, p. 397–403.

Maliva, R. G., and R. Siever, 1988a, Diagenetic replacement controlled by force of crystallization: Geology, v. 16, p. 688–691.

———1988b, Mechanisms and controls of silicification of fossils in limestone: Jour. Geology, v. 96, p. 387–398.

———1988c, Pre-Cenozoic nodular cherts: evidence for opal-CT precursors and direct quartz replacement: Am. Jour. Science, v. 288, p. 798–809.

———1989, Chertification histories of some Late Mesozoic and Middle Paleozoic platform carbonates: Sedimentology, v. 36, p. 907–926.

Marshall, J. F., 1983, Submarine cementation in a high-energy platform reef: One Tree Reef, southern Great Barrier Reef: Jour. Sed. Petrology, v. 53, p. 1133–1149.

Marshall, J. R., 1987, Clastic particles: Van Nostrand Reinhold, New York, 346 p.

Martens, C. S., and R. C. Harris, 1970, Inhibition of apatite precipitation in marine environment by magnesium ions: Geochim. et Cosmochim. Acta, v. 34, p. 621–625.

Martin, R. F., 1974, Controls of ordering and subsolidus phase relations in the alkali feldspars, *in* MacKenzie, W. S., and J. Zussman, eds., The feldspars: Manchester University Press, Manchester, p. 313–336.

Matalucci, R. V., J. W. Shelton, and M. Abdel-Hady, 1969, Grain orientation in Vicksburg Loess: Jour. Sed. Petrology, v. 39, p. 969–979.

Matlack, K. S., D. W. Houseknecht, and K. R. Applin, 1989, Emplacement of clay into sand by infiltration: Jour. Sed. Petrology, v. 59, p. 77–87.

Matter, A., and K. Ramseyer, 1985, Cathodoluminescence microscopy as a tool for provenance studies of sandstones, *in* Zuffa, G. G., ed., Provenance of arenites: D. Reidel, Dordrecht, p. 191–211.

Mawson, Sir D., 1929, Some South Australian limestones in the process of forming: Quaternary Jour. Geol. Society, v. 85, p. 613–623.

Maynard, J. B., 1984, Composition of plagioclase feldspar in modern deep-sea sands: relationship to tectonic setting: Sedimentology, v. 31, p. 493–502.

Maynard, J. B., R. Valloni, and H.-S. Yu, 1982, Composition of modern deep-sea sands from arc-related basins, *in* Legget, J. E., ed., Trench-forearc geology: Sedimentation and tectonics on modern and ancient active plate margins: Geol. Soc. London Spec. Pub. 10, Blackwell, Oxford, p. 551–561.

Mazzullo, J., 1987, Origin of grain shape types in the St. Peter Sandstone: Determination by Fourier shape analysis and scanning electron microscopy: *in* Marshall, J. R., ed., Clastic particles: Van Nostrand Reinhold, New York, p. 302–313.

Mazzullo, J., and J. B. Anderson, 1987, Grain shape and surface texture analysis of till and glacial-marine sand grains from the Weddell and Ross seas, Antarctica, *in* Marshall, J. R., ed., Clastic particles: Van Nostrand Reinhold, New York, p. 314–327.

Mazzullo, J. M., and R. Ehrlich, 1983, Grain-shape variation in the St. Peter Sandstone: A record of eolian and fluvial sedimentation of an Early Paleozoic cratonic sheet sand: Jour. Sed. Petrology, v. 53, p. 105–120.

Mazzullo, J., R. Ehrlich, and M. A. Hemming, 1984, Provenance and areal distribution of Late Pleistocene and Holocene quartz sand on the southern New England continental shelf: Jour. Sed. Petrology, v. 54, p. 1335–1348.

Mazzullo, J., R. Ehrlich, and O. H. Pilkey, 1982, Local and distal origin of sands in the Hatteras abyssal plain: Marine Geology, v. 48, p. 75–88.

Mazzullo, J., and S. K. Kennedy, 1985, Automated measurement of the nominal sectional diameters of individual sedimentary particles: Jour. Sed. Petrology, v. 55, p. 593–595.

Mazzullo, J., P. Leschak, and D. Prusak, 1988, Sources and distribution of Late Quaternary silt in the surfi-cial sediment of the northeastern continental shelf of the United States: Marine Geology, v. 78, p. 241–254.

Mazzullo, J., and S. Magenheimer, 1987, The original shapes of quartz sand grains: Jour. Sed. Petrology, v. 57, p. 479–487.

Mazzullo, J., D. Sims, and D. Cunningham, 1986, The effects of shape sorting and abrasion upon the shapes of fine quartz sand grains: Jour. Sed. Petrology, v. 56, p. 45–56.

Mazzullo, J., and K. D. Withers, 1984, Sources, distribution, and mixing of Late Pleistocene and Holocene sands on the south Texas continental shelf: Jour. Sed. Petrology, v. 54, p. 1319–1334.

McBride, E. F., 1963, A classification of common sandstones: Jour. Sed. Petrology, v. 33, p. 664–669.

———1984, Diagenetic processes that affect provenance determinations in sandstone, *in* Zuffa, G. G., ed., Provenance of arenites: D. Reidel, Dordrecht, Holland, p. 95–113.

McBride, E. F., and R. L. Folk, 1977, The Caballos Novaculite revisited: Part II: Chert and shale members and synthesis: Jour. Sed. Petrology, v. 47, p. 1261–1286.

McBride, E. F., and M. D. Picard, 1987, Downstream changes in sand composition, roundness, and gravel size in a short-headed, high-gradient stream, northwestern Italy: Jour. Sed. Petrology, v. 57, p. 1018–1026.

McBride, E. F., R. G. Shepard, and R. A. Crawley, 1975, Origin of parallel, near-horizontal laminae by migration of bed forms in a small flume: Jour. Sed. Petrology, v. 45, p. 132–139.

McCabe, P. J., 1984, Depositional environments of coal and coal-bearing strata, *in* Rahmani, R. A., and R. M. Flores, eds., Sedimentology of coal and coal-bearing sequences: Internat. Assoc. Sedimentologists Spec. Pub. 7, p. 13–42.

McCabe, P. J., and C. M. Jones, 1977, Formation of reactivation surfaces within superimposed deltas and bedforms: Jour. Sed. Petrology, v. 47, p. 707–715.

McCave, I. N., R. J. Bryant, H. F. Cook, and C. A. Coughanowr, 1986, Evaluation of a laser-diffraction-size analyzer for use with natural sediments: Jour. Sed. Petrology, v. 56, p. 561–564.

McConchie, D. M., and D. W. Lewis, 1980, Varieties of glauconite in late Cretaceous and early Tertiary rocks of the South Island of New Zealand, and a new proposal for classification: New Zealand Jour. Geol. Geophysics, v. 23, p. 413–437.

McGowen, J. H., 1970, Gum Hollow Fan Delta, Nueces Bay, Texas: Bur. Econ. Geology, The University of Texas at Austin, Rept. Inv. 69, 91 p.

McHardy, W. J., and A. C. Birnie, 1987, Scanning electron microscopy, *in* Wilson, M. J., ed., A handbook of determinative methods in clay mineralogy: Blackie, Glasgow and London (Chapman and Hall, New York), p. 174–208.

McKee, E. D., 1965, Experiments on ripple lamination, *in* Middleton, G. V., ed., Primary sedimentary structures and their hydrodynamic interpretation. Soc. Econ. Paleontologists and Mineralogists Spec. Pub. 12, p. 66–83.

McKee, E. D., J. R. Douglass, and S. Rittenhouse, 1971, Deformation of lee-side laminae in eolian dunes: Geol. Soc. America Bull., v. 82, p. 359–378.

McKee, E. D., and G. W. Weir, 1953, Terminology for stratification and cross-stratification in sedimentary rocks: Geol. Soc. America Bull., v. 64, p. 381–390.

McKelvey, V. E., 1973, Abundance and distribution of phosphorus in the lithosphere, *in* Environmental phosphorus handbook: John Wiley & Sons, New York, p. 13–31.

McKelvey, V. E., J. S. Williams, R. P. Sheldon, E. R. Cressman, and T. M. Channey, 1959, The Phosphoria, Park City and Shedhorn formations in the Western Phosphate Field: U.S. Geol. Survey Prof. Paper 313-A.

Mckenzie, F. T., and R. Gees, 1971, Quartz synthesis at earth-surface conditions: Science, v. 173, p. 533–535.

McKenzie, J., 1981, Holocene dolomitization of calcium carbonate sediments from the coastal sabkhas of Abu Dhabi, U.A.E.: a stable isotope study: Jour. Geology, v. 89, p. 185–198.

McKenzie, J. A., K. J. Hsü, and J. F. Schneider, 1980, Movement of subsurface waters under the Sabkha, Abu Dhabi, UAE, and its relation to evaporative dolomite genesis, *in* Zenger, D. H., J. B. Dunham, and R. L. Ethington, eds., Concepts and models of dolomitization: Soc. Econ. Paleontologists and Mineralogists Spec. Pub. 28, p. 11–30.

Mellon, G. B., 1964, Discriminatory analysis of calcite- and silicate-cemented phases of the Mountain Park Sandstone: Jour. Geology, v. 72, p. 786–809.

Merkel, R. H., 1979, Well log formation evaluation: AAPG Course Note Series, No. 4, Am. Assoc. Petroleum Geologists, Tulsa, 82 p.

Mero, J. L., 1965, The mineral resources of the sea: Elsevier, New York, 312 p.

Meyer, W. J., 1974, Carbonate cement stratigraphy of the Lake Valley Formation (Mississippian) Sacramento Mountains, New Mexico: Jour. Sed. Petrology, v. 44, p. 837–861.

Meyer, W. J., and K. C. Lohmann, 1985, Isotope geochemistry of regionally extensive calcite cement zones and marine components in Mississippian limestones, New Mexico, in Schneidermann, N., and P. M. Harris, eds., Carbonate cements: Soc. Econ. Paleontologists and Mineralogists Spec. Pub. 36, p. 223–239.

Meyers, P. A., and R. M. Mitterer, eds., 1986, Deep ocean black shales: organic geochemistry and paleooceanographic setting: Marine Geology, v. 70, p. 1–174.

Mezzadri, G., and E. Saccani, 1989, Heavy mineral distribution in Late Quaternary sediments of the southern Aegean Sea: Implications for provenance and sediment dispersal in sedimentary basins at active margins: Jour. Sed. Petrology, v. 59, p. 412–422.

Miall, A. D., 1977, A review of braided-river depositional environments: Earth Science Rev., v. 13, p. 1–62.

——— 1978, Lithofacies types and vertical profile models in braided river deposits: A summary, in Miall, A. D., ed., Fluvial sedimentology: Canadian Soc. Petroleum Geologists Mem. 5, p. 597–604.

——— 1990, Principles of sedimentary basin analysis, 2nd ed.: Springer-Verlag, New York, Berlin, 668 p.

Middleburg, J. J., G. J. de Lange, and R. Kreulen, 1990, Dolomite formation in anoxic sediments of Kau Bay, Indonesia: Geology, v. 18, p. 399–402.

Middleton, G. V., 1976, Hydraulic interpretation of sand size distributions: Jour. Geology, v. 94, p. 405–426.

Middleton, L. T., and A. P. Trujillo, 1984, Sedimentology and depositional setting of the Upper Proterozoic Scanlan Conglomerate, central Arizona, in Koster, E. H., and R. J. Steel, eds.,

Sedimentology of gravels and conglomerates: Canadian Soc. Petroleum Geologists Mem. 10, p. 189–201.

Miller, J. 1988, Microscopical techniques: I. Slices, slides, stains and peels, in Tucker, M., ed., Techniques in sedimentology: Blackwell, p. 86–107.

Milliken, K. L., 1988, Loss of provenance information through subsurface diagenesis in Plio-Pleistocene sandstones, northern Gulf of Mexico: Jour. Sed. Petrology, v. 58, p. 992–1002.

——— 1989, Petrography and composition of authigenic feldspars, Oligocene Frio Formation, south Texas: Jour. Sed. Petrology, v. 59, p. 361–374.

Milliken, K. L., L. S. Land, and R. G. Loucks, 1981, History of burial diagenesis determined from isotopic geochemistry, Frio Formation, Brazoria County, Texas: Am. Assoc. Petroleum Geologists Bull., v. 65, p. 1397–1413.

Milliken, K. L., and L. E. Mack, 1990, Subsurface dissolution of heavy minerals, Frio Formation sandstones of the ancestral Rio Grande Province, south Texas: Sed. Geology, v. 68, p. 187–199.

Milliken, K. L., E. F. McBride, and L. S. Land, 1989, Numerical assessment of dissolution versus replacement in the subsurface destruction of detrital feldspars, Oligocene Frio Formation, south Texas: Jour. Sed. Petrology, v. 59, p. 740–757.

Milliman, J. D., 1972, Petrology of the sand fraction of sediments, northern New Jersey to southern Florida: U.S. Geol. Survey Prof. Paper 529-J, 40 p.

——— 1974, Marine carbonates: Springer-Verlag, New York, 375 p.

Mills, H. H., 1984, Clast orientation in Mount St. Helens debris-flow deposits, North Fork Toutle River, Washington: Jour. Sed. Petrology, v. 54, p. 626–634.

Milner, H. B., 1926, Supplement to introduction to sedimentary petrography: Murby, London, 157 p.

——— 1962, Sedimentary petrography, 4th ed., v. I, II: George Allen & Unwin, London, 715 p.

Minnis, M. M., 1984, An automatic point-counting method for mineralogic assessment: Am. Assoc. Petroleum Geologists Bull., v. 68, p. 744–752.

Mitchell, A. H. G., and H. G. Reading, 1986, Sedimentation and tectonics, in Reading, H. G., ed., Sedimentary environments and facies, 2nd ed., Blackwell, p. 471–519.

Molnia, B. F., and J. R. Hein, 1982, Clay mineralogy of a glacially dominated subarctic continental shelf: Northeast Gulf of Alaska: Jour. Sed. Petrology, v. 52, p. 515–527.

Monicard, R. P., 1980, Properties of reservoir rocks: Core analyses: Gulf Pub. Co., Houston (Editions Technip, Paris), 168 p.

Monty, C. L. V., 1976, The origin and development of cryptalgal fabrics, in Walter, M. R., ed., Stromatolites: Elsevier, Amsterdam, p. 193–249.

Moon, C. F., and C. W. Hurst, 1984, Fabric of muds and shales: an overview, in Stow, D. A. V., and D. J. W. Piper, eds., Fine-grained sediments: deep-water processes and facies: Geol. Soc. Spec. Pub. 15, Blackwell, Oxford, p. 579–593.

Moore, C. H., 1985, Upper Jurassic subsurface cements: A case history, in Schneidermann N., and P. M. Harris, eds., Carbonate cements: Soc. Econ. Paleontologists and Mineralogists Spec. Pub. 36, p. 291–308.

——— 1989, Carbonate diagenesis and porosity: Developments in sedimentology 46, Elsevier, Amsterdam, 338 p.

Moore, C. H., and Y. Druckman, 1981, Burial diagenesis and porosity evolution, upper Jurassic Smackover, Arkansas and Louisiana: Am. Assoc. Petroleum Geologists Bull., v. 65, p. 597–628.

Moore, D. M., 1978, A sample of the Purington Shale prepared as a geochemical standard: Jour. Sed. Petrology, v. 48, p. 995–998.

Moore, J. G., R. L. Phillips, R. W. Grigg, D. W. Peterson, and D. A. Swanson, 1973, Flow of lava into the sea, 1969–1971, Kilauea Volcano, Hawaii: Geol. Soc. America Bull., v. 84, p. 537–546.

Morad, S., 1984, Diagenetic matrix in Proterozoic graywackes from Sweden: Jour. Sed. Petrology, v. 54, p. 1157–1168.

Morad, S., and A. A. Aldahan, 1987, Diagenetic replacement of feldspars by quartz in sandstones: Jour. Sed. Petrology, v. 57, p. 488–493.

Morad, S., M. Bergan, R. Knarud, and J. P. Nystuen, 1990, Albitization of detrital plagioclase in Triassic reservoir sandstones from the Snorre Field, Norwegian North Sea: Jour. Sed. Petrology, v. 60, p. 411–425.

Morad, S., R. Marfil, and J. A. De La Peña, 1989, Diagenetic K-feldspar pseudomorphs in the Triassic

Buntsandstein sandstones of the Iberian Range, Spain: Sedimentology, v. 36, p. 635–650.

Moraes, M. A. S., and L. F. De Ros, 1990, Infiltrated clays in fluvial Jurassic sandstones of Recôncavo Basin, northeastern Brazil: Jour. Sed. Petrology, v. 60, p. 809–819.

Morey, G. W., R. O. Fournier, and J. J. Rowe, 1962, The solubility of quartz in water in the temperature interval from 25°C to 300°C: Geochim. et Cosmochim. Acta, v. 26, p. 1029–1043.

——— 1964, The solubility of amorphous silica at 25°C: Jour. Geophys. Research, v. 69, p. 1995–2002.

Morris, S. W., 1987, Diagenesis of the Permo-Triassic Ivishak Formation of the Sadlerochit Group, Prudhoe Bay Field, Alaska: M.S. Thesis, University of Oregon, 198 p.

Morrow, D. W., and B. D. Ricketts, 1988, Experimental investigation of sulfate inhibition of dolomite and its mineral analogues, in Shukla, V., and P. A. Baker, eds., Sedimentology and geochemistry of dolostones: Soc. Econ. Paleontologists and Mineralogists Spec. Pub. 43, p. 25–38.

Morse, J. W., 1985, Kinetic control of morphology, composition, and mineralogy of abiotic sedimentary carbonates: Discussion: Jour. Sed. Petrology, v. 55, p. 919–920.

Morton, A. C., 1985a, Heavy minerals in provenance studies, in Zuffa, G. G., ed., Provenance of arenites: D. Reidel, Dordrecht, Holland, p. 249–277.

——— 1985b, A new approach to provenance studies: electron microprobe analysis of detrital garnets from Middle Jurassic sandstones of the northern North Sea: Sedimentology, v. 32, p. 553–566.

Moseley, F., 1981, Methods in field geology: W. H. Freeman and Co., Oxford and San Francisco, 211 p.

Mount, J., 1985, Mixed siliciclastic and carbonate sediments: A proposed first-order textural compositional classification: Sedimentology, v. 32, p. 435–442.

Mucci, A., and J. W. Morse, 1983, The incorporation of Mg^{+2} and Sr^{+2} into calcite overgrowths: influences of growth rate and solution composition: Geochim. et Cosmochim. Acta, v. 47, p. 217–233.

Mueller, G., 1967, Methods in sedimentary petrology: E. Schweizer-bart'sche Verlagsbuchhandlung, Stuttgart, 283 p.

Muerdter, D. R., J. P. Dauphin, and G. Steele, 1981, An interactive computerized system for grain size analysis of silt using electro-resistance: Jour. Sed. Petrology, v. 51, p. 647–650.

Müller, D. W., J. A. McKenzie, and P. A. Mueller, 1990, Abu Dhabi sabkha, Persian Gulf, revisited: Application of strontium isotopes to test an early dolomitization model: Geology, v. 18, p. 618–621.

Müller, G., 1967, Diagenesis in argillaceous sediments, in Larsen, G., and G. V. Chilingar, eds., Diagenesis in sediments: Elsevier, Amsterdam, p. 127–177.

Mullins, H. T., A. C. Neumann, R. J. Wilber, and M. R. Boardman, 1980, Nodular carbonate sediment on Bahamian slopes: possible precursors to nodular limestones: Jour. Sed. Petrology, v. 50, p. 117–131.

Mumpton, F. A., 1981, Natural zeolites, in Mumpton, F. A., ed., Mineralogy and geology of natural zeolites: Mineralogical Soc. America Rev. Mineralogy, v. 4, p. 1–17.

Murray, J. W., and E. E. Mackintosh, 1968, Occurrence of interstratified glauconite-montmorillonoid pellets, Queen Charlotte Sound, British Columbia: Canadian Jour. Earth Science, v. 5, p. 243–247.

Murray, R. C., 1964, Preservation of primary structures and fabrics in dolomite, in Imbrie, J., and N. Newell, eds., Approaches in paleoecology: John Wiley & Sons, New York, p. 388–403.

Mustard, P. S., and J. A. Donaldson, 1987, Early Proterozoic ice-proximal glaciomarine deposition: The lower Gowganda Formation at Cobalt, Ontario, Canada: Geol. Soc. America Bull., v. 98, p. 373–387.

Nadler, A., and M. Magaritz, 1980, Studies of marine solution basins—isotopic and compositional changes during evaporation, in Nissenbaum, A., ed., Hypersaline brines and evaporitic environments: Elsevier, Amsterdam, p. 115–129.

Naidu, A. S., D. C. Burrell, and D. W. Hood, 1971, Clay mineral composition and geological significance of some Beauford Sea sediments: Jour. Sed. Petrology, v. 41, p. 691–694.

Nanz, R. H., Jr., 1954, Genesis of Oligocene sandstone reservoir, Seeligson Field, Jim Wells and Kleberg counties, Texas: Am. Assoc. Petroleum Geologists Bull., v. 38, p. 96–117.

Nathan, Y., 1984, The mineralogy and geochemistry of phosphorites, in Nriagu, J. O., and P. B. Moore, eds., Phosphate minerals: Springer-Verlag, Berlin, p. 275–291.

Naylor, M. A., 1980, The origin of inverse grading in muddy debris flow deposits—a review: Jour. Sed. Petrology, v. 50, p. 1111–1116.

Nelson, B. K., and D. J. DePaolo, 1988, Comparison of isotopic and petrographic provenance indicators in sediments from Tertiary continental basins of New Mexico: Jour. Sed. Petrology, v. 58, p. 348–357.

Nelson, C. H., 1982, Modern shallow-water graded sand layers from storm surges, Bering shelf: A mimic of Bouma sequences and turbidite systems: Jour. Sed. Petrology, v. 52, p. 537–545.

Nemec, W., S. J. Porebski, and R. J. Steel, 1980, Texture and structure of resedimented conglomerates: examples from Ksiaz Formation (Famennian-Tournaisian), southwestern Poland: Sedimentology, v. 27, p. 519–538.

Nemec, W., and R. J. Steel, 1984, Alluvial and coastal conglomerates: Their significant features and some comments on gravelly mass-flow deposits, in Koster, E. H., and R. J. Steel, eds., Sedimentology of gravels and conglomerates: Canadian Soc. Petroleum Geologists Mem. 10, p. 1–31.

Nesse, W. D., 1986, Introduction to optical mineralogy: Oxford University Press, New York, 325 p.

Neumann, A. C., J. W. Kofoed, and G. H. Keller, 1977, Lithoherms in the Straits of Florida: Geology, v. 5, p. 4–11.

Neumann, A. C., and L. S. Land, 1975, Lime mud deposition and calcareous algae in the Bight of Abaco, Bahamas: a budget: Jour. Sed. Petrology, v. 45, p. 763–786.

Newell, N. D., J. K. Rigby, A. G. Fischer, A. J. Whitman, J. E. Hickox, and J. S. Bradley, 1953, The Permian reef complex of the Guadalupe Mountains region, Texas and New Mexico: W. H. Freeman, San Francisco, 236 p.

Newell, N. D., J. K. Rigby, A. J. Whitman, and J. S. Bradley, 1951, Shoal-water geology and environments, eastern Andros Island, Bahamas: Am. Mus. Nat. History Bull., v. 97, p. 1–29.

Newman, A. C. D., and G. Brown, 1987, The chemical constitution of clays, in Newman, A. C. D., ed., Chemistry of clays and clay minerals:

Mineralog. Soc. Mon. 6, Longman Scientific & Technical, Essex, UK, p. 1–128.

Noble, J. P. A., and K. D. M. Howells, 1974, Early marine lithification of the nodular limestones in the Silurian of New Brunswick: Sedimentology, v. 21, p. 597–609.

Notholt, A. J. G., R. P. Sheldon, and D. F. Davidson, 1989a, Preface to Phosphate deposits of the world: Cambridge University Press, Cambridge, p. xxiii–xxv.

———1989b, Phosphate deposits of the world: Cambridge University Press, Cambridge, 566 p.

Nuegebauer, J., 1978, Micritization of crinoids by diagenetic dissolution: Sedimentology, v. 25, p. 267–283.

Nuhfer, E. B., 1981, Mudrock fabrics and their significance—Discussion: Jour. Sed. Petrology, v. 51, p. 1027–1029.

O'Brien, N. R., 1981, SEM study of shale fabric—a review: Scanning Electron Microscopy/1981/I, SEM Inc., p. 569–575.

———1987, The effects of bioturbation on the fabric of shale: Jour. Sed. Petrology, v. 57, p. 449–455.

———1989, Origin of lamination in Middle and Upper Devonian black shales, New York State: Northeastern Geology, v. 11, p. 159–165.

———1990, Significance of lamination in Toarcian (Lower Jurassic) shales from Yorkshire, Great Britain: Sed. Geology, v. 67, p. 25–34.

O'Brien, N. R., and R. M. Slatt, 1990, Argillaceous rock atlas: Springer-Verlag, New York, 141 p.

Odin, G. S., and A. Matter, 1981, De glauconiarum origine: Sedimentology, v. 28, p. 611–642.

Odom, I. E., 1975, Feldspar-grain size relations in Cambrian arenites, Upper Mississippi Valley: Jour. Sed. Petrology, v. 45, p. 636–650.

———1976, Microstructures, mineralogy, and chemistry of Cambrian glauconitic pellets and glauconite, central U.S.A.: Clays and Clay Minerals, v. 24, p. 232–238.

Odom, I. E., T. W. Doe, and R. H. Dott, Jr., 1976, Nature of feldspar-grain size relations in some quartz-rich sandstones: Jour. Sed. Petrology, v. 46, p. 862–870.

Odom, I. E., T. N. Willand, and R. J. Lassin, 1979, Paragenesis of diagenetic minerals in the St. Peter Sandstone (Ordovician), Wisconsin and Illinois, *in* Scholle, P. A., and P. R.

Schluger, eds., Aspects of diagenesis: Soc. Econ. Paleontologists and Mineralogists Spec. Pub. 26, p. 425–443.

Ogihara, S., and A. Iijima, 1989, Clinoptilolite to heulandite transformation in burial diagenesis, *in* Jacobs, P. A., and R. A. Van Santen, eds., Zeolites: facts, figures, future: Elsevier, Amsterdam, p. 491–500.

Ogren, D. E., and C. J. Waag, 1986, Orientation of cobble and boulder beach clasts: Sed. Geology, v. 47, p. 69–76.

Ojakangas, R. W., 1963, Petrology and sedimentation of the Upper Cambrian Lamotte Sandstone in Missouri: Jour. Sed. Petrology, v. 33, p. 860–873.

Okada, H., 1966, Non-greywacke "turbidite" sandstones in the Welsh geosyncline: Sedimentology, v. 7, p. 211–232.

———1971, Classification of sandstone: Analysis and proposal: Jour. Geology, v. 79, p. 509–525.

Ondrick, C. W., and J. C. Griffiths, 1969, Frequency distribution of elements in Rensselaer Graywacke, Troy, New York: Geol. Soc. America Bull., v. 80, p. 509–518.

Orford, J. D., and W. B. Whalley, 1983, The use of the fractal dimension to characterize irregular-shaped particles: Sedimentology, v. 30, p. 655–668.

———1987, The quantitative description of highly irregular sedimentary particles: The use of the fractal dimension, *in* Marshall, J. R., ed., Clastic particles: Van Nostrand Reinhold, New York, p. 267–280.

Owen, M. R., 1987, Hafnium content of detrital zircons, a new tool for provenance study: Jour. Sed. Petrology, v. 57, p. 824–830.

Owen, M. R., and A. V. Carozzi, 1986, Southern provenance of upper Jackfork Sandstone, southern Ouachita Mountains: Cathodoluminescence petrography: Geol. Soc. America Bull., v. 97, p. 110–115.

Owens, J. P., and N. F. Sohl, 1973, Glauconites from the New Jersey–Maryland coastal plain: Their K/Ar ages and applications in stratigraphic studies: Geol. Soc. America Bull., v. 84, p. 2811–2838.

Packer, B. M., and R. V. Ingersoll, 1986, Provenance and petrology of Deep Sea Drilling Project sands and sandstones from the Japan and Mariana forearc and backarc regions: Sed. Geology, v. 51, p. 5–28.

Pandalai, H. S., and S. Basumallick, 1984, Packing in a clastic sediment: concepts and measurements: Sed. Geology, v. 39, p. 87–93.

Paola, C., 1988, Subsidence and gravel transport in alluvial basins, *in* Kleinspehn, K. L., and C. Paola, eds., New perspectives in basin analysis: Springer-Verlag, New York, p. 231–243.

Park, W.-C., and E. H. Schot, 1968, Stylolitization in carbonate rocks, *in* Müller, G., and G. M. Friedman, eds., Carbonate sedimentology in central Europe: Springer-Verlag, New York, p. 66–74.

Parkash, B., and G. V. Middleton, 1970, Downcurrent textural changes in Ordovician turbidite greywackes: Sedimentology, v. 14, p. 259–293.

Passega, R., 1964, Grain size representation by CM patterns as a geological tool: Jour. Sed. Petrology, v. 34, p. 830–847.

———1977, Significance of CM diagrams of sediments deposited by suspensions: Sedimentology, v. 24, p. 723–733.

Paterson, E., and R. Swaffield, 1987, Thermal analysis, *in* Wilson, M. J., ed., A handbook of determinative methods in clay mineralogy: Blackie, Glasgow and London (Chapman and Hall, New York), p. 99–132.

Patterson, R. J., and D. J. J. Kinsman, 1982, Formation of diagenetic dolomite in coastal sabkha along Arabian (Persian) Gulf: Am. Assoc. Petroleum Geologists Bull., v. 66, p. 28–43.

Payne, T. G., 1942, Stratigraphical analysis and environmental reconstruction: Am. Assoc. Petroleum Geologists Bull., v. 26, p. 1697–1770.

Pederson, G. K., and F. Surlyk, 1977, Dish structures in Eocene volcanic ash layers, Denmark: Sedimentology, v. 24, p. 581–590.

Pederson, T. F., and S. E. Calvert, 1990, Anoxia vs. productivity: What controls the formation of organic-carbon rich sediments of sedimentary rocks? Am. Assoc. Petroleum Geologists Bull., v. 74, p. 454–466.

Pedley, H. M., 1990, Classification and environmental models of cool freshwater tufas: Sed. Geology, v. 68, p. 143–154.

Perry, E. C., Jr., 1983, Oxygen isotope geochemistry of iron-formation, *in* Trendall, A. F., and R. C. Morris, eds., Iron-formation: facts and problems: Elsevier, Amsterdam, p. 359–371.

Peryt, T. M., 1983a, Classification of coated grains, *in* Peryt, T. M., ed., Coated grains: Springer-Verlag, Berlin, p. 3–6.

———— Peryt, T. M., ed., 1983b, Coated grains: Springer-Verlag, Berlin, 655 p.

Peryt, T. M., and M. Magaritz, 1990, Genesis of evaporite-associated platform dolomites: Case study of the Main Dolomite (Zechstein, Upper Permian), Leba elevation, northern Poland: Sedimentology, v. 37, p. 745–761.

Peterman, Z. E., R. G. Coleman, and C. M. Bunker, 1981, Provenance of Eocene graywackes of the Fluornoy Formation near Agness, Oregon—a geochemical approach: Geology, v. 9, p. 81–86.

Peterson, C. D., P. D. Komar, and K. F. Scheidegger, 1985, Distribution, geometry, and origin of heavy mineral placer deposits on Oregon beaches: Jour. Sed. Petrology, v. 56, p. 67–77.

Peterson, M. N., and C. C. von der Borch, 1965, Chert: Modern inorganic deposition in a carbonate-precipitating locality: Science, v. 149, p. 1501–1503.

Pettijohn, F. J., 1941, Persistence of heavy minerals and geologic age: Jour. Geology, v. 49, p. 610–625.

———— 1963, Chemical composition of sandstones—excluding carbonate and volcanic sands, *in* Data of geochemistry, 6th ed.: U.S. Geol. Survey Prof. Paper 440S, 19 p.

———— 1975, Sedimentary rocks, 3rd ed.: Harper & Row, New York, 628 p.

Pettijohn, F. J., and P. E. Potter, 1964, Atlas and glossary of primary sedimentary structures: Springer-Verlag, New York, 370 p.

Pettijohn, F. J., P. E. Potter, and R. Siever, 1987, Sand and sandstone, 2nd ed., Springer-Verlag, New York, 553 p.

Phillips, J. O., 1986, Petrology of the Late Proterozoic(?)—Early Cambrian Arumbera Sandstone and the Late Proterozoic Qandong Conglomerate, east-central Amadeus Basin, central Australia: M.S. Thesis, Utah State University, 402 p.

Phillips, R. L., 1983, Late Miocene tidal shelf sedimentation, Santa Cruz Mountains, California, *in* Larue, D. K., and R. J. Steel, eds., Cenozoic marine sedimentation, Pacific margin, U.S.A.: Pacific Sect., Soc. Economic Paleontologists and Mineralogists, p. 45–61.

———— 1984, Depositional features of Late Miocene, marine cross-bedded conglomerates, California, *in* Koster, E. H., and R. J. Steel, eds., Sedimentology of gravels and conglomerates: Canadian Soc. Petroleum Geologists Mem. 10, p. 345–358.

Picard, M. D., 1971, Classification of fine-grained sedimentary rocks: Jour. Sed. Petrology, v. 41, p. 179–195.

———— 1977, Stratigraphic analysis of the Navajo Sandstone: A discussion: Jour. Sed. Petrology, v. 47, p. 475–483.

Pierson, S. J., 1983, Geological well log analysis, 3rd ed.: Gulf Pub. Co., Houston, 424 p.

Pilkey, O. H., 1963, Heavy minerals of the U.S. South Atlantic continental shelf and slope: Geol. Soc. America Bull., v. 74, p. 641–648.

Pittman, E. D., 1963, The use of zoned plagioclase as an indicator of provenance: Jour. Sed. Petrology, v. 33, p. 380–386.

———— 1969, Destruction of plagioclase twins by stream transport: Jour. Sed. Petrology, v. 39, p. 1432–1437.

———— 1970, Plagioclase feldspar as an indicator of provenance in sedimentary rocks: Jour. Sed. Petrology, v. 40, p. 591–598.

———— 1979, Recent advances in sandstone diagenesis: Ann. Rev. Earth and Planetary Science, v. 7, p. 39–62.

Plummer, P. S., and V. A. Gostin, 1981, Shrinkage cracks: Desiccation or synaeresis? Jour. Sed. Petrology, v. 51, p. 1147–1156.

Plymate, T. G., and L. J. Suttner, 1983, Evaluation of optical and X-ray technique for determining source-rock controlled variation in detrital potassium feldspars: Jour. Sed. Petrology, v. 53, p. 509–519.

Popeno, W. P., 1941, The Trabuco and Baker conglomerates of the Santa Ana Mountains: Jour. Geology, v. 49, p. 738–752.

Poppe, L. J., A. H. Eliason, and J. J. Fredricks, 1985, APSAS—an automated particle size analysis system: U.S. Geol. Survey Circ. 963, 77 p.

Porrenga, D. H., 1965, Chamosite in Recent sediments of the Niger and Orinoco deltas: Geologie en Mijnbouw, v. 44, p. 400–403.

———— 1966, Clay minerals in recent sediments of the Niger Delta, *in* Bailey, S. W., ed., Clays and clay minerals, 14th National Conference on Clays and Clay Minerals: Pergamon, p. 221–233.

———— 1967, Glauconite and chamosite as depth indicators in the marine environment: Marine Geology, v. 5, p. 495–501.

Potter, P. E., 1978, Petrology and chemistry of modern big river sands: Jour. Geology, v. 86, p. 423–449.

———— 1986, South America and a few grains of sand: Part I—Beach sands: Jour. Geology, v. 94, p. 301–319.

Potter, P. E., J. B. Maynard, and W. A. Pryor, 1980, Sedimentology of shale: Springer-Verlag, New York, 306 p.

Potter, P. E., and F. J. Pettijohn, 1977, Paleocurrents and basin analysis, 2nd ed.: Springer-Verlag, Berlin, 413 p.

Potter, P. E., and W. A. Pryor, 1961, Dispersal centers of Paleozoic and later clastics of the Upper Mississippi Valley and adjacent area: Geol. Soc. America Bull., v. 72, p. 1195–1250.

Powell, R. D., 1983, Glacial-marine sedimentation processes and lithofacies of temperate tidewater glaciers, Glacier Bay, Alaska, *in* Molnia, B. F., ed., Glacial-marine sedimentation: Plenum Press, New York, p. 185–232.

Powers, M. C., 1953, A new roundness scale for sedimentary particles: Jour. Sed. Petrology, v. 23, p. 117–119.

Prezbindowski, D. R., 1985, Burial cementation—is it important? A case study, Stuart City Trend, south-central Texas, *in* Schneidermann, N., and P. M. Harris, eds., Carbonate cements, Soc. Econ. Paleontologists and Mineralogists, Spec. Pub. 36, p. 241–264.

Prusak, D., and J. Mazzullo, 1987, Sources and provinces of Late Pleistocene and Holocene sand and silt on the Mid-Atlantic continental shelf: Jour. Sed. Petrology, v. 57, p. 278–287.

Purdy, E. G., 1963, Recent calcium carbonate facies of the Great Bahama Bank: Jour. Geology, v. 71, p. 334–355, 472–497.

Pye, K., and D. H. Krinsley, 1984, Petrographic examination of sedimentary rocks in the SEM using backscattered electron detectors: Jour. Sed. Petrology, v. 84, p. 877–888.

Radke, B. M., and R. L. Mathis, 1980, On the formation and occurrence of saddle dolomite: Jour. Sed. Petrology, v. 50, p. 1149–1168.

Rahmani, R. A., and R. M. Flores, eds., 1984, Sedimentology of coal and coal-bearing sequences: Internat.

Assoc. Sedimentologists Spec. Pub. 7, 412 p.

Ramos, A., and A. Sopeña, 1983, Gravel bars in low-sinuosity streams (Permian and Triassic, central Spain), *in* Collinson, J. D., and J. Lewin, eds., Modern and ancient fluvial systems: Internat. Assoc. Sedimentologists Spec. Pub. 6, p. 301–312.

Ramseyer, K., and J. R. Boles, 1986, Mixed-layer illite/smectite minerals in Tertiary sandstones and shales, San Joaquin basin, California: Clays and Clay Minerals, v. 34, p. 115–124.

Rand, B., and I. E. Melton, 1977, Particle interactions in aqueous kaolinite suspensions: Jour. Colloid and Interface Science, v. 60, p. 308–320.

Ranganathan, V., and R. S. Tye, 1984, Petrography, diagenesis, and facies controls on porosity in Shannon Sandstone, Hartzog Draw Field, Wyoming: Am. Assoc. Petroleum Geologists Bull., v. 70, p. 56–69.

Rapson, J. E., 1965, Petrography and derivation of Jurassic-Cretaceous clastic rocks, southern Rocky Mountains, Canada: Am. Assoc. Petroleum Geologists Bull., v. 49, p. 1426–1452.

Raup, O. B., 1966, Clay mineralogy of Pennsylvanian redbeds and associated rocks flanking ancestral Front Range of central Colorado: Am. Assoc. Petroleum Geologists Bull., v. 50, p. 251–268.

Rautman, C. A., and R. H. Dott, Jr., 1977, Dish structures formed by fluid escape in Jurassic shallow marine sandstones: Jour. Sed. Petrology, v. 47, p. 101–106.

Reading, H. G., 1982, Sedimentary basins and global tectonics: Proc. Geol. Assoc., v. 93, p. 321–350.

Reading, H. G., ed., 1986, Sedimentary environments and facies, 2nd ed.: Blackwell, Oxford, 615 p.

Reddy, M. M., and K. K. Wang, 1980, Crystallization of calcium carbonate in the presence of metal ions. I. Inhibition by magnesium ions at pH 8.8 and 25°C: Jour. Crystal Growth, v. 50, p. 470–480.

Reed, W. R., R. LeFever, and G. J. Moir, 1975, Depositional environmental interpretation from settling-velocity (psi) distributions: Geol. Soc. America Bull., v. 86, p. 1321–1328.

Reeder, R. J., 1983, Crystal chemistry of the rhombohedral carbonates, *in* Reeder, R. J., ed., Carbonates: Mineralogy and chemistry: Mineralog. Soc. America Rev. Mineralogy, v. 11, p. 1–47.

Reeder, R. J., and J. L. Prosky, 1986, Compositional sector zoning in dolomite: Jour. Sed. Petrology, v. 56, p. 237–247.

Rees, A. I., 1965, The use of anisotropy of magnetic susceptibility in the estimation of sedimentary fabric: Sedimentology, v. 14, p. 257–271.

———— 1968, The production of preferred orientation in a concentrated dispersion of elongated and flattened grains: Jour. Geology, v. 76, p. 457–465.

———— 1979, The orientation of grains in a sheared dispersion: Tectonophysics, v. 55, p. 275–287.

———— 1983, Experiments on the production of transverse grain alignment in a sheared dispersion: Sedimentology, v. 30, p. 437–448.

Rees, A. I., and W. A. Woodall, 1975, The magnetic fabric of some laboratory-deposited sediments: Earth and Planetary Sci. Letters, v. 25, p. 121–130.

Reicke, H. H., III, and G. V. Chilingarian, 1974, Compaction of argillaceous sediments: Elsevier, Amsterdam, 424 p.

Reid, R. P., I. G. Macintyre, and N. P. James, 1990, Internal precipitation of microcrystalline carbonate: a fundamental problem for sedimentologists: Sed. Geology, v. 68, p. 163–170.

Reineck, H. E., and I. B. Singh, 1980, Depositional sedimentary environments, 2nd ed.: Springer-Verlag, Berlin, 549 p.

Renne, P. R., T. A. Becker, and S. M. Swapp, 1990, $^{40}Ar/^{39}Ar$ laserprobe dating of detrital micas from the Montgomery Creek Formation, northern California: clues to provenance, tectonics, and weathering processes: Geology, v. 18, p. 563–566.

Ribbe, P. H., 1983, Aluminum-silicon order in feldspars: domain textures and diffraction patterns, *in* Ribb, P. H., ed., Feldspar mineralogy, 2nd ed.: Mineralog. Soc. America Rev. Mineralogy, v. 2, p. 21–55.

Richter, D. K., 1983a, Classification of coated grains: Discussion, *in* Peryt, T. M., ed., Coated grains: Springer-Verlag, p. 7–8.

———— 1983b, Calcareous ooids: a synopsis, *in* Peryt, T., ed., Coated grains, Springer-Verlag, Berlin, p. 72–99.

Richter, D. K., and H. Besenecker, 1983, Subrecent high-Sr ooids from hot springs near Tekke Ilica (Tur-

key), *in* Peryt, T., ed., Coated grains, Springer-Verlag, Berlin, p. 154–162.

Richter, D. K., and H. Füchtbauer, 1978, Ferroan calcite replacement indicates former magnesian calcite skeletons: Sedimentology, v. 25, p. 843–861.

Ricken, W., 1987, The carbonate compaction law: A new tool: Sedimentology, v. 34, p. 571–584.

Rigby, D., and B. D. Batts, 1986, The isotopic composition of nitrogen in Australian coals and oil shales: Geochem. Geology (Isotope Geoscience Section), v. 58, p. 273–282.

Riggs, S. R., 1980, Intraclast and pellet phosphorite sedimentation in the Miocene of Florida: Jour. Geol. Soc. (London), v. 137, p. 741–748.

Rittenhouse, G., 1943, Transportation and deposition of heavy minerals: Geol. Soc. America Bull., v. 54, p. 1725–1730, 1739–1780.

Rittman, A., 1962, Volcanoes and their activity: John Wiley & Sons, New York, 305 p.

Robb, L. J., D. W. Davis, and S. L. Kamo, 1990, U-Pb ages on single detrital zircon grains from the Witwatersrand Basin, South Africa: constraints on the ages of sedimentation and on the evolution of granites adjacent to the basin: Jour. Geology, v. 98, p. 311–328.

Robin, P. F., 1978, Pressure solution at grain-to-grain contacts: Geochim. et Cosmochim. Acta, v. 42, p. 1383–1389.

Roedder, E., 1979, Fluid inclusion evidence on the environments of sedimentary diagenesis, a review, *in* Scholle, P. A., and P. R. Schluger, eds., Aspects of diagenesis: Soc. Econ. Paleontologists and Mineralogists Spec. Pub. 26, p. 89–107.

Rohrlich, V., N. B. Price, and S. E. Calvert, 1969, Chamosite in the Recent sediments of Loch Etive, Scotland: Jour. Sed. Petrology, v. 39, p. 624–631.

Roman, S., 1974, Palynoplanktological analysis of some Black Sea cores, *in* Degens, E. T., and D. A. Ross, eds., The Black Sea—Geology, chemistry, and biology: Am. Assoc. Petroleum Geologists Mem. 20, p. 396–410.

Ronov, A. B., 1983, The Earth's sedimentary shell: Am. Geol. Institute Reprint Ser.: V, American Geological Institute, Falls Church, Va., 80 p.

Ronov, A. B., E. E. Khain, A. N. Balukhovsky, and K. B. Seslavinsky, 1980, Quantitative analysis of Phanerozoic sedimentation: Sed. Geology, v. 25, p. 311–325.

Ronov, A. B., and A. A. Migdisov, 1971, Geochemical history of the crystalline basement and sedimentary cover of the Russian and North American platforms: Sedimentology, v. 16, p. 137–185.

Roser, B. P., and R. J. Korsch, 1985, Plate tectonics and geochemical composition of sandstones: A discussion: Jour. Geology, v. 93, p. 81–84.

———— 1986, Determination of tectonic setting of sandstone-mudstone suites using SiO_2 content and K_2O/Na_2O ratio: Jour. Geology, v. 94, p. 635–650.

———— 1988, Provenance signatures of sandstone-mudstone suites determined using discriminant function analysis of major-element data: Chem. Geology, v. 67, p. 119–139.

Rottman, C. J. F., 1970, Physical parameters and interrelationships of modern beach sands, Pleistocene terrace sands, and Eocene sandstones from Cape Arago, Oregon: A study combining the evolution of grain morphology in the zone of surf action and local aspects of the present and past depositional environment: Ph.D. Dissertation, University of Oregon, 238 p.

Rubin, D. M., 1987, Crossbedding, bedforms, and paleocurrents: Soc. Econ. Paleontologists and Mineralogists, Concepts in Sedimentology and Paleontology, v. 1, 187 p.

Ruppel, S. C., and H. S. Cander, 1988, Dolomitization of shallow-water carbonates by seawater and seawater-derived brines: San Andres Formation (Guadalupian), West Texas, *in* Shukla, V., and P. A. Baker, eds., Sedimentology and geochemistry of dolostones, Soc. Econ. Paleontologists and Mineralogists Spec. Pub. 43, p. 245–262.

Russell, J. D., 1987, Infrared methods, *in* Wilson, M. J., ed., A handbook of determinative methods in clay mineralogy: Blackie, Glasgow and London (Chapman and Hall, New York), p. 133–173.

Russell, P. L., 1990, Oil shales of the world: Their origin, occurrence and exploitation: Pergamon Press, Oxford, 736 p.

Rust, B. R., 1966, Late Cretaceous paleogeography near Wheeler Gorge, Ventura County, California: Am. Assoc. Petroleum Geologists Bull., v. 50, p. 1384–1398.

———— 1972, Pebble orientation in fluvial sediments: Jour. Sed. Petrology, v. 42, p. 384–388.

———— 1977, Mass flow deposits in a Quaternary succession near Ottawa, Canada: diagnostic criteria for subaqueous outwash: Canadian Jour. Earth Science, v. 14, p. 175–184.

———— 1978, Depositional models for braided alluvium, *in* Miall, A. D., ed., Fluvial sedimentology: Canadian Soc. Petroleum Geologists Mem. 5, p. 605–625.

Rust, B. R., and R. Romanelli, 1975, Late Quaternary subaqueous outwash deposits near Ottawa, Canada, *in* Jopling, A. V., and B. C. Macdonald, eds., Glaciofluvial and glaciolacustrine sedimentation: Soc. Econ. Paleontologists and Mineralogists Spec. Pub. 23, p. 177–192.

Ryer, T. A., and A. W. Langer, 1980, Thickness change involved in the peat-to-coal transformation for a bituminous coal of Cretaceous age in central Utah: Jour. Sed. Petrology, v. 50, p. 987–992.

Sagoe, K.-M. O., and G. S. Visher, 1977, Population breaks in grain-size distributions of sand—A theoretical model: Jour. Sed. Petrology, v. 47, p. 285–310.

Sahu, B. K., 1982, Multigroup discrimination of river, beach, and dune sands using roundness statistics: Jour. Sed. Petrology, v. 52, p. 779–784.

Saigal, G. C., S. Morad, K. Bjørlykke, P. K. Egeberg, and P. Aagaard, 1988, Diagenetic albitization of detrital K-feldspars in Jurassic, Lower Cretaceous and Tertiary clastic reservoir rocks from offshore Norway, I. Textures and origin: Jour. Sed. Petrology, v. 58, p. 1003–1013.

Saigal, G. C., and E. K. Walton, 1988, On the occurrence of displacive calcite in Lower Old Red Sandstones of Carnoustie, eastern Scotland: Jour. Sed. Petrology, v. 58, p. 131–135.

Saller, A. H., 1984, Petrologic and geochemical constraints on the origin of subsurface dolomite: An example of dolomitization by normal seawater, Eniwetok Atoll: Geology, v. 12, p. 217–220.

Sandberg, P. A., 1983, An oscillating trend in Phanerozoic non-skeletal carbonate mineralogy: Nature, v. 305, p. 19–22.

Sanderson, I. D., 1984, Recognition and significance of inherited quartz overgrowths in quartz arenites: Jour. Sed. Petrology, v. 54, p. 473–487.

Sass, E., and A. Bein, 1988, Dolomites and salinity: a comparative geochemical study, *in* Shukla, V., and P. A. Baker, eds., Sedimentology and geochemistry of dolostones: Soc. Econ. Paleontologists and Mineralogists Spec. Pub. 43, 223–233.

Sass, E., and A. Katz, 1982, The origin of platform dolomites: Am. Jour. Science, v. 282, p. 1184–1213.

Sassen, R., C. H. Moore, and F. C. Meendsen, 1987, Distribution of hydrocarbon source potential in the Jurassic Smackover Formation: Organic Geochemistry, v. 11, p. 379–383.

Savin, S. M., and H. W. Yeh, 1981, Stable isotopes in ocean sediments, *in* Emiliani, C., ed., The sea, v. 7: Wiley-Interscience, New York, p. 1521–1554.

Schäfer, A., and T. Teyssen, 1987, Size, shape and orientation of grains in sands and sandstones image analysis applied to rock thin-sections: Sed. Geology, v. 52, p. 251–271.

Schermerhorn, L. J. G., 1974, Late Precambrian mictites: glacial and/or nonglacial? Am. Jour. Science, v. 274, p. 673–824.

Schieber, J., 1987, Small scale sedimentary iron deposits in a mid-Proterozoic basin: viability of iron supply by rivers, *in* Appel, P. W. U., and G. L. LaBerge, eds., Precambrian iron-formations: Theophrastus Pub., S.A., Athens, Greece, p. 267–295.

Schlanger, S. O., and R. G. Douglas, 1974, The pelagic ooze-chalk-limestone transition and its implications for marine stratigraphy, *in* Hsü, K. J., and H. C. Jenkyns, eds., Pelagic sediments on land and under the sea: Internat. Assoc. Sedimentologists Spec. Pub. 1, p. 117–148.

Schlanger, W., and N. P. James, 1978, Low-magnesian calcite forming on the deep-sea floor, Tongue of the Ocean, Bahamas: Sedimentology, v. 25, p. 675–702.

Schlater, J. G., and P. A. F. Christie, 1980, Continental stretching: an explanation of the post-mid-Cretaceous subsidence of the central North Sea Basin: Jour. Geophys. Research, v. 85, p. 3711–3739.

Schlee, J., 1957, Fluvial gravel fabric: Jour. Sed. Petrology, v. 27, p. 162–176.

Schmalz, R. F., 1967, Kinetics and diagenesis in carbonate sediments: Jour. Sed. Petrology, v. 37, p. 60–68.

Schmid, R., 1981, Descriptive nomenclature and classification of pyroclastic deposits and fragments: Recommendations of the IUGS Subcommission on the Systematics of Igneous Rocks: Geology, v. 9, p. 41–43.

Schmidt, V., and D. A. McDonald, 1979a, The role of secondary porosity in the course of sandstone diagenesis, *in* Scholle, P. A., and P. R. Schluger, eds., Aspects of diagenesis: Soc. Econ. Paleontologists and Mineralogists Spec. Pub. 26, p. 175–207.

———1979b, Texture and recognition of secondary porosity in sandstones, *in* Scholle, P. A., and P. R. Schluger, eds., Aspects of diagenesis: Soc. Econ. Paleontologists and Mineralogists Spec. Pub. 26, p. 209–225.

Schmoker, J. W., 1984, Empirical relation between carbonate porosity and thermal maturity: an approach to regional porosity prediction: Am. Assoc. Petroleum Geologists Bull., v. 68, p. 1697–1703.

Schmoker, J. W., and R. B. Halley, 1982, Carbonate porosity versus depth: a predictable relation for south Florida: Am Assoc. Petroleum Geologists Bull., v. 66, p. 2561–2570.

Schneidermann, N., and P. M. Harris, 1985, Carbonate cements: Soc. Econ. Paleontologists and Mineralogists Spec. Pub. 36, 379 p.

Scholle, P. A., 1977, Chalk diagenesis and its relation to petroleum exploration—oil from chalks, a modern miracle?: Am. Assoc. Petroleum Geologists Bull., v. 61, p. 982–1009.

———1978, Carbonate rock constituents, textures, cements, and porosities: Am. Assoc. Petroleum Geologists Mem. 27, 241 p.

———1979, Constituents, textures, cements, and porosities of sandstones and associated rocks: Am. Assoc. Petroleum Geologists Mem. 28, 201 p.

Scholle, P. A., D. G. Bebout, and C. H. Moore, eds., 1983, Carbonate depositional environments: Am. Assoc. Petroleum Geologists Mem. 33, 708 p.

Scholle, P. A., and R. B. Halley, 1985, Burial diagenesis: out of sight, out of mind, *in* Schneidermann, N., and P. M. Harris, eds., Carbonate cements: Soc. Econ. Paleontologists and Mineralogists Spec. Pub. 36, p. 309–334.

Schopf, J. M., 1956, A definition of coal: Econ. Geology, v. 51, p. 521–527.

Schreiber, B. C., 1982, Environments of subaqueous gypsum deposition, *in* Dean, W. E., and B. C. Schreiber, eds., Marine evaporites: Soc. Econ. Paleontologists and Mineralogists Short Course 4, p. 43–73.

———1988a, Introduction, *in* Schreiber, B. C., ed., Evaporites and hydrocarbons: Columbia University Press, New York, p. 1–10.

———1988b, Evaporites and hydrocarbons: Columbia University Press, New York, 475 p.

———1988c, Subaqueous evaporite deposition, *in* Schreiber, B. C., ed., Evaporites and hydrocarbons: Columbia University Press, New York, p. 182–255.

Schreiber, B. C., and K. J. Hsü, 1980, Evaporites, *in* Hobson, G. D., ed., Developments in petroleum geology—2.: Applied Science Publishers, Barking, Essex, Great Britain, p. 87–138.

Schreiber, B. C., M. E. Tucker, and R. Till, l986, Arid shorelines and evaporites, *in* Reading, H. G., ed., Sedimentary environments and facies: Blackwell, p. 189–228.

Schroeder, J. H., 1972, Calcified filaments of an endolithic alga in Recent Bermuda Reefs: Neues Jahrb. Geologie u. Paläontologie Abh., v. 12, p. 16–33.

Schubel, K. A., and B. M. Simonson, 1990, Petrography and diagenesis of cherts from Lake Magadi, Kenya: Jour. Sed. Petrology, v. 60, p. 761–776.

Schultz, A. W., 1984, Subaerial debris-flow deposition in the upper Paleozoic Cutler Formation, western Colorado: Jour. Sed. Petrology, v. 54, p. 759–772.

Schwab, F. L., 1970, Origin of the Antietam Formation (Late Precambrian[?]-Lower Cambrian) central Virginia: Jour. Sed. Petrology, v. 40, p. 354–366.

———1975, Framework mineralogy and chemical composition of continental margin-type sandstone: Geology, v. 3, p. 487–490.

———1981, Evolution of the western continental margin, French-Italian Alps: Sandstone mineralogy as an index of plate tectonic setting: Jour. Geology, v. 89, p. 349–368.

Schwarcz, H. P., and K. C. Shane, 1969, Measurement of particle shape by Fourier analysis: Sedimentology, v. 13, p. 213–231.

Scoffin, T. P., 1987, Carbonate sediments and rocks: Blackie, Glasgow and London, 274 p.

Scotford, D. M., 1965, Petrology of the Cincinnatian Series shales and environmental implications: Geol. Soc. America Bull., v. 76, p. 193–222.

Searl, A., 1989, Pedogenic columnar calcite from the Oolite Group (Lower Carboniferous), south Wales: Sed. Geology, v. 62, p. 47–58.

Sedimentary Petrology Seminar, 1965, Gravel fabric in Wolf Run: Sedimentology, v. 4, p. 273–283.

Sedimentation Seminar, 1981, Comparison of methods of size analysis for sands of the Amazon and Solimões rivers, Brazil and Peru: Sedimentology, v. 28, p. 123–128.

Seiders, V. M., and C. D. Blome, 1988, Implications of upper Mesozoic conglomerate for suspect terrane in western California and adjacent areas: Geol. Soc. America Bull., v. 100, p. 374–391.

Seilacher, A., 1964, Biogenic sedimentary structures, *in* Imbrie, J., and N. D. Newell, eds., Approaches to paleoecology: John Wiley & Sons, New York, p. 296–315.

———1968, Origin and diagenesis of the Oriskany Sandstone (Lower Devonian, Appalachians) as reflected in its shell fossils; *in* Müller, G., and G. M. Friedman, eds., Recent developments in carbonate sedimentation in central Europe: Springer-Verlag, New York, p. 175–185.

Selley, R. C., 1978, Ancient sedimentary environments, 2nd ed.: Cornell Univ. Press, Ithaca, N.Y., 287 p.

———1985, Elements of petroleum geology: W. H. Freeman, New York, 449 p.

Sellwood, B. W., 1986, Shallow-marine carbonate environments, *in* Reading, H. G., ed., Sedimentary environments and facies, 2nd ed.: Blackwell, Oxford, p. 283–342.

Sellwood, B. W., ed., 1989, Zoned carbonate cements: techniques, applications and implications: Sed. Geology, v. 65, nos. 3/4, p. 205–377.

Sellwood, B. W., T. J. Shepherd, M. R. Evans, and B. James, 1989, Origin of late cements in oolitic reservoir facies: a fluid inclusion and isotopic study (Mid-Jurassic, southern England): Sed. Geology, v. 61, p. 223–237.

Sestini, G., 1970, Flysch facies and turbidite sedimentology: Sed. Geology, v. 4, p. 559–597.

Shackleton, N. J., 1967, Oxygen isotope analysis and paleotemperatures reassessed: Nature, v. 215, p. 15–17.

Shanks, W. C., W. E. Seyfried, W. C. Meyer, and T. J. O'Neil, 1976, Mineralogy of oil shale, *in* Yen, T. F., and G. V. Chilingarian, eds., Oil shale: Elsevier, Amsterdam, p. 81–102.

Shanmugam, G., 1984, Types of porosity in sandstones and their significance in interpreting provenance, *in* Zuffa, G. G., ed., Provenance of arenites: D. Reidel, Dordrecht, Holland, p. 115–137.

Shannon, P. M., 1978, The petrology of some Lower Paleozoic greywackes from southeast Ireland: A clue to the origin of matrix: Jour. Sed. Petrology, v. 48, p. 1185–1192.

Sharp, W. E., and G. C. Kennedy, 1965, The solution alteration of carbonate rocks, the effects of temperature and pressure: Jour. Geology, v. 73, p. 391–403.

Shaw, D. B., and C. E. Weaver, 1965, The mineralogical composition of shales: Jour. Sed. Petrology, v. 35, p. 213–222.

Shaw, D. M., 1956, Geochemistry of pelitic rocks. Part III: Major elements and general geochemistry: Geol. Soc. America Bull., v. 67, p. 919–934.

Shearman, D. J., 1982, Evaporites of coastal sabkhas. *in* Dean, W. E., and B. C. Schreiber, eds., Marine evaporites: Soc. Econ. Paleontologists and Mineralogists Short Course 4, p. 6–42.

Shearman, D. J., and A. J. Smith, 1985, Ikaite, the parent mineral of jarrowite-type pseudomorphs: Proc. Geol. Assoc., v. 96, p. 305–314.

Shearman, D. J., J. Twyman, and M. Z. Karmi, 1970, The diagenesis of oolites: Proc. Geol. Assoc., v. 81, p. 561–575.

Sheldon, R. P., 1989, Phosphorite deposits of the Phosphoria Formation, western United States, *in* Notholt, A. J. G., R. P. Sheldon, and D. F. Davidson, eds., Phosphate deposits of the world: Cambridge Univ. Press, Cambridge, p. 53–61.

Shelton, J. W., H. R. Burman, and R. L. Noble, 1974, Directional features in braided-meandering-stream deposits, Cimarron River, north-central Oklahoma: Jour. Sed. Petrology, v. 44, p. 1114–1117.

Shimkus, K. M., and E. S. Trimons, 1974, Modern sedimentation in the Black Sea, *in* Degens, E. T., and D. A. Ross, eds., The Black Sea—Geology, chemistry and biology: Am. Assoc. Petroleum Geologists Mem. 20, p. 249–278.

Shinn, E. A., 1969, Submarine lithification of Holocene carbonate sediments in the Persian Gulf: Sedimentology, v. 12, p. 109–144.

Shinn, E. A., and D. M. Robbin, 1983, Mechanical and chemical compaction in fine-grained shallow-water lime-stones: Jour. Sed. Petrology, v. 53, p. 595–618.

Shinn, E. A., R. P. Steinen, B. H. Lidz, and P. K. Swart, 1989, Whitings, a sedimentologic dilemma: Jour. Sed. Petrology, v. 59, p. 147–161.

Shipp, R. C., 1984, Bedforms and depositional sedimentary structures of a barred nearshore system, eastern Long Island, New York: Marine Geology, v. 60, p. 235–259.

Shukla, V., 1986, Epigenetic dolomitization and the origin of xenotopic dolomite texture: Jour. Sed. Petrology, v. 56, p. 733–734.

Sibley, D. F., 1990, Unstable to stable transformations during dolomitization: Jour. Geology, v. 98, p. 739–748.

Sibley, D. F., and H. Blatt, 1976, Intergranular pressure solution and cementation of the Tuscarora orthoquartzite: Jour. Sed. Petrology, v. 46, p. 881–896.

Sibley, D. F., and J. M. Gregg, 1987, Classification of dolomite rock textures: Jour. Sed. Petrology, v. 57, p. 967–975.

Sibley, D. F., and K. J. Pentony, 1978, Provenance variations in turbidite sediments, Sea of Japan: Jour. Sed. Petrology, v. 48, p. 1241–1248.

Siever, R., 1957, The silica budget in the sedimentary cycle: Am. Mineralogist, v. 42, p. 821–841.

———— 1962, Silica solubility, 0°C–200°C, and the diagenesis of siliceous sediments: Jour. Geology, v. 70, p. 127–150.

———— 1983, Evolution of chert at active and passive continental margins, *in* Iijima, A., J. R. Hein, and R. Siever, eds., Siliceous deposits in the Pacific region: Developments in Sedimentology 36, Elsevier, Amsterdam, p. 7–24.

Siever, R., and M. Kastner, 1972, Shale petrology by electron microprobe: Pyrite-chlorite relations: Jour. Sed. Petrology, v. 42, p. 350–355.

Simms, M., 1984, Dolomitization by groundwater-flow systems in carbonate platforms: Gulf Coast Assoc. Geol. Soc. Trans., v. 34, p. 411–420.

Simons, D. B., and E. V. Richardson, 1961, Forms of bed roughness in alluvial channels: Am. Soc. Civil Engineers Proc., Jour. Hydraulics Div., v. 87 (HY3), p. 87–105.

Simonson, B. M., 1985, Sedimentological constraints on the origins of Precambrian iron-formations: Geol. Soc. America Bull., v. 96, p. 244–252.

Simonson, B. M., and A. D. T. Goode, 1989, First discovery of ferruginous chert arenites in the early Precambrian Hamersley Group of Western Australia: Geology, v. 17, p. 269–272.

Simpson, S., 1975, Classification of trace fossils, *in* Frey, R. W., ed., The study of trace fossils: Springer-Verlag, New York, p. 39–54.

Sindowski, F. K. H., 1949, Results and problems of heavy mineral analysis in Germany; a review of sedimentary-petrological papers, 1936–1948: Jour. Sed. Petrology, v. 19, p. 3–25.

Singer, A., 1980, The paleoclimatic interpretation of clay minerals in soils and weathering profiles: Earth Science Rev., v. 15, p. 303–326.

———— 1984, The paleoclimatic interpretation of clay minerals in sediments—a review: Earth Science Rev., v. 21, p. 251–293.

Singer, J. K., J. B. Anderson, M. T. Ledbetter, I. N. McCave, K. P. N. Jones, and R. Wright, 1988, An assessment of analytical techniques for the size analysis of fine-grained sediments: Jour. Sed. Petrology, v. 58, p. 534–543.

Singer, A., and G. Müller, 1983, Diagenesis in argillaceous sediments, *in* Larsen, G., and G. V. Chilingar, eds., Diagenesis in sediments and sedimentary rocks, 2: Elsevier, Amsterdam, p. 115–212.

Sippel, R. F., 1968, Sandstone petrology, evidence from luminescence petrography: Jour. Sed. Petrology, v. 38, p. 530–554.

———— 1971, Quartz grain orientations—1 (the photometric method): Jour. Sed. Petrology, v. 41, p. 38–58.

Slansky, M., 1986, Geology of sedimentary phosphates: North Oxford Academic Pub., Essex, Great Britain, 210 p.

Sloane, R. L., and T. F. Kell, 1966, The fabric of mechanically compacted kaolinite: Clays and Clay Minerals, v. 14, p. 289–296.

Sloss, L. L., and D. E. Feray, 1948, Microstylolites in sandstone: Jour. Sed. Petrology, v. 18, p. 3–13.

Smalley, I. J., 1964a, A method for describing the packing texture of clastic sediments: Nature, v. 203, p. 281–284.

———— 1964b, Representation of packing in a clastic sediment: Am. Jour. Sci., v. 262, p. 242–248.

Smalley, I. J., and J. G. Cabrera,

1969, Particle association in compacted kaolinite: Nature, v. 222, p. 80–81.

Smith, J. V., 1974, Feldspar minerals, v. 1 and 2: Springer-Verlag, New York.

Smith, N. D., 1974, Sedimentology and bar formation in the upper Kicking Horse river, a braided outwash stream: Jour. Geology, v. 82, p. 205–224.

Smith, S. A., 1990, The sedimentology and accretionary styles of an ancient gravel-bed stream: The Budleigh Salterton Pebble Beds (Lower Triassic), southwest England: Sed. Geology, v. 67, p. 199–220.

Smosna, R., 1987, Compositional maturity of limestones—a review: Sed. Geology, v. 51, p. 137–146.

——1989, Compaction law for Cretaceous sandstones of Alaska's North Slope: Jour. Sed. Petrology, v. 59, p. 572–584.

Sneed, E. D., and R. L. Folk, 1958, Pebbles in the lower Colorado River, Texas, a study in particle morphogenesis: Jour. Geology, v. 66, p. 114–150.

Sonnenfeld, P., and C. G. St.C. Kendall, conveners, 1989, Marine evaporites: genesis, alteration, and associated deposits: Penrose Conference Rept., Geology, v. 17, p. 573–574.

Sorby, H. C., 1880, On the structure and origin of non-calcareous stratified rocks: Geol. Soc. London Proc., v. 36, p. 46–92.

Soudry, D., and P. N. Southgate, 1989, Ultrastructure of a middle Cambrian primary nonpelletal phosphorite and its early transformation into phosphate vadoids: Georgina Basin, Australia: Jour. Sed. Petrology, v. 59, p. 53–64.

Southard, J. B., and L. A. Boguchwal, 1990, Bed configurations in steady unidirectional water flows. Part 2. Synthesis of flume data: Jour. Sed. Petrology, v. 60, p. 658–679.

Spears, D. A., 1976, The fissility of some Carboniferous shales: Sedimentology, v. 23, p. 721–725.

——1980, Toward a classification of shales: Jour. Geol. Soc. London, v. 137, p. 125–129.

Speer, J. A., 1983, Crystal chemistry and phase relations of orthorhombic carbonates, *in* Reeder, R. J., ed., Carbonates: Mineralogy and chemistry: Mineralog. Soc. America Rev. Mineralogy, v. 11, p. 145–190.

Spotts, J. H., 1964, Grain orientation and imbrication in Miocene turbidity current sandstones, California: Jour. Sed. Petrology, v. 34, p. 229–253.

Stablein, N. K., III, and E. C. Dapples, 1977, Feldspars of the Tunnel City Group (Cambrian), western Wisconsin: Jour. Sed. Petrology, v. 47, p. 1512–1538.

Stach, E., M.-Th. Mackowsky, M. Teichmüller, G. H. Taylor, D. Chandra, and R. Teichmüller, 1982, Coal petrology, 3rd ed.: Gebrüder Borntraeger, Berlin-Stuttgart, 535 p.

Stattegger, K., 1987, Heavy minerals and provenance of sand: modeling of lithologic end members from river sands of northern Austria and from sandstones of the Austroalpine Gosau Formation (Late Cretaceous): Jour. Sed. Petrology, v. 57, p. 301–310.

Stauffer, P. H., 1967, Grain-flow deposits and their implications, Santa Ynez Mountains, California: Jour. Sed. Petrology, v. 37, p. 487–508.

Stein, R., 1985, Rapid grain-size analyses of clay and silt fraction by Sedigraph 5000D: Comparison with Coulter counter and Atterberg methods: Jour. Sed. Petrology, v. 55, p. 590–593.

Stephenson, L. P., 1977, Porosity dependence on temperature: limits on maximum possible effect: Am. Assoc. Petroleum Geologists Bull., v. 61, p. 407–415.

Stewart, F. H., 1963, Marine evaporites, *in* Fleischer, M., ed., Data of geochemistry: U.S. Geol. Survey Prof. Paper 440-Y, 54 p.

Stewart, R. J., 1976, Turbidites of the Aleutian abyssal plain: mineralogy, provenance, and constraints for Cenozoic motion of the Pacific Plate: Geol. Soc. America Bull., v. 87, p. 793–808.

Stoesser, R. K., and C. H. Moore, 1983, Chemical constraints and origins of four groups of Gulf Coast reservoir fluids: Am. Assoc. Petroleum Geologists Bull., v. 67, p. 896–906.

Stoffers, P., and G. Müller, 1962, Clay mineralogy of Black Sea sediments: Sedimentology, v. 18, p. 113–121.

Stokes, S., C. S. Nelson, and T. R. Healy, 1989, Textural procedures for the environmental discrimination of late Neogene coastal sand deposits, southwest Auckland, New Zealand: Sed. Geology, v. 61, p. 135–150.

Stopes, M. C., 1919, On the four visible ingredients in banded bituminous coal. Studies in the composition of coal: Royal Soc. London Proc., Ser. B., v. 90, p. 470–487.

——1935, On the petrology of banded bituminous coal: Fuel, London, v. 14, p. 4–13.

Stow, D. A. V., and A. J. Bowen, 1978, Origin of lamination in deep sea, fine-grained sediments: Nature, v. 274, p. 324–328.

——1980, A physical model for the transport and sorting of fine-grained sediment by turbidity currents: Sedimentology, v. 27, p. 31–46.

Stow, D. A. V., and D. J. W. Piper, 1984, Deep-water fine-grained sediments: History, methodology and terminology, *in* Stow, D. A. V., and D. J. W. Piper, eds., Fine-grained sediments: Geol. Soc. Spec. Pub. 15, Blackwell, Oxford, p. 3–14.

Strasser, A., E. Davaud, and Y. Jedouri, 1989, Carbonate cements in Holocene beachrock: examples from Bahiret el Biban, southeastern Tunisia: Sed. Geology, v. 62, p. 89–100.

Sudo, T., S. Shimoda, H. Yotsumoto, and S. Aita, 1981, Electron micrographs of clay minerals: Elsevier, Amsterdam, 203 p.

Suess, E., 1909, Das Antlitz der Erde: Leipzig, Freytag, v. 3, pt. 2, 789 p.

Suess, E., W. Balzer, K.-F. Hesse, P. J. Müller, P. J. Ungerer, and G. Wefer, 1982, Calcium carbonate hexahydrate from organic-rich sediments of the Antarctic shelf: Precursor of Glendonites: Science, v. 1216, p. 1128–1131.

Sugisaki, R., 1984, Relation between chemical composition and sedimentation rate of Pacific ocean-floor sediments deposited since the middle Cretaceous: Basic evidence for chemical constraints on depositional environments of ancient sediments: Jour. Geology, v. 92, p. 235–260.

Summerhayes, C. P., and N. J. Shackleton, eds., 1986, North Atlantic palaeoceanography: Geol. Soc. Spec. Pub. 21, Blackwell, Oxford, 473 p.

Sun, S. Q., and V. P. Wright, 1989, Peloidal fabrics in Upper Jurassic reefal limestones, Weald Basin, southern England: Sed. Geology, v. 65, p. 165–181.

Surdam, R. C., 1981, Zeolites in closed hydrologic systems, *in* Mumpton, F. A., ed., Mineralogy and geology of natural zeolites: Mineralog. Soc. America, Rev. Mineralogy, v. 4, p. 65–91.

Surdam, R. C., S. W. Boese, and L. J. Crossey, 1984, The chemistry of secondary porosity, *in* McDonald, D. A., and R. C. Surdam, eds., Clastic diagenesis: Am. Assoc. Petroleum Geologists Mem. 37, p. 127–151.

Surdam, R. C., and J. R. Boles, 1979, Diagenesis of volcanic sandstones, *in* Scholle, P. A., and P. R. Schluger, eds., Aspects of diagenesis: Soc. Econ. Paleontologists and Mineralogists Spec. Pub. 26, p. 227–242.

Surdam, R. C., L. J. Crossey, E. S. Hagen, and H. P. Heasler, 1989, Organic-inorganic interactions and sandstone diagenesis: Am. Assoc. Petroleum Geologists Bull., v. 73, p. 1–23.

Surdam, R. C., T. L. Dunn, H. P. Heasler, and D. B. MacGowan, 1989, Porosity evolution in sandstone/shale systems, *in* Hutcheon, I. E., ed., Burial diagenesis: v. 15, Mineralog. Assoc. Canada Short Course Handb., p. 61–134.

Suttner L. J., 1969, Stratigraphic and petrographic analysis of Upper Jurassic-Lower Cretaceous Morrison and Kootenai formations, southwest Montana: Am. Assoc. Petroleum Geologists Bull., v. 53, p. 1391–1410.

Suttner, L. J., and A. Basu, 1977, Structural state of detrital alkali feldspars: Sedimentology, v. 24, p. 63–74.

——— 1985, The effects of grain size on detrital modes: A test of the Gazzi-Dickinson point-counting method: Jour. Sed. Petrology, v. 55, p. 616–617.

Suttner, L. J., A. Basu, and G. H. Mack, 1981, Climate and the origin of quartz arenites: Jour. Sed. Petrology, v. 51, p. 1235–1246.

Suttner, L. J., and P. K. Dutta, 1986, Alluvial sandstone composition and paleoclimate, I. Framework mineralogy: Jour. Sed. Petrology, v. 56, p. 329–345.

Suttner, L. J., and R. K. Leninger, 1972, Comparison of trace element content of plutonic, volcanic, and metamorphic quartz from southwestern Montana: Geol. Soc. America Bull., v. 83, p. 1855–1862.

Sutton, R. G., and T. L. Lewis, 1966, Regional patterns of cross-laminae and convolution in a single bed: Jour. Sed. Petrology, v. 36, p. 225–229.

Swett, K., and H. H. Knoll, 1989, Marine pisolites from upper Proterozoic carbonates of east Greenland and Spitsbergen: Sedimentology, v. 36, p. 75–93.

Swift, D. J. P., 1971, Grain mounts, *in* Carver, R. E., ed., Procedures in sedimentary petrology: John Wiley & Sons, New York, p. 499–510.

Swift, D. J. P., and D. Nummedal, 1987, Hummocky cross-stratification, tropical hurricanes and intense winter storms: Discussion: Sedimentology, v. 34, p. 338–344.

Swift, D. J. P., J. R. Schubel, and R. E. Sheldon, 1972, Size analysis of fine-grained suspended sediments: A review: Jour. Sed. Petrology, v. 42, p. 122–134.

Swineford, A., 1955, Petrography of upper Permian rocks in south-central Kansas: Kansas State Geol. Survey Bull. III, 179 p.

Taira, A., and B. R. Lienert, 1979, The comparative reliability of magnetic, photometric, and microscopic methods of determining the orientations of sedimentary grains: Jour. Sed. Petrology, v. 49, p. 759–772.

Taira, A., and P. A. Scholle, 1979, Discrimination of depositional environments using settling tube data: Jour. Sed. Petrology, v. 49, p. 787–800.

Tanner, W. F., 1983, Hydrodynamic origin of the Gaussian grain size distribution, *in* Turner, W. F., ed., Nearshore sedimentology: Proceedings of 6th Symposium on coastal sedimentology.

Taylor, J. M., 1950, Pore-space reduction in sandstones: Am. Assoc. Petroleum Geologists Bull., v. 34, p. 701–716.

Teichert, C., 1970, Oolite, oolith, ooid: Discussion: Am. Assoc. Petroleum Geologists Bull., v. 54, p. 1748–1749.

Telford, R. W., M. Lyons, J. D. Orford, W. B. Whalley, and D. Q. M. Fay, 1987, A low-cost, microcomputer-based image analyzing system for characterization of particle outline morphology, *in* Marshall, J. R., ed., Clastic particles: Van Nostrand Reinhold, New York, p. 281–289.

Textoris, D. A., 1971, Grain-size measurements in thin-section, *in* Carver, R. E., ed., Procedures in sedimentary petrology: John Wiley & Sons, New York, p. 95–107.

Thiel, G. A., 1935, Sedimentary and petrographic analysis of the St. Peter Sandstone: Geol. Soc. America Bull., v. 46, p. 559–614.

Tickell, F. G., 1965, The techniques of sedimentary mineralogy: Elsevier, Amsterdam, 220 p.

Till, R., 1974, Statistical methods for the earth scientist: an introduction: John Wiley & Sons, New York, 154 p.

Till, R., and D. A. Spears, 1969, The determination of quartz in sedimentary rocks using an x-ray diffraction method: Clays and Clay Minerals, v. 17, p. 323–327.

Ting, F. T. C., 1982, Coal macerals, *in* Meyer, R. A., ed., Coal structure: Academic Press, New York, p. 7–49.

Tissot, B. P., and D. H. Welte, 1984, Petroleum formation and occurrence, 2nd ed.: Springer-Verlag, Berlin, 699 p.

Toby, A. C., 1962, Characteristic patterns of plagioclase twinning: Norsk geol. tidsskr., v. 42, p. 264–271.

Tolstov, G. P., 1976, Fourier series: Dover, New York, 336 p.

Tourtelet, H. A., 1960, Origin and use of the word "shale": Am. Jour. Sc., Bradley volume, v. 258-A, p. 335–343.

Trendall, A. F., 1983, Introduction, *in* Trendall, A. F., and R. C. Morris, eds., Iron-formation: Facts and problems: Elsevier, Amsterdam, p. 1–12.

Trendall, A. F., and J. G. Blockley, 1970, The iron formations of the Precambrian Hamersley Group, Western Australia: Geol. Survey Western Australia Bull., v. 119, p. 1–136.

Trevena, A. S., and W. P. Nash, 1979, Chemistry and provenance of detrital feldspars: Geology, v. 7, p. 475–478.

——— 1981, An electron microprobe study of detrital feldspar: Jour. Sed. Petrology, v. 51, p. 137–150.

Trurnit, P., 1968, Analysis of pressure-solution contacts and classification of pressure solution phenomena, *in* Müller, G., and G. M. Friedman, eds., Carbonate sedimentology in central Europe: Springer-Verlag, New York, p. 75–84.

Tucker, M. E., 1982, Field description of sedimentary rocks: Open University Press and Halsted Press (John Wiley & Sons), New York, Toronto, 112 p.

Tucker, M., ed., 1988, Techniques in sedimentology: Blackwell, Oxford, 394 p.

Tucker, M. E., and V. P. Wright, 1990, Carbonate sedimentology: Blackwell, Oxford, 482 p.

Tucker, R. W., and H. L. Vacher, 1980, Effectiveness of discriminating beach, dune, and river sands by moments and the cumulative weight percentages: Jour. Sed. Petrology, v. 50, p. 165–172.

Turekian, K. H., and K. H. Wedepohl, 1961, Distribution of elements in some major units of the earth's crust: Geol. Soc. America Bull., v. 72, p. 175–192.

Turner, F. J., 1951, Observations on twinning of plagioclase in metamorphic rocks: Am. Mineralogist, v. 36, p. 581–589.

Turner, F. J., and L. E. Weiss, 1963, Structural analysis of metamorphic tectonites: McGraw-Hill, New York, 545 p.

Turner, J. V., 1982, Kinetic fractionation of carbon-13 during calcium carbonate precipitation: Geochim. et Cosmochim. Acta, v. 46, p. 1183–1192.

Tuttle, O. F., 1952, Origin of the contrasting mineralogy of extrusive and plutonic salic rocks: Jour. Geology, v. 60, p. 107–124.

Tweto, O., and T. S. Lovering, 1977, Geology of the Minturn 15-minute Quadrangle, Eagle and Summit counties, Colorado: U.S. Geol. Survey Prof. Paper 956, 96 p.

Tyler, S. A., E. S. Berghoorn, and L. P. Barrett, 1957, Anthracitic coal from Precambrian upper Huronian black shale of the Iron River district, northern Michigan: Geol. Soc. America Bull., v. 68, p. 1293–1304.

Unrug, R., 1957, Recent transport and sedimentation of gravels in the Dunajec Valley (western Carpathians): Acta Geol. Polon., v. 7, p. 217–257.

Vahrenkamp, V. C., and P. K. Swart, 1990, New distribution coefficients for the incorporation of strontium into dolomite and its implications for the formation of ancient dolomites: Geology, v. 18, p. 387–391.

Valloni, R., 1985, Reading provenance from modern marine sands, in Zuffa, G. G., ed., Provenance of arenites: D. Reidel, Dordrecht, p. 309–332.

Valloni, R., and J. B. Maynard, 1981, Detrital modes of recent deep-sea sands and their relation to tectonic setting: A first approximation: Sedimentology, v. 28, p. 75–83.

Van Andel, Tj. H., 1959, Reflections on the interpretation of heavy mineral analyses: Jour. Sed. Petrology, v. 29, p. 153–163.

Van Andel, Tj. H., and D. M. Poole, 1960, Sources of recent sediments in the northern Gulf of Mexico: Jour. Sed. Petrology, v. 30, p. 91–122.

Van de Kamp, P. C., B. E. Leake, and A. Senior, 1976, The petrology and geochemistry of some Californian arkoses with application to identifying gneisses of metasedimentary origin: Jour. Geology, v. 84, p. 195–212.

Van Houten, F. B., 1982, Phanerozoic oolitic ironstones—Geologic record and facies models: Ann. Rev. Earth and Planetary Science Letters, v. 10, p. 441–457.

Van Houten, F. B., and M. E. Purucker, 1984, Glauconitic peloids and chamositic ooids—favorable factors, constraints, and problems: Earth Science Rev., v. 20, p. 211–243.

van Olphen, H., 1977, An introduction to clay colloid chemistry: John Wiley & Sons, New York.

Vandenberghe, N., 1975, An evaluation of CM patterns for grain-size studies of fine grained sediments: Sedimentology, v. 22, p. 615–622.

Vavra, C. L., 1989, Mineral reactions and controls on zeolite-facies alterations in sandstones of the central Transantarctic Mountains, Antarctica: Jour. Sed. Petrology, v. 59, p. 688–703.

Veizer, J., 1983, Trace elements and isotopes in sedimentary rocks, in Reeder, R. J., ed., Carbonates: mineralogy and chemistry: Mineralog. Soc. America Rev. Mineralogy, v. 11, p. 265–299.

Veizer, J., and J. Hoefs, 1976, The nature of O^{18}/O^{16} and C^{13}/C^{12} secular trends in sedimentary carbonate rocks: Geochim. et Cosmochim. Acta, v. 40, p. 1387–1395.

Velde, B., 1983, Diagenetic reactions in clays, in Parker, A., and B. W. Sellwood, eds., Sediment diagenesis, D. Reidel, Dordrecht, Holland, p. 215–268.

————1985, Clay minerals: A physico-chemical explanation of their occurrence: Elsevier, Amsterdam, 428 p.

Vincent, P., 1986, Differentiation of modern beach and coastal dune sands—a logistic regression approach using parameters of the hyperbolic function: Sed. Geology, v. 49, p. 167–176.

Vinopal, R. J., and A. H. Coogan, 1978, Effect of particle shape on the packing of carbonate sands and gravels: Jour. Sed. Petrology, v. 48, p. 7–24.

Visher, G. S., 1969, Grain size distributions and depositional processes: Jour. Sed. Petrology, v. 39, p. 1074–1106.

Von Der Borch, C. C., and J. B. Jones, 1976, Spherular modern dolomite from the Coorong area, South Australia: Sedimentology, v. 23, p. 587–591.

Waag, C. J., and D. E. Ogren, 1984, Shape evolution and fabric in a boulder beach, Monument Cove, Maine: Jour. Sed. Petrology, v. 54, p. 98–102.

Wadell, H., 1932, Volume, shape and roundness of rock particles: Jour. Geology, v. 40., p. 443–451.

Wagoner, J. L., and J. L. Younker, 1982, Characterization of alluvial sources in the Owens Valley of eastern California using Fourier shape analysis: Jour. Sed. Petrology, v. 52, p. 209–214.

Walderhaug, O. 1990, A fluid inclusion study of quartz-cemented sandstones from offshore mid-Norway: Jour. Sed. Petrology, v. 60, 203–210.

Walkden, G. M., and J. R. Berry, 1984, Syntaxial overgrowths in muddy crinoidal limestones: Cathodoluminescence sheds new light on an old problem: Sedimentology, v. 31, p. 251–267.

Walker, G. P. L., 1973, Explosive volcanic eruptions—a new classification scheme: Geol. Rundschau, v. 62, p. 431–446.

Walker, R. G., 1975a, Conglomerate: sedimentary structure and facies models, in Harms, J. C., J. B. Southard, D. R. Spearing, and R. G. Walker, eds., Depositional environments as interpreted from primary sedimentary structures and stratification sequences: Soc. Econ. Paleontologists and Mineralogists Short Course Notes 2, p. 133–161.

————1975b, Generalized facies models for resedimented conglomerates of turbidite association: Geol. Soc. America Bull., v. 86, p. 737–748.

————1977, Deposition of upper Mesozoic resedimented conglomerates and associated turbidites in southwestern Oregon: Geol. Soc. America Bull., v. 88, p. 273–285.

————1984, Turbidites and associated coarse clastic deposits, in Walker, R. G., ed., Facies models, 2nd ed., Geoscience Canada Reprint Ser. 1, p. 171–188.

Walker, T. R., 1962, Reversible nature of chert-carbonate replacement in sedimentary rocks: Geol. Soc. America Bull., v. 73, p. 237–242.

————1967, Formation of red beds in modern and ancient deserts. Geol. Soc. America Bull., v. 78, p. 353–368.

————1974, Formation of red beds in moist tropical climates: A hypothesis: Geol. Soc. America Bull., v. 85, p. 633–638.

————1984, Diagenetic albitization of potassium feldspars in arkosic sandstones: Jour. Sed. Petrology, v. 54, p. 3–16.

Walker, T. R., B. Waugh, and

A. J. Grone, 1978, Diagenesis in first-cycle desert alluvium of Cenozoic age, southwestern United States and northwestern Mexico: Geol. Soc. America Bull., v. 89, p. 19–32.

Wallace, M. W., 1987, The role of internal erosion and sedimentation in the formation of stromatactis mudstones and associated lithologies: Jour. Sed. Petrology, v. 57, p. 695–700.

Walton, A. W., 1986, Recognition and significance of inherited quartz overgrowths in quartz arenites: Discussion: Jour. Sed. Petrology, v. 56, p. 317–318.

Wanless, H. R., 1979, Limestone response to stress: Pressure solution and dolomitization: Jour. Sed. Petrology, v. 49, p. 437–462.

Waples, D. W., 1980, Time and temperature in petroleum formation: application of Lepatin's method to petroleum exploration: Am. Assoc. Petroleum Geologists Bull., v. 64, p. 916–926.

———1983, Reappraisal of anoxia and organic richness, with emphasis on Cretaceous of North America: Am. Assoc. Petroleum Geologists Bull., v. 67, p. 963–978.

Ward, C. R., ed., 1984, Coal geology and coal technology: Blackwell, Melbourne, 345 p.

Wardlaw, N., A. Oldlershaw, and M. Stout, 1978, Transformation of aragonite to calcite in a marine gastropod: Canadian Jour. Earth Science, v. 15, p. 1861–1866.

Warren, J. K., 1989, Evaporite sedimentology: Prentice Hall, Englewood Cliffs, N.J., 285 p.

———1990, Sedimentology and mineralogy of dolomitic Coorong lakes, South Australia: Jour. Sed. Petrology, v. 60, p. 843–858.

Warren, J. K., and G. C. St. C. Kendall, 1985, Comparison of marine sabkhas (subaerial) and salina (subaqueous) evaporites: Modern and ancient: Am. Assoc. Petroleum Geologists Bull., v. 69, p. 1013–1023.

Waugh, B., 1978, Authigenic K-feldspars in British Permo-Triassic sandstones: Jour. Geol. Soc. London, v. 135, p. 51–56.

Weaver, M., and S. W. Wise, Jr., 1974, Opaline sediments of the southeastern coastal plain and horizon A: Biogenic origin: Science, v. 184, p. 899–901.

Wells, N. A., 1984, Sheet debris flow and sheetflood conglomerates in Cretaceous cool-marine alluvial fans, South Orkney Islands, Antarctica, in Ko-

ster, E. H., and R. J. Steel, eds., Sedimentology of gravels and conglomerates: Canadian Soc. Petroleum Geologists Mem. 10, p. 133–145.

Wentworth, C. K., 1922, A scale of grade and class terms for clastic sediments: Jour. Geology, v. 30, p. 377–392.

Wentworth, C. K., and H. Williams, 1932, The classification and terminology of the pyroclastic rocks: Natl. Research Council Bull., v. 89, p. 19–53.

Wescott, W. A., and F. G. Ethridge, 1983, Eocene fan delta-submarine fan deposition in the Wagwater Trough, east-central Jamaica: Sedimentology, v. 30, p. 235–245.

Wetzel, A., 1989, Influence of heat flow on ooze/chalk cementation: Quantification from consolidation parameters in DSDP sites 504 and 505 sediments: Jour. Sed. Petrology, v. 59, p. 539–547.

Whalley, W. B., and J. D. Orford, 1986, Practical methods for analysing and quantifying two-dimensional images, in Sieveking, C. De C., and M. B. Hart, eds., The scientific study of flint and chert, Cambridge University Press, Cambridge, UK, p. 235–242.

Whetton, J. T., and J. W. Hawkins, 1970, Diagenetic origin of greywacke matrix minerals: Sedimentology, v. 15, p. 347–361.

Whisonant, R. C., 1987. Paleocurrent and petrographic analysis of imbricate intraclasts in shallow-marine carbonates, Upper Cambrian, southwestern Virginia: Jour. Sed. Petrology, v. 57, p. 983–994.

Whitaker, F. F., and P. L. Smart, 1990, Active circulation of saline ground water in carbonate platforms: Evidence from the Great Bahama Bank: Geology, v. 18, p. 200–203.

White, D. E., 1965, Saline waters of sedimentary rocks, in Young, A. and J. E. Galley, eds., Fluids in subsurface environments: Am. Assoc. Petroleum Geologists Mem. 4, p. 342–366.

White, D. E., J. D. Hem, and G. A. Waring, 1963, Data of geochemistry, Chapter F, Chemical composition of subsurface waters: U.S. Geol. Survey Prof. Paper 440-F, 67 p.

White, S. H., H. F. Shaw, and J. H. Huggett, 1984, The use of backscattered electron imaging for the petrographic study of sandstones and shales: Jour. Sed. Petrology, v. 54, p. 487–495.

Wiegman, J., C. H. Horte, and

G. Kranz, 1982, Determination of the complete mineral composition of clays, in van Olphen, H., and F. Veniale, eds., Internat. Clay Conference 1981, Developments in Sedimentology 35: Elsevier, Amsterdam, p. 365–372.

Wiesnet, D. R., 1961, Composition, grain size, roundness, and sphericity of the Potsdam Sandstone (Cambrian), in northeastern New York: Jour. Sed. Petrology, v. 31, p. 5–14.

Wilkinson, B. H., 1979, Biomineralization, paleooceanography, and the evolution of calcareous marine organisms: Geology, v. 7, p. 524–527.

Willey, J. D. 1974, The effect of pressure on the solubility of amorphous silica in seawater at 0°C: Marine Chemistry, v. 2, p. 239–250.

Williams, H., and A. R. McBirney, 1979, Volcanology: Freeman, Cooper & Co., San Francisco, 397 p.

Williams, H., F. J. Turner, and C. M. Gilbert, 1982, Petrography: An introduction to the study of rocks in thin sections, 2nd ed.: W. H. Freeman, San Francisco, 626 p.

Williams, L. A., and D. A. Crerar, 1985, Silica diagenesis, II. General mechanisms: Jour. Sed. Petrology, v. 55, p. 312–321.

Williams, L. A., G. A. Parks, and D. A. Crerar, 1985, Silica diagenesis, I. Solubility controls: Jour. Sed. Petrology, v. 55, p. 301–311.

Wilson, J. C., and E. F. McBride, 1988, Compaction and porosity evaluation of Pliocene sandstones, Ventura Basin, California: Am. Assoc. Petroleum Geologists Bull., v. 72, p. 664–681.

Wilson, J. E., 1975, Carbonate facies in geologic history: Springer-Verlag, New York, 471 p.

Wilson, M. D., 1970, Upper Cretaceous-Paleocene synorogenic conglomerates of southwestern Montana: Am. Assoc. Petroleum Geologists Bull., v. 54, p. 1843–1867.

Wilson, M. D., and E. D. Pittman, 1977, Authigenic clays in sandstones: recognition and influence on reservoir properties and paleoenvironmental analysis: Jour. Sed. Petrology, v. 47, p. 3–31.

Wilson, M. J., ed., 1987a, A handbook of determinative methods in clay mineralogy: Blackie, Glasgow and London (Chapman and Hall, New York), 308 p.

Wilson, M. J., 1987b, X-ray powder diffraction methods, in Wilson, M. J., ed., A handbook of determinative methods in clay mineralogy: Blackie, Glasgow and London (Chapman and Hall, New York), p. 26–98.

Windley, B. F., 1977, The evolving continents: John Wiley & Sons, New York, 385 p.

Winklemolen, A. M., 1972, Dielectric anisotropy and grain orientation: Am. Assoc. Petroleum Geologists Bull., v. 56, p. 2150–2159.

——— 1982, Critical remarks on grain parameters, with special emphasis on shape: Sedimentology, v. 29, p. 255–265.

Winland, H. D., and R. K. Matthews, 1974, Origin and significance of grapestone, Bahama Island: Jour. Sed. Petrology, v. 44, p. 921–927.

Wise, S. W., Jr., and K. R. Kelts, 1972, Inferred diagenetic history of a weakly silicified deep sea chalk: Gulf Coast Assoc. Geol. Soc. Trans., v. 22, p. 177–203.

Wohletz, K. H., 1983, Mechanisms of hydrovolcanic pyroclast formation: grain-size, scanning electron microscopy, and experimental studies: Jour. Volcanology and Geothermal Research, v. 17, p. 31–63.

Wolf, K. H., and G. V. Chilingarian, 1976, Diagenesis of sandstones and compaction, *in* Chilingarian, G. V., and K. H. Wolf, eds., Compaction of coarse-grained sediments, II: Elsevier, Amsterdam, p. 69–444.

Wolff, R. G., and N. K. Huber, 1973, The Copper Harbor Conglomerate (Middle Keweenawan) on Isle Royale, Michigan, and its regional implications:

U. S. Geol. Survey, Prof. Paper 754-B, p. B1–B15.

Woodland, B. G., 1964, The nature and origin of cone-in-cone structure: Fieldiana: Geology, v. 13, p. 185–305.

Wright, J. V., A. L. Smith, and S. Self, 1980, A working terminology of pyroclastic deposits: Jour. Volcanology and Geothermal Research, v. 8, p. 315–336.

Wright, T. L., and D. B. Stewart, 1968, X-ray and optical study of alkali feldspars: I: Determination of structural state from refined unit-cell parameters and 2 V: Am. Mineralogist, v. 53, p. 38–87.

Wyatt, M., 1954, Zircons as provenance indicators: Am. Mineralogist, v. 39, p. 983–990.

Wyllie, P. J., 1976, The way the Earth works: John Wiley & Sons, New York, 296 p.

Wyrwoll, K.-H., and G. K. Smyth, 1985, On using the log-hyperbolic distribution to describe the textural characteristics of eolian sediments: Jour. Sed. Petrology, v. 55, p. 471–478.

Yaalon, D. H., 1962, Mineral composition of the average shale: Clay Min. Bull., v. 5, p. 31–36.

Yemel'yanov, Ye. M., 1975, Organic carbon in Atlantic sediments: Doklady Acad. Sci. USSR (English translation), v. 220, p. 220–223.

Yen, T. F., and G. V. Chilingar, 1976, Introduction to oil shales, *in* Yen, T. F., and G. V. Chilingarian, eds., Oil shale: Elsevier, Amsterdam, p. 1–12.

Yen, T. F., and G. V. Chilingarian, eds., 1976, Introduction to oil shales: Elsevier, Amsterdam, 292 p.

Yim, W. W.-S., A. J. W. Gleadow, and J. C. van Moort, 1985, Fission track dating of alluvial zircons and heavy mineral provenance in northeast Tasmania: Jour. Geol. Soc. London, v. 142, p. 351–356.

Young, S. W., 1976, Petrographic textures of detrital polycrystalline quartz as an aid to interpreting crystalline source rocks: Jour. Sed. Petrology, v. 46, p. 595–603.

Young, S. W., A. Basu, G. Mack, N. Darnell, and L. J. Suttner, 1975, Use of size-composition trends in Holocene soil and fluvial sand for paleoclimatic interpretation: Proc. IXth Internat. Cong. Sedimentation, Th. 1, Nice, France.

Zenger, D. H., 1989, Dolomite abundance and stratigraphic age: Constraints on rates and mechanisms of Phanerozoic dolostone formation—Discussion: Jour. Sed. Petrology, v. 59, p. 162–164.

Zimmerle, W., and L. C. Bonham, 1962, Rapid methods for dimensional grain orientation measurements: Jour. Sed. Petrology, v. 32, p. 751–763.

Zingg, Th., 1935, Beiträge zur Schotteranalyse: Schweiz. Mineralog. Petrog. Mitt., v. 15, p. 39–140.

Zuffa, G. G., 1980, Hybrid arenites: their composition and classification: Jour. Sed. Petrology, v. 50, p. 21–29.

Author Index

Subject Index

Aggregate grains (carbonate), 434, 435, 436
Albite, 128, 134, 136, 378, 391, 397, 400, 406, 407
Albitization, 393, 396, 397, 399, 400
Amphiboles, 128, 138, 139, 179, 329
Analcime, 391, 404, 405, 406, 407
Andalusite, 138, 139
Anhydrite, 128, 171, 280, 292, 378, 390, 397, 400, 416, 468, 469, 539, 554, 567, 569, 570–75, 585
Ankerite, 139, 280, 372, 378, 381, 415, 485, 603, 606
Anorthoclase, 128, 134, 135
Anoxic conditions, 268, 272, 273, 296, 297, 364, 369, 644, 645, 649
Apatite, 128, 138, 139, 141, 169, 187, 280, 554, 621, 626
Aragonite, 128, 378, 379, 382, 392, 400, 401, 412, 413, 415, 416, 435, 440, 441, 442, 467, 470, 471, 472, 473, 496, 498, 499, 515, 517, 518, 519, 522, 524–30, 532, 534, 536, 537, 552, 553, 555
Arkose, 164, 165, 184, 185
Aulacogens, 13, 15, 16, 17
Authigenic minerals, 146–53

Bacterial activity, 365, 368, 369, 373, 379, 380, 502, 644
Bafflestone, 449, 450, 468
Barite, 128, 139, 171, 280, 378, 390
Basin analysis, 30
Basins, 13, 15–23
Beachrock, 521, 529
Bindstone, 449, 450, 468
Bioherms, 460, 461, 463, 468
Biotite 128, 136, 137, 138, 139, 178, 179, 187, 188, 392
Bioturbation, 172, 175, 194, 356, 357, 522

Birdseye structure, 456, 457
Bitumin, 633, 646, 650, 658
Böhm lamellae, 129, 130
Boring organisms, 517, 525
Bouma sequences, 83–86, 94, 195
Boundstone, 449, 454, 469
Breccias, 213
Burial-history curves, 386, 408, 401

Calcite, 128, 169, 171, 174, 179, 180, 181, 187, 207, 280, 292, 369, 372, 376, 378, 379, 381, 382, 392, 396, 397, 400, 401, 412, 413, 414, 416, 435, 440, 467, 470, 472, 473, 476, 486, 496, 501, 515, 518, 522, 524, 526–29, 536, 546, 552, 553, 555, 603, 606, 622
   magnesian, 413, 416, 435, 442, 467, 470, 471, 472, 473, 515, 517–19, 524–30, 532, 534, 536, 537, 552, 553
Calcitization, 515, 526
Calcium carbonate precipitation, 441, 470–73
Caliche, 379, 476–79
Carbonaceous sedimentary rocks, 631–59
Carbonate
   grains, 423–37
   minerals, 412–16
   solubility, 518
Carbonate compaction law, 560, 561
Carboxylic acid, 374
Cardhouse structures, 266
Caries texture, 556
Catagenesis, 655, 656
Cathodoluminescence, 29, 126, 169, 354, 412, 461, 492, 511, 545, 546, 547, 550, 551, 554
Cement stratigraphy, 550

Cements
   carbonate rocks, 169, 171, 174, 178, 179, 180, 181, 187, 189, 522, 523, 524, 530, 531, 532, 546–51
   characteristics of, 153, 154, 155, 382
   siliciclastic rocks, 378, 382–90
Chalcedony, 128, 129, 132, 147, 383, 416, 583, 584, 585, 591
Chamosite, 149–53, 365, 602, 606
Chemical compaction, 360, 394, 541, 560
Chert, 129, 133, 141, 144, 163, 167, 169, 178, 185, 186, 187, 188, 190, 196, 207, 208, 371, 383, 389, 392, 396, 401, 554, 620, 622
Cherts, 565, 566, 583–98
   bedded, 586–88
   chemical composition of, 586
   deposition of, 589–94
   diagenesis of, 594
   mineralogy of, 584
   nodular, 588, 589, 594–96, 598
   precipitation of, 591–94
   texture of, 584
   varieties of, 586
Chickenwire structure, 571
Chlorite, 128, 136, 137, 138, 179, 180, 187, 280, 365, 369, 376, 378, 402, 404, 406
Clarain, 639, 640
Clasts, stability of, 347
Clay minerals, 128, 138, 155, 156, 172, 182, 207, 279, 280–88, 365, 369, 370, 376, 378, 401, 392, 396, 586
   abundance of, 288, 289
   classification of, 284, 287, 288
   structures of, 281–86
Clay particles
   charge on, 266
   domains of, 268, 269, 270, 272